A Textbook of Physical Chemistry

Volume 1

States of Matter and Ions in Solution
(SI Units)

Sixth Edition

A Textbook of Physical Chemistry

Volume 1 : States of Matter and Ions in Solution
Volume 2 : Thermodynamics and Chemical Equilibrium
Volume 3 : Applications of Thermodynamics
Volume 4 : Quantum Chemistry, Molecular Spectroscopy and Molecular Symmetry
Volume 5 : Dynamics of Chemical Reactions, Statistical Thermodynamics, Macromolecules, and Irreversible Processes
Volume 6 : Computational Aspects in Physical Chemistry
Volume 7 : Practical Aspects in Physical Chemistry

A Textbook of Physical Chemistry

Volume 1

States of Matter and Ions in Solution
(SI Units)

Sixth Edition

K L KAPOOR

Former Associate Professor
Hindu College
University of Delhi
New Delhi

McGraw Hill Education (India) Private Limited

Published by McGraw Hill Education (India) Private Limited,
Registered Office: Anjana Complex No: 5/90 A, Butt Road, St. Thomas Mount, Chennai – 600016, Tamil Nadu, India.

A Textbook of Physical Chemistry, Volume 1

Eighth reprint 2024

Print Edition:
ISBN-13: 978-93-5316-518-5
ISBN-10: 93-5316-518-0

E-Book Edition:
ISBN-13: 978-93-5316-525-3
ISBN-10: 93-5316-525-3

09 10 11 12 13 RAJ 28 27 26 25 24

Typeset at The Composers, 260, C.A. Apt,, Paschim Vihar, New Delhi 110 063, and Printed and bound in India at Rajkamal Electric Press, Plot No. 2, Phase-IV, Kundli, Haryana

Cover Printer: Rajkamal Electric Press
Cover Designer: Creative Designer
Cover Image Source: Thinkstock

Visit us at: www.mheducation.co.in
Toll free in India: 1800 103 5875
Write to us at: info.india@mheducation.com
CIN: U80302TN2010PTC111532

To the memory of
my parents

Preface

Most of the textbooks on Physical Chemistry include qualitative description of the isothermal variation of compression factor with pressure for real gases. These also include the qualitative explanation of these isotherms by van der Waals equation of state. It is worth exploring these isotherms quantitatively for better understanding of the variations. In view of this, I venture to present the quantitative accounts of these variations based on van der Waals, Berthelot, Dieterici and Redlich-Kwong equations of state. In general, the variations can be explained quantitatively including the variation at Boyle temperature. It is found that the compression factor at Boyle temperature varies linearly with square of pressure in the lower pressure region.

Since the reduced equation of state is applicable to all gases irrespective of their nature, the variations of compression factor versus reduced pressure at different reduced temperatures is worth considering. A procedure to display these variations for the four equations of state mentioned above is described in this revised edition. The same isotherms were obtained irrespective of reference gas chosen to draw the isotherms. It is also found that the variation of compression factor is linear with respect to square of reduced pressure at reduced Boyle temperature.

Salient Features

- Strictly in accordance with latest IUPAC recommendations and SI units being adopted throughout the text
- Detailed discussion on Compression Factor of a Gas in Terms of Reduced Variables as chapter-end annexure
- Perfect blend of both theoretical and application-based concepts
- Comprehensive explanation of all equations of state-van der Waals, Berthelot, Dieterici and Redlich Kwong
- Extensive chapter end numericals including Revisionary Problems, Try Yourself Problems and Numerical Problems

K L Kapoor

Publisher's Note

McGraw-Hill Education (India) invites suggestions and comments from you, all of which can be sent to *info.india@mheducation.com* (kindly mention the title and author name in the subject line).

Piracy-related issues may also be reported.

Preface

Preface to the Fifth Edition

In recent years, the teaching curriculum of Physical Chemistry in many Indian universities has been restructured with a greater emphasis on a theoretical and conceptual methodology and the applications of the underlying basic concepts and principles. This shift in the emphasis, as I have observed, has unduly frightened undergraduates whose performance in Physical Chemistry has been otherwise generally far from satisfactory. This poor performance is partly because of the non-availability of a comprehensive textbook which also lays adequate stress on the logical deduction and solution of numericals and related problems. Naturally, the students find themselves unduly constrained when they are forced to refer to various books to collect the necessary reading material.

It is primarily to help these students that I have ventured to present a textbook which provides a systematic and comprehensive coverage of the theory as well as of the illustration of the applications thereof.

The present volumes grew out of more than a decade of classroom teaching through lecture notes and assignments prepared for my students of BSc (General) and BSc (Honours). The schematic structure of the book is assigned to cover the major topics of Physical Chemistry in six different volumes. **Volume 1** discusses the states of matter and ions in solutions. It comprises five chapters on the gaseous state, physical properties of liquids, solid state, ionic equilibria and conductance. **Volume 2** describes the basic principles of thermodynamics and chemical equilibrium in seven chapters, viz., introduction and mathematical background, zeroth and first laws of thermodynamics, thermochemistry, second law of thermodynamics, criteria for equilibrium and *A* and *G* functions, systems of variable composition, and thermodynamics of chemical reactions. **Volume 3** seeks to present the applications of thermodynamics to the equilibria between phases, colligative properties, phase rule, solutions, phase diagrams of one-, two- and three-component systems, and electrochemical cells. **Volume 4** deals with quantum chemistry, molecular spectroscopy and applications of molecular symmetry. It focuses on atomic structure, chemical bonding, electrical and magnetic properties, molecular spectroscopy and applications of molecular symmetry. **Volume 5** covers dynamics of chemical reactions, statistical and irreversible thermodynamics, and macromolecules in six chapters, viz., adsorption, chemical kinetics, photochemistry, statistical thermodynamics, macromolecules and introduction to irreversible processes. **Volume 6** describes computational aspects in physical chemistry in three chapters, viz., synopsis of commonly used statements in BASIC language, list of programs, and projects.

The study of Physical Chemistry is incomplete if students confine themselves to the ambit of theoretical discussions of the subject. They must grasp the practical significance of the basic theory in all its ramifications and develop a clear perspective to appreciate various problems and how they can be solved.

It is here that these volumes merit mention. Apart from having a lucid style and simplicity of expression, each has a wealth of carefully selected examples and solved illustrations. Further, three types of problems with different objectives in view are listed at the end of each chapter: (1) Revisionary Problems, (2) Try Yourself Problems, and (3) Numerical Problems. Under *Revisionary Problems*, only those problems pertaining to the text are included which should afford an opportunity to the students in self-evaluation. In *Try Yourself Problems*, the problems related to the text but not highlighted therein are provided. Such problems will help students extend their knowledge of the chapter to closely related problems. Finally, unsolved *Numerical Problems* are pieced together for students to practice.

Though the volumes are written on the basis of the syllabi prescribed for undergraduate courses of the University of Delhi, they will also prove useful to students of other universities, since the content of physical chemistry remains the same everywhere. In general, the SI units (*Systeme International d' unite's*), along with some of the common non-SI units such as atm, mmHg, etc., have been used in the books.

Salient Features

- Comprehensive coverage given to gaseous state, physical properties of liquids, the solid state, physical properties of liquids, ionic equilibria and conductance
- Emphasis given to applications and principles
- Explanation of equations in the form of solved problems and numericals
- IUPAC recommendations and SI units have been adopted throughout
- Rich and illustrious pedagogy

Acknowledgements

I wish to acknowledge my greatest indebtedness to my teacher, late Prof. R P Mitra, who instilled in me the spirit of scientific inquiry. I also record my sense of appreciation to my students and colleagues at Hindu College, University of Delhi, for their comments, constructive criticism and valuable suggestions towards improvement of the book. I am grateful to late Dr Mohan Katyal (St. Stephen's College), and late Prof. V R Shastri (Ujjain University) for the numerous suggestions in improving the book. I would like to thank Sh. M M Jain, Hans Raj College, for his encouragement during the course of publication of the book.

I wish to extend my appreciation to the students and teachers of Delhi University for the constructive suggestions in bringing out this edition of the book. I also wish to thank my children, Saurabh-Urvashi and Surabhi-Jugnu, for many useful suggestions in improving the presentation of the book.

Finally, my special thanks go to my wife, Pratima, for her encouragement, patience and understanding.

Feedback Request

The author takes the entire responsibility for any error or ambiguity, in fact or opinion, that may have found its way into this book. Comments and criticism from readers will, therefore, be highly appreciated and incorporated in subsequent editions.

K L Kapoor

Publisher's Note

McGraw-Hill Education (India) invites suggestions and comments from you, all of which can be sent to *info.india@mheducation.com* (kindly mention the title and author name in the subject line).

Piracy-related issues may also be reported.

Contents

Preface *vii*
Preface to the Fifth Edition *ix*

1. GASEOUS STATE **1**

1.1 The Three States of Matter *1*
1.2 Experimentally Derived Gas Laws *2*
1.3 Equation of State *5*
1.4 Application of Equation of State *7*
1.5 Concepts of Partial Pressure and Partial Volume *12*
1.6 The Kinetic Gas Equation *14*
1.7 Some Derivation from the Kinetic Gas Equation *17*
1.8 Real Gases *20*
1.9 Van Der Waals Equation of State for a Real Gas *23*
1.10 Other Equations of State for Real Gases *29*
1.11 Reduction of Van Der Waals Equation to Virial Equation *31*
1.12 Critical Constants *35*
1.13 Continuity of State *39*
1.14 Isotherms of Van Der Waals Equation *40*
1.15 The Law of Corresponding States *44*
1.16 Maxwellian Distribution of Molecular Speeds *52*
1.17 Derivation of Some Expressions from the Maxwell Distribution *56*
1.18 Barometric Distribution Law *62*
1.19 Molecular Collisions in a Gas *68*
1.20 Viscosity *73*
1.21 Self Diffusion and Effusion Processes *77*
1.22 Thermal Conductivity *81*
1.23 Electrical Conductivity *83*
1.24 Law of Equipartition of Energy *84*
1.25 Degress of Freedom and the Average Energy of a Molecule *86*
1.26 Heat Capacities *88*
1.27 Comparison of Theoretical and Experimental Heat Capacities *90*
1.28 Miscellaneous Problems *93*
Revisionary Problems *96*
Try Yourself Problems *100*
Numerical Problems *103*
Annexure I: Quantitative Explanation of Variation of Compression Factor with Pressure of a Gas *107*
Annexure II: Variation of Compression Factor of a Gas in terms of Reduced Variables *133*

2. PHYSICAL PROPERTIES OF LIQUIDS **144**

2.1 Introduction *144*
2.2 Vapour Pressure *144*
2.3 Viscosity *153*
2.4 Surface Tension *158*
Revisionary Problems 168
Numerical Problems 169

3. THE SOLID STATE **171**

3.1 Introduction *171*
3.2 Faces, Edges and Interfacial Angle of a Crystal *171*
3.3 Haüy's Idea and Space Lattice *172*
3.4 Crystal Systems *173*
3.5 The Fourteen Bravais Lattices *176*
3.6 Point Groups *180*
3.7 Space Groups *181*
3.8 Crystal Cleavage and Development of its Faces *182*
3.9 Designation of Planes *184*
3.10 Interplanar Distances for Cubic Systems *187*
3.11 Diffraction of Electromagnetic Radiation *189*
3.12 Diffraction of X-rays by Crystals *191*
3.13 Bragg's Equation *192*
3.14 Powder Method *195*
3.15 Diffraction Pattern of a Cubic System *197*
3.16 Crystal Structure of Sodium Chloride *202*
3.17 Crystal Structure of Potassium Chloride *205*
3.18 Density of Cubic Crystals *206*
3.19 Classification of Crystals Based on Bond Type *210*
3.20 General Discussion on Structure of Liquids *242*
Revisionary Problems 247
Try Yourself Problems 251
Numerical Problems 251
Annexure I: Symmetry Elements and Symmetry Operations *254*
Annexure II: Supplementary Materials *260*

4. IONIC EQUILIBRIA **273**

4.1 The Dissolution Process *273*
4.2 Classification of Substances *276*
4.3 The Arrhenius Theory of Dissociation *277*
4.4 Effect of Dilution on Degree of Dissociation *279*
4.5 Dissociation of Pure Water *280*
4.6 The pH-Scale *282*
4.7 Classification of Acids and Bases in Water *284*

4.8 Exact Treatment for Ionization of a Monoprotic Acid *284*
4.9 Exact Treatment for Ionization of a Base *295*
4.10 Exact Treatment for Ionization of a Diprotic Acid *298*
4.11 Dissociation of Polyprotic Acid *305*
4.12 Solutions of Salts in Water: Hydrolysis *308*
4.13 Exact Treatment of Hydrolysis of Salt Formed from a Weak Acid and a Strong Base *310*
4.14 Exact Treatment of Hydrolysis of Salt Formed from a Strong Acid and a Weak Base *316*
4.15 Exact Treatment of Hydrolysis of Salt Formed from a Weak Acid and a Weak Base *319*
4.16 Exact Treatment of Hydrolysis of Salt Involving Weak Conjugate Cation and an Amphiprotic Anion *324*
4.17 Exact Treatment of Hydrolysis of Salt Involving Strong Conjugate Cation and Amphiprotic Anion *327*
4.18 Hydrolysis of a Salt Containing Multivalent Cation or Anion *330*
4.19 Exact Treatment of a Mixture of a Weak Acid (HA) and Salt of its Conjugate Base (MA) *336*
4.20 Exact Treatment of a Mixture of a Weak Base (BOH) and Salt of its Conjugate Acid (BA) *339*
4.21 Exact Treatment of a Mixture of a Strong Acid and a Weak Acid *341*
4.22 Exact Treatment of a Mixture of Two Weak Acids *343*
4.23 Buffer Solutions *348*
4.24 Acid-Base Indicators *355*
4.25 Titration of a Strong Monoprotic Acid with a Strong Base *361*
4.26 Titration of a Weak Monoprotic Acid with a Strong Base *364*
4.27 Titration of a Weak Base with a Strong Monoprotic Acid *367*
4.28 General Treatment of Titration of an Acid with a Strong Base *371*
4.29 Titration of a Dibasic Acid with a Strong Base *374*
4.30 General Treatment of Titration of a Diprotic Acid with a Strong Base *381*
4.31 Titration of Sodium Carbonate Solution with Hydrochloric Acid *383*
4.32 Solubility Product *387*
4.33 Equilibria Involving Complex Ions *403*
4.34 Amphoterism *417*
4.35 Some Concepts of Acids and Bases *420*
4.36 Acid and Base Strengths and Structure *430*
Revisionary Problems 433
Try Yourself Problems 440
Numerical Problems 443

5. CONDUCTANCE **448**

5.1 Introduction *448*
5.2 Metallic Conduction *448*
5.3 Electrolytic Conduction *449*
5.4 Electrolysis *450*

5.5 Conductance of Solutions *452*
5.6 Equivalent and Molar Conductivities *455*
5.7 Variation of Conductivity and Molar Conductivity with Concentration *459*
5.8 Conductivity at High Electric Fields and High Frequencies *465*
5.9 Kohlrausch's Law of Independent Migration of Ions *466*
5.10 Values of Limiting Ionic Molar Conductivities *467*
5.11 Transport Numbers *474*
5.12 Ionic Speed and Ionic Mobility *497*
5.13 The Walden's Rule *501*
5.14 Application of Conductance Measurements *502*
Revisionary Problems *511*
Try Yourself Problems *512*
Numerical Problems *513*

Appendix I: A Note on Changing Concepts in Physical Chemistry *517*
Appendix II: Units and Conversion Factors *543*

Index *545*

1 Gaseous State

1.1 THE THREE STATES OF MATTER

Introduction

In order to determine experimentally the properties of substances, we deal with the aggregates of molecules as they occur in nature. It is the aggregations of molecules which come within the scope of human experience that constitute what is known as matter. The various kinds of substances that make up matter can be divided roughly into three categories, namely, gases, liquids and solids. These are called the three states of matter. These states can be considered to arise as a result of competition between two opposing molecular forces, namely, the forces of attraction which tend to hold the molecules together, and the disruptive forces due to the thermal energy of molecules.

Gaseous State

If the thermal energy is much greater than the forces of attraction, then we have matter in its gaseous state. Molecules in the gaseous state move with very large speeds and the forces of attraction amongst them are not sufficient to bind the molecules at one place, with the result that the molecules move practically independent of one another. Because of this feature, gases are characterized by marked sensitivity of volume change with change in temperature and pressure. There exists no boundary surface and, therefore, gases tend to fill completely any available space, resulting in no fixed volume to the gaseous state.

Liquid State

If the forces of attraction are greater than the thermal energy, we have matter in the liquid state. Molecules in the liquid state too have kinetic energy but they cannot go very far away because of the larger forces of attraction amongst them. Due to this feature, liquids have definite volume, but no definite shape. They take the shape of the vessel in which they are placed. In general, liquids are more dense and less compressible than gases.

Solid State

If the forces of attraction between molecules are much greater than the thermal energy, the positions of the molecules remain fixed and we have matter in the solid state. The molecules in the solid state, therefore, do not possess any translational energy, but have only vibrational energy since they can vibrate about their mean positions. Extremely large forces of attraction exist amongst them. That is why solids differ markedly from liquids and gases in respect of size, shape and volume. Solids, in general, have definite size, shape and volume.

Comments on the Gaseous System

Of all the three states of molecular aggregation, only the gaseous state allows a comparatively simple quantitative description. We are generally concerned with the relations among four properties, namely, mass, pressure, volume and temperature. A system is in a definite state (or condition) when all the properties of the system have definite values. It is not necessary to specify each and every property of the matter as these are interrelated. The relationship which connects the above four variables is known as the *equation of state* of the system. For gases, only three of these must be specified to describe the state, the fourth automatically has a fixed value and can be calculated from the equation of state established from the experimental behaviour of the system.

1.2 EXPERIMENTALLY DERIVED GASEOUS LAWS

Boyle's Law

At constant temperature, the volume of a definite mass of a gas is inversely proportional to its pressure, that is,

$$V \propto \frac{1}{p} \quad \text{i.e.} \quad V = \frac{K}{p} \quad \text{or} \quad pV = K \tag{1.2.1}$$

where K is a constant whose value depends upon (i) nature of the gas, (ii) temperature of the gas, and (iii) mass of the gas. For a given mass of a gas at constant temperature, Boyle's law gives

$$p_1V_1 = p_2V_2 \tag{1.2.2}$$

where V_1 and V_2 are volumes at pressures p_1 and p_2, respectively.

Graphical Representation

Equation (1.2.1) can be represented graphically by plotting pressures as ordinates and the corresponding volumes as abscissae (Fig. 1.2.1). The nature of the curve is a rectangular hyperbola. The general term *isothermal* or *isotherm* (meaning at constant temperature) is used to describe these plots.

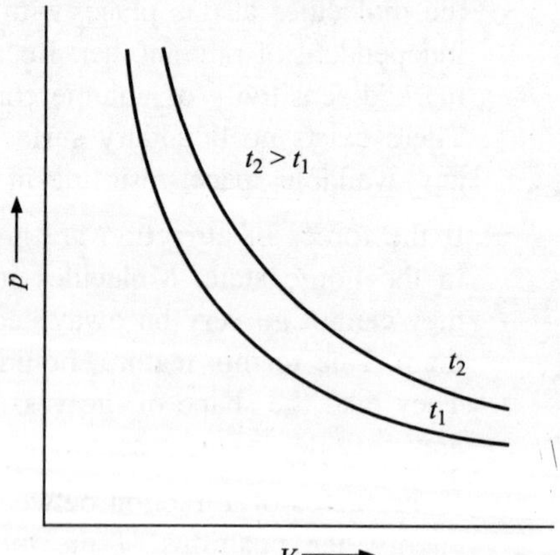

Fig. 1.2.1 A typical variation of pressure of a gas with volume

Charles Law

Charles made measurements of the volume of a fixed mass of a gas at various temperatures under the condition of constant pressure and found that *the volume of a fixed mass of a gas at constant pressure is a linear function of its Celsius temperature.* This can be expressed as

$$V_t = a + bt \tag{1.2.3}$$

where t is Celsius temperature and a and b are constants.

Graphical Representation

Equation (1.2.3) has been plotted in Fig. 1.2.2. The intercept on the vertical axis is a and it is equal to V_0, the volume at 0 °C. The slope of the plot is the derivative

$$b = \left(\frac{\partial V_t}{\partial t}\right)_p$$

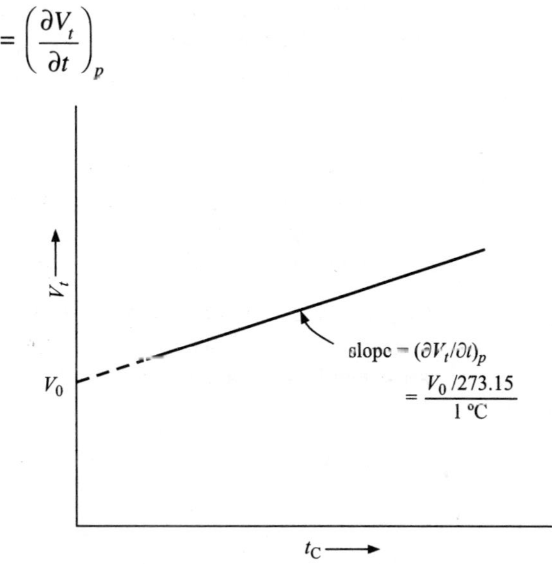

Fig. 1.2.2 A typical variation of volume of a gas with temperature expressed in °C

Alternative Form of Charles Law

Experimental data shows that for each Celsius degree rise in temperature, the volume of a gas expands 1/273.15 of its volume at 0 °C. If V_0 is the volume of a gas at 0 °C, then b is given by

$$b = \left(\frac{V_0/273.15}{1\,°C}\right)$$

With this, Eq. (1.2.3) becomes

$$V_t = V_0 + \left(\frac{V_0/2.73.15}{1\,°C}\right)t$$

or
$$V_t = V_0\left(1 + \frac{t/°C}{273.15}\right) = V_0\left(\frac{273.15 + t/°C}{273.15}\right) \qquad (1.2.4)$$

or
$$V_T = V_0\,\frac{(T/\mathrm{K})}{273.15} \qquad (T \text{ is kelvin temperature})^\dagger$$

† It is convenient to use the absolute temperature scale on which temperatures are measured in kelvin (K). A reading on this scale is obtained by adding 273.15 to the Celsius value. Temperature on the kelvin scale is denoted by T. Thus

$$T/\mathrm{K} = 273.15 + t/°C$$

i.e. $$V_T = \left(\frac{V_0}{273.15\ \mathrm{K}}\right)T$$

Since V_0, the volume of the gas at 0 °C, has a constant value at a given pressure, the above relation can be expressed as

$$V = K_2T \qquad (1.2.5)$$

where K_2 is a constant whose value depends upon the nature, mass and pressure of the gas.

Equation (1.2.5) is an alternative form of Charles law according to which *the volume of a given mass of a gas at constant pressure is directly proportional to its kelvin temperature.*

Graphical Representation

A typical variation of volume of a gas with change in its kelvin temperature is shown in Fig. 1.2.3. The general term *isobar*, which means at constant pressure, is assigned to these plots.

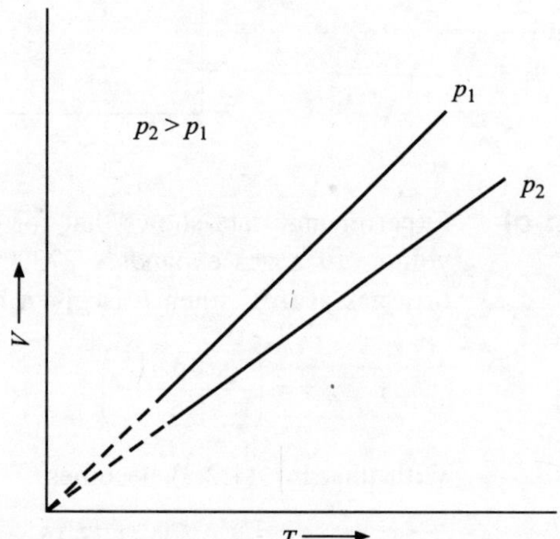

Fig. 1.2.3 Variation of volume of a gas with kelvin temperature

Comment on Zero Kelvin

Since volume is directly proportional to kelvin temperature, the volume of a gas should theoretically be zero at kelvin zero. However, gases liquefy and then solidify before this low temperature is reached. In fact, no substance exists as a gas at a temperature near kelvin zero, though the straight-line plots can be extrapolated to zero volume. The temperature that corresponds to zero volume is −273.15 °C.

Gay-Lussac's Law: Dependence of Pressure on Temperature

An expression similar to volume dependence of gas on temperature has been derived for the pressure dependence also. The pressure of a given mass of a gas at constant volume varies linearly with Celsius temperature.

$$p_t = a + bt \qquad (1.2.6)$$

where $a = p_0$ and $b = (\partial p_t/\partial t)_V$. The value of the latter can be determined experimentally and is found to be (p_0/273.15 °C). Thus, Eq. (1.2.6) modifies to

$$p_t = p_0 + \left(\frac{p_0}{273.15}\right)(t/°\mathrm{C})$$

Therefore $$p_t = p_0\left(\frac{273.15 + (t/°\mathrm{C})}{273.15}\right) = \left(\frac{p_0}{273.15\ \mathrm{K}}\right)T$$

or $$p_t \propto T \tag{1.2.7}$$

that is, *the pressure of a given mass of a gas at constant volume is directly proportional to its kelvin temperature.*

Graphical Representations

Equations (1.2.6) and (1.2.7) are shown graphically in Figs. 1.2.4 and 1.2.5, respectively. The general term *isochor* (meaning at constant volume) is given to the plots of Fig. 1.2.5.

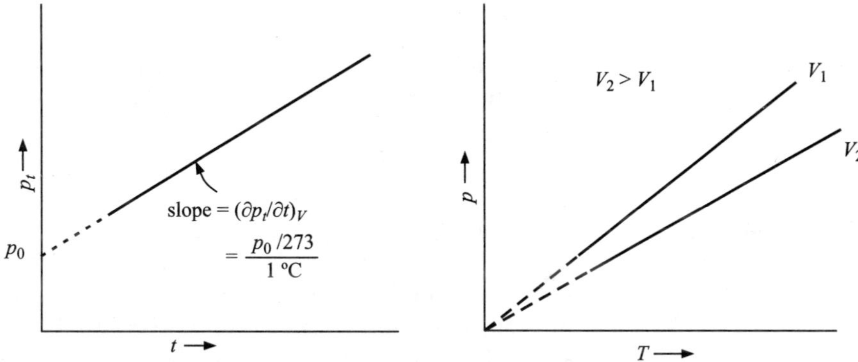

Fig. 1.2.4 A typical variation of pressure of a gas with temperature expressed in °C

Fig. 1.2.5 A typical variation of pressure of a gas with kelvin temperature

Graham's Law of Diffusion

The phenomenon of diffusion may be described as the tendency for any substance to spread uniformly throughout the space available to it. Diffusion through fine pores is called effusion.

According to Graham's law of diffusion, *the rate of diffusion (or effusion) of a gas is inversely proportional to the square root of its density or molar mass.* If r_1 and r_2 are the rates of diffusion of two gases under identical conditions, whose densities under the given conditions are ρ_1 and ρ_2, respectively, then from Graham's law,

$$\frac{r_1}{r_2} = \sqrt{\frac{\rho_2}{\rho_1}} \quad \text{or} \quad \frac{r_1}{r_2} = \sqrt{\frac{M_2}{M_1}} \tag{1.2.8}$$

where M_1 and M_2 are the respective molar masses of the two gases.

1.3 EQUATION OF STATE

Derivation of Equation of State

The results of the laws of Boyle and Charles can be combined into an expression which represents the relationship between pressure, volume and temperature of a given mass of a gas; such an expression is described as an *equation of state.*

Suppose the gas is in the initial state with volume V_1, pressure p_1 and temperature T_1. We then change the state of the gas to a volume V_2, pressure p_2 and temperature T_2. Let us carry out this change in two steps.

(i) First we change the pressure from p_1 to p_2 keeping the temperature T_1 constant. The resultant volume V_r as given by Boyle's law is

$$V_r = \frac{p_1 V_1}{p_2}$$

(ii) Next, temperature is changed from T_1 to T_2, keeping the pressure p_2 constant. The final volume V_2 as given by Charles law is

$$\frac{V_2}{T_2} = \frac{V_r}{T_1} \quad \text{i.e.} \quad V_2 = \frac{V_r T_2}{T_1} = \frac{(p_1 V_1 / p_2) T_2}{T_1}$$

or
$$\frac{p_1 V_1}{T_1} = \frac{p_2 V_2}{T_2} \tag{1.3.1}$$

It follows that no matter how we change the state of the given amount of a gas, the ratio pV/T always remains constant, i.e.

$$\frac{pV}{T} = K$$

Universal Gas Constant

The value of K depends on the amount of gas in the system. Since V is an extensive property (which is mass dependent), its value at constant p and T is proportional to the amount of the gas present in the system. Then K must also be proportional to the amount of gas because p and T are intensive properties (which have no mass dependence). We can express this by writing $K = nR$, in which n is the amount of gas in a given volume of gas and R is independent of all variables and is, therefore, a universal constant. We thus have the general gas law

$$pV = nRT \tag{1.3.2}$$

Physical Significance of Gas Constant *R*

The universal gas constant as given by Eq. (1.3.2) is $R = pV/nT$. Thus, it has the units of (pressure × volume) divided by (amount of gas × temperature). Now the dimensions of pressure and volume are,

$$\text{Pressure} = (\text{force/area}) = (\text{force/length}^2) = \text{force} \times \text{length}^{-2}$$

$$\text{Volume} = \text{length}^3$$

Thus
$$R = \frac{(\text{force} \times \text{length}^{-2})(\text{length}^3)}{(\text{amount of gas})(\text{kelvin})} = \frac{(\text{force} \times \text{length})}{(\text{amount of gas})(\text{kelvin})}$$

$$= \frac{\text{work (or energy)}}{(\text{amount of gas})(\text{kelvin})}$$

Thus, the dimensions of R are energy per mole per kelvin and hence it represents the amount of work (or energy) that can be obtained from one mole of a gas when its temperature is raised by one kelvin.

1.4 APPLICATION OF EQUATION OF STATE

Concept of an Ideal Gas

So far, we have assumed that all gases obey the gas laws under all conditions of temperature and pressure; however, for real gases this is not true. Real gases obey these laws only under limited conditions of low pressures and high temperatures. They exhibit deviations from the gaseous laws and these deviations are greater when the temperature and pressure are close to the conditions at which the gas can be condensed into a liquid. Thus Boyle's law, Charles law, and the equation of state derived from these two laws may be regarded as approximations for real gases and are expected to be applicable only at relatively low pressures and moderately high temperatures. It is, nevertheless, very useful to postulate a hypothetical ideal gas, defined as a gas to which the laws of Boyle and Charles are strictly applicable under all conditions of temperatures and pressures. It is for this reason that Eq. (1.3.2) is commonly referred to as the *ideal gas equation*. Real gases attain ideal behaviour only at very low pressures and very high temperatures.

Characteristics of an Ideal Gas

Since Eq. (1.3.2) is not applicable to real gases, the evaluation of the universal gas constant R cannot be done directly by utilizing the pressure, volume, and temperature data of real gases. Equation (1.3.2) is strictly applicable only for ideal gases and thus if the pressure and volume of one mole of an ideal gas were known at a definite temperature, it would be a simple matter to evaluate R from Eq. (1.3.2). However, as no gas behaves ideally, this procedure would appear to be ruled out. But we know from experiments that gases approach ideal behaviour as the pressure is decreased. Hence, the extrapolation method ($p \to 0$) on the data of real gases can be utilized to determine the corresponding properties of an ideal gas. The data obtained in this manner, after extrapolation, should be independent of the characteristics of the actual gas employed for the experiment.

By measuring the volumes of one mole of a real gas at different pressures and constant temperature, a graph between pV and p can be drawn. On extrapolating this graph to zero pressure to correct for departure from ideal behaviour it is possible to determine the value of pV which is expected to be applicable to one mole of an ideal gas. Since this value of pV is expected to be independent of the nature of the gas, the same value of $(pV)_{p\to 0}$ would be obtained irrespective of the gas employed for this purpose. In other words, the graphs of pV versus p of different gases must yield the same value of $(pV)_{p\to 0}$. In fact, it is found to be so, as is evident from Fig. 1.4.1. The value of $(pV)_{p\to 0}$ at 273.15 K is found to be 22.711 dm^3 bar. Thus if p = 1 bar, then V = 22.711 dm^3, that is, the volume occupied by one mole of an ideal gas at standard temperature (273.15 K) and pressure (1 bar) is 22.711 dm^3.

Value of Gas Constant in SI Units

The value of R in SI units can be worked out as follows.

$$R = \frac{pV}{nT} = \frac{(1\ \text{bar})\ (22.711\ \text{dm}^3)}{(1\ \text{mol})\ (273.15\ \text{K})} = 0.083\,14\ \text{bar dm}^3\ \text{K}^{-1}\ \text{mol}^{-1}$$

Since 10^2 kPa = 1 bar, the value of R expressed in kPa dm^3 K^{-1} mol^{-1} will be

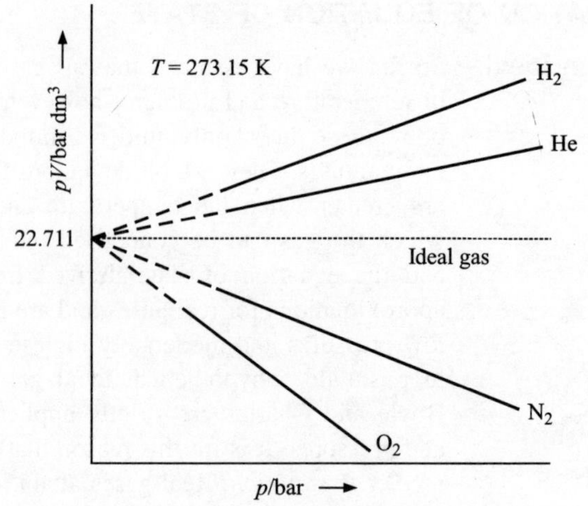

Fig. 1.4.1 Plots of pV versus p of a few gases

$$R = 0.083\,14\ (10^2\ \text{kPa})\ \text{dm}^3\ \text{K}^{-1}\ \text{mol}^{-1}$$
$$= 8.314\ \text{kPa dm}^3\ \text{K}^{-1}\ \text{mol}^{-1} \equiv 8.314\ \text{MPa cm}^3\ \text{K}^{-1}\ \text{mol}^{-1}$$
$$\equiv 8.314\ \text{Pa m}^3\ \text{K}^{-1}\ \text{mol}^{-1} \equiv 8.314\ \text{J K}^{-1}\ \text{mol}^{-1}$$

Example 1.4.1 Determine the value of gas constant R when pressure is expressed in Torr and volume in dm^3.

Solution By definition, 1.013 25 bar = 760 Torr. Hence

$$R = \frac{pV}{nT} = \frac{\left\{(1\ \text{bar})\left(\dfrac{760\ \text{Torr}}{1.013\,25\ \text{bar}}\right)\right\}(22.711\ \text{dm}^3)}{(1\ \text{mol})\,(273.15\ \text{K})}$$
$$= 62.36\ \text{Torr dm}^3\ \text{K}^{-1}\ \text{mol}^{-1}$$

Example 1.4.2 Derive the value of R when (a) pressure is expressed in atmospheres, volume in cm^3 and (b) p in dyn m^{-2} and V in mm^3.

Solution Since pV = 22.711 dm^3 bar, the volume of an ideal gas at 1 atm (= 1.013 25 bar) will be

$$V = \frac{22.711\ \text{dm}^3\ \text{bar}}{1.013\,25\ \text{bar}} = 22.414\ \text{dm}^3$$

(a) *p in* atm *and V in* cm^3

$$R = \frac{pV}{nT} = \frac{(1\ \text{atm})\,(22\,414\ \text{cm}^3)}{(1\ \text{mol})\,(273.15\ \text{K})} = 82.06\ \text{atm cm}^3\ \text{K}^{-1}\ \text{mol}^{-1}$$

(b) *p in* dyn m^{-2} *and V in* mm^3

$$p = 1\ \text{atm} \equiv 1.013\,2\times10^6\ \text{dyn cm}^{-2} \equiv 1.013\,2\times10^{10}\ \text{dyn m}^{-2}$$
$$V = 22\,414\ \text{cm}^3 \equiv 22\,414\times10^3\ \text{mm}^3$$
$$R = \frac{pV}{nT} = \frac{(1.013\,2\times10^{10}\ \text{dyn m}^{-2})\,(22\,414\times10^3\ \text{mm}^3)}{(1\ \text{mol})\,(273.15\ \text{K})}$$
$$= 8.314\times10^{14}\ (\text{dyn m}^{-2})\,(\text{mm}^3)\ \text{K}^{-1}\ \text{mol}^{-1}$$

Avogadro's Law

According to Avogadro's law, *equal number of molecules of different gases under identical conditions of temperature and pressure occupy the same volume.*

When this law is applied to real gases, it is found that the law does not hold good at ordinary temperatures and pressures. However, when the measurements are made at low pressures, deviations from the law become less and thus, like other gaseous laws, Avogadro's law may be regarded as an approximation which is expected to be applicable only under conditions of low pressures and high temperatures. Strictly speaking, this law would be applicable only for ideal gases.

The fact that Avogadro's law is applicable to real gases at very low pressures and high temperatures indicates that the volume occupied by different gases having the same number of molecules under identical conditions of temperature and pressure is independent of the nature of the gaseous molecules. Thus, whether the molecules are heavy (e.g. Br_2) or light (e.g. H_2), gases with equal number of molecules would occupy the same volume. This leads to one of the most important features of gases that the distance between molecules is much larger than the actual dimensions of molecules, since otherwise, Avogadro's law would not have been true.

Avogadro Constant

The facts that the behaviour of a real gas approaches that of an ideal gas as $p \rightarrow 0$ and the volume occupied by one mole of an ideal gas at the specified temperature (273.15 K) and pressure (101.325 kPa) has a fixed value (22.414 dm^3) indicate that the number of molecules contained in one mole of any real gas should be a constant quantity. This physical quantity has a value of $6.022 \times 10^{23}\ \mathrm{mol}^{-1}$ and is known as *Avogadro constant.*

Equation of State in Terms of Numbers of Molecules

The amount of gas containing N number of molecules is given by

$$n = \frac{N}{N_\mathrm{A}}$$

With this, Eq. (1.3.2) becomes

$$pV = nRT = \frac{N}{N_\mathrm{A}} RT \tag{1.4.1}$$

Avogadro's law follows directly from the Eq. (1.4.1). We have

$$V = \left(\frac{RT}{pN_\mathrm{A}}\right) N$$

For a fixed condition of pressure and temperature, a gas will have fixed volume for a fixed number of gaseous molecules.

Example 1.4.3 Estimate the number of gaseous molecules left in a volume of 1 mm^3 if it is pumped out to give a vacuum of 10^{-6} mmHg at 298 K.

Solution We are given that

$$V = 1\ \mathrm{mm}^3 = 10^{-6}\ \mathrm{dm}^3$$

$$p = 10^{-6}\ \mathrm{mmHg} = (10^{-6}\ \mathrm{mmHg})\left(\frac{101.235\ \mathrm{kPa}}{760\ \mathrm{mmHg}}\right) = 1.333 \times 10^{-7}\ \mathrm{kPa}$$

$$\text{Amount of the gas, } n = \frac{pV}{RT} = \frac{(1.333\times10^{-7}\ \text{kPa})\,(10^{-6}\ \text{dm}^3)}{(8.314\ \text{kPa dm}^3\ \text{K}^{-1}\ \text{mol}^{-1})\,(298\ \text{K})}$$

$$= 5.38\times10^{-17}\ \text{mol}$$

Hence, number of molecules

$$N = n\,N_A = (5.38\times10^{-17}\ \text{mol})\,(6.022\times10^{23}\ \text{mol}^{-1}) = 3.240\times10^{7}$$

Equation of State in Terms of Mass of a Gas

For a gas of mass m, the amount of gas is given by

$$n = \frac{m}{M}$$

where M is the molar mass of the gas. With this, Eq. (1.3.2) becomes

$$pV = nRT = \left(\frac{m}{M}\right)RT \qquad (1.4.2)$$

Example 1.4.4

When 2 g of gaseous substance A is introduced into an initially evacuated flask kept at 25 °C, the pressure is found to be 101.325 kPa. The flask is evacuated and 3 g of B is introduced. The pressure is found to be 50.662 5 kPa at 25 °C. Calculate the ratio M_A/M_B.

Solution

From the ideal gas equation, we have

$$pV = nRT = \left(\frac{m}{M}\right)RT \quad \text{or} \quad M = m\frac{RT}{pV}$$

Hence $\quad M_A = (2\ \text{g})\dfrac{RT}{(101.325\ \text{kPa})\,V} \quad$ and $\quad M_B = (3\ \text{g})\dfrac{RT}{(50.662\,5\ \text{kPa})\,V}$

Thus, $\quad \dfrac{M_A}{M_B} = \dfrac{2\times0.5}{3} = \dfrac{1}{3}$

Example 1.4.5

A certain mixture of helium and argon weighing 5.0 g occupies a volume of 10 dm³ at 25 °C and 101.325 kPa. What is the composition of the mixture in mass percentage?

Solution

Given that $m_{mix} = 5.0$ g; $\quad V = 10$ dm³; $\quad T = 25$ °C ≡ 298.15 K; $\quad p = 101.325$ kPa

Let the mass of He be x. Therefore

$$\text{Amount of He} = \frac{m}{M} = \frac{x}{(4.0\ \text{g mol}^{-1})}$$

$$\text{Amount of Ar} = \frac{m}{M} = \frac{5.0\ \text{g} - x}{(39.95\ \text{g mol}^{-1})}$$

$$\text{Total amount of gases} = \frac{pV}{RT} = \frac{(101.325\ \text{kPa})\,(10\ \text{dm}^3)}{(8.314\ \text{kPa dm}^3\ \text{K}^{-1}\ \text{mol}^{-1})\,(298.15\ \text{K})}$$

$$= 0.409\ \text{mol}$$

Hence $\quad \left(\dfrac{x}{4.0\ \text{g mol}^{-1}}\right) + \left(\dfrac{5.0\ \text{g} - x}{39.95\ \text{g mol}^{-1}}\right) = 0.409\ \text{mol}$

Solving for x, we get

$$x = 1.262\ \text{g}$$

$$\text{Mass per cent of He} = \frac{1.262\ \text{g}}{5.0\ \text{g}} \times 100 = 25.24$$

$$\text{Mass per cent of Ar} = 100 - 25.24 = 74.76$$

Example 1.4.6 A flask of 2 dm^3 capacity contains O_2 at 101.325 kPa and 300 K. The gas pressure is reduced to 0.10 Pa by attaching the flask to a pump. Assuming ideal behaviour, answer the following:

(i) What will be the volume of the gas which is left behind?
(ii) What amount of O_2 and the corresponding number of molecules are left behind in the flask?
(iii) If now 2 g of N_2 is introduced, what will be the pressure of the flask?

Solution Given that $V_1 = 2\ \text{dm}^3$, $p_1 = 101.325$ kPa, $p_2 = 0.10$ Pa, $T = 300$ K

We have the following results.

(i) The volume of O_2 left behind will be the same, i.e. 2 dm^3.
(ii) The amount of O_2 left behind is given by

$$n = \frac{p_2V_1}{RT} = \frac{(10^{-4}\ \text{kPa})\,(2\ \text{dm}^3)}{(8.314\ \text{kPa dm}^3\ \text{K}^{-1}\ \text{mol}^{-1})\,(300\ \text{K})} = 8.019 \times 10^{-8}\ \text{mol}$$

$$N = nN_A = (8.109 \times 10^{-8}\ \text{mol})\,(6.022 \times 10^{23}\ \text{mol}^{-1})$$

$$= 4.88 \times 10^{16}$$

(iii) $2\ \text{g of}\ N_2 = \dfrac{1}{14}\ \text{mol}$

$$\text{Total amount of gases in flask} = \frac{1}{14}\ \text{mol} + 8.019 \times 10^{-8}\ \text{mol} \simeq \frac{1}{14}\ \text{mol}$$

Thus, the pressure of the flask is given by

$$p = \frac{nRT}{V} = \frac{(1\ \text{mol}/14)\,(8.314\ \text{kPa dm}^3\ \text{K}^{-1}\ \text{mol}^{-1})\,(300\ \text{K})}{(2\ \text{dm}^3)} = 89.08\ \text{kPa}$$

Example 1.4.7 Two flasks of equal volume connected by a narrow tube (of negligible volume) are at 300 K and contain 0.70 mol of H_2 gas at 50.662 5 kPa pressure. One of the flasks is then immersed into a bath kept at 400 K, while the other remains at 300 K. Calculate the final pressure and the amount of H_2 in each flask.

Solution The final pressure in both the flasks will be the same, since both of them are connected with each other. Let n_1 be the amount of the gas in flask 1 (T_1 = 300 K) and n_2 in the flask 2 (T_2 = 400 K);

For flask 1, $pV = n_1RT_1$ For flask 2, $pV = n_2RT_2$

Therefore, $n_1T_1 = n_2T_2$ i.e. $\dfrac{n_1}{n_2} = \dfrac{T_2}{T_1} = \dfrac{400\ \text{K}}{300\ \text{K}} = \dfrac{4}{3}$

But $n_1 + n_2 = 0.7$ mol

Hence $n_1 = 0.4$ mol at 300 K $n_2 = 0.3$ mol at 400 K

Volume of each flask is

$$V = \frac{nRT}{p} = \frac{(0.35\ \text{mol})\,(8.314\ \text{kPa dm}^3\ \text{K}^{-1}\ \text{mol}^{-1})\,(300\ \text{K})}{(50.662\ 5\ \text{kPa})} = 17.23\ \text{dm}^3$$

Final pressure is

$$p_f = \frac{n_1RT_1}{V} = \frac{(0.4\ \text{mol})(8.314\ \text{kPa dm}^3\ \text{K}^{-1}\ \text{mol}^{-1})(300\ \text{K})}{(17.23\ \text{dm}^3)} = 57.90\ \text{kPa}$$

1.5 CONCEPTS OF PARTIAL PRESSURE AND PARTIAL VOLUME

Definition of Partial Pressure

The relation between the total pressure of a mixture of gases and the pressures of the individual gases was expressed by Dalton in the forms of law of partial pressures. The partial pressure of a gas in a mixture is defined as the pressure which the gas would exert if it is allowed to occupy the whole volume of the mixture at the same temperature.

Definition of Dalton's Law

According to Dalton's law of partial pressures, *the total pressure of a mixture of gases is equal to the sum of the partial pressures of the constituent gases.*

Partial Pressures in a Gaseous Mixture

Let a mixture of gases have the amount n_1 of the first gas, n_2 of the second gas, and so on. Let the corresponding partial pressures be $p_1, p_2, \ldots$. The total pressure is given by

$$p_{\text{total}} = p_1 + p_2 + \cdots$$

If the gases present in the mixture behave ideally, then, it is possible to write separately for each gas,

$$p_1V = n_1RT \tag{1.5.1a}$$

$$p_2V = n_2RT \tag{1.5.1b}$$

$$\cdots\cdots$$

Hence $\quad (p_1 + p_2 + \cdots)\,V = (n_1 + n_2 + \cdots)RT$

i.e. $\quad p_{\text{total}}\,V = n_{\text{total}}\,RT \qquad (1.5.2)$

where n_{total} is the total amount of gases in the mixture. Dividing Eqs (1.5.la) and (1.5.1b) by Eq. (1.5.2), we get

$$p_1 = \frac{n_1}{n_{\text{total}}}\,p_{\text{total}} = x_1p_{\text{total}} \tag{1.5.3a}$$

$$p_2 = \frac{n_2}{n_{\text{total}}}\,p_{\text{total}} = x_2p_{\text{total}} \tag{1.5.3b}$$

Definition of Amount (Mole) Fraction

The fractions n_1/n_{total}, n_2/n_{total}, are called the *amount* (*mole*) *fractions* of the respective gases. The amount fraction of a constituent in any mixture (gaseous, liquid or solid) is defined as the amount (or number of molecules) of that constituent divided by the total amount (or number of molecules) of constituents in the mixture. If xs are given, it is possible to calculate partial pressures by using Eqs (1.5.3).

Partial Volumes: Amagat's Law

The partial volume of a gas in a mixture is defined as the volume which the gas would occupy if it were present alone in a container at temperature T and pressure p of the mixture. According to the ideal gas equation, this is given by

$$V_1 = n_1\left(\frac{RT}{p}\right) \tag{1.5.4a}$$

$$V_2 = n_2\left(\frac{RT}{p}\right) \tag{1.5.4b}$$

$\cdots$

Adding, we get

$$V_1 + V_2 + \cdots = (n_1 + n_2 + \cdots)\frac{RT}{P} = n_{\text{total}}\left(\frac{RT}{p}\right)$$

From the ideal gas equation

$$n_{\text{total}}\left(\frac{RT}{p}\right) = V_{\text{total}} \tag{1.5.5}$$

we have $V_1 + V_2 + \cdots = V_{\text{total}}$

which is Amagat's law of partial volumes according to which *the total volume of a mixture of gases is equal to the sum of the partial volumes of the constituent gases.*

Dividing Eqs (1.5.4) by Eq. (1.5.5), we get

$$V_i = x_i V_{\text{total}} \qquad i = 1, 2, \ldots \tag{1.5.6}$$

Example 1.5.1 The following reaction is carried out at 101.325 kPa and 383 K,

$$2CH_4 + 3O_2 \rightarrow 2CO + 4H_2O$$

with the initial amounts of CH_4 and O_2 as 0.01 mol and 0.03 mol, respectively. All reactants and products are gaseous at 383 K. A short while after completion of the reaction, the flask is cooled to 283 K at which temperature H_2O is completely condensed. Calculate:

(i) The volume of the flask.
(ii) Total pressure and partial pressures of various species after the reaction at 383 K and 283 K.
(iii) The number of molecules of the various substances before and after the reaction.

Solution The reaction is $2CH_4 + 3O_2 \rightarrow 2CO + 4H_2O$

Amount/mol	$2CH_4$	$3O_2$	$2CO$	$4H_2O$	*Temp.*
In the beginning	0.01	0.03	0	0	383 K
At the end	0.0	0.015	0.01	0.02	383 K
	0.0	0.015	0.01	condensed	283 K

(i) Volume of the flask

$$V = \frac{nRT}{p} = \frac{(0.04\ \text{mol})\,(8.314\ \text{kPa dm}^3\ \text{K}^{-1}\ \text{mol}^{-1})\,(383\ \text{K})}{(101.325\ \text{kPa})} = 1.257\ \text{dm}^3$$

(ii) $p(\text{total, 383 K}) = \left(\frac{0.045 \text{ mol}}{0.040 \text{ mol}}\right)(101.325 \text{ kPa}) = 113.99 \text{ kPa}$

$p(\text{total, 283 K}) = \left(\frac{0.025 \text{ mol}}{0.040 \text{ mol}}\right)\left(\frac{283 \text{ K}}{383 \text{ K}}\right)(101.325 \text{ kPa}) = 46.81 \text{ kPa}$

$p(CH_4, 383 \text{ K}) = 0$

$p(O_2, 383 \text{ K}) = \left(\frac{0.015 \text{ mol}}{0.045 \text{ mol}}\right)(113.99 \text{ kPa}) = 38.00 \text{ kPa}$

$p(CO, 383 \text{ K}) = \left(\frac{0.01 \text{ mol}}{0.045 \text{ mol}}\right)(113.99 \text{ kPa}) = 25.33 \text{ kPa}$

$p(H_2O, 383 \text{ K}) = \left(\frac{0.02 \text{ mol}}{0.045 \text{ mol}}\right)(113.99 \text{ kPa}) = 50.66 \text{ kPa}$

$p(CH_4, 283 \text{ K}) = 0$

$p(O_2, 283 \text{ K}) = \left(\frac{0.015 \text{ mol}}{0.025 \text{ mol}}\right)(46.81 \text{ kPa}) = 28.09 \text{ kPa}$

$p(CO, 283 \text{ K}) = \left(\frac{0.01 \text{ mol}}{0.025 \text{ mol}}\right)(46.81 \text{ kPa}) = 18.72 \text{ kPa}$

(iii) Number of molecules before the reaction are

$N(CH_4) = (0.01 \text{ mol})\ (6.022 \times 10^{23} \text{ mol}^{-1}) = 6.022 \times 10^{21}$

$N(O_2) = (0.03 \text{ mol})\ (6.022 \times 10^{23} \text{ mol}^{-1}) = 1.807 \times 10^{22}$

Number of molecules after the reaction are

$N(CH_4) = 0$

$N(O_2) = (0.015 \text{ mol})\ (6.022 \times 10^{23} \text{ mol}^{-1}) = 9.033 \times 10^{21}$

$N(CO) = (0.01 \text{ mol})\ (6.022 \times 10^{23} \text{ mol}^{-1}) = 6.022 \times 10^{21}$

$N(H_2O) = (0.02 \text{ mol})\ (6.022 \times 10^{23} \text{ mol}^{-1}) = 1.204 \times 10^{22}$

1.6 THE KINETIC GAS EQUATION

Postulates of an Ideal Gas

After knowing the experimental gas laws, it is of interest to develop a theoretical model based on the structure of gases, which can correlate the experimental facts. Fortunately, such a theory has been developed (known as the kinetic theory of gases) and based upon certain essential postulates (which are supposed to be applicable to an ideal gas) it is possible to derive an expression (known as the kinetic gas equation) from where all these gas laws can be derived. The essential postulates are:

- A gas consists of a large number of very small spherical tiny particles, which may be identified with the molecules. The molecules of a given gas are completely identical in size, shape and mass.

- The volume occupied by the molecules is negligible in comparison to the total volume of the gas.
- The molecules are in rapid motion which is completely random. During their motion, they collide with one another and with the sides of the vessel. The pressure of the gas is due to the collisions of molecules with the sides of the vessel.
- The molecules are perfectly elastic, i.e. there occurs no loss of energy when they collide with one another and with the sides of the vessel.
- The laws of classical mechanics, in particular Newton's second law of motion, are applicable to the molecules in motion.
- There is no force of attraction or repulsion amongst the molecules, i.e. they are moving independent of one another.
- At any instant, a given molecule can have energy ranging from a small value to a very large value, but the average kinetic energy remains constant for a given temperature, i.e. the average kinetic energy is proportional to the absolute temperature of the gas.

Derivation of the Kinetic Gas Equation

Imagine a cube of edge-length l, containing N molecules, each having a mass of m. Molecules are moving at random in all directions, with speed covering a considerable range of values.

The velocity u of any molecule may be resolved into three-component velocities designated as u_x, u_y and u_z. These are in the three directions at right angles to each other and parallel to the sides of the cube as shown in Fig. 1.6.1. The component velocities are related by the expression

$$u^2 = u_x^2 + u_y^2 + u_z^2 \qquad (1.6.1)$$

Considering the x-component motion of a molecule, we will have

Momentum of the molecule before collision with the side ABCD = mu_x.

Momentum of the molecule after collision with the side ABCD = $-mu_x$.

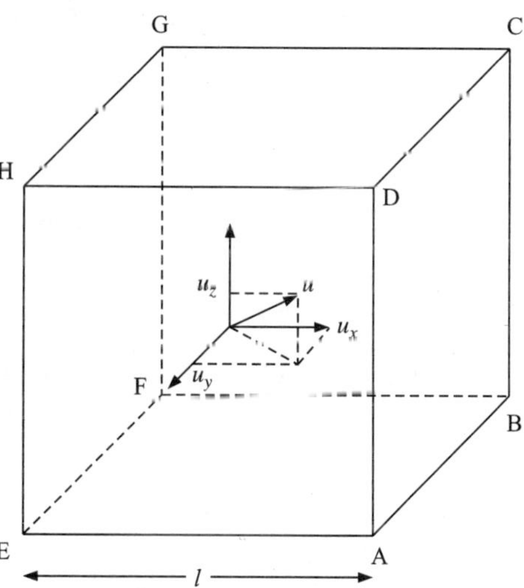

Fig. 1.6.1 Molecular velocity and its components

Change of momentum of the molecule in a single collision with the side ABCD $= |2mu_x|$.

Since l is the edge length of the cube, the molecule has to travel a distance $2l$ to arrive back at the wall ABCD. The number of collisions per unit time with the wall ABCD will be equal to $u_x/2l$.

The total change of momentum per unit time due to such impacts is

$$2mu_x\left(\frac{u_x}{2l}\right) = \frac{mu_x^2}{l}$$

According to Newton's second law of motion

$$\begin{aligned}\text{Force} &= \text{mass} \times \text{acceleration}\\ &= \text{mass} \times \frac{\mathrm{d(velocity)}}{\mathrm{d}t} = \frac{\mathrm{d}}{\mathrm{d}t}(\text{mass} \times \text{velocity})\\ &= \frac{\mathrm{d}}{\mathrm{d}t}(\text{momentum}) = \text{rate of change of momentum}\end{aligned}$$

Hence, total force due to impacts of a single molecule with the wall ABCD of the vessel is mu_x^2/l.

The area of the wall is l^2. Hence, the pressure exerted due to the collision of x-component velocity of a single molecule with the side ABCD is

$$p_x = \frac{mu_x^2/l}{l^2} = \frac{mu_x^2}{V} \tag{1.6.2}$$

where V is the volume of the vessel.

Since each molecule will exert similar pressure, the total pressure exerted on the wall ABCD will be

$$p = \sum_{i=1}^{N} p_{ix} = \frac{m}{V}\sum_{i=1}^{N} u_{ix}^2 \tag{1.6.3}$$

Defining the mean square speed as

$$\overline{u_x^2} = \frac{1}{N}\sum_{i=1}^{N} u_{ix}^2 \tag{1.6.4}$$

we can write Eq. (1.6.3) as

$$p = \frac{mN}{V}\overline{u_x^2} \tag{1.6.5}$$

Since the directions x, y and z are equivalent, we will also have

$$\overline{u_x^2} = \overline{u_y^2} = \overline{u_z^2} \tag{1.6.6}$$

But from Eq. (1.6.1), we will have

$$\overline{u^2} = \overline{u_x^2} + \overline{u_y^2} + \overline{u_z^2} \tag{1.6.7}$$

From Eqs (1.6.6) and (1.6.7), we can write

$$\overline{u_x^2} = \overline{u_y^2} = \overline{u_z^2} = \frac{1}{3}\overline{u^2} \tag{1.6.8}$$

Substituting this in Eq. (1.6.5), we get

$$p = \frac{mN}{V}\left(\frac{1}{3}\overline{u^2}\right) \quad \text{or} \quad pV = \frac{1}{3}mN\overline{u^2} \tag{1.6.9}$$

Example 1.6.1 Calculate the pressure exerted by 10^{23} gas particles each of mass 10^{-22} g in a container of volume 1 dm^3. The root mean square speed is 10^5 cm s^{-1}.

Solution From the given data, we have

$$N = 10^{23}; \quad m = 10^{-22}\text{ g} = 10^{-25}\text{ kg}; \quad V = 1\text{ dm}^3 = 10^{-3}\text{ m}^3$$

$$\sqrt{\overline{u^2}} = 10^5\text{ cm s}^{-1} = 10^3\text{ m s}^{-1}$$

Therefore, from the kinetic gas equation

$$p = \frac{1}{3}\frac{mN\overline{u^2}}{V}$$

we have $$p = \frac{1}{3}\frac{(10^{-25}\text{ kg})(10^{23})(10^3\text{ m s}^{-1})^2}{(10^{-3}\text{ m}^3)}$$

$$= \frac{1}{3}(10^7)\text{ kg m}^{-1}\text{ s}^{-2} = \frac{1}{3}\times 10^7\text{ Pa}$$

1.7 SOME DERIVATIONS FROM THE KINETIC GAS EQUATION

Kinetic Gas Equation Involving Kelvin Temperature

The kinetic gas equation (1.6.9) can be used to derive the various gaseous laws and to define expressions for some useful quantities such as the root mean square speed and the average kinetic energy. Before deriving these, it is helpful to write this equation in the following form:

One of the postulates of the kinetic theory of gases is

$$\text{Average kinetic energy} \propto T$$

i.e. $$\frac{1}{2}m\overline{u^2} \propto T \quad \text{or} \quad \frac{1}{2}m\overline{u^2} = KT$$

where K is the proportionality constant. Introducing this in Eq. (1.6.9) we have

$$pV = \frac{1}{3}mN\overline{u^2} = \frac{2}{3}N\left(\frac{1}{2}m\overline{u^2}\right) = \frac{2}{3}NKT \tag{1.7.1}$$

Now, we proceed to derive the various gaseous laws from Eq. (1.7.1).

Boyle's Law

The essential conditions for Boyle's law to be applicable are:

(i) Temperature (T) should remain constant.
(ii) Mass of the gas should remain constant. In other words, the total number of molecules (N) remains unchanged.

Under these conditions, Eq. (1.7.1) yields

$$pV = \text{constant} \qquad \text{or} \qquad p \propto \frac{1}{V}$$

which is the expression for Boyle's law.

Charles Law

In this case:

(i) Pressure (p) remains fixed.
(ii) Mass of the gas remains unchanged, i.e. N is constant.

With these conditions, Eq. (1.7.1) yields

$$V = \left(\frac{2}{3}\frac{NK}{p}\right)T \qquad \text{i.e.} \qquad V = (\text{constant})\ T \qquad \text{or} \qquad V \propto T$$

as required by Charles law.

Avogadro's Law

It states that *under similar conditions of pressure and temperature, equal volume of all gases contain equal number of molecules.* Considering two gases, we have

$$p_1V_1 = \frac{2}{3}N_1KT_1 \quad \text{and} \quad p_2V_2 = \frac{2}{3}N_2KT_2$$

Since $p_1 = p_2$ and $T_1 = T_2$, therefore

$$\frac{p_1V_1}{p_2V_2} = \frac{(2/3)N_1KT_1}{(2/3)N_2KT_2} \quad \Rightarrow \quad \frac{V_1}{V_2} = \frac{N_1}{N_2}$$

If volumes are identical, obviously $N_1 = N_2$.

Graham's Law of Effusion

The rate of diffusion or effusion can be assumed to be directly proportional to the root mean square speed (or any other average speed). Thus

$$\frac{r_1}{r_2} = \sqrt{\frac{\overline{u_1^2}}{\overline{u_2^2}}}$$

From Eq. (1.6.9), we have

$$\overline{u^2} = \frac{3pV}{mN}$$

For 1 mol of an ideal gas

$$pV = RT$$

and $\quad N = N_A \qquad$ (N_A is Avogadro constant)

With these, the above equation becomes

$$\overline{u^2} = \frac{3RT}{mN_A} = \frac{3RT}{M} \tag{1.7.2}$$

where M is the molar mass of the gas.

Thus, $$\frac{r_1}{r_2} = \sqrt{\frac{\overline{u_1^2}}{\overline{u_2^2}}} = \sqrt{\frac{3RT/M_1}{3RT/M_2}} = \sqrt{\frac{M_2}{M_1}}$$

which is Graham's law of effusion.

Root Mean Square Speed

Root mean square (rms) speed is defined as the square root of the average of the squares of speeds, i.e.

$$\sqrt{\overline{u^2}} = \sqrt{\frac{u_1^2 + u_2^2 + \cdots + u_N^2}{N}}$$

According to Eq. (1.7.2), this is given as

$$\sqrt{\overline{u^2}} = \sqrt{\frac{3RT}{M}} \tag{1.7.3}$$

or $$\sqrt{\overline{u^2}} = \sqrt{\frac{3pV_m}{M}} \tag{1.7.4}$$

Thus, rms speed is directly proportional to the square root of temperature and inversely proportional to the square root of molar mass. Hence, at a given temperature lighter molecules (say H_2, He) move faster than the heavier molecules (say O_2, N_2). There is no effect of change of pressure or volume on the rms speed since, at a given temperature, pV_m = constant.

Example 1.7.1

A bulb of capacity 1 dm³ contains 1.03 × 10²³ gaseous hydrogen molecules and the pressure exerted by these molecules is 101.325 kPa. Calculate the average square molecular speed and the temperature.

Solution

We have $V = 1\ \text{dm}^3$, $N = 1.03 \times 10^{23}$, $p = 101.325$ kPa

$$n = \frac{N}{N_A} = \frac{1.03 \times 10^{23}}{(6.022 \times 10^{23}\ \text{mol}^{-1})} = 0.171\ \text{mol}$$

$$T = \frac{pV}{nR} = \frac{(101.325\ \text{kPa})\,(1\ \text{dm}^3)}{(0.171\ \text{mol})\,(8.314\ \text{kPa dm}^3\ \text{K}^{-1}\ \text{mol}^{-1})} = 71.27\ \text{K}$$

$$\overline{u^2} = \frac{3RT}{M} = \frac{3(8.314\ \text{J K}^{-1}\ \text{mol}^{-1})\,(71.27\ \text{K})}{(2.0 \times 10^{-3}\ \text{kg mol}^{-1})}$$

$$= 8.888 \times 10^5\ \text{J kg}^{-1} \equiv 8.888 \times 10^5 (\text{m s}^{-1})^2$$

Average Kinetic Energy

The average kinetic energy $(\overline{\text{KE}})$ is defined as

$$\overline{\text{KE}} = \frac{1}{2} m\overline{u^2}$$

According to Eq. (1.7.1), this is given as

$$\overline{\text{KE}} = \frac{3}{2}\frac{pV}{N}$$

For 1 mole of an ideal gas

$$pV = RT \qquad \text{and} \qquad N = N_\text{A}$$

With these, the above equation becomes

$$\overline{\text{KE}} = \frac{3}{2}\frac{RT}{N_\text{A}} = \frac{3}{2}kT \tag{1.7.5}$$

where $k = R/N_\text{A}$ and is known as the Boltzmann constant. Its value is given by

$$k = \frac{R}{N_\text{A}} = \frac{8.314 \text{ J K}^{-1} \text{ mol}^{-1}}{6.022 \times 10^{23} \text{ mol}^{-1}} = 1.380\,6 \times 10^{-23} \text{ J K}^{-1}$$

The total kinetic energy for 1 mole of the gas is

$$E_\text{total} = N_\text{A}(\overline{\text{KE}}) = \frac{3}{2}RT \tag{1.7.6}$$

Example 1.7.2 For a gas containing 10^{23} molecules (each having mass 10^{-22} g) in a volume of 1 dm^3, calculate the total kinetic energy of molecules if their root mean square speed is 10^5 cm s^{-1}. What will be its temperature?

Solution Total kinetic energy

$$= N\left(\frac{1}{2}m\overline{u^2}\right) = (10^{23})\left\{\frac{1}{2}(10^{-25} \text{ kg})(10^3 \text{ m s}^{-1})^2\right\}$$

$$= 0.5 \times 10^4 \text{ kg m}^2 \text{ s}^{-2} = 0.5 \times 10^4 \text{ J}$$

Total kinetic energy is also equal to $N(3/2)kT$. Thus

$$N\left(\frac{3}{2}\right)kT = 0.5 \times 10^4 \text{ J}$$

Hence $$T = \frac{2}{3}\frac{(0.5 \times 10^4 \text{ J})}{Nk} = \frac{2}{3}\frac{(0.5 \times 10^4 \text{ J})}{(10^{23})(1.380\,6 \times 10^{-23} \text{ J K}^{-1})}$$

$$= 2\,414 \text{ K}$$

Example 1.7.3 Calculate the total kinetic energy of 0.5 mol of an ideal gas at 273 K.

Solution Total Kinetic energy

$$= n\left(\frac{3}{2}RT\right) = (0.5 \text{ mol})\left\{\frac{3}{2}(8.314 \text{ J K}^{-1} \text{ mol}^{-1})(273 \text{ K})\right\} = 1\,702 \text{ J}$$

1.8 REAL GASES

Deviation from Ideal Behaviour Real gases do not obey the ideal gas laws exactly under all conditions of temperature and pressure. Experiments show that at low pressure and moderately high temperatures, gases obey the laws of Boyle, Charles and Avogadro approximately, but as the pressure is increased or the temperature is decreased, a marked departure from ideal behaviour is observed. Figure 1.8.1 shows, for example, the type of deviation that occurs in Boyle's law for H_2 at room temperature.

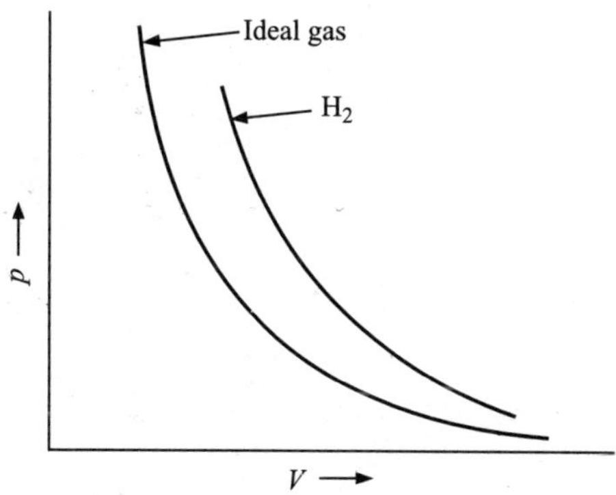

Fig. 1.8.1 Plot of *p* versus *V* of hydrogen, as compared to that of an ideal gas

The curve for the real gas has a tendency to coincide with that of an ideal gas at low pressures when the volume is large. At higher pressures, however, deviations are observed.

Compression Factor

The deviations can be displayed more clearly, by plotting the ratio of the observed molar volume V_m to the ideal molar volume $V_{m,ideal}$ (= RT/p) as a function of pressure at constant temperature. This ratio is called the compression factor Z and can be expressed as

$$Z = \frac{V_m}{V_{m,ideal}} = \frac{p}{RT}V_m \tag{1.8.1}$$

Example 1.8.1

At 273.15 K and under a pressure of 10.132 5 MPa, the compression factor of O_2 is 0.927. Calculate the mass of O_2 necessary to fill a gas cylinder of 100 dm^3 capacity under the given conditions.

Solution

From the given data, we have

$$T = 273.15\text{ K},\quad Z = 0.927,\quad p = 10.132\ 5\text{ MPa}$$

Thus, the molar volume of O_2 is

$$V_m = \frac{ZRT}{p} = \frac{(0.927)(8.314\text{ MPa cm}^3\text{ K}^{-1}\text{ mol}^{-1})(273.15\text{ K})}{10.132\ 5\text{ MPa}}$$

$$= 207.77\text{ cm}^3\text{ mol}^{-1}$$

The mass of this molar volume will be equal to the molar mass of oxygen, i.e. 207.77 cm^3 weighs 0.032 kg. Thus, the mass of oxygen required to fill a gas cylinder of 100 dm^3 (i.e. 10^5 cm^3) under the given condition is

$$\left(\frac{0.032\text{ kg}}{207.77\text{ cm}^3}\right)(10^5\text{ cm}^3) = 15.40\text{ kg}$$

Example 1.8.2 The compression factor ($Z = pV/nRT$) for N_2 at 223 K and 81.06 MPa is 1.95 and at 373 K and 20.265 MPa it is 1.10. A certain mass of N_2 occupies a volume of 1.0 dm^3 at 223 K and 81.06 MPa. Calculate the amount of the gas and the volume occupied by the same quantity of N_2 at 373 K and 20.265 MPa.

Solution For T = 223 K, p = 81.06 MPa, Z = 1.95 and V = 1.0 dm^3 = 10^3 cm^3, we have

$$n = \frac{pV}{ZRT} = \frac{(81.06\ \text{MPa})(10^3\ \text{cm}^3)}{(1.95)(8.314\ \text{MPa cm}^3\ \text{K}^{-1}\ \text{mol}^{-1})(223\ \text{K})} = 22.42\ \text{mol}$$

Now at T = 373 K, p = 20.265 MPa, Z = 1.10, the volume occupied would be

$$V = \frac{ZnRT}{p} = \frac{(1.10)(22.42\ \text{mol})(8.314\ \text{MPa cm}^3\ \text{K}^{-1}\ \text{mol}^{-1})(373\ \text{K})}{(20.265\ \text{MPa})}$$

$$= 3\,774.0\ \text{cm}^3 = 3.774\ \text{dm}^3$$

Plots of Compression Factor versus Pressure

For an ideal gas, Z = 1 and is independent of pressure and temperature. For a real gas, $Z = f(T, p)$, a function of both temperature and pressure. Figure 1.8.2 shows a graph between Z and p for some gases at 273.15 K, the pressure range in this graph is very large. It can be noted that:

(1) Z is always greater than 1 for H_2.

(2) For N_2, $Z < 1$ in the lower pressure range and is greater than 1 at higher pressures. It decreases with increase of pressure in the lower pressure region, passes through a minimum at some pressure and then increases continuously with pressure in the higher pressure region.

(3) For CO_2, there is a large dip in the beginning. In fact, for gases which are easily liquefied, Z dips sharply below the ideal line in the low pressure region.

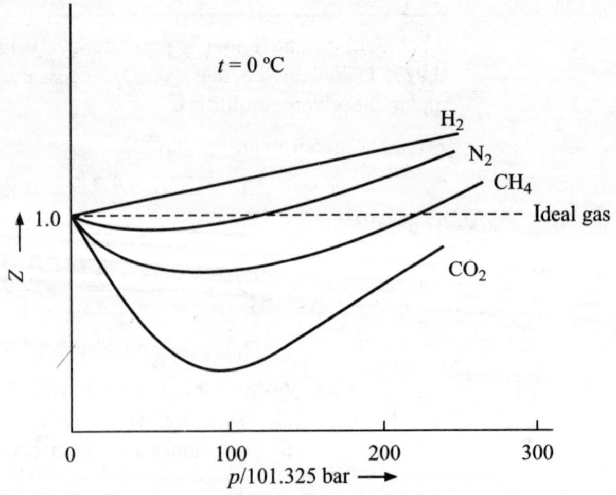

Fig. 1.8.2 Plots of *Z* versus *p* of a few gases

Figure 1.8.2 gives an impression that the nature of deviations depend upon the nature of the gas. In fact, it is not so. The determining factor is the temperature relative to the critical temperature (*see* p. 36) of the particular gas; near the critical temperature, the pV curves are like those for CO_2, but when far away, the curves are like those for H_2 (Fig. 1.8.3).

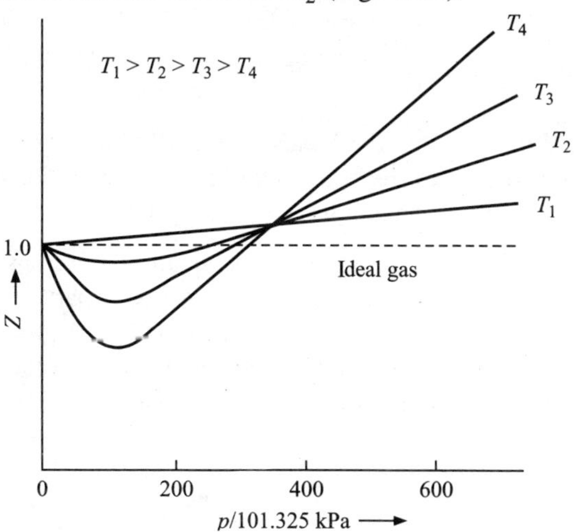

Fig. 1.8.3 Plots of Z versus p of a single gas at various temperatures

Provided the pressure is of the order of 1 bar or less, and the temperature is not too near the point of liquefaction, the observed deviations from the ideal gas laws are not more than a few per cent. Under these conditions, therefore, the equation $pV = nRT$ and related expressions may be used.

1.9 VAN DER WAALS EQUATION OF STATE FOR A REAL GAS

Causes of Deviations from Ideal Behaviour

The ideal gas laws can be derived from the kinetic theory of gases which is based on the following two important assumptions:

(i) The volume occupied by the molecules is negligible in comparison to the total volume of the gas.

(ii) The molecules exert no forces of attraction upon one another.

It is because neither of these assumptions can be regarded as applicable to real gases that the latter show departure from the ideal behaviour.

Evidence for Molecular Volume

The molecules of a gas, however, do occupy a certain volume as can be seen from the fact that gases can be liquefied and solidified at low temperatures and high pressures. On decreasing the temperature of a gas, the thermal energy of molecules is decreased and the effect of applying high pressure is to bring the molecules closer to one another, thereby increasing the forces of attraction amongst them. Both these factors favour liquefaction and solidification. In the solid state, however, there is a considerable resistance to any further attempt at compression. It is, therefore, apparent that the molecules of a gas must have an appreciable volume, which is probably of the same order as that occupied by the same number of molecules in the solid state.

Evidence for Molecular Attractions

The molecules in gases also have weak forces of attraction (called van der Waals attraction) amongst themselves, as otherwise, the gases could never be liquefied and solidified. This is also supported by the fact that when a compressed gas is passed through a porous plug of silk or cotton in adiabatic condition, the emerging gas is found to be cooler than the entering gas provided the temperature of the gas is less than its inversion temperature (Joule-Thomson effect).† This is because on expansion, some work has to be done against the internal forces of attraction, which requires energy. This energy comes from the system itself.

Derivation of Van der Waals Equation

Van der Waals was the first to introduce systematically the correction terms due to the above two invalid assumptions in the ideal gas equation $p_iV_i = nRT$. His corrections are given below.

Correction for Volume

V_i in the ideal gas equation represents an ideal volume where the molecules can move freely. In real gases, a part of the total volume is, however, occupied by the molecules of the gas. Hence, the free volume V_i is the total volume V minus the volume occupied by the molecules. If b represents the *effective volume* occupied by the molecules of 1 mole of a gas, then for the amount n of the gas V_i is given by

$$V_i = V - nb \tag{1.9.1}$$

where b is called the *excluded volume* or *co-volume*. The numerical value of b is four times the actual volume occupied by the gas molecules. This can be shown as follows.

Expression of Excluded Volume

If we consider only bimolecular collisions, then the volume occupied by the sphere of radius $2r$ represents the excluded volume per pair of molecules as shown in Fig. 1.9.1.

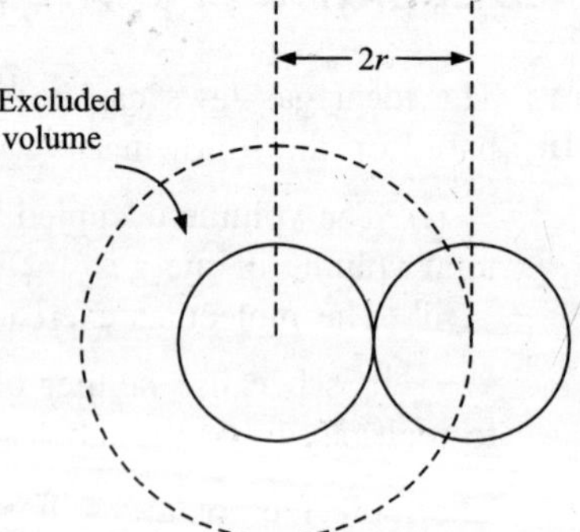

Fig. 1.9.1 Excluded volume per pair of molecules

Thus, excluded volume per pair of molecules

$$= \frac{4}{3}\pi(2r)^3 = 8\left(\frac{4}{3}\pi r^3\right)$$

† See Section 2.9 of Volume 2 of this series of book

Excluded volume per molecule

$$= \frac{1}{2}\left[8\left(\frac{4}{3}\pi r^3\right)\right] = 4\left(\frac{4}{3}\pi r^3\right) = 4 \text{ (volume occupied by a molecule)}$$

Since b represents excluded volume per mole of the gas, it is obvious that

$$b = N_A\left[4\left(\frac{4}{3}\pi r^3\right)\right] \qquad (1.9.2)$$

Correction for Forces of Attraction

Consider a molecule A in the bulk of a vessel as shown in Fig. 1.9.2. This molecule is surrounded by other molecules in a symmetrical manner, with the result that this molecule on the whole experiences no net force of attraction.

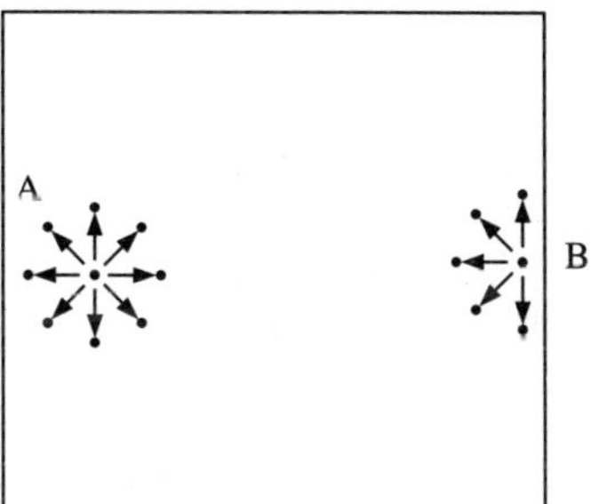

Fig. 1.9.2 Arrangement of molecules within and near the surface of a vessel

Now, consider a molecule B near the side of the vessel, which is about to strike one of its sides, thus contributing towards the total pressure of the gas. There are molecules only on one side of the vessel, i.e. towards its centre, with the result that this molecule experiences a net force of attraction towards the centre of the vessel. This results in decreasing the velocity of the molecule, and hence its momentum. Thus, the molecule does not contribute as much force as it would have, had there been no forces of attraction. Thus, the pressure of a real gas would be smaller than the corresponding pressure of an ideal gas, i.e.

$$p_i = p + \text{correction term} \qquad (1.9.3)$$

This correction term depends upon two factors:

(i) *The number of molecules per unit volume of the vessel* Larger the number, larger the net force of attraction with which the molecule B is dragged behind. This results in a greater decrease in the velocity of the molecule B and hence a greater decrease in the rate of change of momentum. Consequently, the correction term also has a large value. If n is the amount of the gas present in the volume V of the container, the number of molecules per unit volume of the container is given as

$$N' = \frac{nN_A}{V} \quad \text{or} \quad N' \propto \frac{n}{V}$$

Thus, the correction term is given as:

$$\text{Correction term} \propto \frac{n}{V} \qquad (1.9.4a)$$

(ii) *The number of molecules striking the side of the vessel per unit time* Larger this number, larger the decrease in the rate of change of momentum. Consequently, the correction term also has a larger value. Now, the number of molecules striking the side of vessel in a unit time also depends upon the number of molecules present in unit volume of the container, and hence in the present case:

$$\text{Correction term} \propto \frac{n}{V} \tag{1.9.4b}$$

Taking both these factors together, we have

$$\text{Correction term} \propto \left(\frac{n}{V}\right)\left(\frac{n}{V}\right)$$

or $$\text{Correction term} = a\frac{n^2}{V^2} \tag{1.9.5}$$

where a is the proportionality constant and is a measure of the forces of attraction between the molecules. Thus

$$p_\mathrm{i} = p + a\frac{n^2}{V^2} \tag{1.9.6}$$

The unit of the term an^2/V^2 is the same as that of the pressure. Thus, the SI unit of a is Pa m^6 mol^{-2}. It may be conveniently expressed in kPa dm^6 mol^{-2}.

Expression of Van der Waals Equation of State

When the expressions as given by Eqs (1.9.1) and (1.9.6) are substituted in the ideal gas equation $p_\mathrm{i}V_\mathrm{i} = nRT$, we get

$$\left(p + \frac{n^2a}{V^2}\right)(V - nb) = nRT \tag{1.9.7}$$

This equation is applicable to real gases and is known as the *van der Waals equation.*

Values of Van der Waals Constants

The constants a and b in van der Waals equation are called van der Waals constants and their values depend upon the nature of the gas (Table 1.9.1). They

Table 1.9.1 Van der Waals Constants

Gas	a / kPa dm^6 mol^{-2}	b / dm^3 mol^{-1}	*Gas*	a / kPa dm^6 mol^{-2}	b / dm^3 mol^{-1}
H_2	21.764	0.026 61	CH_4	228.285	0.042 78
He	3.457	0.023 70	C_2H_6	556.173	0.063 80
N_2	140.842	0.039 13	C_3H_8	877.880	0.084 45
O_2	137.802	0.031 83	C_4H_{10}(n)	1 466.173	0.122 6
Cl_2	657.903	0.056 22	C_4H_{10}(iso)	1 304.053	0.114 2
NO	135.776	0.027 89	C_5H_{12}(n)	1 926.188	0.146 0
NO_2	535.401	0.044 24	CO	150.468	0.039 85
H_2O	553.639	0.030 49	CO_2	363.959	0.042 67

are characteristics of the gas. The values of these constants are determined by the critical constants of the gas. Actually, the so-called constants vary to some extent with temperature and this shows that the van der Waals equation is not a complete solution of the behaviour of real gases.

Example 1.9.1 Calculate the pressure exerted by 22 g of carbon dioxide in 0.5 dm^3 at 298.15 K using: (a) the ideal gas law and (b) van der Waals equation. Given:

$$a = 363.76 \text{ kPa dm}^6 \text{ mol}^{-2} \quad \text{and} \quad b = 42.67 \text{ cm}^3 \text{ mol}^{-1}$$

Solution

$$\text{Amount of } CO_2 = \frac{22 \text{ g}}{44 \text{ g mol}^{-1}} = 0.5 \text{ mol}$$

$$V = 0.5 \text{ dm}^3 \qquad T = 298.15 \text{ K}$$

$$a = 363.76 \text{ kPa dm}^6 \text{ mol}^{-2} \qquad b = 42.67 \text{ cm}^3 \text{ mol}^{-1} = 0.042\ 67 \text{ dm}^3 \text{ mol}^{-1}$$

(a) From the ideal gas law, $p = nRT/V$, we have

$$p = \frac{(0.5 \text{ mol})(8.314 \text{ kPa dm}^3 \text{ K}^{-1} \text{ mol}^{-1})(298.15 \text{ K})}{(0.5 \text{ dm}^3)} = 2.479 \times 10^3 \text{ kPa}$$

(b) From the van der Waals equation, $p = \dfrac{nRT}{V - nb} - \dfrac{n^2 a}{V^2}$, we have

$$p = \frac{(0.5 \text{ mol})(8.314 \text{ kPa dm}^3 \text{ K}^{-1} \text{ mol}^{-1})(298.15 \text{ K})}{0.5 \text{ dm}^3 - (0.5 \text{ mol})(0.042\,67 \text{ dm}^3 \text{ mol}^{-1})} - \frac{(0.5 \text{ mol})^2 (363.76 \text{ kPa dm}^6 \text{ mol}^{-2})}{(0.5 \text{ dm}^3)^2}$$

$$= 2\ 589.31 \text{ kPa} - 363.76 \text{ kPa} = 2\ 225.55 \text{ kPa}$$

Example 1.9.2 Two van der Waals gases have the same value of b but different a values. Which of these would occupy greater volume under identical conditions? If the gases have the same a value but different values of b which would be more compressible?

Solution If two gases have same value of b but different values of a, then the gas having a larger value a will occupy lesser volume. This is because the gas with a larger value of a will have a larger force of attraction, and hence lesser distance between its molecules.

If two gases have the same value of a but different values of b, then the smaller the value of b, larger will be the compressibility because the gas with the smaller value of b will occupy lesser volume and hence will be more compressible.

Example 1.9.3 Calculate molecular diameter d of helium from its van der Waals constant b (b = 24 cm^3 mol^{-1}).

Solution

Since $b = 4 \times$ volume occupied by the molecules in 1 mole of a gas

or

$$b = 4N_A\left(\frac{4}{3}\pi r^3\right)$$

therefore

$$r = \left(\frac{3b}{16N_A\pi}\right)^{1/3} = \left\{\frac{3 \times 24 \text{ cm}^3 \text{ mol}^{-1}}{16(6.022 \times 10^{23} \text{ mol}^{-1})(3.14)}\right\}^{1/3}$$

$$= 1.335 \times 10^{-8} \text{ cm} = 133.5 \text{ pm}$$

$$d = 2r = 267 \text{ pm}$$

Example 1.9.4 The molar volume of helium at 10.132 5 MPa and 273 K is 0.011 075 of its molar volume at 101.325 kPa at 273 K. Calculate the radius of helium atom. The value of a may be neglected.

Solution The van der Waals equation after neglecting a reduces to

$$p(V_m - b) = RT$$

Substituting the given data, we have

at 101.325 kPa:

$$(101.325\ \text{kPa})\ (V_m - b) = (8.314\ \text{kPa dm}^3\ \text{K}^{-1}\ \text{mol}^{-1})\ (273\ \text{K})$$

i.e. $$V_m - b = 22.4\ \text{dm}^3\ \text{mol}^{-1} = 22.4 \times 10^3\ \text{cm}^3\ \text{mol}^{-1} \qquad (1)$$

at 10.132 5 MPa:

$$(10.132\ 5\ \text{MPa})\ (0.011\ 075\ V_m - b) = (8.314\ \text{MPa cm}^3\ \text{K}^{-1}\ \text{mol}^{-1})\ (273\ \text{K})$$

i.e. $$0.011\ 075\ V_m - b = 224.00\ \text{cm}^3\ \text{mol}^{-1} \qquad (2)$$

Multiplying Eq. (1) by 0.011 075 and then subtracting Eq. (2) from it, we get

$$b - 0.011\ 075\ b = (248.08 - 224)\ \text{cm}^3\ \text{mol}^{-1}$$

$$b = \frac{24.08\ \text{cm}^3\ \text{mol}^{-1}}{0.988\ 925} = 24.35\ \text{cm}^3\ \text{mol}^{-1}$$

Since $$b = \left(\frac{4}{3}\pi r^3\right)(4N_A)$$

we have $$r = \left(\frac{3b}{16\pi N_A}\right)^{1/3} = \left(\frac{3 \times 24.35\ \text{cm}^3\ \text{mol}^{-1}}{16 \times 3.14 \times 6.022 \times 10^{23}\ \text{mol}^{-1}}\right)^{1/3}$$

$$= 1.34 \times 10^{-8}\ \text{cm} = 134\ \text{pm}$$

Applicability of the van der Waals Equation

Since the van der Waals equation is applicable to real gases, it is worth considering how far this equation can explain the experimental behaviour of real gases, as represented by Fig. 1.8.2[†]. The van der Waals equation for 1 mole of a gas is

$$\left(p + \frac{a}{V_m^2}\right)(V_m - b) = RT \qquad (1.9.8)$$

At low pressure When pressure is low, the volume is sufficiently large and b can be ignored in comparison to V_m in Eq. (1.9.8). Thus, we have

$$\left(p + \frac{a}{V_m^2}\right)V_m = RT \quad \text{or} \quad pV_m + \frac{a}{V_m} = RT$$

or $$Z = 1 - \frac{a}{V_m RT} \qquad (1.9.9)$$

From the above equation it is clear that in the low pressure region, Z is less than 1. On increasing the pressure in this region, the value of the term $(a/V_m RT)$ increases as V is inversely proportional to p. Consequently, Z decreases with increase of p.

[†]See Annexure I at the end of this chapter for quantitative explanation of Fig. 1.8.2 by van der Waals, Berthelot, Dieterici and Redlich-Kwong equations of state.

At high pressure When p is large, V_m will be small and one cannot ignore b in comparison to V_m. However, the term a/V_m^2 may be considered negligible in comparison to p in Eq. (1.9.8). Thus,

$$p(V_m - b) = RT$$

or $$Z = 1 + \frac{pb}{RT} \qquad (1.9.10)$$

Here Z is greater than 1 and it increases linearly with pressure. This explains the nature of the graph in the high pressure region.

At high temperature and low pressure If temperature is high, V_m will also be sufficiently large and thus the term a/V_m^2 will be negligibly small. At this stage, b may also be negligible in comparison to V_m. Under these conditions, Eq. (1.9.8) reduces to an ideal gas equation of state:

$$pV_m = RT$$

Hydrogen and helium The value of a is extremely small for these gases as they are difficult to liquefy. Thus, we have the equation of state as $p(V_m - b) = RT$, obtained from the van der Waals equation by ignoring the term a/V_m^2. Hence, Z is always greater than 1 and it increases with increase of p.

The van der Waals equation is a distinct improvement over the ideal gas law. It gives qualitative reasons for the deviations from ideal behaviour. However, the generality of the equation is lost as it contains two constants, the values of which depend upon the nature of the gas.

1.10 OTHER EQUATIONS OF STATE FOR REAL GASES

Berthelot's Equation The van der Waals equation is one of the many equations of state suggested in order to account for the behaviour of real gases. There are two other simple equations of state which involve just two arbitrary constants. The first of these, due to Berthelot, is

$$\left(p + \frac{n^2 a}{TV^2}\right)(V - nb) = nRT \qquad (1.10.1)$$

where a and b are constants called the *Berthelot's constants* (different from van der Waals constants) and are characteristics of the gas.

Dieterici's Equation The second equation, due to Dieterici, is

$$\{p \exp(na/VRT)\}\ (V - nb) = nRT \qquad (1.10.2)$$

Redlich-Kwong Equation The third equation, due to Redlich-Kwong, is

$$\left(p + \frac{n^2 a}{T^{1/2}\, V(V + b)}\right)(V - nb) = nRT$$

where a and b are constants called the Redlich-Kwong constants.

Virial Equation

All these three equations of state can be expressed approximately in one common form, called the *virial equation of state*, which has the following form for 1 mole of a gas

$$Z = \frac{pV_m}{RT} = 1 + B\frac{1}{V_m} + C\frac{1}{V_m^2} + D\frac{1}{V_m^3} + \cdots \qquad (1.10.3)$$

where B, C, . . . are temperature dependent constants known as second, third, etc., virial coefficients. These coefficients must be evaluated experimentally at each different temperature.

The second virial coefficient B may be obtained from the experimental data. Rearranging the virial equation, we get

$$V_m\left(\frac{pV_m}{RT} - 1\right) = B + \frac{C}{V_m} + \cdots$$

Thus, extrapolating the graph between $V_m\{(pV_m/RT) - 1\}$ versus $1/V_m$ to $1/V_m = 0$ gives the value of B, i.e.

$$B = \lim_{V_m \to \infty} V_m\left[\frac{pV_m}{RT} - 1\right] \qquad (1.10.4)$$

The third virial coefficient C would be the slope of this plot if there were no higher terms in Eq. (1.10.3). These further terms cause the plot to be curved so that C must be evaluated from the initial slope.

Physical Significance of the Constant *B*

The second virial coefficient B has the unit of volume and may be considered to be an excluded molar volume as can be shown by using statistical mechanics. The term B can be expressed in terms of intermolecular attraction by the equation

$$B = 2\pi N_A \int_0^\infty \{1 - \exp(-\Phi/kT)\}r^2\, dr \qquad (1.10.5)$$

where r is the intermolecular distance, k is the Boltzmann constant and Φ is the potential of molecular interaction. As the simplest example of the use of this equation, we consider a gas made up of rigid spherical molecules of diameter d. We assume that the molecules do not interact unless they touch one another and thus $\Phi = 0$ if $r > d$. The molecules cannot penetrate one another as they are rigid; thus $\Phi = \infty$ if $r < d$. Therefore, we have

$$B = 2\pi N_A \int_0^d r^2\, dr = \frac{2}{3}\pi N_A d^3 = 4N_A\left(\frac{4}{3}\pi r^3\right) \qquad (1.10.6)$$

Hence B is the product of Avogadro constant and the volume excluded per molecule.

In general, the numerical values of the virial coefficients decrease very sharply with higher powers of volume.

Problem 1.10.1

Show that at low densities, the van der Waals equation

$$\left(p + \frac{a}{V_m^2}\right)(V_m - b) = RT$$

and the Dieterici's equation

$$p(V_m - b) = RT \exp(-a/RTV_m)$$

give essentially the same value of p.

Solution

At low densities, volume of the gas is large, therefore b may be ignored in comparison to V_m. Moreover, the term a/RTV_m will have small value and thus the term $\exp(-a/RTV_m)$ can be expanded as

$$\exp(-a/RTV_m) \simeq 1 - \frac{a}{RTV_m}$$

Thus, under these approximations we can write van der Waals equation as

$$\left(p + \frac{a}{V_m^2}\right)(V_m) = RT \quad \text{or} \quad p = \frac{RT}{V_m} - \frac{a}{V_m^2}$$

and Dieterici's equation as

$$p = \frac{RT}{V_m}\left(1 - \frac{a}{RTV_m}\right) = \frac{RT}{V_m} - \frac{a}{V_m^2}$$

Thus, we see that both van der Waals equation and Dieterici's equation reduce to the same expression of p at low densities.

1.11 REDUCTION OF VAN DER WAALS EQUATION TO VIRIAL EQUATION

Virial Equation in Volume

The van der Waals equation of state for 1 mole of a gas is

$$\left(p + \frac{a}{V_m^2}\right)(V_m - b) = RT \quad \text{or} \quad p = \frac{RT}{V_m - b} - \frac{a}{V_m^2}$$

Multiplying both sides by V_m/RT, we get

$$\frac{pV_m}{RT} = \frac{V_m}{V_m - b} - \frac{a}{V_m RT} \quad \text{or} \quad Z = \left(1 - \frac{b}{V_m}\right)^{-1} - \frac{a}{V_m RT}$$

In the low pressure region, V_m is large and $b/V_m << 1$. Thus, the expression $(1 - b/V_m)^{-1}$ can be expanded into a power series in b/V_m:

$$\left(1 - \frac{b}{V_m}\right)^{-1} = 1 + \frac{b}{V_m} + \left(\frac{b}{V_m}\right)^2 + \left(\frac{b}{V_m}\right)^3 + \cdots$$

Substituting this in the expression for Z, we get

$$Z = 1 + \left(b - \frac{a}{RT}\right)\frac{1}{V_m} + \left(\frac{b}{V_m}\right)^2 + \cdots \qquad (1.11.1)$$

Thus for the second virial coefficient, we have

$$B = b - \frac{a}{RT}$$

Third virial coefficient $C = b^2$, and so on.

Virial Equation in Pressure

An alternative form of the virial equation of state involves the expression of Z in terms of a power series in p, i.e.

$$Z = 1 + A_1 p + A_2 p^2 + \cdots \tag{1.11.2}$$

The expressions for A_1 and A_2 can be derived as follows:

$$Z = 1 + \left(b - \frac{a}{RT}\right)\frac{1}{V_m} + \left(\frac{b}{V_m}\right)^2 + \cdots$$

Since $Z = pV_m/RT$, therefore, $1/V_m = p/RTZ$. Hence

$$Z = 1 + \left(b - \frac{a}{RT}\right)\frac{p}{RTZ} + b^2\left(\frac{p}{RTZ}\right)^2 + \cdots \tag{1.11.3}$$

Comparing Eqs (1.11.2) and (1.11.3), we get

$$1 + A_1 p + A_2 p^2 + \cdots = 1 + \left(b - \frac{a}{RT}\right)\frac{p}{RTZ} + \left(\frac{b}{RTZ}\right)^2 p^2 + \cdots$$

or

$$A_1 p + A_2 p^2 + \cdots = \frac{1}{RTZ}\left(b - \frac{a}{RT}\right)p + \left(\frac{b}{RT}\right)^2 \frac{p^2}{Z^2} + \cdots$$

Dividing by p, we get

$$A_1 + A_2 p + \cdots = \frac{1}{RTZ}\left(b - \frac{a}{RT}\right) + \left(\frac{b}{RT}\right)^2 \frac{p}{Z^2} + \cdots$$

In the limiting state of zero pressure, $Z = 1$ and this equation becomes

$$A_1 = \frac{1}{RT}\left(b - \frac{a}{RT}\right) \tag{1.11.4}$$

which is the required expression for A_1. Thus

$$A_1 + A_2 p + \cdots = A_1\left(\frac{1}{Z}\right) + \left(\frac{b}{RT}\right)^2 \frac{p}{Z^2} + \cdots$$

We repeat the procedure by subtracting A_1 from both sides of this equation, dividing by p and taking the limiting value at zero pressure. Note that $(Z - 1)/p = A_1$ at zero pressure. Then

$$A_2 = \left(\frac{b}{RT}\right)^2 - A_1^2 = \frac{a}{(RT)^3}\left(2b - \frac{a}{RT}\right) \tag{1.11.5}$$

Thus, the expression for Z correct up to the third coefficient is

$$Z = 1 + \frac{1}{RT}\left(b - \frac{a}{RT}\right)p + \frac{a}{(RT)^3}\left(2b - \frac{a}{RT}\right)p^2 + \cdots \qquad (1.11.6)$$

The correct coefficient for p could have been obtained by simply replacing $1/V_m$ in Eq. (1.11.1) by the ideal value; however, this would yield incorrect values of the coefficients of higher powers of pressures.

The slope of Z versus p curve is obtained by differentiating the above virial equation in Z with respect to pressure, keeping the temperature constant, i.e.

$$\left(\frac{\partial Z}{\partial p}\right)_T = \frac{1}{RT}\left(b - \frac{a}{RT}\right) + \frac{2a}{(RT)^3}\left(2b - \frac{a}{RT}\right)p + \cdots \qquad (1.11.7)$$

At $p = 0$, all higher terms drop out and this derivative simply reduces to

$$\left(\frac{\partial Z}{\partial p}\right)_T = \frac{1}{RT}\left(b - \frac{a}{RT}\right); \qquad (p = 0) \qquad (1.11.8)$$

Comment on the Plots of Compression Factor versus Pressure

The derivative in Eq. (1.11.8) is the initial slope of the plot of Z versus p (Fig. 1.8.2). Now if $b > a/RT$, the initial slope is positive and the size effect (i.e. b factor) will dominate the behaviour of the gas. However, if $b < a/RT$, the initial slope is negative and the effect of the attractive forces (i.e. a factor) will dominate. Thus, the van der Waals equation, which includes both the effects of size and of intermolecular forces, can interpret both the positive and negative slopes of the Z versus p plots. In interpreting Fig. 1.8.2, we can say that at 0 °C, the effect of attractive forces dominate the behaviour of methane and carbon dioxide, while the molecular size effect dominates the behaviour of hydrogen.

While interpreting Fig. 1.8.3 (graph of Z versus p of the same gas at different temperatures), we can say that if the temperature is low enough, the term a/RT will be larger than b and so the initial slope of Z versus p will be negative. As the temperature rises, a/RT becomes smaller. At a sufficiently high temperature it becomes less than b, and the initial slope of Z versus p curve turns positive.

Boyle Temperature

At some intermediate temperature T_B, called *Boyle temperature*, the initial slope is zero. This is obtained from Eq. (1.11.8) by putting $b - a/RT_B = 0$, which yields

$$T_B = \frac{a}{Rb} \qquad (1.11.9)$$

At the Boyle temperature, the Z versus p line of an ideal gas is tangent to that of a real gas when p approaches zero. The latter rises above the ideal gas line only very slowly. In Eq. (1.11.6) the second term is zero at T_B and the remaining terms are small until the pressure becomes very high. Thus, at the Boyle temperature, the real gas behaves ideally over a wide range of pressure, because the effects of the size of molecules and intermolecular forces roughly compensate each other.

The Boyle temperature of some gases are given below:

$T_B(H_2) = -156\ °C$ $\qquad$ $T_B(N_2) = 59\ °C$

$T_B(He) = -249\ °C$ $\qquad$ $T_B(CH_4) = 224\ °C$

$T_B(NH_3) = 587\ °C$

Thus we can see that for H_2 and He, the temperature of 0 °C is above their respective Boyle temperatures and so they have Z values greater than unity. The other gases at 0 °C are below their respective Boyle temperatures and so they have Z values less than unity in the low pressure range.

Example 1.11.1 Given that $Z = 1.000\ 54$ at 273.15 K and 101.325 kPa pressure and the Boyle temperature of the gas is 107 K, estimate the values of a and b.

Solution We are given that

$$Z = 1.000\ 54, \quad T = 273.15\ \text{K}, \quad p = 101.325\ \text{kPa}, \quad \text{and} \quad T_B = 107\ \text{K}$$

From the expression

$$Z = 1 + \frac{p}{RTZ}\left(b - \frac{a}{RT}\right)$$

we have $\quad b = (Z-1)\dfrac{RTZ}{p} + \dfrac{a}{RT}$

At Boyle temperature, $T_B = a/Rb$ so that $a = RbT_B$. Therefore,

$$b = (Z-1)\frac{RTZ}{p} + \frac{RbT_B}{RT}$$

On rearranging, we get

$$b = \left(\frac{T}{T - T_B}\right)\left(\frac{Z-1}{p}\right)RTZ$$

Substituting the values, we have

$$b = \left(\frac{273.15\ \text{K}}{273.15\ \text{K} - 107\ \text{K}}\right)\left(\frac{1.000\ 54 - 1}{101.325\ \text{kPa}}\right)(8.314\ \text{kPa dm}^3\ \text{K}^{-1}\ \text{mol}^{-1})$$
$$\times (273.15\ \text{K})(1.000\ 54)$$
$$= 0.019\ 9\ \text{dm}^3\ \text{mol}^{-1}$$

Also $\quad a = RbT_B = (8.314\ \text{kPa dm}^3\ \text{K}^{-1}\ \text{mol}^{-1})(0.019\ 9\ \text{dm}^3\ \text{mol}^{-1})(107\ \text{K})$

$$= 17.703\ \text{kPa dm}^6\ \text{mol}^{-2}$$

Problem 1.11.1 Express Berthelot, Dieterici and Redlich-Kwong equations in the form of virial equation of state and derive the expressions for the second virial coefficient of these equations. Also derive the expressions for Boyle temperature.

Solution ***Berthelot's equation***

$$p = \frac{RT}{V_m - b} - \frac{a}{TV_m^2}$$

Multiplying by V_m/RT, we get

$$Z = \frac{pV_m}{RT} = \frac{V_m}{V_m - b} - \frac{a}{V_m RT^2} = \left(1 - \frac{b}{V_m}\right)^{-1} - \frac{a}{V_m RT^2}$$

$$= 1 + \frac{b}{V_m} + \left(\frac{b}{V_m}\right)^2 + \cdots - \frac{a}{V_m RT^2}$$

$$= 1 + \frac{1}{V_m}\left(b - \frac{a}{RT^2}\right) + \left(\frac{b}{V_m}\right)^2 + \cdots$$

The second virial coefficient and Boyle temperature are

$$B = \left(b - \frac{a}{RT^2}\right); \qquad T_B = \left(\frac{a}{Rb}\right)^{1/2}$$

Dieterici's equation

$$p = \frac{RT}{(V_m - b)} \exp(-a/V_m RT)$$

Therefore,

$$Z = \frac{pV_m}{RT} = \frac{V_m}{V_m - b} \exp(-a/V_m RT) = \left(1 - \frac{b}{V_m}\right)^{-1} \exp(-a/V_m RT)$$

$$= \left[1 + \frac{b}{V_m} + \left(\frac{b}{V_m}\right)^2 + \cdots\right]\left[1 - \frac{a}{V_m RT} + \cdots\right] = 1 + \frac{1}{V_m}\left(b - \frac{a}{RT}\right) + \cdots$$

Thus, the second virial coefficient is $(b - a/RT)$ and Boyle temperature T_B is a/Rb.

Redlich-Kwong equation

$$p = \frac{RT}{V_m - b} - \frac{a}{T^{1/2} V_m (V_m + b)}$$

Therefore,

$$Z = \frac{pV_m}{RT} = \frac{V_m}{V_m - b} - \frac{a}{RT^{3/2}(V_m + b)}$$

$$= \left(1 - \frac{b}{V_m}\right)^{-1} - \frac{a}{V_m RT^{3/2}}\left(1 + \frac{b}{V_m}\right)^{-1}$$

$$= \left[1 + \left(\frac{b}{V_m}\right) + \left(\frac{b}{V_m}\right)^2 + \cdots\right] - \frac{a}{V_m RT^{3/2}}\left[1 - \left(\frac{b}{V_m}\right) + \left(\frac{b}{V_m}\right)^2 - \cdots\right]$$

$$= 1 + \frac{1}{V_m}\left(b - \frac{a}{RT^{3/2}}\right) + \cdots$$

At Boyle temperature

$$b - \frac{a}{RT_B^{3/2}} = 0 \quad \Rightarrow \quad T_B = \left(\frac{a}{Rb}\right)^{2/3}$$

Note: The values of the constants a and b of a gas following the given equation of state are determined through their respective expressions in terms critical constants of the gas.

1.12 CRITICAL CONSTANTS

Andrews Isotherms

In 1869, Thomas Andrews carried out an experiment in which p–V relations of carbon dioxide gas were measured at various temperatures. The types of isotherms obtained are shown in Fig. 1.12.1. Other real gases also show the same types of isotherms.

We observe from Fig. 1.12.1 the following:

(1) At high temperatures, such as T_4, the isotherms look like those of an ideal gas.

(2) At low temperatures, the curves have altogether different appearances. Consider, for example, a typical curve abcd. As the pressure increases, the volume of the gas decreases (curve a to b). At point b liquefaction commences and the volume decreases rapidly as the gas is converted to a liquid with a much higher density. This conversion takes place at constant pressure p. At the point c, liquefaction is complete and thus the line cd represents the variation of V with p of the liquid state. The steepness of the line cd is evidence of the fact that the liquid cannot be easily compressed. Thus, we note that ab represents the gaseous state, bc, liquid and vapour in equilibrium, and cd shows the liquid state only.

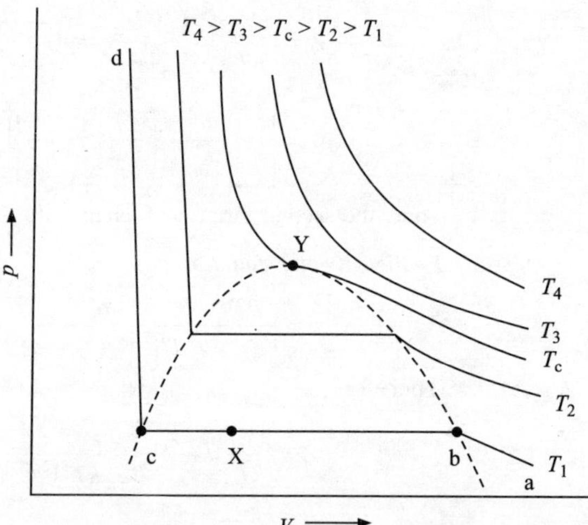

Fig. 1.12.1 Andrew's graphs of p versus V

Liquefaction commences at b and is complete at c. At a point between b and c, say X, the ratio of liquid to gas is equal to bX/cX. The pressure corresponding to the line bc is known as *vapour pressure* of the liquid.

(3) At still higher temperatures we get a similar type of curve as discussed in (2) above, except that the width of the horizontal portion is reduced; the pressure corresponding to this portion being higher than at lower temperatures.

(4) At temperature T_c the horizontal portion is reduced to a mere point. At temperatures higher than T_c there is no indication of liquefaction at all.

Thus for every gas, there is a limit of temperature above which it cannot be liquefied, no matter what the pressure is.

Definitions of Critical Constants

Critical temperature T_c is the maximum temperature at which a gas can be liquefied, i.e. the temperature above which a liquid cannot exist.

Critical pressure p_c is the minimum pressure required to cause liquefaction at the temperature T_c.

Critical volume V_c is the volume occupied by one mole of a gas at critical temperature T_c and critical pressure p_c.

Characteristics of Critical Isotherm

The point Y in Fig. 1.12.1 represents the gas in its critical state. At this point the temperature, pressure and volume have critical values. These three are known as *critical constants*. The isotherm corresponding to the temperature T_c is known as critical isotherm. To the left of the point Y on this isotherm, we have the liquid state whereas to the right, we have the gaseous state. Thus, at point Y transition from liquid to gaseous state (or vice versa) takes place and thus it is not possible to state whether the substance is in the gaseous form or in the liquid form. In fact, both the states become indistinguishable at the critical point. The surface of separation between liquid and gas disappears. At this point the various physical properties such as density, refractive index, etc., have identical values for both the states.

Experimental Determination of Critical Constants

One of the methods to determine the critical constants T_c, p_c and V_c of a gas is to use the procedure of Andrews, i.e. to plot p–V isotherms for a number of temperatures. Other methods, which are much simpler are also available.

Determination of T_c and p_c

The two variables T_c and p_c can be determined based on the principle that at T_c and p_c, the densities of liquid and vapour become identical and the surface of separation—the meniscus between them—disappears.

Bulb A in Fig. 1.12.2, where the substance is placed, is attached to a mercury manometer B.

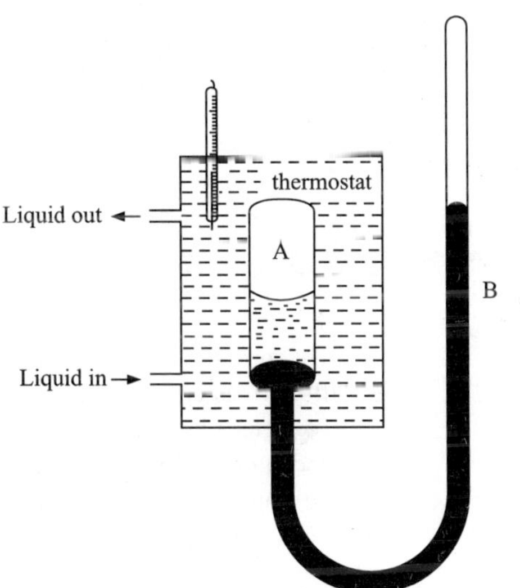

Fig. 1.12.2 Apparatus used for the determination of critical temperature and pressure

We adopt the following procedure for obtaining critical values of temperature and pressure of the gas under examination.

(i) The bulb is cooled so that the surface of separation between the liquid and vapour is clearly visible.

(ii) Temperature of the thermostat is gradually increased until the meniscus just disappears. The temperature and the corresponding pressure are noted down.

(iii) The bulb is cooled slowly and again the temperature and pressure at which the surface of separation just reappears are noted down.

The mean values of the above two temperatures and pressures give the critical temperature and pressure, respectively.

Determination of V_c

The determination of V_c may be carried out by using the law of rectilinear diameter (given by L. Cailletal and E. Mathias). According to this law, the mean value of the densities of any substance in the states of liquid and of saturated vapour at the same temperature is a linear function of the temperature. These densities where liquid and vapour are in equilibrium are known as *orthobaric densities*.

Mathematically, we can write

$$\frac{1}{2}(\rho_1 + \rho_v) = a + bt$$

where a and b are constants. If a graph is drawn between density and temperature; one obtains a graph of the type shown in Fig. 1.12.3.

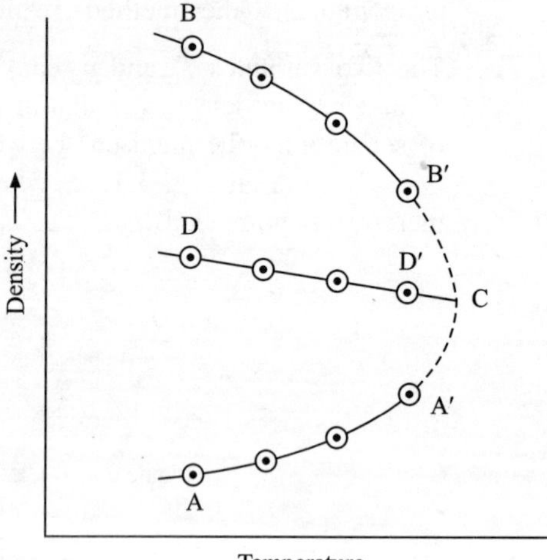

Fig. 1.12.3 Plot of orthobaric densities versus temperature. AA′, BB′ and DD′ represent densities of vapour, liquid and means, respectively

The point C, obtained by extrapolating lines AA′, BB′ and DD′, gives the critical density.

To determine the densities, a known mass of the liquid is sealed in a graduated tube and heated to a particular temperature. The volumes V_1 and V_v of liquid and vapour, respectively, are then read off from the graduation. If ρ_1 and ρ_v are the densities of liquid and vapour, respectively, then

$$m = V_1\rho_1 + V_v\rho_v$$

The experiment is repeated with another mass and from the resulting equations the values of ρ_1 and ρ_v are obtained.

Values of Critical Constants

Table 1.12.1 records the critical constants of a few substances.

Table 1.12.1 Critical Constants of a few Substances

Gas	Critical Constants			Gas	Critical Constants		
	$\frac{p_c}{\text{kPa}}$	$\frac{V_c}{\text{dm}^3\ \text{mol}^{-1}}$	$\frac{T_c}{\text{K}}$		$\frac{p_c}{\text{kPa}}$	$\frac{V_c}{\text{dm}^3\ \text{mol}^{-1}}$	$\frac{T_c}{\text{K}}$
He	228.99	0.057 8	5.3	C_2H_6	4 883.87	0.139	305.5
H_2	1 296.96	0.065 0	33.3	C_3H_8	4 265.78	0.195	370.0
Ne	2 624.32	0.041 7	44.5	*n*-Butane	3 647.7	0.250	426
Ar	4 863.60	0.075 2	151	Ethylene	5 116.92	0.126	282.8
Xe	5 865.70	0.120 2	289.81	Acetylene	6 241.62	0.113	308.6
N_2	3 394.39	0.090 1	126.1	H_2O	22 058.45	0.056 6	647.3
O_2	5 035.85	0.074 4	154.4	NH_3	11 368.67	0.072 0	405.5
CH_4	4 640.69	0.099 0	190.7	CH_3OH	7 971.24	0.118	513.2
CO_2	7 376.46	0.094 2	304.2				

1.13 CONTINUITY OF STATE

In Fig. 1.12.1 end-points of the horizontal lines have been connected with a dotted line. This portion, known as the *surface of discontinuity*, separates the liquid state on one side and the gas on the other. Within this curve the liquid and the gas coexist. Because of this coexistence curve, it is possible to distinguish between the two states of matter, namely, gas and liquid. However, in practice, this is not always true because it is possible to convert matter from one state into another without any sharp discontinuity. This can be done as shown in Fig. 1.13.1.

(i) Increase the temperature of the gas keeping volume constant. The pressure rises along AL.

(ii) Having reached L, the pressure is kept constant and the gas is cooled; this decreases the volume along the line LD.

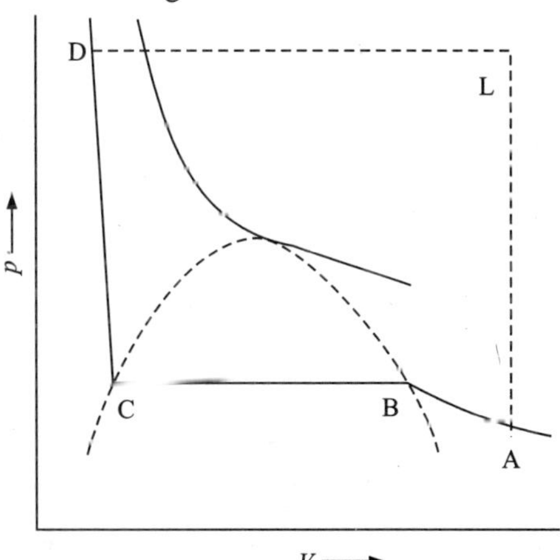

Fig. 1.13.1 Schematic representation of the continuity of state

Thus, we have passed from A to D without the gradual change as it occurs along the line BC, i.e. condensation in the usual sense of the term did not occur. Point D could be said to represent a highly compressed gaseous state of the substance. Whether we refer to the state in the region of point D as liquid state or as highly compressed gaseous state depends purely upon which of the two viewpoints happens to be convenient at the moment. Thus, in the absence of the surface of discontinuity, there is no way of distinguishing between liquid and gas.

1.14 ISOTHERMS OF VAN DER WAALS EQUATION

For one mole of a gas, the van der Waals equation

$$\left(p + \frac{a}{V_m^2}\right)(V_m - b) = RT$$

can be written as

$$V_m^3 - \left(b + \frac{RT}{p}\right)V_m^2 + \frac{a}{p}V_m - \frac{ab}{p} = 0 \qquad (1.14.1)$$

This equation has three roots in V_m for given values of a, b, p and T. It is found that either all the three roots are real or one is real and the other two are imaginary (Fig. 1.14.1).

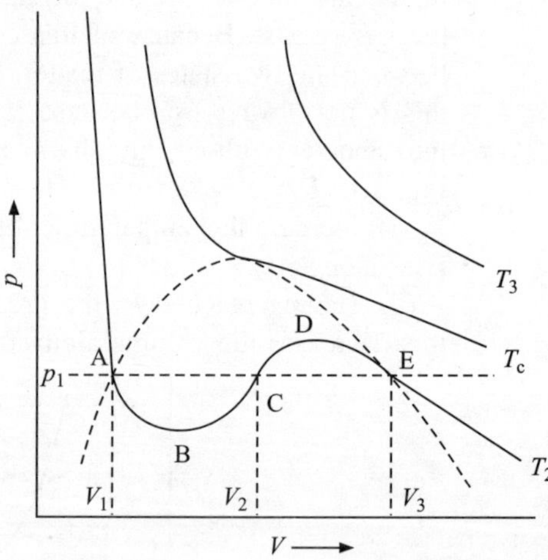

Fig. 1.14.1 Van der Waals isotherms

Main Characteristics of Van der Waals Isotherms

The main characteristics of Fig. 1.14.1 are given in the following:

- At higher temperatures such as T_3 and in the higher volume region, the isotherms look much like the isotherms for a real gas.
- At a temperature lower than T_c the isotherm exhibits a maximum and a minimum. For certain values of pressure, the equation gives three roots of volume, e.g., V_1, V_2 and V_3 at pressure p_1. The sections AB and ED of the van der Waals curve at T_2 can be realized experimentally. ED represents supersaturated (or supercooled) vapour and AB represents superheated liquid. Both these states are metastable. These are realized only when the volume is changed very slowly. These states are unstable in the sense that slight

disturbances are sufficient to cause the system to revert spontaneously into the stable state with the two phases present in equilibrium.

- The section BCD of the van der Waals isotherm cannot be realized experimentally. In this region the slope of the p–V curve is positive. Increasing (decreasing) the volume of such a system would increase (decrease) the pressure. The line BCD also represents the system in the metastable state.
- At the end points of the horizontal line AE, the conversion of gas into liquid or vice versa has just commenced, the system will have the same value of Gibbs function at the points A and E. Thus,

$$\Delta G_{\mathrm{A}\to\mathrm{E}} = 0$$

Since $\mathrm{d}G = V\mathrm{d}p$ at constant temperature, this leads to the fact that

$$\Delta G_{\mathrm{A}\to\mathrm{E}} = \int_{\mathrm{A}}^{\mathrm{E}} V\,\mathrm{d}p = 0$$

Consequently, the line AE which divides the curly curve of van der Waals isotherm within the discontinuity region into two portions is so placed that the area ABCA and CDEC have the same value but of opposite sign so that their sum is equal to zero.

Isotherm at Critical Temperature

On increasing the temperature, the three roots such as A, C and E, become closer to one another and ultimately at critical temperature, they become identical. Thus, the cubic equation in V_{m} can be written as

$$(V_{\mathrm{m}} - V')(V_{\mathrm{m}} - V'')(V_{\mathrm{m}} - V''') = 0$$

which at the critical point (where $V' = V'' = V''' = V_{\mathrm{c}}$) becomes

$$(V_{\mathrm{m}} - V_{\mathrm{c}})^3 = 0$$

Expressions of Critical Constants in Terms of Van der Waals Constants

Expanding the above expression, we obtain

$$V_{\mathrm{m}}^3 - V_{\mathrm{c}}^3 - 3V_{\mathrm{c}}V_{\mathrm{m}}^2 + 3V_{\mathrm{c}}^2V_{\mathrm{m}} = 0 \qquad (1.14.2)$$

Equation (1.14.1) with $p = p_{\mathrm{c}}$ and $T = T_{\mathrm{c}}$ and Eq. (1.14.2) must be completely identical and coefficients of the individual powers of V_{m} must be the same in both. Setting the corresponding coefficients equal, we obtain the following three equations:

$$3V_{\mathrm{c}} = b + \frac{RT_{\mathrm{c}}}{p_{\mathrm{c}}}; \qquad 3V_{\mathrm{c}}^2 = \frac{a}{p_{\mathrm{c}}}; \qquad V_{\mathrm{c}}^3 = \frac{ab}{p_{\mathrm{c}}} \qquad (1.14.3)$$

Solving these equations for p_{c}, V_{c} and T_{c} in terms of a, b and R, we get

$$V_{\mathrm{c}} = 3b; \qquad p_{\mathrm{c}} = a/27b^2; \qquad T_{\mathrm{c}} = 8a/27Rb \qquad (1.14.4)$$

Thus, if the values of p_{c}, V_{c} and T_{c} for a given gas are known, it is possible to calculate the values of a, b and R from the equations

$$b = \frac{V_{\mathrm{c}}}{3}, \qquad a = 3p_{\mathrm{c}}V_{\mathrm{c}}^2, \qquad R = \frac{8p_{\mathrm{c}}V_{\mathrm{c}}}{3T_{\mathrm{c}}} \qquad (1.14.5)$$

Since experimentally it is difficult to determine V_{c} accurately, it would be better if a and b could be obtained from p_{c} and T_{c} only. Thus

$$b = \frac{1}{3}\left(\frac{3RT_c}{8p_c}\right) = \frac{RT_c}{8p_c}; \qquad a = 3p_cV_c^2 = 3p_c\left(\frac{3RT_c}{8p_c}\right)^2 = \frac{27(RT_c)^2}{64p_c}$$

(Note that for an ideal gas $a = 0$ since there exists no forces of attraction between its molecules. Thus, for such a gas, $T_c = 0$ K. Since the essential condition for a gas to liquefy is to cool it up to or below its critical temperature, it is obvious that an ideal gas cannot be liquefied as it is not possible to attain 0 K.)

Comment on the Value of Gas Constant at the Critical State

The value of pV/RT at critical state is

$$\frac{p_cV_c}{RT_c} = \frac{3}{8} = 0.375 \qquad (1.14.6)$$

If we compare this value with the experimental values (Table 1.14.1), it is found that the agreement is very poor. The reason behind this poor agreement is that the van der Waals equation is not accurate enough to predict the behaviour of a gas near its critical state.

Table 1.14.1 Order of Observed Values of p_cV_c/RT_c for Some Gases

Type of molecules	*Examples*	*Value of p_cV_c/RT_c*
Molecules having small symmetrical nature (nonpolar and slightly polarizable)	He, Ne, Ar, O_2, CH_4	Very close to 0.29
Molecules having polarity or polarizability	Cl_2, CS_2, CCl_4, C_2H_4	About 0.26 or 0.27
Molecules having hydrogen bonding	NH_3, H_2O, CH_3OH	0.22 to 0.24

Alternative Way of Expressing Critical Constants in Terms of Van Der Waals Constants

Another way of expressing critical constants in terms of a, b and R is to use the condition of maximum slope at the critical state. The slope of p–V curve is negative before and after the critical point and has a maximum value of zero at the critical point (the point of inflection). The condition of slope being equal to zero is given by

$$\left(\frac{\partial p}{\partial V}\right)_T = 0 \qquad (1.14.7)$$

and the condition that this slope has a maximum value is

$$\frac{\partial}{\partial V}\left\{\left(\frac{\partial p}{\partial V}\right)_T\right\}_T = 0 \qquad (1.14.8)$$

From the van der Waals equation

$$p = \frac{RT}{V_m - b} - \frac{a}{V_m^2}$$

we get $\left(\frac{\partial p}{\partial V_m}\right)_T = \frac{-RT}{(V_m - b)^2} + \frac{2a}{V_m^3}$ and $\left(\frac{\partial^2 p}{\partial V_m^2}\right)_T = \frac{2RT}{(V_m - b)^3} - \frac{6a}{V_m^4}$

Hence at the critical point,

$$-\frac{RT_c}{(V_c-b)^2}+\frac{2a}{V_c^3}=0 \tag{1.14.9}$$

and
$$\frac{2RT_c}{(V_c-b)^3}-\frac{6a}{V_c^4}=0 \tag{1.14.10}$$

Solving Eqs (1.14.9) and (1.14.10) for V_c and T_c, we get

$$V_c = 3b \quad \text{and} \quad T_c=\frac{8a}{27Rb} \tag{1.14.11}$$

Substituting the values of V_c and T_c in the van der Waals equation, we have

$$p_c=\frac{RT_c}{V_c-b}-\frac{a}{V_c^2}=\frac{R(8a/27Rb)}{(3b-b)}-\frac{a}{(3b)^2}$$

$$=\frac{4a}{27b^2}-\frac{a}{9b^2}=\frac{a}{27b^2} \tag{1.14.12}$$

Example 1.14.1

The critical constants for water are 647 K, 22.09 MPa and 0.056 6 $dm^3\ mol^{-1}$. Calculate the values of a, b and R and explain the abnormal value of R.

Solution

We have $T_c = 647$ K; $p_c = 22.09$ MPa $= 22.09\times10^3$ kPa; $V_c = 0.056\,6\ dm^3\ mol^{-1}$

Thus
$$b=\frac{V_c}{3}=\frac{0.056\,6\ dm^3\ mol^{-1}}{3}=0.018\,9\ dm^3\ mol^{-1}$$

$$a=3p_cV_c^2=3(22.09\times10^3\ kPa)(0.056\,6\ dm^3\ mol^{-1})^2$$
$$=212.3\ kPa\ dm^6\ mol^{-2}$$

$$R=\frac{8}{3}\frac{p_cV_c}{T_c}=\frac{8}{3}\left(\frac{(22\,090\ kPa)(0.056\,6\ dm^3\ mol^{-1})}{647\ K}\right)=5.153\ J\ K^{-1}\ mol^{-1}$$

The value of R is 8.314 J K^{-1} mol^{-1}. The experimentally determined value is much different from this value. This is due to the fact that simple van der Waals attraction does not exist at the critical state.

Example 1.14.2

The critical temperature and pressure for NO are 177 K and 6.485 MPa, respectively, and for CCl_4 these are 550 K and 4.56 MPa, respectively. Which gas (i) has smaller value for the van der Waals constant b; (ii) has smaller value of constant a; (iii) has larger critical volume; and (iv) is most nearly ideal in behaviour at 300 K and 1.013 MPa?

Solution

We have $T_c(NO) = 177$ K $\qquad T_c(CCl_4) = 550$ K

$p_c(NO) = 6.485$ MPa $\qquad p_c(CCl_4) = 4.56$ MPa

(i) Since $\dfrac{p_c}{T_c}=\dfrac{a/27b^2}{8a/27Rb}=\dfrac{R}{8b}$

therefore, $b = \dfrac{T_c R}{8p_c}$

Thus, $b(\text{NO}) = \dfrac{(177\ \text{K})(8.314\ \text{MPa cm}^3\ \text{K}^{-1}\ \text{mol}^{-1})}{(8)(6.485\ \text{MPa})} = 28.36\ \text{cm}^3\ \text{mol}^{-1}$

and $b(\text{CCl}_4) = \dfrac{(550\ \text{K})(8.314\ \text{MPa cm}^3\ \text{K}^{-1}\ \text{mol}^{-1})}{(8)(4.56\ \text{MPa})} = 125.35\ \text{cm}^3\ \text{mol}^{-1}$

Hence $b(\text{NO}) < b(\text{CCl}_4)$

(ii) Since $a = 27p_c b^2$

therefore $a(\text{NO}) = (27)(6.485\ \text{MPa})(28.36\ \text{cm}^3\ \text{mol}^{-1})^2$

$= 140\,827\ \text{MPa cm}^6\ \text{mol}^{-2} \equiv 140.827\ \text{kPa dm}^6\ \text{mol}^{-2}$

$a(\text{CCl}_4) = (27)(4.56\ \text{MPa})(125.35\ \text{cm}^3\ \text{mol}^{-1})^2$

$= 1\,934\,538\ \text{MPa cm}^6\ \text{mol}^{-2} \equiv 1\,934.538\ \text{kPa dm}^6\ \text{mol}^{-2}$

Hence $a(\text{NO}) < a(\text{CCl}_4)$

(iii) Since $V_c = 3b$

therefore, $V_c(\text{NO}) = 3\,(28.36\ \text{cm}^3\ \text{mol}^{-1}) = 85.08\ \text{cm}^3\ \text{mol}^{-1}$

$V_c(\text{CCl}_4) = 3\,(125.35\ \text{cm}^3\ \text{mol}^{-1}) = 376.05\ \text{cm}^3\ \text{mol}^{-1}$

Hence $V_c(\text{NO}) < V_c(\text{CCl}_4)$

(iv) NO is more ideal in behaviour at 300 K and 1.013 MPa, because its critical temperature is less than 300 K, whereas for CCl_4 the corresponding critical temperature is greater than 300 K.

1.15 THE LAW OF CORRESPONDING STATES

Reduced Equation of State

Van der Waals equation can be written in a form which does not contain any constant characteristics of individual gases. Such an equation will, therefore, be applicable to all gases. In order to obtain this equation, we define reduced pressure, reduced temperature and reduced volume as follows:

$$p_r = \frac{p}{p_c}, \qquad T_r = \frac{T}{T_c} \qquad \text{and} \qquad V_r = \frac{V_m}{V_c} \tag{1.15.1}$$

Thus $p = p_r p_c$, $\quad T = T_r T_c \quad$ and $\quad V_m = V_r V_c$

Substituting these expressions in the van der Waals equation

$$\left(p + \frac{a}{V_m^2}\right)(V_m - b) = RT$$

we obtain $\left(p_r p_c + \dfrac{a}{V_r^2 V_c^2}\right)(V_r V_c - b) = R T_r T_c$

Replacing p_c, V_c and T_c in terms of a, b and R, we get

$$\left\{p_r \frac{a}{27b^2} + \frac{a}{V_r^2 (3b)^2}\right\}\{V_r(3b) - b\} = RT_r\left(\frac{8a}{27Rb}\right)$$

i.e. $$(p_r + 3/V_r^2)(3V_r - 1) = 8T_r \tag{1.15.2}$$

Equation (1.15.2), known as the *reduced equation of state*, does not contain any constant which is characteristic of a gas and is thus applicable to all gases. According to it, *if two gases have the same values of reduced pressure and reduced temperature, they will have the same reduced volume.* Thus, they correspond to each other. This statement is known as the law of corresponding states.

In actual practice, the above reduced equation of state is not directly used. One makes use of the graphs between compression factor Z and the reduced pressure at different reduced temperatures. The same graphs are applicable to all gases. This can be seen from the following consideration.

Since $Z = pV_m/RT$, writing this in reduced terms gives

$$Z = \frac{pV_m}{RT} = \frac{(p_r p_c)(V_r V_c)}{R(T_r T_c)} = \frac{p_c V_c}{T_c}\left(\frac{p_r V_r}{T_r}\right) = \frac{3}{8}\frac{p_r V_r}{T_r} \tag{1.15.3}$$ †

According to the law of corresponding states, if two gases have the same reduced temperature and reduced pressure they will have the same reduced volume. Thus, the right hand side of Eq. (1.15.3) is independent of the nature of gas and hence the value of Z is same for all gases.

Comment

Any equation of state which involves only two constants in addition to R can be written in terms of reduced variables only.

Problem 1.15.1

Express Berthelot's constants in terms of critical constants and obtain its reduced equation of state.

Solution

The Berthelot's equation of state is

$$p = \frac{RT}{V_m - b} - \frac{a}{V_m^2 T} \tag{1.15.4}$$

Thus $$\left(\frac{\partial p}{\partial V_m}\right)_T = -\frac{RT}{(V_m - b)^2} + \frac{2a}{V_m^3 T} \quad \text{and} \quad \left(\frac{\partial^2 p}{\partial V_m^2}\right)_T = \frac{2RT}{(V_m - b)^3} - \frac{6a}{V_m^4 T}$$

At critical state

$$\left(\frac{\partial p}{\partial V_m}\right)_T = 0 \quad \text{and} \quad \left(\frac{\partial^2 p}{\partial V_m^2}\right)_T = 0$$

---

†See Annexure II at the end of this chapter for the variation of compression factor with reduced pressure at a constant reduced temperature including reduced Boyle temperature.

Thus $$\frac{RT_c}{(V_c-b)^2}=\frac{2a}{V_c^3T_c} \tag{1.15.5}$$

and $$\frac{2RT_c}{(V_c-b)^3}=\frac{6a}{V_c^4T_c} \tag{1.15.6}$$

Dividing Eq. (1.15.5) by Eq. (1.15.6), we get

$$\frac{RT_c}{(V_c-b)^2}\frac{(V_c-b)^3}{2RT_c}=\frac{2a}{V_c^3T_c}\frac{V_c^4T_c}{6a}$$

i.e. $$\frac{V_c-b}{2}=\frac{V_c}{3}$$

or $$3V_c-3b=2V_c \quad \text{or} \quad V_c=3b \quad \text{or} \quad b=\frac{V_c}{3} \tag{1.15.7}$$

From Eq. (1.15.5)

$$a=\frac{RT_c}{(V_c-b)^2}\left(\frac{V_c^3T_c}{2}\right)$$

Eliminating b, we get

$$a=\frac{RT_c}{\left(\frac{2}{3}V_c\right)^2}\left(\frac{V_c^3T_c}{2}\right)=\frac{9RT_c^2V_c}{8} \tag{1.15.8}$$

The value of p_cV_c/RT_c of this equation can be obtained as follows:

At critical state, Eq. (1.15.4) modifies to

$$p_c=\frac{RT_c}{V_c-b}-\frac{a}{V_c^2T_c}$$

Multiplying by V_c/RT_c, we get

$$Z_c=\frac{p_cV_c}{RT_c}=\frac{V_c}{V_c-b}-\frac{a}{V_cT_c^2R}$$

Substituting b and a in the above expression in terms of V_c and T_c from Eq. (1.15.7) and Eq. (1.15.8), we get

$$Z_c=\frac{3b}{3b-b}-\frac{9RT_c^2V_c}{8V_cT_c^2R}=\frac{3}{2}-\frac{9}{8}=\frac{3}{8}=0.375 \tag{1.15.9}$$

Reduced form of Berthelot's equation of state Replacing p, V and T in the Berthelot's equation of state

$$\left(p+\frac{a}{V_m^2T}\right)(V_m-b)=RT$$

in terms of p_c, V_c and T_c, respectively, we get

$$\left(p_rp_c+\frac{a}{V_r^2V_c^2T_rT_c}\right)(V_rV_c-b)=RT_rT_c$$

Further, we replace a and b in terms of V_c and T_c and get

$$\left(p_r p_c + \frac{9RT_c^2 V_c}{8V_r^2 V_c^2 T_r T_c}\right)\left(V_r V_c - \frac{V_c}{3}\right) = RT_r T_c$$

or
$$\left(p_r p_c + \frac{9RT_c}{8V_r^2 T_r V_c}\right)\left(V_r - \frac{1}{3}\right)V_c = RT_r T_c$$

or
$$\left(p_r + \frac{9}{8V_r^2 T_r}\frac{RT_c}{p_c V_c}\right)\left(V_r - \frac{1}{3}\right) = T_r \frac{RT_c}{p_c V_c}$$

Substituting the value of $(RT_c/p_c V_c)$ from Eq. (1.15.9), we get

$$\left(p_r + \frac{3}{V_r^2 T_r}\right)\left(V_r - \frac{1}{3}\right) = \frac{8}{3}T_r \tag{1.15.10}$$

which is the reduced form of the Berthelot's equation of state.

Problem 1.15.2

Express Dieterici's constants in terms of critical constants and obtain its reduced equation of state.

Solution

The Dieterici's equation of state is

$$p = \left(\frac{RT}{V_m - b}\right)\exp(-a/V_m RT) \tag{1.15.11}$$

Differentiating with respect to V_m at constant temperature, we get

$$\left(\frac{\partial p}{\partial V_m}\right)_T = -\frac{RT}{(V_m - b)^2}\exp(-a/V_m RT) + \frac{RT}{(V_m - b)}\frac{a}{V_m^2 RT}\exp(-a/V_m RT)$$

$$= \left(\frac{RT}{V_m - b}\right)\exp(-a/V_m RT)\left[-\frac{1}{(V_m - b)} + \frac{a}{V_m^2 RT}\right]$$

$$= p\left[\frac{a}{V_m^2 RT} - \frac{1}{(V_m - b)}\right] \tag{1.15.12}$$

At the critical state,

$$\left(\frac{\partial p}{\partial V_m}\right)_T = 0$$

Therefore $\dfrac{a}{V_c^2 RT_c} - \dfrac{1}{(V_c - b)} = 0$ or $a = \dfrac{V_c^2 RT_c}{V_c - b}$ (1.15.13)

Differentiation of Eq. (1.15.12) with respect of V_m at constant temperature gives

$$\left(\frac{\partial^2 p}{\partial V_m^2}\right)_T = \left(\frac{\partial p}{\partial V_m}\right)_T\left[\frac{a}{V_m^2 RT} - \frac{1}{(V_m - b)}\right] + p\left[\frac{-2a}{V_m^3 RT} + \frac{1}{(V_m - b)^2}\right]$$

$$= p\left[\frac{a}{V_m^2 RT} - \frac{1}{V_m + b}\right]^2 + p\left[\frac{-2a}{V_m^3 RT} + \frac{1}{(V_m - b)^2}\right]$$

This must be zero at the critical state. Hence

$$p_c\left[\frac{a}{V_c^2RT}-\frac{1}{V_c-b}\right]^2+p_c\left[-\frac{2a}{V_c^3RT_c}+\frac{1}{(V_c-b)^2}\right]=0$$

Since the first term is zero because $(\partial p/\partial V_m)_T = 0$, we can write

$$-\frac{2a}{V_c^3RT_c}+\frac{1}{(V_c-b)^2}=0$$

This gives $$a=\frac{V_c^3RT_c}{2(V_c-b)^2} \qquad (1.15.14)$$

Dividing Eq. (1.15.14) by Eq. (1.15.13), we get

$$1=\frac{V_c}{2(V_c-b)} \quad \text{i.e.} \quad 2(V_c-b)=V_c \quad \text{or} \quad b=\frac{V_c}{2} \qquad (1.15.15)$$

Substituting Eq. (1.15.15) in Eq. (1.15.13), we get

$$a=\frac{(2b)^2RT_c}{(2b-b)}=4bRT_c=2V_cRT_c \qquad (1.15.16)$$

At critical state, Dieterici's equation becomes

$$p_c=\frac{RT_c}{(V_c-b)}\exp\left(-\frac{a}{V_cRT_c}\right)=\frac{R\left(\dfrac{a}{4bR}\right)}{(2b-b)}\exp\left(-\frac{a}{(2b)R(a/4bR)}\right)$$

which gives $$p_c=\frac{a}{4b^2e^2} \qquad (1.15.17)$$

The value of p_cV_c/RT_c can be obtained by substituting a and b from Eqs (1.15.15) and (1.15.16) into Eq. (1.15.17). Thus, we have

$$p_c=\frac{2V_cRT_c}{4(V_c/2)^2e^2} \quad \text{i.e.} \quad \frac{p_cV_c}{RT_c}=\frac{2}{e^2}=0.270\,7 \qquad (1.15.18)$$

Reduced form of Dieterici's equation of state Substituting p, V and T in terms of reduced pressure, volume and temperature, respectively, in Dieterici's equation of state

$$\{p\exp(a/V_mRT)\}(V_m-b)=RT$$

we get $$\{p_rp_c\exp(a/V_rV_cRT_rT_c)\}(V_rV_c-b)=RT_rT_c$$

Substituting a and b in terms of p_c, V_c and T_c, we get

$$\{p_r p_c \exp(2V_c RT_c / V_r V_c RT_r T_c)\}\left(V_r V_c - \frac{V_c}{2}\right) = RT_r T_c$$

that is $$p_r \exp(2/V_r T_r)\left(V_r - \frac{1}{2}\right) = RT_r \frac{T_c}{p_c V_c}$$

or $$p_r \exp(2/V_r T_r)\left(V_r - \frac{1}{2}\right) = \frac{e^2 T_r}{2} \tag{1.15.19}$$

which is the required reduced form of Dieterici's equation of state.

Problem 1.15.3 Express Redlich-Kwong constants in terms of critical constants and obtain its reduced equation of state.

Solution The Redlich-Kwong equation

$$p = \frac{RT}{V_m - b} - \frac{a}{T^{1/2} V_m (V_m + b)} \tag{1.15.20}$$

may be written as

$$p(V_m - b) = RT - \frac{a}{T^{1/2}}\left[\frac{V_m - b}{V_m(V_m + b)}\right]$$

$$= RT - \frac{a}{T^{1/2}}\left(-\frac{1}{V_m} + \frac{2}{V_m + b}\right) \tag{1.15.21}$$

Differentiating this equation with respect to V_m at constant T, we get

$$\left(\frac{\partial p}{\partial V_m}\right)_T (V_m - b) + p = -\frac{a}{T^{1/2}}\left[\frac{1}{V_m^2} - \frac{2}{(V_m + b)^2}\right] \tag{1.15.22}$$

At critical state, $(\partial p/\partial V_m)_T = 0$. Hence

$$p_c = -\frac{a}{T_c^{1/2}}\left[\frac{1}{V_c^2} - \frac{2}{(V_c + b)^2}\right] \tag{1.15.23}$$

Differentiating Eq. (1.15.22) with respect to V_m at constant T, we get

$$\left(\frac{\partial^2 p}{\partial V_m^2}\right)_T (V_m - b) + \left(\frac{\partial p}{\partial V_m}\right)_T + \left(\frac{\partial p}{\partial V_m}\right)_T = -\frac{a}{T^{1/2}}\left[-\frac{2}{V_m^3} + \frac{4}{(V_m + b)^3}\right] \tag{1.15.24}$$

Since $(\partial^2 p/\partial V_m^2)_T = 0$ at the critical state, Eq. (1.15.24) gives

$$-\frac{2}{V_c^3} + \frac{4}{(V_c + b)^3} = 0$$

Thus $$\frac{V_c + b}{V_c} = 2^{1/3} \quad \text{i.e.} \quad b = (2^{1/3} - 1)V_c \tag{1.15.25}$$

Substituting $V_c + b$ from Eq. (1.15.25) into Eq. (1.15.23), we get

$$p_c = -\frac{a}{T_c^{1/2}}\left(\frac{1}{V_c^2} - \frac{2^{1/3}}{V_c^2}\right) = \frac{a}{V_c^2 T_c^{1/2}}(2^{1/3}-1)$$

This gives $a = p_c T_c^{1/2} V_c^2/(2^{1/3}-1)$ (1.15.26)

The value of $Z_c = \left(\frac{p_c V_c}{RT_c}\right)$ may be computed as follows.

The Redlich-Kwong equation at the critical state is

$$p_c = \frac{RT_c}{V_c - b} - \frac{a}{T_c^{1/2} V_c (V_c + b)} \tag{1.15.27}$$

Substituting the expressions of a and b in terms of critical constants, we get

$$p_c = \frac{RT_c}{V_c - (2^{1/3}-1)V_c} - \frac{p_c T_c^{1/2} V_c^2/(2^{1/3}-1)}{T_c^{1/2} V_c (2^{1/3} V_c)}$$

Cancelling the common terms and rearranging the resultant expression, we get

$$p_c + \frac{p_c}{2^{1/3}(2^{1/3}-1)} = \frac{RT_c}{2^{1/3}(2^{2/3}-1)V_c}$$

or
$$p_c\left[\frac{2^{1/3}(2^{1/3}-1)+1}{2^{1/3}(2^{1/3}-1)}\right] = \frac{RT_c}{2^{1/3}(2^{2/3}-1)V_c}$$

or
$$\frac{p_c V_c}{RT_c} = \frac{1}{(2^{2/3}-1)}\frac{(2^{1/3}-1)}{(2^{2/3}-2^{1/3}+1)} = \frac{1}{3} \tag{1.15.28}$$

The constants a and b in terms of T_c and p_c are given by the expressions

$$b = (2^{1/3} - 1)V_c = (2^{1/3} - 1)(RT_c/3p_c) \tag{1.15.29}$$

$$a = \frac{p_c T_c^{1/2} V_c^2}{(2^{1/3}-1)} = \frac{T_c^{1/2}(p_c^2 V_c^2)}{(2^{1/3}-1)p_c} = \frac{T_c^{1/2}(RT_c/3)^2}{(2^{1/3}-1)p_c} = \frac{R^2 T_c^{5/2}}{9(2^{1/3}-1)p_c} \tag{1.15.30}$$

The values of constants a and b are obtained from their expressions in terms of p_c and T_c of a gas (Eqs 1.15.29 and 1.15.30).

Reduced Redlich-Kwong Equation

The Redlich-Kwong equation is

$$p = \frac{RT}{V_m - b} - \frac{a}{T^{1/2} V_m (V_m + b)} \tag{1.15.20}$$

Expressing p, V_m and T in terms of reduced variables, we get

$$p_r p_c = \frac{RT_r T_c}{V_r V_c - b} - \frac{a}{(T_r T_c)^{1/2} V_r V_c (V_r V_c + b)} \tag{1.15.31}$$

Substituting a and b in terms of critical constants, we get

$$p_r p_c = \frac{RT_r T_c}{V_r V_c - (2^{1/3}-1)V_c} - \frac{R^2 T_c^{5/2}/9(2^{1/3}-1)p_c}{(T_r T_c)^{1/2} V_r V_c [V_r V_c + (2^{1/3}-1)V_c]}$$

$$p_r = \frac{RT_c}{p_c V_c} \frac{T_r}{V_r - (2^{1/3}-1)} - \frac{(RT_c/p_c V_c)^2/9(2^{1/3}-1)}{T_r^{1/2} V_r [V_r + (2^{1/3}-1)]}$$

$$= \frac{3T_r}{V_r - (2^{1/3}-1)} - \frac{1/(2^{1/3}-1)}{T_r^{1/2} V_r [V_r + (2^{1/3}-1)]} \tag{1.15.32}$$

Equation (1.15.32) is the reduced Redlich-Kwong equation.

Example 1.15.1

Calculate the volume occupied by 2.0 mol of N_2 at 200 K and 10.132 5 MPa pressure if $p_c V_c / RT_c = 3/8$ and $p_r V_r / T_r = 2.21$.

Solution

Since $\quad \dfrac{p_c V_c}{RT_c} = \dfrac{3}{8} \quad$ and $\quad \dfrac{p_r V_r}{T_r} = 2.21$

therefore $\quad \dfrac{p_c V_c}{RT_c} \times \dfrac{p_r V_r}{T_r} = \dfrac{3}{8} \times 2.21 \quad$ or $\quad \dfrac{pV_m}{RT} = \dfrac{3}{8} \times 2.21$

or

$$V_m = \left(\frac{3 \times 2.21}{8}\right)\frac{RT}{p}$$

$$= \left(\frac{3}{8} \times 2.21\right)\frac{(8.314 \text{ MPa cm}^3 \text{ K}^{-1} \text{ mol}^{-1})(200 \text{ K})}{(10.132\,5 \text{ MPa})} = 136.0 \text{ cm}^3 \text{ mol}^{-1}$$

Volume of 2 mol of N_2 = (2.0 mol) (136.0 cm^3 mol^{-1}) = 272.0 cm^3

Problem 1.15.4

Compare the values of $p_c V_c / RT_c$ of van der Waals, Berthelot, Dieterici and Redlich-Kwong equations of state with the experimental values given in Table 1.14.1 and suggest which one is the most appropriate to be used at the critical state of gases.

Solution

The values of $p_c V_c / RT_c$ for the three equations of state are

van der Waals: $\quad \dfrac{p_c V_c}{RT_c} = \dfrac{3}{8} = 0.375 \quad$ (Eq. 1.14.6)

Berthelot's: $\quad \dfrac{p_c V_c}{RT_c} = \dfrac{3}{8} = 0.375 \quad$ (Eq. 1.15.9)

Dieterici's: $\quad \dfrac{p_c V_c}{RT_c} = \dfrac{2}{e^2} = 0.270\,7 \quad$ (Eq. 1.15.18)

Redlich-Kwong: $\quad \dfrac{p_c V_c}{RT_c} = \dfrac{1}{3}$

Comparing these with the values given in Table 1.14.1 reveals that the Dieterici's equation of state is the most appropriate to be used at the critical state of gases.

1.16 MAXWELLIAN DISTRIBUTION OF MOLECULAR SPEEDS

Distribution of Molecular Speeds

The speed of a molecule of a gas changes continuously as a result of collisions with other molecules and with the walls of the container. Thus, the net result is that one cannot speak of the speed of an individual molecule; rather, one must consider the statistical average of the speeds of the whole collection of gas molecules. Since the observed properties such as pressure, volume and temperature of an isolated gaseous sample do not change with time, it is expected that the same is also true in the case of distribution of molecular speeds. That is, the fraction of total number of molecules having speeds between any definite range must be constant, even though the speeds of individual molecules may be changing as a result of molecular collisions.

Law of Distribution of Molecular Speeds

The manner in which the molecules of a gas are distributed over the possible speed ranges, from zero to very high values, was first investigated by J.C. Maxwell using the theory of probability. His results are expressed as the law of distribution of molecular speeds, one form of which is

$$\mathrm{d}N_u = 4\pi N\left(\frac{M}{2\pi RT}\right)^{3/2} \exp(-Mu^2/2RT)\, u^2\mathrm{d}u$$

$$= 4\pi N\left(\frac{m}{2\pi kT}\right)^{3/2} \exp(-mu^2/2kT)\, u^2\mathrm{d}u \tag{1.16.1}$$

This expression gives the number of molecules $\mathrm{d}N_u$ having speeds between u and $u + \mathrm{d}u$ in terms of the total number N of molecules present, molar mass M (or mass of a single molecule m) of the gas and its temperature. According to this expression the fraction $\mathrm{d}N_u/N$ of molecules having speeds between u and $u + \mathrm{d}u$ for a gas of molar mass M depends only on temperature. Thus, for a given temperature, this fraction has a constant value.

Consequence of Distribution of Speeds

A direct consequence of the distribution of speeds is that *the average kinetic energy of a gas is also constant for a given temperature*. Qualitatively, this may be verified as follows:

The average kinetic energy is defined as

$$\overline{\mathrm{KE}} = \frac{1}{N}\left(\frac{1}{2}mu_1^2 + \frac{1}{2}mu_2^2 + \cdots + \frac{1}{2}mu_N^2\right)$$

$$= \frac{1}{2N}m(u_1^2 + u_2^2 + \cdots + u_N^2) = \frac{1}{2}m\overline{u^2}$$

Alternatively, it may be defined as

$$\overline{\mathrm{KE}} = \frac{1}{N}\left(\frac{1}{2}m\sum_i \mathrm{d}N_i\, u_i^2\right) = \frac{1}{2}m\left(\sum_i \frac{\mathrm{d}N_i}{N}u_i^2\right) \tag{1.16.2}$$

where $\mathrm{d}N_i$ is the number of molecules having speed equal to u_i. Since the fraction $\mathrm{d}N_i/N$ having speed u_i (or more precisely between speed range u_i and $u_i + \mathrm{d}u$) is constant at a given temperature, the right side of Eq. (1.16.2) has a constant value. Thus, the average kinetic energy has a constant value at a given temperature. This is, in fact, one of the assumptions of the kinetic theory of gases.

Plots of Maxwell Distribution of Speeds

The Maxwell distribution of speeds is customarily plotted with the fraction $(1/N)\ (\mathrm{d}N_u/\mathrm{d}u)$ as the ordinate and u as the abscissa. The term $(1/N) \times (\mathrm{d}N_u/\mathrm{d}u) = (1/\mathrm{d}u)\ (\mathrm{d}N_u/N)$ gives the fraction of molecules in the speed range of u to $u + \mathrm{d}u$ per unit interval of speed. Roughly speaking, this gives the probability of finding a molecule with a speed between u and (u + 1 m/s). The distribution at two temperatures is shown in Fig. 1.16.1.

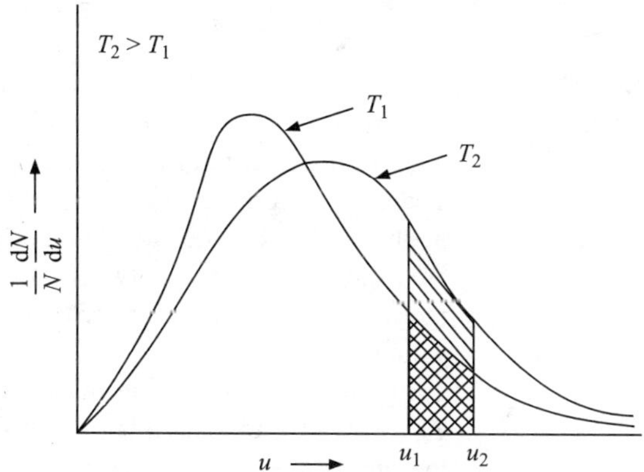

Fig. 1.16.1 Plots of $(1/N)\ (\mathrm{d}N_u/\mathrm{d}u)$ versus u

The curve at any temperature is parabolic near the origin, since the factor u^2 is dominant in this region, the exponential function being approximately equal to unity. At high values of u, however, the exponential factor dominates the behaviour of the function, causing it to decrease rapidly in value.

Most Probable Speed

As a consequence of the contrasting behaviour of the two factors, the product function passes through a maximum at a speed known as the *most probable speed*, u_{mp}. Thus, the most probable speed is the speed possessed by the maximum fraction of the molecules. The expression of u_{mp} can be derived mathematically using the condition of maxima by setting the first derivative of $(1/N)(\mathrm{d}N_u/\mathrm{d}u)$ with respect to speed equal to zero. Thus

$$\frac{\mathrm{d}\left(\frac{1}{N}\frac{\mathrm{d}N_u}{\mathrm{d}u}\right)}{\mathrm{d}u} = 4\pi\left(\frac{M}{2\pi RT}\right)^{3/2}\left[2u\exp(-Mu^2/2RT) + u^2\left(-\frac{2Mu}{2RT}\right)\exp(-Mu^2/2RT)\right]$$

Hence at u_{mp}, we have

$$4\pi\left(\frac{M}{2\pi RT}\right)^{3/2} u_{\mathrm{mp}}\exp(-Mu_{\mathrm{mp}}^2/2RT)\left[2-\frac{Mu_{\mathrm{mp}}^2}{RT}\right] = 0 \qquad (1.16.3)$$

In fact, any of the terms u_{mp}, $\exp(-Mu_{mp}^2/2RT)$ and $(2 - Mu_{mp}^2/RT)$ being equal to zero will make the whole Eq. (1.16.3) equal to zero. The first two correspond to the minimum fraction of molecules having speeds zero and infinity, respectively. The third term gives

$$2 - \frac{Mu_{mp}^2}{RT} = 0 \quad \text{or} \quad u_{mp}^2 = \frac{2RT}{M}$$

or
$$u_{mp} = \sqrt{\frac{2RT}{M}} = \sqrt{\frac{2kT}{m}} \tag{1.16.4}$$

General Comments on the Distribution of Speeds

- The fractions of molecules having either very low speeds or very high speeds are small in number (see Fig. 1.16.1). The majority of molecules have speeds which cluster around u_{mp} in the middle of the range of u.
- The total area under the curve is a measure of the total number of molecules in the collection, i.e. the number of molecules comprising all speeds. The area under the curve between any two speeds, for example, u_1 and u_2, is a measure of the total number of molecules having speeds between these two values.

  Figure 1.16.1 gives the distribution of speeds at two temperatures T_1 and T_2. Since the total number of molecules is the same at both temperatures, the area under both the curves will have the same value.
- Since increase in temperature causes an increase in the kinetic energy of molecules, it follows that the fraction of molecules having lower speed range decreases whereas the fraction of molecules having higher speed range increases on increasing the temperature of the gas. This is primarily because of the $\exp(-Mu^2/2RT)$ factor in Eq. (1.16.1). This is evident in Fig. 1.16.1. Also, the curve at the higher temperature T_2 has its u_{mp} shifted to a higher value compared with that for T_1, whereas the corresponding fraction of molecules has decreased. But at the same time, the curve near u_{mp} has become broader at the higher temperature indicating that more molecules possess speeds near to the most probable speed. In general, the distribution of speeds is wider at a higher temperature than at a lower temperature.
- The speed distribution also depends on the mass of the molecule. At the same temperature a heavy gas has a narrower distribution of speeds than a light gas. It can be seen from Eq. (1.16.1) that the distribution, in general, depends upon the value of M/T. Thus, the distribution will be the same for a gas of molar mass $2M$ at temperature $2T$ since the ratio remains the same. For example, the distribution of O_2 molecules at temperature T will be the same as those of SO_2 molecules at temperature $2T$.

Verification of Maxwell Distribution Law: Molecular Beam Method

In this method, the gaseous molecules after passing through a pinhole in the source S are collimated by the slits and then pass through one of the openings between the cogs in the cogwheel C_1 (Fig. 1.16.2). The cogwheels C_1 and C_2 are mounted on the same axle which is rapidly rotated. The molecule which has passed through C_1 will pass through C_2 only when the time required for the particle to come from C_1 to C_2 is the same as that required for the cogwheel C_2

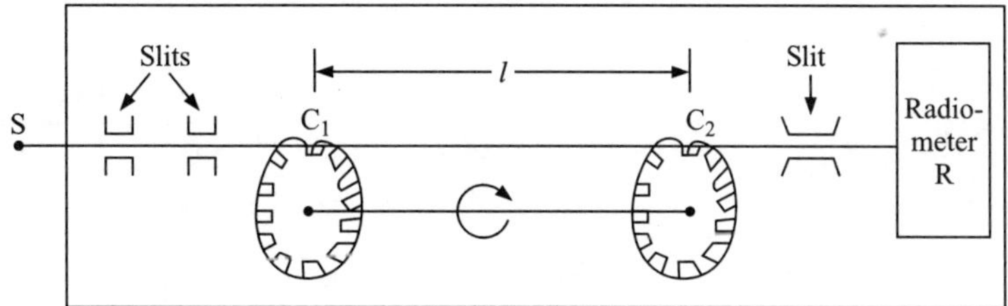

Fig. 1.16.2 Verification of Maxwell distribution. The entire apparatus is kept in highly evacuated chamber.

to rotate once, i.e. when C_2 is exactly in the same position as C_1 was when the molecule passed through it.† If ν is the frequency of rotation of cogwheels, then the time required for one rotation is

$$t = \frac{1}{\nu}$$

Let l bc the distance between the two cogwheels. The speed of the molecule which covers this distance in time t is

$$u = \frac{l}{t} = l\nu$$

The molecules with the above speed after passing through the cogwheel C_2 are admitted in R where their number is determined from the radiometer deflection. By changing the frequency of rotation of the cogwheels the number of molecules having different speeds can be determined.

Example 1.16.1

Calculate the fraction of N_2 molecules at 101.325 kPa and 300 K whose speeds are in the range of $u_{mp} - 0.005\,u_{mp}$ to $u_{mp} + 0.005\,u_{mp}$.

Solution

Most probable speed,

$$u_{mp} = \sqrt{\frac{2RT}{M}} = \left\{\frac{2(8.314\ \text{J K}^{-1}\ \text{mol}^{-1})(300\ \text{K})}{(0.028\ \text{kg mol}^{-1})}\right\}^{1/2}$$

$$= 422.09\ \text{J}^{1/2}\ \text{kg}^{-1/2} \equiv 422.09\ \text{m s}^{-1}$$

Since $\quad 0.005\,u_{mp} = (0.005)(422.09\ \text{m s}^{-1}) = 2.11\ \text{m s}^{-1}$

therefore $\quad du = (u_{mp} + 0.005\,u_{mp}) - (u_{mp} - 0.005\,u_{mp}) = 4.22\ \text{m s}^{-1}$

Maxwell distribution law states

$$\frac{dN}{N} = 4\pi\left(\frac{M}{2\pi RT}\right)^{3/2} u^2 \exp(-Mu^2/2RT)\,du$$

† Since the speeds of the molecules are large, this time may be made equal to that of the cogweel to be displaced to the next opening.

Thus, we have

$$\left(\frac{M}{2\pi RT}\right)^{3/2} = \left\{\frac{(0.028\ \text{kg mol}^{-1})}{2\times 3.14\times(8.314\ \text{J K}^{-1}\ \text{mol}^{-1})\,(300\ \text{K})}\right\}^{3/2}$$

$$= 2.390\times 10^{-9}\ \text{kg}^{3/2}\ \text{J}^{-3/2} \equiv 2.390\times 10^{-9}\ \text{m}^{-3}\ \text{s}^{3}$$

$$\exp\left(-\frac{Mu^2}{2RT}\right) = \exp\left\{-\frac{(0.028\ \text{kg mol}^{-1})\,(422.09\ \text{m s}^{-1})^2}{2(8.314\ \text{J K}^{-1}\ \text{mol}^{-1})\,(300\ \text{K})}\right\}$$

$$= \exp(-1.000) = 0.367\,9$$

Hence $$\frac{\text{d}N}{N} = 4\times 3.14\times(2.390\times 10^{-9}\ \text{m}^{-3}\ \text{s}^{3})\,(422.09\ \text{m s}^{-1})^2(0.367\,9)\,(4.22\ \text{m s}^{-1})$$

$$= 8.303\times 10^{-3}$$

Example 1.16.2 What is the ratio of the number of molecules having speeds in the range of $2u_{\text{mp}}$ and $2u_{\text{mp}} + du$ to the number of molecules having speeds in the range of u_{mp} and $u_{\text{mp}} + du$?

Solution If $\text{d}N_1$ is the number of molecules in the speed range u_{mp} to $u_{\text{mp}} + du$ and $\text{d}N_2$ is the corresponding number in the speed range $2u_{\text{mp}}$ to $2u_{\text{mp}} + du$, then according to the Maxwell distribution, we have

$$\frac{\text{d}N_1}{N} = 4\pi\left(\frac{M}{2\pi RT}\right)^{3/2} u_{\text{mp}}^2 \exp(-Mu_{\text{mp}}^2/2RT)\,\text{d}u$$

and $$\frac{\text{d}N_2}{N} = 4\pi\left(\frac{M}{2\pi RT}\right)^{3/2} (2u_{\text{mp}})^2 \exp(-4Mu_{\text{mp}}^2/2RT)\,\text{d}u$$

Therefore $$\frac{\text{d}N_2}{\text{d}N_1} = 4\frac{\exp(-4Mu_{\text{mp}}^2/2RT)}{\exp(-Mu_{\text{mp}}^2/2RT)} = 4\exp(-3Mu_{\text{mp}}^2/2RT)$$

Now, since $u_{\text{mp}}^2 = \dfrac{2RT}{M}$, therefore $\dfrac{\text{d}N_2}{\text{d}N_1} = 4\,\text{e}^{-3} = 0.199$

1.17 DERIVATION OF SOME EXPRESSIONS FROM THE MAXWELL DISTRIBUTION

Maxwell distribution expression (Eq. 1.16.1) can be used to derive expressions for average speed, root mean square speed, average kinetic energy and the fraction of molecules possessing kinetic energies greater than some specified energy.

Average Speed The average value of speeds is given by the relation

$$\bar{u} = \frac{u_1 + u_2 + \cdots + u_N}{N} = \frac{1}{N}\sum_i u_i \qquad (1.17.1)$$

Equation (1.17.1) can be written in the form

$$\bar{u} = \frac{1}{N}\int_0^\infty u\,\mathrm{d}N_u = \int_0^\infty u\frac{\mathrm{d}N_u}{N} \tag{1.17.2}$$

where $\mathrm{d}N_u$ is the number of molecules having speed u. The summation of different speeds is replaced by integration since all types of speed ranging from zero to infinity are involved.

Substituting $\mathrm{d}N_u/N$ from Eq. (1.16.1) in Eq. (1.17.2), we get

$$\bar{u} = 4\pi\left(\frac{M}{2\pi RT}\right)^{3/2}\int_0^\infty u^3\exp(-Mu^2/2RT)\,\mathrm{d}u$$

which on integration yields

$$\bar{u} = 4\pi\left(\frac{M}{2\pi RT}\right)^{3/2}\left\{2\left(\frac{RT}{M}\right)^2\right\} = \sqrt{\frac{8RT}{\pi M}} = \sqrt{\frac{8kT}{\pi m}} \tag{1.17.3}$$

Root Mean Square Speed

The mean square speed is given by

$$\overline{u^2} = \frac{u_1^2 + u_2^2 + \cdots + u_N^2}{N} = \frac{1}{N}\sum_i u_i^2 = \frac{1}{N}\int_0^\infty u^2\,\mathrm{d}N_u \tag{1.17.4}$$

Using Eq. (1.16.1), we get

$$\overline{u^2} = 4\pi\left(\frac{M}{2\pi RT}\right)^{3/2}\int_0^\infty u^4\exp(-Mu^2/2RT)\,\mathrm{d}u$$

which on integration yields

$$\overline{u^2} = 4\pi\left(\frac{M}{2\pi RT}\right)^{3/2}\left\{\left(\frac{RT}{M}\right)^{5/2}\frac{3}{\sqrt{2}}\sqrt{\pi}\right\} = 3\frac{RT}{M}$$

Thus, $$u_{\mathrm{rms}} = \sqrt{\overline{u^2}} = \sqrt{\frac{3RT}{M}} = \sqrt{\frac{3kT}{m}} \tag{1.17.5}$$

Example 1.17.1

Arrange root mean square, most probable and average speeds in the order of decreasing value. Discuss the effects of temperature and pressure on these speeds.

Solution

From Eqs (1.16.4), (1.17.3) and (1.17.5), we find that

$$u_{\mathrm{mp}} = \sqrt{\frac{2RT}{M}};\quad \bar{u} = \sqrt{\frac{8RT}{\pi M}};\quad \sqrt{\overline{u^2}} = \sqrt{\frac{3RT}{M}}$$

Therefore

$$\frac{\sqrt{\overline{u^2}}}{\bar{u}} = \frac{\sqrt{3RT/M}}{\sqrt{8RT/\pi M}} = \sqrt{\frac{3\pi}{8}} = \sqrt{\frac{3}{8/\pi}} = \frac{1.732}{1.596}$$

$$\frac{\sqrt{\overline{u^2}}}{u_{\mathrm{mp}}} = \frac{\sqrt{3RT/M}}{\sqrt{2RT/M}} = \sqrt{\frac{3}{2}} = \frac{1.732}{1.414}$$

Hence $\sqrt{\overline{u^2}} : \bar{u} : u_{mp} :: 1.732 : 1.596 : 1.414$

From this, it follows that

$$\sqrt{\overline{u^2}} > \bar{u} > u_{mp}$$

It may be concluded that all the three speeds are directly proportional to the square root of absolute temperature and are independent of pressure of gas.

Example 1.17.2 For O_2 gas molecules, the root means square speed at T_1, the average speed at T_2 and most probable speed at T_3 are all equal to 1.5×10^3 m s^{-1}. Calculate T_1, T_2 and T_3.

Solution We can calculate T_1, T_2 and T_3 as follows.

$$\sqrt{\overline{u_1}} = \sqrt{\frac{3RT_1}{M}} = 1.5 \times 10^3 \text{ m s}^{-1}$$

which gives $T_1 = (1.5 \times 10^3 \text{ m s}^{-1})^2 \dfrac{M}{3R} = \dfrac{(1.5 \times 10^3 \text{ m s}^{-1})^2 (0.032 \text{ kg mol}^{-1})}{3(8.314 \text{ J K}^{-1} \text{ mol}^{-1})}$

$= 2\,887$ K

$$u_{av} = \sqrt{\frac{8RT_2}{\pi M}} = 1.5 \times 10^3 \text{ m s}^{-1}$$

which gives $T_2 = (1.5 \times 10^3 \text{ m s}^{-1})^2 \left(\dfrac{\pi M}{8R}\right) = \dfrac{(1.5 \times 10^3 \text{ m s}^{-1})^2 (3.14 \times 0.032 \text{ kg mol}^{-1})}{8(8.314 \text{ J K}^{-1} \text{ mol}^{-1})}$

$= 3\,399$ K

$$u_{mp} = \sqrt{\frac{2RT_3}{M}} = 1.5 \times 10^3 \text{ m s}^{-1}$$

which gives $T_3 = \dfrac{(1.5 \times 10^3 \text{ m s}^{-1})^2 (M)}{2R} = \dfrac{(1.5 \times 10^3 \text{ m s}^{-1})^2 (0.032 \text{ kg mol}^{-1})}{2(8.314 \text{ J K}^{-1} \text{ mol}^{-1})}$

$= 4\,330$ K

Example 1.17.3 Calculate the temperature at which the average speed of H_2 equals that of O_2 at 320 K.

Solution We have $\bar{u}(O_2) = \sqrt{\dfrac{8RT}{\pi M}} = \left\{\dfrac{8R(320 \text{ K})}{\pi(0.032 \text{ kg mol}^{-1})}\right\}^{1/2}$

$$\bar{u}(H_2) = \sqrt{\frac{8RT}{\pi M}} = \left\{\frac{8RT}{\pi(0.002 \text{ kg mol}^{-1})}\right\}^{1/2}$$

Since $\bar{u}(O_2) = \bar{u}(H_2)$

therefore $\dfrac{8R(320 \text{ K})}{\pi(0.032 \text{ kg mol}^{-1})} = \dfrac{8RT}{\pi(0.002 \text{ kg mol}^{-1})}$

which gives $T = 20$ K

Example 1.17.4

Calculate the root mean square, average and most probable speeds of H_2 molecules. The density of the gas at 101.325 kPa is 0.09 g dm^{-3} ($\equiv$ 0.09 kg m^{-3}). Assume ideal behaviour.

Solution

The three speeds can be calculated as follows:

$$\sqrt{\overline{u^2}} = \sqrt{\frac{3RT}{M}} = \sqrt{\frac{3pV_{\mathrm{m}}}{M}} = \sqrt{\frac{3p}{\rho}} = \left\{\frac{3(101.325 \times 10^3\ \mathrm{Pa})}{(0.09\ \mathrm{kg\ m^{-3}})}\right\}^{1/2} = 1\,838\ \mathrm{m\ s^{-1}}$$

$$\bar{u} = \sqrt{\frac{8RT}{\pi M}} = \sqrt{\frac{8pV_{\mathrm{m}}}{\pi M}} = \sqrt{\frac{8p}{\pi\rho}} = \left\{\frac{8(101.325 \times 10^3\ \mathrm{Pa})}{3.14(0.09\ \mathrm{kg\ m^{-3}})}\right\}^{1/2} = 1\,694\ \mathrm{m\ s^{-1}}$$

$$u_{\mathrm{mp}} = \sqrt{\frac{2RT}{M}} = \sqrt{\frac{2pV_{\mathrm{m}}}{M}} = \sqrt{\frac{2p}{\rho}} = \left\{\frac{2(101.325 \times 10^3\ \mathrm{Pa})}{(0.09\ \mathrm{kg\ m^{-3}})}\right\}^{1/2} = 1\,501\ \mathrm{m\ s^{-1}}$$

Average Kinetic Energy

The average kinetic energy is given by

$$\bar{\varepsilon} = \frac{1}{N}\left(\frac{1}{2}m\,u_1^2 + \frac{1}{2}m\,u_2^2 + \cdots + \frac{1}{2}m\,u_N^2\right)$$

$$= \frac{1}{N}\frac{m}{2}\sum_i u_i^2 = \frac{1}{N}\frac{m}{2}\int_0^\infty u^2\,\mathrm{d}N_u$$

Substituting $\mathrm{d}N_u/N$ from Eq. (1.16.1) and intergrating the resultant expression, we have

$$\bar{\varepsilon} = \frac{1}{2}m\left(\frac{3kT}{m}\right) = \frac{3}{2}kT \tag{1.17.6}$$

Expression of Energy Distribution

The Maxwell distribution of speeds (Eq. 1.16.1) can be converted into energy distribution by substituting

$$\varepsilon = \frac{1}{2}mu^2$$

which gives

$$u = \left(\frac{2}{m}\right)^{1/2} \varepsilon^{1/2}$$

Differentiating, we have

$$\mathrm{d}u = \left(\frac{1}{2m}\right)^{1/2} \varepsilon^{-1/2}\,\mathrm{d}\varepsilon$$

The energy range $d\varepsilon$ corresponds to the speed range du, and so the number of particles dN_u having speeds between u and $u + du$ corresponds to the number of particles dN_ε having energies between ε and $\varepsilon + d\varepsilon$. Replacing u and du in Eq. (1.16.1) in terms of ε and $d\varepsilon$, we have

$$dN_\varepsilon = 2\pi N\left(\frac{1}{\pi kT}\right)^{3/2} \varepsilon^{1/2} \exp(-\varepsilon/kT)\, d\varepsilon \qquad (1.17.7)$$

Figure 1.17.1 shows the plot of $(1/N)\,(dN_\varepsilon/d\varepsilon)$ versus ε. Shape of this curve is different from that of the speed distribution curve. The energy distribution has a vertical tangent at the origin and thus it rises much more rapidly than the speed distribution curve which starts with a horizontal tangent. After passing the maximum, the energy distribution falls off more gently than does the speed distribution. As usual, the distribution is broadened at higher temperatures. Thus, a greater proportion of the molecules possess higher energies.

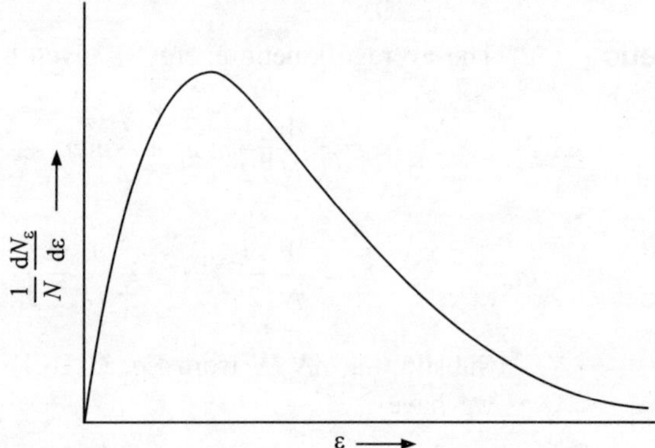

Fig. 1.17.1 Plot of $(1/N)\,(dN_\varepsilon/d\varepsilon)$ versus ε

Fraction of Molecules Possessing Kinetic Energies Greater than some Specified Energy

The fraction of molecules having energies greater than ε' is given by

$$\frac{N(\varepsilon')}{N} = \int_{\varepsilon'}^{\infty} \frac{dN_\varepsilon}{N} \qquad (1.17.8)$$

Substituting dN_ε/N from Eq. (1.17.7) in the above expression, we have

$$\frac{N(\varepsilon')}{N} = 2\pi\left(\frac{1}{\pi kT}\right)^{3/2} \int_{\varepsilon'}^{\infty} \varepsilon^{1/2} \exp(-\varepsilon/kT)\, d\varepsilon$$

which on integration yields

$$\frac{N(\varepsilon')}{N} = 2\left(\frac{\varepsilon'}{\pi kT}\right)^{3/2} \exp(-\varepsilon'/kT), \qquad (\varepsilon' >> kT) \qquad (1.17.9)$$

Equation (1.17.9) describes how the fraction of molecules having kinetic energies greater than ε' varies with temperature. Due to the exponential dependence, this fraction varies quite rapidly with temperature, particularly at low temperatures. This equation is often required in describing many concepts of physical chemistry. For example, in the study of effect of temperature on reaction rates, we require the fraction of molecules having energies equal to or greater than some minimum energy (known as threshold energy). It is known that only those molecules which have energies equal to or greater than the threshold energy can react chemically. Since this fraction increases with temperature, the rate of a chemical reaction also increases with temperature.

Example 1.17.5

Calculate the fraction of N_2 molecules at 101.325 kPa and 300 K whose kinetic energies are in the range of $\bar{\varepsilon} - 0.005\,\bar{\varepsilon}$ and $\bar{\varepsilon} + 0.005\,\bar{\varepsilon}$.

Solution

The average kinetic energy at 300 K is

$$\bar{\varepsilon} = \frac{3}{2}kT = \frac{3}{2}(1.380\,6 \times 10^{-23}\ \text{J K}^{-1})(300\ \text{K}) = 6.213 \times 10^{-21}\ \text{J}$$

Now $$\mathrm{d}\varepsilon = (\bar{\varepsilon} + 0.005\,\bar{\varepsilon}) - (\bar{\varepsilon} - 0.005\,\bar{\varepsilon}) = 0.01\,\bar{\varepsilon} = 6.213 \times 10^{-23}\ \text{J}$$

Equation for energy distribution is

$$\frac{\mathrm{d}N_\varepsilon}{N} = 2\pi\left(\frac{1}{\pi kT}\right)^{3/2} \varepsilon^{1/2} \exp(-\varepsilon/kT)\,\mathrm{d}\varepsilon$$

Therefore, we have

$$\left(\frac{1}{\pi kT}\right)^{3/2} = \left(\frac{1}{3.14 \times (1.380\,6 \times 10^{-23}\ \text{J K}^{-1})(300\ \text{K})}\right)^{3/2}$$
$$= 6.742 \times 10^{29}\ \text{J}^{-3/2}$$

$$\exp(-\varepsilon/kT) = \exp\left(-\frac{6.213 \times 10^{-21}\ \text{J}}{(1.380\,6 \times 10^{-23}\ \text{J K}^{-1})(300\ \text{K})}\right) = 0.223$$

Thus $$\frac{\mathrm{d}N_\varepsilon}{N} = 2 \times 3.14 \times (6.742 \times 10^{29}\ \text{J}^{-3/2})(6.213 \times 10^{-21}\text{J})^{1/2}$$
$$\times (0.223)(6.213 \times 10^{-23}\ \text{J}) = 4.624 \times 10^{-3}$$

Example 1.17.6

Calculate the number of molecules in one mole of an ideal gas that have energies greater than four times the average thermal energy at 25 °C and 50 °C.

Solution

The average thermal energy $\bar{\varepsilon}$ is given as

$$\bar{\varepsilon} = \frac{3}{2}kT$$

The expression which gives the fraction of molecules having energies greater than ε is given as

$$\frac{N(\varepsilon)}{N_A} = 2\left(\frac{\varepsilon}{\pi kT}\right)^{1/2} \exp(-\varepsilon/kT)$$

Now, $$\varepsilon = 4\bar{\varepsilon} = 4\left(\frac{3}{2}kT\right) = 6\,kT$$

Substituting this in the above expression, we have

$$\frac{N(\varepsilon)}{N_A} = 2\left(\frac{6kT}{\pi kT}\right)^{1/2} \exp(-6kT/kT) = 2\left(\frac{6}{\pi}\right)^{1/2} \exp(-6)$$

$$= 2\left(\frac{6}{3.14}\right)^{1/2} (0.002\,479) = 6.85 \times 10^{-3}$$

Since $N_A = 6.022 \times 10^{23}\ \text{mol}^{-1}$, therefore

$$N(\varepsilon) = (6.022 \times 10^{23}\ \text{mol}^{-1})(6.85 \times 10^{-3}) = 4.13 \times 10^{21}\ \text{mol}^{-1}$$

Since the fraction $N(\varepsilon)/N_A$ is independent of temperature, the same number of molecules will have energies greater than four times the average thermal energy at the given temperatures of 25 °C and 50 °C.

1.18 BAROMETRIC DISTRIBUTION LAW

Derivation of Distribution Law

Barometric distribution law deals with the variation of pressure with height produced as a result of the influence of the gravitational field. The pressure at any height h is determined by the total weight of fluid in the column of unit area of cross-section above that height.

Let F_h be equal to weight of fluid in the column of area of cross-section A above height h and $(F_h + dF)$ be equal to weight of fluid in the column above height $(h + dh)$. Thus, $|dF|$ is equal to the weight of fluid in the column between heights h and $(h + dh)$ (Fig. 1.18.1).

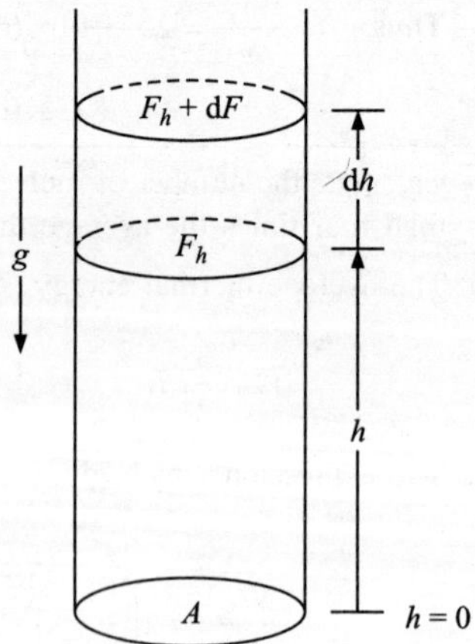

Fig. 1.18.1 Column of fluid in a gravitational field

Now,

$$\text{Weight of the fluid between heights } h \text{ and } (h + dh) = \rho(A\ dh)g$$

where ρ is the density of the fluid between the heights h and $(h + \mathrm{d}h)$. Since the weight of the fluid above any particular height decreases with increase in the height, it is obvious that $\mathrm{d}F$ and $\mathrm{d}h$ will have opposite signs. Thus

$$-\,\mathrm{d}F = \rho(A\,\mathrm{d}h)\,g \qquad \text{or} \qquad \mathrm{d}F = -\,\rho g A\,\mathrm{d}h$$

Now, since $\mathrm{d}p = \mathrm{d}F/A$, therefore $\mathrm{d}p = -\,\rho g\,\mathrm{d}h$ (1.18.1)

For Liquids

Since density ρ for liquids is independent of pressure, therefore, at height h the pressure is given by

$$\int_{p_0}^{p_h} \mathrm{d}p = \int_0^h -\rho g\,\mathrm{d}h$$

that is, $$p_h - p_0 = -\,\rho g h \tag{1.18.2}$$

where p_0 is the pressure at the bottom of the column and p_h is the pressure at height h above the bottom.

For Gases

In case of gases, density is a function of pressure. We can assume the gas to be ideal, such that

or $$pV = nRT = \frac{m}{M}RT$$

Therefore $$\frac{pM}{RT} = \frac{m}{V} = \rho$$

With this, Eq. (1.18.1) modifies to

$$\mathrm{d}p = -\frac{pM}{RT}g\,\mathrm{d}h$$

Separating variables, it yields

$$\frac{\mathrm{d}p}{p} = -\left(\frac{Mg}{RT}\right)\mathrm{d}h \tag{1.18.3}$$

which on integration gives

$$\ln(p/p^\circ) = -\frac{Mgh}{RT} + I \tag{1.18.4}$$

where p° is the standard unit pressure. Since at $h = 0$, $p = p_0$, therefore, the constant of integration I is given as

$$I = \ln(p_0/p^\circ)$$

Thus, Eq. (1.18.4) becomes

$$\ln\left(\frac{p}{p_0}\right) = -\frac{Mgh}{RT} \quad \text{or} \quad p = p_0 \exp(-Mgh/RT) \tag{1.18.5}$$

Equation (1.18.5) is known as the barometric distribution law.

Alternative Forms of Barometric Distribution Law

Two alternative forms of Eq. (1.18.5) are also commonly used. These are described in the following.

In terms of density of gas The density of a gas is directly proportional to its pressure. Thus, we have

$$\rho = \rho_0 \exp(-Mgh/RT) \tag{1.18.6}$$

In terms of number of particles per unit of volume

Since $pV = nRT$, therefore

$$pV = \frac{N}{N_A} RT \quad \text{or} \quad pN_A = \frac{N}{V} RT = N'RT$$

or $p \propto N'$

that is, the pressure of gas is directly proportional to the number of particles per unit volume. Thus, we have

$$N' = N'_0 \exp(-Mgh/RT) \quad \text{or} \quad N' = N'_0 \exp(-mgh/kT) \tag{1.18.7}$$

Parallelism with Boltzmann Distribution Law

Equation (1.18.7) is, in fact, the expression of Boltzmann distribution law, which basically deals with the distribution of molecules in various energy levels with respect to some reference level. The general form of the expression is

$$\frac{N_2}{N_1} = \exp(-\Delta E/RT) \tag{1.18.8}$$

where N_1 is the number of molecules in the reference level and N_2 is the corresponding number in a level whose energy differs from that of a reference level by an amount ΔE.

Effect of Temperature on Distribution

According to Eq. (1.18.5) the value of p/p_0 at a given height will be larger at a higher temperature than that at a lower temperature. Thus, the variation in pressure with height is less gradual at higher temperatures; if the temperature were infinite, the pressure would be the same everywhere in the column. Figure 1.18.2 shows the variation of p/p_0 with h at three different temperatures where $T_3 > T_2 > T_1$.

Effect of Height on Distribution

In Eq. (1.18.4), if $\ln(p/p^\circ)$ is plotted against h, one gets a straight line, with the slope = $-Mg/RT$ as shown in Fig. 1.18.3.

According to Eq. (1.18.5), the variation of p/p_0 with height is an exponential decay (Fig. 1.18.4). One of the characteristics of an exponential decay is that the relative decrease in pressure $-\mathrm{d}p/p$ is directly proportional to $\mathrm{d}h$ (Eq. 1.18.3). Thus, if the pressure is decreased to half (i.e. $p/p_0 = 1/2$, say at a distance x from the ground level), then at a distance x further from this, the ratio p_1/p will again be half. Thus $p_1/p = 1/2$ or $p_1 = p/2 = (1/2)\,(p_0/2)$, i.e. $p_1 = (1/4)p_0$. Thus, at a distance of $2x$ from the ground level the value of pressure will be one-fourth of the pressure at ground level and so on.

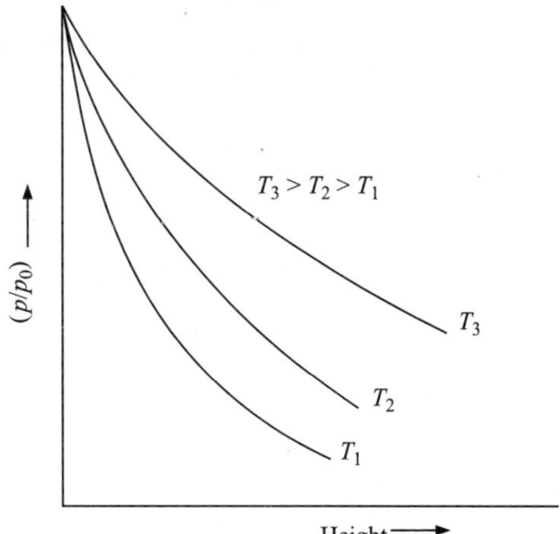

Fig. 1.18.2 Plot of p/p_0 with height at different temperatures

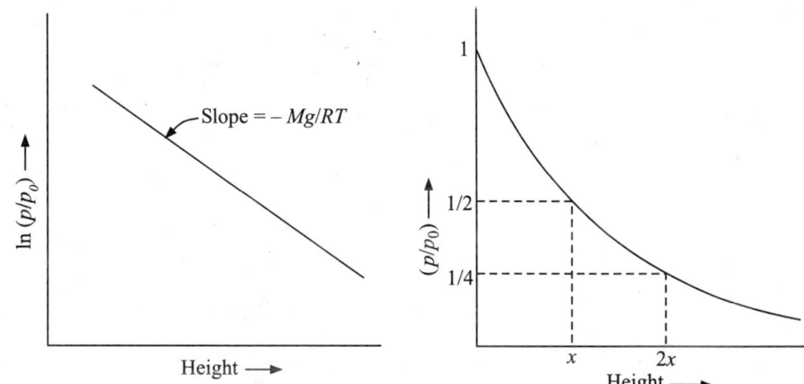

Fig. 1.18.3 Plot of ln (p/p_0) versus height

Fig. 1.18.4 Plot of p/p_0 versus height

Effect of Molar Mass on Distribution

The relative decrease in pressure $(-dp/p)$ is directly proportional to the molar mass of the gas (Eq. 1.18.3). Thus at a specified temperature, the relative decrease is larger for a gas having a higher molar mass than for a gas having a lower molar mass. For a gaseous mixture, as in the case of air, each of the gases obeys the distribution law independently of the others, so that

$$p_i = (p_0)_i \exp(-M_i gh/RT)$$

where $(p_0)_i$ is the partial vapour pressure of the constituent i at $h = 0$ and p_i is the corresponding pressure at the height h. The total pressure at any height will be the sum of the partial pressures at that height. It follows that the partial pressures of the lighter gases, such as H_2, He and Ne, decrease less rapidly with height than do those of the heavier gases, such as O_2 and N_2. Thus, the composition of air at very great heights is quite different from that at the ground level, the percentage of the lighter gases being more at greater heights.

Example 1.18.1 Determine the molar mass of a gas if its pressure is to fall to one-half of its value in a vertical distance of one kilometer at 298 K.

Solution From the given data, we write

$$p = p_0/2, \quad h = 10^3\ \text{m} \quad \text{and} \quad g = 9.807\ \text{m s}^{-2}$$

Substituting these values in the expression

$$p = p_0 \exp(-Mgh/RT)$$

i.e.

$$M = -\frac{RT}{gh} \ln \frac{p}{p_0}$$

we get

$$M = -\frac{(8.314\ \text{J K}^{-1}\text{mol}^{-1})(298\ \text{K})}{(9.807\ \text{m s}^{-2})(10^3\ \text{m})} \ln\left(\frac{1}{2}\right) = 175.1\ \text{g mol}^{-1}$$

Example 1.18.2 Calculate the pressure of a barometer on an aeroplane which is at an altitude of 10 km. Assume the pressure to be 101.325 kPa at sea level and the mean temperature 243 K. Use the average molar mass of air (80% N_2 and 20% O_2).

Solution From the given data, we have

$$p_0 = 101.325\ \text{kPa}, \quad h = 10\ \text{km} = 10^4\ \text{m}, \quad T = 243\ \text{K}$$

$$M = 0.8\,(0.028\ \text{kg mol}^{-1}) + 0.2\,(0.032\ \text{kg mol}^{-1}) = 0.028\,8\ \text{kg mol}^{-1}$$

Substituting these values in the expression

$$p = p_0 \exp(-\,Mgh/RT)$$

we get

$$p = (101.325\ \text{kPa}) \exp\left\{-\,\frac{(0.028\,8\ \text{kg mol}^{-1})(9.807\ \text{m s}^{-2})(10^4\text{m})}{(8.314\ \text{J K}^{-1}\ \text{mol}^{-1})(243\ \text{K})}\right\}$$

$$= (101.325\ \text{kPa})\,(0.247) = 25.036\ \text{kPa}$$

Problem 1.18.1 If the compression factor of a gas is $Z(p, T)$, we write equation of state as $pV/nRT = Z$. Show how this will affect the equation for the distribution of the gas in a gravitational field. From the differential equation for the distribution, show that if $Z > 1$, the distribution is broader for a real gas than for an ideal gas and converse is true if $Z < 1$. If $Z = 1 + bp$ where b is a function of temperature, integrate the equation and evaluate the constant of integration to obtain the explicit form of the distribution function.

Solution We have $\mathrm{d}p = -\,g\rho\,\mathrm{d}h$

Since $pV = ZnRT = Z\dfrac{m}{M}RT$, we obtain $\dfrac{pM}{ZRT} = \dfrac{m}{V} = \rho$

Therefore $\quad -\dfrac{dp}{p} = \dfrac{Mg}{ZRT}\,dh$

For $Z > 1$ the term Mg/ZRT for a real gas will be less than the corresponding term for an ideal gas, hence the distribution will be broader for the real gas. If $Z < 1$, this term is greater for a real gas than for an ideal gas and hence the distribution of a real gas is not as broad as that of an ideal gas.

If $Z = 1 + bp$, we have

$$dp = -\frac{p}{(1+bp)}\frac{Mg}{RT}\,dh \quad \text{or} \quad dp\left(\frac{1+bp}{p}\right) = -\frac{Mg}{RT}\,dh$$

or
$$\frac{dp}{p} + b\,dp = -\frac{Mg}{RT}\,dh \quad \text{or} \quad \frac{d(p/p^\circ)}{(p/p^\circ)} + b\,dp = -\frac{Mg}{RT}\,dh$$

where p° is standard unit pressure. Integrating this, we have

$$\ln(p/p^\circ) + bp = -\frac{Mgh}{RT} + C$$

The value of constant C is obtained by substituting $p = p_0$ at $h = 0$. Thus

$$C = \ln(p_0/p^\circ) + bp_0$$

The resultant expression becomes

$$\ln\frac{p}{p_0} + b(p - p_0) = -\frac{Mgh}{RT}$$

Problem 1.18.2 The temperature of air decreases linearly with altitude in accordance with the equation $T = T_0 - ah$, where a is a constant, h is altitude, T_0 is temperature at ground level and T is temperature at altitude h. Derive a modified form of the barometric equation taking into account this temperature dependence on altitude.

Solution We have $\quad dp = -\rho g\,dh$

that is $\quad dp = -\dfrac{pM}{RT}g\,dh$

or
$$\int\frac{dp}{p} = \frac{Mg}{R}\int\frac{dh}{T}$$

Now, since $T = T_0 - ah$, we can write

$$dT = -a\,dh \quad \text{or} \quad dh = dT/(-a)$$

Replacing dh in terms of dT in the above equation, we get

$$\int_{p_0}^{p} \frac{dp}{p} = -\frac{Mg}{R}\int_{T_0}^{T} \frac{1}{(-a)}\frac{dT}{T} = \frac{Mg}{aR}\int_{T_0}^{T}\frac{dT}{T}$$

or $$\ln\frac{p}{p_0} = \frac{Mg}{aR}\ln\frac{T}{T_0}$$

In terms of height, we have

$$\int_{p_0}^{p} \frac{dp}{p} = -\frac{Mg}{R}\int_{0}^{h}\frac{dh}{(T_0 - ah)}$$

i.e. $$\ln\frac{p}{p_0} = \frac{Mg}{Ra}\ln\left(\frac{T_0 - ah}{T_0}\right)$$

1.19 MOLECULAR COLLISIONS IN A GAS

While considering collisions of the molecules among themselves we assume the molecules to be rigid, non-interacting, and spherical with diameter σ. It is also assumed that all the molecules move with the same average speed $\overline{u}$.

Collision Cross-Section of Molecules

Two identical molecules of diameter σ will just touch each other when the distance separating their centres is σ. Thus, a moving molecule will collide with other molecules whose centres come within a distance of σ from its centre. The quantity $(\pi\sigma^2)$ is called the *collision cross-section* for the rigid spherical molecule. From Fig. 1.19.1 it is obvious that this collision cross-section is an area of an imaginary sphere of radius σ around the molecule within which the centre of another molecule cannot penetrate.

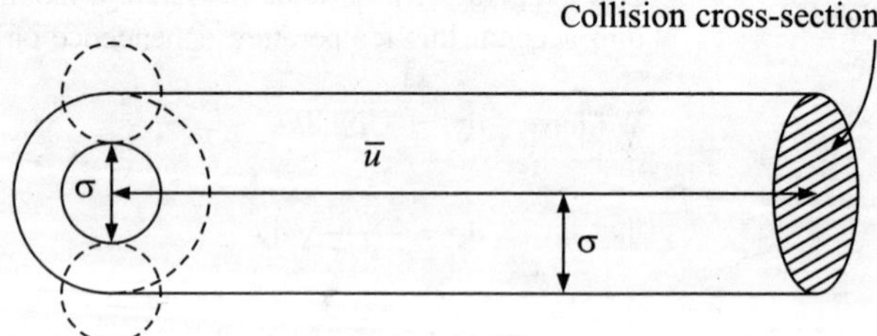

Fig. 1.19.1 Collision cross-section of a molecule

Expression of Number of Collisions

The volume swept by a single molecule in unit time is

$$V = (\pi\sigma^2)\,\overline{u}$$

If N^* is the number of molecules per unit volume, then the number of molecules within the volume V is

$$N = VN^* = (\pi\sigma^2\overline{u})\,N^*$$

Hence, the number of collisions made by a single molecule in unit time will be

$$Z_1 = N = (\pi\sigma^2\bar{u})\, N^*$$

So far, it is assumed that only one molecule is moving and all the others are stationary. In practice, however, this is not true. In order to account for the movements of all molecules, one must consider the average velocity along the line of centres of two colliding molecules instead of the average velocity of a single molecule. If it is assumed that, on an average, molecules collide while approaching each other perpendicularly, then the average velocity along their centres is $\sqrt{2}\bar{u}$ as shown below.

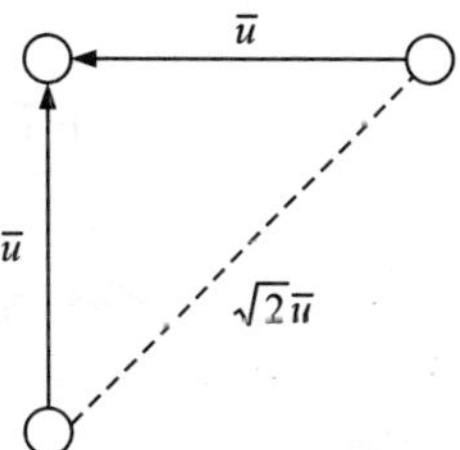

Number of collisions made by a single molecule with other molecules per unit time is given by

$$Z_1 = \pi\sigma^2(\bar{u}_{\text{rel}})N^* = \sqrt{2}\pi\sigma^2\bar{u}\; N^* \tag{1.19.1}$$

The total number of bimolecular collisions Z_{11} per unit volume per unit time is given by

$$Z_{11} = \frac{1}{2}(Z_1\; N^*)$$

or

$$Z_{11} = \frac{1}{2}(\sqrt{2}\pi\sigma^2\bar{u}\; N^*)N^* = \frac{1}{\sqrt{2}}\pi\sigma^2\bar{u}\; N^{*2} \tag{1.19.2}$$

(Note that the division by two is essential since the simple multiplication of Z_1 by N^* would count every collision twice.)

If the collisions involve two unlike molecules, then the number of collisions Z_{12} per unit volume per unit time is given as

$$Z_{12} = \pi\sigma_{12}^2\left(\sqrt{\frac{8kT}{\pi\mu}}\right)N_1N_2 \tag{1.19.3}$$

where N_1 and N_2 are the number of molecules per unit volume of the two types of molecules, σ_{12} is the average diameter of the two molecules (i.e. $\sigma_{12} = (\sigma_1 + \sigma_2)/2$) and μ is the reduced mass such that

$$\frac{1}{\mu} = \frac{1}{m_1} + \frac{1}{m_2}$$

The Mean Free Path The mean free path is *the average distance travelled by a molecule between two successive collisions*. We can express it as follows:

$$\lambda = \frac{\text{Average distance travelled per unit time}}{\text{No. of collisions made by a single molecule per unit time}}$$

$$= \frac{\bar{u}}{Z_1}$$

Substituting Z_1 from Eq. (1.19.1), we get

$$\lambda = \frac{\bar{u}}{\sqrt{2}\pi\sigma^2\bar{u}N^*} = \frac{1}{\sqrt{2}\pi\sigma^2 N^*} \tag{1.19.4}$$

Example 1.19.1 Calculate the values of σ, λ, Z_1 and Z_{11} for oxygen at 298.15 K at the pressure of 101.325 kPa, given van der Waals constant $b = 3.183 \times 10^{-2}$ dm^3 mol^{-1}.

Solution Number of molecules per unit volume,

$$N^* = \frac{p}{kT} = \frac{(101.325 \times 10^3 \text{ Pa})}{(1.380\,6 \times 10^{-23} \text{ J K}^{-1})(298.15 \text{ K})} = 246.157 \times 10^{23} \text{ m}^{-3}$$

The van der Waals constant b is

$$b = 4N_{\text{A}}\left(\frac{4}{3}\pi r^3\right)$$

Thus $$r = \left(\frac{3b}{16\pi N_{\text{A}}}\right)^{1/3} = \left(\frac{3 \times 3.183 \times 10^{-2} \text{ dm}^3 \text{ mol}^{-1}}{16 \times 3.14 \times 6.022 \times 10^{23} \text{ mol}^{-1}}\right)^{1/3}$$

$$= 1.467 \times 10^{-9} \text{ dm}$$

Therefore

$$\sigma = 2r = 2.934 \times 10^{-9} \text{ dm} = 2.934 \times 10^{-10} \text{ m}$$

Average speed,

$$\bar{u} = \sqrt{\frac{8RT}{\pi M}} = \left\{\frac{8(8.314 \text{ J K}^{-1} \text{ mol}^{-1})(298.15 \text{ K})}{(3.14)(0.032 \text{ kg mol}^{-1})}\right\}^{1/2} = 444.25 \text{ m s}^{-1}$$

Mean free path,

$$\lambda = \frac{1}{\sqrt{2}\pi\sigma^2 N^*}$$

$$= \frac{1}{(1.414)(3.14)(2.934 \times 10^{-10} \text{ m})^2(246.157 \times 10^{23} \text{ m}^{-3})} = 1.06 \times 10^{-7} \text{ m}$$

$$Z_1 = \sqrt{2}\pi\sigma^2\bar{u}\,N^*$$
$$= (1.414)\,(3.14)\,(2.934 \times 10^{-10}\text{ m})^2\,(444.25\text{ m s}^{-1})\,(246.157 \times 10^{23}\text{ m}^{-3})$$
$$= 4.18 \times 10^9\text{ s}^{-1}$$

$$Z_{11} = \frac{1}{2}Z_1 N^* = \frac{1}{2}(4.18 \times 10^9\text{ s}^{-1})(246.157 \times 10^{23}\text{ m}^{-3})$$
$$= 5.144 \times 10^{34}\text{ m}^{-3}\text{ s}^{-1}$$

Example 1.19.2

Calculate λ, Z_1 and Z_{11} for oxygen at 298 K and 10^{-3} mmHg. Given: $\sigma = 3.61 \times 10^{-8}$ cm.

Solution

From the given data, we have

$$N^* = \frac{p}{kT} = \frac{(10^{-3}\text{ mmHg})(133.322\text{ Pa/1 mmHg})}{(1.380\,6 \times 10^{-23}\text{ J K}^{-1})(298\text{ K})} = 3.24 \times 10^{19}\text{ m}^{-3}$$

$$\bar{u} = \sqrt{\frac{8RT}{\pi M}} = \left(\frac{8(8.314\text{ J K}^{-1}\text{ mol}^{-1})(298\text{ K})}{(3.14)(0.032\text{ kg mol}^{-1})}\right)^{1/2} = 444.1\text{ m s}^{-1}$$

Thus
$$Z_1 = \sqrt{2}\pi\sigma^2\bar{u}\,N^*$$
$$= (1.414)\,(3.14)\,(3.61 \times 10^{-10}\text{ m})^2\,(444.1\text{ m s}^{-1})(3.24 \times 10^{19}\text{ m}^{-3})$$
$$= 8.326 \times 10^3\text{ s}^{-1}$$

$$Z_{11} = \frac{1}{2}Z_1 N^* = \frac{1}{2}(8.326 \times 10^3\text{ s}^{-1})\,(3.24 \times 10^{19}\text{ m}^{-3})$$
$$= 13.488 \times 10^{22}\text{ m}^{-3}\text{ s}^{-1}$$

$$\lambda = \frac{1}{\sqrt{2}\pi\sigma^2 N^*} = \frac{1}{(1.414)(3.14)(3.61 \times 10^{-10}\text{m})^2(3.24 \times 10^{19}\text{m}^{-3})}$$
$$= 5.334 \times 10^{-2}\text{ m} = 5.334\text{ cm}$$

Example 1.19.3

The mean free path of the molecule of a certain gas at 300 K is 2.6×10^{-5} m. The collision diameter of the molecule is 0.26 nm. Calculate pressure of the gas and the number of molecules per unit volume of the gas.

Solution

Here $\lambda = 2.6 \times 10^{-5}\text{m}$, $\sigma = 0.26 \times 10^{-9}\text{m}$, $T = 300\text{ K}$

Since
$$\lambda = \frac{1}{\sqrt{2}\pi\sigma^2 N^*}$$

we get
$$N^* = \frac{1}{\sqrt{2}\pi\sigma^2\lambda} = \frac{1}{(1.414)(3.14)(0.26 \times 10^{-9}\text{m})^2(2.6 \times 10^{-5}\text{m})}$$
$$= 1.281 \times 10^{23}\text{m}^{-3}$$

Now,
$$p = N^*kT = (1.281 \times 10^{23}\text{ m}^{-3})\,(1.380\,6 \times 10^{-23}\text{ J K}^{-1})\,(300\text{ K})$$
$$= 5.306 \times 10^2\text{ J m}^{-3} \equiv 5.306 \times 10^2\text{ Pa}$$

Effect of Temperature and Pressure on Mean Free Path and Molecular Collisions

In order to discuss the effect of temperature and pressure on λ, Z_1 and Z_{11}, first we express the pressure and temperature dependence of N^* and $\bar{u}$. This can be done as follows.

Dependence of N^* on p and T

According to the ideal gas equation, we have

$$pV = nRT = \frac{N'}{N_A}RT \quad \text{or} \quad p = \left(\frac{N'}{V}\right)\left(\frac{R}{N_A}\right)T = \left(\frac{N'}{V}\right)kT$$

Thus, the number of molecules per unit volume is given by

$$N^* = \frac{N'}{V} = \frac{p}{kT} \tag{1.19.5}$$

Thus, $$N^* \propto \frac{p}{T} \tag{1.19.6}$$

Dependence of $\bar{u}$ on T

The average speed $\bar{u}$ is given as

$$\bar{u} = \sqrt{\frac{8RT}{\pi M}} = \sqrt{\frac{8kT}{\pi m}}$$

Thus, $$\bar{u} \propto \sqrt{T} \tag{1.19.7}$$

Effects on Mean Free Path

Now from Eq. (1.19.4), we get

$$\lambda = \frac{1}{\sqrt{2}\pi\sigma^2 N^*}$$

Thus, $$\lambda \propto \frac{1}{N^*}$$

Employing Eq. (1.19.6), we get

$$\lambda \propto \frac{T}{p} \tag{1.19.8}$$

Thus, $\lambda \propto T$ provided p is held constant

and $\lambda \propto \frac{1}{p}$ provided T is held constant.

Since, according to Gay Lussac's law, $p \propto T$ at constant volume, therefore, Eq. (1.19.8) under these conditions modifies to

$$\lambda \propto \text{(constant)}$$

that is, there will be no effect of changing T or p on λ if the volume of the gas is kept constant.

Effects of p and T on Molecular Collisions

From Eq. (1.19.1), we have

$$Z_1 = \sqrt{2}\pi\sigma^2 \, \bar{u} \, N^*$$

or $$Z_1 \propto \bar{u} \, N^*$$

Employing Eq. (1.19.6) and Eq. (1.19.7), this modifies to

$$Z_1 \propto (\sqrt{T})\left(\frac{p}{T}\right) \quad \text{or} \quad Z_1 \propto \frac{p}{\sqrt{T}} \tag{1.19.9}$$

Thus, $Z_1 \propto p$ when temperature is held constant and $Z_1 \propto 1/\sqrt{T}$ when pressure is held constant.

The effect of changing p or T at constant volume can be described by making use of Gay-Lussac's law in Eq. (1.19.9). Thus, we have

$$Z_1 \propto \frac{T}{\sqrt{T}} \propto \sqrt{T} \qquad \text{(volume constant)}$$

and $$Z_1 \propto \frac{p}{\sqrt{p}} \propto \sqrt{p} \qquad \text{(volume constant)}$$

From Eq. (1.19.2), we have

$$Z_{11} = \frac{1}{\sqrt{2}}\pi\sigma^2\bar{u} \, N^{*2} \quad \text{or} \quad Z_{11} \propto \bar{u} \, N^{*2}$$

Employing Eq. (1.19.6) and Eq. (1.19.7), this modifies to

$$Z_{11} \propto (\sqrt{T})\left(\frac{p}{T}\right)^2 \quad \text{or} \quad Z_{11} \propto \frac{p^2}{T^{3/2}}$$

Thus, $Z_{11} \propto p^2$ provided temperature is held constant and $Z_{11} \propto T^{-3/2}$ provided pressure is held constant.

The effect of changing p or T at constant volume can again be described by making use of Gay Lussac's law. Thus

$$Z_{11} \propto \frac{p^2}{T^{3/2}} \propto \frac{T^2}{T^{3/2}} \propto T^{1/2} \qquad \text{(volume constant)}$$

and $$Z_{11} \propto \frac{p^2}{T^{3/2}} \propto \frac{p^2}{p^{3/2}} \propto p^{1/2} \qquad \text{(volume constant)}$$

1.20 VISCOSITY

Introduction

The internal friction which opposes the relative motion of adjacent layers of a fluid is known as viscosity.

Viscosity of Gases

In a laminar flow of a fluid in a cylindrical tube, layers just touching the sides of the tube are stationary, and velocities of the adjacent layers increase towards the centre of the tube, the layer in the centre of the tube has a maximum velocity. There thus exists a velocity gradient amongst different layers of a liquid.

In case of gases, because of their continuous movement, there occurs an interchange of molecules between two layers, with the result that a fraction of momentum of one layer is passed over to the other layer. The net effect is to decrease the relative rate of movement of one layer with respect to the other. In order to maintain a uniform velocity gradient, one has to apply a force along the direction of movement of the layers. This applied force is a measure of internal friction or viscosity of the fluid.

Viscosity of Liquids

In case of liquids, this internal friction arises because of intermolecular attractions. Molecules in a slower moving layer try to decrease the velocity of the molecules in a faster moving layer and vice versa, with the result that some tangential force is required to maintain a uniform flow.

Definition of Coefficient of Viscosity

The tangential force F required to maintain uniform velocity of layers will depend upon two factors, viz.,

(i) Area A of contact between the two adjacent layers
(ii) Velocity gradient $\mathrm{d}u/\mathrm{d}z$

Thus,

$$F \propto A\frac{\mathrm{d}u}{\mathrm{d}z} \quad \text{that is} \quad F = \eta A\frac{\mathrm{d}u}{\mathrm{d}z} \tag{1.20.1}$$

where η is known as the *coefficient of viscosity* (or simply viscosity). It is the tangential force that must be applied in order to maintain a velocity difference of unity between two parallel layers unit distance apart and having unit area of contact. SI unit of the coefficient of viscosity is N $\mathrm{m^{-2}}$s. In CGS units, it has the unit of dyn $\mathrm{cm^{-2}}$ s and is known as the poise unit.

Expression of Viscosity of Gases

Viscosity in case of a gas arises because of transfer of momentum across the layers of the gas. Consider a layer P–P′ at a height z, (Fig. 1.20.1), moving with a velocity u_z. Let the velocity gradient be $\mathrm{d}u/\mathrm{d}z$. Let us consider the molecules entering and leaving this layer. We assume:

(i) That the flow velocity u_z is very small as compared with the mean gas velocity $\overline{u}$.

(ii) That the only molecules reaching P–P′ are those which, on an average, have just made their last collision at a distance λ from the height z.

(iii) The number of molecules passing downwards or upwards through a unit area per unit time = $N^*\overline{u}/4$, where N^* is the number of molecules per unit volume and $\overline{u}$ is the average speed of gaseous molecules.[†]

[†] Assuming statistical motion of molecules, the number of molecules will come out to be $N^*\overline{u}/6$ as one sixth of the molecules will be moving along each of the positive and negative directions of x-, y-, and z-axes. However, the explicit expression of $N^*\overline{u}/4$ can be derived statistically by using velocity-component distribution function (Eq. 4.14.12 of Volume 5 of this series of book).

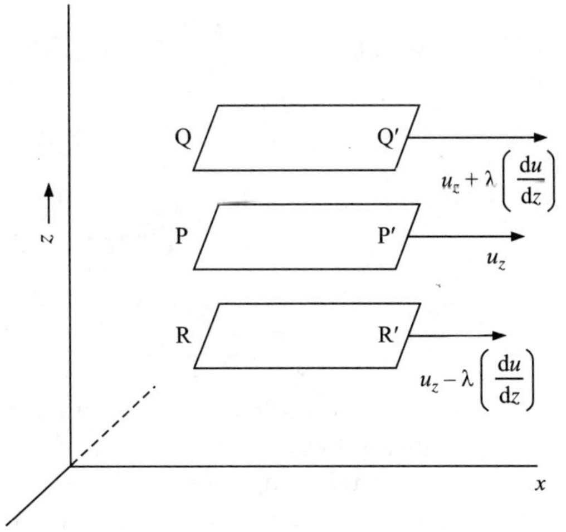

Fig. 1.20.1 **Display of velocity gradient with the height of layer**

In the plane P–P′, the amount of horizontal momentum coming up through a unit area per unit time is

$$(mu)\uparrow = m\left(\frac{1}{4}N^*\bar{u}\right)\left(u_z - \lambda\frac{du}{dz}\right)$$

and the amount of horizontal momentum coming down

$$(mu)\downarrow = m\left(\frac{1}{4}N^*\bar{u}\right)\left(u_z + \lambda\frac{du}{dz}\right)$$

The net downward flow of x momentum in unit time

$$(mu)\downarrow - (mu)\uparrow = \frac{1}{2}N^*\bar{u}\, m\lambda\frac{du}{dz}$$

Since momentum transfer in unit time is numerically equal to the force, therefore, the force acting in the x-direction on a unit area of the layer is

$$F = \frac{1}{2}mN^*\bar{u}\lambda\frac{du}{dz}$$

Comparing this with Eq. (1.20.1) with A = unit area, we get

$$\eta = \frac{1}{2}mN^*\bar{u}\lambda = \frac{1}{2}\rho\bar{u}\lambda \qquad (1.20.2)^{\dagger}$$

where ρ is the density of the medium.

† The more rigorous calculations based on the hard sphere model gives the expression of viscosity as $\eta = (5\pi/32)\, mN^*\bar{u}\,\lambda$.

Effect of p and T on Viscosity of Gases

We have $\bar{u} = \sqrt{\dfrac{8kT}{\pi m}}$ and $\lambda = \dfrac{1}{\sqrt{2}\pi\sigma^2 N^*}$

Substituting these in Eq. (1.20.2), we have

$$\eta = \frac{1}{2} mN^* \left(\sqrt{\frac{8kT}{\pi m}}\right)\left(\frac{1}{\sqrt{2}\pi\sigma^2 N^*}\right)$$

or $$\eta = \frac{(mkT)^{1/2}}{\pi^{3/2}\sigma^2} \tag{1.20.3}$$

According to this equation, η is independent of pressure. Experimentally this is found to be true. When the pressure is so low that the mean free path becomes comparable with the dimensions of the apparatus, the collisions of molecules are primarily with the walls, and under such circumstances, Eq. (1.20.2) is not applicable.

Equation (1.20.3) shows that η should also be independent of density of gas and this is in agreement with the experimental observations. This equation also suggests that $\eta \propto T^{1/2}$, but a somewhat larger exponent more like $T^{0.7}$ is observed for real gases, partly due to the fact that the cross-sectional diameter becomes smaller at higher temperatures due to increased penetration of the potential energy barrier by gas molecules of higher velocities. In contrast with the viscosity of ideal gases, the viscosity of liquids decreases with the rise in temperature.

Example 1.20.1 The van der Waals constant b for n-heptane is 0.265 4 dm^3 mol^{-1}. Estimate the coefficient of viscosity of this gas at 298 K. Calculate σ from b assuming molecules to be spherical.

Solution We are given that $b = 0.265\,4\ \text{dm}^3\ \text{mol}^{-1}$

Since $$b = 4N_A\left(\frac{4}{3}\pi r^3\right)$$

we get $$r = \left(\frac{3b}{16N_A\pi}\right)^{1/3} = \left\{\frac{3(0.265\,4\ \text{dm}^3\ \text{mol}^{-1})}{16(6.022\times 10^{23}\ \text{mol}^{-1})(3.14)}\right\}^{1/3}$$

$$= 0.297\,4\times 10^{-8}\ \text{dm}$$

Therefore $\sigma = 2r = 0.594\,8\times 10^{-8}\ \text{dm}$

Molar mass of n-heptane = 100 g mol^{-1} = 0.1 kg mol^{-1}

Thus $$\eta = \frac{\sqrt{mkT}}{\pi^{3/2}\sigma^2} = \frac{\{(0.1\ \text{kg}/6.022\times 10^{23})(1.380\,6\times 10^{-23}\ \text{J K}^{-1})(298\ \text{K})\}^{1/2}}{(3.14)^{3/2}(0.594\,8\times 10^{-9}\ \text{m})^2}$$

$$= \frac{(2.613\,8\times 10^{-23}\ \text{J}^{1/2}\ \text{kg}^{1/2})}{(1.968\,5\times 10^{-18}\text{m}^2)} = 1.328\times 10^{-5}\ \text{J}^{1/2}\ \text{kg}^{1/2}\ \text{m}^{-2}$$

$$= 1.328\times 10^{-5}\ \text{N m}^{-2}\ \text{s} = 1.328\times 10^{-4}\text{dyn cm}^{-2}\ \text{s} = 1.328\times 10^{-4}\ \text{poise}$$

Example 1.20.2 Calculate the number of collisions per square metre per second of O_2 molecules with a wall at a pressure of 101.325 kPa and temperature 298 K.

Solution Number of collisions per unit area per unit time = $\frac{1}{4}N^* u_{av}$

Now $$N^* = \frac{p}{kT} = \frac{101.325 \times 10^3 \text{ N m}^{-2}}{(1.380\,6 \times 10^{-23} \text{ J K}^{-1})(298 \text{ K})} = 2.463 \times 10^{25} \text{ m}^{-3}$$

$$\bar{u} = \sqrt{\frac{8RT}{\pi M}} = \left\{\frac{8(8.314 \text{ J K}^{-1} \text{ mol}^{-1})(298 \text{ K})}{3.14(0.032 \text{ kg mol}^{-1})}\right\}^{1/2} = 444.1 \text{ m s}^{-1}$$

Hence, Number of collisions

$$= \frac{1}{4}N^* \bar{u} = \frac{1}{4}(2.463 \times 10^{25} \text{ m}^{-3})(444.1 \text{ m s}^{-1}) = 2.734 \times 10^{27} \text{ m}^{-2} \text{ s}^{-1}$$

1.21 SELF DIFFUSION AND EFFUSION PROCESSES

Introduction to Self Diffusion

Movement of molecules from high-density region to low-density region constitutes the phenomenon of self diffusion. Experimentally, the self diffusion is given by Fick's law:

$$J_z = -D \frac{\mathrm{d}c}{\mathrm{d}z} \tag{1.21.1}$$

where J_z is the net flow of matter per unit area per unit time and D is the coefficient of diffusion.

Expression of Coefficient of Diffusion

Consider a unit area in the gaseous layer at height z (Fig. 1.21.1). The number of molecules per unit volume at height z relative to zero level is given by

$$N = N_0 + \left(\frac{\partial N}{\partial z}\right) z \tag{1.21.2}$$

where N_0 is the number of molecules per unit volume at $z = 0$.

Also consider two layers at the heights $z - \lambda$ and $z + \lambda$ respectively, where λ is the mean free path of gaseous molecules. Only those molecules will pass through the layer at the height z who have their last collision at a distance λ from this layer. Hence, the number of molecules passing downward through a unit area at the height z per unit time is given by

$$N\downarrow = \frac{1}{4}N^*_{z+\lambda}\,\bar{u} = \frac{1}{4}\left[N_0 + \left(\frac{\partial N}{\partial z}\right)(z+\lambda)\right]\bar{u}$$

Similarly, the number of molecules coming up from the lower layer is

$$N\uparrow = \frac{1}{4}N^*_{z-\lambda}\,\bar{u} = \frac{1}{4}\left[N_0 + \left(\frac{\partial N}{\partial z}\right)(z-\lambda)\right]\bar{u}$$

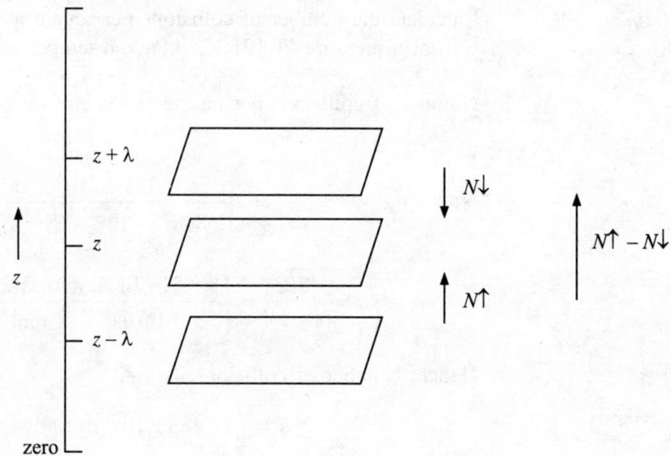

Fig. 1.21.1 Process of self diffusion

The net flow of the molecules in the upward direction is

$$N\uparrow - N\downarrow = \frac{1}{4}\bar{u}\left[\left\{N_0 + \left(\frac{\partial N}{\partial z}\right)(z-\lambda)\right\} - \left\{N_0 + \left(\frac{\partial N}{\partial z}\right)(z+\lambda)\right\}\right]$$

$$= -\frac{1}{2}\bar{u}\lambda\left(\frac{\partial N}{\partial z}\right) \tag{1.21.3}$$

Comparing Eq. (1.21.3) with Eq. (1.21.1), we get

$$D = \frac{1}{2}\bar{u}\lambda \tag{1.21.4}$$

Effect of p and T on the Diffusion Coefficient

We have

$$\bar{u} = \sqrt{\frac{8kT}{\pi m}} \tag{1.21.5}$$

$$\lambda = \frac{1}{\sqrt{2}\pi\sigma^2 N^*} = \frac{kT}{\sqrt{2}\pi\sigma^2 p} \tag{1.21.6}$$

Substituting Eqs (1.21.5) and (1.21.6) in Eq. (1.21.4), we get

$$D = \frac{1}{2}\left(\sqrt{\frac{8kT}{\pi m}}\right)\left(\frac{kT}{\sqrt{2}\pi\sigma^2 p}\right) \tag{1.21.7}$$

Hence, the coefficient of diffusion depends

(i) inversely on pressure at constant temperature, and
(ii) directly on $T^{3/2}$ at constant pressure.

Effusion of Gases

The process of effusion involves the escaping of gas molecules through a tiny hole.

Suppose a gas enclosed in a container is placed in vacuum. Let the wall of the container has a tiny hole of diameter much lesser than the mean free path of gaseous molecules. A molecule that strikes the tiny hole undergoes effusion and thus is escaped from the container. We know that the number of molecules striking per unit area of the wall of a container in a unit time is given by

$$N = \frac{1}{4}\bar{u}\, N^* \tag{1.21.8}$$

where $\bar{u}(=\sqrt{8RT/\pi M})$ is the average speed of gaseous molecules and $N^*(=p/kT)$ is its number density (i.e. number of molecules per unit volume). If A_{hole} is the area of tiny hole, then

Number of molecules striking the tiny hole in a unit time is given by

$$N' = NA_{\text{hole}} = \left(\frac{1}{4}\bar{u}\, N^*\right)A_{\text{hole}} \tag{1.21.9}$$

Hence, the rate of effusion becomes

$$\begin{aligned} r_{\text{effusion}} &= -\frac{\mathrm{d}N}{\mathrm{d}t} = N' \\ &= \left(\frac{1}{4}\bar{u}\, N^*\right)A_{\text{hole}} = \frac{1}{4}\left(\frac{8RT}{\pi M}\right)^{1/2}\left(\frac{p}{kT}\right)A_{\text{hole}} \\ &= \frac{1}{4}\left(\frac{8RT}{\pi M}\right)^{1/2}\left(\frac{N_{\text{A}}p}{RT}\right)A_{\text{hole}} = \frac{pN_{\text{A}}A_{\text{hole}}}{\sqrt{2\pi MRT}} \end{aligned} \tag{1.21.10}$$

Thus, the rate of effusion of a gas is inversely proportional to the square root of its molar mass, which is, Graham's law of effusion. It is also directly proportional to the pressure of the gas.

Knudsen Method for the Measurement of Vapour Pressure

Equation (1.21.10) forms the basis of determining the vapour pressure of a liquid or solid. The substance is taken in a closed container of thin wall with a tiny hole through which the vapours of the substance can escape in vacuum. The pressure of vapours inside the container remains constant as these are in equilibrium with the solid or liquid. Consequently, the rate at which vapours escape through the hole is equal to the rate of evaporation of the substance in the container. The effusion leads to the loss of mass of the substance in the container. The rate of loss of mass of the substance is measured. We will have

$$-\frac{\mathrm{d}m}{\mathrm{d}t} \equiv -\frac{\mathrm{d}}{\mathrm{d}t}\left(N\frac{M}{N_{\text{A}}}\right) \equiv -\frac{M}{N_{\text{A}}}\frac{\mathrm{d}N}{\mathrm{d}t}$$

Substituting dN/dt from Eq. (1.21.10), we get

$$-\frac{\mathrm{d}m}{\mathrm{d}t} = \frac{M}{N_A}\left[\frac{pN_A A_{\text{hole}}}{\sqrt{2\pi MRT}}\right]$$

or
$$p = \frac{1}{A_{\text{hole}}}\left(-\frac{\mathrm{d}m}{\mathrm{d}t}\right)\left(\frac{2\pi RT}{M}\right)^{1/2} \tag{1.21.11}$$

Since p remains constant, dm/dt may be replaced by $\Delta m/\Delta t$, where Δm is the loss of mass of the substance in a small interval of time, Δt.

To account for the size of the hole, an empirical parameter C (known as Clausing factor) is introduced in Eq. (1.21.11) to give

$$p = \frac{C}{A_{\text{hole}}}\left(-\frac{\Delta m}{\Delta t}\right)\left(\frac{2\pi RT}{M}\right)^{1/2} \tag{1.21.12}$$

The value of $C = 1$ for $l/r \to 0$ and 1.95 for $l/r \to 2$, where l is the length and r is the radius of the hole.

Example 1.21.1 Solid Sc exists in equilibrium with its vapours at 1 690 K in a container with a tiny hole of diameter 1.763 mm. The container is placed in vacuum. A loss of 10.5 mg was recorded in 49.5 min. Determine the vapour pressure of Sc at 1 690 K. The molar mass of Sc is 44.96 g mol^{-1}.

Solution We have

$$-\frac{\Delta m}{\Delta t} = \frac{10.5\times 10^{-6}\ \text{kg}}{49.5\times 60\ \text{s}} = 3.535\times 10^{-9}\ \text{kg s}^{-1}$$

$$A_{\text{hole}} = \pi r^2 = \pi(d/2)^2 = (3.14)(1.763\times 10^{-3}\text{m}/2)^2$$
$$= 2.44\times 10^{-6}\text{m}^2$$

$$\sqrt{\frac{M}{2\pi RT}} = \left[\frac{(44.96\times 10^{-3}\ \text{kg mol}^{-1})}{2\,(3.14)\,(8.314\ \text{J K}^{-1}\ \text{mol}^{-1})\,(1\,690\ \text{K})}\right]^{1/2}$$
$$= 7.14\times 10^{-4}\ \text{m}^{-1}\ \text{s}$$

$$p = \frac{-(\Delta m/\Delta t)}{A_{\text{hole}}\sqrt{M/2\pi RT}} = \frac{(3.535\times 10^{-9}\ \text{kg s}^{-1})}{(2.44\times 10^{-6}\ \text{m}^2)(7.14\times 10^{-4}\ \text{m}^{-1}\ \text{s})}$$
$$= 2.03\ \text{Pa}$$

1.22 THERMAL CONDUCTIVITY

Coefficient of Thermal Conductivity

If there exists a temperature gradient in a gaseous system, energy flows from a high-temperature region to a low-temperature region. The heat flowing through a unit area of a layer per unit time along the temperature gradient $\partial T/\partial z$ is given by

$$Q = -\kappa\left(\frac{\partial T}{\partial z}\right) \qquad (1.22.1)$$

where κ is known as the coefficient of thermal conductivity. It has the unit of J K^{-1} m^{-1} s^{-1}.

Expression of Thermal Conductivity

Consider a unit area in a horizontal plane at the height z of the gaseous system (Fig. 1.22.1). Only those molecules are able to pass through this area that have suffered the last collision at a distance λ (mean free path) from the considered plane. Thus, we consider two layers at the distance λ above and below the layer at height z, respectively (Fig. 1.22.1).

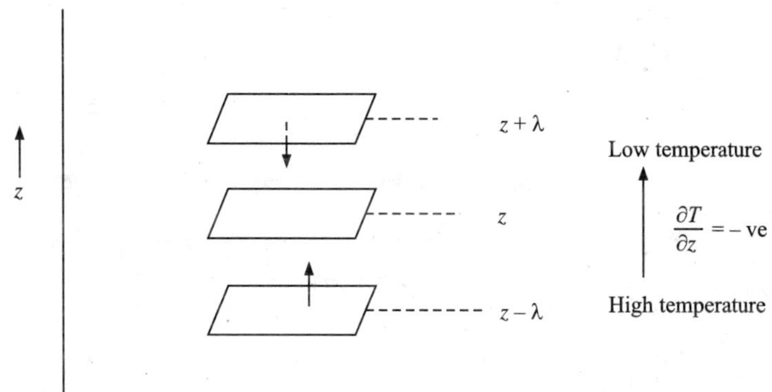

Fig. 1.22.1 Process of thermal conductivity

If $\bar{\varepsilon}_z$ is the average energy of molecules in the layer at height z, then Average energy transferred per unit area per unit time from the plane below the considered plane is

$$E\uparrow = \left(\frac{N^*}{4}\bar{u}\right)\bar{\varepsilon}_{z-\lambda} = \left(\frac{N^*}{4}\bar{u}\right)\left[\bar{\varepsilon}_z - \left(\frac{\partial E}{\partial z}\right)\lambda\right]$$

Similarly, from the plane above the considered plane the average energy transferred is

$$E\downarrow = \left(\frac{N^*}{4}\bar{u}\right)\bar{\varepsilon}_{z+\lambda} = \left(\frac{N^*}{4}\bar{u}\right)\left[\bar{\varepsilon}_z + \left(\frac{\partial E}{\partial z}\right)\lambda\right]$$

The net flow of energy from below the considered plane per unit area per unit time is

$$
\begin{aligned}
Q &= E\uparrow - E\downarrow \\
&= \left(\frac{N^*}{4}\bar{u}\right)\left[\left\{\bar{\varepsilon}_z - \left(\frac{\partial E}{\partial z}\right)\lambda\right\} - \left\{\bar{\varepsilon}_z + \left(\frac{\partial E}{\partial z}\right)\lambda\right\}\right] \\
&= -\frac{N^*}{2}\bar{u}\lambda\left(\frac{\partial E}{\partial z}\right) = -\frac{N^*}{2}\bar{u}\lambda\left(\frac{\partial E}{\partial T}\right)\left(\frac{\partial T}{\partial z}\right) \\
&= -\frac{1}{2}N^*\bar{u}\lambda c\left(\frac{\partial T}{\partial z}\right) \qquad (1.22.2)
\end{aligned}
$$

where c is the average heat capacity per gaseous molecule. Comparing Eq. (1.22.2) with Eq. (1.22.1), we get

$$\kappa = \frac{1}{2}N^*\bar{u}\lambda c \qquad (1.22.3)^*$$

Effect of *T* and *p* on the Coefficient of Thermal Conductivity

We have

$$\bar{u} = \sqrt{\frac{8kT}{\pi m}}, \quad \text{and} \quad \lambda = \frac{1}{\sqrt{2}\pi\sigma^2 N^*}$$

Substituting these in Eq. (1.22.3), we get

$$
\begin{aligned}
\kappa &= \frac{1}{2}N^*\left(\frac{8kT}{\pi m}\right)^{1/2}\left(\frac{1}{\sqrt{2}\pi\sigma^2 N^*}\right)c \\
&= \frac{(kT)^{1/2}c}{\pi^{3/2}m^{1/2}\sigma^2} \qquad (1.22.4)^\dagger
\end{aligned}
$$

Hence, the coefficient of thermal conductivity is

(i) independent of pressure of the gas, and
(ii) directly proportional to the square root of temperature.

Though the phenomenon of thermal conductivity requires the presence of temperature gradient, yet temperature appears in the expression of coefficient of thermal conductivity due to the average velocity of gaseous molecules. Equation (1.22.4) is valid if the temperature difference is not large, so that the average value of temperature may used to compute the value of average velocity.

---

* The factor 1/2 in Eq. (1.22.3) is not to be trusted too much in this simplified calculation.

† It may be mentioned that the collision diameter σ also varies with the temperature as discussed in Section 1.19. As a result κ increases somewhat more rapidly with increasing temperature than is given by Eq. (1.22.4).

Example 1.22.1

Two parallel plates 0.5 cm apart are maintained at 299 K and 301 K. The space between the two plates is filled with hydrogen. If σ = 190 pm and C_V = (5/2) R, Calculate the flow of heat between the two plates.

Solution

The expression of flow of heat per unit area per unit time is

$$Q = -\kappa\left(\frac{\partial T}{\partial z}\right)$$

$$= -\frac{(kT)^{1/2}c}{\pi^{3/2}m^{1/2}\sigma^2}\left(\frac{\partial T}{\partial z}\right)$$

$$= -\frac{[(1.38\times10^{-23}\ \text{J K}^{-1})(300\ \text{K})]^{1/2}\left[\frac{5}{2}\left(\frac{8.314\ \text{J K}^{-1}\ \text{mol}^{-1}}{6.022\times10^{23}\ \text{mol}^{-1}}\right)\right]}{(3.14)^{3/2}\left(\frac{0.002\ \text{kg mol}^{-1}}{6.022\times10^{23}\ \text{mol}^{-1}}\right)^{1/2}(190\times10^{-12}\ \text{m})^2}\times\left(\frac{-2\ \text{K}}{0.5\times10^{-2}\ \text{m}}\right)$$

$$= 76.75\ \text{J m}^{-2}\ \text{s}^{-1}$$

Comment on Ratio κ/η

Dividing Eq. (1.22.4) by Eq. (1.20.3), we get

$$\frac{\kappa}{\eta} = \frac{(kT)^{1/2}c/\pi^{3/2}m^{1/2}\sigma^2}{(mkT)^{1/2}/\pi^{3/2}\sigma^2} = \frac{c}{m} = \frac{C_V}{M}$$

or
$$\frac{\kappa/\eta}{C_V/M} = 1 \qquad (1.22.5)$$

Experimentally, the above ratio is found to be in the range between 1.3 and 2.5. This discrepancy may be accounted for as the effects due to the distribution of molecular velocities have not been considered in deriving Eqs (1.22.4) and (1.22.5). The faster molecules cross a given plane more frequently than the slower ones. In thermal conductivity, the faster molecules carry more kinetic energy but not the greater average x component of momentum. Thus, the ratio of Eq. (1.22.5) is increased from the value of one.

1.23 ELECTRICAL CONDUCTIVITY

Definition

If a gaseous system containing charged particles is subjected to an electric field E there occurs migration of charged particles from the high-potential region to the low-potential region. The generated electric density (i.e. the average charge crossing per unit area per unit time) is given by

$$j_z = \sigma_{\text{el}}E \qquad (1.23.1)$$

where σ_{el} is known as electrical conductivity.

Expression of Electrical Conductivity

If $\bar{v}_z$ is the average component of velocity of charged particles under the influence of electric field in z-direction, then the number of charged particles crossing per unit area (perpendicular to the z-direction) per unit time is given by

$$N' = N^* \bar{v}_z \tag{1.23.2}$$

where N^* is the number of molecules per unit volume of the gaseous system. If each particle carries a charge e, then the electric density current is given by

$$j_z = N^* \bar{v}_z e \tag{1.23.3}$$

The drift velocity of the charged particle may be obtained by using Newton's second law of motion. This gives

$$m\frac{\mathrm{d}v_z}{\mathrm{d}t} = eE \quad \text{or} \quad \mathrm{d}v_z = \frac{eE}{m}\mathrm{d}t$$

which on integrating gives

$$v_z = \frac{eE}{m}t + \text{constant}$$

If it be assumed that after each collision, the particles restore to thermal equilibrium, the value of constant may be taken to be zero. Hence,

$$v_z = \frac{eE}{m}t \tag{1.23.4}$$

The average value of v_z is obtained by taking the average time between the collisions. Let it be represented by τ. Hence

$$\bar{v}_z = \frac{eE}{m}\tau \tag{1.23.5}$$

With this, Eq. (1.23.3) becomes

$$j_z = N^* e\left(\frac{eE}{m}\tau\right) = \left(\frac{N^* e^2}{m}\tau\right)E \tag{1.23.6}$$

Comparing Eq. (1.23.6) with Eq. (1.23.1), we get

$$\sigma_{\text{el}} = \frac{N^* e^2}{m}\tau \tag{1.23.7}$$

1.24 LAW OF EQUIPARTITION OF ENERGY

Classical Law of Equipartition of Energy

The average energy of a molecule can be calculated with the help of the classical law of equipartition of energy. The latter may be stated as follows:

If the energy of a molecule can be written in the form of a sum of terms, each of which is proportional to the square of a velocity component (or to the square of a position coordinate), then each of these square terms contributes (1/2)kT to the average energy.

The above law can be derived by evaluating the average value of the x-component of average kinetic energy with the help of Maxwell distribution law.[†] But here we follow a simple method based on the conceptual analysis of the average kinetic energy of molecules. According to Eq. (1.7.5), the latter is given as

$$\overline{\mathrm{KE}} = \bar{\varepsilon} = \frac{1}{2}m\overline{u^2} = \frac{3}{2}kT \qquad (1.24.1)$$

Now the mean square velocity can be written in terms of its components as

$$\overline{u^2} = \overline{u_x^2} + \overline{u_y^2} + \overline{u_z^2}$$

Since the designation of the components is arbitrary, it follows that

$$\overline{u_x^2} = \overline{u_y^2} = \overline{u_z^2}$$

Thus, Eq. (1.25.1) can be written as

$$\bar{\varepsilon} = \frac{1}{2}m\overline{u^2} = \frac{1}{2}m(\overline{u_x^2} + \overline{u_y^2} + \overline{u_z^2}) = \frac{3}{2}m\overline{u_x^2} = \frac{3}{2}kT$$

which gives

$$\frac{1}{2}m\overline{u_x^2} = \frac{1}{2}kT \qquad (1.24.2)$$

Similarly,

$$\frac{1}{2}m\overline{u_y^2} = \frac{1}{2}kT \qquad (1.24.3)$$

$$\frac{1}{2}m\overline{u_z^2} = \frac{1}{2}kT \qquad (1.24.4)$$

Thus, the average total kinetic energy can be divided into three components, each of which is proportional to the square of the velocity component and thus contributes $(1/2)kT$ to the average energy.

Average Energy of Different Modes of Motion

A polyatomic molecule, besides having translational motion, also has rotational and vibrational motions. The average energy stored in these motions can be calculated using the law of equipartition of energy as shown in the following.

Translational Motion The translational energy of a gas molecule is

$$\bar{\varepsilon}_{\mathrm{trans}} = \frac{1}{2}mu^2 = \frac{1}{2}mu_x^2 + \frac{1}{2}mu_y^2 + \frac{1}{2}mu_z^2$$

Since each term is proportional to the square of a velocity component, each contributes $(1/2)kT$ to the average energy. Thus, the average contribution of translational energy is

$$\bar{\varepsilon}_{\mathrm{trans}} = \frac{1}{2}kT + \frac{1}{2}kT + \frac{1}{2}kT = \frac{3}{2}kT \qquad (1.24.5)$$

Rotational Motion In order to describe the orientation of a linear molecule in space, we need to specify two angles about two axes while for a nonlinear molecule, three angles are required. Motion in these coordinates corresponds to rotation about two axes

[†] See Section 4.14 of Vol. 5 of this series of the book.

for a linear molecule and about three axes for a nonlinear molecule in space. The equation for the energy of rotation is given by

$$\varepsilon_{\text{rot}} = \frac{1}{2}I_x\omega_x^2 + \frac{1}{2}I_y\omega_y^2 \qquad \text{(linear molecule)}$$

$$\varepsilon_{\text{rot}} = \frac{1}{2}I_x\omega_x^2 + \frac{1}{2}I_y\omega_y^2 + \frac{1}{2}I_z\omega_z^2 \qquad \text{(nonlinear molecule)}$$

where ω_x, ω_y, ω_z are angular velocities and I_x, I_y, I_z are moments of inertia about the *x*-, *y*-, and *z*-axes, respectively. Since each term in the above expressions is proportional to the square of the velocity component, it contributes on an average $(1/2)kT$ towards the average rotational energy. Thus,

Average rotational energy of a linear molecule

$$= \frac{1}{2}kT + \frac{1}{2}kT = kT \tag{1.24.6}$$

Average rotational energy of a nonlinear molecule

$$= \frac{1}{2}kT + \frac{1}{2}kT + \frac{1}{2}kT = \frac{3}{2}kT \tag{1.24.7}$$

Vibrational Motion

If the vibrational motion is assumed to be harmonic, then the energy of each vibrational mode is given as

$$\varepsilon_{\text{vib}} = \text{kinetic energy} + \text{potential energy}$$

$$= \frac{1}{2}\mu\left(\frac{dr}{dr}\right)^2 + \frac{1}{2}k_{\text{f}}(r - r_0)^2$$

where μ is the reduced mass, k_{f} is the force constant, r_0 is the equilibrium value of the coordinate *r*, and d*r*/d*t* is the change of internuclear distance with time. Since both the terms in the above expression contain square of either the velocity or the coordinate, it follows that each will contribute $(1/2)kT$ towards the total average vibrational energy. Thus, the average energy stored in a vibrational motion is

$$\varepsilon_{\text{vib}} = \frac{1}{2}kT + \frac{1}{2}kT = kT \tag{1.24.8}$$

1.25 DEGREES OF FREEDOM AND THE AVERAGE ENERGY OF A MOLECULE

A system consisting of *N* particles can be described by specifying three coordinates for each particle or a total of 3*N* coordinates. Even when these particles are undergoing motions, however complicated they may be, the system at any instant can be described by stating these 3*N* coordinates. Since any one of these components can vary by any amount, the system is said to possess 3*N* independent components of motion or degrees of freedom. If *N* particles are bound together to form a molecule, then the 3*N* coordinates and components of motion are conveniently chosen as follows.

Translational Motion The translational motion of a molecule as a whole can be described by the motion of its centre of mass. Since three coordinates are required to describe the position of centre of mass, it follows that the energy (kinetic only) stored in the translational motion as described by Eq. (1.24.5) is

$$\bar{\varepsilon}_{\text{trans}} = \frac{1}{2}kT + \frac{1}{2}kT + \frac{1}{2}kT = \frac{3}{2}kT \qquad (1.25.1)$$

Rotational Motion ***Linear molecule*** Since a linear molecule can rotate around two axes, the energy stored in the rotational motion as described by Eq. (1.24.6) is

$$\bar{\varepsilon}_{\text{rot}} = \frac{1}{2}kT + \frac{1}{2}kT = kT \qquad (1.25.2)$$

Nonlinear molecule Since a nonlinear molecule can rotate around three axes, the energy stored in this motion as described by Eq. (1.24.7) is

$$\bar{\varepsilon}_{\text{rot}} = \frac{1}{2}kT + \frac{1}{2}kT + \frac{1}{2}kT = \frac{3}{2}kT \qquad (1.25.3)$$

Thus a linear molecule contributes, on an average, kT and a nonlinear molecule $(3/2)kT$ towards the total average energy due to rotational motion.

Vibrational Motion The remaining $(3N - 5)$ coordinates for a linear molecule and $(3N - 6)$ coordinates for a nonlinear molecule, describe the bond distances and bond angles within the molecule. Motion in these coordinates corresponds to the vibrations (stretching or bending) of the molecule. Thus, linear molecules have $(3N - 5)$ and nonlinear molecules have $(3N - 6)$ vibrational modes. Vibrational energy stored in each of the vibrational mode as given by Eq. (1.24.8) is

$$\bar{\varepsilon}_{\text{vib}} = \frac{1}{2}kT + \frac{1}{2}kT = kT \qquad (1.25.4)$$

Thus, each vibrational mode contributes on an average kT towards the total average energy. Thus, linear molecules contribute $(3N - 5)kT$ and nonlinear molecules contribute $(3N - 6)kT$ towards the total average energy due to their vibrational motions.

Total Energy of Molecules The total average energy contribution due to all the three modes is:

For linear molecule

$$\bar{\varepsilon}_{\text{linear}} = \left(\frac{3}{2}kT\right)_{\text{trans}} + \left(\frac{2}{2}kT\right)_{\text{rot}} + \{(3N-5)kT)\}_{\text{vib}} \qquad (1.25.5)$$

For nonlinear molecule

$$\bar{\varepsilon}_{\text{nonlinear}} = \left(\frac{3}{2}kT\right)_{\text{trans}} + \left(\frac{3}{2}kT\right)_{\text{rot}} + \{(3N-6)kT)\}_{\text{vib}} \qquad (1.25.6)$$

The total average energy stored by the molecules in one mole of the gas is

$$\overline{E}_{\text{linear}} = N_A\, \overline{\varepsilon}_{\text{linear}} = \frac{3}{2}RT + RT + (3N-5)RT \tag{1.25.7}$$

$$\overline{E}_{\text{nonlinear}} = N_A\, \overline{\varepsilon}_{\text{nonlinear}} = \frac{3}{2}RT + \frac{3}{2}RT + (3N-6)RT \tag{1.25.8}$$

Table 1.26.1 describes the average energies of molecules of different gases.

Table 1.26.1 Average Energies of Gaseous Molecules

Gas	*N*	$\overline{\varepsilon}$	$\overline{E}$
Monatomic	1	(3/2) kT	(3/2) RT
Diatomic	2	(7/2) kT	(7/2) RT
Triatomic	3		
Linear		(13/2) kT	(13/2) RT
Nonlinear		6 kT	6 RT

1.26 HEAT CAPACITIES

Definition of Heat Capacities

The change in internal energy of a gaseous system is given by

$$\Delta U = q + w \tag{1.26.1}$$

where q is the heat absorbed (or released) by the system and w is the work of expansion (or compression) of the system.[†]

The work done due to expansion at constant pressure (Fig. 1.26.1) is given by

$$w = -F \times l = -p\,(A \times l)$$

or $\quad w = -\,p\,\Delta V$

With this, Eq. (1.26.1) becomes

$$q = \Delta U + p\,\Delta V \tag{1.26.2}$$

If heat flows at constant volume condition, then $w = 0$ and the entire heat is utilized in increasing the internal energy of the gas, i.e.

$$q_V = \Delta U \quad \text{or} \quad \mathrm{d}q_V = \mathrm{d}U \tag{1.26.3}$$

and if the heat flows at constant pressure condition, then it is utilized both in increasing the internal energy and in doing the work of expansion, i.e.

$$q_p = \Delta U + p\,\Delta V \quad \text{or} \quad \mathrm{d}q_p = \mathrm{d}U + p\,\mathrm{d}V \tag{1.26.4}$$

[†] Heat is assigned a positive value if it is absorbed by the system, work is assigned a positive value if it is done on the system.

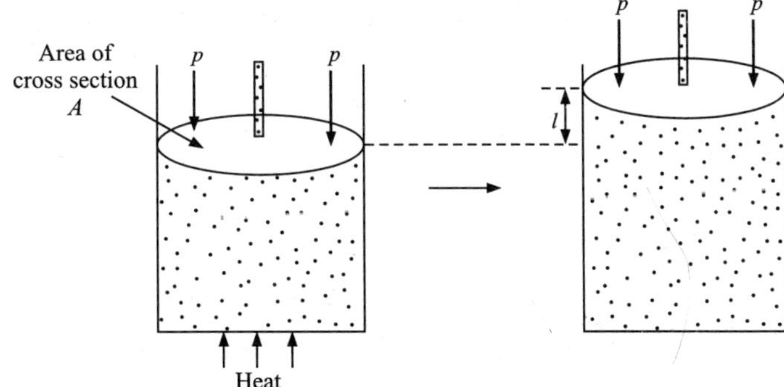

Fig. 1.26.1 Work done on account of expansion of a gas

Heat capacity of a system is defined as the amount of heat required to increase its temperature by 1 degree Celsius. Thus one can have two heat capacities, (i) heat capacity C_V at constant volume condition and (ii) heat capacity C_p at constant pressure condition. These are represented as:

$$C_V = \left(\frac{\partial q}{\partial T}\right)_V = \left(\frac{\partial U}{\partial T}\right)_V \tag{1.26.5}$$

$$C_p = \left(\frac{\partial q}{\partial T}\right)_p = \left(\frac{\partial U}{\partial T}\right)_p + p\left(\frac{\partial V}{\partial T}\right)_p \tag{1.26.6}$$

Difference in Heat Capacities of an Ideal Gas

If the gas is assumed to be ideal, then

$$pV = nRT, \quad p(\partial V/\partial T)_p = nR$$

Since for an ideal gas, the internal energy U depends only on T, we will have

$$(\partial U/\partial T)_V = (\partial U/\partial T)_p$$

Thus, for an ideal gas it follows that

$$C_p = C_V + nR \tag{1.26.7}$$

For one mole of the gas

$$C_{p,\text{m}} = C_{V,\text{m}} + R \tag{1.26.8}$$

Molar Heat Capacities of Gases

Molar heat capacity at constant volume of a gas can be obtained by differentiating the molar energy with respect to temperature. Thus, from Eqs (1.25.7) and (1.25.8), and Table 1.26.1, we have

Monatomic gases

$$C_{V,\text{m}} = \left(\frac{\partial U}{\partial T}\right)_V = \frac{3}{2}R; \qquad C_{p,\text{m}} = \frac{5}{2}R$$

$$\gamma = \frac{C_{p,\text{m}}}{C_{V,\text{m}}} = \frac{5}{3} \simeq 1.667$$

Polyatomic gases

(Linear) $C_{V,\mathrm{m}} = \left(\dfrac{\partial U}{\partial T}\right)_V = \dfrac{3}{2}R + R + (3N-5)R$ (1.26.9)

(Nonlinear) $C_{V,\mathrm{m}} = \left(\dfrac{\partial U}{\partial T}\right)_V = \dfrac{3}{2}R + \dfrac{3}{2}R + (3N-6)R$ (1.26.10)

The molar heat capacities for diatomic and triatomic molecules are as follows:

Diatomic molecule $N = 2$. Thus

$$C_{V,\mathrm{m}} = \frac{3}{2}R + R + R = \frac{7}{2}R; \qquad C_{p,\mathrm{m}} = \frac{9}{2}R$$

$$\gamma = \frac{C_{p,\mathrm{m}}}{C_{V,\mathrm{m}}} = \frac{9}{7} \simeq 1.286 \tag{1.26.11}$$

Triatomic molecule $N = 3$. Thus

Linear $C_{V,\mathrm{m}} = \dfrac{3}{2}R + R + 4R = \dfrac{13}{2}R; \qquad C_{p,\mathrm{m}} = \dfrac{15}{2}R$ (1.26.12)

$$\gamma = \frac{C_{p,\mathrm{m}}}{C_{V,\mathrm{m}}} = \frac{15}{13} \simeq 1.154$$

Nonlinear $C_{V,\mathrm{m}} = \dfrac{3}{2}R + \dfrac{3}{2}R + 3R = 6R; \qquad C_{p,\mathrm{m}} = 7R$ (1.26.13)

$$\gamma = \frac{C_{p,\mathrm{m}}}{C_{V,\mathrm{m}}} = \frac{7}{6} \simeq 1.167$$

1.27 COMPARISON OF THEORETICAL AND EXPERIMENTAL HEAT CAPACITIES

Comparison Between Experimental and Theoretical Values

Comparison of the theoretical and experimental heat capacities reveals the following facts:

(i) Agreement of monatomic gases is perfect, i.e. $C_{V,\mathrm{m}}/R = 1.5$. This value is independent of temperature over a wide range.

(ii) For polyatomic gases, two points of disagreement are found for the observed heat capacities: (a) they are always substantially lower than the predicted values, and (b) they noticeably depend on the temperature. For example, in case of diatomic molecules, we have

$$\frac{C_{V,\mathrm{m}}}{R}\text{(calculated)} = \frac{3}{2} + \frac{2}{2} + 1 = \frac{7}{2} = 3.5$$

$$\frac{C_{V,\mathrm{m}}}{R}\text{ (observed) lies in the range 2.5 to 3.5}$$

The equipartition principle is a law of classical physics and these discrepancies of polyatomic molecules indicate that classical mechanics is not adequate to describe the molecular properties. In fact, one should use quantum mechanics for this purpose.

Quantization of Energies

It is known that all forms of energies are quantized. The permissible values of energies in a molecule and the order of energy difference between them are given below:

$$E_{\text{trans}} = n^2\left(\frac{h^2}{8ml^2}\right), \qquad n = 0, 1, 2, \ldots \qquad \Delta E_{\text{trans}} \simeq 10^{-29}\,\text{J} \quad (1.27.1)$$

$$E_{\text{rot}} = J(J+1)\left(\frac{h^2}{8\pi^2 I}\right), \qquad J = 0, 1, 2, \ldots \qquad \Delta E_{\text{rot}} \simeq 10^{-24}\,\text{J} \quad (1.27.2)$$

$$E_{\text{vib}} = \left(\upsilon + \frac{1}{2}\right)\frac{h}{2\pi}\sqrt{\frac{k_{\text{f}}}{\mu}}, \qquad \upsilon = 0, 1, 2, \ldots \qquad \Delta E_{\text{vib}} \simeq 10^{-20}\,\text{J} \quad (1.27.3)$$

where n, J, υ are translational, rotational and vibrational quantum numbers, respectively, h is the Planck's constant, I is the moment of inertia, μ is the reduced mass, k_{f} is the force constant, l is length of the vessel and m is the mass of the atom.

Distribution of Molecules in Energy Levels

The distribution of the molecules in the energy levels is given by the Boltzmann distribution law

$$N_i = N \exp(-\Delta\varepsilon_i/kT)$$

where

N_i is the number of molecules in the ith level

N is the number of molecules in the lowest quantum level

$\Delta\varepsilon_i$ is the energy difference of the ith level from that of the lowest quantum level

k is the Boltzmann constant.

The relative population N_i/N in any given level depends upon the ratio $\Delta\varepsilon_i/kT$. It can be shown that at room temperature ($kT \simeq 10^{-21}$ J), many of the lower translational and rotational levels are populated (since $\Delta\varepsilon_i/kT < 1$), whereas only the lowest vibrational level is thickly populated (since $\Delta\varepsilon_i/kT > 1$) and very few molecules are present in the second vibrational level.

Contributions from Vibrational Energies

It can also be shown that on raising the temperature of the gas by 1 °C, the number of molecules in the lowest vibrational level remains unaltered, i.e. no energy is utilized in promoting these molecules to the higher levels. Hence, it does not contribute towards the heat capacity. Thus at lower temperature, the contributions to the heat capacity come only through translational and rotational motions giving the value of $C_{V,\text{m}}$ as

$$\frac{C_{V,m}}{R}(\text{lower temperature}) = \frac{3}{2} + \frac{2}{2} = 2.5$$

which agrees with the experimental values of many of the diatomic molecules.

The contribution to the heat capacity from vibrational level comes only when the temperature is high, since only then the molecules can be excited from the lowest level to the higher levels, thereby increasing the vibrational energy. Thus, the discrepancy between the theoretical value and the experimental value is only due to vibrational motion. Since the difference between translational levels is too small to be detected experimentally, translational energy, for all practical purposes, is considered to be continuous. Since the difference of energies between translational levels and rotational levels are much less than the thermal energy kT, one can use classical mechanics without introducing much error in heat capacities. The contribution from vibrational levels has to be considered from the view point of quantum mechanics, since here $\Delta E_{\text{vib}} > kT$.

We can calculate the average value of vibrational energy using the expression

$$\bar{\varepsilon}_{\text{vib}} = \frac{\sum\limits_{\upsilon=0}^{\upsilon=\infty} \varepsilon_\upsilon N_\upsilon}{N} \tag{1.27.4}$$

On substituting quantum mechanical expressions for ε_υ and n_υ, we will get

$$\bar{\varepsilon}_{\text{vib}} = \frac{h\nu}{2} + \frac{h\nu}{(e^{h\nu/kT} - 1)} \tag{1.27.5}$$

We observe that the average energy is made up of zero point energy $h\nu/2$ which is the lowest energy possible for the molecule, plus a term which depends on temperature. At very low temperature $h\nu/kT >> 1$, hence $e^{h\nu/kT} >> 1$ so that the second term is very small and thus

$$\bar{\varepsilon} = \frac{1}{2}h\nu \tag{1.27.6}$$

Effectively, all the molecules are in the lowest quantum state with $\upsilon = 0$. At very high temperature where $h\nu/kT << 1$, we may expand the exponential function as $e^{h\nu/kT} \approx 1 + h\nu/kT$ and thus $e^{h\nu/kT} - 1 \approx h\nu/kT$ and we have

$$\bar{\varepsilon} = \frac{h\nu}{2} + \frac{h\nu}{h\nu/kT} = \frac{h\nu}{2} + kT \tag{1.27.7}$$

Thus $\quad C_V = k$

or $\quad C_{V,\text{m}} = R$ for 1 mol of gas $\quad$ (1.27.8)

Thus, it is only at high temperatures that the vibrational heat capacity attains the classical value of R per mole. Substituting $\theta = h\nu/k$ in Eq. (1.27.5), we get

$$\varepsilon_{\text{vib}} = \frac{h\nu}{2} + \frac{k\theta}{e^{\theta/T} - 1}$$

Thus $$C_V(\text{vib}) = \left(\frac{\partial \varepsilon_{\text{vib}}}{\partial T}\right) = \left(\frac{\theta}{T}\right)^2 \frac{k\,e^{\theta/T}}{(e^{\theta/T}-1)^2}$$

or $$\frac{C_V(\text{vib})}{k} = \left(\frac{\theta}{T}\right)^2 \frac{e^{\theta/T}}{(e^{\theta/T}-1)^2} \quad \text{(for 1 molecule)}$$

or $$\frac{C_{V,\text{m}}(\text{vib})}{R} = \left(\frac{\theta}{T}\right)^2 \frac{e^{\theta/T}}{(e^{\theta/T}-1)^2} \quad \text{(for 1 mol of the gas)} \qquad (1.27.9)$$

A plot of $C_{V,\text{m}}(\text{vib})/R$ versus T/θ gives a graph of the type shown in Fig. 1.27.1.

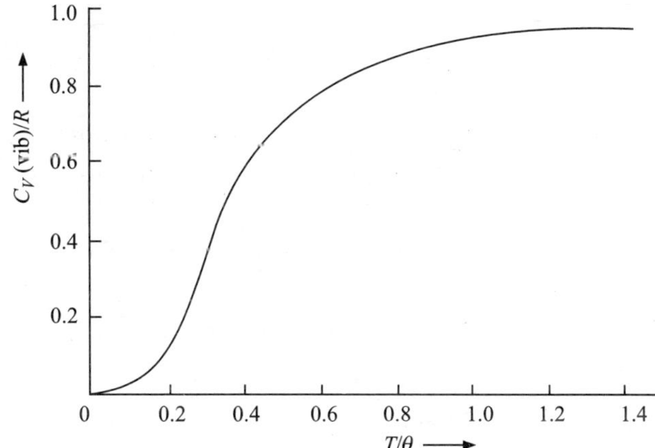

Fig. 1.27.1 Variation of $C_{V,\text{m}}(\text{vib})/R$ with T/θ

1.28 MISCELLANEOUS PROBLEMS

1. Two flasks A and B have equal volumes. A is maintained at 300 K and B at 600 K. While A contains H_2 gas, B has an equal mass of CH_4 gas. Assuming ideal behaviour for both the gases, find the following and establish your answers quantitatively.

(a) Flask containing greater number of molecules.
(b) Flask in which the pressure is greater.
(c) Flask in which the molecules are moving faster.
(d) Flask with greater number of collisions with the walls.
(e) Flask with greater mean free path of molecules. (The collision diameter of CH_4 may be assumed to be twice as that of H_2.)
(f) Flask in which the viscosity of gas is greater.
(g) Flask with greater molar kinetic energy (KE).
(h) Flask in which the total kinetic energy is greater.
(i) Flask in which Z_1 and Z_{11} are greater.
(j) Flask in which the compression factor is greater.

Solution

(a) Let n_A be the amount of H_2 and n_B be the amount of CH_4

We have

$$n_A = \frac{m}{2\ \text{g mol}^{-1}} \qquad n_B = \frac{m}{16\ \text{g mol}^{-1}}$$

where m is the mass of the gas in each of the two flasks A and B. We can write

$$\frac{n_A}{n_B} = 8$$

Now, the number of molecules in these flasks are given as

$$N_A = \frac{m}{2 \text{ g mol}^{-1}} N, \quad N_B = \frac{m}{16 \text{ g mol}^{-1}} N; \qquad \text{where } N \text{ is Avogadro constant}$$

Therefore, $\dfrac{N_A}{N_B} = 8$

(b) Since $p_A V = n_A R T_A$ and $p_B V = n_B R T_B$

therefore, $\dfrac{p_A}{p_B} = \dfrac{n_A T_A}{n_B T_B} = \left(\dfrac{n_A}{n_B}\right)\left(\dfrac{T_A}{T_B}\right) = 8 \times \dfrac{300}{600} = 4$

(c) Since $(u_{av})_A = \sqrt{\dfrac{8RT_A}{\pi M_A}} = \sqrt{\dfrac{8R(300 \text{ K})}{\pi(0.002 \text{ kg mol}^{-1})}}$

$$(u_{av})_B = \sqrt{\frac{8RT_B}{\pi M_B}} = \sqrt{\frac{8R(600 \text{ K})}{\pi(0.016 \text{ kg mol}^{-1})}}$$

therefore, $\dfrac{(u_{av})_A}{(u_{av})_B} = \sqrt{\dfrac{300}{2}} \times \sqrt{\dfrac{16}{600}} = 2$

(d) Number of collisions per unit area per unit time, $X = \left(\dfrac{1}{4}\right)(N^* u_{av})$

Therefore $X_A = \dfrac{1}{4} N_A^* (u_{av})_A \qquad X_B = \dfrac{1}{4} N_B^* (u_{av})_B$

Hence $\dfrac{X_A}{X_B} = \dfrac{N_A^*(u_{av})_A}{N_B^*(u_{av})_B} = 8 \times 2 = 16$

(e) Since $\lambda_A = \dfrac{kT_A}{\sqrt{2}\pi\sigma_A^2 p_A}$ and $\lambda_B = \dfrac{kT_B}{\sqrt{2}\pi\sigma_B^2 p_B}$

therefore, $\dfrac{\lambda_A}{\lambda_B} = \left(\dfrac{T_A}{T_B}\right)\left(\dfrac{\sigma_B^2}{\sigma_A^2}\right)\left(\dfrac{p_B}{p_A}\right) = \left(\dfrac{300}{600}\right)\left(\dfrac{2}{1}\right)^2\left(\dfrac{1}{4}\right) = \dfrac{1}{2}$

(f) Since $\eta_A = \dfrac{(m_A k T_A)^{1/2}}{\pi^{3/2}\sigma_A^2}$ and $\eta_B = \dfrac{(m_B k T_B)^{1/2}}{\pi^{3/2}\sigma_B^2}$

therefore $\dfrac{\eta_A}{\eta_B} = \left(\dfrac{m_A}{m_B}\right)^{1/2}\left(\dfrac{T_A}{T_B}\right)^{1/2}\left(\dfrac{\sigma_B}{\sigma_A}\right)^2 = \left(\dfrac{2}{16}\right)^{1/2}\left(\dfrac{300}{600}\right)^{1/2}\left(\dfrac{2}{1}\right)^2 = 1$

(g) Since KE per mole $= \dfrac{3}{2}RT$, therefore

$$\text{KE per mole of A} = \frac{3}{2}RT_A \qquad \text{KE per mole of B} = \frac{3}{2}RT_B$$

Hence $\dfrac{\text{KE/mol of A}}{\text{KE/mol of B}} = \dfrac{T_A}{T_B} = \dfrac{1}{2}$

(h) Since Total KE = $\overline{\text{KE}}$ × number of molecules, therefore

$$(\text{Total KE})_A = \frac{3}{2}kT_AN_A \qquad (\text{Total KE})_B = \frac{3}{2}kT_BN_B$$

and hence $\dfrac{(\text{Total KE})_A}{(\text{Total KE})_B} = \dfrac{T_AN_A}{T_BN_B} = \dfrac{1}{2} \times 8 = 4$

(i) Since $Z_1 = \dfrac{u_{av}}{\lambda}$, therefore

$$Z_{1A} = \frac{(u_{av})_A}{\lambda_A} \qquad \text{and} \qquad Z_{1B} = \frac{(u_{av})_B}{\lambda_B}$$

Hence, $\dfrac{Z_{1A}}{Z_{1B}} = \left(\dfrac{(u_{av})_A}{(u_{av})_B}\right)\left(\dfrac{\lambda_B}{\lambda_A}\right) = 2 \times 2 = 4$

Since $Z_{11} = \dfrac{N^*}{2}Z_1$, therefore

$$\frac{Z_{11A}}{Z_{11B}} = \frac{(N_A^*Z_{1A}/2)}{(N_B^*Z_{1B}/2)} = \left(\frac{N_A^*}{N_B^*}\right)\left(\frac{Z_{1A}}{Z_{1B}}\right) = 8 \times 4 = 32$$

(j) Both the gases will have the same compression factor, i.e. 1, since both of them are ideal gases.

---

2. Correct the following statements giving reasons, if necessary.

Note Correction, if necessary, is provided immediately after each statement.

(i) The product of pressure and volume of a fixed amount of a gas is independent of temperature.
This statement is incorrect. According to the Boyle's law the product of pressure and volume depends upon temperature.

(ii) Rise in the compression factor with increasing pressure is due to *a* or *b*.
This is due to the factor *b*. Since

$$\left(p + \frac{a}{V_m^2}\right)(V_m - b) = RT$$

Ignoring a/V_m^2, we get

$$pV_m - pb = RT \quad \text{or} \quad Z = 1 + b\frac{p}{RT}$$

(iii) A gas can be liquefied at, and (a) $T = T_c$ and $p < p_c$, and (b) $T < T_c$ and $p = p_c$.
Statement (a) is incorrect since p should be equal to or greater than p_c. Statement (b) is correct.

(iv) The gas gets heated if its temperature is less than its inversion temperature in Joule–Thomson effect.
This statement is incorrect. Heating is observed when $T > T_i$.

(v) It is possible to liquefy an ideal gas.
No, since there exist no forces of attraction amongst the molecules of an ideal gas.

(vi) All molecules in a gas are moving with the same speed.
No, movements of molecules in a gas follow Maxwell's distribution law.

(vii) Average speed of molecules of a gas in a container moving only in one dimension will be zero.
Yes, since otherwise, with time all the molecules will be collected in one direction,

(viii) The fraction of molecules having speeds in the range of u to $u + du$ of a gas of molar mass M at temperature T is the same as that of the gas of molar mass $2M$ at temperature $T/2$.
Incorrect, since

$$\frac{1}{N}\frac{dN}{du} = 4\pi\left(\frac{M}{2\pi RT}\right)^{3/2} u^2 \exp(-Mu^2/2RT)$$

This will be so if temperature of the gas of molar mass $2M$ is $2T$.

REVISIONARY PROBLEMS

1.1 (a) Experimentally, it is known that gas laws are not applicable to real gases under all conditions of temperature and pressure. Under what conditions are these laws obeyed by real gases?
(b) What is an ideal gas? Does it exist in nature?
(c) The equation $pV = nRT$ is strictly applicable to an ideal gas, How would you determine experimentally the volume occupied by such a gas at STP?
(d) Discuss the nature of gas constant R. Derive its value when
(i) p is expressed in atm and V in litres,
(ii) p is expressed in dyn/cm^2 and V in cm^3.
(e) Does the value of R depend upon the nature of the gas?
(f) What is the Boltzmann constant? Derive its value in joule kelvin^{-1}.

1.2 (a) Starting from the postulates of kinetic theory of gases, derive the kinetic gas equation $pV = mN\overline{u^2}/3$.
(b) Deduce the following gas laws from the kinetic gas equation.
(i) Boyle's law; (ii) Charles law; (iii) Graham's law of effusion, and (iv) Avogadro's hypothesis.
(c) Starting from the kinetic gas equation deduce the expressions for (i) root mean square speed, and (ii) average kinetic energy.
(d) Why is the simplest kinetic theory applicable only to ideal gases?

1.3 (a) Explain the factors which led van der Waals to modify the ideal gas equation $pV = nRT$, and hence derive the van der Waals equation of state

$$\left(p + \frac{n^2a}{V^2}\right)(V - nb) = nRT$$

(b) What are the units of van der Waals constants a and b? Do they have the same values for all real gases?
(c) How would you determine the values of a and b for a given gas experimentally? Do these constants depend upon temperature of the gas?
(d) It is stated that the excluded volume b is approximately four times the actual volume occupied by the molecules of one mole of a gas. How would you derive this conclusion?
(e) From the given value of b for a real gas, how would you estimate the molecular diameter of a molecule of a given gas?

(f) Explain how the van der Waals equation of state accounts for the behaviour of real gases.

(g) Two van der Waals gases have the same value of b but different a values. Which of these would occupy greater volume under identical conditions? If the gases have the same a value but different b values, which of the two would be more compressible?

1.4 (a) What is the virial equation of state? Express the following equations of states in the form of virial equation of state:

(i) Van der Waals equation

$$\left(p + \frac{a}{V_m^2}\right)(V_m - b) = RT$$

(ii) Berthelot's equation

$$\left(p + \frac{a}{TV_m^2}\right)(V_m - b) = RT$$

(iii) Dieterici's equation

$$p\{\exp(a/V_m RT)\}\,(V_m - b) = RT$$

(b) Derive expressions for the second virial coefficient of the above equations of states.

1.5 (a) What do you understand by the term 'compression factor Z' of a gas?

(b) The slope of Z versus p curve of a real gas at a given temperature can be obtained by differentiating the virial equation of state in p with respect to pressure keeping the temperature constant. Show that this slope is given by

$$\left(\frac{\partial Z}{\partial p}\right)_T = \left(\frac{1}{RT}\right)\left(b - \frac{a}{RT}\right) + \frac{2}{(RT)^3}\left(2b - \frac{a}{RT}\right)p + \cdots$$

(c) What will be the value of this slope when $p = 0$?

(d) Show that the initial slope of Z versus p curve (i.e. slope at $p = 0$) will have the following characteristics:

(i) Slope is positive if $b > a/RT$, i.e. the size effect dominates the behaviour of the gas.

(ii) Slope is negative if $b < a/RT$, i.e. the attractive forces dominate the behaviour of the gas.

(iii) Slope is zero if $b = a/RT$.

1.6 (a) What is Boyle temperature? Show that at this temperature the second virial coefficient is zero. From this condition, find the expressions of Boyle temperature for van der Waals, Berthelot's and Dieterici's equations of states.

(b) Explain why at Boyle temperature the van der Waals gases behave ideally over a wide range of pressures. Also explain why the initial slope of the compression factor $Z(= pV/RT)$ versus p curves is (i) positive if a gas is above its Boyle temperature, and (ii) negative if it is below its Boyle temperature.

1.7 In the liquefaction of gases, what would you regard the change from gas to liquid—a continuous transition or a discontinuous one and why?

1.8 (a) Differentiate between the Andrew's isotherms and the van der Waals isotherms for carbon dioxide.

(b) Define the terms critical temperature, critical pressure and critical volume. How are these quantities determined experimentally?

(c) At the critical state, all the three roots of volume of van der Waals equation are real and identical. Using this rule, derive the relations between van der Waals constants and critical constants and vice versa.

1.9 (a) The slope of critical isotherm has a maximum value of zero at the critical state. Using the mathematical condition of maxima, derive the relations between critical constants and (i) van der Waals constants, (ii) Berthelot's constants, and (iii) Dieterici's constants, (b) Calculate the value of compression factor at the critical point for the above three equations of states. Compare these values with experimental values and comment.

1.10 What is the law of corresponding states? Derive the reduced equation of state for (i) van der Waals equation of state, and (ii) Dieterici's equation of state.

1.11 (a) State Maxwell's law for the distribution of speeds among molecules of a gas. How does a change in temperature or pressure of a gas influence this distribution?
(b) Give an experimental method which can demonstrate the distribution of speeds among molecules.
(c) Following the Maxwell distribution of molecular speeds, derive the expressions for (i) root mean square speed, (ii) average speed, and (iii) most probable speed. Arrange these three speeds in order of increasing magnitude.
(d) Following the Maxwell distribution of molecular speeds, derive the expressions for (i) average kinetic energy, and (ii) fraction of molecules having kinetic energy exceeding a specified value of ε'.

1.12 (a) Derive the barometric formula $p = p_0 \exp(-Mgh/RT)$. Under what assumptions is the above expression applicable?
(b) Explain how would you construct a straight-line graph relating altitude to air pressure?
(c) Show that the height at which the atmospheric pressure is reduced to half its value is given by

$$h = \frac{0.6909RT}{Mg}$$

(d) Explain what effects temperature and molar mass of the gas have on the variation of pressure with height?

1.13 (a) If the compression factor of a gas is $Z(p, T)$, the equation of state may be written as $pV/nRT = Z$. Show how this affects the equation for the distribution of the gas in a gravity field. From the differential equation for the distribution, show that if $Z > 1$, the distribution is broader for a real gas than for an ideal gas and the converse is true if $Z < 1$. If $Z = 1 + bp$, where b is a function of temperature, integrate the equation and evaluate the constant of integration to obtain explicit form of the distribution function.
(b) The temperature of air decreases linearly with altitude in accordance with the equation $T = T_0 - ah$, where a is constant, h is altitude, T_0 is temperature at the ground level and T is temperature at an altitude h. Derive a modified form of the barometric equation taking this temperature dependence into account.

1.14 Explain the terms σ, λ, Z_1 and Z_{11}. Discuss the effects of temperature and pressure on these terms.

1.15 What is the viscosity of a fluid? Discuss how exactly this arises in liquids and gases? Derive the expression of the coefficient of viscosity of a gas in terms of average speed, mean free path and number of molecules per unit volume. Discuss the effects of temperature and pressure on the viscosity of gases and compare them with those of liquids.

1.16 (a) Show that the rate of effusion of a gas through a fine hole in a container placed in vacuum is given by

$$r = \frac{pN_A A_{hole}}{\sqrt{2\pi MRT}}$$

where the various symbols have their usual meanings.
(b) Describe Knudsen method to determine the vapour pressure of a liquid or solid.

1.17 Define Fick's law. Show that the coefficient of self diffusion is given by

$$D = \frac{1}{2}\bar{u}\lambda$$

What are the effects of pressure and temperature on the coefficient of self diffusion?

1.18 (a) Show that the heat flowing per unit area per unit time in the gaseous system having temperature gradient $\partial T/\partial z$ is given by

$$Q = -\frac{1}{2}N^*\bar{u}\,\lambda c\left(\frac{\partial T}{\partial z}\right)$$

where the various symbols have their usual meanings.
(b) What are the effects of temperature and pressure on the coefficient of thermal conductivity?

1.19 Show that the net charge flowing per unit area per unit time in a system containing charged particles under the influence of electric field E given by

$$j_z = \left(\frac{N^* e^2}{m}\tau\right)E$$

where the various symbol have their usual meanings.

1.20 Define the following terms:

(a) Degrees of freedom. (b) Principle of equipartition of energy.

1.21 (a) How much average energy per mole is associated with (i) translational, (ii) rotational, and (iii) vibrational degrees of freedom of motion? Calculate the values of total average energies and the corresponding values of molar heat capacities of a mon-, di- and triatomic molecules (linear as well as nonlinear).
(b) Explain, why polyatomic gases at low temperature have lower values of C_p and C_V compared with the theoretical values? Why do these values approach the theoretical values at very high temperatures?
(c) A triatomic molecule has the value of $C_V = 24.9$ J K^{-1} mol^{-1} at low temperatures. Comment upon the geometry of the molecule.

TRY YOURSELF PROBLEMS

1.1 In an attempt to derive the formula $pV = KT$, where K is a constant, by combining the laws of Boyle and Charles, the following argument was proposed:

(i) $pV = k_1 \Rightarrow V = k_1/p$ (ii) $\dfrac{V}{T} = k_2 \Rightarrow V = k_2T$

(iii) $\dfrac{k_1}{p} = k_2T \Rightarrow \dfrac{k_1}{k_2} = pT \Rightarrow pT = \text{constant}$

The conclusion that pT = constant is obviously incorrect. What is the error? At what steps has it been introduced? Give a correct derivation.

1.2 Derive the value of R when
(a) Pressure is expressed in atm and volume in mm^3.
(b) p in cmHg and V in cm^3.
(c) p in dynes/metre^2 and V in cubic millimetres.

1.3 Explain whether a gas would approach ideal behaviour or deviate from it under the following conditions:

(i) It is compressed to a small volume at a constant temperature.
(ii) Temperature is raised keeping the volume constant.
(iii) More gas is introduced into the same volume at the same temperature.

1.4 At the same T and p, which of the gases will have higher average kinetic energy per mole: H_2, O_2, CH_4 and SF_6?

1.5 Briefly explain the following statements:
(a) An ideal gas is not expected to show any cooling on free expansion.
(b) The mean free path of molecules in a gas increases and the number of collisions per unit time decreases with the lowering of pressure.
(c) Ideal gas does not exist but is a useful concept.
(d) CO and N_2 have the same speed distribution at the same temperature.

1.6 Briefly answer the following:
(a) Three different containers of equal volume at 27 °C contain H_2, N_2, and O_2, respectively. Which one of the gases has the maximum average speed of its molecules?
(b) If u_x, u_y, and u_z denote the components of the velocity of a molecule of a gas in a container at rest, show qualitatively that the averages

$$\bar{u}_x = \bar{u}_y = \bar{u}_z = 0; \qquad \overline{u_x^2} = \overline{u_y^2} = \overline{u_z^2} \neq 0; \qquad \overline{u^2} = 3\overline{u_x^2}$$

(c) Which of the following gases will have higher average energy per molecule at the same temperature? Give reasons.
(i) CO_2, and (ii) H_2O vapour.

1.7 The deviation of a real gas from an ideal gas can be seen from its compression factor Z. Show that in a gas obeying van der Waals equation, Z is given by

$$Z = 1 + \frac{bp}{RT} - \frac{a}{RTV_\text{m}} + \frac{ab}{RTV_\text{m}^2}$$

(a) What conclusions do you draw from this equation concerning the behaviour of gases at low and high pressures? Do they agree with experiments?
(b) How does this equation explain the fact that the real gases are more nearly ideal, the lower the pressure, and the higher the temperature.

(c) Explain the variation of compression factor with pressure of N_2 and H_2 at (i) the same temperature and (ii) at different temperatures.

1.8 Show that at low densities the van der Waals equation

$$\left(p + \frac{a}{V_m^2}\right)(V_m - b) = RT$$

and the Dieterici's equation

$$p = \frac{RT}{V_m - b}\exp(-a/V_m RT)$$

give essentially the same value of p.

1.9 (a) A certain gas has the equation of state

$$p = \frac{RT}{V_m - bT} - \frac{a}{V_m^3}$$

where a and b are constants characteristic of the gas. Show that

$$a = \frac{4RT_c V_c^2}{3}, \qquad b = \frac{V_c}{2T_c}$$

where V_c and T_c are critical constants.

(b) Obtain the reduced form of the above equation of state.

1.10 The acceleration due to gravity g varies in magnitude with distance r from the centre of the earth in accordance with the formula $g = GM_E/r^2$, where G is the gravitational constant and M_E is the mass of the earth. Derive a modified form of the barometric equation that takes into account this variation of g with altitude h above the earth's surface.

1.11 Explain why the viscosity of a gas (i) increases with increasing temperature, (ii) depends on the diameter of the molecules, and (iii) is independent of pressure at constant temperature? Why does this independence fail at very low pressures?

1.12 Arrange in order of increasing value, the most probable speed, root mean square speed and average speed of molecules of a gas. Would you expect the difference between these three to increase, decrease or remain unchanged with increasing temperature?

1.13 Derive the explicit formulas for the coefficient of thermal expansion $\alpha(=(\partial V/\partial T)_p/V)$ and the isothermal compressibility $\kappa(= -(\partial V/\partial p)_T/V)$ for gases obeying (i) ideal gas equation, and (ii) van der Waals equation of state.

1.14 Correct the following statements, giving reasons, where necessary.

(i) The product of pressure and volume of a fixed amount of a gas is independent of temperature.

(ii) Rise in the compression factor with increasing pressure is due to a or b or both.

(iii) A gas can be liquefied at

$$T = T_c \text{ and } p < p_c \quad \text{and} \quad T < T_c \text{ and } p = p_c$$

(iv) The gas gets heated if its temperature is less than its inversion temperature in Joule–Thomson effect.

(v) It is possible to liquefy an ideal gas.

(vi) All molecules in a gas move with the same speed.

(vii) Average speed of molecules of a gas in a container at rest moving in any one direction will be zero.

(viii) The fraction of molecules having speed in the range of u to $u + du$ of a gas of molar mass M at temperature T is the same as that of the gas of molar mass $2M$ at temperature $T/2$.

1.15 Two flasks A and B have equal volumes. Flask A containing H_2 gas is maintained at 300 K while B containing an equal mass of C_2H_6 gas is maintained at 900 K. Assuming ideal behaviour for both the gases, answer the following:

(i) Which flask contains the largest number of molecules? How many times more?

(ii) In which flask, and by how many times, is the pressure greater?

(iii) In which flask, and by how many times, are the molecules moving faster?

(iv) In which flask, and by how many times, is the number of collisions with the wall greater?
(*Hint*: Number of collisions with the wall = $N\bar{u}/4$.)

(v) In which flask, and by how many times, is the mean free path of molecules greater? (The collision diameter of C_2H_6 gas is four times that of H_2 gas.)

(vi) In which flask, and by how many times, is the viscosity greater?

(vii) In which flask, and by how many times, is the kinetic energy per mole greater?

(vii) In which flask, and by how many times, is the total kinetic energy greater?

(ix) In which flask, and by how many times, are Z_1 and Z_{11} greater?

(x) In which flask, and by how many times, is the compression factor greater?

1.16 Can a van der Waals gas with $a = 0$ or $b = 0$ be liquefied? Is the equation $p(V - nb) = nRT$ suitable near the critical point? Explain mathematically.

1.17 Estimate the increase in energy per Celsius degree rise in temperature for:

(a) One mole of a gas consisting of rigid linear polyatomic molecules.
(b) One mole of a solid containing rigid nonlinear polyatomic molecules, if each molecule is free to rotate and if each molecule is bound to its position in the solid by forces that are proportional to the displacement from its equilibrium position.

1.18 Show that:

(i) Average kinetic energy of molecules is independent of the molecular mass.

(ii) The distribution of energy amongst the molecules is independent of the molecular mass.

(iii) The most probable kinetic energy is $kT/2$.

1.19 At Boyle temperature, the initial slope of pV versus p is zero, i.e.

$$\left(\frac{\partial(pV)}{\partial p}\right)_{T_B} = p\left(\frac{\partial V}{\partial p}\right)_{T_B} + V = 0. \quad \text{Thus} \quad \left(\frac{\partial V}{\partial p}\right)_{T_B} = -\frac{V}{p} = -\frac{V^2}{RT_B}$$

Starting from the van der Waals equation of state and using the above equation, show that the Boyle temperature is given by

$$T_B = \frac{a}{Rb}$$

(*Hint*: Write the van der Waals equation in the open form, neglect the term ab/V^2. Differentiate the resultant expression with respect to p at constant T and substitute $(\partial V/\partial p)_{TB} = -V^2RT_B$.)

NUMERICAL PROBLEMS

Equation of State for an Ideal Gas

1.1 A sample of gas weighing 0.289 g occupies a volume of 50 cm^3 at 76 cmHg and 25 °C. What is the molar mass of the gas? (Ans. 141.33 g mol^{-1})

1.2 What is the density of helium at 0 °C and 100 mmHg pressure? (Ans. 2.35 × 10^{-2} g dm^{-3})

1.3 The density of gas X is 1 g cm^{-3} at –135 °C and 110.6 kPa. What is the density at STP? (Ans. 0.463 g cm^{-3})

1.4 The density of a mixture of CH_4 and C_2H_6 at 100 °C and a pressure of 93. 326 kPa is 0.699 g dm^{-3}. What is the composition of this mixture: (a) by volume, (b) by mass; (c) in mole per cent? (d) What is the partial pressure of each of the components? (Ans. (a) 484 cm^3, 516 cm^3, (b) 0.2313 g, 0.4627 g, (c) 48.42, 51.48 (d) 339 mmHg, 361 mmHg)

1.5 A mixture of H_2 and N_2 weighing 0.116 g is collected over water at 50 °C and occupies a volume of 275 cm^3 when the total pressure is 101.325 kPa. Calculate the composition of the dry mixture in mole per cent. The vapour pressure of water at 50 °C is 12.332 kPa. (Ans. 58.73, 41.27)

1.6 A mixture of CH_4 and C_3H_8 occupied a volume of 100 cm^3 at 25 °C and 50.662 5 kPa. Enough O_2 was added to make the pressure 74.660 5 kPa when the volume was 400 cm^3. The mixture was then ignited and after the temperature returned to 25 °C, the volume was 300 cm^3 at 64.928 kPa. Find the composition of the mixture. (Assume that the combustion was complete. Ignore the volume of liquid water formed and take the vapour pressure of water at 25 °C to be 3.20 kPa.) (Ans. $n(CH_4)$ = 0.001 557 2 mol, $n(C_3H_8)$ = 0.000 488 8 mol)

1.7 A 1 dm^3 gas bulb contains 1.03 × 10^{23} H_2 molecules. If the pressure exerted by these molecules is 6.34 mmHg, what must be the average squared molecular speed? What must be the temperature? (Ans. 7.414 × 10^7 (cm $s^{-1})^2$, 0.60 K)

1.8 A good vacuum pump will bring a tight system down to 10^{-10} mmHg. How many gas molecules remain per cm^3 at this pressure and at a temperature of 300 K? (Ans. 3.22 × 10^7)

1.9 Calculate the root mean square speed of oxygen molecules having kinetic energy of 8.368 kJ mol^{-1}. At what temperature would the molecules have this value of root mean square speed? (Ans. 723.2 ms^{-1}, 671 K)

1.10 Calculate the average and total kinetic energies of 0.6 mol of an ideal gas at 0 °C. (Ans. 3.405 kJ, 2.043 kJ)

1.11 Three moles of helium are contained in a volume of 20 dm^3 at a pressure of 506.625 kPa. Calculate (a) the average molecular kinetic energy, and (b) the root mean square speed of the molecules. (Ans. (a) 8.40 × 10^{-21} J, (b) 1.59 × 10^5 cm s^{-1})

Van der Waals Equation

1.12 The density of mercury vapours is 13.6 g/cm^3. Estimate the van der waals constant *b* value. (Ans. 58.997 cm^3)

1.13 Using the van der Waals equation, calculate the pressure exerted by 20.0 g of carbon dioxide in 1 dm^3 vessel at 25 °C and compare this with the ideal gas value. (Ans. 1.074 MPa, 1.125 MPa)

1.14 Using the van der Waals equation, find the temperature at which 3 mol of SO_2 will occupy a volume of 10 dm^3 at a pressure of 1.52 MPa. Given: a = 678.88 dm^6 kPa mol^{-2} and b = 5.6 × 10^{-2} dm^3 mol^{-1} (Ans. 623.3 K)

1.15 The density of solid nitrogen at 0 °C and 303.975 MPa is 0.835 g cm^{-3}. Calculate the average distance between the centres of the molecules. How does this compare with the molecular diameter calculated from van der Waals constant b = 39.1 cm^3 mol^{-1}? Given: N_2 crystallizes as primitive cubic lattice. (Ans. 382 pm, 314 pm from *b*)

Critical Constants

1.16 (a) The orthobaric densities of CCl_4 liquid and vapour at a series of temperatures have the following values:

Temperature/°C	150	250	270	280
Density, ρ_l/g cm^{-3}	1.321 5	0.998 0	0.866 6	0.763 4
Density, ρ_v/g cm^{-3}	0.030 4	0.175 4	0.271 0	0.359 7

Given that the critical temperature of CCl_4 is 283.1 °C, find the critical volume of CCl_4. (Ans. 276 cm^3 mol^{-1})

(b) From the value of V_c obtained above calculate the value of van der Waals constant b and the molecular diameter of CCl_4 (assuming the latter to be spherical).

(c) Since $p_cV_c/RT_c = 3/8$ for a van der Waals gas, what would be the value of p_c for CCl_4?

(d) Estimate the value of van der Waals constant a from the relation $p_c = a/27b^2$ or $T_c = 8a/27Rb$ or $a = 3p_cV_c^2$.

1.17 The van der Waals constants for HC1 are a = 371.843 kPa dm^6 mol^{-2} and b = 40.8 cm^3 mol^{-1}. Find the critical constants of this substance. (Ans. 8.273 MPa, 122.4 cm^3, 325 K)

1.18 The van der Waals constants for gases A, B and C are as follows:

Gas	a/dm^6 kPa mol^{-2}	b/dm^3 mol^{-1}
A	405.3	0.027
B	1 215.9	0.030
C	607.95	0.032

Which gas has (i) the highest critical temperature, (ii) the largest molecular volume, and (iii) most ideal behaviour around STP? (Ans. (i) B, (ii) C, (iii)) A)

Maxwell Distribution of Speeds

1.19 What fraction of molecules in hydrogen gas have a kinetic energy within $\overline{KE}$ ± 10 per cent of $\overline{KE}$ at 25 °C and at 500 °C? What fraction of molecules in mercury vapour have the above range of kinetic energy at 25 °C? (Ans. 0.092 4; same fraction in all cases)

1.20 (a) Calculate the root mean square, arithmetic mean and the most probable speeds for oxygen molecules at 25 °C. At what temperatures would these speeds be possessed by butane molecules?

(b) The escape speed, defined as the speed at which molecules can escape the earth's gravitational field, is approximately 1.1 × 10^6 cm s^{-1}. At what temperature approximately will the root mean square speed for H_2 be equal to its escape speed? (Ans. (a) 481.95 m s^{-1}, 444.14 m s^{-1}, 393.51 m s^{-1} (b) 9.703 × 10^3 K)

Barometric Distribution

1.21 Calculate the pressure at an altitude of 800 km above the earth assuming that the atmosphere is isothermal with a temperature of 0 °C and that the average molar mass is 28.8 g mol^{-1} (80% N_2 + 20% O_2), independent of height. (Ans. 6.237 × 10^{-39} Pa)

1.22 If in Problem 1.21, the composition of air changes with height, what would have been the pressure at an altitude of 800 km? (Ans. 7.91 × 10^{-38} Pa)

Molecular Collisions

1.23 Calculate the mean free path in CO_2 at 27 °C and a pressure of 10^{-6} mmHg. (Take σ to be 460 pm.) (Ans. 3.307 × 10^3 cm)

1.24 (a) Calculate the number of bimolecular collisions per second per cubic centimeter in argon at a pressure of 101.325 kPa and a temperature of 100 °C if the collision diameter is 364 pm.

(b) What would be the collision rate if the argon pressure was doubled and the temperature was reduced to 50 °C?

(Ans. (a) 5.07 × 10^{28} cm^{-1} s^{-1}, (b) 2.52 × 10^{29} cm^{-1} s^{-1})

1.25 The number of collisions per unit time between unlike molecules A and B per unit volume of gas is

$$Z_{12} = \pi N_A N_B \left(\frac{d_A + d_B}{2}\right)^2 \sqrt{\frac{8kT}{\pi\mu}}$$

where the reduced mass $\mu = m_A m_B/(m_A + m_B)$. In an equimolar mixture of H_2 and I_2 at 500 K and 101.325 kPa pressure, calculate the number of collisions per second per cm^3 between H_2 and H_2, H_2 and I_2, I_2 and I_2. For H_2 take d = 218 pm and for I_2, d = 376 pm.

(Ans. 5.24×10^{28} cm^{-3} s^{-1}, 1.38×10^{29} cm^{-3} s^{-1}, 1.38×10^{28} cm^{-3} s^{-1})

1.26 The collision diameter for helium is 207 pm and for methane, 414 pm. How does the mean free path for He compare with that for CH_4 under the same conditions?

(Ans. 4)

1.27 Calculate the value of σ, λ, Z_1 and Z_{11} for nitrogen molecules at 25 °C and at pressure of 10^{-3} mmHg, given that b for nitrogen is 39.1 cm^3 mol^{-1}.

(Ans. 314 pm, 7.05 cm, 6734 s^{-1}, 1.09×10^{17} cm^{-1} s^{-1})

What is the effect on the number of collisions of (i) doubling the absolute temperature at constant pressure, and (ii) doubling the pressure at constant temperature?

(Ans. (i) 0.353 fold decrease, (ii) 4 fold increase)

Viscosity of Gases

1.28 Neon and mercury vapours have very nearly equal van der Waals b parameters, which implies equal atomic volumes and radii. Would you expect any difference in the viscosities of these gases (at the same temperature)?

(Ans. Yes $\eta(\text{Hg})/\eta(\text{Ne}) = \{m(\text{Hg})/m(\text{Ne})\}^{1/2}$)

1.29 At 280 K the viscosity coefficient of CH_4 is 105 micropoise. Calculate its collision diameter, and (ii) the number of bimolecular collisions per second per cubic centimetre at atmospheric pressure. (Ans. (i) 4.164×10^{-8} cm, (ii) 1.619×10^{29})

1.30 The van der Waals constant b of N_2O is 0.044 2 dm^3 mol^{-1}. Estimate the viscosity of N_2O at 25 °C. (Ans. 2.908×10^{-5} N m^{-2} s)

Knudsen Method

1.31 A solid exists in equilibrium with its vapours at 400 K in an effusion cell having a circular hole of diameter 2.50 mm. If the solid mass is decreased by 104 mg in 2 h, determine its vapour pressure. Given: Molar mass of solid = 260 g mol^{-1}.

(Ans. 0.835 Pa)

1.32 A solid exists in equilibrium with its vapours at 500 K in an effusion cell having a circular hole of diameter 3.04 mm. How much mass of the solid is decreased in 10 h if its molar mass is 350 g mol^{-1} and vapour pressure of 0.25 Pa ?

(Ans. 0.239 g)

Heat Capacities

1.33 How many Celsius degrees would the temperature of 1 mole of liquid water be raised by the addition of an amount of energy equal to the translational kinetic energy at 25 °C of 1 mol of water vapour? (Ans. 49 °C)

1.34 The heat capacities at constant volume (C_V) of C_2N_2 and $AsCl_3$ in gaseous state are 77.82 and 75.31 J K^{-1} mol^{-1} respectively. Assume that all degrees of freedom contribute to the heat capacity. On the basis of this assumption, what can you say about the structure of these molecules? (Ans. C_2N_2 linear, $AsCl_3$ nonlinear)

1.35 Estimate the molar heat capacity $C_{V,m}$ of the following molecules at low and high temperatures: CCl_4, F_2, O_3, HCN and HCl.

1.36 The specific heat capacity of iodine vapour at high temperatures is 0.115 J g^{-1} K^{-1}. Is iodine a monatomic or diatomic gas?

Miscellaneous Numericals

1.37 From the relations between the variables for two gases, given below on the left what can you conclude regarding the variables on the right?

Given	Inference (> or < or =)
(a) Equal p, V, T, $m_1 > m_2$	$\overline{\text{KE}}_1$ and $\overline{\text{KE}}_2$
(b) Equal p, V, T, $m_1 > m_2$	N_1 and N_2
(c) Equal p, V, $N_1 > N_2$	T_1 and T_2
(d) Equal T, N, $p_1 > p_2$, $m_1 > m_2$	V_1 and V_2
(e) Equal V, N, $\overline{\text{KE}}$; $m_1 > m_2$	p_1 and p_2
(f) Equal p, $T_1 = 2T_2$, $\sigma_2 = 2\sigma_1$	λ_1 and λ_2
(g) Equal T, p, $m_1 = 2m_2$, $\sigma_1 = 2\sigma_2$	$Z_{1(1)}$ and $Z_{1(2)}$
(h) Equal T, p, $m_1 = 2m_2$, $\sigma_2 = 2\ \sigma_1$	$Z_{11(1)}$ and $Z_{11(2)}$
(i) Equal T, p, $m_1 = 2m_2$, $\sigma_1 = 2\sigma_2$	η_1 and η_2
(j) Equal van der Waals gas constant a, $b_1 > b_2$	$T_{B(1)}$ and $T_{B(2)}$

(Ans. $\overline{\text{KE}}_1 = \overline{\text{KE}}_2$, $N_1 = N_2$, $T_1 < T_2$, $V_1 < V_2$, $p_1 = p_2$, $\lambda_1 > \lambda_2$, $Z_{1(1)} > Z_{1(2)}$, $Z_{11(1)} > Z_{11(2)}$, $\eta_1 < \eta_2$ and $T_{B(1)} < T_{B(2)}$)

1.37 The molar mass of a real gas can be determined by utilizing the ideal gas equation in the form

$$pV = nRT = \frac{m}{M}RT \quad \text{i.e.} \quad M = \frac{m}{V}\frac{RT}{p} = \frac{\rho}{p}RT$$

where ρ is the density of the gas at pressure p and temperature T. However, as real gases do not behave ideally, the ratio ρ/p varies with the pressure. To overcome this difficulty, one can make use of the extrapolation method to determine the limiting value of $(\rho/p)_{p=0}$ from the graph of (ρ/p) versus p. This is known as the method of limiting densities. Following this method, determine the molar mass of nitrous oxide from the data given below:

p/101.325 kPa	1	2/3	1/2	1/3
ρ/g cm^{-3} at 0 °C	1.9804	1.3164	0.9861	0.6565

Annexure I Quantitative Explanation of Variation of Compression Factor with Pressure (Fig. 1.8.2) of a Gas

Van der Waals Equation of State

The van der Waals equation of state is

$$\left(p+\frac{a}{V_m^2}\right)(V_m-b)=RT \tag{AI.1}$$

This gives $p=\dfrac{RT}{V_m-b}-\dfrac{a}{V_m^2}$

Multiplying throughout by V_m/RT, we get

$$\begin{aligned} Z=\frac{pV_m}{RT}&=\frac{V_m}{V_m-b}-\frac{a}{V_mRT}\\ &=1+\left(\frac{V_m}{V_m-b}-1\right)-\frac{a}{V_mRT}\\ &=1+\frac{b}{V_m-b}-\frac{a}{V_mRT} \end{aligned} \tag{AI.2}$$

Negative Deviation from Ideality

For negative deviation, $Z < 1$. This is observed when

$$\frac{a}{V_mRT}>\frac{b}{V_m-b} \tag{AI.3a}$$

Under the condition of low pressure (i.e. large volume), we can approximate $V_m - b \simeq V_m$. Hence,

$$\frac{a}{RT}>b \tag{AI.3b}$$

Equation (AI.3b) implies that the attractive force (represented by the constant a) predominates over repulsive force (represented by the constant b).

Positive Deviation from Ideality

For positive deviation, $Z > 1$. This is observed when

$$\frac{a}{V_mRT}<\frac{b}{V_m-b} \tag{AI.4a}$$

Under the condition of low pressure (i.e. large volume), we can approximate $V_m - b \simeq V_m$. Hence,

$$\frac{a}{RT} < b \tag{AI.4b}$$

Equation (AI.4b) implies that the repulsive force predominates over attractive force.

Ideal Behaviour

For ideal behaviour, $Z = 1$. This is observed when

$$\frac{a}{V_m RT} = \frac{b}{V_m - b} \tag{AI.5a}$$

Under the condition of low pressure (or large volume), we can approximate $V_m - b \simeq V_m$. Hence,

$$\frac{a}{RT} = b \tag{AI.5b}$$

Equation (AI.5b) implies that the attractive force compensates the repulsive force.

Boyle Temperature

From Eq. (AI.5b), we write

$$T_B = \frac{a}{bR} \tag{AI.6}$$

This equation provides the expression of Boyle temperature at which the gas is expected to behave ideally over a wide range of low pressure.

Pressure Corresponding to Minimum Value of Z

At minimum value of Z, we will have

$$\left(\frac{\partial Z}{\partial p}\right)_T = \left(\frac{\partial Z}{\partial V_m}\right)_T \left(\frac{\partial V_m}{\partial p}\right)_T = 0$$

Since $(\partial V_m/\partial p)_T$ is not zero, we evaluate $(\partial Z/\partial V_m)_T$ and set it to zero. From Eq. (AI.2), we have

$$\left(\frac{\partial Z}{\partial V_m}\right)_T = -\frac{b}{(V_m - b)^2} + \frac{a}{V_m^2 RT} = 0 \tag{AI.7}$$

This gives

$$\frac{b}{(V_m - b)^2} = \frac{a}{V_m^2 RT} \qquad \text{i.e.} \qquad \frac{V_m - b}{V_m} = \sqrt{\frac{bRT}{a}}$$

Solving for V_m, we get

$$V_m = \frac{b}{1-\sqrt{bRT/a}} \tag{AI.8}$$

Knowing V_m, the corresponding pressure of the gas may be calculated by using Eq. (AI.1)

Illustration AI.1

For CO_2(g), a = 365.8 kPa dm^6 mol^{-2} and b = 0.0429 dm^3 mol^{-1}, calculate the pressure at which the compression factor at 300 K has a minimum value. Also, calculate the minimum value of compression factor. The gas follows van der Waals equation.

Solution

We have

$$\sqrt{\frac{bRT}{a}} = \left[\frac{(0.0429 \text{ dm}^3 \text{ mol}^{-1})(8.314 \text{ kPa dm}^3 \text{ K}^{-1} \text{ mol}^{-1})(300 \text{ K})}{(365.8 \text{ kPa dm}^6 \text{ mol}^{-2})}\right]^{1/2}$$

$$= 0.541$$

$$V_m = \frac{b}{1-\sqrt{bRT/a}} = \frac{0.0429 \text{ dm}^3 \text{ mol}^{-1}}{(1-0.541)} = 0.0935 \text{ dm}^3 \text{ mol}^{-1}$$

For pressure, we calculate

$$\frac{RT}{V_m - b} = \frac{(8.314 \text{ kPa dm}^3 \text{ K}^{-1} \text{ mol}^{-1})(300 \text{ K})}{(0.0935-0.0429) \text{ dm}^3 \text{ mol}^{-1}} = 4.929 \times 10^4 \text{ kPa}$$

$$\frac{a}{V_m^2} = \frac{(365.8 \text{ kPa dm}^6 \text{ mol}^{-2})}{(0.0935 \text{ dm}^3 \text{ mol}^{-1})^2} = 4.184 \times 10^4 \text{ kPa}$$

Hence, $$p = \frac{RT}{V_m - b} - \frac{a}{V_m^2} = (4.929 - 4.184) \times 10^4 \text{ kPa}$$

$$= 7.397 \times 10^3 \text{ kPa} \equiv 73.0 \text{ atm}$$

For computing Z_{min}, we have

$$\frac{a}{V_m RT} = \frac{(365.8 \text{ kPa dm}^6 \text{ mol}^{-2})}{(0.0935 \text{ dm}^3 \text{ mol}^{-1})(8.314 \text{ kPa dm}^3 \text{ K}^{-1} \text{ mol}^{-1})(300 \text{ K})}$$

$$= 1.569$$

$$\frac{b}{V_m - b} = \frac{(0.0429 \text{ dm}^3 \text{ mol}^{-1})}{(0.0935-0.0429)\text{dm}^3 \text{ mol}^{-1}} = 0.848$$

Hence $$Z = 1 + \frac{b}{V_m - b} - \frac{a}{V_m RT} = 1 + 0.848 - 1.569$$
$$= 0.28$$

Pressure Corresponding to Ideal Behaviour

For ideal behaviour, Eq. (AI.5a) gives

$$V_m = \frac{ab}{a - bRT} \tag{AI.9}$$

Pressure corresponding to this value of V_m is

$$p = \frac{RT}{V_m} = \frac{RT(a - bRT)}{ab} \tag{AI.10}$$

Illustration AI.2

Calculate the pressure of CO_2 at which gas at 300 K behaves as an ideal gas. The gas follows van der Waals equation. (Given a = 365.8 kPa dm^6 mol^{-2} and b = 0.0429 dm^3 mol^{-1}.)

Solution

For calculating pressure, we calculate

$$RT = (8.314 \text{ kPa dm}^3 \text{ K}^{-1} \text{ mol}^{-1})(300 \text{ K}) = 2494.2 \text{ kPa dm}^3 \text{ mol}^{-1}$$
$$b(RT) = (0.0429 \text{ dm}^3 \text{ mol}^{-1})(2494.2 \text{ kPa dm}^3 \text{ mol}^{-1})$$
$$= 106.9 \text{ kPa dm}^6 \text{ mol}^{-2}$$
$$ab = (365.8 \text{ kPa dm}^6 \text{ mol}^{-2})(0.0429 \text{ dm}^3 \text{ mol}^{-1})$$
$$= 15.69 \text{ kPa dm}^9 \text{ mol}^{-3}$$

Hence, $$V_m = \frac{ab}{a - bRT} = \frac{(15.69 \text{ kPa dm}^9 \text{ mol}^{-3})}{(365.8 - 106.9)\text{kPa dm}^6 \text{ mol}^{-2}}$$
$$= 0.0606 \text{ dm}^3 \text{ mol}^{-1}$$

$$p = \frac{RT}{V_m} = \frac{(8.314 \text{ kPa dm}^3 \text{ K}^{-1} \text{ mol}^{-1})(300 \text{ K})}{(0.0606 \text{ dm}^3 \text{ mol}^{-1})}$$
$$= 41158.4 \text{ kPa} \equiv 406.2 \text{ atm}$$

Illustration AI.3

If p is the pressure of 1 mol of CO_2(g) at which it has a minimum value of Z, show that the slope of Z versus p has (i) a negative value in the range p = 0 to p and (ii) a positive value in the range p to p_{ideal}. The gas follows van der Waals equation.

Solution

The value of V_m at which CO_2(g) has a minimum value of Z is V_m = 0.0935 dm^3 mol^{-1} (see, Example AI.1).

(i) For the pressure less than p at which Z has a minimum value, the volume V will be larger than V_m.

Taking an example of 0.2 dm^3 mol^{-1}, we have

$$\frac{b}{(V_m - b)^2} = \frac{0.0429\ \text{dm}^3\ \text{mol}^{-1}}{\{(0.2 - 0.0429)\text{dm}^3\ \text{mol}^{-1}\}^2} = 1.738\ \text{dm}^{-3}\ \text{mol}$$

$$\frac{a}{V_m^2 RT} = \frac{(365.8\ \text{kPa dm}^6\ \text{mol}^{-2})}{(0.2\ \text{mol dm}^{-3})^2(8.314\ \text{kPa dm}^3\ \text{K}^{-1}\ \text{mol}^{-1})(300\ \text{K})}$$

$$= 3.665\ \text{dm}^{-3}\ \text{mol}$$

Now $$\left(\frac{\partial Z}{\partial V_m}\right)_T = -\frac{b}{(V_m - b)^2} + \frac{a}{V_m^2 RT} \qquad \text{(Eq. AI.7)}$$

$$= (-1.738 + 3.665)\ \text{dm}^{-3}\ \text{mol}$$
$$= 2.593\ \text{dm}^3\ \text{mol}^{-1}$$

Finally, $$\left(\frac{\partial Z}{\partial p}\right)_T = \left(\frac{\partial Z}{\partial V_m}\right)_T \left(\frac{\partial V_m}{\partial p}\right)_T = (+\text{ve})(-\text{ve}) = -\text{ve}$$

that is, the slope of the plot of Z versus p is negative in the range $p = 0$ to p.

(ii) For the pressure in between p and p_{ideal}, the volume of CO_2(g) will be lesser than V_m.
Taking an example of V_m = 0.0729 dm^3 mol^{-1}, we have

$$\frac{b}{(V_m - b)^2} = \frac{(0.0429\ \text{dm}^3\ \text{mol}^{-1})}{\{(0.0729 - 0.0429)\ \text{dm}^3\ \text{mol}^{-1}\}^2} = 47.67\ \text{dm}^{-3}\ \text{mol}$$

$$\frac{a}{V_m^2 RT} = \frac{(365.8\ \text{kPa dm}^6\ \text{mol}^{-2})}{(0.0729\ \text{dm}^3\ \text{mol}^{-1})^2(8.314\ \text{kPa dm}^3\ \text{K}^{-1}\ \text{mol}^{-1})(300\ \text{K})}$$

$$= 27.60\ \text{dm}^{-3}\ \text{mol}^{-1}$$

Now $$\left(\frac{\partial Z}{\partial V_m}\right)_T = -\frac{b}{(V_m - b)^2} + \frac{a}{V_m^2 RT}$$

$$= (-47.67 + 27.60)\ \text{dm}^{-3}\ \text{mol}$$
$$= -20.07\ \text{dm}^3\ \text{mol}^{-1}$$

Since $$\left(\frac{\partial Z}{\partial p}\right)_T = \left(\frac{\partial Z}{\partial V_m}\right)_T \left(\frac{\partial V_m}{\partial p}\right)_T = (-\text{ve})(-\text{ve}) = +\text{ve}$$

that is, the slope is positive in the range p to p_{ideal}. It is true for any pressure greater than p.

Display of *Z* versus *p* Plots

Figure AI.1 displays the above discussed characteristics of the plot of compression factor versus pressure for CO_2(g) at 300 K along with those expected at 400 K, 500 K and 600 K.

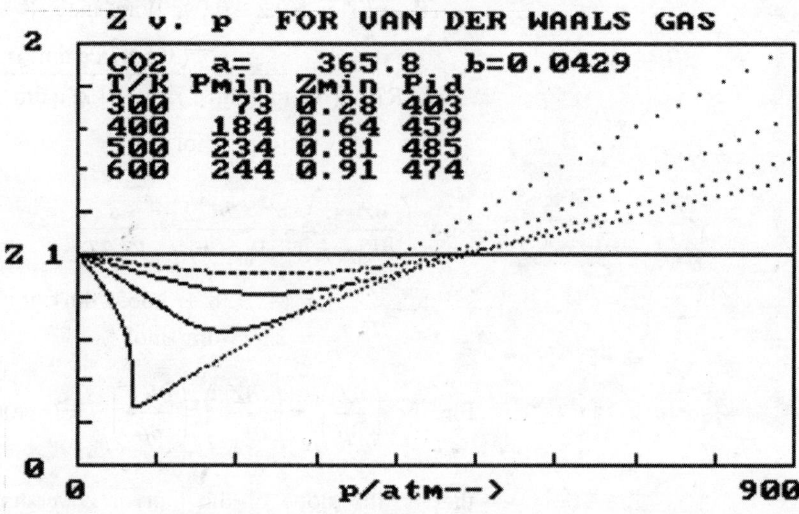

Fig. AI.1

Compression Factor at Boyle Temperature

Since a real gas behaves ideally over a wide range of low pressure (or large volume) at Boyle temperature, we may replace V_m in Eq. (AI.2) by RT_B/p. This gives

$$Z = 1 + \frac{b}{(RT_B/p) - b} - \frac{a}{(RT_B/p)RT_B}$$

$$= 1 + \frac{pb}{RT_B - pb} - \frac{ap}{(RT_B)^2}$$

Since $T_B = a/Rb$ (Eq. AI.6), the above expression becomes

$$Z = 1 + \frac{pb}{(a/b) - pb} - \frac{ap}{(a/b)^2}$$

$$= 1 + \frac{(b^2/a)p}{1-(b^2/a)p} - \left(\frac{b^2}{a}\right)p = 1 + \left(\frac{b^2}{a}p\right)\left[\frac{1}{1-(b^2/a)p} - 1\right]$$

$$= 1 + \frac{(b^2/a)^2 p^2}{1-(b^2/a)p} \qquad \text{(AI.11)}$$

The value of b^2/a for CO_2(g) is

$$\frac{b^2}{a} = \frac{(0.0429\ \text{dm}^3\ \text{mol}^{-1})^2}{(365.8\ \text{kPa dm}^6\ \text{mol}^{-2})} = 5.03 \times 10^{-6}\ \text{kPa}^{-1}$$

Since b^2/a has a small value, Eq. (AI.11) in the lower pressure region may be approximated as

$$Z \simeq 1 + \left(\frac{b^2}{a}\right)^2 p^2 \tag{AI.12}$$

Thus, the plot of Z versus p^2 at T_B in the lower pressure region is linear with slope = $(b^2/a)^2$ and intercept = 1.

(**Note:** The coefficient (b^4/a^2) may be identified with the third virial coefficient in the expression of Z in terms of p.)

The value of slope for CO_2(g) is

$$\left(\frac{b^2}{a}\right)^2 = (5.03 \times 10^{-6}\ \text{kPa}^{-1})^2 = 2.53 \times 10^{-11}\ \text{kPa}^{-2}$$

$$\equiv 2.60 \times 10^{-7}\ \text{atm}^{-2}$$

For such a small value of the slope, Z is expected to increase to a very small extent with increase in the value of p^2. In fact, the value Z remains close to one over a wide range of low pressure.

If we write Eq. (AI.2) as

$$Z = 1 + \left(\frac{b^4}{a^2} p\right) p \tag{AI.13}$$

it is expected that Z varies nonlinearly when plotted against p. The slope of this plot varies with p with the characteristic of slope $\to 0$ as $p \to 0$. This implies that the plot of Z versus p meets tangentially with the straight-line plot of $Z(= 1)$ versus p for an ideal gas as the pressure of a real gas approaches zero.

Display of Z versus p and p^2 Graphs at T_B

Figure AI.2 displays the plot of Z versus p and Z versus p^2 for CO_2(g) at its Boyle temperature.

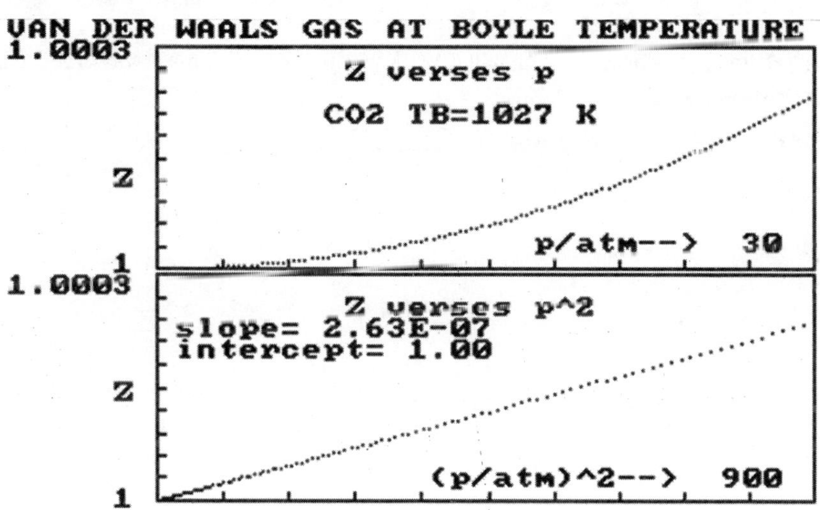

Fig. AI.2

Compression Factor in Terms of Reduced Pressure

For $CO_2(g)$, the variation of Z with p as given by Eq. (AI.12) is

$$Z = 1 + \left(\frac{b^2}{a}\right)^2 p^2 = 1 + (2.53 \times 10^{-11}\ \text{kPa}^{-2})p^2$$

Multiplying and dividing the second term on the right side by p_c^2 of $CO_2(g)$, we get

$$Z = 1 + \{(2.53 \times 10^{-11}\ \text{kPa}^{-2})\, p_c^2\}(p/p_c)^2$$

For $CO_2(g)$, $p_c = 7376.46$ kPa. Hence

$$\begin{aligned} Z &= 1 + \{(2.53 \times 10^{-11}\ \text{kPa}^{-2})(7376.46\ \text{kPa})^2\}\, p_r^2 \\ &= 1 + (1.376 \times 10^{-3})p_r^2 \end{aligned} \qquad \text{(AI.14)}$$

Berthelot Equation of State

The Berthelot equation of state is

$$\left(p + \frac{a}{TV_m^2}\right)(V_m - b) = RT \qquad \text{(AI.15)}$$

This gives $p = \dfrac{RT}{V_m - b} - \dfrac{a}{TV_m^2}$

Multiplying throughout by V_m/RT, we get

$$\begin{aligned} Z = \frac{pV_m}{RT} &= \frac{V_m}{V_m - b} - \frac{a}{RT^2V_m} = 1 + \left(\frac{V_m}{V_m - b} - 1\right) - \frac{a}{RT^2V_m} \\ &= 1 + \frac{b}{V_m - b} - \frac{a}{RT^2V_m} \end{aligned} \qquad \text{(AI.16)}$$

Negative Deviation for Ideality

For negative deviation, $Z < 1$. This is observed when

$$\frac{a}{RT^2V_m} > \frac{b}{V_m - b} \qquad \text{(AI.17a)}$$

Under the condition of low pressure (or large volume), we can approximate $V_m - b \simeq V_m$. Hence

$$\frac{a}{RT^2} > b \qquad \text{(AI.17b)}$$

Equation (AI.17b) implies that the attractive force (represented by the constant a) predominates over repulsive force (represented by the constant b).

Positive Deviation from Ideality

For positive deviation, $Z > 1$. This is observed when

$$\frac{a}{RT^2 V_m} < \frac{b}{V_m - b} \tag{AI.18a}$$

Under the condition of low pressure (i.e. large volume), Eq. (AI.18a) may be approximated as

$$\frac{a}{RT^2} < b \tag{AI.18b}$$

Equation (AI.18b) implies that the repulsive force predominates over attractive force.

Ideal Behaviour

For ideal behaviour, $Z = 1$. This is observed when

$$\frac{a}{RT^2 V_m} = \frac{b}{V_m - b} \tag{AI.19a}$$

Under the condition of low pressure (i.e. high volume), Eq. (AI.19a) may be approximated as

$$\frac{a}{RT^2} = b \tag{AI.19b}$$

Equation (AI.19b) implies that the attractive force compensates the repulsive force.

Boyle Temperature

From Eq. (AI.19b), we write

$$T_B = \left(\frac{a}{Rb}\right)^{1/2} \tag{AI.20}$$

Equation (AI.20) provides the expression of Boyle temperature at which the gas is expected to behave ideally over a wide range of low pressure.

Pressure Corresponding to Minimum Value of Z

At the minimum value of Z, we will have $(\partial Z/\partial V_m)_T = 0$. From Eq. (AI.16), we get

$$\left(\frac{\partial Z}{\partial V_m}\right)_T = -\frac{b}{(V_m - b)^2} + \frac{a}{RT^2 V_m^2} = 0 \tag{AI.21}$$

Solving Eq. (AI.21) for V_m, we get

$$V_m = \frac{b}{1 - \sqrt{bRT^2/a}} \tag{AI.22}$$

Knowing V_m, the corresponding pressure may be calculated by using Eq. (AI.15).

Illustration AI.3 For $CO_2(g)$, $p_c = 7376.46$ kPa and $T_c = 304.2$ K. Calculate the pressure at which $CO_2(g)$ at 300 K has a minimum value of compression factor (Z). What is the value of Z? The gas follows Berthelot equation of state.

Solution For Berthelot gas,

$$a = \frac{27}{64}\frac{R^2T_c^3}{p_c} = \frac{27}{64}\frac{(8.314\ \text{kPa dm}^3\ \text{K}^{-1}\ \text{mol}^{-1})^2(304.2\ \text{K})^3}{(7376.46\ \text{kPa})}$$

$$= 111\,284.2\ \text{kPa dm}^6\ \text{mol}^{-2}$$

$$b = \frac{RT_c}{8p_c} = \frac{(8.314\ \text{kPa dm}^3\ \text{K}^{-1}\ \text{mol}^{-1})\,(304.2\ \text{K})}{8\,(7376.46\ \text{kPa})}$$

$$= 0.0429\ \text{dm}^3$$

For calculating V_m at which Z has a minimum value, we evaluate

$$\sqrt{\frac{bRT^2}{a}} = \left[\frac{(0.0429\,\text{dm}^3\ \text{mol}^{-1})(8.314\,\text{kPa dm}^3\ \text{K}^{-1}\ \text{mol}^{-1})(300\,\text{K})^2}{111\,284.2\,\text{kPa dm}^6\ \text{K}^{-1}\ \text{mol}^{-2}}\right]^{1/2}$$

$$= 0.537$$

Hence, $$V_m = \frac{b}{1-\sqrt{bRT^2/a}} = \frac{0.0429\ \text{dm}^3\ \text{mol}^{-1}}{(1-0.537)} = 0.0927\ \text{dm}^3\ \text{mol}^{-1}$$

For pressure, we calculate

$$\frac{RT}{V_m - b} = \frac{(8.314\ \text{kPa dm}^3\ \text{K}^{-1}\ \text{mol}^{-1})(300\ \text{K})}{(0.0927-0.0429)\ \text{dm}^3\ \text{mol}^{-1}} = 50084.34\ \text{kPa}$$

$$\frac{a}{TV_m^2} = \frac{(111284.2\ \text{kPa dm}^6\ \text{K mol}^{-2})}{(300\ \text{K})(0.0927\ \text{dm}^3\ \text{mol}^{-1})^2} = 43167.09\ \text{kPa}$$

Hence, $$p = \frac{RT}{V_m - b} - \frac{a}{TV_m^2} = (50084.34 - 43167.09)\ \text{kPa}$$

$$= 6917.25\ \text{kPa} \equiv 68.27\ \text{atm}$$

For computing Z_{min}, we have

$$\frac{b}{V_m - b} = \frac{(0.0429\ \text{dm}^3\ \text{mol}^{-1})}{(0.0927-0.0429)\ \text{dm}^3\ \text{mol}^{-1}} = 0.86$$

$$\frac{a}{RT^2V_m} = \frac{(111\,284.2\ \text{kPa dm}^6\ \text{K mol}^{-2})}{(8.314\ \text{kPa dm}^3\ \text{K}^{-1}\ \text{mol}^{-1})\,(300\ \text{K})^2(0.0927\ \text{dm}^3\ \text{mol}^{-1})}$$

$$= 1.60$$

Hence, $$Z = 1 + \frac{b}{V_m - b} - \frac{a}{RT^2V_m} = 1 + 0.86 - 1.60 = 0.26$$

Pressure Corresponding to Ideal Behaviour

For ideal behaviour, V_m as determined from Eq. (AI.19a) is

$$V_m = \frac{ab}{a - bRT^2} = \frac{b}{1 - (bRT^2/a)} \tag{AI.23}$$

The pressure corresponding to this volume is

$$p = \frac{RT}{V_m} = \frac{RT(a - bRT^2)}{ab} \tag{AI.24}$$

Illustration AI.4

Calculate the pressure at which CO_2(g) at 300 K behaves ideally. The gas follows Berthelot equation of state. (Given: a = 111 284.2 kPa dm^6 K mol^{-2} and b = 0.0429 dm^3 mol^{-1}.)

Solution

From illustration AI.3, we have

$$\frac{bRT^2}{a} = (0.537)^2 = 0.2884;$$

$$\text{Hence, } V_m = \frac{0.0429 \text{ dm}^3 \text{ mol}^{-1}}{1 - 0.2884} = 0.0603 \text{ dm}^3 \text{ mol}^{-1}$$

$$\text{Hence, } \quad p = \frac{RT}{V_m} = \frac{(8.314 \text{ kPa dm}^3 \text{ K}^{-1} \text{ mol}^{-1})(300 \text{ K})}{(0.0603 \text{ dm}^3 \text{ mol}^{-1})}$$

$$= 41363.2 \text{ kPa} \equiv 408.2 \text{ atm}$$

Graph of Z versus p

Figure AI.3 displays the above characteristics of the plot of Z versus p for CO_2(g) at 300 K along with those expected at 400 K, 500 K and 600 K.

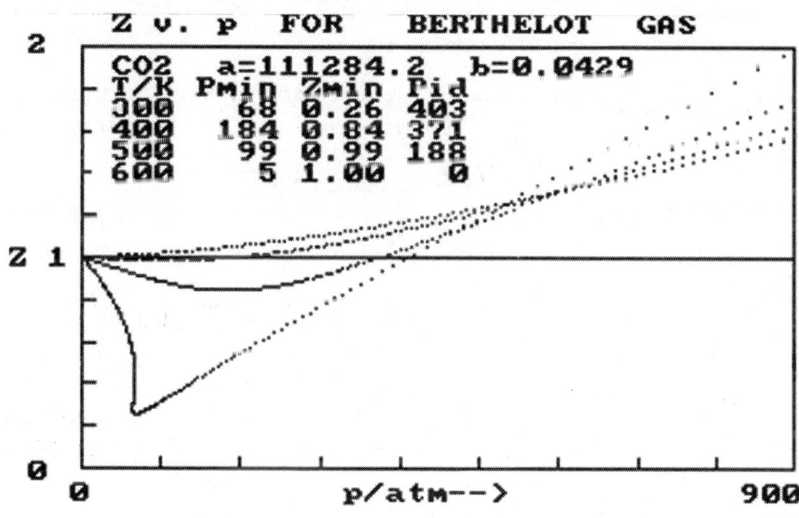

Fig. AI.3

Compression Factor at Boyle Temperature

Since a real gas behaves ideally over a wide range of low pressure (or larger volume) at Boyle temperature, we may replace V_m in Eq. (AI.15) by RT_B/p. This gives

$$Z = 1 + \frac{b}{(RT_B/p) - b} - \frac{a}{RT_B^2(RT_B/p)}$$

$$= 1 + \frac{pb}{RT_B - bp} - \frac{ap}{R^2T_B^3} \qquad \text{(AI.25)}$$

Since $T_B = (a/Rb)^{1/2}$, we get

$$Z = 1 + \frac{pb}{R(a/Rb)^{1/2} - pb} - \frac{ap}{R^2(a/Rb)^{3/2}}$$

$$= 1 + \frac{pb^{3/2}}{(Ra)^{1/2}}\left[\frac{1}{1 - b^{3/2}p/(Ra)^{1/2}} - 1\right]$$

$$= 1 + \frac{(b^3/Ra)p^2}{1 - \{b^{3/2}/(Ra)^{1/2}\}p}$$

Let $x = b^{3/2}/(Ra)^{1/2}$. With this, the above expression becomes

$$Z = 1 + \frac{x^2p^2}{1 - xp} \qquad \text{(AI.26)}$$

The value of $x (= b^{3/2}/\sqrt{Ra})$ for CO_2(g) is

$$x = \frac{b^{3/2}}{\sqrt{Ra}} = \frac{(0.0429\,\text{dm}^3\,\text{mol}^{-1})^{3/2}}{\{(8.314\,\text{kPa dm}^3\,\text{K}^{-1}\,\text{mol}^{-1})(111\,284.2\,\text{kPa dm}^6\,\text{K mol}^{-2})\}^{1/2}}$$

$$= 9.24 \times 10^{-6}\ \text{kPa}^{-1}$$

In the lower pressure region $xp \ll 1$ and thus may be ignored in comparison to 1 in Eq. (AI.26). with this, Eq. (AI.26) may be approximated as

$$Z = 1 + x^2p^2 \qquad \text{(AI.27)}$$

The plot of Z versus p^2 at T_B of a gas is linear with slope equal to x^2 ($= b^3/Ra$). The value of slope for CO_2(g) at 300 K is

$$\frac{b^3}{Ra} = x^2 = (9.24 \times 10^{-6}\ \text{kPa}^{-1})^2 = 8.54 \times 10^{-11}\ \text{kPa}^{-2}$$

$$\equiv 8.77 \times 10^{-7}\ \text{atm}^{-2}$$

For such a small value of slope, Z increases to a very small extent with increase in the value of p^2. In fact, the value of Z remains close to 1 over a wide range of lower pressure.

When Eq. (AI.27) is written as

$$Z = 1 + (x^2 p)p$$

we find that Z varies nonlinearly when plotted against p. The slope of this plot varies with p with the characteristic of slope $\rightarrow 0$ as $p \rightarrow 0$. This implies that the plot of Z versus p meets tangentially with the straight-line plot of Z (= 1) versus p for an ideal gas as the pressure of real gas approaches zero.

Graphical Display of *Z* versus *p* and p^2

Figure AI.4 displays the plots of Z versus p and Z versus p^2 for CO_2(g) at its Boyle temperature.

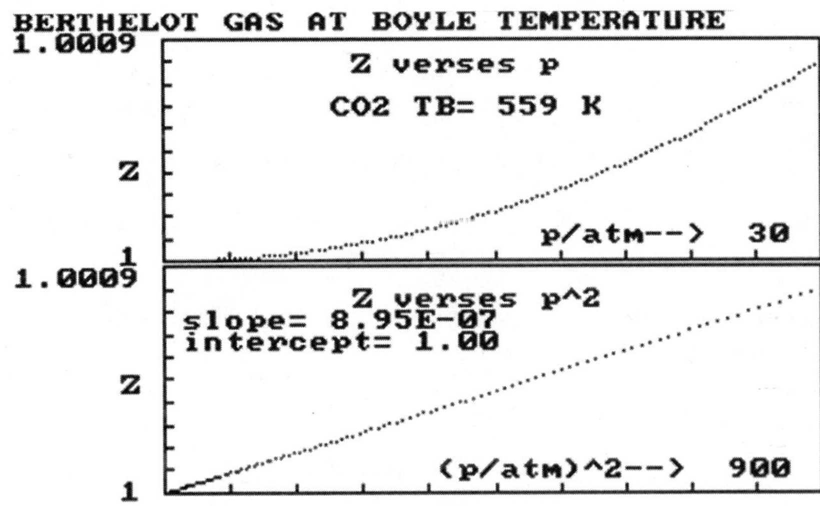

Fig. AI.4

Compression Factor in Terms of Reduced Pressure

For CO_2(g), the variation of Z with p^2 as given by Eq. (AI.27) is

$$Z = 1 + x^2 p^2 = 1 + (8.54 \times 10^{-11}\ \text{kPa}^{-2})p^2$$

Dividing and multiplying the second term on the right side by p_c, we get

$$Z = 1 + \{(8.54 \times 10^{-11}\ \text{kPa}^{-2})\, p_c^2\,\}(p/p_c)^2$$

For CO_2(g), p_c = 7376.46 kPa. Hence

$$Z = 1 + \{(8.54 \times 10^{-11}\ \text{kPa}^{-2})(7376.46\ \text{kPa})^2\}\, p_r^2$$

$$\equiv 1 + (4.65 \times 10^{-3})\, p_r^2 \qquad \text{(AI.28)}$$

Dieterici Equation of State

The Dieterici equation of state is

$$p = \left(\frac{RT}{V_m - b}\right) e^{-a/V_m RT} \qquad \text{(AI.29)}$$

Multiplying throughout by V_m/RT, we get

$$Z = \frac{pV_m}{RT} = \left(\frac{V_m}{V_m - b}\right) e^{-a/V_m RT} = \left(\frac{1}{1 - b/V_m}\right) e^{-a/V_m RT} \tag{AI.30}$$

Negative Deviation from Ideality

For negative deviation, $Z < 1$. This is observed when

$$\left(\frac{1}{1 - b/V_m}\right) e^{-a/V_m RT} < 1 \quad \text{i.e.} \quad e^{-a/V_m RT} < \left(1 - \frac{b}{V_m}\right) \tag{AI.31a}$$

Under the condition of low pressure (i.e. large volume), the exponential factor may be approximated as

$$e^{-a/V_m RT} = 1 - \frac{a}{V_m RT}$$

with this, Eq. (AI.31a) becomes

$$\left(1 - \frac{a}{V_m RT}\right) < \left(1 - \frac{b}{V_m}\right) \quad \text{i.e.} \quad \frac{a}{RT} > b \tag{AI.31b}$$

Positive Deviation from Ideality

For positive deviation, $Z > 1$. This is observed when

$$\left(\frac{1}{1 - b/V_m}\right) e^{-a/V_m RT} > 1 \quad \text{i.e.} \quad e^{-a/V_m RT} > \left(1 - \frac{b}{V_m}\right) \tag{AI.32a}$$

Under the condition of low pressure (i.e. large volume), Eq. (AI.32a) may be written as

$$\left(1 - \frac{a}{V_m RT}\right) > \left(1 - \frac{b}{V_m}\right) \quad \text{i.e.} \quad \frac{a}{RT} < b \tag{AI.32b}$$

Ideal Behaviour

For ideal behaviour, $Z = 1$. This is observed when

$$\left(\frac{1}{1 - b/V_m}\right) e^{-a/V_m RT} = 1 \quad \text{i.e.} \quad e^{-a/V_m RT} = \left(1 - \frac{b}{V_m}\right) \tag{AI.33a}$$

Under the condition of low pressure (or large volume), Eq. (AI.33a) may be written as

$$\left(1 - \frac{a}{V_m RT}\right) = \left(1 - \frac{b}{V_m}\right) \quad \text{i.e.} \quad \frac{a}{RT} = b \tag{AI.33b}$$

Boyle Temperature

Equation (AI.33b) provides

$$T_B = \frac{a}{Rb} \tag{AI.34}$$

Equation (AI.34) is the expression of Boyle temperature at which the gas is expected to behave ideally over a wide range of pressure.

Pressure Corresponding to Minimum Value of Z

At minimum value of Z, we set

$$\left(\frac{\partial Z}{\partial V_\mathrm{m}}\right)_T = 0$$

that is $$\left(\frac{\partial Z}{\partial V_\mathrm{m}}\right)_T = \left[\left\{\frac{1}{V_\mathrm{m}-b} - \frac{V_\mathrm{m}}{(V_\mathrm{m}-b)^2}\right\} + \left(\frac{V_\mathrm{m}}{V_\mathrm{m}-b}\right)\left(\frac{a}{V_\mathrm{m}^2 RT}\right)\right] \mathrm{e}^{-a/V_\mathrm{m}RT} = 0$$

This gives

$$\frac{1}{V_\mathrm{m}-b} - \frac{V_\mathrm{m}}{(V_\mathrm{m}-b)^2} + \frac{a}{V_\mathrm{m}(V_\mathrm{m}-b)RT} = 0$$

i.e. $$\frac{(V_\mathrm{m}-b)V_\mathrm{m}RT - V_\mathrm{m}^2 RT + a(V_\mathrm{m}-b)}{(V_\mathrm{m}-b)^2 V_\mathrm{m}RT} = 0$$

i.e. $$(V_\mathrm{m}^2 - V_\mathrm{m}b - V_\mathrm{m}^2)RT + a(V_\mathrm{m}-b) = 0$$

This gives

$$-V_\mathrm{m}bRT + a(V_\mathrm{m} - b) = 0$$

Solving for V_m, we get

$$V_\mathrm{m} = \frac{ab}{a - bRT} \qquad \text{(AI.35)}$$

Substituting this value in Eq. (AI.29) provides the value of p.

Illustration AI.5

The critical constants p_c and T_c for CO_2(g) are: p_c = 7376.46 kPa and T_c = 304.2 K. Calculate the pressure at which CO_2(g) at 300 K has a minimum value of compression factor (Z). What is the value of Z? The gas follows Dieterici equation of state.

Solution

For Dieterici equation of state

$$a = 2V_\mathrm{c}RT_\mathrm{c} = 2\left(\frac{2}{\mathrm{e}^2}\frac{RT_\mathrm{c}}{p_\mathrm{c}}\right)RT_\mathrm{c} = \frac{4}{\mathrm{e}^2}\frac{(RT_\mathrm{c})^2}{p_\mathrm{c}}$$

$$b = \frac{V_\mathrm{c}}{2} = \frac{1}{2}\left(\frac{2}{\mathrm{e}^2}\frac{RT_\mathrm{c}}{p_\mathrm{c}}\right) = \frac{1}{\mathrm{e}^2}\frac{RT_\mathrm{c}}{p_\mathrm{c}}$$

Hence, $$a = \frac{4(8.314\ \mathrm{kPa\ dm^3\ K^{-1}\ mol^{-1}})^2(304.2\ \mathrm{K})^2}{(2.718)^2(7376.46\ \mathrm{kPa})}$$

$$= 469.52\ \mathrm{kPa\ dm^6\ mol^{-2}}$$

$$b = \frac{1}{(2.718)^2}\frac{(8.314\ \mathrm{kPa\ dm^3\ K^{-1}\ mol^{-1}})(304.2\ \mathrm{K})}{(7376.46\ \mathrm{kPa})}$$

$$= 0.0464\ \mathrm{dm^3\ mol^{-1}}$$

For calculating V_m, we evaluate

$$ab = (469.52 \text{ kPa dm}^6 \text{ mol}^{-2})(0.0464 \text{ dm}^3 \text{ mol}^{-1}) = 21.79 \text{ kPa dm}^9 \text{ mol}^{-3}$$

$$bRT = (0.0464 \text{ dm}^3 \text{ mol}^{-1})(8.314 \text{ kPa dm}^3 \text{ K}^{-1} \text{ mol}^{-1})(300 \text{ K}) = 115.73 \text{ kPa dm}^6 \text{ mol}^{-2}$$

Hence, $$V_m = \frac{ab}{a - bRT} = \frac{21.79 \text{ kPa dm}^9 \text{ mol}^{-3}}{(469.52 - 115.73) \text{ kPa dm}^6 \text{ mol}^{-2}} = 0.0616 \text{ dm}^3 \text{ mol}^{-1}$$

For calculating p, we evaluate

$$V_mRT = (0.0616 \text{ dm}^3 \text{ mol}^{-1})(8.314 \text{ kPa dm}^3 \text{ K}^{-1} \text{ mol}^{-1})(300 \text{ K}) = 153.64 \text{ kPa dm}^6 \text{ mol}^{-2}$$

Hence, $$p = \frac{RT}{(V_m - b)} e^{-a/V_mRT} = \left[\frac{(8.314 \text{ kPa dm}^3 \text{ K}^{-1} \text{ mol}^{-1})(300 \text{ K})}{(0.0616 - 0.0464) \text{ dm}^3 \text{ mol}^{-1}}\right] e^{-469.52/153.64}$$

$$= (164\,092.1 \text{ kPa})(0.047) = 7712.33 \text{ kPa} \equiv 76.1 \text{ atm}$$

The minimum value of Z is

$$Z = \left(\frac{V_m}{V_m - b}\right) e^{-a/V_mRT} = \left[\frac{0.0616 \text{ dm}^3 \text{ mol}^{-1}}{(0.0616 - 0.0464)\text{dm}^3 \text{ mol}^{-1}}\right](0.047) = (4.053)(0.047) = 0.19$$

Pressure Corresponding to Ideal Behaviour

For ideal behaviour, V_m as given by Eq. (AI.33a) is

$$V_m = \frac{b}{1 - e^{-a/V_mRT}} \tag{AI.36}$$

The value of V_m may be determined by the method of successive approximation Once the value of V_m is known, the value of p may be computed by using the expression $p = RT/V_m$.

Illustration AI.6

Calculate the pressure of CO_2(g) at which it behaves ideally. (Given: a = 469.52 kPa dm^6 mol^{-2} and b = 0.0464 dm^3 mol^{-1}). The gas follows Dietrici equation of state.

Solution

The volume of CO_2(g) behaving as an ideal gas is

$$V_m = \frac{b}{1 - e^{-a/V_m RT}} \qquad \text{(Eq. AI.26)}$$

The value of a/RT at 300 K is

$$\frac{a}{RT} = \frac{(469.52 \text{ kPa dm}^6 \text{ mol}^{-2})}{(8.314 \text{ kPa dm}^3 \text{ K}^{-1} \text{ mol}^{-1})(300 \text{ K})}$$
$$= 0.1882 \text{ dm}^3 \text{ mol}^{-1}$$

With this, the expression of V_m becomes

$$V_m = \frac{(0.0464 \text{ dm}^3 \text{ mol}^{-1})}{1 - e^{-(0.1882 \text{ dm}^3 \text{ mol}^{-1})/V_m}}$$

We start with V_m = 0.05 mol dm^{-3}. Substituting this value on right side of the above expression provides new value of V_m. This is again substituted on the right side to provide the new value of V_m. This procedure is repeated till the two successive values of V_m are very close to each other. The following results were obtained.

Cycle	V_m/dm^3
0	0.05
1	0.04750
2	0.04730
3	0.04730

The pressure corresponding to V_m = 0.0473 dm^3 mol^{-1} is

$$p = \frac{RT}{V_m} = \frac{(8.314 \text{ kPa dm}^3 \text{ K}^{-1} \text{ mol}^{-1})(300 \text{ K})}{(0.0473 \text{ dm}^3 \text{ mol}^{-1})}$$
$$= 52731.5 \text{ kPa} = 520.4 \text{ atm}$$

Display of Z versus p

Figure AI.5 displays plot of Z versus p for CO_2(g) at 300 K along with the plots of 400 K, 500 K and 600 K.

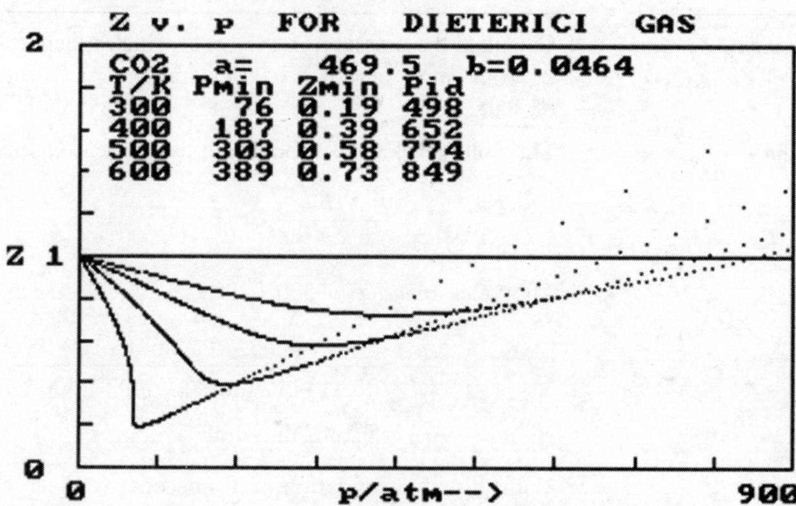

Fig. AI.5

Compression Factor at Boyle Temperature

Since a real gas behaves ideally over a wide range of low pressure (or larger volume) at Boyle temperature, we may replace V_m in Eq. (AI.30) by RT_B/p as per ideal-gas equation of state. This gives

$$Z = \left[\frac{RT_B/p}{(RT_B/p) - b}\right] e^{-ap/R^2T_B^2} = \left(\frac{1}{1 - pb/RT_B}\right) e^{-ap/R^2T_B^2} \tag{AI.37}$$

Since $T_B = a/Rb$, the above expression becomes

$$Z = \frac{1}{(1 - pb^2/a)} e^{-pb^2/a} \tag{AI.38}$$

The value of b^2/a for CO_2(g) is

$$\frac{b^2}{a} = \frac{(0.0464 \text{ dm}^3 \text{ mol}^{-1})^2}{(469.52 \text{ kPa dm}^6 \text{ mol}^{-2})} = 4.59 \times 10^{-6} \text{ kPa}^{-1}$$

Since b^2/a is very small, the exponential term in Eq. (AI.38) in the low pressure region may be expanded as

$$e^{-pb^2/a} = 1 - \frac{pb^2}{a} + \frac{1}{2}\left(\frac{pb^2}{a}\right)^2$$

With this, Eq. (AI.38) becomes

$$Z = \frac{1}{1 - pb^2/a}\left[1 - \frac{pb^2}{a} + \frac{1}{2}\left(\frac{pb^2}{a}\right)^2\right]$$

$$= 1 + \frac{1}{2}\frac{(pb^2/a)^2}{(1 - pb^2/a)} \tag{AI.39}$$

Since b^2/a is very small, Eq. (AI.39) in the lower pressure region may be approximated as

$$Z = 1 + \frac{1}{2}\left(\frac{b^2}{a}\right)^2 p^2 \qquad \text{(AI.40)}$$

Thus, the plot of Z versus p^2 at T_B in the lower pressure region is linear with

$$\text{slope} = \frac{1}{2}\left(\frac{b^2}{a}\right)^2 \text{ and intercept} = 1.$$

The value of slope for CO_2(g) is

$$\frac{1}{2}\left(\frac{b^2}{a}\right)^2 = \frac{1}{2}(4.59 \times 10^{-6}\ \text{kPa}^{-1})^2 = 1.04 \times 10^{-11}\ \text{kPa}^{-2}$$

$$\equiv 1.07 \times 10^{-7}\ \text{atm}^{-2}$$

For such a small value of slope, Z increases to a very small extent with increase in the value of p^2. In fact, the value of Z remains close to one over a wide range of lower pressure.

When Eq. (AI.40) is written as

$$Z = 1 + \left(\frac{b^4}{2a^2}p\right)p \qquad \text{(AI.41)}$$

we find that Z varies nonlinearly when plotted against p with the characteristic of slope $\rightarrow$ 0 when $p \rightarrow 0$. This implies that the plot of Z varies p meets tangentially with the straight-line plot of Z (= 1) versus p for an ideal gas as the pressure of real gas approaches zero.

Display of Z versus p and p^2 Plots

Figure AI.6 displays the plot of Z versus p and Z versus p^2 for CO_2(g) at its Boyle temperature.

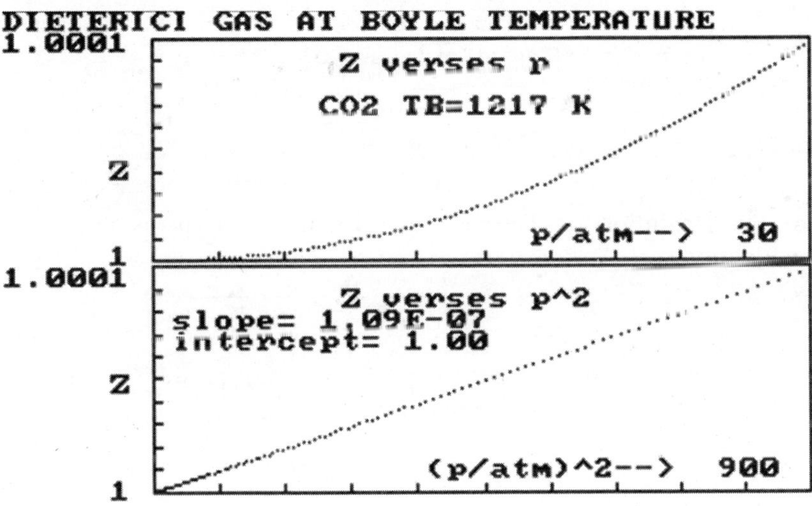

Fig. AI.6

Z versus Reduced Pressure

The variation of Z with p^2 for $CO_2(g)$ as given by Eq. (AI.41) is

$$Z = 1 + \frac{1}{2}\left(\frac{b^2}{a}\right)^2 p^2 = 1 + (1.04 \times 10^{-11}\ \text{kPa}^{-2})p^2$$

Multiplying and dividing the second term on right side by p_c, we get

$$Z = 1 + \{(1.04 \times 14^{-11}\ \text{kPa}^{-2})p_c^{\,2}\}(p/p_c)^2$$

For $CO_2(g)$, p_c = 7376.46 kPa. Hence

$$Z = 1 + \{(1.04 \times 10^{-11}\ \text{kPa}^{-2})(7376.46\ \text{kPa})^2\}\ p_r^2$$

$$= 1 + (5.66 \times 10^{-4})\ p_r^2 \qquad \text{(AI.42)}$$

Redlich-Kwong Equation of State

The Redlich-Kwong equation of state is

$$p = \frac{RT}{V_m - b} - \frac{a}{T^{1/2}V_m(V_m + b)} \qquad \text{(AI.43)}$$

Multiplying throughout by V_m/RT, we get

$$Z = \frac{pV_m}{RT} = \frac{V_m}{V_m - b} - \frac{a}{RT^{3/2}(V_m + b)}$$

$$= 1 + \frac{b}{V_m - b} - \frac{a}{RT^{3/2}(V_m + b)} \qquad \text{(AI.44)}$$

Negative Deviation from Ideality

For negative deviation, $Z < 1$. This is observed when

$$\frac{a}{RT^{3/2}(V_m + b)} > \frac{b}{V_m - b} \qquad \text{(AI.45a)}$$

Under the condition of low pressure (i.e. large volume), Eq. (AI.45a) may be approximated as

$$\frac{a}{RT^{3/2}} > b \qquad \text{(AI.45b)}$$

Positive Deviation for Ideality

For positive deviation, $Z > 1$. This is observed when

$$\frac{a}{RT^{3/2}(V_m + b)} < \frac{b}{V_m - b} \qquad \text{(AI.46a)}$$

Under the condition of low pressure (i.e. large volume), Eq. (AI.49a) may be approximated as

$$\frac{a}{RT^{3/2}} < b \qquad \text{(AI.46b)}$$

Ideal Behaviour

For ideal behaviour, $Z = 1$. This is observed when

$$\frac{a}{RT^{3/2}(V_{\mathrm{m}}+b)}=\frac{b}{V_{\mathrm{m}}-b} \tag{AI.47a}$$

Under the condition of low pressure (i.e. large volume), Eq. (AI.47a) may be approximated as

$$\frac{a}{RT^{3/2}}=b \tag{AI.47b}$$

Boyle Temperature

Equation (AI.47b) may be rearranged to give

$$T_{\mathrm{B}}=\left(\frac{a}{Rb}\right)^{2/3} \tag{AI.48}$$

Equation (AI.48) provides the expression of Boyle temperature at which the gas is expected to behave ideally over a wide range of low pressure.

Pressure Corresponding to Minimum Value of Z

At minimum value of Z, we get

$$\left(\frac{\partial Z}{\partial V_{\mathrm{m}}}\right)_T=0$$

that is $$\left(\frac{\partial Z}{\partial V_{\mathrm{m}}}\right)_T=-\frac{b}{(V_{\mathrm{m}}-b)^2}+\frac{a}{RT^{3/2}(V_{\mathrm{m}}+b)^2}=0$$

This gives

$$\frac{V_{\mathrm{m}}+b}{V_{\mathrm{m}}-b}=\left(\frac{a}{RT^{3/2}b}\right)^{1/2} \tag{AI.49}$$

Let $$x=\left(\frac{a}{RT^{3/2}b}\right)^{1/2} \tag{AI.50}$$

With this, Eq. (AI.49) becomes

$$\frac{V_{\mathrm{m}}+b}{V_{\mathrm{m}}-b}=x$$

This gives

$$V_{\mathrm{m}}=\frac{b(x+1)}{(x-1)} \tag{AI.51}$$

Knowing V_{m}, the value of pressure may be evaluated by using Eq. (AI.43).

Illustration AI.7

For CO_2(g), p_c = 7376.46 kPa and T_c = 304.2 K. Calculate the pressure at which CO_2(g) at 300 K has a minimum value of compression factor (Z). Also, calculate the value of Z. The gas follows Redlich-Kwong equation of state.

Solution For Redlich-Kwong equation, we have

$$a = \frac{R^2 T_c^{5/2}}{9(2^{1/3}-1)p_c} \quad \text{and} \quad b = \frac{(2^{1/3}-1)RT_c}{3p_c}$$

Hence,
$$a = \frac{(8.314\ \text{kPa dm}^3\ \text{K}^{-1}\ \text{mol}^{-1})^2(304.2\ \text{K})^{5/2}}{9(2^{1/3}-1)(7376.46\ \text{kPa})}$$

$$= 6465.3\ \text{kPa dm}^6\ \text{K}^{1/2}\ \text{mol}^{-2}$$

$$b = \frac{(2^{1/3}-1)(8.314\ \text{kPa dm}^3\ \text{K}^{-1}\ \text{mol}^{-1})(304.2\ \text{K})}{3(7376.46\ \text{kPa})}$$

$$= 0.0297\ \text{dm}^3\ \text{mol}^{-1}$$

Now
$$x = \left(\frac{a}{RT^{3/2}b}\right)^{1/2}$$

$$= \left[\frac{(6465.3\ \text{kPa dm}^6\ \text{K}^{1/2}\ \text{mol}^{-2})}{(8.314\ \text{kPa dm}^3\ \text{K}^{-1}\ \text{mol}^{-1})(300\ \text{K})^{3/2}(0.0297\ \text{dm}^3\ \text{mol}^{-1})}\right]^{1/2}$$

$$= 2.245$$

Hence
$$V_m = \frac{b(x+1)}{(x-1)} = \frac{(0.0297\ \text{dm}^3\ \text{mol}^{-1})(3.245)}{(1.245)}$$

$$= 0.0774\ \text{dm}^3\ \text{mol}^{-1}$$

$$V_m - b = (0.0774 - 0.0297)\ \text{dm}^3\ \text{mol}^{-1} = 0.0477\ \text{dm}^3\ \text{mol}^{-1}$$

$$V_m + b = (0.0774 + 0.0297)\ \text{dm}^3\ \text{mol}^{-1} = 0.1071\ \text{dm}^3\ \text{mol}^{-1}$$

To calculate pressure, we evaluate

$$\frac{RT}{V_m - b} = \frac{(8.314\ \text{kPa dm}^3\ \text{K}^{-1}\ \text{mol}^{-2})(300\ \text{K})}{(0.0477\ \text{dm}^3\ \text{mol}^{-1})} = 52289.3\ \text{kPa}$$

$$\frac{a}{T^{1/2}V_m(V_m+b)} = \frac{(6465.3\ \text{kPa dm}^6\ \text{K}^{1/2}\ \text{mol}^{-2})}{(300\ \text{K})^{1/2}(0.0774\ \text{dm}^3\ \text{mol}^{-1})(0.1071\ \text{dm}^3\ \text{mol}^{-1})}$$

$$= 45029.6\ \text{kPa}$$

Hence,
$$p = \frac{RT}{V_m - b} - \frac{a}{T^{1/2}V_m(V_m+b)} = (52289.3 - 45029.6)\ \text{kPa}$$

$$= 7259.7\ \text{kPa} \equiv 71.65\ \text{atm}$$

For calculating Z, we evaluate

$$\frac{b}{V_m - b} = \frac{(0.0297\ \text{dm}^3\ \text{mol}^{-1})}{(0.0477\ \text{dm}^3\ \text{mol}^{-1})} = 0.623$$

$$\frac{a}{RT^{3/2}(V_m+b)} = \frac{(6465.3\ \text{kPa dm}^6\ \text{K}^{1/2}\ \text{mol}^{-2})}{(8.314\ \text{kPa dm}^3\ \text{K}^{-1}\ \text{mol}^{-1})(300\ \text{K})^{3/2}(0.1071\ \text{dm}^3\ \text{mol}^{-1})}$$

$$= 1.397$$

Hence, $$Z = 1 + \frac{b}{V_m - b} - \frac{a}{RT^{3/2}(V_m + b)}$$

$$Z = 1 + 0.623 - 1.397 = 0.23$$

Pressure Corresponding to Ideal Behaviour

For ideal behaviour, the expression V_m derived from Eq. (AI.47a) is

$$\frac{V_m + b}{V_m - b} = \frac{a}{bRT^{3/2}}$$

Since $(a/RT^{3/2}\ b)^{1/2} = x$, we write the above expression as

$$\frac{V_m + b}{V_m - b} = x^2$$

This gives

$$V_m = \frac{b(x^2+1)}{x^2-1} \tag{AI.52}$$

Knowing V_m, the pressure can be calculated by using the expression $p = RT/V_m$.

Illustration AI.8

For $CO_2(g)$, the Redlich-Kwong constants are: $a = 6465.3$ kPa dm^6 K$^{1/2}$ mol^{-2} and $b = 0.0297$ dm^3 mol^{-1}. Calculate the pressure at which the gas is expected to behave ideally.

Solution

From the value of x derived in illustration AI.7, we have

$$x^2 = \left(\frac{a}{RT^{3/2}b}\right) = (2.245)^2 = 5.04$$

Hence, $$V_m = \frac{b(x^2+1)}{(x^2-1)} = \frac{(0.0297\ \text{dm}^3\ \text{mol}^{-1})(6.04)}{(4.04)}$$

$$= 0.0444\ \text{dm}^3\ \text{mol}^{-1}$$

Finally, $$p = \frac{RT}{V_m} = \frac{(8.314\ \text{kPa dm}^3\ \text{K}^{-1}\ \text{mol}^{-1})(300\ \text{K})}{(0.0444\ \text{dm}^3\ \text{mol}^{-1})}$$

$$= 56175.7\ \text{kPa} \equiv 554.4\ \text{atm}$$

Display of Z versus p Plots

Figure AI.7 displays the plot of Z versus p for CO_2(g) at 300 K along with the plots and 400 K, 500 K and 600 K.

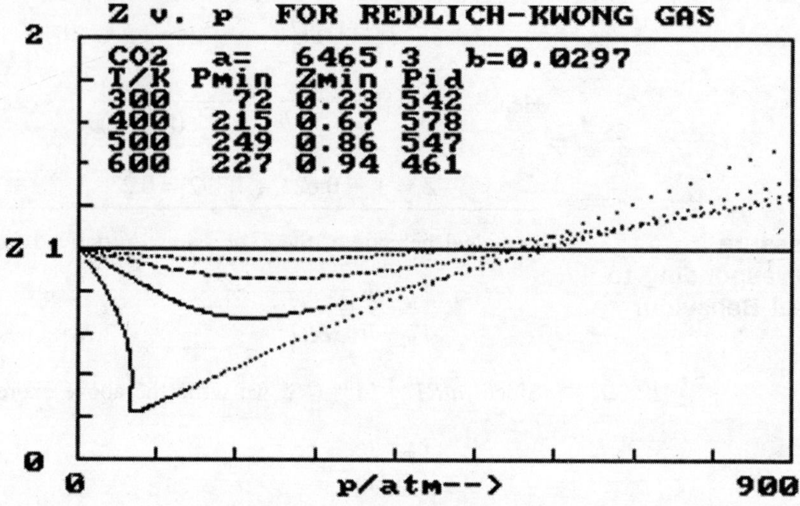

Fig. AI.7

Compression Factor at Boyle Temperature

Since a real gas behaves ideally over a wide range of low pressure (or larger volume) at Boyle temperature, we may replace V_m in Eq. (AI.44) by RT_B/p. This gives

$$Z = 1 + \frac{b}{(RT_B/p) - b} - \frac{a}{RT_B^{3/2}\{(RT_B/p) + b\}}$$

Since $T_B = (a/Rb)^{2/3}$, we may write the above expression as

$$Z = 1 + \frac{b}{(RT_B/p) - b} - \frac{b}{(RT_B/p) + b}$$

$$= 1 + \frac{1}{(RT_B/pb) - 1} - \frac{1}{(RT_B/pb) + 1}$$

$$= 1 + \frac{2}{(RT_B/pb)^2 - 1} = 1 + \frac{2(pb/RT_B)^2}{1 - (pb/RT_B)^2} \qquad \text{(AI.53)}$$

Let $y = b/RT_B$. With this the above expression becomes

$$Z = 1 + \frac{2(yp)^2}{1 - (yp)^2} \qquad \text{(AI.54)}$$

The expression of y is

$$y = \frac{b}{RT_B} = \frac{b}{R(a/Rb)^{2/3}} = \left(\frac{b^5}{Ra^2}\right)^{1/3} \qquad \text{(AI.55)}$$

The value of y for CO_2(g) is

$$y = \left[\frac{(0.0297\ \text{dm}^3\ \text{mol}^{-1})^5}{(8.314\ \text{kPa dm}^3\ \text{K}^{-1}\ \text{mol}^{-1})(6465.3\ \text{kPa dm}^6\ \text{K}^{1/2}\ \text{mol}^{-2})^2}\right]^{1/3}$$

$$= 4.05 \times 10^{-6}\ \text{kPa}^{-1}$$

Since y is very small, Eq. (AI.54) in the lower pressure region may be written as

$$Z = 1 + (2y^2)p^2 \qquad \text{(AI.56)}$$

Thus, the plot of Z versus p^2 at T_B in the lower pressure region is linear. With slope = $2y^2$ and intercept = 1.

The value of slope is

$$2y^2 = 2(4.05 \times 10^{-6}\ \text{kPa}^{-1})^2 = 3.28 \times 10^{-11}\ \text{kPa}^{-2}$$
$$\equiv 3.37 \times 10^{-7}\ \text{atm}^{-2}$$

For such a small value of slope, the compression factor Z increases to a very small extent with increase in the value of p^2. In fact, the value of Z remains close to 1 over wide range of pressures.

When Eq. (AI.56) is written as

$$Z = 1 + (2y^2p)p \qquad \text{(AI.57)}$$

we find that Z varies nonlinearly when plotted against p. The slope of this plot varies with p with the characteristic of slope → 0 as p → 0. This implies that the plot of Z versus p meets tangentially with the straight-line plot of Z (= 1) versus p for an ideal gas as the pressure of real gas approaches zero.

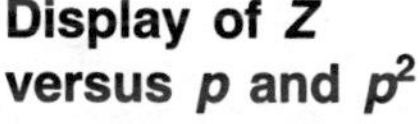

Display of Z versus p and p^2

Figure AI.8 displays the plot of Z versus p and Z versus p^2 for CO_2(g) at its Boyle temperature

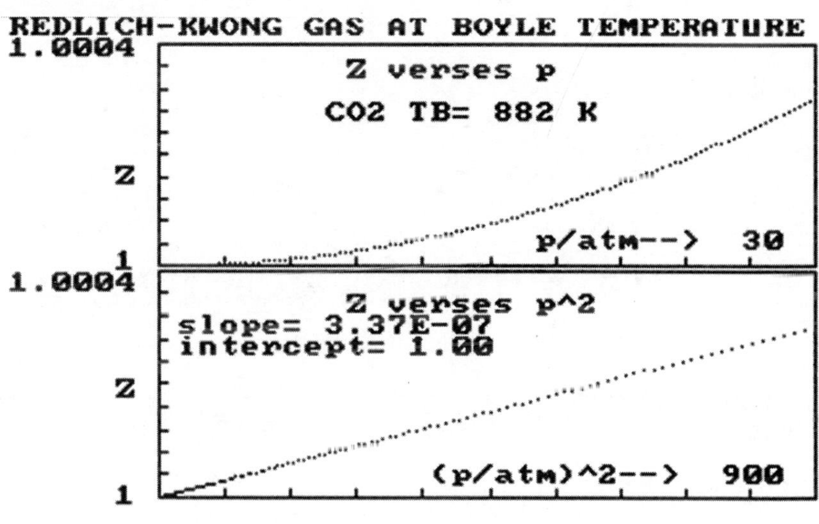

Fig. AI.8

Z versus Reduced Pressure

The variation of Z with p^2 for $CO_2(g)$ as given by (AI.56) is

$$Z = 1 + (2y^2)p^2 = 1 + (3.28 \times 10^{-11}\ \text{kPa}^{-2})p^2 \qquad \text{(AI.58)}$$

Multiplying and dividing second term on right side by p_c, we get

$$Z = 1 + \{(3.28 \times 10^{-11}\ \text{kPa}^{-2})\ p_c^2\}(p/p_c)^2$$

For $CO_2(g)$, p_c = 7376.46 kPa. Hence,

$$Z = 1 + \{(3.28 \times 10^{-11}\ \text{kPa}^{-2})(7376.46\ \text{kPa})^2\}\ p_r^2$$

$$= 1 + (1.79 \times 10^{-3})\ p_r^2 \qquad \text{(AI.59)}$$

Annexure II Variation of Compression Factor of a Gas in terms of Reduced Variables

The expression of compression factor in terms of reduced variables may be derived as shown in the following

$$Z = \frac{pV_m}{RT} = \frac{(p_r p_c)(V_r V_c)}{R(T_r T_c)} = \left(\frac{p_c V_c}{RT_c}\right)\left(\frac{p_r V_r}{T_r}\right) = Z_c\left(\frac{p_r V_r}{T_r}\right)$$

where Z_c takes the following values.

For van der Waals equation	$Z_c = (3/8)$
For Berthelot equation	$Z_c = (3/8)$
For Dieterici equation	$Z_c = (2/e^2)$
For Redlich-Kwong equation	$Z_c = (1/3)$

The expression Z in terms of reduce variables may be conveniently derived starting from the reduced equation of state.

Plots of Compression Factor versus Reduced Pressure at a Constant Reduced Temperature

The following procedure may be adopted to draw a plot of compression factor (Z) versus reduced pressure (p_r) at a given reduced temperature (T_r).

- Let the plot of Z versus p_r be drawn at $T_r = 1$.
- Start with initial pressure of 1 atm and temperature T (= $T_r T_c$, where T_c is the critical temperature of a chosen gas, may be $CO_2(g)$).
- Calculate V_m of an ideal gas at $p = 1$ atm and temperature T. Convert the calculated V_m into V_r by using the expression $V_r = V_m/V_c$.
- Using this value of V_r along with the given T_r, calculate p_r by using the appropriate reduced equation of state.
- Compute $Z = Z_c(p_r V_r/T_r)$ where Z_c is equal to $p_c V_c/RT_c$ of the chosen equation of state. It has the following values.

Van der Waals	$Z_c = (3/8)$
Berthelot	$Z_c = (3/8)$
Dieterici	$Z_c = (2/e^2)$
Redlich-Kwong	$Z_c = (1/3)$

- Plot Z against the calculated value of p_r.
- Increase the value of p with a suitable increment and follow the above procedure to calculate p_r and Z. Continue plotting Z versus p_r.
- The entire procedure may be repeated upto a suitable maximum value of p.
- Repeat the entire procedure for the new value of T_r including reduced Boyle temperature.

Van der Waals Equation

The reduced van der Waals equation is

$$\left(p_\mathrm{r} + \frac{3}{V_\mathrm{r}^2}\right)(3V_\mathrm{r} - 1) = 8T_\mathrm{r}$$

i.e. $$3p_\mathrm{r}V_\mathrm{r} - p_\mathrm{r} + \frac{9}{V_\mathrm{r}} - \frac{3}{V_\mathrm{r}^2} = 8T_\mathrm{r}$$

Dividing throughout by $8T_\mathrm{r}$, we get

$$\frac{3}{8}\left(\frac{p_\mathrm{r}V_\mathrm{r}}{T_\mathrm{r}}\right) - \frac{p_\mathrm{r}}{8T_\mathrm{r}} + \frac{9}{8}\frac{1}{V_\mathrm{r}T_\mathrm{r}} - \frac{3}{8}\frac{1}{V_\mathrm{r}^2T_\mathrm{r}} = 1$$

This gives

$$Z = 1 + \frac{p_\mathrm{r}}{8T_\mathrm{r}} - \frac{9}{8}\frac{1}{V_\mathrm{r}T_\mathrm{r}} + \frac{3}{8}\frac{1}{V_\mathrm{r}^2T_\mathrm{r}} \qquad \text{(AII.1)}$$

Display of Z versus p_r

Figure AII.1 displays the plot of Z versus p_r at different T_r. These plots have been drawn taking CO_2 and CH_4 as reference gases.

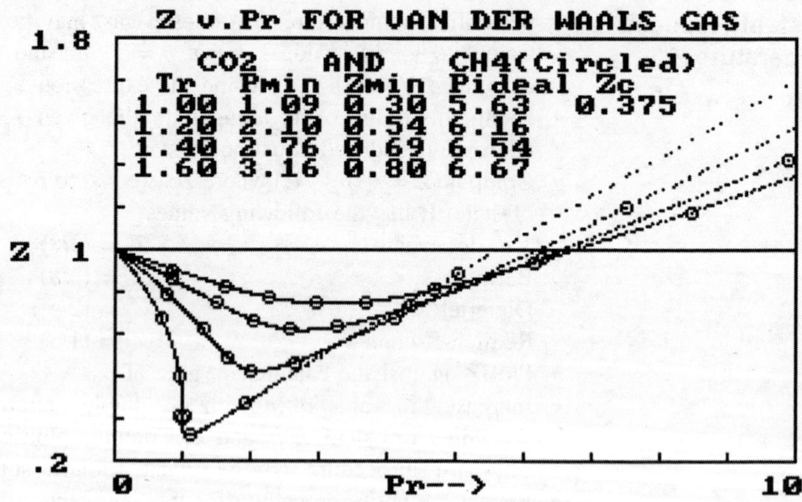

Fig. AII.1

Expression of Z in terms of Reduced Boyle Temperature

The expression of Boyle temperature is

$$T_\mathrm{B} = \frac{a}{bR}$$

Replacing a and b in terms of critical constants, we get

$$T_B = \frac{3p_c V_c^2}{(V_c/3)R} = 9\left(\frac{p_c V_c}{R}\right) = 9\left(\frac{3}{8}T_c\right) = \frac{27}{8}T_c$$

The expression of reduced Boyle temperature is

$$T_{r(B)} = \frac{T_B}{T_c} = \frac{27}{8} \tag{AII.2}$$

Substituting this expression of $T_{r(B)}$ in place of T_r in Eq. (AII.1), we get

$$Z = 1 + \frac{p_r}{27} - \frac{1}{3V_r} + \frac{1}{9V_r^2} \tag{AII.3}$$

Since at T_B, the gas behaves ideally over a wide range of low pressure, we may express V_r in terms of p_r as described in the following.

$$V_r = \frac{V}{V_c} = \frac{RT/p}{V_c} = \frac{R(T_r T_c)}{(p_r p_c)V_c} = \frac{T_r}{p_r}\left(\frac{RT_c}{p_c V_c}\right) = \frac{8T_r}{3p_r}$$

At reduced Boyle temperature, this becomes

$$V_r = \frac{8(27/8)}{3p_r} = \frac{9}{p_r} \tag{AII.4}$$

Substituting Eq. (AII.4) in Eq. (AII.3), we get

$$\begin{aligned} Z &= 1 + \frac{p_r}{27} - \frac{1}{3(9/p_r)} + \frac{1}{9(9/p_r)^2} \\ &= 1 + \frac{1}{(27)^2} p_r^2 = 1 + (1.372 \times 10^{-3}) p_r^2 \end{aligned} \tag{AII.5}$$

that is, Z varies linearly with p_r^2 with slope = $1/27^2$ and intercept = 1.

When Eq. (AII.5) is written as

$$Z = 1 + \left(\frac{1}{27^2} p_r\right) p_r \tag{AII.6}$$

it follows that Z varies nonlinearly with p_r with a characteristic of slope $\rightarrow 0$ as $p_r \rightarrow 0$. This implies that the plot of Z versus p_r meets tangentially with the straight-line plot of Z (= 1) versus p_r for an ideal gas as reduced pressure of the gas approaches zero.

Display of Z versus p_r and p_r^2 at $T_{r(B)}$

Figure AII.2 displays Z versus p_r and Z versus p_r^2 at reduced Boyle temperature. These plots are drawn by taking CO_2 and CH_4 as the reference gases.

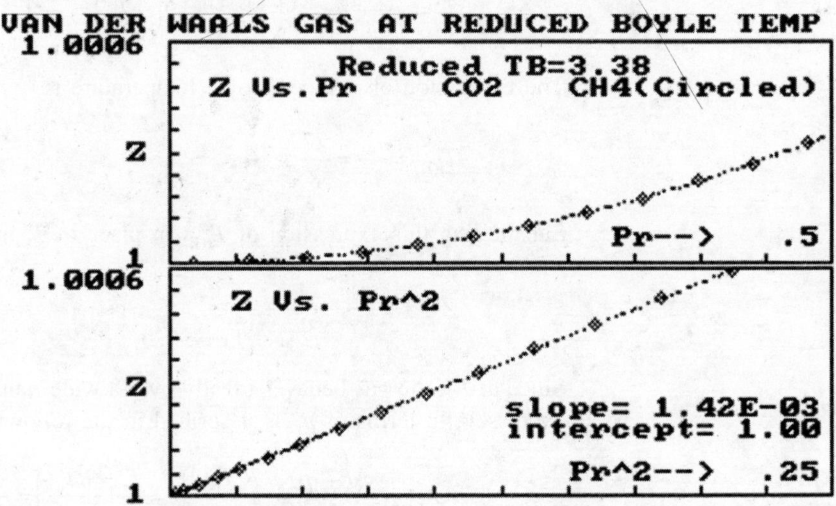

Fig. AII.2

Berthelot Equation of State

The reduced Berthelot equation is

$$\left(p_r + \frac{3}{V_r^2 T_r}\right)(3V_r - 1) = 8T_r$$

i.e.
$$3p_r V_r - p_r + \frac{9}{V_r T_r} - \frac{3}{V_r^2 T_r} = 8T_r$$

Dividing throughout by $8T_r$, we get

$$\frac{3}{8}\left(\frac{p_r V_r}{T_r}\right) - \frac{p_r}{8T_r} + \frac{9}{8}\frac{1}{V_r T_r^2} - \frac{3}{8}\frac{1}{V_r^2 T_r^2} = 1$$

This gives

$$Z = 1 + \frac{p_r}{8T_r} - \frac{9}{8}\frac{1}{V_r T_r^2} + \frac{3}{8}\frac{1}{V_r^2 T_r} \tag{AII.7}$$

Display of Z versus p_r

Figure AII.3 displays Z versus p_r at different reduced temperatures. These plots were drawn by taking CO_2 and CH_4 as reference gases.

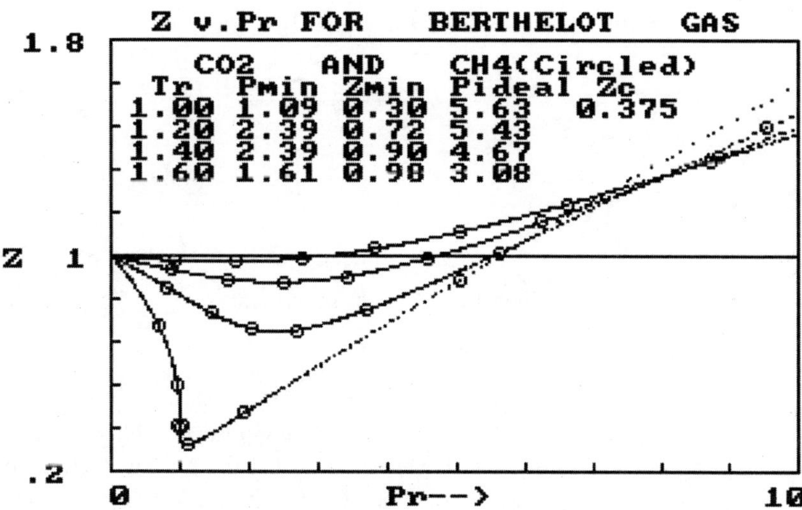

Fig. AII.3

Expression of Z in terms of Reduced Boyle Temperature

The expression of Boyle temperature is

$$T_B = \left(\frac{a}{bR}\right)^{1/2}$$

Replacing a and b in terms of critical constants, we get

$$T_B = \left(\frac{9RT_c^2V_c/8}{(V_c/3)R}\right)^{1/2} = \left(\frac{27}{8}\right)^{1/2} T_c$$

Thus $$T_{r(B)} = \frac{T_B}{T_c} = \left(\frac{27}{8}\right)^{1/2} \qquad \text{(AII.8)}$$

Substituting this expression of $T_{r(B)}$ in place of T_r in Eq. (AII.7), we get

$$Z = 1 + \frac{p_r}{6\sqrt{6}} - \frac{1}{3V_r} + \frac{1}{9V_r^2} \qquad \text{(AII.9)}$$

Since at Boyle temperature, the gas behaves ideally over a wide range of low pressure, we may express V_r in terms of p_r as described in the following.

$$V_r = \frac{V}{V_c} = \frac{RT/p}{V_c} = \frac{R(T_rT_c)}{(p_rp_c)V_c} = \left(\frac{RT_c}{p_cV_c}\right)\left(\frac{T_r}{p_r}\right) = \frac{8}{3}\frac{T_r}{p_r}$$

At reduced Boyle temperature, this becomes

$$V_r = \frac{8}{3}\frac{(27/8)^{1/2}}{p_r} = \frac{2\sqrt{6}}{p_r} \qquad \text{(AII.10)}$$

Substituting Eq. (AII.10) in Eq. (AII.9), we get

$$Z = 1 + \frac{p_r}{6\sqrt{6}} - \frac{p_r}{6\sqrt{6}} + \frac{p_r^2}{216}$$

i.e.
$$Z = 1 + \left(\frac{1}{216}\right) p_r^2 = 1 + (4.63 \times 10^{-3}) p_r^2 \qquad \text{(AII.11)}$$

that is Z varies linearly with p_r^2 with slope = 1/216 and intercept = 1. When Eq. (AII.11) is written as

$$Z = 1 + \left(\frac{1}{216} p_r\right) p_r \qquad \text{(AII.12)}$$

it follows that Z varies nonlinearly with p_r with a characteristic of slope $\rightarrow 0$ as $p_r \rightarrow 0$. This implies that the plot of Z versus p_r meets tangentially with the straight-line plot of Z (= 1) versus p_r for an ideal gas as reduced pressure of the gas approaches zero.

Display of Z versus p_r and p_r^2 at $T_{r(B)}$

Figure AII.4 displays the plot of Z versus p_r and Z versus p_r^2 at reduced Boyle temperature. These have been drawn taking CO_2 and CH_4 as reference gases.

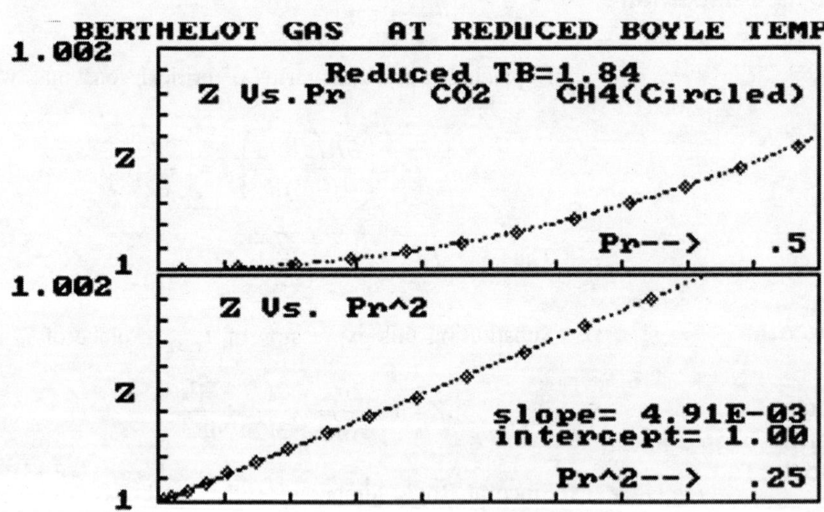

Fig. AII.4

Dieterici Equation of State

The reduced Dieterici equation is

$$p_r e^{2/V_r T_r}\left(V_r - \frac{1}{2}\right) = \frac{e^2}{2} T_r$$

that is $$p_r V_r - \frac{p_r}{2} = \left(\frac{e^2}{2} T_r\right) e^{-2/V_r T_r}$$

Dividing throughout by $(e^2/2)T_r$, we get

$$\frac{2}{e^2}\left(\frac{p_r V_r}{T_r}\right) - \frac{p_r}{e^2 T_r} = e^{-2/V_r T_r}$$

or $$Z = e^{-2/V_r T_r} + \frac{p_r}{e^2 T_r} \qquad \text{(AII.13)}$$

Display of Z versus p_r

Figure AII.5 displays the plot of Z versus p_r at different reduced temperatures. These plots are drawn taking CO_2 and CH_4 as reference gases.

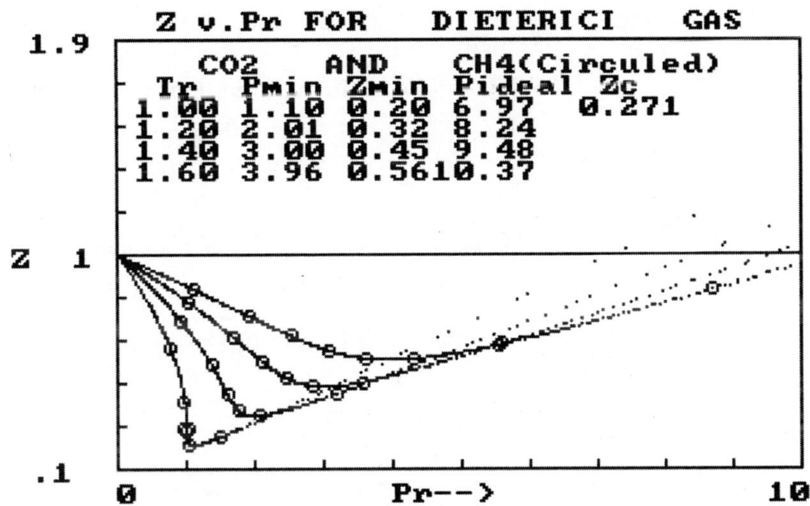

Fig. AII.5

Expression of Z in terms of Reduced Boyle Temperature

The expression of Boyle temperature is

$$T_B = \frac{a}{Rb}$$

Replacing a and b in terms of critical constants, we get

$$T_B = \frac{2RV_c T_c}{R(V_c/2)} = 4T_c$$

The expression of reduced Boyle temperature is

$$T_{r(B)} = \frac{T_B}{T_c} = 4 \qquad \text{(AII.14)}$$

Substituting this expression of $T_{r(B)}$ in place of T_r in Eq. (AII.13), we get

$$Z = e^{-1/2V_r} + \frac{p_r}{4e^2} \tag{AII.15}$$

Since at Boyle temperature, the gas behaves ideally over a wide range of low pressure, we may express V_r in terms of p_r as described in the following.

$$V_r = \frac{V}{V_c} = \frac{RT/p}{V_c} = \frac{R(T_r T_c)}{(p_r p_c)V_c} = \left(\frac{RT_c}{p_c V_c}\right)\frac{T_r}{p_r} = \frac{e^2}{2}\frac{T_r}{p_r}$$

At reduced Boyle temperature, this becomes

$$V_r = \left(\frac{e^2}{2}\right)\left(\frac{4}{p_r}\right) = \frac{2e^2}{p_r} \tag{AII.16}$$

Substituting Eq. (AII.16) in Eq. (AII.15), we get

$$Z = e^{-p_r/4e^2} + \frac{p_r}{4e^2} \tag{AII.17}$$

In the low pressure region, the exponential term may be expanded as

$$e^{-p_r/4e^2} = 1 - \frac{p_r}{4e^2} + \frac{1}{2}\left(\frac{p_r}{4e^2}\right)^2$$

With this, Eq. (AII.17) becomes

$$Z = 1 - \frac{p_r}{4e^2} + \frac{1}{2}\left(\frac{p_r}{4e^2}\right)^2 + \frac{p_r}{4e^4}$$

$$= 1 + \left(\frac{1}{32e^2}\right)p_r^2 = 1 + (5.72 \times 10^{-4})p_r^2 \tag{AII.18}$$

that is Z varies linearly with p_r^2 with slope = $1/32e^2$ and intercept = 1
When Eq. (AII.18) is written as

$$Z = 1 + \left(\frac{1}{32e^4}p_r\right)p_r$$

it follows that Z varies non linearly with p_r with a characteristic of slope of Z versus p_r meets tangentially with the straight-line plot of Z (= 1) versus p_r for an ideal gas as reduced pressure of the gas approaches zero.

Display of Z versus p_r and p_r^2 at $T_{r(B)}$

Figure AII.6 displays the plot of Z versus p_r and Z versus p_r^2 at reduced Boyle temperature. These are drawn by taking CO_2 and CH_4 as reference gases.

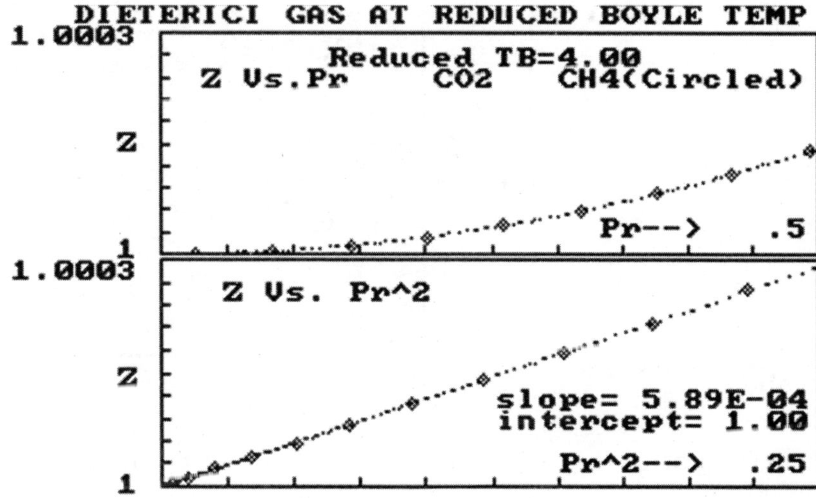

Fig. AII.6

Redlich-Kwong Equation

The reduced Redlich-Kwong equation is

$$p_r = \frac{3T_r}{V_r - x} - \frac{1}{xT_r^{1/2}V_r(V_r + x)}; \qquad \text{(where } x = 2^{1/3} - 1)$$

Multiplying throughout the above expression by $V_r/3T_r$, we get

$$\frac{1}{3}\left(\frac{p_r V_r}{T_r}\right) = \frac{V_r}{V_r - x} - \frac{1}{3xT_r^{3/2}(V_r + x)}$$

i.e.
$$Z = 1 + \frac{x}{V_r - x} - \frac{1}{3xT_r^{3/2}(V_r + x)} \qquad \text{(AII.19)}$$

Display of Z versus p_r

Figure AII.7 displays the plot of Z versus p_r at different reduced temperatures. These are drawn by taking CH_4 and CO_2 as reference gases.

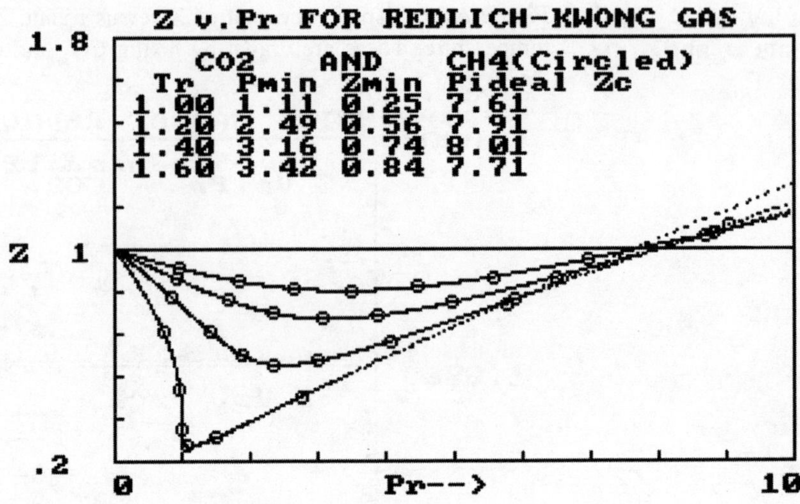

Fig. AII.7

Expression of Z in terms of Reduced Boyle Temperature

The expression of Boyle temperature is

$$T_{\mathrm{B}}^{3/2} = \frac{a}{bR}$$

Replacing a and b in terms of critical constants, we get

$$T_{\mathrm{B}}^{3/2} = \frac{R^2 T_{\mathrm{c}}^{5/2}/9xp_{\mathrm{c}}}{R(xRT_{\mathrm{c}}/3p_{\mathrm{c}})} = \left(\frac{1}{3x^2}\right) T_{\mathrm{c}}^{3/2}$$

The expression of reduced Boyle temperature is

$$T_{\mathrm{r(B)}}^{3/2} = \left(\frac{T_{\mathrm{B}}}{T_{\mathrm{c}}}\right)^{3/2} = \frac{1}{3x^2} \tag{AII.20}$$

Substituting this expression of $T_{\mathrm{r(B)}}$ in place of T_{r} in Eq. (AII.19), we get

$$Z = 1 + \frac{x}{V_{\mathrm{r}} - x} - \frac{x}{V_{\mathrm{r}} + x} = 1 + \frac{2x^2}{V_{\mathrm{r}}^2 - x^2} \tag{AII.21}$$

Since at T_{B}, the gas behaves ideally over a wide range of low pressures, we may express V_{r} in terms of p_{r} as described in the following.

$$V_{\mathrm{r}} = \frac{V}{V_{\mathrm{c}}} = \frac{RT/p}{V_{\mathrm{c}}} = \frac{R(T_{\mathrm{r}}T_{\mathrm{c}})}{(p_{\mathrm{r}}p_{\mathrm{c}})V_{\mathrm{c}}} = \left(\frac{RT_{\mathrm{c}}}{p_{\mathrm{c}}V_{\mathrm{c}}}\right)\left(\frac{T_{\mathrm{r}}}{p_{\mathrm{r}}}\right) = 3\left(\frac{T_{\mathrm{r}}}{p_{\mathrm{r}}}\right)$$

At reduced Boyle temperature, this becomes

$$V_{\mathrm{r}} = \left(\frac{3}{p_{\mathrm{r}}}\right)\left(\frac{1}{3x^2}\right)^{2/3} = \frac{y}{p_{\mathrm{r}}}; \qquad \text{(where } y = 3/(3x^2)^{2/3}\text{)} \tag{AII.22}$$

Substituting Eq. (AII.22) in Eq. (AII.21), we get

$$Z = 1 + \frac{2x^2}{(y/p_r)^2 - x^2} = 1 + \frac{2x^2 p_r^2}{y^2 - x^2 p_r^2} \tag{AII.23}$$

The values of x^2 and y^2 are:

$$x^2 = (2^{1/3} - 1)^2 = (0.26)^2 = 0.0676$$

$$y^2 = \left(\frac{3}{(3x^2)^{2/3}}\right)^2 = \left(\frac{3}{(3 \times 0.0676)^{2/3}}\right)^2 = \left(\frac{3}{0.345}\right)^2 = 75.61$$

With these, Eq. (AII.23) becomes

$$Z = 1 + \frac{(0.1352)p_r^2}{75.61 - (0.0676)p_r^2} \tag{AII.24}$$

Neglecting (0.0676) p_r^2 in comparison to 75.61, we get

$$Z = 1 + \left(\frac{0.1352}{75.61}\right)p_r^2 = 1 + (1.79 \times 10^{-3})p_r^2 \tag{AII.25}$$

that is, Z varies linearly with p_r^2 with slope = 0.00179 and intercept = 1.

When Eq. (AII.25) is written as

$$Z = 1 + (0.00179\ p_r)p_r$$

it follows that Z varies nonlinearly with p_r with a characteristic of slope $\rightarrow$ 0 as $p_r \rightarrow 0$. This implies that the plot of Z versus p_r meet tangentially with the straight-line plot of Z (= 1) versus p_r for an ideal gas as reduced pressure of the gas approaches zero.

Display of Z versus p_r and p_r^2 at $T_{r(B)}$

Figure AII.8 displays the plot of Z versus p_r and Z versus p_r^2 at reduced Boyle temperature. These are drawn by taking CH_4 and CO_2 as the reference gases.

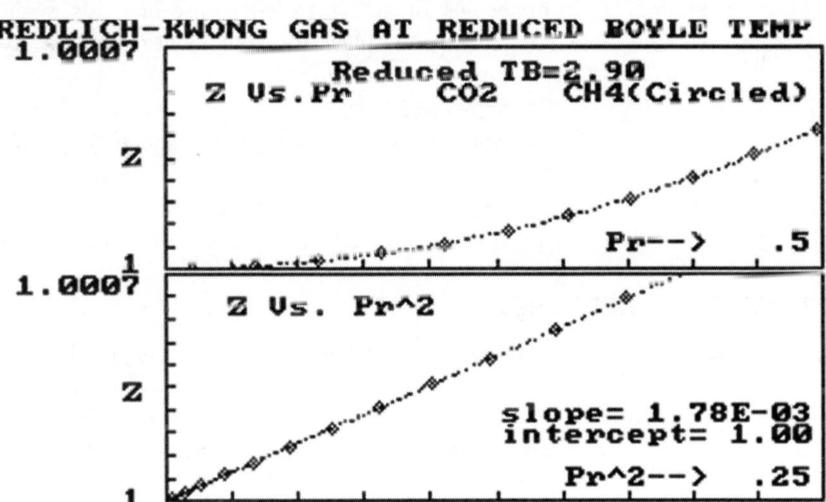

Fig. AII.8

2 Physical Properties of Liquids

2.1 INTRODUCTION

Characteristics of Liquid State

In general, liquids can be obtained from gases by cooling the latter below their respective critical temperatures, followed by the treatment of high pressure. The effect of cooling is to decrease the thermal energies of molecules and the effect of high pressure is to decrease the volume of the system so as to allow the molecules to come closer, thereby increasing the forces of attraction amongst them. Alternatively, liquids can be obtained by heating solids up to or beyond their melting points. In solids, molecules do not possess any translational energy but possess only vibrational energy. The forces of attractions amongst them are very strong. The effect of heating solids is to impart sufficient energy to molecules so that they can overcome these strong forces of attractions. Thus, we see that the properties of liquids lie in between those of solids and gases. For example, liquids are less compressible than gases but a little more compressible than solids. They are less dense than solids but more dense than gases. The two important properties of liquids, namely, fixed volume but no fixed shape, arise mainly because of the following two facts:

(i) The energies binding the molecules are larger than their average thermal energy.

(ii) These binding energies are not strong enough to stop the motion of the molecules altogether, as is the case in solids, with the result that molecules can move from one place to another but cannot escape from the liquid unless they are present at the surface.

In this chapter, we will discuss only three properties of liquids, namely, (i) vapour pressure, (ii) viscosity, and (iii) surface tension. The origin of these properties in liquids is basically due to the existence of strong intermolecular attractions. The structural aspect of liquids is discussed in the next chapter as it requires some knowledge about the arrangement of the molecules in solids.

2.2 VAPOUR PRESSURE

Introduction

Suppose a beaker containing a liquid is placed in an evacuated vessel (Fig. 2.2.1). Let the latter be connected to a manometer so that any pressure that is developed in the free space could be measured. After some time, it is found that the manometer records a constant pressure. This pressure is known as the vapour pressure of the liquid.

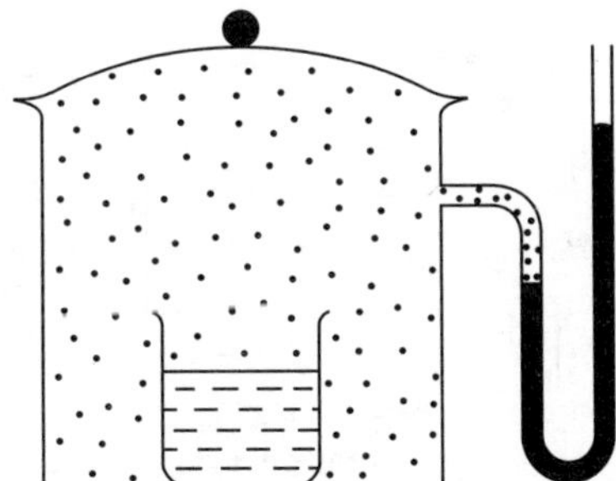

Fig. 2.2.1 Vapour pressure of a liquid

Origin of Vapour Pressure

Before making an attempt to understand how the vapour pressure arises, we will have to consider the following two facts:

(i) The molecules of a liquid, like those of a gas, have different kinetic energies. The distribution of speeds amongst molecules follows the Maxwell-Boltzmann distribution.

Figure 2.2.2 shows such a distribution at two different temperatures. The most important point to be noted in these distribution curves is that the fraction of molecules having higher speeds increases with the increase in temperature (shown by the area marked by tilted lines).

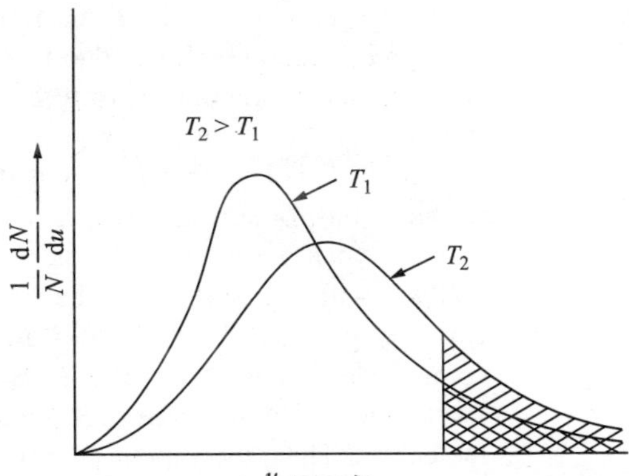

Fig. 2.2.2 Distribution of speeds of molecules

(ii) If we consider a molecule in the bulk of a liquid, it will be surrounded by other molecules in a symmetrical manner. Thus, the forces of attraction on this molecule by the molecules present on one side are completely balanced by the molecules present on the opposite side. Hence, the net force of attraction experienced by this molecule will, on the whole, be zero. It will move as if there exists no force of attraction on it. However, the situation is altogether different at the surface of the liquid (Fig. 2.2.3). There are larger number of molecules towards the liquid side of a molecule than towards the open space above it, with the result that this molecule experiences a net force of attraction in the downward direction.

The forces of attraction between the molecules of a liquid are of a stronger nature and are larger than the average thermal energy of the molecules. However, because of the Maxwell-Boltzmann distribution, some of the molecules can

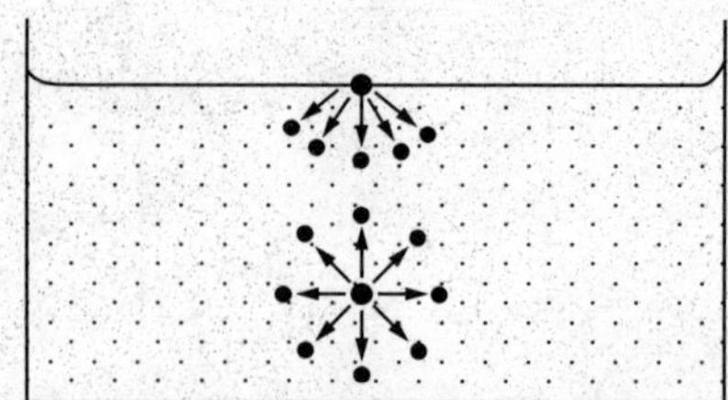

Fig. 2.2.3 Arrangement of molecules within and at the surface of a liquid

have thermal energies equal to or greater than the characteristic energy which is just sufficient to overcome the forces of attraction. If such a molecule happens to be at the surface, it will overcome the net downward forces of attraction and will immediately leave the surface and escape to the empty space above. If the space above the surface is an open one, then the molecules will continue to escape resulting in the phenomenon of evaporation. Since molecules of higher thermal energies are leaving the surface of the liquid, it follows, therefore, that the average thermal energy of molecules in the liquid will decrease. Consequently, the temperature of the liquid is reduced and hence cooling is observed.

Definition of Vapour Pressure

If the space 'above the liquid is a closed one, then the molecules escaping from the surface of the liquid (referred to as vapour molecules) will go on collecting in the empty space. After some time it is observed that a constant pressure is registered. This pressure is due to vapour molecules of the liquid and hence it is known as the vapour pressure of the liquid. Since this pressure is constant, it follows that there must be a constant number of molecules in the space above the liquid. This can be true only if the molecules in the space are also returning to the liquid; otherwise, the pressure in the space would continue to increase. In fact, when a vapour molecule with a comparatively smaller thermal energy collides with the surface of the liquid, it sticks to the latter. Thus, there is a two-way process; the molecules are leaving the liquid and are simultaneously coming back to it. We get a state of dynamic equilibrium when the rate of evaporation of liquid molecules is equal to the rate of condensation of the vapour molecules. Thus, the vapour pressure of a liquid may be defined as the *pressure of the vapour in equilibrium with the liquid.* It is the same constant pressure which is required to convert a gas into a liquid when the former is below its critical temperature.

Effect of Temperature on Vapour Pressure

On raising the temperature, more and more molecules of a liquid will have energies equal to or greater than the critical energy which is just sufficient to overcome the forces of attraction between the molecules. As a result, larger number of molecules can leave the surface of the liquid which will consequently have higher vapour pressure. Thus, *the vapour pressure of a liquid increases with the increase in temperature.* The variation of vapour pressure of liquids with temperature is of the type shown in Fig. 2.2.4 with the highest and lowest limits correspond to critical point and triple point, respectively.

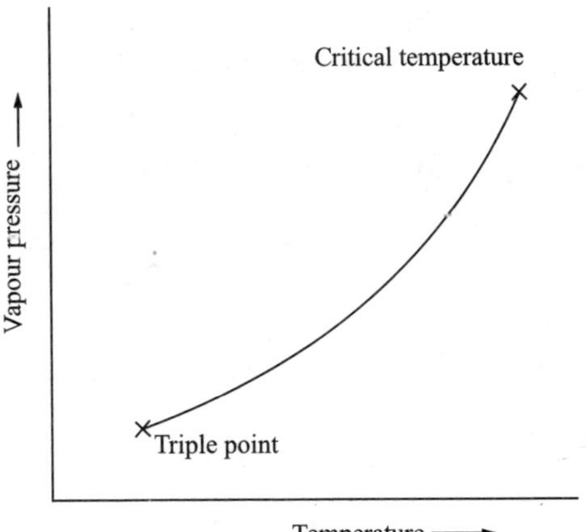

Fig. 2.2.4 Variation of vapour pressure with temperature

Clapeyron Equation

The quantitative variation of vapour pressure with temperature is given by *Clapeyron equation,* according to which

$$\frac{\mathrm{d}p}{\mathrm{d}T} = \frac{\Delta_{\mathrm{vap}}H_{\mathrm{m}}}{T(V_{\mathrm{m,v}} - V_{\mathrm{m,l}})} \tag{2.2.1}$$

where $\mathrm{d}p/\mathrm{d}T$ is the rate of change of vapour pressure with temperature, $\Delta_{\mathrm{vap}}H_{\mathrm{m}}$ is the molar enthalpy of vaporization of the liquid, and $V_{\mathrm{m,v}}$ and $V_{\mathrm{m,l}}$ are the molar volumes of vapour and liquid, respectively. Equation (2.2.1) represents the slope of the plot of vapour pressure against the absolute temperature. This equation can be used to determine the molar enthalpy of vaporization of a liquid if the rate of change of its vapour pressure with temperature is known. Alternatively, if the value of $\Delta_{\mathrm{vap}}H_{\mathrm{m}}$ of the liquid is known, the rate of change of its vapour pressure with temperature can be calculated.

Clausius-Clapeyron Equation

Equation (2.2.1) can be simplified under the following approximations as suggested by R. Clausius.

(i) The molar volume $V_{\mathrm{m,l}}$ of a liquid may be considered to be negligible in comparison to the molar volume of its vapour. This will be true provided the temperature is not near the critical temperature.

(ii) The liquid's vapour may be assumed to obey the ideal gas laws so that

$$V_{\mathrm{m,v}} = \frac{RT}{p}$$

With these two approximations, Eq. (2.2.1) modifies to

$$\frac{\mathrm{d}p}{\mathrm{d}T} = \frac{\Delta_{\mathrm{vap}}H_{\mathrm{m}}}{RT^2/p} \quad \text{or} \quad \frac{\mathrm{d}p}{p} = \frac{\Delta_{\mathrm{vap}}H_{\mathrm{m}}}{R}\left(\frac{\mathrm{d}T}{T^2}\right)$$

or
$$\mathrm{d}\ln(p/p^\circ) = \frac{\Delta_{\mathrm{vap}}H_{\mathrm{m}}}{R}\left(\frac{\mathrm{d}T}{T^2}\right) \tag{2.2.2}$$

where $p°$ is the standard unit pressure. Equation (2.2.2) is known as the *Clausius-Clapeyron equation.*

Plot of Vapour Pressure versus Temperature

If the molar enthalpy of vaporization can be regarded as constant, Eq. (2.2.2) may be readily integrated to give

$$\mathrm{d}\ \ln(p/p°) = \frac{\Delta_{\mathrm{vap}}H_{\mathrm{m}}}{R} \int \frac{\mathrm{d}T}{T^2}$$

or
$$\ln(p/p°) = -\frac{\Delta_{\mathrm{vap}}H_{\mathrm{m}}}{RT} + I \qquad (2.2.3)$$

where I is the constant of integration. According to this relation a graph between $\ln(p/p°)$ and $1/T$ will give a straight line with a slope of $-\Delta_{\mathrm{vap}}H_{\mathrm{m}}/R$ as shown in Fig. 2.2.5.

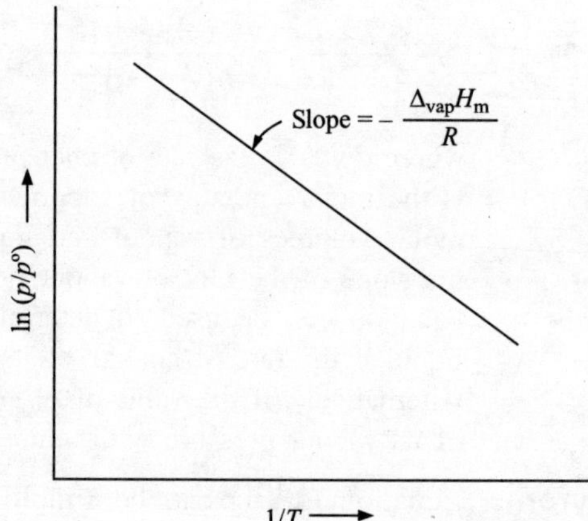

Fig. 2.2.5 Plot of ln ($p/p°$) versus $1/T$

Correlating Vapour Pressures at two Temperatures

If p_1 and p_2 are the vapour pressures at T_1 and T_2, respectively, then Eq. (2.2.2) can be integrated within the limits to give

$$\int_{p_1}^{p_2} \mathrm{d}\ \ln\left(\frac{p}{p°}\right) = \frac{\Delta_{\mathrm{vap}}H_{\mathrm{m}}}{R} \int_{T_1}^{T_2} \frac{\mathrm{d}T}{T^2}$$

or
$$\ln\frac{p_2}{p_1} = -\frac{\Delta_{\mathrm{vap}}H_{\mathrm{m}}}{R}\left(\frac{1}{T_2} - \frac{1}{T_1}\right)$$

or
$$\log\frac{p_2}{p_1} = -\frac{\Delta_{\mathrm{vap}}H_{\mathrm{m}}}{2.303\,R}\left(\frac{1}{T_2} - \frac{1}{T_1}\right) \qquad (2.2.4)$$

Example 2.2.1 The enthalpy of vaporization of cyclohexane (C_6H_{12}) at its boiling point 80.75 °C is 385.15 J g^{-1}. The densities of the liquid and vapour at this temperature are 0.719 9 g cm^{-3} and 0.002 9 g cm^{-3}, respectively, (a) Calculate the value of dp/dT. (b) Estimate the boiling point at 740 mmHg. (c) If it is to be distilled at 25 °C, to what value must the pressure be reduced?

Solution Given that

Molar mass of cyclohexane, M = 84 g mol^{-1}

Enthalpy of vaporization per mole of cyclohexane

$$\Delta_{vap}H_m = (358.15\ \mathrm{J\ g^{-1}})(84\ \mathrm{g\ mol^{-1}}) = 30\ 084.6\ \mathrm{J\ mol^{-1}}$$

Molar volume of liquid,

$$V_{m,1} = \frac{M}{\rho_1} = \frac{(84\ \mathrm{g\ mol^{-1}})}{(0.719\ 9\ \mathrm{g\ cm^{-3}})} = 177\ \mathrm{cm^3\ mol^{-1}}$$

Molar volume of vapour,

$$V_{m,v} = \frac{M}{\rho_v} = \frac{(84\ \mathrm{g\ mol^{-1}})}{(0.002\ 9\ \mathrm{g\ cm^{-3}})} = 28\ 966\ \mathrm{cm^3\ mol^{-1}}$$

(a) Now $$\frac{dp}{dT} = \frac{\Delta_{vap}H_m}{T(V_{m,v} - V_{m,1})} = \frac{(30\ 084.6\ \mathrm{J\ mol^{-1}})}{(353.9\ \mathrm{K})(28\ 966 - 117)\ \mathrm{cm^3\ mol^{-1}}}$$

$$= 2.947 \times 10^{-3}\ \mathrm{J\ cm^{-3}\ K^{-1}} \equiv 2.947 \times 10^{3}\ \mathrm{N\ m^{-2}\ K^{-1}} \equiv 2\ 947\ \mathrm{Pa\ K^{-1}}$$

(b) $$dp = 740\ \mathrm{mmHg} - 760\ \mathrm{mmHg} = -\ 20\ \mathrm{mmHg}$$

$$= -\ (20\ \mathrm{mmHg})\left(\frac{133.322\ \mathrm{Pa}}{1\ \mathrm{mmHg}}\right) = -\ 2\ 666.44\ \mathrm{Pa}$$

Thus, from dp/dT = 2 947 Pa K^{-1}, we get

$$dT = \frac{dp}{2\ 947\ \mathrm{Pa\ K^{-1}}} = \frac{-2\ 666.44\ \mathrm{Pa}}{2\ 947\ \mathrm{Pa\ K^{-1}}} = -\ 0.905\ \mathrm{K}$$

Hence, boiling point at this pressure = 80.75 °C – 0.905 °C ≈ 79.85 °C

(c) Making use of Clausius-Clapeyron equation,

$$\log \frac{p_2}{p_1} = \frac{\Delta_{vap}H_m}{2.303R}\left(\frac{1}{T_1} - \frac{1}{T_2}\right)$$

we `get $$\log\left(\frac{p_2}{\mathrm{mmHg}}\right)\ \log(760) = \frac{(30\ 084.6\ \mathrm{J\ mol^{-1}})}{2.303(8.314\ \mathrm{J\ K^{-1}\ mol^{-1}})}\left(\frac{1}{353.9\ \mathrm{K}} - \frac{1}{298.15\ \mathrm{K}}\right)$$

$$\log(p_2/\mathrm{mmHg}) = 2.880\ 8 - 0.830\ 4 = 2.050\ 4$$

or $$p_2 = 112.3\ \mathrm{mmHg}$$

Vapour Pressure and Boiling Point

If the pressure above a liquid is adjusted to a definite value, it is possible to raise the temperature so that the vapour pressure becomes equal to the external pressure. At this stage, bubbles of vapour are formed within the liquid which rise upwards and escape from the surface. We then say that the liquid has started boiling and the corresponding temperature is known as the *boiling point* of the liquid. Thus, the boiling point of a liquid can be defined as the *temperature at which its vapour pressure is equal to the external pressure.* If the latter is 1.013 25 bar (=1 atm), the boiling point is known as the normal boiling point. Obviously, the boiling point of a liquid can be varied by varying the external pressure. Increase in the external pressure will increase the boiling point of a liquid and decrease in external pressure will decrease its boiling point. It is evident from this that a plot of the variation of vapour pressure with temperature also represents the variation of boiling point of a liquid with the external pressure.

The Effect of Impurities on Boiling Point

The vapour pressure of a liquid invariably decreases with nonvolatile impurities. This is because of the fact that the surface will now be occupied by the molecules of both solute and solvent. Thus, there will be lesser number of molecules of the solvent as compared to those present in the pure liquid. Consequently, lesser number of molecules of the solvent will get a chance to escape from the surface of the liquid, resulting in lower vapour pressure. Obviously, boiling point of the liquid will rise as it will now require more heating to make its vapour pressure equal to the external pressure.

Measurement of Vapour Pressure

The methods generally employed for the measurement of vapour pressure fall into three categories: (i) the static method, (ii) the dynamic method, and (iii) the gas saturation method.

The Static Method

In this method, vapour pressure of a given liquid is measured by a manometer attached to the open space above the liquid placed in a closed vessel. One of the simplest procedures is to employ two barometric tubes. A small amount of the liquid is introduced in one of the tubes until the space above the mercury is saturated with vapour as shown by a small quantity of liquid remaining on the surface of mercury. The difference in the levels of mercury in the two tubes gives the vapour pressure of the liquid. An alternative form of the above method is due to A. Smith and A.W.C. Menzies and is known as the *isoteniscope method.* In this method, the liquid is introduced in the bulb A and in the attached U-tube as shown in Fig. 2.2.6. This system, which is known as the isoteniscope, is connected to a manometer M and to a vessel B which has a large volume in order to smooth out minor pressure fluctuations. The isoteniscope is put in a thermostat to keep the temperature constant. First of all, the apparatus is evacuated until the liquid begins to boil and the air between the liquid in the bulb and the U-tube is removed and contains only the vapour of the liquid. Now air is introduced slowly until the levels of liquid are same in both the limbs of the U-tube. At this stage, the pressure above the liquid in A (vapour pressure) is equal to pressure in B. The latter can be read with the help of a manometer (Fig. 2.2.6).

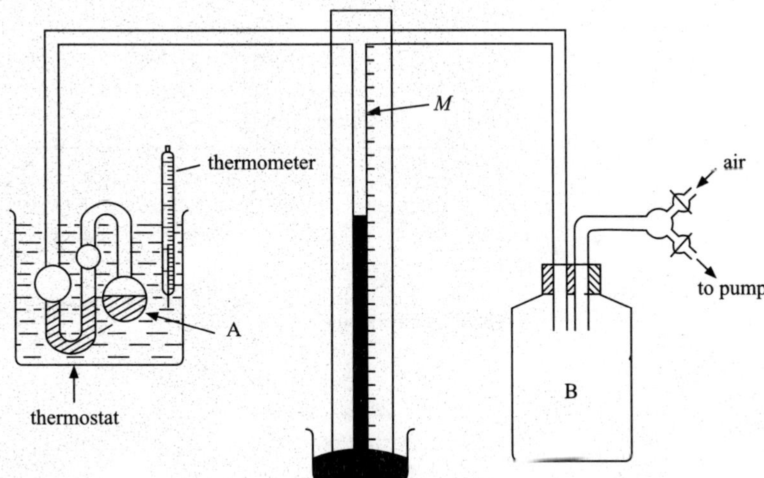

Fig. 2.2.6 The isoteniscope method for the measurement of vapour pressure

The Dynamic Method

In this method, the external pressure is kept constant and temperature of the liquid is raised till it starts boiling. By definition, the vapour pressure of the liquid at this temperature will be equal to the external pressure. In recording the temperature, the thermometer should be placed in the vapour phase and not in the liquid so as to avoid the superfluous temperature recording, which may be caused due to the superheating of the liquid.

The Gas-Saturation Method

In this method, a known mass of the liquid, whose vapour pressure is to be determined, is taken in a closed vessel and a known volume of dry air (or any gas which does not interact with the liquid) is passed repeatedly through it, keeping its pressure P constant. This way, the air carries along with it some of the molecules of the liquid and thus there occurs a loss in mass of the liquid. The gas is continuously passed till no further loss in mass occurs. At this stage, the air is saturated with the vapour of the liquid and the partial pressure p of the vapour in the mixture is equal to the vapour pressure of the liquid. If n_1 and n_2 are the amounts of vapour and dry air, respectively, then

$$p = \left(\frac{n_1}{n_1 + n_2}\right) P$$

If m is the loss in mass of the liquid and M is its molar mass, then

$$n_1 = \frac{m}{M}$$

Assuming air to be an ideal gas, we have

$$n_2 = \frac{PV}{RT}$$

Thus $$p = \frac{(m/M)P}{\left(\frac{m}{M} + \frac{PV}{RT}\right)} = \frac{mPRT}{mRT + PVM} \tag{2.2.5}$$

Thus knowing m (the loss in mass of the liquid), V (the volume of air to start with), P (the pressure at which the gas is being passed), M (the molar mass of the liquid), the vapour pressure p of the liquid at the given temperature T can be calculated from Eq. (2.2.5).

Example 2.2.2 A volume of 10.5 dm^3 of nitrogen measured at 30 °C and 102.66 kPa pressure is bubbled through a saturator containing bromobenzene (C_6H_5Br) at 40 °C. The mass of the saturator and its contents is reduced by 0.856 g. Calculate the vapour pressure of bromobenzene at 40 °C.

Solution

Amount of nitrogen, $n_2 = \dfrac{pV}{RT} = \dfrac{(102.66\ \text{kPa})(10.5\ \text{dm}^3)}{(8.314\ \text{kPa dm}^3\ \text{K}^{-1}\ \text{mol}^{-1})(303.15\ \text{K})}$

$= 0.427\,7$ mol

Pressure of nitrogen gas at 40 °C, $p = (102.66\ \text{kPa})\left(\dfrac{313.15\ \text{K}}{303.15\ \text{K}}\right) = 106.05\ \text{kPa}$

Amount of bromobenzene lost, $n_1 = \dfrac{m}{M} = \dfrac{0.856\ \text{g}}{(156.9\ \text{g mol}^{-1})} = 0.005\,5\ \text{mol}$

Total amount in the mixture, $(n_1 + n_2) = 0.005\,5\ \text{mol} + 0.427\,7\ \text{mol} = 0.433\,2\ \text{mol}$

Mole fraction of bromobenzene in the mixture, $x = \dfrac{n_1}{n_1 + n_2} = \dfrac{0.005\,5\ \text{mol}}{0.433\,2\ \text{mol}}$

Vapour pressure of bromobenzene = partial pressure of bromobenzene in the mixture

$$= x\,p = \frac{(0.005\,5)}{(0.433\,2)} \times 106.05\ \text{kPa} = 1.347\ \text{kPa}$$

Empirical Relations Relating the Normal Boiling Point

It is of interest to mention here some of the empirical relations between the normal boiling point of a liquid and its other physical properties. Two such relations are given below.

For most liquids, the ratio of normal boiling point T_b and the critical temperature T_c is found to be 0.66.

$$\frac{T_b}{T_c} \simeq 0.66$$

Trouton's rule The molar enthalpy of vaporization divided by the normal boiling point has an approximately constant value of 10.5R, i.e.,

$$\frac{\Delta_{vap}H_m}{T_b} \approx 10.5\,R$$

This relation holds more closely for non-associated substances of molar mass of about 100 g mol^{-1} and having not too high boiling points. For liquids of low boiling points, e.g., H_2 and He, the ratio is much less than 10.5R and for the associated liquids, such as water and alcohols, the value is greater.

2.3 VISCOSITY

Origin of Viscosity

It is a common experience of daily life that different liquids flow with different speeds. For example, water flows with greater speed than glycerol. Obviously, some sort of an internal friction is operating which checks the flow of liquids and which varies from liquid to liquid. This internal friction in the case of liquids is primarily due to forces of attraction between the molecules. If we have a laminar flow of a liquid in a tube, then the velocity of the layer just in touch with the side of the tube is zero and it increases as we proceed towards the centre of the tube (Fig. 2.3.1). Thus, there exists what is known as the *velocity gradient* between different layers of the liquid.

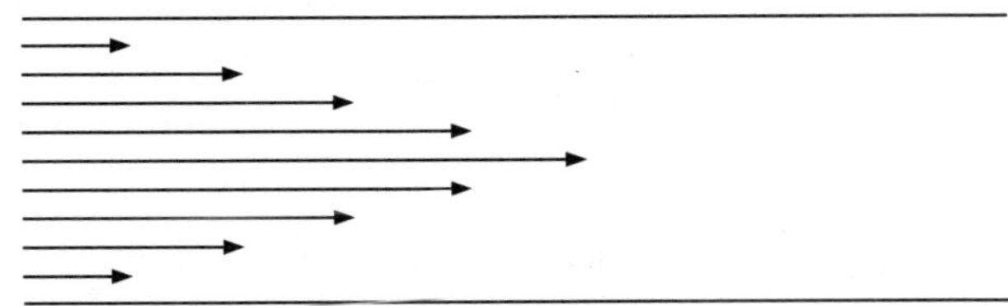

Fig. 2.3.1 Laminar flow of liquid in a tube

Due to greater intermolecular attractions amongst molecules of a liquid, the molecules moving in any one layer will tend to impede the movement of the molecules in the adjacent faster moving layer. As a result, the velocity of molecules in the faster layer decreases. Unless this decrease is prevented by applying a force along the layer in the forward direction, the velocity of the faster moving layer will go on decreasing and ultimately it will become zero. At this stage the liquid will stop flowing.

Definition of Coefficient of Viscosity

The internal friction which resists the flow of a liquid can be measured in terms of the tangential force which is needed to keep the speeds of different layers constant. This force F depends upon the following factors:

(i) It is directly proportional to A, the area of contact of two adjacent layers. Larger the area of contact between the two layers, larger the effect of intermolecular attractions and hence larger the decrease in speeds. Consequently, larger force is required to maintain the speeds constant.

(ii) It is directly proportional to $\mathrm{d}u$, the velocity difference between two adjacent layes. Larger the velocity difference, larger the force required to maintain the constant velocity difference.

(iii) It is inversely proportional to $\mathrm{d}x$, the distance between the two adjacent layers. Larger the distance, lesser the effect of intermolecular attractions and thus lesser the decrease in speed. Consequently, lesser force will be required to maintain the speeds of different layers.

Taking these factors together, we express F in the form,

$$F \propto A\frac{\mathrm{d}u}{\mathrm{d}x}$$

Removing the proportionality sign, we have

$$F = \eta A\frac{\mathrm{d}u}{\mathrm{d}x} \quad \text{or} \quad \eta = \frac{F}{A(\mathrm{d}u/\mathrm{d}x)} \tag{2.3.1}$$

The constant η is known as the *coefficient of viscosity* of a liquid or simply *viscosity* of a liquid. It may be defined as *the force per unit area required to maintain a velocity difference of unity between two parallel layers of liquid unit distance apart.* The unit of viscosity in CGS units is dyn cm^{-2} s and is known as the poise unit. The unit of viscosity in SI units is N m^{-2} s (= Pa s). The unit viscosity in this case is equal to a unit force (1 N) required to maintain a velocity difference of unity (1 m s^{-1}) between two adjacent layers having unit area of contact (1 m^2) and being unit distance (1 m) apart. The viscosities of most of liquids are small in magnitude. Therefore, these are usually expressed in the units of centipoise (10^{-2} poise) and millipoise (10^{-3} poise). The SI equivalent of the poise unit may be derived as follows.

Unit of viscosity in CGS units = dyn cm^{-2} s
Unit of viscosity in SI units = N m^{-2} s = Pa s

$$1 \text{ poise} = 1 \text{ dyn cm}^{-2}\text{ s} = 1\ (10^{-5}\text{ N})\ (10^{-2}\text{ m})^{-2}\text{ s} = 10^{-1}\text{ N m}^{-2}\text{ s}$$

Dimension of η The dimension of viscosity η can be worked out as follows:

$$\eta = \frac{F}{A\dfrac{du}{dx}} = \frac{\text{Mass} \times \text{Acceleration}}{\text{Area} \times \dfrac{\text{Velocity difference}}{\text{Distance}}} = \frac{m(l\,t^{-2})}{l^2\left(\dfrac{(l\,t^{-1})}{l}\right)} = m\,l^{-1}t^{-1} \quad (2.3.2)$$

Measurement of Viscosity

The following two methods are commonly employed for the determination of viscosity of a liquid.

Ostwald's Viscometer Method

This method is based on the Poiseuille's equation. If the volume v of a liquid requires the time t to flow through a capillary tube of radius r and length l under pressure head p, then its coefficient of viscosity as given by Poiseuille's equation is

$$\eta = \frac{\pi r^4 pt}{8\,lv} \quad (2.3.3)$$

The apparatus used, known as Ostwald's viscometer, is shown in Fig. 2.3.2. It consists of two bulbs A and B connected through a fine capillary tube. The bulb A is of bigger size whereas the bulb B is of smaller size. A known volume of liquid is taken in the bulb A and the liquid is sucked into the bulb B. The time taken in seconds for the liquid to flow from the mark a to mark b is noted down.

In order to avoid measurements of l, v and other factors in Eq. (2.3.3), one employs the relative method. Here, using the same viscometer, the experiment is repeated by taking same volume of a liquid whose viscosity is known. Water is usually employed for this purpose. If t_1 and t_w are the times taken by the liquid and water to flow from mark a to mark b, respectively, we have

$$\eta_1 = \frac{\pi r^4 p_1 t_1}{8lv} \quad \text{and} \quad \eta_w = \frac{\pi r^4 p_w t_w}{8lv}$$

Dividing, we get

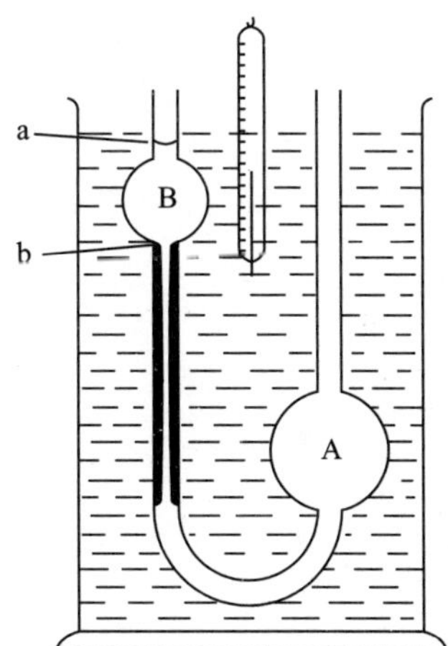

Fig. 2.3.2 Ostwald viscometer

$$\frac{\eta_1}{\eta_w} = \frac{p_1 t_1}{p_w t_w} \tag{2.3.4}$$

The pressure head because of which the liquid flows through the capillary tube is numerically equal to the weight of the liquid in the bulb B. Though this pressure head changes during the flow of the liquid, yet at any instant it may be considered to be proportional to the density of the liquid. Thus Eq. (2.3.4) modifies to

$$\eta_1 = \left(\frac{\rho_1 t_1}{\rho_w t_w}\right)\eta_w \tag{2.3.5}$$

Hence if the factors on the right side of Eq. (2.3.5) are known, the value of η_1 can be calculated.

Stokes Falling Sphere Method

If a steel ball is allowed to fall through a liquid, after some time it acquires a constant velocity. This happens when the force responsible for its downward motion (i.e. weight of the ball in the liquid) becomes equal to the frictional forces (due to the viscosity of the liquid) acting in the opposite direction. The uniform rate of streamlined fall of a sphere of radius r in a liquid is related to its viscosity through the Stokes relation;

$$u = \frac{F}{6\pi\eta r} \tag{2.3.6}$$

where F is the force acting downwards and is equal to the weight of the ball in air minus the weight of the displaced liquid, i.e.

$$F = \left(\frac{4}{3}\pi r^3\right)\rho_s\, g - \left(\frac{4}{3}\pi r^3\right)\rho_1 g = \frac{4}{3}\pi r^3(\rho_s - \rho_1)g$$

Substituting this in Eq. (2.3.6) and rearranging, we have

$$\eta = \frac{F}{6\pi ur} = \frac{\frac{4}{3}\pi r^3(\rho_s - \rho_1)g}{6\pi ur} = \frac{2r^2(\rho_s - \rho_1)g}{9u} \tag{2.3.7}$$

The apparatus used in this method is shown in Fig. 2.3.3.

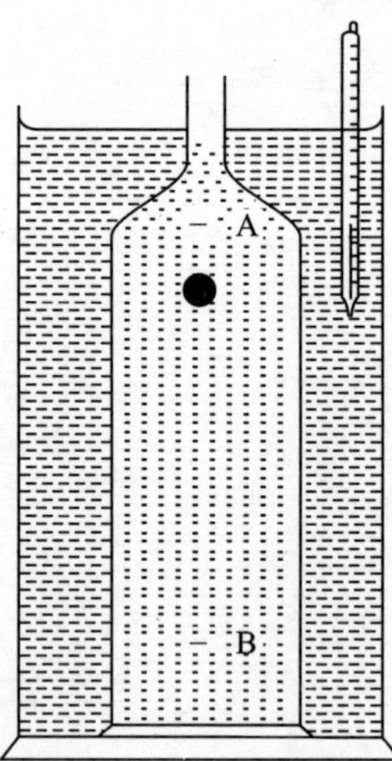

Fig. 2.3.3 Stokes falling sphere method

The given liquid is taken in a tall cylindrical vessel and its temperature is kept constant by surrounding the vessel with a thermostat. A small stainless steel ball is introduced from the top of the liquid and the time taken by it to travel between two marks A and B on the cylinder with a uniform velocity is noted down.

Thus in Eq. (2.3.7), $u = l/t$, where l is the distance between the two marks A and B. If we know the radius of the ball and densities of the ball and the liquid, the coefficient of viscosity of the liquid as given by Eq. (2.3.7) becomes

$$\eta = \frac{2r^2(\rho_s - \rho_1)g}{9(l/t)} \tag{2.3.8}$$

If the experiment is repeated with a liquid (known as the reference liquid) of known viscosity then r and l in Eq. (2.3.8) can be eliminated. Thus

$$\frac{\eta_1}{\eta_2} = \frac{(\rho_s - \rho_1)t_1}{(\rho_s - \rho_2)t_2} \tag{2.3.9}$$

where η_1 and η_2 are the viscosities, ρ_1 and ρ_2 are the densities, and t_1 and t_2 are the times required for the steel ball to travel through the given liquid and the reference liquid, respectively. Thus by knowing η_2 and determining values of t_1 and t_2 experimentally, the value of η_1 can be calculated by using Eq. (2.3.9.)

A General Comment on the Viscosity of Liquids

The coefficients of viscosity at 20 °C of some of the common liquids are given in Table 2.3.1.

It can be concluded from these values, that, in general, the coefficients of viscosity of associated liquids are larger than those of non-associated liquids. For example, $\eta(C_2H_5OH) > \eta(C_2H_5OC_2H_5)$. This is largely due to the hydrogen bonding between the molecules of a liquid. Glycerol and ethylene glycol have comparatively larger values of coefficients of viscosity since they have more than one hydroxyl group. Because of these hydroxyl groups, a network of hydrogen bonds is formed between the molecules which extends throughout the liquid, thus offering greater resistance to flow.

Table 2.3.1 Coefficients of Viscosity of Some Common Liquids at 20 °C

Substance	*Coefficient of viscosity* (η/poise) × 10^3 or (η/N m^{-2} s) × 10^4	*Substance*	*Coefficient of viscosity* (η/poise) × 10^3 or (η/N m^{-2} s) × 10^4
Acetic acid	12.29	Ethyl ether	2.33
n-Butyl alcohol	29.5	Acetone	3.29
Ethyl alcohol	12.0	Carbon tetrachloride	9.68
Methyl alcohol	5.92	Chloroform	5.63
Water	10.02	Benzene	6.47
Nitrobenzene	20.10	Toluene	5.90
Ethylene glycol	199		
Glycerol	8500		

Fluidity of Liquids

The flow of a liquid can be expressed in terms of fluidity ϕ, which is reciprocal of the viscosity, i.e.

$$\phi = \frac{1}{\eta}$$

The fluidity (or the viscosity) of a liquid depends on various factors, some of which are:

(i) Heavier and larger molecules flow less easily than the lighter and smaller molecules.

(ii) Spherical molecules offer less resistance to flow than the plate-like molecules. Molecules with flexible chains offer a very high resistance to flow because of entangling of side chains.

(iii) Impurities invariably increase the viscosity of a liquid. The presence of lyophilic colloids, in particular, enhance the viscosity of a liquid to a large extent.

Effect of Temperature on Viscosity

The viscosity of a liquid decreases with increase in temperature and this decrease is roughly of the order of 2 per cent per Celsius degree. This is due to the fact that on raising the temperature of a liquid, the average thermal energy of its molecules is increased and thus the effect of intermolecular attractions is decreased. Before a molecule can take part in liquid flow, it is expected that it should have sufficient energy to overcome the forces of intermolecular attraction due to the surrounding molecules. This energy is known as the *activation energy for viscous flow*. It is known that the number of molecules having this minimum energy or greater than this, increases in proportion to the Boltzmann factor exp($-E/RT$) and hence the resistance to flow of the viscosity may be expected to decrease in a reciprocal manner, i.e.,

$$\eta \propto \exp(E/RT) \quad \text{or} \quad \eta = A\exp(E/RT)$$

or

$$\ln(\eta/\eta^{\circ}) = \ln(A/\eta^{\circ}) + \frac{E}{RT}$$

where η° is the standard unit viscosity. Thus, $\ln(\eta/\eta^\circ)$ varies linearly with $(1/T)$ as shown in Fig. 2.3.4. From the slope, the activation energy for viscous flow can be calculated.

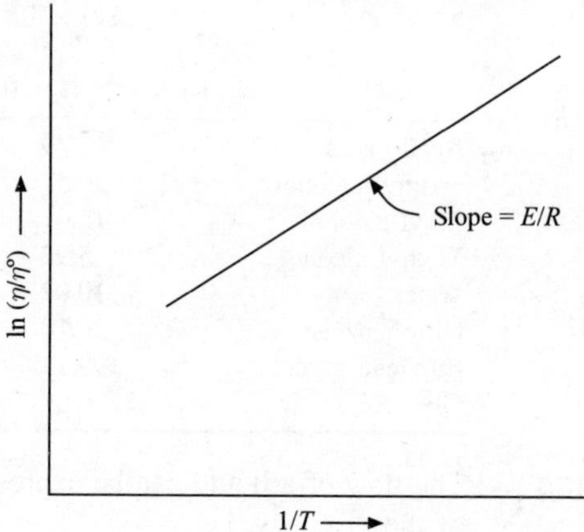

Fig. 2.3.4 Plot of ln (η/η°) versus $1/T$

Viscosities of Mixtures

The fluidity of an ideal mixture which involves similar types of liquids, such as a mixture of benzene and toluene, is best represented by the following equations:

$$\phi = x_A\phi_A + x_B\phi_B \qquad \text{(Bingham's equation)}$$

$$\log(\phi/\phi^\circ) = x_A \log(\phi_A/\phi^\circ) + x_B \log(\phi_B/\phi^\circ) \qquad \text{(Kendall's equation)}$$

where ϕ° is the standard unit fluidity. These equations are not applicable to non-ideal solutions which involve dissimilar type of liquids. If the constituents of a solution strongly interact with each other then the observed fluidity is less than the calculated value, i.e., a considerable increase in viscosity occurs in such a case. One of the examples is the water and ethanol mixture. In this mixture, a strong interaction between the molecules of water and ethanol takes place due to the hydrogen bondings. If the mixture involves a polar and a nonpolar liquid such as alcohol and benzene then the observed fluidities are higher than the predicted fluidities as obtained by the use of either Bingham or Kendall equation.

2.4 SURFACE TENSION

Introduction

It was seen during the study of vapour pressure that the molecules at the surface of a liquid experience a net inward pull because of the larger number of molecules towards the liquid side than towards the vapour side. There is a tendency on the part of surface molecules to go in the bulk of the liquid. The surface of liquid is therefore in some sort of tension and it tends to contract to the smallest possible area in order to have the minimum number of molecules at the surface. It is for this reason that the surface of a liquid is spherically curved, since the surface area is minimum for a given volume in the case of a sphere.

Surface Energy

If the area of the surface is to be extended then one has to bring more molecules from the bulk of a liquid to its surface. This will require expenditure of some energy because work has to be done in bringing molecules from the bulk against the inward attractive forces. The amount of work done in increasing the area by unity is known as the *surface energy*. Larger the forces of attraction amongst the molecules of a liquid, larger the net inward pull, and thus, larger work will be involved in increasing the area. Hence, the surface energy can be used to define these net inward pulls.

Surface Tension

It is customary to define these inward pulls in terms of the surface tension. The latter is defined as *the force acting along the surface of a liquid at right angle to any line of unit length*. The surface tension is equal, both numerically and dimensionally, to the surface energy. This can be proved as follows:

Equivalency of Surface Tension and Surface Energy

Consider a line of unit length (say, 1 cm) anywhere on the surface of a liquid (Fig. 2.4.1).

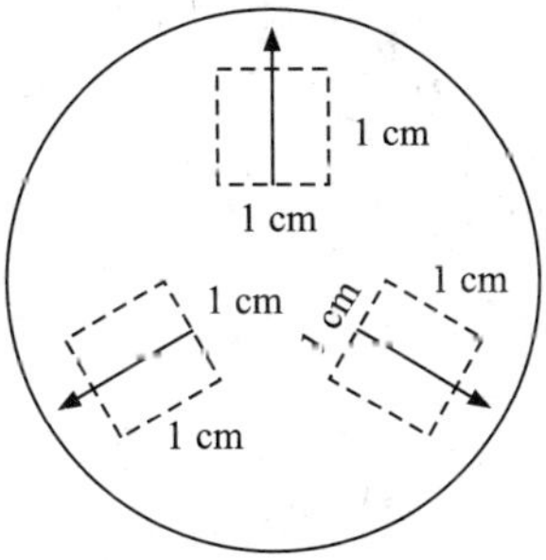

Fig. 2.4.1 Representation of surface tension

The surface area of the liquid can be increased by pulling this line perpendicularly by a force acting along the surface of the liquid. Let this line be extended by unity so as to enclose a unit area of the surface. Then

$$\text{Surface energy} = \frac{\text{Amount of work done}}{\text{Amount of area extended}} = \frac{\text{Force} \times \text{Distance}}{\text{Area}}$$

Hence in CGS units, we have

$$\text{Surface energy} = \frac{\text{dyn} \times \text{cm}}{\text{cm}^2} = \frac{\text{dyn}}{\text{cm}} = \text{surface tension}$$

and in SI units, we have

$$\text{Surface energy} = \frac{\text{N} \times \text{m}}{\text{m}^2} = \text{N m}^{-1}$$

Obviously, 1 dyn $\text{cm}^{-1} = (10^{-5}\ \text{N})\ (10^{-2}\ \text{m})^{-1} = 10^{-3}\ \text{N m}^{-1}$

The surface tension is represented by the symbol γ.

Dimension of surface tension This can be determined as follows.

$$\gamma = \frac{\text{Force}}{\text{Distance}} = \frac{\text{Mass} \times \text{Acceleration}}{\text{Distance}} = \frac{m\, l\, t^{-2}}{l} = m\, t^{-2}$$

Formation of Bubbles

The formation of a bubble is basically due to surface tension. It is obvious that the total pressure acting on a concave side must be larger than the pressure acting on the convex side. Therefore, the pressure inside a bubble must be larger than the external pressure. If this excess of pressure is not balanced by any other force, then the bubble will not be stable and will collapse immediately. Since stable bubbles are formed, this excessive pressure must be balanced by certain forces. These forces are due to surface tension and are called the *cementing forces*.

Expression of Excessive Pressure

Let a gas bubble in a liquid be cut into two hemispheres by an imaginary plane (Fig. 2.4.2). There will be a tendency for the two halves to be driven apart by a force equal to the inward pressure multiplied by the surface area of the hemispheres. This will be prevented by the external forces due to atmospheric pressure and by the cementing forces acting along the circumference of the circle. For a stable bubble, the total disruptive forces must be equal to the binding forces.

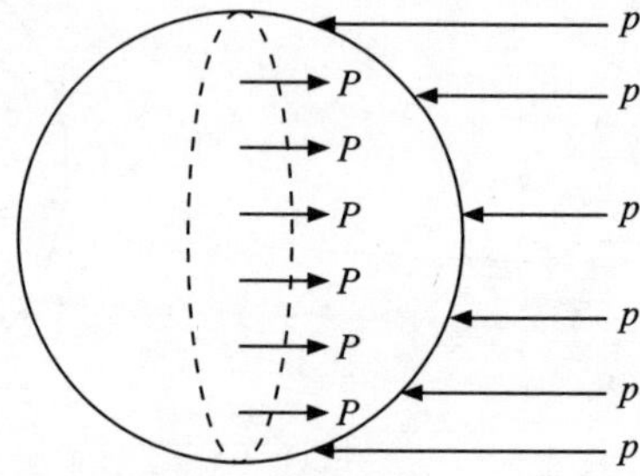

Fig. 2.4.2 Formation of a bubble

We have

$$\text{Disruptive forces} = P(\pi r^2)$$

$$\text{Binding forces} = p(\pi r^2) + \gamma(2\pi r)$$

Thus, for a stable bubble,

$$P(\pi r^2) = p(\pi r^2) + \gamma(2\pi r) \quad \text{i.e.} \quad \gamma(2\pi r) = (P - p)\pi r^2$$

or $$(P - p) = \frac{2\gamma}{r}$$

The excessive pressure inside the spherical bubble is thus inversely related to its radius. From this, it follows that the formation and maintenance of smaller bubbles will need greater values of excess pressure than the larger ones. This is, in fact, in agreement with the daily experience of blowing a balloon. It is difficult to blow a balloon initially but becomes easier later on as the excess pressure to be generated inside the balloon is inversely related to its radius.

Phenomenon of Capillary Action

When a capillary tube is dipped in a liquid, there occurs either a rise or a fall of liquid in the tube. This phenomenon is known as *capillary action* and is basically due to surface tension of the liquid. If the forces of attraction between the molecules of a liquid and those of the solid surface of the tube are greater than those existing amongst the molecules of the liquid, then the liquid has a tendency to spread on the solid surface and its meniscus in the tube is concave upwards. Such types of liquids are known as *wetting liquids* and they rise in the capillary tube. The angle of contact, which is measured within the liquid from the side of the tube to the tangent drawn at the meniscus touching the surface of the tube, in this case, is less than 90°. This is shown in Fig. 2.4.3a. If the cohesive forces in the liquid are greater than the solid-liquid attraction forces, or if there occurs repulsion between the molecules of the liquid and those of the solid surface, the liquid detaches from the surface of the solid. The meniscus of such a liquid in the tube is convex upwards and its level falls within the tube. The angle of contact, in this case, is greater than 90° as shown in Fig. 2.4.3b.

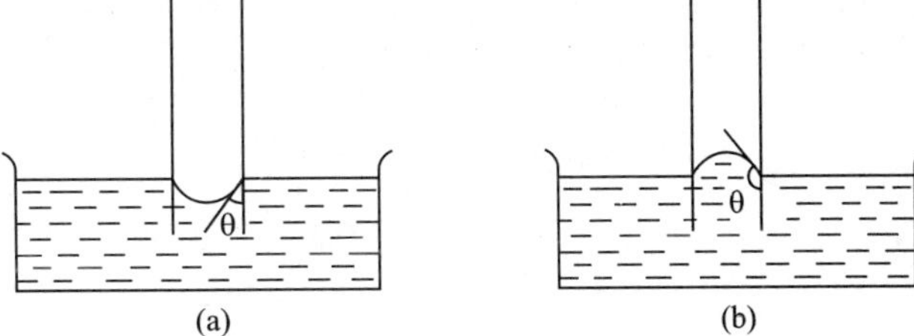

Fig. 2.4.3 The angle of contact of the meniscus (a) wetting liquid, < 90° and (b) non-wetting liquid, > 90°

Expression of Surface Tension

As mentioned above, the rise and fall of a liquid in a capillary tube is due to surface tension. Take, for example, the case of a wetting liquid. The surface tension forces act all around the capillary tube in the direction shown in Fig. 2.4.4. The liquid rises in the tube because of these upward forces. It continues to rise till the vertical component of the lifting force becomes equal to the weight of the liquid in the capillary tube. Thus,

$$\text{Lifting force} = (\gamma \cos \theta)\,(2\pi r_c)$$

$$\text{Weight of the liquid in the capillary tube} = \{(\pi r_c^2)h\}\rho g$$

At equilibrium, lifting force is equal to the downward force due to weight of the liquid in the capillary tube. Therefore,

$$(\gamma \cos \theta)\,(2\pi r_c) = \pi r_c^2\, h\rho g$$

or

$$\gamma = \frac{1}{2} h\rho g \frac{r_c}{\cos \theta} \tag{2.4.1}$$

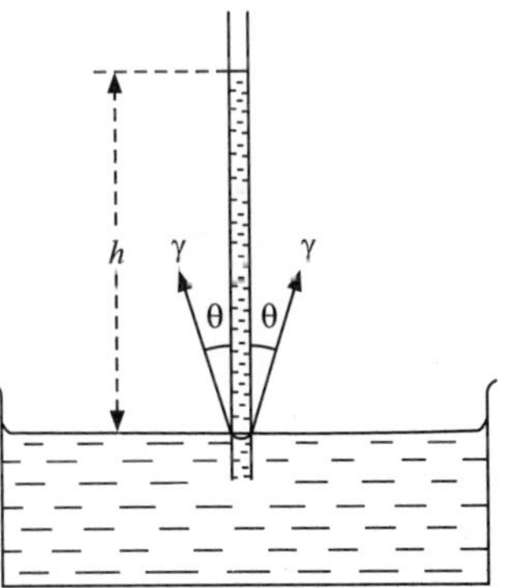

Fig. 2.4.4 Rise of a liquid in a capillary tube

For most of the wetting liquids, θ is very very small and thus $\cos\theta \simeq 1$. Therefore, the value of γ for such liquids is

$$\gamma = \frac{1}{2}h\rho g r_c \tag{2.4.2}$$

For non-wetting liquids, Eq. (2.4.1) is still applicable. Here h denotes depression of the liquid level.

Measurement of Surface Tension

Method Using Capillary Action

The phenomenon of capillary action can be used to determine surface tension of a liquid. However, to avoid the necessity of determining r, we use the relative method in which the value of h is also determined for a liquid whose surface tension is known. Thus

$$\gamma_1 = \frac{1}{2}h_1\rho_1 g r, \quad \text{and} \quad \gamma_2 = \frac{1}{2}h_2\rho_2 g r$$

$$\text{Hence,} \quad \frac{\gamma_1}{\gamma_2} = \frac{h_1\rho_1}{h_2\rho_2} \quad \text{or} \quad \gamma_1 = \left(\frac{h_1\rho_1}{h_2\rho_2}\right)\gamma_2 \tag{2.4.3}$$

Hence, for known values of γ_2, h_1, ρ_1, h_2 and ρ_2, the value of γ_1 can be calculated. The accuracy of this method can be further increased by the use of two capillary tubes of different radii as shown in Fig. 2.4.5. We have

$$\gamma = \frac{1}{2}h_1\rho g r_1 \quad \text{or} \quad \frac{\gamma}{r_1} = \frac{1}{2}h_1\rho g$$

$$\text{and} \quad \gamma = \frac{1}{2}h_2\rho g r_2 \quad \text{or} \quad \frac{\gamma}{r_2} = \frac{1}{2}h_2\rho g$$

Subtracting, we have

$$\gamma\left(\frac{1}{r_1} - \frac{1}{r_2}\right) = \frac{1}{2}\rho g(h_1 - h_2) = \frac{1}{2}\rho g\,\Delta h$$

$$\text{or} \quad \gamma = \frac{\rho g \Delta h}{2\left(\dfrac{1}{r_1} - \dfrac{1}{r_2}\right)}$$

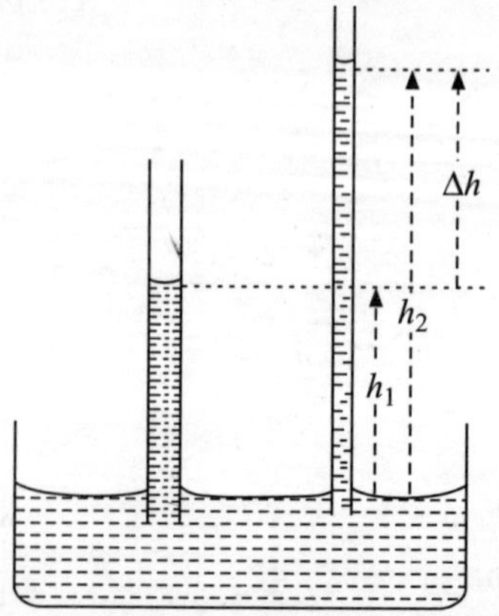

Δh
h_2
h_1

Fig. 2.4.5 Capillary rise method for measurement of surface tension

If the same pair of capillaries, after proper cleaning, are dipped in two different liquids one after the other, we will have

$$\gamma_A = \frac{\rho_A g \Delta h_A}{2\left(\frac{1}{r_1} - \frac{1}{r_2}\right)} \quad \text{and} \quad \gamma_B = \frac{\rho_B g \Delta h_B}{2\left(\frac{1}{r_1} - \frac{1}{r_2}\right)}$$

Dividing, we get

$$\gamma_A = \frac{\rho_A}{\rho_B}\frac{\Delta h_A}{\Delta h_B}\gamma_B \qquad (2.4.4)$$

Thus, knowing the values of various terms on the right side of Eq. (2.4.4), the value of surface tension of the liquid A can be determined.

A Laboratory Method

The other method which is commonly employed in laboratories is to use a *stalagmometer* (Fig. 2.4.6).

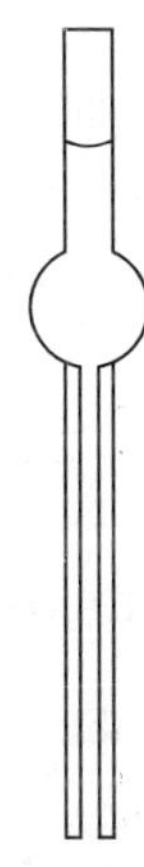

Fig. 2.4.6 The stalagmometer

A stalagmometer consists of a bulb attached to a fine capillary with a sharp edge. The liquid is allowed to fall drop by drop through the capillary tube. As the size of a liquid drop grows, its weight goes on increasing. It remains attached to the edge because of the forces of surface tension which act around the circumference of the tube and in the upward direction. When the downward force, due to weight of the liquid, becomes infinitesimally greater than the upward forces of surface tension, the drop detaches from the apparatus and falls down. Thus, the weight of the drop can be correlated with the surface tension of the liquid:

$$\text{Upward force} = (2\pi r)\,\gamma$$

$$\text{Downward force} = mg$$

Thus, $mg = (2\pi r)\,\gamma$

In order to avoid measurement of r, we use the relative method in which the experiment is repeated with a liquid whose surface tension is known. Usually water is employed for this purpose. Thus,

$$m_1 g = 2\pi r\gamma_1 \qquad \text{and} \qquad m_{\rm w} g = 2\pi r\gamma_{\rm w}$$

Dividing and rearranging, we get

$$\gamma_1 = \left(\frac{m_1}{m_{\rm w}}\right)\gamma_{\rm w} \tag{2.4.5}$$

The values of m_1 and $m_{\rm w}$ can be determined following either the drop weight method or the drop number method.

Drop Weight Method In the drop weight method, the mass of about 20 to 30 drops of each liquid is found accurately, and then the mass of a single drop is calculated.

Drop Number Method In the drop number method, the number of drops obtained for the same volume, say V, of each liquid is counted. Then the mass of a single drop is determined as follows:

$$v_1 = \frac{m_1}{\rho_1} \qquad \text{and also} \qquad v_1 = \frac{V}{n_1}$$

Therefore, $m_1 = v_1\rho_1 = \dfrac{V\rho_1}{n_1}$

Similarly, $m_2 = \dfrac{V\rho_2}{n_2}$

and thus,
$$\frac{\gamma_1}{\gamma_2} = \frac{m_1}{m_2} = \frac{\rho_1 n_2}{\rho_2 n_1} \tag{2.4.6}$$

Values of Surface Tensions of a few Liquids Table 2.4.1 records the surface tension values at 20 °C for some liquids.

Table 2.4.1 Surface Tensions of Some Common Liquids at 20 °C

Substance	*Surface tension* γ/dyn cm^{-1} or $\gamma \times 10^3$/N m^{-1}	*Substance*	*Surface tension* γ/dyn cm^{-1} or $\gamma \times 10^3$/N m^{-1}
Water	72.8	Acetic acid	27.42
Benzene	28.87	Methyl alcohol	22.55
Toluene	28.53	Ethyl alcohol	22.30
Carbon disulphide	32.25	Ethyl ether	17.05
Chloroform	27.2	Nitrobenzene	43.35
Carbon tetra-		*n*-Butyl alcohol	24.52
chloride	26.75	*n*-Propyl alcohol	23.75
Acetone	23.32	*o*-Xylene	30.03
Ethyl acetate	23.75	*p*-Xylene	28.31
Methyl acetate	24.8	Chlorobenzene	33.25

Effect of Temperature on Surface Tension Since the effects of forces of attraction between the molecules of a liquid decrease with increase in temperature, it follows, therefore, that the surface tension decreases with increase in temperature.

Eötvös Equation R. von Eötvös proposed the following relation between surface tension and temperature.

$$\gamma(Mv)^{2/3} = a - kt \tag{2.4.7}$$

where a and k are constants and v is the specific volume of the liquid. The factor Mv represents the molar volume of the liquid. If the volume is considered to be spherical, the area of the sphere will be proportional to $(Mv)^{2/3}$. Since γ represents the surface energy per unit area, the factor $\gamma(Mv)^{2/3}$, therefore, represents some sort of surface energy which according to Eötvös, varies linearly with temperature. The value of surface energy will be zero at the critical temperature since at this temperature the surface of separation between a liquid and its vapour disappears. This fact can be utilized in deriving the value of a in terms of k. At critical temperature

$$\gamma(Mv)^{2/3} = 0 = a - kt_c$$

Therefore

$$a = kt_c$$

Ramsay and Shields Modification W. Ramsay and J. Shields made a study of Eötvös equation and found that the experimental results could be better expressed by the relationship:

$$\gamma(Mv)^{2/3} = k(t_c - t - 6) \tag{2.4.8}$$

According to this equation, the surface energy becomes zero when

$$t = t_c - 6$$

that is, it becomes zero at a temperature 6 °C below the critical temperature. This is also supported by experimental observations that for some liquids, the meniscus disappears at a temperature a few degrees before the critical temperature. For normal and non-associated liquids the constant k is found to have a value of 2.12. For associated liquids such as water and alcohol, k is found to have a value less than 2.12.

Interfacial Tension

Consider two immiscible liquids in contact with each other. The molecules at the surface of either of these liquids will experience an unbalanced force of attraction in the same way as the molecules at the surface of a liquid experience when they have the liquid on one side and the vapour on the other. These unbalanced forces at the surface of separation between the two immiscible liquids give rise to interfacial tension, which can be defined in the same way as surface tension. It is thus, the force acting along the surface of a liquid perpendicular to any line of unit length. According to Antonoff's rule, the interfacial tension γ_{AB} between two liquids A and B is equal to the difference of the respective surface tensions γ_A and γ_B, i.e.

$$\gamma_{AB} = |\gamma_A - \gamma_B| \tag{2.4.9}$$

Obviously, the interfacial tension will be smaller than the larger of the two surface tensions. This is because the attraction across the interface between the molecules of one liquid and those of the other, tends to reduce the inward pull of the molecules in the surface by those of the same kind. The numerical value of the interfacial tension can be determined experimentally by following either the capillary rise method or the drop weight method. The interfacial tension plays a very important role in determining the properties of emulsions, detergents, etc.

Surface Tension of Mixtures

When two liquids whose surface tensions do not differ very much are mixed in various proportions, the surface tension of the resultant mixture varies linearly with the mole fraction of either of the liquids. Ideal solutions follow this behaviour; an example of which is the benzene-toluene mixture. If the surface tension of liquids differ very much, the addition of even a small amount of the liquid of lower surface tension in the liquid of higher surface tension decreases the surface tension of the latter to a large extent. This is because such a liquid has a tendency to accumulate at the surface and thus the surface has a relatively richer constituent of the lower surface tension as compared to the bulk of the mixture. In such cases, the variation of surface tension is given by the following expression, provided the solution is not very dilute:

$$\frac{\mathrm{d}\gamma}{\mathrm{d}\ln c} = -X\gamma^* \tag{2.4.10}$$

In this expression, X is a constant and γ^* is the surface tension of the pure liquid. According to this, the surface tension of the mixture varies linearly with the logarithm of the concentration of the added liquid. A few examples are alcohol-water, acetone-water, and acetic acid-water mixtures.

Surface Active Materials

Certain substances like soap, salts of higher sulphonic acids and of higher amines, in water decrease the surface tension of the latter to a large extent and thus are called *surface active materials* (due to their tendency to accumulate at the surface.). These substances act as detergents and they have the characteristics of lowering the interfacial tension between water and grease and thus render wetting of the latter easier. This also explains why these substances are used to clean garments. We can now explain the cleansing action of soap as follows.

Cleansing Action of Soap

Most of the dirt or dust sticks on to grease or oily materials which gather on clothes. As grease is not readily wetted by water, it is difficult to clean the garment with water alone. If a soap solution is used, the intefacial tension between water and grease decreases and thus facilitating the mixing of the two. This decrease of interfacial tension is due to higher concentration of soap molecules at the surface. The orientations of these molecules at the surface is specific, —COO groups pointing towards the surface of water and hydrocarbon chains pointing outwardly. These hydrocarbon chains act as a solvent for the grease and thus the latter is detached from the cloth along with the dust.

Gibbs Adsorption Isotherm

In general, the variation of surface tension of a solvent with the addition of solute is given by the Gibbs adsorption isotherm:

$$\Gamma_2 = -\frac{c}{RT}\frac{\mathrm{d}\gamma}{\mathrm{d}c} \tag{2.4.11}$$

where Γ_2 is the excessive concentration of the solute at the surface relative to that existing in the bulk of the solution. According to Eq. (2.4.11), $\mathrm{d}\gamma$ is positive if Γ_2 is negative, and vice versa. For solutes like acetone, soaps, and detergents, Γ_2 is positive and hence there occurs a decrease in surface tension of water with addition of such solutes. The solutes like NaCl, KCl, etc., have more concentration in the bulk as compared to that existing at the surface (i.e. Γ_2 is negative) and thus there occurs an increase in the surface tension of water when such electrolytes are added in this solvent.

The Parachor

D.B. Macleod observed that the surface tension of a liquid was related to the orthobaric densities ρ_l and ρ_v of liquid and saturated vapour, respectively, through the equation

$$\frac{\gamma^{1/4}}{\rho_l - \rho_v} = C$$

where C is a constant, whose value depends upon the nature of the liquid. If the above equation is multiplied by M, the molar mass of the liquid, then we have

$$\frac{M\gamma^{1/4}}{\rho_l - \rho_v} = MC = \text{constant} = [P]$$

If the liquid is not very near to its critical temperature, then ρ_l is much greater than ρ_v, and the approximate form of the above expression becomes

$$\left(\frac{M}{\rho_l}\right)\gamma^{1/4} = [P] \quad \text{or} \quad V_m\gamma^{1/4} = [P]$$

Thus, if the surface tension is unity, the value of the constant $[P]$ is equal to the molar volume of the liquid. The symbol $[P]$ is known as *parachor*, meaning *comparative volume, since the comparison of the parachors of different liquids is equivalent to a comparison of molar volumes under the condition of equal (unit) surface tension.*

Parachor has been found to be largely an *additive* property and partly a constitutive property. It is possible to evaluate the parachor equivalents of various atoms and for various structural units (Table 2.4.2).

Table 2.4.2 Vogel's Values of Atomic and Structural Parachor

Unit	*Parachor* ($[P]/[P^\circ]$)	*Unit*	*Parachor* ($[P]/[P^\circ]$)†
C	8.6	—OH	30.2
H	15.7	—COOH	73.7
O	19.8	—NO_2	73.8
S	48.2	Single bond	0
Cl	55.2	Double bond	19.9
Br	68.8	Triple bond	40.6
I	90.3	3-membered ring	12.3
>CO	44.4	4-membered ring	10.0
		5-membered ring	14.6
		6-membered ring	1.4

$^\dagger[P^\circ] = 1\ \text{cm}^3\ \text{mol}^{-1}\ (\text{dyn cm}^{-1})^{1/4} \equiv 10^{-6}\ \text{m}^3\ \text{mol}^{-1}\ (10^{-3}\ \text{N m}^{-1})^{1/4}$

With the help of these values, we can calculate the parachor values of different isomers of a given compound. The comparison of these values with the experimentally determined values can help in deciding the structure of the given compound.

REVISIONARY PROBLEMS

2.1 Discuss the general characteristics of liquids and compare them with those of gases and solids.

2.2 What is the vapour pressure of a liquid? How does it arise and what is the effect of temperature on it?

2.3 The variation of vapour pressure with temperature can quantitatively be expressed by Clapeyron equation

$$\frac{\mathrm{d}p}{\mathrm{d}T} = \frac{\Delta_{\mathrm{vap}}H_{\mathrm{m}}}{T(V_{\mathrm{m,v}} - V_{\mathrm{m,l}})}$$

Explain the various terms involved in this equation. Discuss the Clausius approximations and hence derive the Clausius-Clapeyron equation.

2.4 Draw a graph indicating:
(i) The variation of vapour pressure with temperature. What are the lower and the upper limits of this graph?
(ii) The variation of ln (p/p°) versus $1/T$. What is its slope?

2.5 Discuss how vapour pressure and boiling point are intimately connected to each other. Explain how a plot of the variation of vapour pressure with temperature can be considered to represent a graph of the variation of boiling point of a liquid with the external pressure.

2.6 Describe the methods commonly employed for the measurement of vapour pressure of a liquid.

2.7 (a) Discuss the effects of nonvolatile impurities on vapour pressure and boiling point of a liquid. (b) What is Trouton's rule?

2.8 What do you understand by the term viscosity of a liquid? How does it arise and what are its units? What is the effect of temperature on the viscosity of a liquid?

2.9 Describe the methods commonly employed for the measurement of viscosity of a liquid.

2.10 Explain why the viscosity of ethyl alcohol is greater than that of ether.

2.11 According to the Arrhenius equation, the variation of viscosity with temperature is given by

$$\ln(\eta/\eta'') = \ln(A/\eta^\circ) + \frac{E}{RT}$$

Explain the various terms involved in this. Give a schematic plot of ln (η/η°) versus $1/T$. What is its slope?

2.12 Discuss the origin of surface tension of a liquid. Show that the surface tension and surface energy have the same dimensions.

2.13 Show that the excessive pressure inside a spherical bubble is given by

$$\Delta p = (P - p) = \frac{2\gamma}{r}$$

where P and p are the pressures inside and outside the bubble, respectively. With the help of this, justify the statement:

'The formation and maintenance of smaller bubbles will need greater values of excess pressure than the larger ones'.

2.14 What is capillary action? Derive the formula

$$\gamma = \frac{1}{2}(\pm h)\rho g r$$

where positive sign is for wetting liquids and negative sign for nonwetting liquids. What is the relation between the angle of contact and the nature of the liquid (both wetting and nonwetting)?

2.15 Discuss the experimental method employed for the determination of surface tension of a liquid.

2.16 Explain the following terms:
(a) Interfacial tension.
(b) Parachor and its importance.
(c) Surface active materials.
(d) Cleansing action of soap.

NUMERICAL PROBLEMS

Vapour Pressure

2.1 The vapour pressure of water at 90 °C is 70.13 kPa, and the mean enthalpy of vaporization between 90 °C and 100 °C may be taken as 2.268 kJ g^{-1}. Calculate the vapour pressure of water at 100 °C. (Ans. 100.79 kPa)

2.2 The vapour pressure of *n*-butyl alcohol is given by the equation

$$\log\ (p/\text{kPa}) = -\frac{2443\ \text{K}}{T} + 8.2617$$

Calculate: (a) $\Delta_{vap}H_m$; and (b) the normal boiling point of alcohol.
(Ans. (a) 46.776 kJ mol^{-1}, (b) 390.5 K)

2.3 The vapour pressure of *n*-propyl alcohol varies with temperature as follows:

Temperature, t/°C	50	60	70	80
Pressure, p/kPa	11.626	19.600	31.864	50.129

(a) Evaluate $\Delta_{vap}H_m$ from a suitable graph of these data.
(b) At what temperature is the vapour pressure 26.665 kPa?
(c) What will be the boiling point if the external pressure is 39.997 kPa?
(Ans. (a) 46.16 kJ mol^{-1}, (b) 66.4 °C, (c) 75 °C)

2.4 At 298 K, liquids X and Y have vapour pressures of 13.332 kPa and 26.665 kPa, respectively, and the corresponding enthalpies of vaporization are 43.57 kJ mol^{-1} and 16.74 kJ mol^{-1}. Calculate the temperature at which both X and Y have the same vapour pressure. (Ans. 318 K)

2.5 At 25 °C, 10 dm^3 of dry air is bubbled slowly through 115.2 g of a pure liquid whose molar mass is 120 g mol^{-1}. The remaining liquid weighed 113.1 g. Assuming the vapour to behave ideally and the volume to be approximately equal to that of the dry air, calculate vapour pressure of the liquid. Given: p_{ext} = 101.325 kPa.
(Ans. 4.16 kPa)

Surface Tension

2.6 The surface tension of water at 20 °C is 72.75 × 10^{-3} N m^{-1}. How high will a column of water rise in a capillary tube with a radius of 0.005 cm? (Ans. 29.7 cm)

2.7 In the determination of the surface tension of a liquid A by the drop number method, equal volumes of A and water gave 60 and 20 drops, respectively. Calculate the surface tension of A if ρ(A) = 0.896 g cm^{-3} and ρ(water) = 0.964 g cm^{-3} Given: $\gamma(H_2O)$ = 72.75 × 10^{-3} N m^{-1}. (Ans. 22.54 × 10^{-3} N m^{-1})

2.8 Benzene has a density of 0.879 g cm^{-3} and has a surface tension of 0.028 88 N m^{-1}. What will be the difference of its heights in two capillaries of diameters 0.10 mm and 0.15 mm, respectively? (Ans. 2.23 cm)

Viscosity

2.9 At 20 °C, pure water with an absolute viscosity of 1.002 × 10^{-3} N m^{-2} s requires 102.2 s to flow through the capillary of an Ostwald viscometer. At 20 °C, toluene requires 68.9 s. If the densities of water and toluene be 0.998 and 0.866 g cm^{-3} respectively, calculate the viscosity of toluene. (Ans. 5.9 × 10^{-4} N m^{-2} s)

2.10 The viscosity of diethyl ether in millipoise is 2.84 at 0 °C, 2.33 at 20 °C and 1.97 at 40 °C. Calculate: (a) activation energy of diethyl ether for viscous flow; (b) its viscosity at 60 °C. (Ans. (a) 6.49 kJ mol^{-1}, (b) 1.70 millipoise)

2.11 A steel ball of density 7.90 g cm^{-3} and 4 mm diameter requires 55 seconds to fall through a distance of 1 metre through a liquid of density 1.10 g cm^{-3}. Calculate the viscosity of the liquid in poise. (Ans. 32.58 poise)

2.12 The viscosity of an oil is 0.05 N m^{-2} s and its density is 0.97 g cm^{-3} at 298 K. How long a given volume of the oil will take to flow through a viscometer if the same volume of water takes 60 s? Given: $\eta(H_2O)$ = 0.000 89 N m^{-2} s. Take $\rho(H_2O)$ = 1.0 g cm^{-3}. (Ans. 3 474 s)

3 The Solid State

3.1 INTRODUCTION

Characteristics of Solids

Solids are characterized by their high density and low compressibility compared with those of the gas phase. The values of these properties for solids indicate that the molecules (or ions) in them are relatively close together. Solids can very easily be distinguished from liquids by their definite shape, considerable mechanical strength and rigidity. These properties are due to the existence of very strong forces of attraction amongst the molecules (or ions) of the solids. It is because of these strong forces that the structural units (atoms, ions, etc.) of the solid do not possess any translatory motion but can have only vibrational motion about their mean positions.

Crystals and Amorphous Solids

Solids can generally be classified into two broad categories: crystals and amorphous substances. The outstanding characteristics of a crystal are its sharp melting point and its flat faces and sharp edges which, in a well developed form, are usually arranged symmetrically. These properties are the result of a high degree of internal order which extends throughout the crystal (a definite pattern constantly repeating in space), i.e. there exists what is known as the *long range order.* The pattern is such that having observed it in some small region of the crystal, it is possible to predict accurately the positions of particles in any region of the crystal, however far it may be from the region under observation. But amorphous solids, such as glass, do not have this ordered arrangement. In many ways they are more closely related to liquids than to crystalline solids and are, therefore, regarded as supercooled liquids with high viscosity. In this chapter, we shall discuss the subject of crystalline structure.

3.2 FACES, EDGES AND INTERFACIAL ANGLE OF A CRYSTAL

Crystals are bound, as stated earlier, by surfaces which are usually planar. These surfaces are called *faces* and where two faces intersect an *edge* is formed. The angle between the normals to the two intersecting faces is the *interfacial angle.*

In general, the shape of a crystalline solid depends upon the conditions (temperature, rate of growth, extent of stirring and impurities) which exist during the period of growth, e.g., NaCl crystallizes in a cubic form from an aqueous solution and in an octahedral form from an aqueous solution containing some urea.

The first quantitative measurements on crystals were made by N. Stensen who measured the angles between the faces of quartz crystals of different shapes. He found that inspite of the differences in shape, the angles between corresponding faces were always the same. Later, when crystals of many other substances were observed, each was found to have characteristic angles between faces. The statement that *the angles between the corresponding faces of various crystals of the same substance are constant* is sometimes called the first law of crystallography.

3.3 HAÜY'S IDEA AND SPACE LATTICE

The Basic Unit of a Crystal

At about the same time as Stensen's work, Robert Hooke speculated that all observed forms of crystals could be pictured as consisting of regular arrangements of small spherical particles. Hooke's idea was extended by Haüy. In 1784, as the result of a fortunate accident, he began a study of the cleavage of calcite crystals and found that regardless of external appearance of crystals, he would obtain the rhombohedral unit in each case. Based on similar cleavage studies of many other crystals, Haüy concluded that the continued cleavage will result in a basic unit, the shape of which was related to the external shape of the crystal. The macroscopic crystal can be generated by stacking this basic unit one upon the other.

Space Lattice

Rather than drawing the entire unit of pattern it is much more convenient to represent the unit of pattern by a point. Each point then represents the position of an atom, ion, molecule or group of ions or molecules. The regular three-dimensional arrangement of the identical points in space gives rise to what is known as a *space lattice.* The definition of a space lattice is strictly a geometrical concept and represents *a three-dimensional translational repetition of the centres of gravity of the units of pattern in the crystal.* This means that if a straight line is drawn through any two points, it will pass at equal intervals through a succession of similar points. Since the arrangement of these points is regular, it implies that the environment around any point is the same as that around any other point in the lattice. It should be kept in mind that the lattice points need not represent the actual atoms in the crystal but represent the spatial arrangement of the units of pattern. These units of pattern can be anything, for example:

(i) In metals or inert gases, each lattice point may represent the position of each atom.

(ii) In a crystal of methane, each lattice point may represent the centre of a methane molecule.

(iii) In ionic crystals, the lattice points may represent ions or ion-pairs. In the former, it is more convenient to consider a lattice of positive ions independently of a lattice of negative ions; then the two can be interlocked to give an adequate representation of the crystal. In the ion-pair case, a point equidistant between a positive and a negative ion can be considered to be a lattice point, thus having a single lattice to represent the ionic crystal.

Unit Cell

The basic idea of Haüy that the crystals can be generated by stacking together the basic units is equivalent to considering the crystal in terms of a lattice. Thus, by connection of the lattice points, a series of parallelepipeds of the type shown in Fig. 3.3.1 can be obtained. Each of these parallelepipeds contains a complete unit of pattern of the crystal. By translation or stacking of the parallelepipeds the entire crystal structure can be generated. Such a parallelepiped can be drawn from any crystal lattice, and is called a *unit cell.*

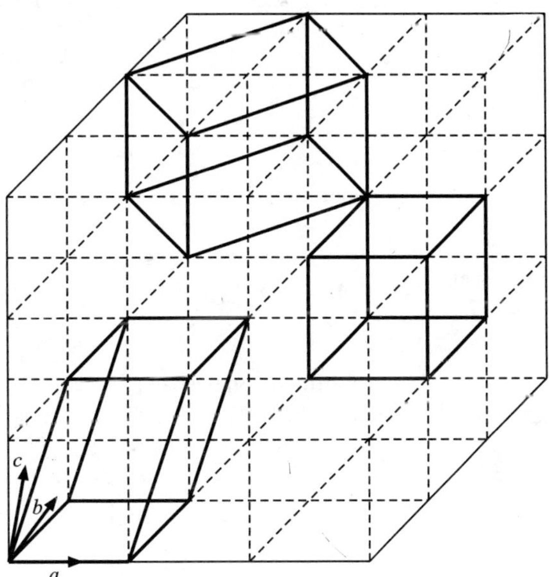

Fig. 3.3.1 Crystal lattice and unit cell

3.4 CRYSTAL SYSTEMS

Unit Cells in a Two-Dimensional Lattice

As stated earlier, in a space lattice or a crystal lattice there exist identical points in space. For example, in a 1-dimensional lattice, we will have a set of points arranged at equal distances along a straight line. The only parameter in this case is the minimum repeat distance. In a 2-dimensional planar lattice, we will have to specify the values of two basis vectors which give the repeat distances along two axes and the angle between these axes. From these parameters a unit pattern or unit cell can be constructed, with which the entire planar lattice can be formed by repeated translations of the unit cell. The values of the two basis vectors in a planar lattice may be same or different, and the angle between them may be 90°, 60° or any value other than these two. These possibilities lead to the existence of just five possible types of planar lattices, each characterized by its unit cell as shown in Fig. 3.4.1.

Primitive and Nonprimitive Unit Cells

It should be noticed that for some lattices the unit cell can be chosen in more than one way. This is shown in Fig. 3.4.1(III). For the same lattice points, the unit cell may be a parallelogram, an equilateral triangle or a regular hexagon with a lattice point in its centre. In the first two unit cells, namely, parallelogram and equilateral triangle, all lattice points lie at the corners of the unit cells. Such unit cells are called *primitive unit cells* and are represented by the symbol P. The third one, i.e. hexagon, in which one lattice point also lies at the centre, is known as *nonprimitive unit cell.*

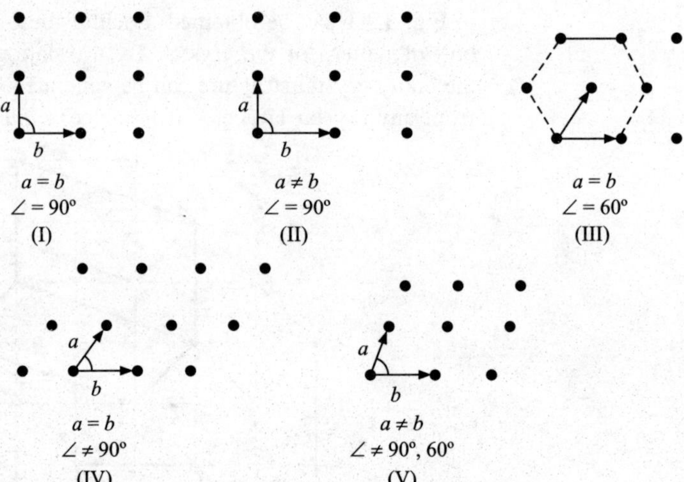

Fig. 3.4.1 Five types of unit cells in planar lattice

Unit Cells in a Three-Dimensional Lattice

In a 3-dimensional space lattice, we need to specify the values of three basis vectors which give the three repeat distances along the three axes and three angles as shown in Fig. 3.4.2.

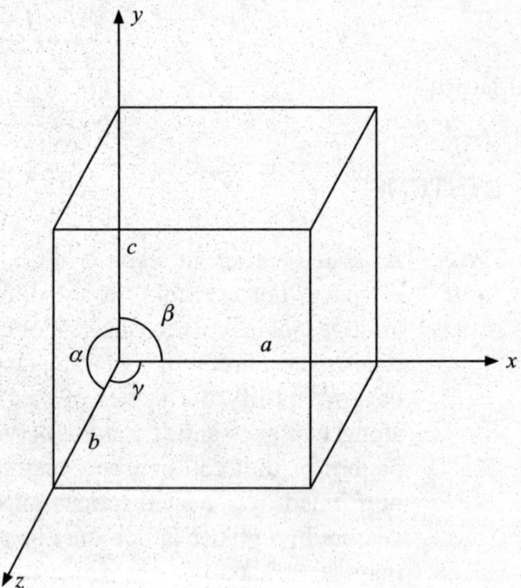

Fig. 3.4.2 Three basis vectors in a 3-dimensional space lattice

Seven Crystal Systems

For a given lattice, there is, in principle, an infinite number of ways in which the three basis vectors *a*, *b* and *c* might be chosen. The choice of the most suitable unit vectors is based on symmetry. If there are no symmetry elements, the three vectors chosen are the shortest ones in three different planes. The morphological study of crystals of different symmetries† showed that they could be classified into seven crystal systems based on the presence of certain rotation axes. The results are listed in Table 3.4.1. The schematic shapes of these primitive (symbol: P) crystal systems are shown in Fig. 3.5.4.

†*See* Annexure I at the end of this chapter for elementary ideas of symmetry.

Table 3.4.1 Seven Crystal Systems

Crystal system	*Minimum symmetry*	*Parallelepiped dimensions*†	*Examples*
Triclinic	1 (or $\bar{1}$)	$a \neq b \neq c$ $\alpha \neq \beta \neq \gamma \neq 90°$	$CuSO_4 \cdot 5H_2O$, $K_2Cr_2O_7$
Monoclinic	2 (or $\bar{2}$)	$a \neq b \neq c$ $\alpha = \beta = 90° \neq \gamma$	S(monoclinic), $CaSO_4 \cdot 2H_2O$, $Na_2SO_4 \cdot 10H_2O$
Orthorhombic	222 (or $\overline{222}$)	$a \neq b \neq c$ $\alpha = \beta = \gamma = 90°$	S(rhombic), $BaSO_4$, KNO_3, K_2SO_4
Trigonal or Rhombohedral	3 (or $\bar{3}$)	$a = b = c$ $\alpha = \beta = \gamma \neq 90°$	$CaCO_3$, calcite
Cubic	Four 3 (or Four $\bar{3}$)	$a = b = c$ $\alpha = \beta = \gamma = 90°$	NaCl, diamond, Alums, CaF_2
Tetragonal	4 (or $\bar{4}$)	$a = b \neq c$ $\alpha = \beta = \gamma = 90°$	TiO_2, Sn(white), $ZrSiO_4$
Hexagonal	6 (or $\bar{6}$)	$a = b \neq c$ $\alpha = \beta = 90°$, $\gamma = 120°$	SiO_2, graphite PbI_2, Mg, ZnO

†Dimensions correspond to maximum possible symmetry of their respective systems.

Absence of Five-Fold Rotation Axis in Crystal Systems

In the crystal classification shown in Table 3.4.1, it can be seen that there is no crystal system in which five-fold rotation axis is present. This is due to the fact that in crystals only those rotation axes are allowed which are consistent with the translational symmetry. According to the latter, the distance between any two points should be equal to a unit vector a or an integral multiple of this vector, i.e. ma. It can be shown that a crystal can never have a five-fold rotation axis as otherwise the condition of translational symmetry is not fulfilled. This may be demonstrated by taking a crystal lattice which has a n-fold rotation axis. It can be shown that n can never be equal to five. The n-fold rotation axis means that if a rotation by an angle $360°/n$ is carried around the rotation axis, then the new arrangement of points will be completely identical to the old arrangement of points, i.e. the new configuration will be indistinguishable from the old configuration. Let this operation be carried out around an axis passing through the point P_2 of the crystal lattice as shown in Fig. 3.4.3 and let the lattice point P_1 be thus thrown into P_3. Similarly, let the lattice point P_2 be thrown into the point P_4 if the rotation operation is carried out around the axis passing through the point P_1 (Fig. 3.4.3).

According to the condition of translational symmetry, the distance between the points P_3 and P_4 must be an integral multiple of the unit vector a. Thus, we have

$$P_3P_4 = ma$$

As can be seen from Fig. 3.4.3, the distance between the points P_3 and P_4 is also given by

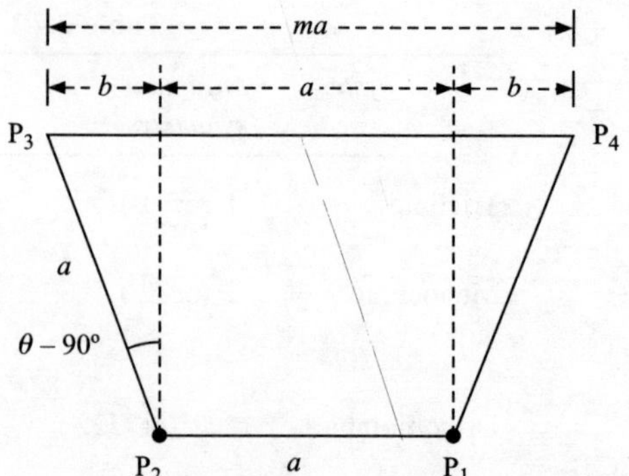

Fig. 3.4.3 Absence of five-fold rotation axis

$$P_3P_4 = a + 2\ b = a + 2\ a \sin (\theta - 90°)$$

Equating this distance to *ma*, we get

$$a(1 - 2 \cos \theta) = ma$$

or $$\cos \theta = \frac{1 - m}{2}$$

By giving an integral value to *m*, the corresponding angle of rotation can be calculated from the above expression. These are given in Table 3.4.2.

Table 3.4.2 The Allowed Rotation Axis in Crystal Systems

m	*cos θ*	*θ*	*Value of n, i.e. order of rotation axis*
0	$\frac{1}{2}$	60°	$\frac{360}{60} = 6$
1	0	90°	$\frac{360}{90} = 4$
2	$-\frac{1}{2}$	120°	$\frac{360}{120} = 3$
3	–1	180°	$\frac{360}{180} = 2$

Higher values cannot be given to *m* as otherwise the magnitude of the value of cos *θ* becomes greater than 1.

It can be seen from the Table 3.4.2 that the five-fold rotation axis is absent Thus, it can be concluded that a crystal can never have five-fold rotation axis if it has to satisfy the translational symmetry.

3.5 THE FOURTEEN BRAVAIS LATTICES

Cubic System

Consider a cube in which a lattice point exists in the centre of each face as well as at the corners. This unit cell can be repeated to give a space lattice. This unit cell, which is known as non-primitive cell, also has the minimum symmetry elements as that of a simple cubic unit cell (four three-fold rotations axes) and thus it belongs to the cubic crystal system. This cell also satisfies the basic requirement of identical environments and is known as the *face-centred cubic system.* Such a type of system is represented by the symbol F.

One can have another unit cell in which a lattice point is at the centre besides being at the corners of the cube. This also has the minimum symmetry elements (four three-fold rotation axes), satisfies the condition of identical environments and hence belongs to the cubic system known as the *body-centred cubic system.* Such a system is represented by the symbol I.

In each of the above cubic lattices, the number of nearest neighbours to any lattice point is different from that of the others. In the primitive space lattice, a given point is surrounded by six nearest neighbours; in a body-centred cubic lattice it is eight and in face-centred, it is twelve.

Tetragonal System

In a tetragonal system, the simplest lattice can be generated by the translation of a primitive unit cell as shown in Fig. 3.5.1. We can now consider the possibility of having body-centred and face-centred tetragonal lattices. By carrying out the appropriate symmetry operation, it is easily shown that the body-centred unit cell is compatible with the symmetry requirements of the tetragonal system. Thus, it results in a different lattice from the primitive one. Coming to the face-centred lattice points, at the very outset, it may be noticed that, unlike the cubic system, all the six faces of a tetragonal are not identical. In fact, we have two different types of faces. Thus, we can have lattice points in one or two or all the three pairs of opposite faces of the tetragonal system.

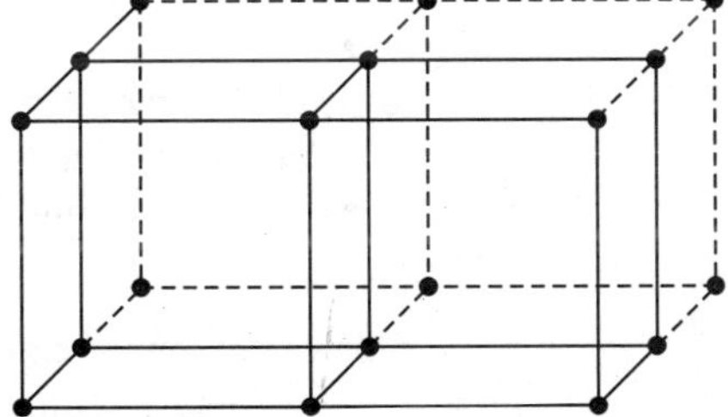

Fig. 3.5.1 Lattice representing a tetragonal unit cell

Consider a tetragonal lattice in which only the end pair of faces is centred (Fig. 3.5.2). However, by rotating the x and y axes by 45°, a new unit cell can be drawn that is primitive, but smaller than the parent primitive tetragonal cell, as shown by heavy lines in Fig. 3.5.2. Hence, it must be concluded that this lattice can still be described by the repetition of a primitive unit cell.

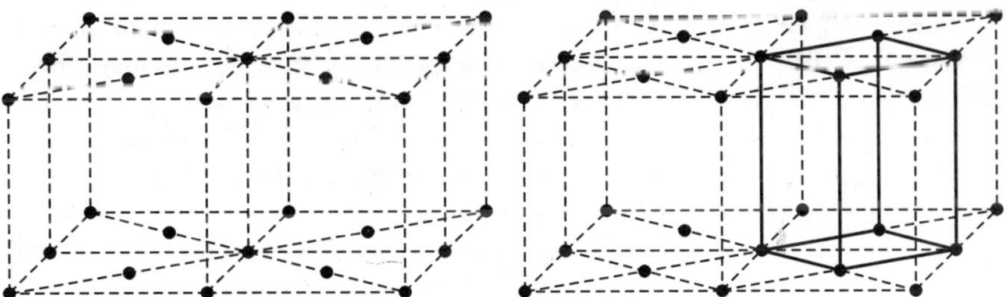

Fig. 3.5.2 Identity of end-paired tetragonal lattice with the primitive lattice

If, instead, lattice points are placed in the centres of the other four faces, the unit cell as shown in Fig. 3.5.3 is obtained. This arrangement is compatible with tetragonal symmetry, but it can be seen by following solid lines that the points labelled as 1 and 2 do not have the same environment. Thus, the basic requirement of a space lattice that all the lattice points have the same environment is not fulfilled. Hence, this arrangement does not give a new type of tetragonal lattice.

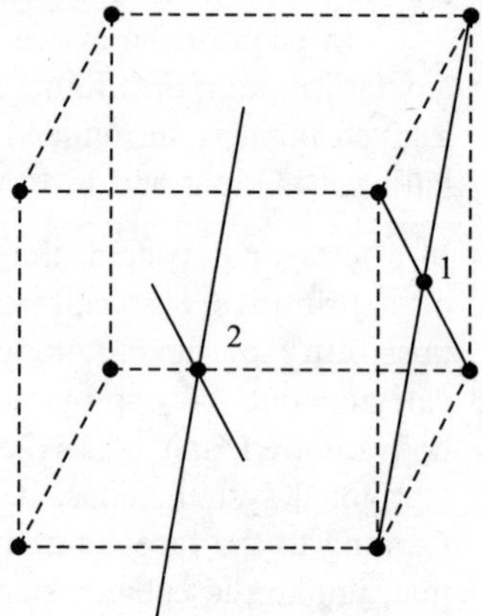

Fig. 3.5.3 Impossibility of the face-centred tetragonal lattice

Finally, we can place lattice points in the centres of all six faces. If this is done and the resultant lattice is carefully studied, it can be shown that the arrangement can be described in terms of a body-centred unit cell. So, again, this does not lead to a new type of tetragonal lattice.

Thus it can be concluded that in a tetragonal crystal system, only two distinct lattices exist: (i) primitive space lattice, and (ii) body-centred space lattice.

Fourteen Bravais Lattices

The same type of reasoning can be extended to the other crystal systems. This was done first by Auguste Bravais in 1848. It can be shown that only *fourteen* independent lattices distributed among the seven crystal systems in which all the points have identical environments can exist. These are known as the *fourteen Bravais lattices.* In contrast to the crystal system, the Bravais lattice type cannot be recognized from the external form of the crystal.

The fourteen Bravais lattices are shown in Fig. 3.5.4. The symbol C is used for the space lattice when one set of the faces in the unit cell is face-centred.

Positions of Lattice Points in Cubic System

Throughout this chapter, the cubic system will be explored in some detail because of its simplicity amongst the fourteen Bravais lattices. In this system, we have three independent Bravais lattices. These are primitive, body-centred and face-centred. All of them have the symmetry of a cube (four three-fold rotation axes). These three cannot be distinguished by macroscopic examination since their lattices are based upon microscopic translations.

A primitive lattice contains one point per unit cell. The position of this point alone can generate the whole of the lattice points since the latter are

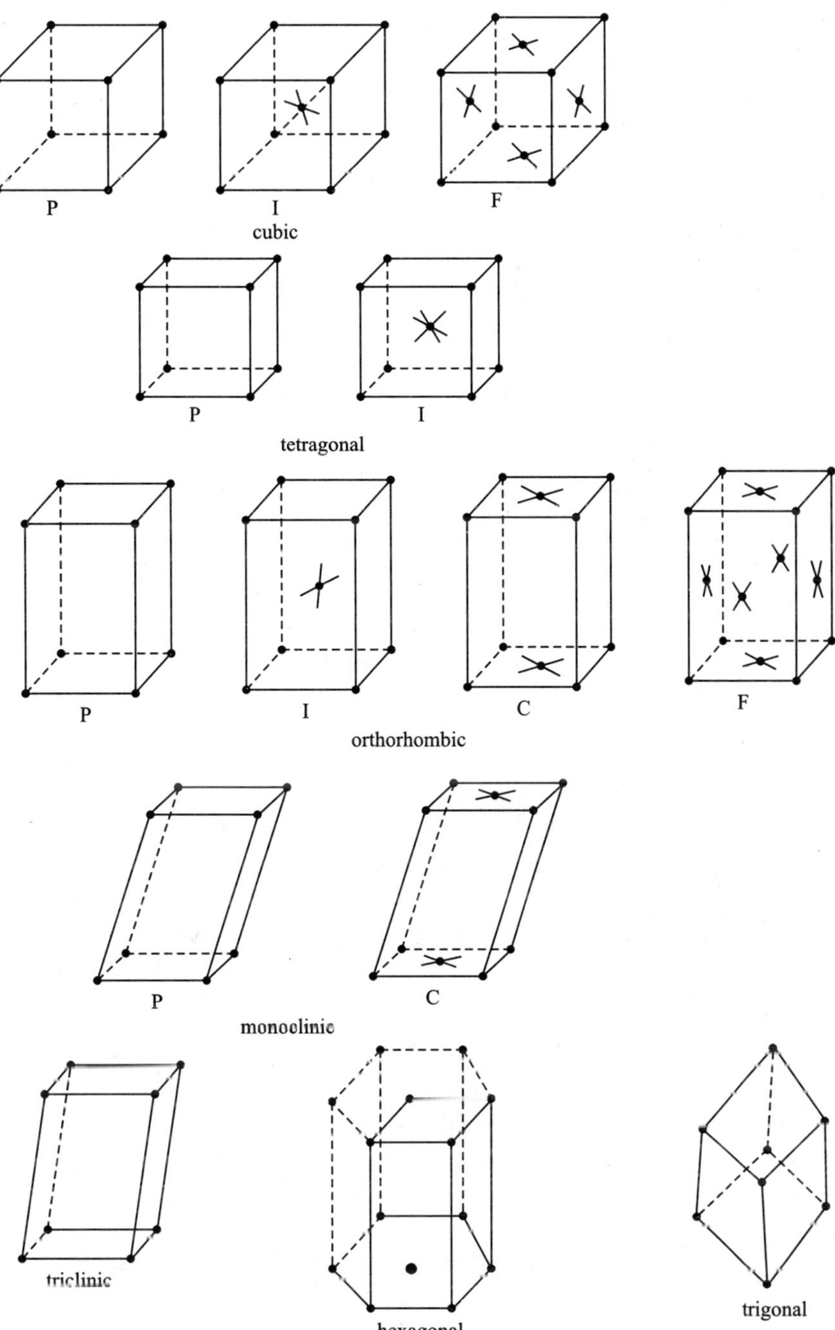

Fig. 3.5.4 Fourteen Bravais lattices

related by translational symmetry. For example, if we assign the coordinates (x, y, z) to one point, then the other points will be at $(x + 1, y, z)$, $(x + 2, y, z)$, ..., $(x, y + 1, z)$, $(x, y + 2, z)$, ..., $(x, y, z + 1)$, $(x, y, z + 2)$, ..., and so on. By convention, the coordinates (x, y, z) are assigned as (0, 0, 0).

A body-centred lattice has two points per unit cell, namely, (x, y, z); $\left(x+\frac{1}{2}, y+\frac{1}{2}, z+\frac{1}{2}\right)$. These will be related to other points by translational symmetry. If we have two lattice points at (0, 0, 0) and $\left(\frac{1}{2}, \frac{1}{2} \frac{1}{2}\right)$, then the other points will be at (1, 0, 0), (2, 0, 0), ..., (0, 1, 0), (0, 2, 0), ..., (0, 0, 1), (0, 0, 2), ..., and $\left(\frac{3}{2}, \frac{1}{2}, \frac{1}{2}\right), \left(\frac{5}{2}, \frac{1}{2}, \frac{1}{2}\right), \ldots, \left(\frac{1}{2}, \frac{3}{2}, \frac{1}{2}\right), \left(\frac{1}{2}, \frac{5}{2}, \frac{1}{2}\right), \ldots, \left(\frac{1}{2}, \frac{1}{2}, \frac{3}{2}\right), \left(\frac{1}{2}, \frac{1}{2}, \frac{5}{2}\right), \ldots$.

A face-centred lattice has four points per unit cell at (x, y, z), $\left(x+\frac{1}{2}, y+\frac{1}{2}, z\right), \left(x, y+\frac{1}{2}, z+\frac{1}{2}\right)$ and $\left(x+\frac{1}{2}, y, z+\frac{1}{2}\right)$. By convention, these are assigned as (0,0,0), $\left(\frac{1}{2}, \frac{1}{2}, 0\right), \left(0, \frac{1}{2}, \frac{1}{2}\right)$ and $\left(\frac{1}{2}, 0, \frac{1}{2}\right)$. The other points of the lattice can be generated by translational symmetry of these four points, i.e. by increasing the coordinates of each by unity each time.

3.6 POINT GROUPS

Point Symmetry

In studying the symmetry of a molecule, we are concerned with the point symmetry. A point is assigned in the molecule and the symmetry with respect to lines and planes passing through this point is considered. Thus when the symmetry operation is carried out, at least one point does not move from its position. It is not necessary that there should be a particle or an atom at this point. In crystals, point symmetry is also found to be of value. But we must realize that in a crystal, symmetry is not restricted to a point. Rather it extends in space throughout the crystal. Consequently, the point symmetry permitted in a crystal is restricted so as to satisfy the requirements of translational symmetry.

Combination of Symmetry Elements

The elements that are generally used to describe point symmetry in a crystal are (i) proper rotation axis, (ii) mirror plane, and (iii) rotation-inversion axis (*see* Annexure I at the end of this chapter). These are not the only choices that can be made. However, all of the point symmetries of a crystal can be expressed in terms of these three elements. For example, a centre of symmetry which exists, in many geometrical structures is not included here. However, it can be shown that a centre of symmetry can be represented by a rotation-inversion axis.

These symmetry elements may be combined in the following ways:

Combinations	*Symbol used**
Rotation axis only	X
Rotation-inversion axis only	$\bar{X}$
Rotation axis normal to a plane of symmetry	$\left(\frac{X}{m}\right)$
Rotation axis with a vertical plane of symmetry	Xm
Rotation-inversion axis with a vertical plane of symmetry	$\bar{X}m$
Rotation axis with a diad axis normal to it	$X2$
Rotation axis with a normal plane and one or more vertical planes	$\left(\frac{X}{m}m\right)$

**X*—Order of principal axis, highest rotation order; *m*—plane of symmetry.

Classification of Crystals

When the seven crystal systems are studied from the viewpoint of the above symmetry combinations, it is found that these can be classified into 32 different groups, known as 32 crystallographic point groups (point group because after carrying out the operations, at least one point does not move). These point groups are distributed among the seven crystal systems as follows:

	Triclinic	*Monoclinic and Orthorhombic*	*Trigonal*	*Tetragonal*	*Hexagonal*	*Cubic*
X	1	2	3	4	6	23
$\bar{X}$	$\bar{1}$	$\bar{2} = m$	$\bar{3}$	$\bar{4}$	$\bar{6}$	$\boxed{\bar{2}\bar{3} = \frac{2}{m}3}$
$\frac{X}{m}$	$\boxed{\frac{1}{m} = \bar{2}}$	$\frac{2}{m}$	$\boxed{\frac{3}{m} = \bar{6}}$	$\frac{4}{m}$	$\frac{6}{m}$	$\frac{2}{m}3 = 3m$
Xm	$\boxed{1m = \bar{2}}$	$2m = mm$	$3m$	$4m = 4mm$	$6m = 6mm$	$\boxed{2m3 - \bar{4}3m}$
$\bar{X}m$	$\boxed{\bar{1}m = \frac{2}{m}}$	$\boxed{\bar{2}m = 2m}$	$\bar{3}m$	$\bar{4}m = \bar{4}2m$	$\bar{6}m = \bar{6}2m$	$\bar{4}3m$
$X2$	$\boxed{12 = 2}$	$22 = 222$	32	42	62	$432 = 43$
$\frac{X}{m}m$	$\boxed{\frac{1}{m}m = 2m}$	$\frac{2}{m}m = mmm$	$\boxed{\frac{3}{m}m = \bar{6}m}$	$\frac{4}{m}m = \frac{4}{m}mm$	$\frac{6}{m}m = \frac{6}{m}mm$	$\frac{4}{m}3m = m3m$

Note: Combinations shown in square borders are the repeated ones.

The vast majority of molecules possess one of the 32 crystallographically allowed point groups.

3.7 SPACE GROUPS

Screw Axis and Glide Plane

In classifying the crystals into different point groups, we consider the operation of symmetry elements about a chosen point, not necessarily a lattice point. However, due to the spatial nature of a crystal lattice, we must also consider the symmetry operations involving translation. Let a given crystal contain a diad axis (Fig. 3.7.1a). There is another way in which the same external symmetry can occur, i.e. a rotation through 180° followed by a translation parallel to the axis as shown in Fig. 3.7.1b. This symmetry operation is called *screw axis*.

Another type of symmetry operation involving translation is glide plane. This is shown in Fig. 3.7.1c. It involves reflection across a mirror plane coupled with a translation parallel to the reflection plane.

Classification of Crystals

The combination of these two translation symmetries (namely, screw axis and glide plane) with point group symmetry elements gives 230 different groups. These are known as *space groups*. Every crystal belongs to one of these space

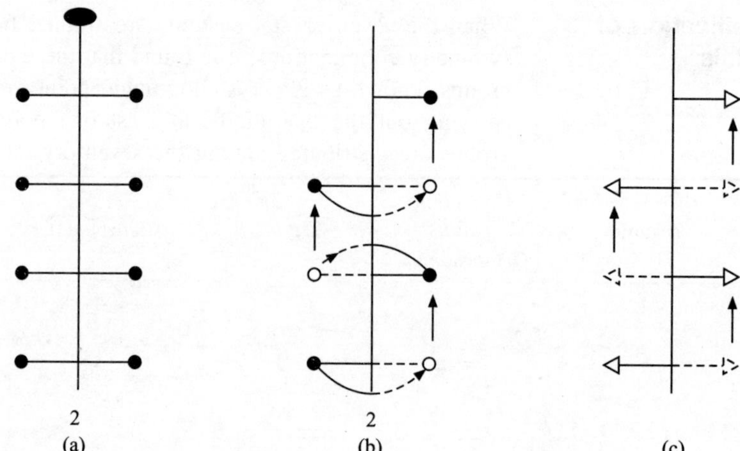

Fig. 3.7.1 (a) Diad axis, (b) Screw axis, and (c) Glide plane

groups which completely specifies it symmetry. Every space group is isomorphic with a point group and morphological examination of the external symmetry of a crystal will yield only the point group. This means that screw axes give the same external symmetry as simple rotation axes and glide planes give the same external symmetry as mirror planes. The actual determination of space groups of crystals can be done with the help of diffraction techniques.

The number of point groups and space groups associated with different seven crystal systems are shown in Table 3.7.1.

Table 3.7.1 Distribution of Bravais Lattices, Point Groups and Space Groups Over the Seven Crystal Systems

Crystal system	*No. of point group*	*No. of space groups*	*No. and nature of groups*	*Bravais lattice*
Triclinic	2	2	1	P
Monoclinic	3	13	2	P, C
Orthorhombic	3	59	4	P, C, I, F
Tetragonal	7	68	2	P, I
Hexagonal	7	27	1	P
Trigonal	5	25	1	P
Cubic	5	36	3	P, I, F
Total	32	230	14	

3.8 CRYSTAL CLEAVAGE AND DEVELOPMENT OF ITS FACES

The development of faces in a crystal can be correlated with the planes that pass through lattice points as shown in Fig. 3.8.1. Theoretically, an infinite number of planes might be drawn through the various lattice points. Any of these planes could represent a crystal face. Inspite of this, only a few faces are actually observed to develop. Qualitatively, this can be explained on the basis of interaction energy involved in atoms, ions or molecules of a crystal. A regular pattern of these units in a crystal results in the maximum interaction and thus minimum energy. Thus, *a crystal face with a large surface density would result into more interaction vis-á-vis minimum energy and hence would show a greater probability to form a crystal face.* It can be seen from Fig. 3.8.1 that the number of lattice sites falling on various planes may differ greatly. The planes parallel to the Cartesian axes have the highest density of points and the density decreases in the order of planes from a to e. The probability of forming a crystal face also decreases in the same order. Thus, we can imagine the development of crystal faces from a statistical point of view, and find that only a few faces with the maximum site density show a high probability of development.

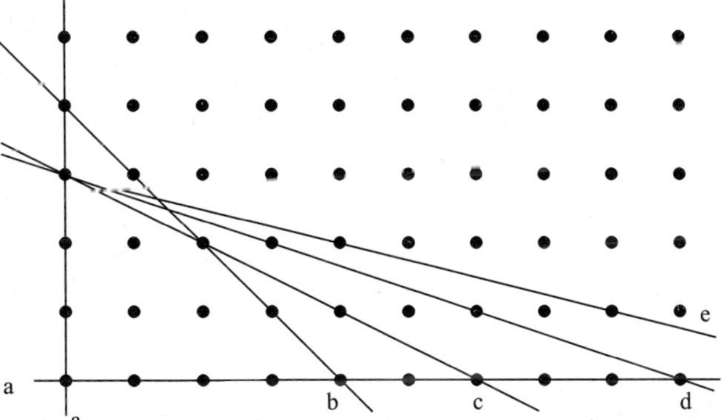

Fig. 3.8.1 Development of faces in a crystal

It can be seen from Fig. 3.8.2 that the planes with the highest site density are also planes with the largest distance of separation. On account of this separation, interatomic interaction between the planes would be minimum. It will thus be expected that the natural fracture will occur parallel to these faces. It also follows that the cleavage planes would correspond to the normally developed crystal faces.

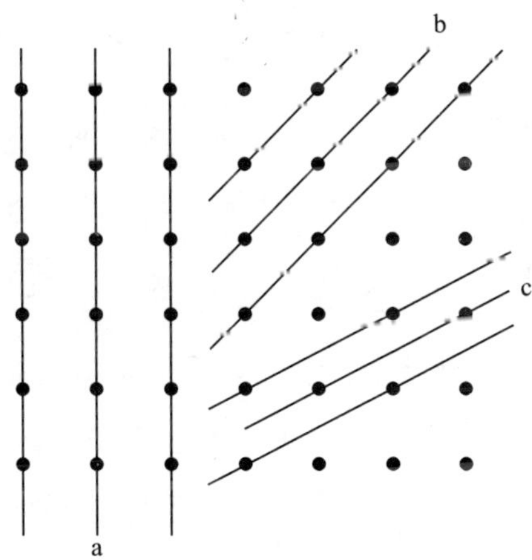

Fig. 3.8.2 Site densities in different planes and their interplanar distances

3.9 DESIGNATION OF PLANES

Weiss Indices

In order to discuss the structure of a crystal, we need to describe the orientation of planes passing through lattice points of the crystal. The orientation of a lattice plane can be described by considering the intercepts of the plane on the three basis vectors of the lattice. According to Haüy, it is possible to choose the unit lengths a, b and c along the three basis vectors such that the ratio of each of the intercepts h', k' and l' to the corresponding unit length is either an integer or a ratio of two integers. This statement is known as the *law of rational indices.* The three ratios are known as the *Weiss indices* of the plane.

Miller Indices

Instead of using Weiss indices, it is more advantageous to use the *Miller indices.* In the latter, we take the reciprocal of the three Weiss indices and then multiply them by the smallest number (if necessary) to make them all integers. The three resultant integers are known as Miller indices and are represented as (hkl). For example, if the intercept ratios are $h'/a = 2$, $k'/b = 1/3$ and $l'/c = 1/2$, the Miller indices of this plane are (164). If a plane is parallel to one of the basis vectors, its intercept on this vector is at infinity and the corresponding Miller index is 0.

Illustrations

The use of Miller indices to specify the orientation of planes parallel to z-axis as shown in Fig. 3.9.1 is illustrated in the following.

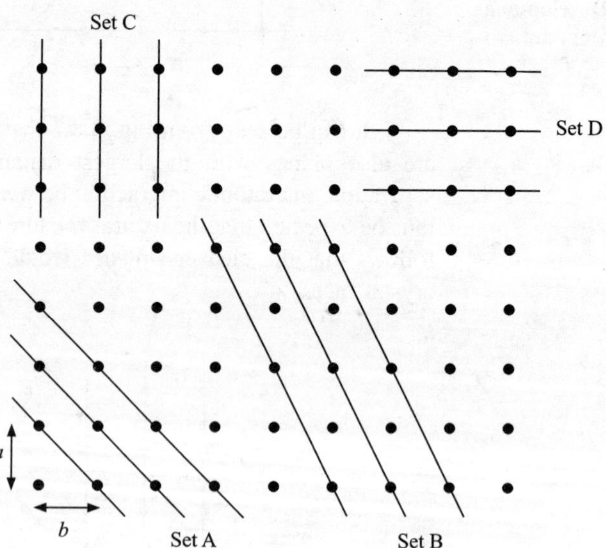

Fig. 3.9.1 Miller indices of various sets of planes parallel to z-axis (only their projections on the x-and y-axes are shown)

Set A The three planes shown in Fig. 3.9.1 have intercepts

(i) a, b, ∞ (ii) $2a$, $2b$, ∞ (iii) $3a$, $3b$, ∞

The intercepts are simple multiples of unit lengths a, b, c of the basis vectors along the crystallographic axes which determine the crystal system.

Miller indices (*hkl*) of such planes are

(i) $\frac{a}{a}, \frac{b}{b}, \frac{c}{\infty},$ i.e. $1, 1, 0 \Rightarrow (110)$

(ii) $\frac{a}{2a}, \frac{b}{2b}, \frac{c}{\infty},$ i.e. $\frac{1}{2}, \frac{1}{2}, 0 \Rightarrow (110)$

(iii) $\frac{a}{3a}, \frac{b}{3b}, \frac{c}{\infty},$ i.e. $\frac{1}{3}, \frac{1}{3}, 0 \Rightarrow (110)$

The Miller indices are therefore the same. *All planes parallel to these planes will have the same Miller indices of* (110). Thus, the *Miller indices actually define a set of parallel planes, one of which passes through the origin, while the others are at a constant distance from each other.*

Set B Intercepts are

(i) $2a, b, \infty$ (ii) $4a, 2b, \infty$ (iii) $6a, 3b, \infty$

and Miller indices (*hkl*) are

(i) $\frac{a}{2a}, \frac{b}{b}, \frac{c}{\infty},$ i.e. $\frac{1}{2}, 1, 0 \Rightarrow (120)$

(ii) $\frac{a}{4a}, \frac{b}{2b}, \frac{c}{\infty},$ i.e. $\frac{1}{4}, \frac{1}{2}, 0 \Rightarrow (120)$

(iii) $\frac{a}{6a}, \frac{b}{3b}, \frac{c}{\infty},$ i.e. $\frac{1}{6}, \frac{1}{3}, 0 \Rightarrow (120)$

Set C Intercepts are

(i) ∞, b, ∞ (ii) $\infty, 2b, \infty$

and Miller indices are (010).

Set D Intercepts are

(i) a, ∞, ∞ (ii) $2a, \infty, \infty$

and Miller indices are (100).

Let us now illustrate the method to calculate the Miller indices of a plane which intercepts all the three crystallographic axes, as shown in Fig. 3.9.2.

The Intercepts of the shown plane with the three axes are

$$h' = 2a; \quad k' = 4b; \quad l' = 3c$$

The intercept ratios are

$$\frac{h'}{a} = 2; \quad \frac{k'}{b} = 4; \quad \frac{l'}{c} = 3$$

Taking the reciprocal of these, we get

$$\frac{a}{h'} = \frac{1}{2}; \quad \frac{b}{k'} = \frac{1}{4}; \quad \frac{c}{l'} = \frac{1}{3}$$

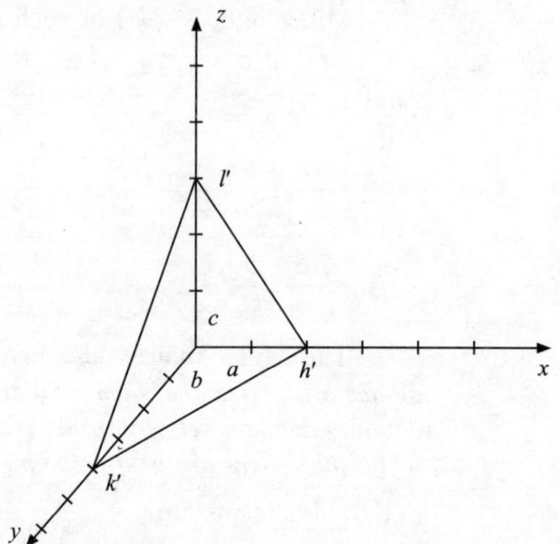

Fig. 3.9.2 Miller indices of a plane intersecting all the three crystallographic axes

Multiplication by the minimum common factor of 12 gives the Miller indices as (634).

It can be seen here that larger the value of Miller index, the smaller is the intercept of that plane on that axis. Thus, (222) plane will have an intercept that is half of the (111) plane.

Lattice Planes in a Cubic System

Some of the lattice planes in the cubic crystals are shown in Fig. 3.9.3.

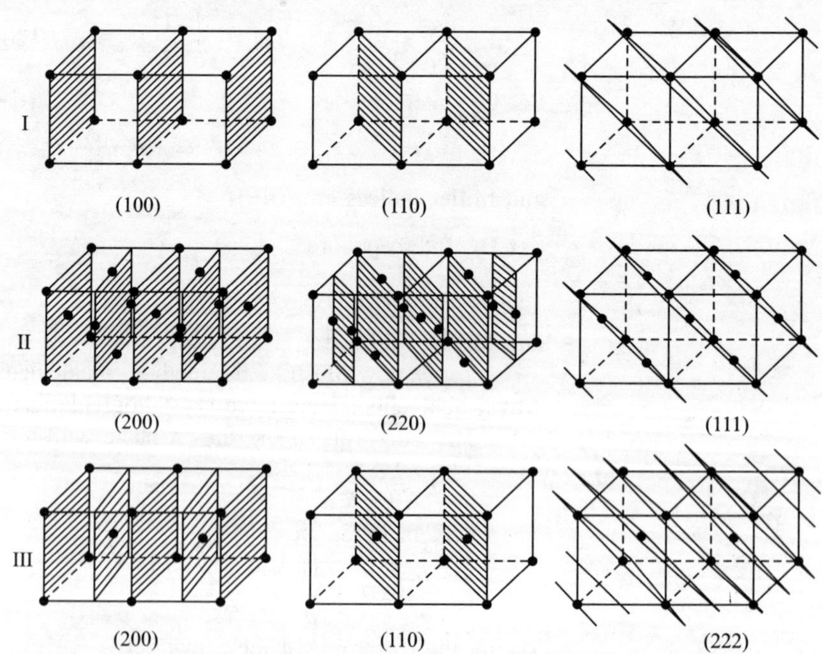

Fig. 3.9.3 Some of the lattice planes of cubic lattice: (I) Primitive cubic lattice, (II) Face-centred cubic lattice, (III) Body-centred cubic lattice

The distances between planes of a crystal can be determined with the help of X-ray diffraction measurements. These can, in turn, be used to determine the system to which the crystal belongs.

3.10 INTERPLANAR DISTANCES FOR CUBIC SYSTEMS

A Few Typical Examples

(1) (100) *planes* Distance between these planes is equal to the length a of the side of a cube, i.e. $d_{100} = a$.

(2) (200) *planes* Distance between these planes is equal to the half length $a/2$ of the side of a cube, i.e. $d_{200} = a/2$.

(3) (110) *planes* Spacing between these planes is one-half of the diagonal of the square base of the cube, i.e. $d_{100} = (\sqrt{a^2 + a^2})/2 = a/\sqrt{2}$

(4) (111) *planes* The entire cross diagonal d of a cube spans three (111) planes. Thus, the distance between the two of each of these planes is $d/3$. Now,

$$d = \sqrt{a^2 + a^2 + a^2} = \sqrt{3}a$$

$$\text{Therefore } d_{111} = \frac{\sqrt{3}a}{3} = \frac{a}{\sqrt{3}}$$

(5) (222) *planes* These planes are in between (111) planes. Thus, the distance between any two such planes is

$$d_{222} = \frac{d_{111}}{2} = \frac{a}{2\sqrt{3}}$$

General Formula

General formula for the interplanar spacing in the cubic system (orthogonal axes) is

$$d_{hkl} = \frac{a}{\sqrt{h^2 + k^2 + l^2}} \qquad (3.10.1)$$

where hkl are Miller indices of the planes and a is edge length of the cube.

Utility of Interplanar Distances

It is obvious that if one had a method for determining d_{hkl} experimentally, the constancy of the quantity $d_{khl}\sqrt{h^2 + k^2 + l^2}$ will not only determine the value of a but will also indicate to what cubic system the given crystal belongs.

The ratios of interplanar distances of different faces in the three cubic lattices (Fig. 3.9.3) are:

Simple cubic lattice $\quad d_{100}: d_{110}: d_{111} = 1 : \dfrac{1}{\sqrt{2}} : \dfrac{1}{\sqrt{3}}$, i.e. $1 : 0.707 : 0.577$

Face-centred cubic lattice $\quad d_{200}: d_{220}: d_{111} = \dfrac{1}{2} : \dfrac{1}{2\sqrt{2}} : \dfrac{1}{\sqrt{3}}$, i.e. $1 : 0.707 : 1.154$

Body-centred cubic lattice $\quad d_{200}: d_{110}: d_{222} = \dfrac{1}{2} : \dfrac{1}{\sqrt{2}} : \dfrac{1}{2\sqrt{3}}$, i.e. $1 : 1.414 : 0.577$

Geometrical Method

The general formula for the interplanar spacing in a system of orthogonal axes can be derived geometrically.† As stated earlier, the Miller indices define the set of parallel planes, one of which passes through the origin. Thus, if a perpendicular

† See Annexure II at the end of this chapter for the more general expression of interplanar distances.

is drawn from the origin to the nearest plane of Miller indices (*hkl*), then this perpendicular distance, represented as d_{hkl}, will be equal to the interplanar spacing between planes of Miller indices (*hkl*) as shown by OD in Fig. 3.10.1.

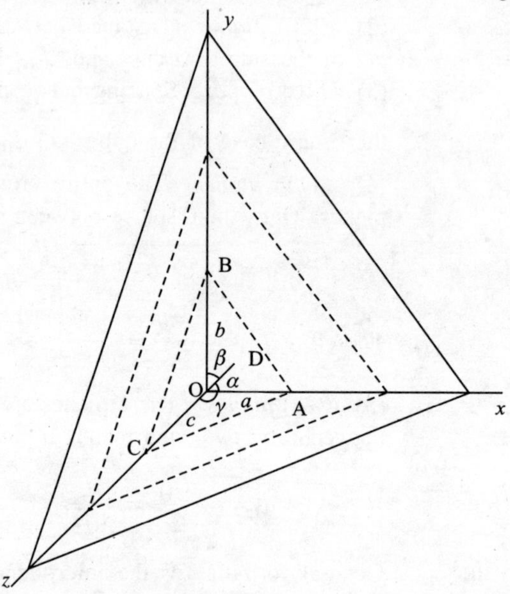

Fig. 3.10.1 Interplanar distance of the planes

If α, β, and γ are the angles which this perpendicular makes with the three axes, then

$$d_{hkl} = \text{OA} \cos \alpha = \left(\frac{a}{h}\right) \cos \alpha, \quad d_{hkl} = \text{OB} \cos \beta = \left(\frac{b}{k}\right) \cos \beta$$

and $$d_{hkl} = \text{OC} \cos \gamma = \left(\frac{a}{l}\right) \cos \gamma$$

or $$\cos \alpha = \left(\frac{h}{a}\right) d_{hkl}, \quad \cos \beta = \left(\frac{k}{b}\right) d_{hkl}, \quad \text{and} \quad \cos \gamma = \left(\frac{l}{c}\right) d_{hkl}$$

Now since $\cos \alpha$, $\cos \beta$ and $\cos \gamma$ are the direction cosines of the perpendicular line, therefore

$$\cos^2 \alpha + \cos^2 \beta + \cos^2 \gamma = 1$$

Substituting their values in terms of d_{hkl}, we have

$$\left(\frac{h}{a}\right)^2 d_{hkl}^2 + \left(\frac{k}{b}\right)^2 d_{hkl}^2 + \left(\frac{l}{c}\right)^2 d_{hkl}^2 = 1$$

or $$\frac{1}{d_{hkl}^2} = \left(\frac{h}{a}\right)^2 + \left(\frac{k}{b}\right)^2 + \left(\frac{l}{c}\right)^2$$

For a cubic system $a = b = c$, we have

$$\frac{1}{d_{hkl}^2} = \frac{h^2 + k^2 + l^2}{a^2} \quad \text{or} \quad d_{hkl} = \frac{a}{\sqrt{h^2 + k^2 + l^2}}$$

For a tetragonal system, $a = b \neq c$, we have

$$\frac{1}{d_{hkl}^2} = \frac{h^2 + k^2}{a^2} + \frac{l^2}{c^2}$$

For an orthorhombic system $a \neq b \neq c$, we have

$$\frac{1}{d_{hkl}^2} = \frac{h^2}{a^2} + \frac{k^2}{b^2} + \frac{l^2}{c^2}$$

3.11 DIFFRACTION OF ELECTROMAGNETIC RADIATION

Characteristics of Waves

Electromagnetic radiation consists of waves that are propagated through space with the velocity of light. It can be characterized by either frequency or wavelength. The former gives the number of cycles through which the wave moves in one second and the latter gives the distance between two points on the wave which have gone through a complete cycle. These two, i.e. frequency and wavelength, are related by the expression

$$\nu = \frac{c}{\lambda}$$

Associated with the wave is the amplitude which is the displacement of the wave in a direction perpendicular to the direction of its propagation. The wave nature can be represented by either a sine wave or a cosine wave as shown in Fig. 3.11.1.

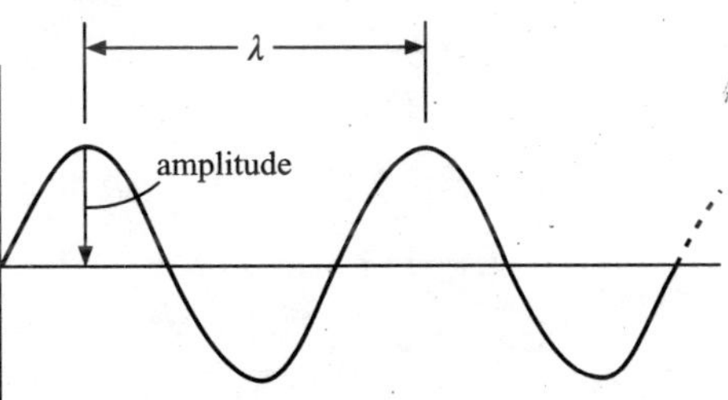

Fig. 3.11.1 Wave propagation

Constructive and Destructive Interferences

Two waves coming from two different sources having the same wavelength and same amplitude reinforce each other when their maxima and minima coincide. On the other hand, they will interfere and exactly cancel each other when their maxima and minima do not coincide. The former is known as *constructive interference* and the latter as *destructive interference.* The amplitude of the resultant wave at any point is obtained by the algebraic sum of the amplitudes of the two individual waves at the point.

This interference phenomenon forms the basis of the diffraction of light by a diffraction grating. A typical diffraction grating consists of a transparent medium (such as glass) on which are ruled a large number of very fine, equidistant, parallel, opaque lines. When light from a monochromatic source is incident perpendicular on the grating, all the clear spaces will act as secondary sources of light and, therefore, will emit light waves in all directions radially outward.

The wavelengths and frequencies of the diffracted light waves are the same as that of the incident light waves. The waves from two apertures will cross at some point beyond the grating. If a screen is placed at this point, a series of bright and dark spots will be observed on the screen. A bright spot is obtained where the waves from adjacent apertures in the grating reinforce one another. A dark spot arises from the destructive interference of waves at the point. Figure 3.11.2 shows the interference of waves from two adjacent apertures of a grating.

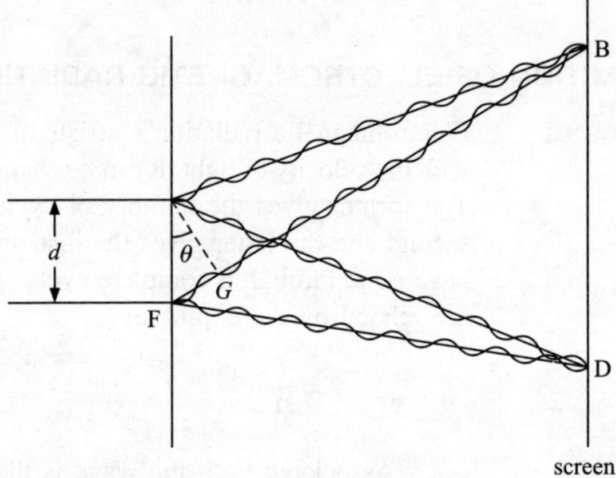

Fig. 3.11.2 Diffraction of light by a diffraction grating (waves from only two adjacent apertures are shown). B—Bright spot. D—Dark spot

Condition for Bright Spot

A bright spot is observed when both the waves are in phase. The essential condition for this to be observed is that the extra distance travelled by one of the waves is an integral multiple of wavelength.

Thus $\quad FG = n\lambda \qquad n = 0, 1, 2, ...$

From Fig. 3.11.2, it follows that

$$FG = d \sin \theta$$

Equating these two, we get

$$n\lambda = d \sin \theta \qquad n = 0, 1, 2, ...$$

Example 3.11.1

A grating with 4,000 lines cm^{-1} is illuminated with Hg green line having a wavelength equal to 546 nm. At what angle will the first and second order diffraction maxima occur?

Solution

The grating spacing, $d = \dfrac{1}{4000\ cm^{-1}} = 2.5 \times 10^{-4} cm = 2.5 \times 10^{-6}\ m$

Therefore, angle of first order diffraction maxima is

$$\sin \theta_1 = \frac{n\lambda}{d} = \frac{(1)(546 \times 10^{-9}\ m)}{2.5 \times 10^{-6}\ m} = 0.218\,4$$

or $\quad \theta_1 = 12.62°$

Angle of second order diffraction maxima is

$$\sin \theta_2 = \frac{(2)(546 \times 10^{-9}\ \text{m})}{2.5 \times 10^{-6}\ \text{m}} = 0.436\,8$$

or $$\theta_2 = 25.9°$$

Example 3.11.2 Calculate the angle at which the first order maxima occurs if an electromagnetic wave of 10 nm is exposed to the above grating.

Solution The angle at which first order maxima is observed is given by

$$\sin \theta = \frac{(1)(10 \times 10^{-9}\ \text{m})}{2.5 \times 10^{-6}\ \text{m}} = 4.0 \times 10^{-3}$$

or $$\theta = 0.23°$$

Comment The calculated angle of first order maxima in Example 3.11.2 is too close to the intense beam which occurs at $\theta = 0$ and thus cannot be determined accurately. In order to obtain measurable separation of maxima, it is essential that the spacing between lines of grating should be approximately equal to the wavelength of the employed electromagnetic radiation.

3.12 DIFFRACTION OF X-RAYS BY CRYSTALS

Laüe Method In 1912, Max von Laüe predicted that since the distances between particles in a crystal are of the same order of magnitude ($\simeq 10^{-8}$ cm) as the wave length of X-rays, the former could be used as a *3-dimensional diffraction grating* and thus if a beam of non-homogeneous X-rays were passed through a crystal, a diffraction pattern would be observed. The experiments carried out on various substances verified Laüe's prediction. The diffraction pattern can be recorded by placing a photographic plate behind the crystal as shown in Fig. 3.12.1. On developing the film, one observes a series of spots arranged in some symmetrical way around the intense central undiffracted beam. The arrangement of these spots (known as Laüe spots) is highly characteristic of the structural arrangement of the crystal. From the position of Laüe spots, it is possible to calculate the size and shape of the unit cell. However, the actual analysis of the Laüe diffraction pattern is highly complicated and difficult.

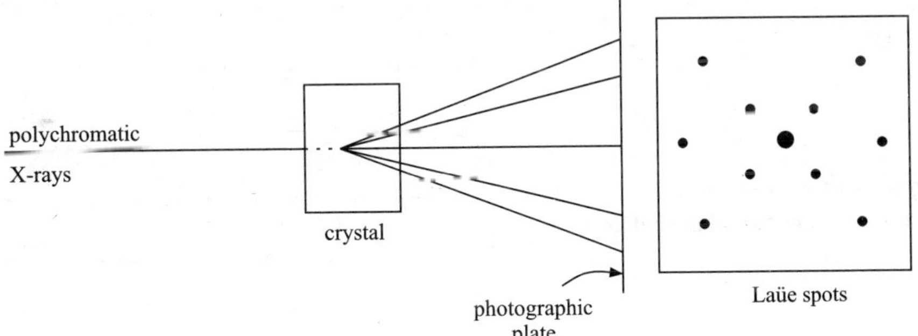

Fig. 3.12.1 Laüe diffraction pattern

3.13 BRAGG'S EQUATION

Crystal as a Reflection Grating

B.H. Bragg and W.L. Bragg were of the opinion that since a crystal is composed of a series of equally spaced atomic planes, it may be employed not only as a transmission grating as in the Laüe method, but also as a *reflection grating*. When X-rays are incident on a crystal face, they penetrate into the crystal and suffer reflections on striking the atoms in successive planes as shown in Fig. 3.13.1. If the reflected waves from successive layers are out of phase, then due to the destructive interference, no diffraction will be observed. If, however, the reflected waves are in phase, then due to the constructive interference, a diffraction spot will be observed. The condition for a reflection to give constructive interference can be derived from Fig. 3.13.1 as follows.

Derivation of Bragg's Equation

Let θ be the incident angle of monochromatic X-rays of wavelength λ with the parallel equidistant planes of atoms with interplanar spacing equal to d. The waves are in phase before striking the planes. Two such waves labelled as wave 1 and wave 2 are shown in Fig. 3.13.1. After the reflection, the two waves will be in phase provided the extra distance travelled by wave 2 is an integral multiple of wavelength λ. This extra distance can be obtained by dropping perpendiculars BG and BH from B on to wave 2. It is obvious from Fig. 3.13.1 that

$$\text{AB} = \text{DG}; \qquad \text{BC} = \text{HF}; \qquad \text{GE} = \text{EH} = d \sin \theta$$

Now the extra distance travelled by wave 2 is given as

$$\begin{aligned} \text{DEF} - \text{ABC} &= (\text{DG} + \text{GE} + \text{EH} + \text{HF}) - (\text{AB} + \text{BC}) \\ &= \text{GE} + \text{EH} \\ &= 2d \sin \theta \end{aligned}$$

Thus, in order to have wave 1 and wave 2 in phase, we must have

$$2d \sin \theta = n\lambda, \qquad n = 1, 2, 3, \ldots \tag{3.13.1}$$

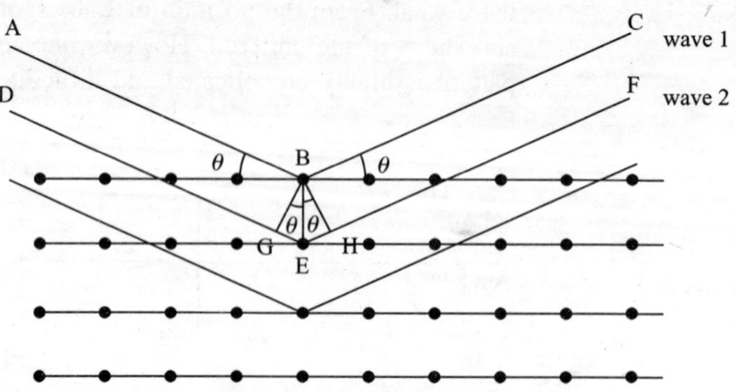

Fig. 3.13.1 Reflection of X-rays from the planes of a crystal

Equation (3.13.1) is known as the Bragg's equation. The constant n gives the order of reflection and is equal to the number of wavelengths in the path difference between waves reflected by adjacent planes. Thus, n can take only integral values subject to the limitation that $\sin\theta$ cannot be greater than one. For fixed values of λ and d, we can have more than one reflection at angles θ_1, θ_2, ... corresponding to the values of n equal to 1, 2, 3,..., respectively. From Eq. (3.13.1), it can be concluded that the higher order reflections will occur at larger values of $\sin\theta$ and hence at larger angles. Experimentally, it is found that the lower order reflections are the most intense and the intensities of higher order reflections decreases rapidly.

Alternative Way of Writing Bragg's Equation

While dealing with X-ray diffraction, it is more convenient to express higher order reflections in terms of the first order reflection from planes of higher Miller indices (hkl). For example, a second order reflection from (111) planes may be considered equivalent to the first order reflection from (222) planes. Similarly a third order reflection from (111) planes may be considered as the first order reflection from (333) planes. This fact can be introduced into the Bragg equation $n\lambda = 2d\sin\theta$ by rewriting it as

$$\lambda = 2\left(\frac{d}{n}\right)\sin\theta = 2\,d_{hkl}\sin\theta$$

where d_{hkl} is the perpendicular distance between adjacent planes having the indices (hkl).

Experimental Set-up

The reflection angles and the intensities of the reflected beams corresponding to these angles can be determined with the help of Bragg X-ray spectrometer, a schematic diagram of which is shown in Fig. 3.13.2. This method consists of the following steps.

(1) The X-rays are generated in tube A by bombarding cathode rays on a suitable target B. The most commonly used target metals and their characteristic wavelengths are *copper* 154.1 pm, *molybdenum* 70.9 pm and *chromium* 229.0 pm.

(2) The generated X-rays are passed through a series of slits and filters (C, D, etc.) in order to get a sharp monochromatic beam. It is then directed to strike the face of a crystal placed on a *graduated turntable* F. The latter may be rotated to any desired angle of incidence.

(3) The reflected beam is passed into an *ionization chamber* G containing SO_2. The reflected X-rays ionize the SO_2 gas in proportion to their intensity. The extent of ionization and hence the intensity of reflection can be determined with the help of an electrometer.

(4) The experiment is repeated at various incident angles of X-rays and the angle at which maximum reflection occurs is determined.

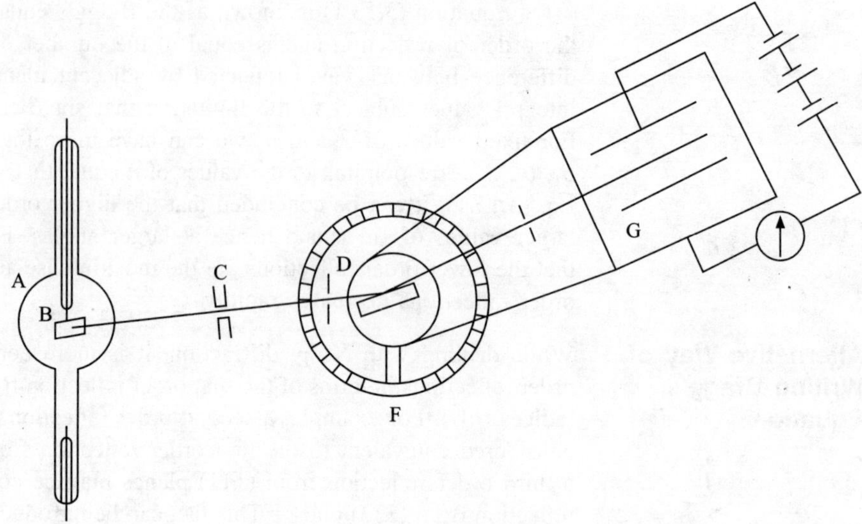

Fig. 3.13.2 Bragg X-ray spectrometer

X-ray Diffraction Pattern

The X-ray diffraction pattern (only schematic) for a crystal of tungsten is shown in Fig. 3.13.3.

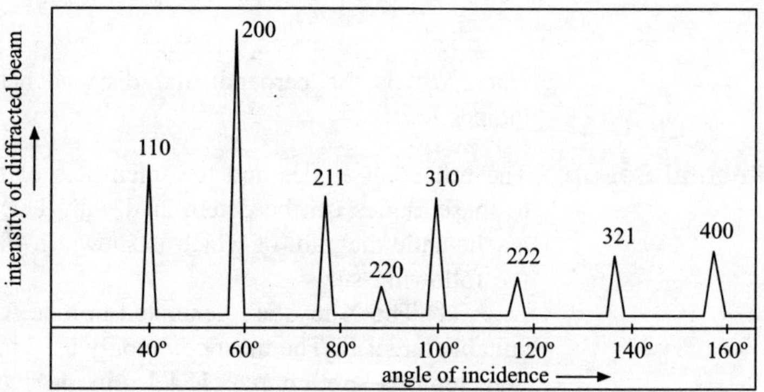

Fig. 3.13.3 X-ray diffraction pattern (only schematic) for a crystal of tungsten

The variation in intensity of the diffracted beam for different sets of planes is due to the variation in density of atoms in these planes. The planes of high atomic density produce better scattering of X-rays which gives a more intense beam.

Scattering Ability of an Atom

If more than one kind of atoms are present in the crystal, the atom with the greater number of electrons has the greater scattering power. For light elements, the scattering power is directly proportional to the number of electrons around the atom. It is because of this reason that the scattering ability of hydrogen atom is very small and thus whatever effect it produces is overshadowed by the effect of neighbouring atoms which contain larger number of electrons. Thus, the positions of hydrogen atoms cannot be deduced from X-ray diffraction. However, their positions can be determined by neutron diffraction since the latter involves the interaction with the nuclei and this interaction is about the same with all nuclei. Similarly, from X-ray diffraction, it is not possible to distinguish between atoms which differ only by the possession of additional electron. Carbon and nitrogen, for example, are indistinguishable because their scattering ability is about the same.

Example 3.13.1

When a certain crystal was studied by the Bragg technique using X-rays of wavelength 229 pm, an X-ray reflection was observed at an angle of 23° 20′. (a) What is the corresponding interplanar spacing? (b) When another X-ray source was used, a reflection was observed at 15° 26′. What was the wavelength of these X-rays?

Solution

(a) Given that

$$\lambda = 229 \text{ pm} \quad \text{and} \quad \theta = 23^\circ 20'$$

Substituting these in the Bragg equation, we have

$$d_{hkl} = \frac{\lambda}{2 \sin \theta} = \frac{229 \text{ pm}}{2 \times \sin (23^\circ 20')} = \frac{229 \text{ pm}}{2 \times 0.396}$$

$$= 289.2 \text{ pm}$$

(b) Now $\theta = 15^\circ 26'$, thus

$$\lambda = 2d_{hkl} \sin \theta = 2(289.2 \text{ pm}) \sin (15^\circ 26')$$

$$= 2(289.2 \text{ pm})(0.266\,2) = 153.9 \text{ pm}$$

3.14 POWDER METHOD

In the Bragg method, one has to use a fairly large crystal with flat faces carefully oriented in a number of specific directions. All crystalline substances are not available in the above needed form. Another method, which is very convenient to handle, was developed independently by Debye and Scherrer, and Hull. In this method, known as the powder method, the given substance is taken in powder form in a thin-walled capillary tube. It is then irradiated with monochromatic X-rays. The particles in the powder act as tiny crystals and are randomly oriented with respect to the incident X-rays. Since the powder contains a very large number of particles, it is possible that some particles will have their (100) planes correctly oriented so that the Bragg equation is satisfied. Some others will have their (110) planes properly oriented, while there may be some with their (111) planes properly oriented and so on. Thus, the powder provides all types of lattice planes for the reflection of X-rays and hence may be considered equivalent to a single crystal rotated not only about one axis, but about all possible axes at once.

Experimental Set-up Consider a set of parallel planes making an angle θ with the incident beam of X-rays as shown in Fig. 3.14.1. The reflected beam from these planes will make an angle 2θ with the unreflected beam. Now, if these planes are rotated around the incident beam, keeping angle θ constant, it is obvious that the diffracted beam will travel over the surface of a cone as shown in Fig. 3.14.2a. Since in the powder, a large number of crystals with all types of orientations are available, the above criterion of a crystal to be rotated around the incident beam is

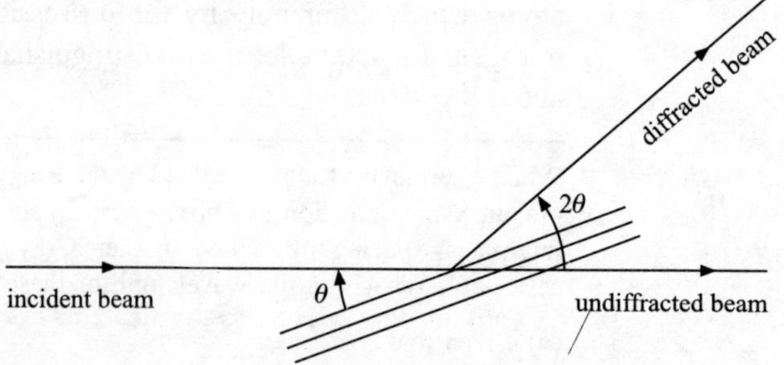

Fig. 3.14.1 Diffraction of X-ray by a set of parallel planes

automatically achieved. Thus, one gets a cone of reflected X-rays corresponding to each and every space lattice planes. If a narrow beam of film is now curved into a cylinder around the capillary tube in such a way that the latter lies along the axis of the cylinder, the diffracted beam will leave impressions on the film in the form of either curved or straight lines which on removing would give the diffraction pattern very similar to the one shown in Fig. 3.14.2b. The big holes in the film correspond to the places where the X-ray beam enters or leaves it.

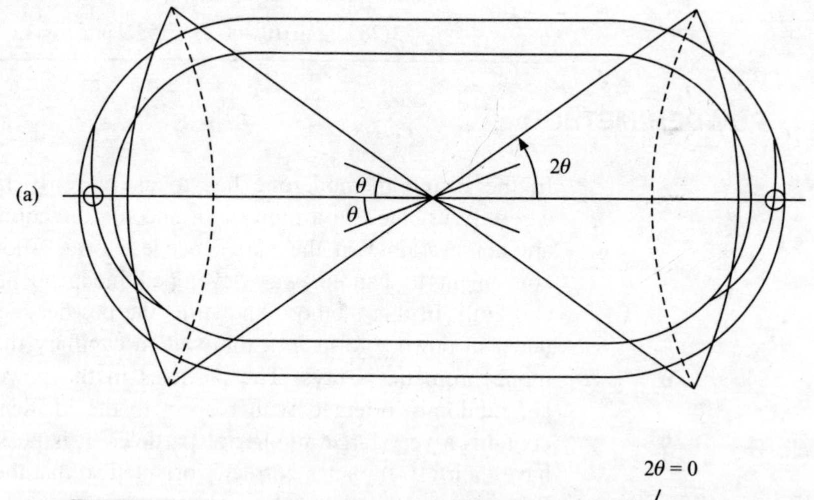

Fig. 3.14.2 (a) X-ray diffraction by a powdered sample. (b) Diffraction impressions on the film

If s is the distance of a diffracted beam from the centre of the hole and r is the distance of the film from the capillary tube, the angle 2θ of the diffracted beam with the unreflected beam can be calculated using the expression $2\theta = s/r$. From this, the incident angle θ of X-rays with the planes can be determined. Thus, knowing the wavelength λ of X-rays, we can calculate the interplanar spacing d_{hkl} using the Bragg equation

$$d_{hkl} = \frac{\lambda}{2 \sin \theta}$$

Criterion for Obtaining Sharp Diffraction Lines

In order to obtain sharp diffraction lines of uniform thickness, the tiny crystals in the powder must have average dimensions of a few microns. If the crystals are too large, relatively few of them will contribute to a diffraction line and thus the diffraction pattern will consist of a discontinuous set of spots. If the crystals are too small, the diffraction lines become broadened. This is due to the fact that as the size of a crystal decreases, the number of its lattice planes and thus the extent of the orderly arrangement also decreases. The powder method is useful for the crystal systems that have only one or two lattice parameters to be determined (cubic, tetragonal, hexagonal and rhombohedral systems).

Example 3.14.1

A powder diffraction pattern for a given substance was obtained using X-rays from a Cu target where λ = 154 pm. The distance from the capillary to the film was 5.0 cm. Diffracted lines were obtained, two of which were at distances 1.2 and 3.4 cm from the undeflected beam. Calculate the spacings for the planes that give rise to these lines.

Solution

Since in the powder method, the diffracted line is observed at angle 2θ from the incident beam, it is obvious that $2\theta = s/r$, where s is the distance of the diffracted beam from the undiffracted one and r is the distance of the film from the capillary tube.

Thus, $$2\theta_1/\text{radian} = \frac{s}{r} = \frac{1.2\ \text{cm}}{5\ \text{cm}}$$

or $$\theta_1 = \frac{1.2}{10}\text{radian} = 0.12\ \text{radian} = (0.12\ \text{radian})\left(\frac{180^\circ}{\pi\ \text{radian}}\right) = 6.9^\circ$$

For second line $\theta_2 = 19.5^\circ$

Taking the sine of θ_1 and θ_2, we get

$$\sin\theta_1 = 0.120 \qquad \text{and} \qquad \sin\theta_2 = 0.334$$

Substituting these in the Bragg equation, we have

$$d_1 = \frac{154\ \text{pm}}{2(0.120)} = 640\ \text{pm} \qquad \text{and} \qquad d_2 = \frac{154\ \text{pm}}{2(0.334)} = 230\ \text{pm}$$

3.15 DIFFRACTION PATTERN OF A CUBIC SYSTEM

Basic Equation

Since for a cubic crystal

$$d_{hkl} = \frac{a}{(h^2 + k^2 + l^2)^{1/2}} \tag{3.15.1}$$

the Bragg's equation

$$\lambda = 2d_{hkl} \sin\theta_{hkl}$$

becomes $$\lambda = 2\frac{a}{(h^2 + k^2 + l^2)^{1/2}} \sin\theta_{hkl}$$

or $$\sin^2\theta_{hkl} = \frac{\lambda^2}{4a^2}(h^2 + k^2 + l^2)$$

or $$\sin^2\theta_{hkl} = K(h^2 + k^2 + l^2) \tag{3.15.2}$$

where $K = \lambda^2/4a^2$. This has a constant value for a given cubic crystal and for a given wavelength λ of X-rays.

Equation (3.15.2) can be used to predict the diffraction patterns for the three types of cubic systems. These are described in the following.

Primitive Cubic Lattice

By assigning consecutive integral values (0, 1, 2, ...) to h, k and l, we can calculate a series of values of d_{hkl} and $\sin^2\theta$ by using Eqs (3.15.1) and (3.15.2), respectively. These are listed in Table 3.15.1

(It can be noted that $\sin^2\theta$ cannot have the value of $7K$ because there is no way in which the integer 7 can be written in the form of $h^2 + k^2 + l^2$. This is also true of the integers 15, 23, 28, etc.)

We will observe diffraction lines at angles listed in Table 3.15.1. Thus, the predicted diffraction pattern consists of a set of six lines which are equally spaced when plotted against $\sin^2\theta$ followed by a gap and then another series of lines. The observation of such a set of diffraction lines shows directly that the crystal under study has a primitive cubic lattice.

Table 3.15.1 Interplanar Distances and the Corresponding Expected Angles of Diffraction for a Primitive Cubic Lattice

hkl	100	110	111	200	210	211	220	221 300	310	311	222	320
d_{hkl}	a	$\frac{a}{\sqrt{2}}$	$\frac{a}{\sqrt{3}}$	$\frac{a}{2}$	$\frac{a}{\sqrt{5}}$	$\frac{a}{\sqrt{6}}$	$\frac{a}{2\sqrt{2}}$	$\frac{a}{3}$	$\frac{a}{\sqrt{10}}$	$\frac{a}{\sqrt{11}}$	$\frac{a}{\sqrt{12}}$	$\frac{a}{\sqrt{13}}$
$\sin^2\theta$	K	$2K$	$3K$	$4K$	$5K$	$6K$	$8K$	$9K$	$10K$	$11K$	$12K$	$13K$

Body-Centred Cubic Lattice

The patterns of lines expected from face-centred and body-centred lattices are different from the pattern of primitive cubic lattice.

It can be seen from Fig. (3.9.3) that in (100) planes of a body-centred lattice, only half of the atoms lie in these planes, the remainder lie in the (200) planes which are located half-way between adjacent (100) planes. As a consequence of this, the X-rays scattered at the Bragg angle for reflection from the (100) planes will be out of phase with those scattered by the (200) planes, with the result that destructive interference will occur, and the diffraction line corresponding to the (100) planes will be absent. On the other hand, at the Bragg angle for reflection from the (200) planes all scattered X-rays will be in phase and thus a strong diffraction line will be observed.

It can be shown, in general, that for a body-centred cubic lattice, all diffraction lines for which $(h + k + l)$ *is an odd integer must be absent.*[†] Thus, we will observe diffraction lines at angles listed in Table 3.15.2.

Face-Centred Cubic Lattice

Again from Fig. 3.9.3, it can be seen that only half the atoms lie in the (100) and the (110) planes. Thus, it is expected that the diffraction from these planes will be absent. On the other hand, all atoms lie in (111), (200) and (220)

[†] See Annexure II at the end of this chapter for generating this criterion from the structure factor.

Table 3.15.2 Angles at which Diffraction Lines are Observed for a Body-Centred Cubic Lattice

hkl	100	110	111	200	210	211	220	221 300	310	311	222	320
d_{hkl}		$\frac{a}{\sqrt{2}}$		$\frac{a}{2}$		$\frac{a}{\sqrt{6}}$	$\frac{a}{2\sqrt{2}}$		$\frac{a}{\sqrt{10}}$		$\frac{a}{\sqrt{12}}$	
$\sin^2\theta$		$2K$		$4K$		$6K$	$8K$		$10K$		$12K$	

planes and thus the corresponding diffraction lines will be observed. It can be shown, in general, that for a face-centred cubic lattice, the strong diffraction lines are observed only from those planes for which the values of h, k and l are either all even or all odd. Thus, one would observe diffraction lines at angles listed in Table 3.15.3.

Table 3.15.3 Angles at which Diffraction Lines are Observed for a Face-Centred Cubic Lattice

hkl	100	110	111	200	210	211	220	221 300	310	311	222
d_{hkl}			$\frac{a}{\sqrt{3}}$	$\frac{a}{2}$			$\frac{a}{2\sqrt{2}}$			$\frac{a}{\sqrt{11}}$	$\frac{a}{\sqrt{12}}$
$\sin^2\theta$			$3K$	$4K$			$8K$			$11K$	$12K$

Summary of Diffraction Pattern for Cubic Systems

The predicted patterns for three types of cubic lattices are also shown in Fig. 3.15.1.

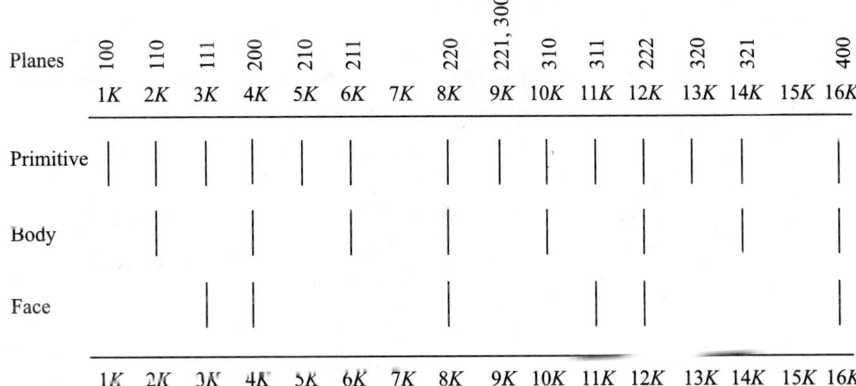

Fig. 3.15.1 A typical X-ray diffraction pattern of a cubic system (The presence of a reflection is indicated by a line)

Importance of Missing Reflections

The difference between the three diffraction patterns clearly indicates the usefulness of missing reflections in distinguishing different lattice types. In general, the search for missing reflections is an important step in the determination of crystal structures. It may be mentioned, however, that the distinction between the primitive and the body-centred cubic systems cannot be made on the basis of the first six lines, since spacing of the lines in the individual diffraction pattern will be same (K in case of primitive and $2K$ in body-centred). If more lines are included only then the distinction can be made since, in the primitive structure, there will be a gap after the sixth line whereas no such gap will be observed in the body-centred cubic system.

Computing Edge Length of Unit Cell

Once the diffraction pattern has been identified, it is then possible to assign each and every line with the correct values of h, k, and l. From the measurement of θ for any one of these lines, the edge length a of the cube can be computed from the equation

$$a = \frac{1}{2 \sin \theta_{hkl}}(h^2 + k^2 + l^2)^{1/2}$$

If the indexing of the lines has been done correctly, the same value of a will be obtained from all values of $\sin \theta_{hkl}$.

Example 3.15.1

Silver is known to be crystallized in cubic form. The Bragg angles, using copper K_α, X-rays with λ = 154.1 pm, for the first six diffraction lines are as follows:

θ	19.08°	22.17°	32.26°	38.74°	40.82°	49.00°

(a) What is the type of cubic crystal formed by silver?
(b) What is the length of a side of the unit cell?
(c) What is the interplanar distance of the planes (111)?

Solution

(a) *Type of cubic crystal*

θ	19.08°	22.17°	32.26°	38.74°	40.82°	49.00°
$\sin \theta$	0.326 8	0.377 3	0.533 8	0.625 7	0.653 6	0.754 7
$\sin^2\theta$	0.106 7	0.142 4	0.284 8	0.391 5	0.427 2	0.569 7
	$3K$	$4K$	$8K$	$11K$	$12K$	$16K$

where K = 0.035 6

Thus, silver crystallizes in the face-centred cubic lattice.

(b) The edge-length of the cube can be calculated from the following expression.

$$a = \frac{\lambda}{2 \sin \theta_{hkl}} \sqrt{(h^2 + k^2 + l^2)}$$

The reflection at 19.08° is due to (111) planes. Hence

$$a = \frac{154.1 \text{ pm}}{2 \times 0.326\,8} \sqrt{1^2 + 1^2 + 1^2} = 408.6 \text{ pm}$$

(c) Distance between (111) planes is

$$d_{111} = \frac{408.6 \text{ pm}}{\sqrt{3}} = 235.9 \text{ pm}$$

Example 3.15.2

Gold has a face-centred cubic lattice with an edge length of 407 pm of the unit cube. Calculate the diffraction pattern when copper X-rays of wavelength 154 pm is used.

Solution

In a face-centred lattice, the reflection of X-rays will occur from the planes for which the Miller indices (hkl) are either all even or all odd. Thus, we will have reflections from (111), (200), (220), (311), (222), (400), etc., planes

For a cubic crystal, we have

$$\sin\theta = \frac{\lambda}{2a}(h^2 + k^2 + l^2)^{1/2}$$

Thus, the diffraction pattern of Au will exhibit reflections at the following angles:

Plane	sin θ	θ
111	$\dfrac{\sqrt{3}\lambda}{2a} = \dfrac{1.732(154\ \text{pm})}{2(407\ \text{pm})} = 0.328\,0$	19° 9′
200	$\dfrac{\lambda}{a} = \dfrac{(154\ \text{pm})}{(407\ \text{pm})} = 0.378\,4$	22° 14′
220	$\dfrac{\sqrt{2}\lambda}{a} = \dfrac{1.414(154\ \text{pm})}{(407\ \text{pm})} = 0.535\,0$	32° 52′
311	$\dfrac{\sqrt{11}\lambda}{2a} = \dfrac{3.317(154\ \text{pm})}{2(407\ \text{pm})} = 0.627\,4$	38° 51′
222	$\dfrac{\sqrt{12}\lambda}{2a} = \dfrac{3.464(154\ \text{pm})}{2(407\ \text{pm})} = 0.655\,4$	40° 51′
400	$\dfrac{\sqrt{16}\lambda}{2a} = \dfrac{4(154\ \text{pm})}{2(407\ \text{pm})} = 0.756\,8$	49° 12′

Example 3.15.3

When an X-ray powder pattern of crystalline copper is obtained using X-rays from copper target (the wavelength of the K_α-line is 154.05 pm), reflections are found at 21.65°, 25.21°, 37.06°, 44.96°, 47.58° and other angles.

(a) What is the type of cubic crystal formed by copper?
(b) What is the length of a side of the unit cell?
(c) Determine the value of Avogadro constant if density of copper is 8.812 g cm^{-3}.
(d) Calculate the radius of copper atom.

Solution

(a) *Type of cubic crystal* Calculating $\sin^2\theta$, we have

Angle	*sin θ*	*$sin^2\theta$*
21.65°	0.368 9	0.136 1
25.21°	0.425 8	0.181 3
37.06°	0.602 4	0.362 9
44.96°	0.706 5	0.499 1
47.58°	0.738 1	0.544 8

Taking the ratios of $\sin^2\theta$, we get

$$0.136\,1 : 0.181\,3 : 0.362\,9 : 0.499\,1 : 0.544\,8$$

or $\quad 3 \times 0.045 : 4 \times 0.045 : 8 \times 0.045 : 11 \times 0.045 : 12 \times 0.045$

that is, $\quad 3K : 4K : 8K : 11K : 12K$, where $K = 0.045$

From these ratios, it is obvious that copper has face-centred cubic crystal.

(b) *Length of the side of the unit cell* Since

$$K = \frac{\lambda^2}{4a^2}$$

therefore,

$$a = \left(\frac{\lambda^2}{4K}\right)^{1/2} = \frac{\lambda}{2}\cdot\frac{1}{\sqrt{K}} = \frac{154.05 \text{ pm}}{2\times\sqrt{0.045}} = \frac{154.05 \text{ pm}}{2\times 0.212\,1} = 363.2 \text{ pm}$$

(c) *Avogadro constant* Since

$$\rho = \frac{nM}{a^3 N_A}, \quad \text{therefore,} \quad N_A = \frac{nM}{a^3 \rho}$$

Substituting the values, we have

$$N_A = \frac{4(63.54 \text{ g mol}^{-1})}{(363.2\times 10^{-10} \text{ cm})^3(8.812 \text{ g cm}^{-3})} = 6.02\times 10^{23} \text{ mol}^{-1}$$

(d) *Radius of Cu atom* In a face-centred cubic lattice, atoms touch one another along the face-diagonal of the cubic unit call.
Therefore,

$$4r = \sqrt{2}a \qquad \text{or} \qquad r = \frac{\sqrt{2}a}{4}$$

Substituting the values, we have

$$r = \frac{(1.414)\,(363.2 \text{ pm})}{4} = 128.3 \text{ pm}$$

3.16 CRYSTAL STRUCTURE OF SODIUM CHLORIDE

The unit lattice of sodium chloride, like the macroscopic crystal, must be a cube and thus the sodium and chloride ions must be arranged in some combination of only three possible space-lattices.

Data from Diffraction Pattern

Schematic representation of the powder pattern of sodium chloride is given in Fig. 3.16.1.

Table 3.16.1 Records values of the angle θ, $\sin\theta$, $\sin^2\theta$ and the relative intensities for some of the lines of NaCl using K_α line from palladium.

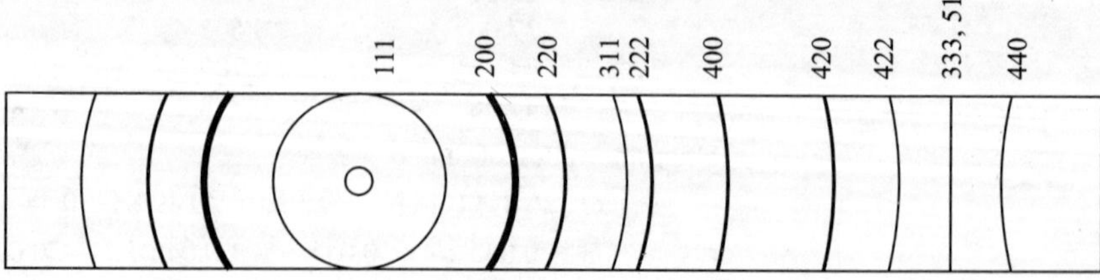

Fig. 3.16.1 Diffraction pattern (only schematic) of sodium chloride

Type of Cubic Lattice

It is obvious from the fifth column of Table 3.16.1 that NaCl crystal must belong to the face-centred cubic crystal. Designation of the plane as given in the first column follows this recognition. This can also be checked by finding out the ratio of the first three planes (200), (220) and (111). This ratio is

Table 3.16.1 Details of Diffraction Pattern of NaCl

Planes	*θ/degree*	*sin θ*	*sin²θ*	*K* = 0.002 73	*Relative Intensities*	
111	5.2	0.091	0.008 2	3*K*	9.0	
200	5.9	0.103	0.010 6	4*K*	100	
220	8.4	0.146	0.021 3	8*K*	50.4	
311	8.7	0.151	0.022 8	11*K*		
222	10.5	0.182	0.033 6	12*K*	33.1	Second order reflection of (111)
333	15.8	0.273	0.074 5	27*K*	0.58	Third order reflection of (111)
444	21.3	0.364	0.132 5	48*K*	2.82	Fourth order reflection of (111)
555	27.1	0.455	0.207 0	75*K*	0.14	Fifth order reflection of (111)
400	11.9	0.208	0.043 2	16*K*	19.90	Second order reflection of (200)
600	18	0.309	0.095 5	36*K*	4.87	Third order reflection of (200)
800	24.3	0.412	0.169 7	64*K*	0.79	Fourth order reflection of (200)
440	17.0	0.292	0.085 3	32*K*	6.10	Second order reflection of (220)
660	26	0.438	0.191 8	72*K*	0.71	Third order reflection of (220)

$$d_{200} : d_{220} : d_{111} = \frac{1}{\sin(5.9°)} : \frac{1}{\sin(8.4°)} : \frac{1}{\sin(5.2°)}$$
$$= \frac{1}{0.103} : \frac{1}{0.146} : \frac{1}{0.090\,6}$$
$$= 1 : 0.705 : 1.137$$

This agrees with the theoretical ratio given in Section 3.10, thus confirming the existence of the face-centred cubic crystal system. The indicated indexing of the diffraction lines also lead to a single value of the edge length of the unit cube. Thus, sodium and chloride ions are arranged in a crystal as a face-centred cubic lattice.

Arrangement of Na^+ and Cl^- Ions

The actual arrangement of Na^+ and Cl^- in this face-centred cubic lattice can be determined by the relative intensities of the reflection maxima for the different orders and planes.

The intensity of a diffracted beam depends upon two factors:

(i) *Number of electrons in the atom* The scattering power of atoms for X-ray depends on the number of electrons in the atom and is roughly proportional to the atomic number.

(ii) *The order of diffraction* Intensity decreases in a definite manner with increase in the order of reflection.

It can be seen from Table 3.16.1 that in the (200) and (220) planes the intensities decrease progressively with order.

n	1	2	3	4
(200)	100	19.90	4.87	0.79
(220)	50.4	6.10	0.71	

This systematic decrease can be accounted for quantitatively by assuming that such planes contain equal numbers of sodium and chloride ions. In the (111) planes, an alternation of intensities is observed.

n	1	2	3	4	5
(111)	9.0	33.1	0.58	2.82	0.14

This can be accounted for by postulating that the planes (111) are composed alternately of sodium and chloride ions.

The possible arrangement which satisfies the above facts is given in Fig. 3.16.2.

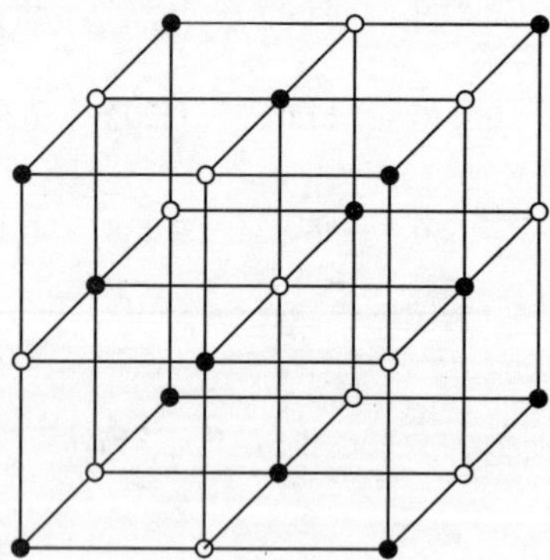

Fig. 3.16.2 Crystal structure of sodium chloride, Na^+ black circles—located at the corners of the cube and at the centre of each of the six faces; Cl^- open circles—located halfway between two sodium ions

The structure shown in Fig. 3.16.2 suggests that it consists essentially of two interpenetrating face-centred cubic lattices, one composed entirely of Na^+ ions and the other of Cl^- ions.

Comment

From the above analysis, it is revealed that the designation 'molecule of sodium chloride' loses a great deal of its definitive meaning. It can hardly be said that any particular chloride ion belongs to any definite sodium ion; rather, each sodium ion is shared equally by six chloride ions, and each chloride ion by six sodium ions. All that can be said is that to each sodium corresponds one-sixth of six chloride ions, so that each sodium has the equivalent of a chloride but not any one ion exclusively.

Example 3.16.1

The first order reflection from (200) planes of NaCl using X-rays of wavelength 58 pm (K_α line from palladium) occurs at an angle of 5.9°. Calculate (i) edge length of the unit cell, (ii) volume of the unit cell, (iii) molar volume, and (iv) density of solid NaCl.

Solution

(i) Using Bragg's law, we have

$$\lambda = 2\, d_{hkl} \sin \theta$$

Thus $$d_{200} = \frac{\lambda}{2 \sin \theta} = \frac{58 \text{ pm}}{2 \sin (5.9°)} = \frac{58 \text{ pm}}{2 \times 0.103} = 281.6 \text{ pm}$$

The edge length of the unit cell = 2 × d_{200} = 563.2 pm

(ii) The volume of unit cell = $(563.2 \text{ pm})^3 = 1.786 \times 10^{-28} \text{ m}^3$.

(iii) In order to calculate molar volume, we must know the number of molecules of NaCl which belong to one unit cell. In one unit cell of NaCl, we have the following arrangements:

(a) Eight Na^+ (or Cl^-) at the end of edges, each of which is shared by eight such unit cells. Thus, contribution from the edges is one Na^+ (or Cl^-).

(b) Six Na^+ (or Cl^-) are present in the centres of the faces, each is shared by two cubes, thus contributing three Na^+ (or Cl^-) ions.

(c) Twelve Cl^- (or Na^+) are present along the edges, each of which is shared by four cubes, thus contributing three Cl^- (or Na^+).

(d) One Cl^- (or Na^+) is present at the centre of the cube.

Thus, we have on an average four Na^+ ions and four Cl^- ions per unit cell, or a total equivalent to four molecules of sodium chloride. The volume allotted to these molecules is the volume of one unit cell. Thus, the molar volume is given by

$$V_m = \frac{\text{Volume of one unit cell}}{4} \times N_A$$

$$= \frac{(1.786 \times 10^{-28} \text{ m}^3)}{4} \times (6.022 \times 10^{23} \text{ mol}^{-1})$$

$$= 2.689 \times 10^{-5} \text{ m}^3 \text{ mol}^{-1}$$

(iv) The density of NaCl is

$$\rho = \frac{M}{V_m} = \frac{(58.443 \text{ g mol}^{-1})}{(2.689 \times 10^{-5} \text{ m}^3 \text{ mol}^{-1})} = 2.173 \times 10^6 \text{ g m}^{-3} = 2.173 \text{ g cm}^{-3}$$

3.17 CRYSTAL STRUCTURE OF POTASSIUM CHLORIDE

The powder pattern of KCl is shown in Fig, 3.17.1a. At first glance the pattern looks like that of a primitive cubic lattice since there are six lines followed by a gap. However, KCl has a face-centred cubic lattice like that of NaCl. The

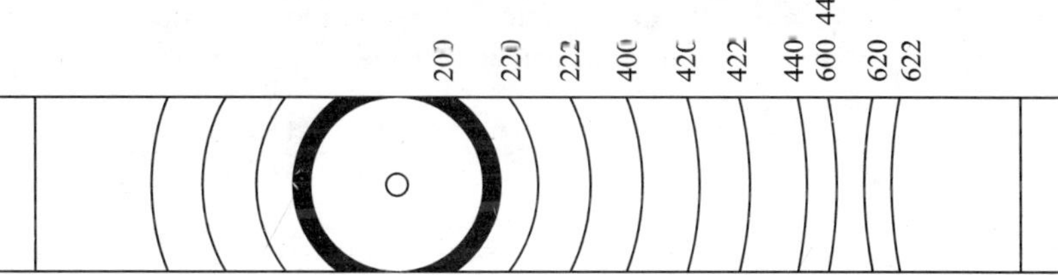

Fig. 3.17.1 (a) Diffraction pattern (only schematic) of potassium chloride

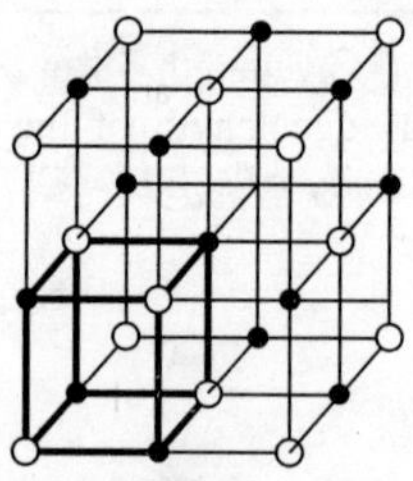

Fig. 3.17.1b

resemblance to a primitive lattice is a consequence of the fact that K^+ ions and Cl^- ions have the same number of electrons and thus have the same scattering power for X-rays. The unit cell of the apparently primitive lattice formed by considering K^+ and Cl^- ions as identical has half the edge length of the actual face-centred unit cell (Fig. 3.17.1b).

Another example exhibiting this type of behaviour is provided by RbBr in the series LiBr, NaBr and RbBr. All these molecules have face-centred lattices. However, X-ray diffraction indicates that RbBr has a simple cubic lattice. This results from the fact that Rb^+ and Br^- have the same number of electrons.

3.18 DENSITY OF CUBIC CRYSTALS

Defining Expression of Density

The density based on the structure can be calculated from the mass contained in a unit cell and its volume. If N is the number of atoms or molecules per unit cubic cell of edge length a, then the mass and volume per unit cell are

$$\text{Mass} = \left(\frac{M}{N_\text{A}}\right)N \qquad \text{Volume} = a^3$$

$$\text{Therefore,} \quad \text{Density} = \frac{\text{mass}}{\text{volume}} = \frac{NM}{a^3 N_\text{A}}$$

The value of N for the three cubic cells can be calculated as follows:

Primitive cubic cell In a primitive cubic cell, atoms are present at the corners of the cube. There are eight corners of a cube and thus eight atoms are present at these corners. Now, any particular corner of the cube is actually shared amongst eight such cubic unit cells placed adjacent to one another. Thus, the contribution of the atom placed at one of the corners to the single cubic unit cell is 1/8. Since there are eight corners of a cube, the number of atoms associated with a single primitive unit cell is 8/8 = 1 (Fig. 3.18.la).

Body-centred cubic cell In a body-centred cubic unit cell, besides atoms being present at the corners, there is one atom in the centre of the cube which belongs exclusively to this cubic unit cell. Therefore, number of atoms per unit cell are two (Fig. 3.18.1b).

Face-centred cubic cell Here, atoms, besides being at the corners, are also present at the centre of the six faces. Each of these atoms is shared between two such unit cells. Thus, their contribution to the unit cell is 6/2 = 3 atoms, making a total of 4 atoms per cubic unit cell (Fig. 3.18.1c).

Illustration

As an example, let us compute the density of silver which has a face-centred cubic unit cell with edge length equal to 408.6 pm. Since $N = 4$, we have

$$\text{Density} = \frac{4\,(107.9 \times 10^{-3}\ \text{kg mol}^{-1})}{(408.6 \times 10^{-12}\ \text{m})^3\,(6.022 \times 10^{23}\ \text{mol}^{-1})}$$

$$= 10.50 \times 10^3\ \text{kg m}^{-3} = 10.50\ \text{g cm}^{-3}$$

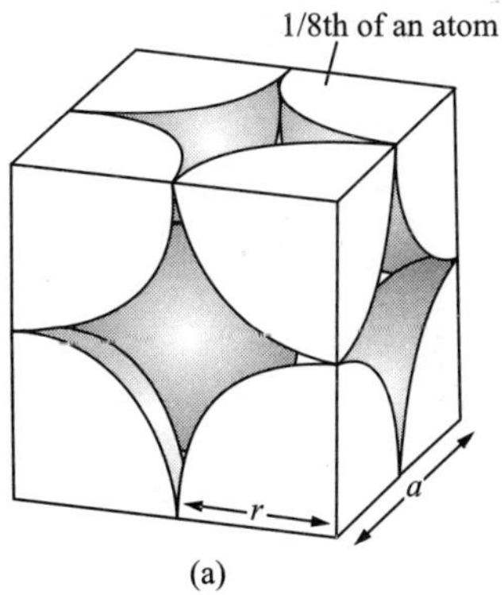

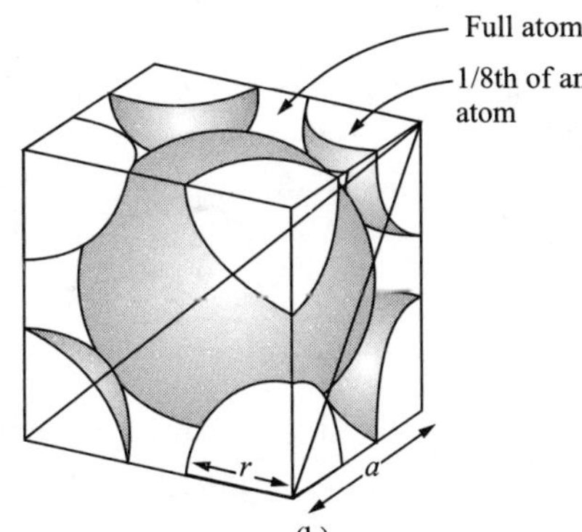

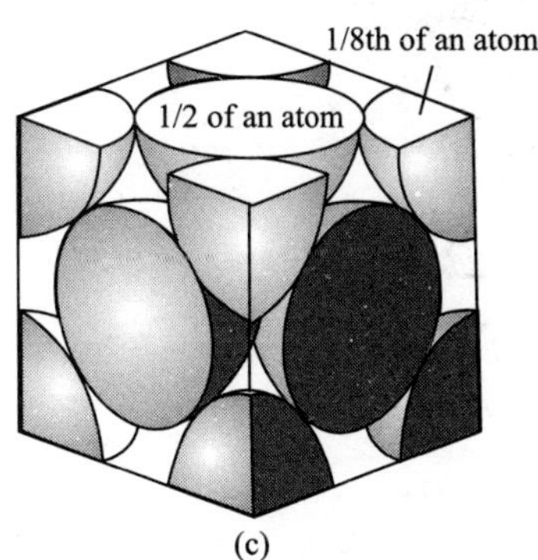

Fig. 3.18.1 Unit cells of cubic systems

Example 3.18.1 Molybdenum forms body-centred cubic crystals whose density is 10.3 g cm^{-3}. Calculate: (a) edge length of the unit cube, (b) distance between the (110) planes and between the (111) planes. Molar mass of Mo = 95.94 g mol^{-1}.

Solution **(a)** Since density of the crystal is

$$\rho = \frac{NM}{a^3 N_A}, \qquad \text{therefore,} \qquad a = \left(\frac{NM}{\rho N_A}\right)^{1/3}$$

Now, the number N of atoms per unit body-centred cubic cell = 2.

Thus, $$a = \left\{\frac{2(95.94\ \text{g mol}^{-1})}{(10.3\ \text{g cm}^{-3})\,(6.022\times10^{23}\ \text{mol}^{-1})}\right\}^{1/3}$$

$$= 3.139\times10^{-8}\ \text{cm} \equiv 313.9\ \text{pm}$$

(b) Since $d_{hkl} = \dfrac{a}{(h^2+k^2+l^2)^{1/2}}$, we get

$$d_{110} = \frac{a}{\sqrt{2}} = \frac{313.9\ \text{pm}}{1.414} = 222.0\ \text{pm}$$

and $$d_{111} = \frac{a}{\sqrt{3}} = \frac{313.9\ \text{pm}}{1.732} = 181.2\ \text{pm}$$

Example 3.18.2 The density of potassium chloride is 1.989 3 g cm^{-3} and the length of a side of the unit cell is 629.082 pm as determined by X-ray diffraction. Calculate the value of Avogadro constant.

Solution Since density, $\rho = \dfrac{NM}{a^3 N_A}$, therefore, $N_A = \dfrac{NM}{a^3 \rho}$

Substituting the values, we have

$$N_A = \frac{4(74.55\ \text{g mol}^{-1})}{(629.082\times10^{-10}\ \text{cm})^3\ (1.986\ 3\ \text{g cm}^{-3})} = 6.03\times10^{23}\ \text{mol}^{-1}$$

Example 3.18.3 Use the data given below to find the type of cubic lattice to which the crystal belongs:

	Fe	V	Pd
a/pm	286	301	388
ρ/g cm^{-3}	7.86	5.96	12.16

Solution Now, $\rho = \dfrac{NM}{a^3 N_A}$, Therefore, $N = \dfrac{\rho a^3 N_A}{M}$

Thus, we have

For Fe $$N = \frac{(7.86\ \text{g cm}^{-3})\,(2.86 \times 10^{-8}\ \text{cm})^3\,(6.022 \times 10^{23}\ \text{mol}^{-1})}{(55.85\ \text{g mol}^{-1})} = 1.983 \simeq 2$$

Hence, the cubic lattice of Fe is *body-centred.*

For V $$N = \frac{(5.96\ \text{g cm}^{-3})\,(3.01 \times 10^{-8}\ \text{cm})^3\,(6.022 \times 10^{23}\ \text{mol}^{-1})}{(50.94\ \text{g mol}^{-1})} = 1.921 \simeq 2$$

Hence, the cubic lattice for V is *body-centred.*

For Pd $$N = \frac{(12.16\ \text{g cm}^{-3})\,(3.88 \times 10^{-8}\ \text{cm})^3\,(6.022 \times 10^{23}\ \text{mol}^{-1})}{(106.4\ \text{g mol}^{-1})} = 4.12 \simeq 4$$

Hence, the cubic lattice for Pd is *face-centred.*

Example 3.18.4 X-ray analysis shows that the unit cell length in NaCl is 562.8 pm. Calculate the density you would expect on this basis. Avogadro constant is $6.022 \times 10^{23}\ \text{mol}^{-1}$.

Solution Since density $\rho = \dfrac{NM}{a^3 N_A}$, therefore, we get

$$\rho = \frac{4(58.5\ \text{g mol}^{-1})}{(5.628 \times 10^{-8}\ \text{cm})^3\,(6.022 \times 10^{23}\ \text{mol}^{-1})} = 2.179\ \text{g cm}^{-3}$$

Example 3.18.5 The X-ray powder pattern for molybdenum has reflections at θ = 20.25°, 29.30°, 36.82°, 43.81°, 50,69°, 58.80°, 66.30° and other larger angles when K_α X-rays from Cu are used (λ = 154 pm).

(a) What is the type of cubic crystal formed by Mo?

(b) What is the length of a side of the unit cell?

(c) What is the density of molybdenum?

Solution **(a)** *Type of cubic crystal* The diffraction pattern is

θ	$\sin\theta$	$\sin^2\theta$				
20.25°	0.346 0	0.119 8	or	1 × 0.1198	or	2 × 0.0599 = 2*K*
29.30°	0.489 4	0.239 5	or	2 × 0.1198	or	4 × 0.0599 = 4*K*
36.82°	0.599 0	0.358 8	or	3 × 0.1198	or	6 × 0.0599 = 6*K*
43.81°	0.692 3	0.479 3	or	4 × 0.1198	or	8 × 0.0599 = 8*K*
50.69°	0.773 6	0.598 4	or	5 × 0.1198	or	10 × 0.0599 = 10*K*
58.80°	0.855 4	0.731 8	or	6 × 0.1198	or	12 × 0.0599 = 12*K*
66.30°	0.915 7	0.838 3	or	7 × 0.1198	or	14 × 0.0599 = 14*K*

The above reflection pattern corresponds to body-centred cubic crystal. The primitive one is eliminated from the fact that the reflection corresponding to $7K$ (which in the present case is $14K$) is present.

(b) *Edge length of unit cell* We have

$$K = \frac{\lambda^2}{4a^2} \qquad \text{Therefore,} \qquad a = \frac{\lambda}{2\sqrt{K}}$$

Substituting the values, we have

$$a = \frac{154 \text{ pm}}{2 \times \sqrt{0.0599}} = \frac{154 \text{ pm}}{2 \times 0.2447} = 314.7 \text{ pm}$$

(c) *Density of molybdenum* We have

Density $\rho = \dfrac{NM}{a^3 N_A}$

Substituting the values, we have

$$\rho = \frac{(2)\,(95.94 \text{ g mol}^{-1})}{(3.147 \times 10^{-8} \text{ cm})^3\,(6.022 \times 10^{23} \text{ mol}^{-1})} = 10.22 \text{ g/cm}^3$$

Example 3.18.6

The unit cell of aluminium (M(Al) = 26.98 g mol^{-1}) is a cube with edge length 405 pm. The density of aluminium is 2.70 g cm^{-3}. What is the structure of aluminium crystals?

Solution

Since density

$$\rho = \frac{NM}{a^3 N_A}, \qquad \text{therefore} \qquad N = \frac{\rho\, a^3 N_A}{M}$$

Substituting the given values, we get

$$N = \frac{(2.70 \text{ g cm}^{-3})\,(4.05 \times 10^{-8} \text{ cm})^3\,(6.022 \times 10^{23} \text{ mol}^{-1})}{(26.98 \text{ g mol}^{-1})} = 4$$

Thus, the unit cell of aluminium is face-centred cubic.

Example 3.18.7

A substance forms face-centred cubic crystals. Its density is 1.984 g cm^{-3} and the length of the edge of the unit is 630 pm. Calculate the molar mass.

Solution

Since $\rho = \dfrac{NM}{a^3 N_A}$, we get $M = \dfrac{\rho a^3 N_A}{N}$

Substituting the values, we have

$$M = \frac{(1.984 \text{ g cm}^{-3})\,(6.30 \times 10^{-8} \text{ cm})^3\,(6.022 \times 10^{23} \text{ mol}^{-1})}{4}$$

$$= 74.69 \text{ g mol}^{-1}$$

Example 3.18.8

From the fact that the length of the side of the unit cell for lithium is 351 pm, calculate its atomic radius. Lithium forms body-centred cubic crystals.

Solution

In body-centred cubic crystals, atoms touch one another along the cross-diagonal of the cube.

Thus $4r = \sqrt{3}a$ or $r = \dfrac{\sqrt{3}a}{4}$

Substituting the value of a, we have

$$r = \frac{\sqrt{3}\,(351\text{ pm})}{4} = 151.9\text{ pm}$$

Example 3.18.9 Magnesium oxide (M = 40.0 g mol^{-1}) is cubic and has a density of 3.620 g/cm^3. An X-ray diffraction diagram of MgO powder has lines at values of sin θ = 0.399, 0.461, 0.652, 0.764, 0.798 and 0.922. (a) Index the pattern and determine the type of cubic structure, (b) Calculate the wavelength of X-rays used. Assume that the number of MgO units per unit cell is the smallest consistent with the structure type.

Solution **(a)** Type of cubic structure

sin θ	0.399	0.461	0.652	0.764	0.798	0.922
sin^2 θ	0.1592	0.2125	0.4250	0.5837	0.6388	0.8500

Obviously the cubic structure is neither primitive nor body-centred since for them the difference between any two successive values of sin^2 θ is constant. The diffraction pattern for the face-centred cubic lattice has sin^2 θ values at $3K$, $4K$, $8K$, $11K$, $12K$, $16K$, ..., corresponding to the planes (111), (200), (220), (311), (222), (400), ... respectively. This pattern is satisfied by the above values of sin^2 θ with K = 0.052 9. Thus, magnesium oxide has a face-centred cubic lattice.

(b) *Wavelength of X-rays used*

Given that ρ = 3.620 g cm^{-3}

For a face-centred cubic lattice, n = 4. Therefore,

$$a = \left(\frac{NM}{\rho N_A}\right)^{1/3} = \left(\frac{4(40\text{ g mol}^{-1})}{(3.62\text{ g cm}^{-3})\,(6.022\times10^{23}\text{ mol}^{-1})}\right)^{1/3}$$
$$= 4.187\times10^{-8}\text{ cm}$$

According to Bragg's equation

$$n\lambda = 2d\sin\theta$$

or
$$\lambda = 2\left(\frac{d}{n}\right)\sin\theta = 2d_{hkl}\sin\theta = 2\left(\frac{a}{\sqrt{h^2+k^2+l^2}}\right)\sin\theta$$

Taking sin θ = 0.399 corresponding to the (111) planes, we have

$$\lambda = \left(\frac{2(4.187\times10^{-8}\text{ cm})}{\sqrt{3}}\right)(0.399) = 1.929\times10^{-8}\text{ cm} = 192.9\text{ pm}$$

3.19 CLASSIFICATION OF CRYSTALS BASED ON BOND TYPE

So far, the classification of crystals was done on the basis of crystal symmetry. This classification does not tell us about the chemical and physical properties of a given crystalline material. In order to understand these properties, a classification based on the bond type is required.

The properties of most of the crystals are found to conform to one of the four general types of chemical bonds, in terms of which it is possible to classify them into four categories, viz., molecular, ionic, covalent and metallic.

Molecular Crystals (or van Der Waals Crystals)

Molecular crystals are those in which the crystalline state is composed of an aggregate of discrete molecules held together by van der Waals forces. If such forces are considered to be nondirectional (e.g., in the nonpolar molecules and molecules with small dipole moments), the structure of the solid should be determined by packing efficiency, thus giving the closest approach of the atoms and therefore the maximum possible interactions. If the molecules are considered spherical, efficient packing can be obtained by stacking them one after the other in a systematic manner (Hooke's proposal).

The packing of the spheres can be done in any one of the following three ways:

Packing in a Simple Cubic Lattice

In a lattice of this type, the spheres are packed in the form of a square array by laying down a base of spheres and then piling upon the base other layers in such a way that each sphere is immediately above the other sphere, as shown in Fig. 3.19.1.

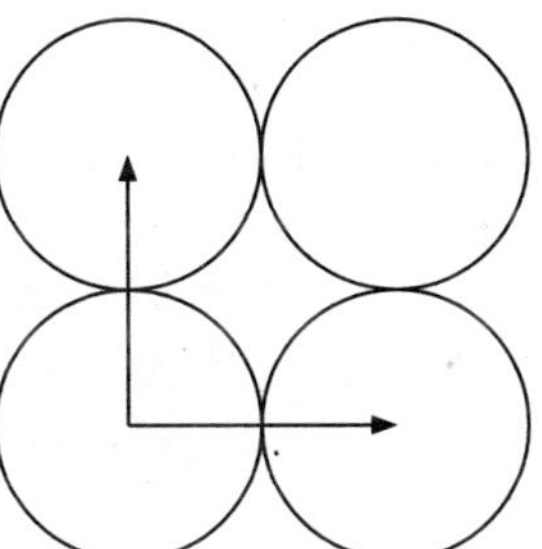

Fig. 3.19.1 Packing in a simple cubic lattice

In this structure, each sphere is in contact with six nearest neighbours (four in the same base, one above and one below). The percentage of occupied volume in this structure can be calculated as follows:

The edge length a of the cube will be twice the radius of the sphere, i.e. $a = 2r$. Since in the primitive cubic lattice, there is only one sphere present in the unit lattice, the volume occupied by the sphere is

$$V = \frac{4}{3}\pi r^3 \quad \text{or} \quad V = \frac{4}{3}\pi\left(\frac{a}{2}\right)^3$$

The fraction of the total volume occupied by the sphere is

$$\phi = \frac{\frac{4}{3}\pi\left(\frac{a}{2}\right)^3}{a^3} = \frac{\pi}{6} = 0.523\,6 \quad \text{or} \quad 52.36 \text{ per cent}$$

Thus, the structure is relatively open since only 52.36% ($\pi/6$) of the total volume is occupied by the spheres. The remainder, i.e. 0.4764 of the total volume is empty space or void volume.

No crystalline element has been found to have this structure.

Packing in a Body-Centred Cubic Lattice

Here the packing consists of a base of spheres, followed by a second layer where each sphere rests in the hollow at the junction of four spheres below it, as shown in Fig. 3.19.2. The third layer then rests on these in arrangement which corresponds exactly to that in the first layer. In this arrangement, spheres are touching one another along the cross diagonal of the cube, making its distance equal to $4r$. This must be equal to $\sqrt{3}a$.

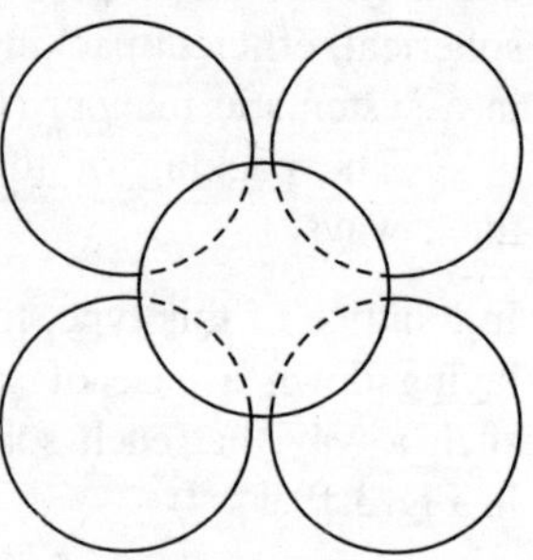

Fig. 3.19.2 Packing in a body-centred cubic lattice

Thus, $\quad 4r = \sqrt{3}a, \quad$ i.e. $\quad r = \dfrac{\sqrt{3}}{4}a$

Volume of the cube $= a^3$

$$\text{Volume of one sphere} = \frac{4}{3}\pi r^3 = \frac{4}{3}\pi\left(\frac{\sqrt{3}}{4}a\right)^3$$

Since there are two spheres in each unit cell, the total volume occupied will be

$$2\left\{\frac{4}{3}\pi\left(\frac{\sqrt{3}}{4}a\right)^3\right\}$$

The fraction of the volume occupied by the spheres

$$\phi = \frac{2\left\{\frac{4}{3}\pi\left(\frac{\sqrt{3}}{4}a\right)^3\right\}}{a^3} = \frac{\sqrt{3}\pi}{8} = 0.6802 \quad \text{or} \quad 68.02 \text{ per cent}$$

In this arrangement each sphere has eight nearest neighbours. Only a few metallic elements are found to have this type of structure.

Closest Packing

In closest packing arrangements, each sphere is in contact with the maximum possible number of nearest neighbours. Figure 3.19.3 shows a closest packed layer of spheres. Each sphere is surrounded by six nearest neighbours lying in the plane, three spheres just above it and three below it, thus making the total number of nearest neighbours equal to twelve.

Fig. 3.19.3 (a) Closest-packed layers of spheres

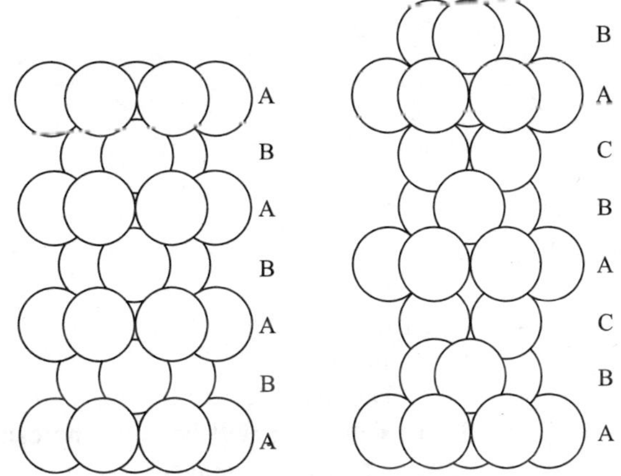

Fig. 3.19.3 (b) Two types of packing

If the spheres are packed in the same plane, then just above these spheres there exist two different types of voids, pointing in different directions as shown in Fig. 3.19.3a. Thus, we can have three different types of locations as shown by A, B and C in Fig. 3.19.3a. Location A is occupied by the spheres while B and C are the two different types of voids. If a second layer of closest-packed spheres is placed upon the first layer, the spheres of the second layer can occupy the region above either the B voids or the C voids. But because of the size of the spheres, both types of voids cannot be occupied simultaneously.

The third layer of closest-packed sphere can be formed in two different ways. If, for example, we choose to place the spheres of the second layer in B sites, one of the available sets of voids for the third layer will be directly above the spheres in the original layer. These are A sites. The other set of voids will be directly above the voids designated by C in the original layer.

Types of Packing

Thus, two types of packing (Fig. 3.19.3b) are possible

ABABA ... or ABCABC ...

We can have many other varieties of patterns such as ABCACB..., ABAC..., etc. But for many of the common substances that form closest-packed structures, one of the above two symmetrical arrangements is observed.

Hexagonal Closest-Packed Structure

The packing ABAB... is known as a *hexagonal closest-packed structure* (HCP). The unit cell of HCP is shown in Fig. 3.19.4.

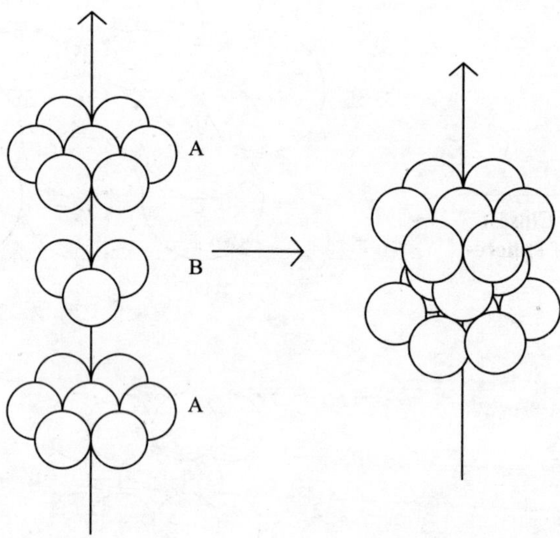

Fig. 3.19.4 Unit cell formed by ABA packing

The fraction of the volume occupied in HCP can be calculated as described in the following.

The distance $C/2$ in Fig. 3.19.5 is the distance between the layers A and B. This distance will be from the centre of a sphere to the plane of the three spheres that are in contact with it.

The spheres a, b and c of the layer A are shown in Fig. 3.19.5a. From this figure, we get

$$\overline{\mathrm{ad}'} = \overline{\mathrm{ab}'}/\cos 30° \quad \text{i.e.} \quad \overline{\mathrm{ad}'} = r/(\sqrt{3}/2) = 2r/\sqrt{3}$$

The centre of sphere d in the layer B is just above the point d′ as shown in Fig. 3.19.5b. From this figure, we get

$$\overline{\mathrm{dd}'} = [(\overline{\mathrm{ad}})^2 - (\overline{\mathrm{ad}'})^2]^{1/2} = [(2r)^2 - (2r/\sqrt{3})^2]^{1/2}$$
$$= (2r)\sqrt{2/3} = 4r/\sqrt{6}$$

This distance is $C/2$. Hence $C = 8r/\sqrt{6}$

Alternatively

The distance $C/2$ can be determined with reference to a face-centred cubic lattice having unit cell length a. In such a lattice, the distance between closest-packed layers (Miller indices 111) is one-third of the body diagonal, i.e. $\sqrt{3}a/3$ (see, Fig. 3.9.3).

Thus, $C/2 = \sqrt{3}a/3 = a/\sqrt{3}$

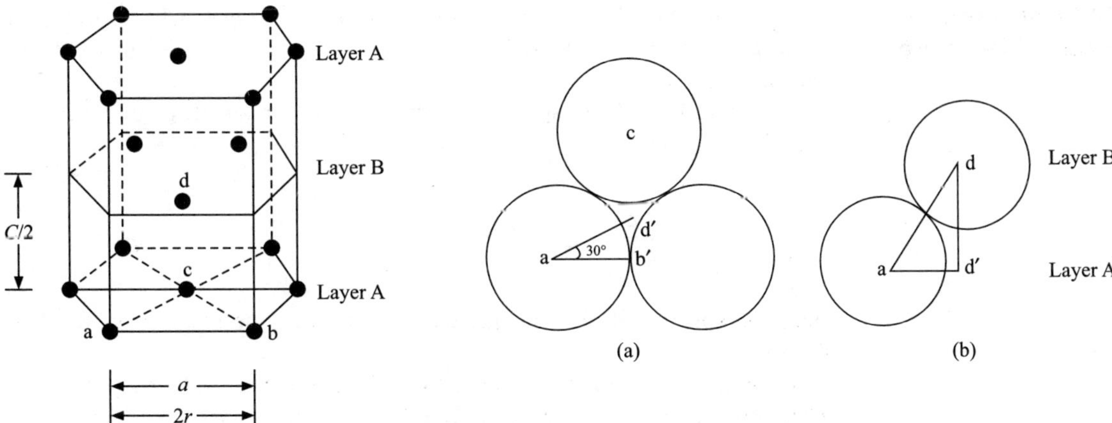

Fig. 3.19.5 Hexagonal closest-packed structure

Now, in the face-centred lattice, spheres touch one another along the face diagonal (see, Fig. 3.18.1c). Thus, we have $a = (4r)/\sqrt{2}$

With this, the distance C becomes

$$C = (2/\sqrt{3})a = (2/\sqrt{3})\,(4r/\sqrt{2}) = 8r/\sqrt{6}$$

The hexagonal base consists of six equilateral triangles, each with side $2r$ and with an altitude of $2r$ sin 60, i.e. $\sqrt{3}r$. Therefore,

$$\text{Area of the base} = 6\left[\frac{1}{2}(\sqrt{3}\,r)\,(2\,r)\right] = 6\sqrt{3}\,r^2$$

$$\text{Volume of the prism} = (6\sqrt{3}\,r^2)\left(\frac{8}{\sqrt{6}}\,r\right) = 24\sqrt{2}\,r^3$$

Number of spheres belonging to this prism

(i) 3 spheres in B layers exclusively belong to this prism.

(ii) 1 from the centre of the base. There are two spheres of this type and each is shared by two prisms.

(iii) 2 from the corners. There are twelve such spheres and each is shared amongst six prisms of this type.

Thus, the total number of spheres is 6.

The fraction of volume of the prism actually occupied by the spheres is

$$\frac{6\left(\frac{4}{3}\pi r^3\right)}{24\sqrt{2}\,r^3} = \frac{\sqrt{2}\,\pi}{6} = 0.740\,5 \quad \text{or} \quad 74.05 \text{ per cent}$$

Example of HCP are Ca, Cd, Cr, Mg and Zn.

Cubical Closest-Packed Structure

The packing ABCABC, ... is a *cubical closest-packing* (CCP) or *face-centred cubic packing* (Fig. 3.19.6). The fraction of volume occupied in CCP can be calculated as follows:

The radius of the sphere in terms of the unit length of the face-centred cube is given by

$$r = \frac{\sqrt{2}\,a}{4}$$

since the sphere will be touching each other along the diagonal of the face of the cube.

In the face-centred cubic lattice, there are four spheres per unit cell. Therefore, fraction of volume occupied by the spheres is

$$\frac{4\left\{\frac{4}{3}\pi\left(\frac{\sqrt{2}\,a}{4}\right)^3\right\}}{a^3} = \frac{\sqrt{2}\pi}{6} = 0.740\,5 \quad \text{or} \quad 74.05 \text{ per cent}$$

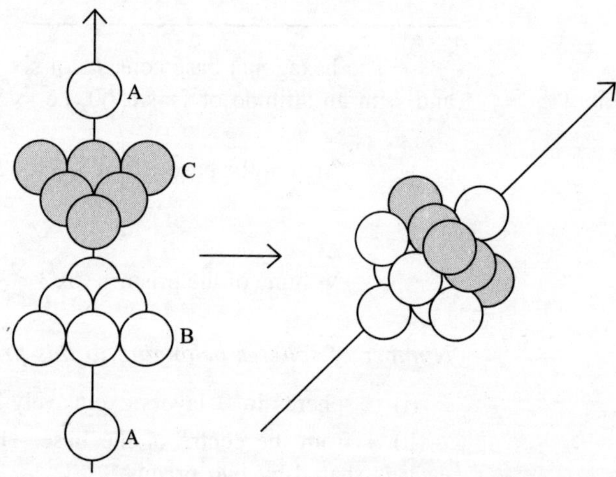

Fig. 3.19.6 Face-centred cubic unit cell formed by ABCA packing

Out of all these packings, HCP and CCP are more common for uniform spheres.

In general, the packing fraction, i.e. fraction of volume occupied, is independent of the radius of the sphere and depends only on the nature of packing. From the values of packing fractions, it follows that the density of a substance in HCP and CCP structures will be more than those in the other two packings.

Example 3.19.1 In a face-centred cubic arrangement of A and B atoms, where A atoms are at the corners of the unit cell and B atoms at the face-centres, one of the A atoms is missing from one corner in each unit cell. What is the simplest formula of the compound?

Solution

Number of atoms A from the corners of unit cell = 7/8
Number of atoms B from the faces of unit cell = 3

Thus, $A : B :: \frac{7}{8} : 3$, so that we can write the formula as $A_{7/8}B_3$.

Hence, the simplest formula is A_7B_{24}.

Example 3.19.2 In a face-centred unit cell with all the positions occupied by A atoms, the body-centred octahedral hole in it is occupied by an atom B of an appropriate size. For such a crystal, calculate the void space per unit volume of unit cell. Also, predict the formula of the compound.

Solution Let a be the edge length of the cube so that

$$4r_A = \sqrt{2}a \quad \text{or} \quad a = 2\sqrt{2}r_A$$

Now, since the atom B is occupied in the body central octahedral hole, it is obvious that

$$2r_A + 2r_B = a \quad \text{or} \quad 2r_A + 2r_B = 2\sqrt{2}r_A$$

or $$2r_B = (2\sqrt{2} - 2)r_A \quad \text{or} \quad \frac{r_B}{r_A} = \sqrt{2} - 1 = 1.44 - 1.0 = 0.414$$

Volume of the cube $= a^3 = 16\sqrt{2}r_A^3$

Since there are 4 atoms of A in the face-centred cube, we will have

$$\text{Volume occupied by A and B} = 4 \times \frac{4}{3}\pi r_A^3 + \frac{4}{3}\pi r_B^3 = \frac{4}{3}\pi(4r_A^3 + r_B^3)$$

Volume occupied per unit volume of unit cell

$$\phi = \frac{\frac{4}{3}\pi(4r_A^3 + r_B^3)}{16\sqrt{2}r_A^3} = \frac{\pi}{12\sqrt{2}}\left\{4 + \left(\frac{r_B}{r_A}\right)^3\right\}$$

$$= \frac{3.14}{12 \times 1.414}\{4 + (0.414)^3\} = \frac{3.14 \times 4.071}{12 \times 1.414} = 0.753\,4$$

Formula of the compound: A_4B

Ionic Crystals

In ionic crystals, basically the structural units are held in position by electrostatic forces. In most of the ionic crystals, cations have only anions as nearest neighbours and vice versa. Earlier we have seen the structure of sodium chloride which consists of the interpenetrating face-centred cubic lattices. Here, each sodium ion is surrounded by six equidistant chloride ions and each chloride ion is, in turn, surrounded by six equidistant sodium ions. The number of nearest neighbours associated with a given ion in a crystal is known as the *coordination number* of the ion.

Lattice Energy

The process of forming an ionic crystal (e.g., NaCl) from the gaseous atoms can be visualized as follows:

$Na(g) \rightarrow Na^+(g) + e^-$	Ionization energy
$Cl(g) + e^- \rightarrow Cl^-(g)$	Electron affinity
$Na^+(g) + Cl^-(g) \rightarrow NaCl(s)$	Lattice energy

The first step involves ionization (easiest for metals); the second step involves electron-capture (favoured for nonmetals): the third involves the interionic attractions and repulsions of ions packed in the lattice, that is, the *lattice energy*.

The last term is determined by the arrangement of ions in a crystal. In general, the crystal structure of ionic crystals is greatly influenced by geometric factors such as size of the ions. The spatial arrangement of ions in ionic crystals is basically different from that in metals. With metals, we are concerned with the packing of spheres of the same size with the same properties; in ionic crystals, though we are still concerned with the packing of spheres, but now the spheres are of different sizes and have different properties.

Common Coordination Numbers

The coordination numbers commonly encountered in ionic crystals are 8, 6 and 4. When a cation has a coordination number of 8, the arrangement is like a body centred cube with the central sphere a cation and the peripheral spheres anions. A coordination number of 6 involves the octahedral configuration, with the anions at the corners of an octahedron and a cation at the centre. A coordination number of 4 involves the tetrahedral arrangement, with the anions at the corners and cation at the centre. These are shown in Fig. 3.19.7.

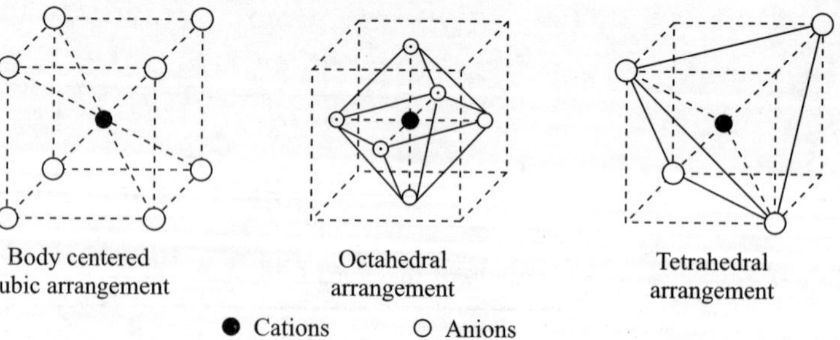

Fig. 3.19.7 Arrangement of ions in crystal lattice

Radius Ratio

A top view of octahedral coordination is shown in Fig. 3.19.8. In this diagram the cation is fairly large so that the anions can touch the cation and still maintain their distances from the other anions. As the cation gets smaller the anions will come close to one another with the corresponding increase in repulsion. This repulsion becomes quite large when the anions just touch one another. Hence, in the case shown in Fig. 3.19.8a the cation has shrunk to its critical size. If the cation gets still smaller, a close approach of anions to the cation will require interpenetration of anionic charge clouds which causes a steep increase in the repulsion energy. Thus, the octahedral configuration becomes unstable when the cation reaches this critical size. On the other hand, the tetrahedral configuration with fewer cation-anions bonds but larger anion-anion distance is the stable form. Thus, the arrangement of the ions in the crystal is determined by the relative sizes of cations and anions. One can calculate the limiting value r_c/r_a, known as radius ratio, from the packing of the ions.

For Octahedral Coordination

From Fig. 3.19.8a, the radius ratio for the octahedral arrangement of ions can be obtained as follows:

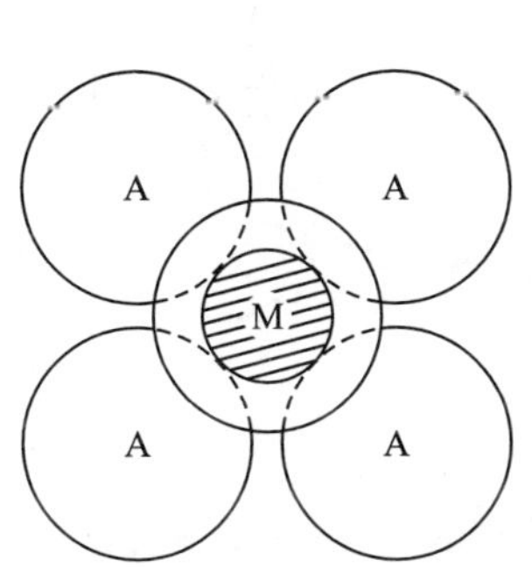

Fig. 3.19.8 A top view of octahedral coordination

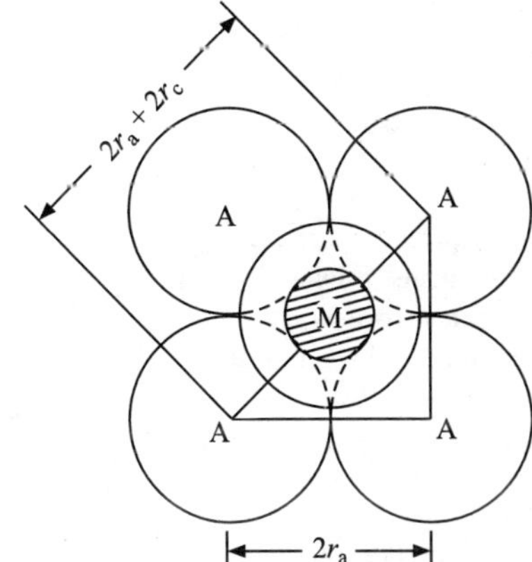

Fig. 3.19.8 (a) Closest approach of anions in octachedral coordination

Draw a right angle triangle passing through the centres of the three anionic spheres. The cation will lie in the middle of the hypotenuse. The sides of the triangle are

$$\text{Base} = 2r_a$$

$$\text{Hypotenuse} = 2r_a + 2r_c$$

It follows that

$$(2r_a + 2r_c)^2 = (2r_a)^2 + (2r_a)^2 \quad \text{or} \quad (2r_a + 2r_c)^2 = 2(2r_a)^2$$

i.e. $$2r_a + 2r_c = 2\sqrt{2}\, r_a$$

$$\frac{r_c}{r_a} = \sqrt{2} - 1 = 0.414$$

This gives the lowest value for the radius ratio that will allow six anions to be arranged octahedrally about the central cation. Whenever the ratio of cation to anion radii falls below 0.414, transition from an octahedral to a tetrahedral configuration is likely to occur.

The critical ratio for the other coordination numbers can be worked out in a similar manner, which we shall discuss in the following.

Eight Coordinate (Body-Centred Cubic Crystal)

A schematic diagram for the body-centred ionic cubic packing is shown in Fig. 3.19.9. It can be seen that each side of the cube has length equal to $2r_a$.

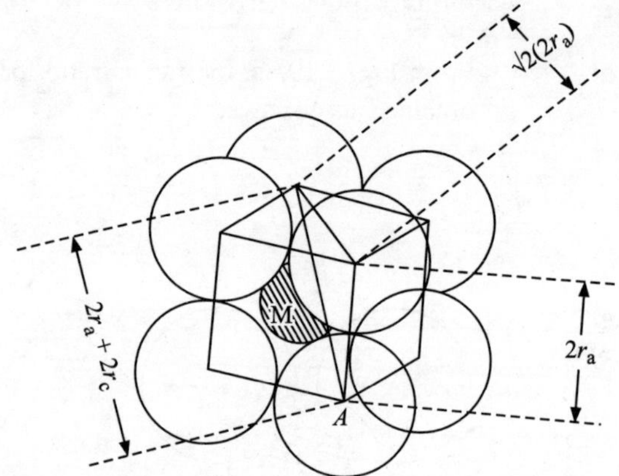

Fig. 3.19.9 Arrangement indicating 8 coordination number in a body-centred cubic crystal

The diagonal across each face of the cube $= \sqrt{2}(2r_a)$.

The cross-diagonal of the cube, which contains the body-centred ions, has length equal to $(2r_c + 2r_a)$.

Using Pythagoras's theorem, we get

$$(2r_c + 2r_a)^2 = (2r_a)^2 + (2\sqrt{2}r_a)^2 \quad \text{or} \quad (2r_c + 2r_a)^2 = 12r_a^2$$

or $$2r_c + 2r_a = 2\sqrt{3}\, r_a$$

or $$\frac{r_c}{r_a} = \sqrt{3} - 1 = 1.732 - 1$$

$$= 0.732$$

Again, it should be noted that this ratio corresponds to the minimum ratio, i.e. it corresponds to the case in which all spheres are in contact. Thus, if r_c were to decrease or r_a to increase, the eight spheres would no longer fit about the central sphere in this configuration without increased overlap of the electron clouds of the ions. This overlap will lead to increased repulsive energy and, therefore, an unstable crystal.

Four Coordinate (Tetrahedral Arrangement)

A tetrahedral arrangement can be drawn within a cube as shown in Fig. 3.19.10. The anions 1 and 3 (or 1 and 2, 2 and 4, 2 and 3, etc.) will touch each other and lie on the face-diagonal, the length of which is $\sqrt{2}a$. Thus

$$r_a + r_a = \sqrt{2}a \quad \text{or} \quad r_a = a/\sqrt{2}$$

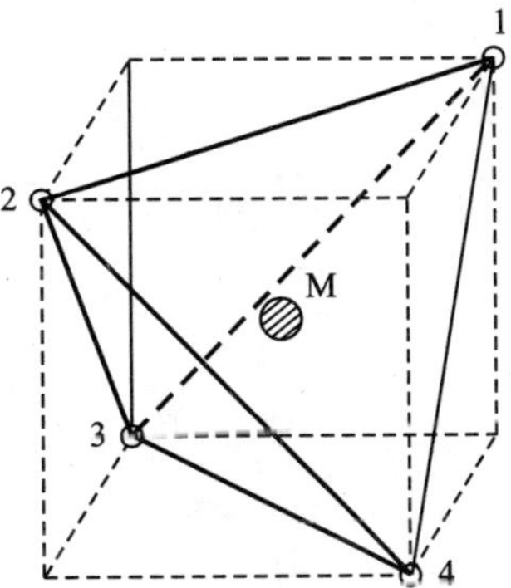

Fig. 3.19.10 Arrangement indicating 4 coordination number in a tetrahedral

The central cation M will be just in between the cross diagonal, the length of which is $\sqrt{3}a$.

Thus $\quad 2(r_a + r_c) = \sqrt{3}a \quad$ or $\quad r_a + r_c = \sqrt{3}a/2$

Dividing this by r_a, we get

$$1 + \frac{r_c}{r_a} = \left(\frac{\sqrt{3}a}{2}\right)\left(\frac{\sqrt{2}}{a}\right) = \sqrt{\frac{3}{2}} = \frac{1.732}{1.414}$$

or $$\frac{r_c}{r_a} = \frac{1.732}{1.414} - 1 = \frac{0.318}{1.414} = 0.225$$

Three Coordinate (Triangular Arrangement)

From Fig. 3.19.11, it is obvious that

$$\text{AB} = \text{AC} = \text{BC} = 2r_a$$

$$\text{CD} = r_c + r_a$$

$$\text{EC} = r_a$$

$$\angle\text{CED} = 90^\circ, \quad \angle\text{ECD} = 30^\circ$$

$$\text{ED} = \frac{r_c + r_a}{2}$$

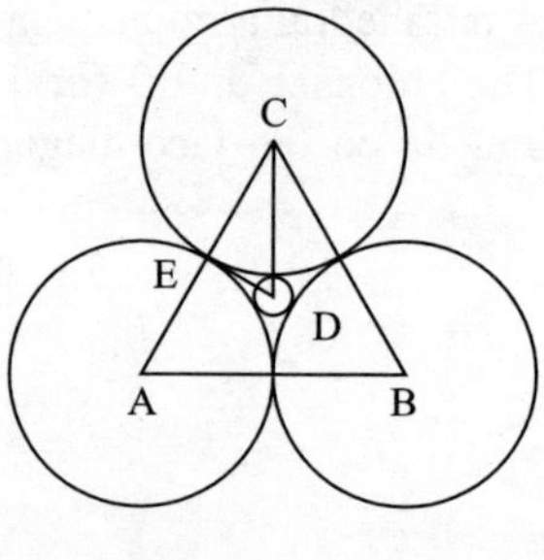

Fig. 3.19.11 Arrangement indicating 3 coordination number in a triangle

Using Pythagoras's theorem in the small triangle CDE, we get

$$CD^2 = EC^2 + ED^2$$

i.e. $$(r_c + r_a)^2 = r_a^2 + \left(\frac{r_c + r_a}{2}\right)^2$$

or $$3\,(r_c + r_a)^2 = 4r_a^2$$

Thus $$r_c + r_a = \frac{2r_a}{\sqrt{3}}$$

and $$\frac{r_c}{r_a} = \frac{2}{\sqrt{3}} \quad 1 = 0.155$$

Summary of Critical Ratio

Thus, the critical ratios are

$$\text{Cubic (8)} \xrightleftharpoons{0.732} \text{Octahedral (6)} \xrightleftharpoons{0.414} \text{Tetrahedral (4)}$$

$$\text{Tetrahedral (4)} \xrightleftharpoons{0.225} \text{Triangular (3)}$$

$$\text{Linear (2)} \xrightleftharpoons[0.155]{} \text{Triangular (3)}$$

These ratios are approximate in number, since in actual calculations, we must also consider the mutual repulsion and attraction of like and unlike charges, respectively, as well as the polarization of one ion by another.

Chlorides, bromides and iodides of Li^+, K^+ and Rb^+ have structures like that of Cl^- (Br^- and I^-) of sodium, i.e. face-centred cubic crystal. But the chloride, bromide and iodide of Cs have the structure of a body-centred cubic crystal. It follows that each Cs^+ ion is surrounded by eight chloride ions and vice versa since the ratio of r_c/r_a increases in going from RbCl to CsCl.

Ionic Crystals and Closest-Packing

It is impossible for both anions and cations to have closest-packed structures but if one of the ions is much larger than the other, it is common for the larger ions alone to approach a closest-packed structure and smaller ions to fit into holes in this structure (Fig. 3.19.12).

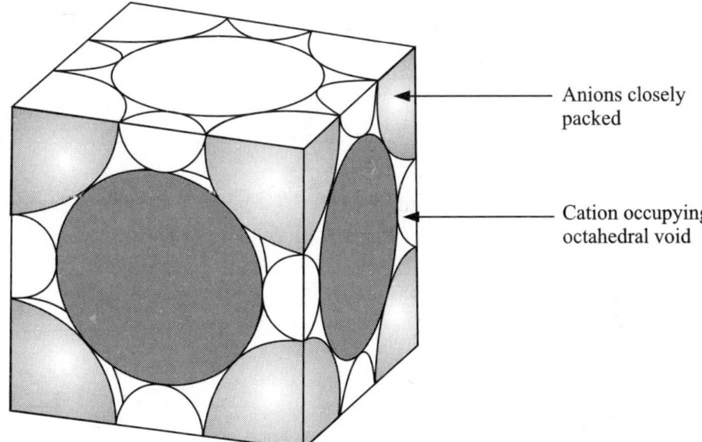

Fig. 3.19.12 Closest Packing in Ionic Crystal

Type of Voids Present in Ionic Crystals

In closest-packing two types of holes are observed: (i) tetrahedral hole, and (ii) octahedral hole.

The tetrahedral hole is formed when a sphere fits into the depression formed by three other spheres closest-packed (forming an equilateral triangle) in two dimensions as shown in Fig. 3.19.13a. In two closest-packed layers, there is such a hole above each atom in the first layer and below each atom in the second layer. In a multilayered closest-packed structure, there is a tetrahedral hole above and below each atom; hence there are twice as many tetrahedral holes as there are closest-packed atoms.

The octahedral hole is formed when the three closest-packed spheres of one layer (forming an equilateral triangle) is put over three closest-packed spheres of the second layer, their positions being inverted with respect to each other (Fig. 3.19.13b).

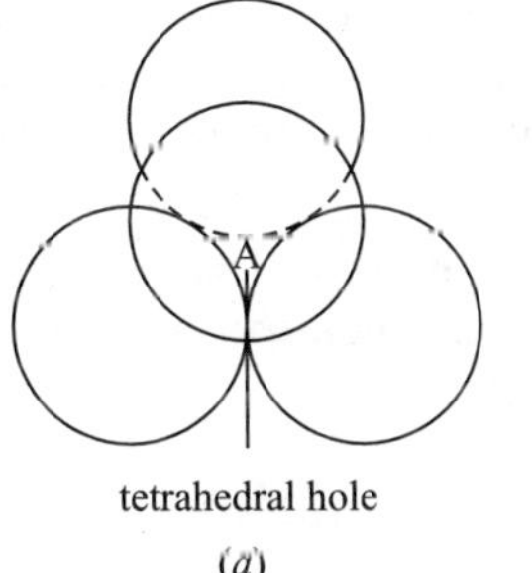

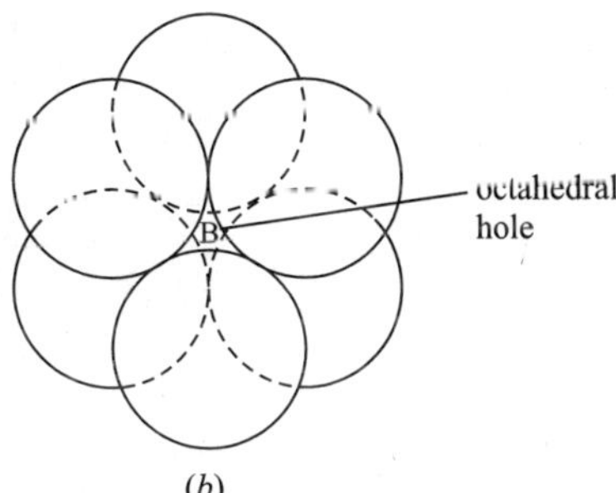

Fig. 3.19.13 (a) Tetrahedral hole shown by position A and (b) octahedral hole shown by position B

In a closest-packed structure, the number of octahedral holes is equal to the number of atoms present and they are located midway between the two closest-packed layers. The positions of these two holes relative to those of the two closest-packed layers are shown below:

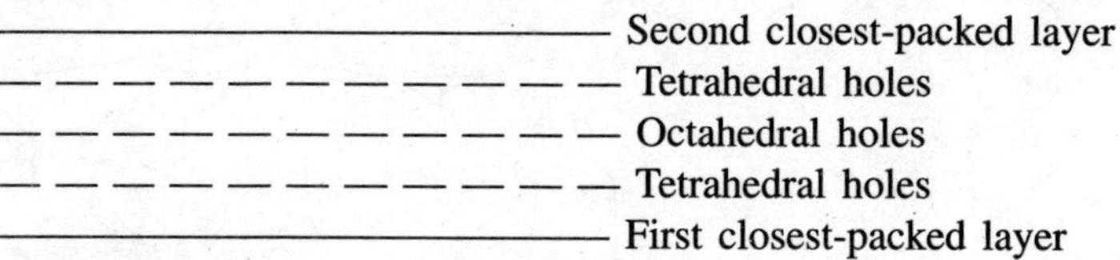

The relation of these holes to the face-centred cubic closest-packing is shown in Figs 3.19.14 and 3.19.15.

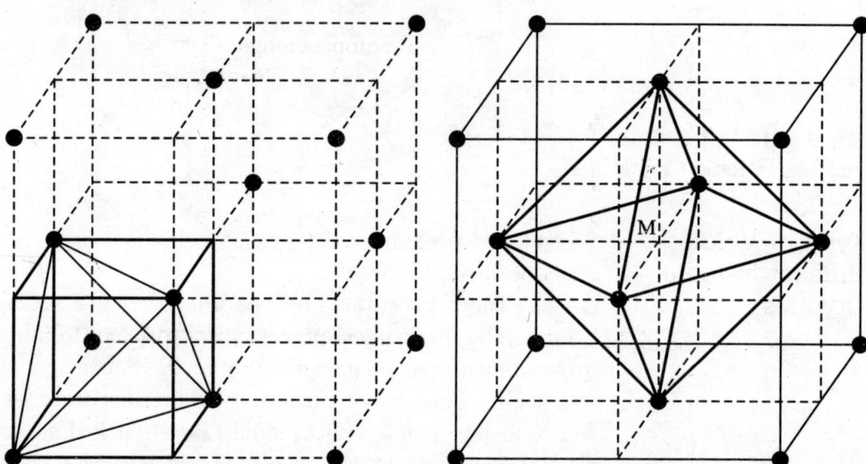

Fig. 3.19.14 Tetrahedral hole in closed-packed atoms

Fig. 3.19.15 Octahedral hole in closed-packed atoms

In Fig. 3.19.14, eight small cubes are shown. The atoms in one small cube actually touch one another. Since these atoms are at alternate corners of the cube they form a tetrahedron with a hole in the centre. The entire unit cell contains eight tetrahedral holes, one in each small cube. Since there are four atoms per cubic unit cell, there will be two tetrahedral holes per closest-packed atom.

The intersections of the lines at M in Fig. 3.19.15 is not occupied by an atom and this is surrounded by six atoms at the corners of an octahedron. This intersection becomes an octahedral hole when the atoms just touch one another. Besides the intersection M, there are twelve more intersections at the centres of the edges. Since each edge is common to four unit cells, there are only three intersections (holes) from edges that belong completely to one unit cell. Thus, in all we have four octahedral holes in one unit cell and hence one octahedral hole per closest-packed atom.

Structure of Common Ionic Compounds

Most of ionic compounds are formed with larger-sized ions (mainly, anions) occupy the lattice positions of a face-centred cubic (FCC) unit cell and smaller-sized ions (mainly, cations) occupy octahederal/terahederal voids in this structure. With this arrangement anions-cations are close to each other.

The common structures of ionic compounds are as follows.

Rock-Salt Structure Anions occupy the lattice positions of the FCC unit cell and cations occupy octahedral voids. When this occurs, cations also occupy lattice points of FCC unit cell. Hence, the unit cell of rock-salt structure contains alternate cations and anions, each occupying a face-centred cubic arrangement. The entire structure may be visualised as the interpenetration of two separate face-centred lattices, one occupied by cations only and the other of anions only (Fig. 3.19.16).

General Formula

The general formula of rock-salt structure is MX. This can be derived either through the concept of coordination number or through the concept of unit cell.

The number of closest cations around any single anion is six and vice versa. Thus, the coordination number in this case is six for cation as well as for anion. Therefore, only 1/6 of a given cation belongs to an anion neighbour and only 1/6 of the anion belongs to the neighbour cation. Hence, the formula is $M_{1/6}X_{1/6}$ or MX. The same formula will be obtained if we consider the number of cations and anions in a unit cell.

Number of cations: 1 central atom wholly within the cell
3 from 12 edges cations, each being shared by 4 unit cells
Total: 4 cations per unit cell

Number of anions: 1 from 8 corners anions, each being shared by 8 unit cells
3 from 6 anions situated in the centre of 6 faces, each being shared by two unit cells
Total: 4 anions per unit cell

Thus, the formula is M_4X_4 or MX.

Examples are MgO, CaO, SrO, BaO and all alkali halides except CsCl, CsBr, and CsI.

Void Space per Unit Volume of Unit Cell

In the rock-salt structure, anion-cation-anion touch each other along the edge of the unit cell. Hence,

$$a = 2(r_c + r_a)$$

There are 4 cations and 4 anions per unit cell. The fraction of volume occupied by ions in a unit cell is

$$\phi = \frac{4\left(\frac{4}{3}\pi r_a^3\right) + 4\left(\frac{4}{3}\pi r_c^3\right)}{\{2(r_a + r_c)\}^3} = \left(\frac{2}{3}\pi\right)\left\{\frac{1 + (r_c/r_a)^3}{(1 + r_c/r_a)^3}\right\}$$

For closest-packed structure (where anions touch each other along the face-diagonal of unit cell, see Fig. 3.19.12), the ratio $r_c/r_a = 0.414$. Hence

$$\phi = \left\{\frac{2}{3}(3.1416)\right\}\left\{\frac{1 + (0.414)^3}{(1 + 0.414)^3}\right\} = 0.793$$

Void volume fraction = 1 – 0.793 = 0.207 per unit volume of unit cell

---

Example 3.19.3 Compute the void space per unit volume of unit cell of sodium chloride structure. Given: $r(Na^+)$ = 95 pm and $r(Cl^-)$ = 181 pm.

Solution We have

$$\text{—} = \frac{95\ \text{pm}}{181\ \text{pm}} = 0.525$$

This ratio is larger than the critical ratio of 0.414 for the octahedral void in the closest-packed structure. Hence, the Cl^- ions are not closely packed but occupy lattice points of FCC unit cell.

Hence

$$\phi = \left(\frac{2}{3}\pi\right)\left\{\frac{1+(r_c/r_a)^3}{1+r_c/r_a)^3}\right\} = 2(3.1416)\left\{\frac{1+(0.525)^3}{(1+0.525)^3}\right\}$$
$$= 0.676$$

Void volume fraction = 1 – 0.676 = 0.324 per unit volume of unit cell.

Body-centred Cubic Structure

In this structure, the cation (or anion) is located at the body centre of primitive cube of anions (cations). The crystal lattice may be considered as the half-way interpenetration of two separate primitive cubic lattices, one occupied by cations and the other one by anions only (Fig. 3.19.16b)

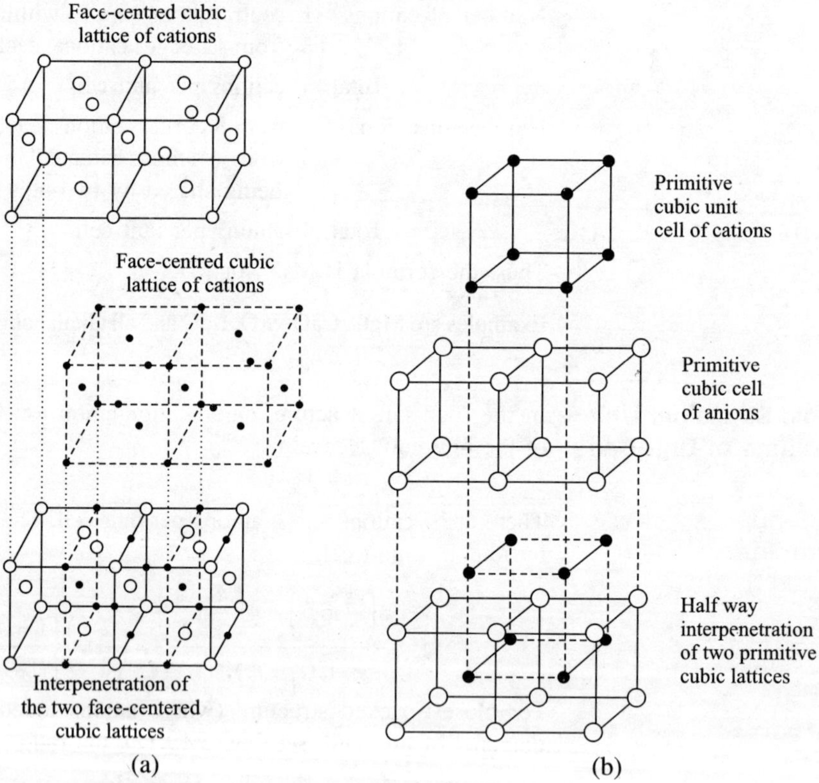

Fig. 3.19.16a Inter-penetration of two face-centred lattices

Fig. 3.19.16b Half-way interpenetration of two primitive lattices

General Formula

Each cation and anion has eight oppositely-charged ions at its next nearest neighbours. Hence, the coordination number of each ion is eight. Thus, the formula is $M_{1/8}X_{1/8}$ i.e. MX. The same formula is obtained if the number of cations and ions in a unit cell are considered.

Considering anions (or cations) occupying lattice points of a primitive unit cell and a (an) cation (or anion) occupying the body centre of the cube, we have

Number of anions (or cations) per unit cell 8(1/8) = 1

Number of cations (or anions) per unit cell = 1

Thus, there is one anion and one cation in a unit cell. Hence, the formula is MX. Examples are CsCl, CsBr and CsI.

Void Space per unit Cell

In this structure anion-cation-anion (or cation-anion-cation) touch each other along the body diagonal of the cube. Hence

$$2(r_c + r_a) = \sqrt{3}\,a$$

Since there is one cation and one anion in a unit cell, the fraction of volume occupied in a unit cell is

$$\phi = \frac{1\left(\frac{4}{3}\pi r_c\right)^2 + 1\left(\frac{4}{3}\pi r_c^3\right)}{\{(2/\sqrt{3})(r_c + r_a)\}^3} = \left(\frac{\sqrt{3}}{2}\pi\right)\left\{\frac{1 + (r_c/r_a)^3}{(1 + r_c/r_a)^3}\right\} \quad \text{or} \quad \left(\frac{\sqrt{3}}{2}\pi\right)\left\{\frac{1 + (r_a/r_c)^3}{(1 + r_a/r_c)^3}\right\}$$

For closest-packed structure, the ratio r_c/r_a (or r_a/r_c) is 0.732. Hence

$$\phi = \left\{\frac{\sqrt{3}}{2}(3.1416)\right\}\left\{\frac{1 + (0.732)^3}{(1 + 0.732)^3}\right\} = 0.729$$

Void volume fraction = 1 – 0.729 = 0.271 per unit volume of unit cell

Example 3.19.4

Compute the void volume of unit cell of CsCl structure. Given: $r(Cs^+)$ = 181 pm and $r(Cl^-)$ = 167 pm

Solution

We have

$$\frac{r_a}{r_c} = \frac{167\ \text{pm}}{181\ \text{pm}} = 0.923$$

This ratio is larger than the critical ratio of 0.732 for the body-centred cubic structure. Hence, the Cs^+ ions are not closely packed.

The fraction of volume occupied per unit volume of unit cell is

$$\phi = \left(\frac{\sqrt{3}}{4}\right)(3.1416)\left\{\frac{1 + (0.923)^3}{(1 + 0.923)^3}\right\} = 0.683$$

Void volume fraction = 1 – 0.683 = 0.317 per unit volume of unit cell.

Anti-fluorite Structure

Anions occupy the lattice positions of the FCC unit cell and cations occupy tetrahedral voids (Fig. 3.19.17)

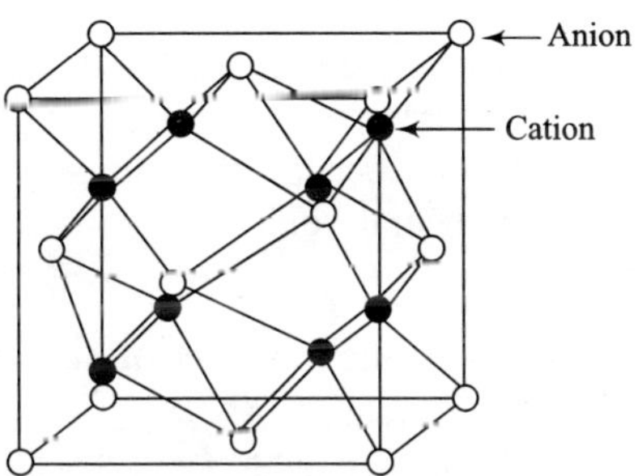

Fig. 3.19.17 Anti-fluorite Structure

General Formula

Coordination number of cations = 4

Coordination number of anions = 8

The formula is $M_{1/4}X_{1/8}$, i.e. M_2X

Examples include Li_2O, Na_2O, K_2O and Rb_2O

Void Space per unit Volume of Unit Cell

Tetrahederal void exists within the smaller cube of edge-length *a*/2 of the entire cubic unit cell (Fig. 3.19.14). Hence, we write

$$2(r_a + r_c) = \sqrt{3}(a/2) \quad \text{i.e.} \quad a = (4/\sqrt{3})(r_a + r_c)$$

There are 4 anions and 8 cations in a unit cell. Hence, the fraction volume occupied in a unit cell is

$$\phi = \frac{4\left(\frac{4}{3}\pi r_a^3\right) + 8\left(\frac{4}{3}\pi r_c^3\right)}{\{(4/\sqrt{3})(r_a + r_c)\}^3} = \left(\frac{\sqrt{3}}{4}\pi\right)\left\{\frac{1 + 2(r_c/r_a)^3}{(1 + r_c/r_a)^3}\right\}$$

For closest-packed structure, $r_c/r_a = 0.225$. Hence,

$$\phi = \left(\frac{\sqrt{3}}{4}\right)(3.1416)\left\{\frac{1 + 2(0.225)^3}{(1 + 0.225)^3}\right\} = 0.757$$

Void volume fraction = 1 – 0.757 = 0.243 per unit volume of unit cell.

Example 3.19.5

Compute the void space per unit volume of unit cell of Na_2O, which crystallizes in anti-fluorite structure. Given: $r(Na^+)$ = 95 pm and $r(O^{2-})$ = 140 pm.

Solution

We have

$$\frac{r_c}{r_a} = \frac{95\ \text{pm}}{140\ \text{pm}} = 0.679$$

This ratio is larger than the critical ratio of 0.225 for the tetrahederal void in the closest-packed structure. Hence, the O^{2-} ions are not closely packed.

The fraction of volume occupied per unit volume of unit cell is

$$\phi = \left(\frac{\sqrt{3}}{4}\right)(3.1416)\left\{\frac{1 + (0.679)^3}{(1 + 0.679)^3}\right\} = 0.377$$

Void volume fraction = 1 – 0.377 = 0.623 per unit volume of unit cell.

Fluorite Structure

Cations occupy the lattice points of the FCC unit cell and anions occupy tetrahedral voids.

General Formula

Coordination number of cations = 8
Coordination number of anions = 4
The formula is $M_{1/8}X_{1/4}$, i.e. MX_2
Examples include UO_2, ThO_2, etc.

Zinc-Blende Unit Cell

Anions occupy the lattice points of a FCC unit cell and half of tetrahedral voids are occupied by cations (Fig. 3.19.18).

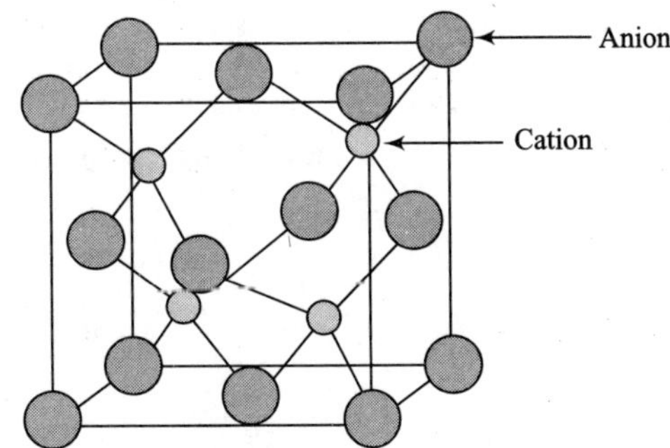

Fig. 3.19.18 Zinc-blende Unit Cell

General Formula

Coordination number of cations = 4
Coordination number of anions = 4
The formula is $M_{1/4}X_{1/4}$, i.e. MX
Example: BeO.

Void Space per Unit Volume of Unit Cell

Tetrahedral void exists within the smaller cube of edge-length $a/2$ of the entire cubic unit cell (Fig. 3.19.14). Hence, we write

$$2(r_a + r_c) = \sqrt{3}\,(a/2) \qquad \text{i.e.} \qquad a = (4/\sqrt{3})(r_a + r_c)$$

There are 4 anions and 4 cations in a unit cell. Hence, the fraction of volume occupied in a unit cell is

$$\phi = \frac{4\left(\frac{4}{3}\pi r_a^3\right) + 4\left(\frac{4}{3}\pi r_c^3\right)}{\{(4/\sqrt{3})(r_a + r_c)\}^3} = \left(\frac{\sqrt{3}}{4}\pi\right)\left\{\frac{1 + (r_c/r_a)^3}{(1 + r_c/r_a)^3}\right\}$$

For closest-packed structure, $r_c/r_a = 0.225$.
Hence,

$$\phi = \left(\frac{\sqrt{3}}{4}\right)(3.1416)\left\{\frac{1 + (0.225)^3}{(1 + 0.225)^3}\right\} = 0.748$$

Void volume fraction = 1 – 0.748 = 0.252 per unit volume of unit cell.

Example 3.19.6

Compute the void volume per unit volume of unit cell of BeO which has a zinc-blende structure. Given: $r(Be^{2+})$ = 45 pm and $r(O^{2-})$ = 140 pm.

Solution

The ratio r_c/r_a of BeO is

$$\frac{r_c}{r_a} = \frac{45\ \text{pm}}{140\ \text{pm}} = 0.321$$

This ratio is larger than the critical ratio of 0.225 for a tetrahederal void, the O^{2-} ions are not closely packed.
The fraction of volume occupied per unit volume of unit cell is

$$\phi = \left(\frac{\sqrt{3}}{4}\right)(3.1416)\left\{\frac{1 + (0.321)^3}{(1 + 0.321)^3}\right\} = 0.610$$

Void volume fraction = 1 – 0.610 = 0.390 per unit volume of unit cell.

Example 3.19.7 The structure of diamond is very similar to that of zinc-blende structure. Carbon atoms form face-centred cubic unit cell with alternate tetrahederal voids occupied by carbon atoms. Compute its void volume per unit volume of unit cell.

Solution The fraction of volume occupied per unit volume of unit cell of zinc-blende structure is

$$\phi = \left(\frac{\sqrt{3}}{4}\pi\right)\left\{\frac{1+(r_c/r_a)^3}{(1+r_c/r_a)^3}\right\}$$

For diamond, $r_c = r_a$. Hence, the fraction of volume occupied per unit volume of unit cell is

$$\phi = \left(\frac{\sqrt{3}}{4}\right)(3.1416)\left(\frac{2}{8}\right) = 0.34$$

Void volume fraction = 1 – 0.34 = 0.66/per unit volume of unit cell.

Corundum Structure

The packing of anions is hexagonal rather than cubic. Not all the octahedral voids are occupied; only 2/3 of these contain cations.

General Formula: M_4X_6 or M_2X_3

Examples: Fe_2O_3, Al_2O_3 and Cr_2O_3

Void Volume per Unit Volume of Unit Cell

Assuming anions form hexagonal closest-packing and cations occupy only two-thirds of octahedral holes, we have

$$\text{Volume of hexagon} = 24\sqrt{2}r_a$$

There are six anions and six octahedral holes per unit cell.

Fraction of volume occupied per unit volume of unit cell is

$$\phi = \frac{6\left(\frac{4}{3}\pi r_a^3\right) + \frac{2}{3}6\left(\frac{4}{3}\pi r_c^3\right)}{24\sqrt{2}r_a^3} = \frac{\pi}{3\sqrt{2}}\left\{1+\frac{2}{3}(0.414)^3\right\}$$

$$= \frac{3.1416}{3\times 1.414}\left\{1+\frac{2}{3}\times 0.070\,96\right\} = \frac{3.1416}{3\times 1.414}\{1+0.0473\} = 0.776$$

Void volume fraction = 0.224 per unit volume of unit cell.

Spinel Structure

A spinel is an important class of oxides containing two types of metal ions with the oxide ions arranged in cubical-closest packing. The normal spinel has one-eighth of the tetrahedral holes occupied by divalent metal ions (A^{2+}) and one-half of the octahedral holes occupied by trivalent metal ions (B^{3+}). In a unit cell, we have

$$\text{Number of divalent metal ions, } A^{2+} = \frac{1}{8}(8) = 1$$

$$\text{Number of trivalent metal ions, } B^{3+} = \frac{1}{2}(4) = 2$$

$$\text{Number of oxide ions, } O^{2-} = 4$$

Hence, formula of normal spinel is AB_2O_4

Examples are $ZnAl_2O_4$, $MgAl_2O_4$ and $ZnFe_2O_4$.

Void Volume per Unit Volume of Unit Cell

Assuming oxides form cubical-closest packing ($4r_O = \sqrt{2}a$), A^{2+} ions occupy one-eighth of tetrahedral holes ($r_A = 0.225\ r_O$) and B^{3+} ions occupy one-half of octahedral holes ($r_B = 0.414\ r_O$). Hence, fraction of volume occupied per unit volume of unit cell is

$$\phi = \frac{4\left(\frac{4}{3}\pi r_O^3\right) + \left(\frac{4}{3}\pi r_A^3\right) + 2\left(\frac{4}{3}\pi r_B^3\right)}{(2\sqrt{2}r_O)^3}$$

$$= \frac{\left(\frac{4}{3}\pi\right)\left[4 + (0.225)^3 + 2(0.414)^3\right]}{(2\sqrt{2})^3}$$

$$= \frac{(4.189)\ (4 + 0.011\,4 + 0.142)}{22.627} = \frac{17.398}{22.627} = 0.77$$

Void volume fraction = 1 – 0.77 = 0.23 per unit volume of unit cell.

Example 3.19.8 Compute void volume per unit volume of unit cell described in Example 3.19.1.

Solution There are (7/8) A atom and 3 B atoms per unit cell. Also

$$2r_A + 2r_B = \sqrt{2}a \quad \text{or} \quad a = \sqrt{2}r_A + \sqrt{2}r_B$$

Volume of unit cell $= a^3 = (\sqrt{2}r_A + \sqrt{2}r_B)^3 = 2\sqrt{2}(r_A + r_B)^3$

Fraction of volume occupied/unit volume of unit cell is given by

$$\phi = \frac{\frac{7}{8}\left(\frac{4}{3}\pi r_A^3\right) + 3\left(\frac{4}{3}\pi r_B^3\right)}{2\sqrt{2}(r_A + r_B)^3} = \frac{\frac{7}{6}\pi r_A^3 + 4\pi r_B^3}{2\sqrt{2}(r_A + r_B)^3}$$

$$= \left[\frac{\frac{7}{6}\pi r_A^3 + 4\pi(0.414r_A)^3}{2\sqrt{2}\{r_A + (0.414r_A)\}^3}\right] = \frac{\pi r_A^3\left\{\frac{7}{6} + 4(0.414)^3\right\}}{2\sqrt{2}r_A^3(1 + 0.414)^3}$$

$$= \frac{3.141\,6(1.166\,7 + 0.283\,8)}{2(1.414)^4} = 0.570$$

Volume unoccupied fraction = 0.430 per unit volume of unit cell.

Example 3.19.9 The ionic radii of the alkali metal ions are Li^+ 68 pm, Na^+ 95 pm, K^+ 133 pm, Rb^+ 148 pm, Cs^+ 169 pm, F^- 136 pm, Cl^- 181 pm, Br^- 195 pm, I^- 216 pm. If the radius ratio rules hold good, predict the structures exactly, give the expected cation coordination number in each case.

Solution The radius ratio of various molecules along with the expected coordination numbers of cations in parenthesis are given below:

	F^-	Cl^-	Br^-	I^-
Li^+	0.5(6)	0.375 7(4)	0.348 7(4)	0.314 8(4)
Na^+	0.698 5(6)	0.524 8(6)	0.487 2(6)	0.439 7(6)
K^+	0.978 1(8)	0.734 8(6)	0.682 2(6)	0.615 7(6)
Rb^+	1.09(8)	0.817 7(8)	0.759 1(8)	0.685 1(6)
Cs^+	1.243(8)	0.933 7(8)	0.866 8(8)	0.782 3(8)

Cohesive Energy of Ionic Crystals

The cohesive energy of an ionic crystal is defined as the energy required to obtain infinitely separated gaseous ions from one mole of an ionic crystal lattice, according to the following reaction:

$$MX(s) \rightarrow M^+(g) + X^-(g)$$

This is negative of the lattice energy which is, by definition, the amount of heat released when one mole of an ionic crystal lattice is formed starting from the requisite number of ions in the gaseous state. The value of the lattice energy (or cohesive energy) is not directly measurable and is, therefore, determined indirectly using the Born-Haber cycle.

The energy of formation of an ionic crystal can be determined experimentally. The equation relating to this is given by

$$M(s) + \frac{1}{2}X_2(g) \rightarrow MX(s) \qquad \Delta_f H$$

This reaction can be obtained by adding the following reactions:

$M(s) \rightarrow M(g)$	$\Delta_{sub}H$	(sublimation energy)
$M(g) \rightarrow M^+(g) + e^-$	$\Delta_{ioniz}H$	(ionization energy)
$\frac{1}{2}X_2(g) \rightarrow X(g)$	$\frac{1}{2}\,\Delta_{diss}H$	(half of dissociation energy)
$X(g) + e^- \rightarrow X^-(g)$	$\Delta_{EA}H$	(electron affinity)
$M^+(g) + X^-(g) \rightarrow MX(s)$	$\Delta_{LE}H$	(lattice energy)

It follows that

$$\Delta_f H = \Delta_{sub}H + \Delta_{ioniz}H + \frac{1}{2}\Delta_{diss}H + \Delta_{EA}H + \Delta_{LE}H$$

Knowing $\Delta_f H$ and other parameters, the lattice energy can be calculated. The cohesive energy will be negative of this lattice energy.

Computation of Lattice Energy from Crystal Geometry

The lattice energy of an ionic compound may be calculated if the crystal geometry and distances between the ions in the crystal are known. An expression for the electrostatic interactions of ions may be derived by treating the charges of the ions as point charges located in the centres of the ions. The expression for potential energy of interaction of two point charges, that is, the energy associated in bringing the two charges together from infinity to a separation distance d, may be derived from Coulomb's law.

$$PE = \frac{Q_1 Q_2}{(4\pi\varepsilon_0)\, d}$$

The potential energy is positive (repulsion) if charges on both the ions have the same sign (either both positive or both negative) indicating that the energy is absorbed when the distance between the charges is decreased. On the other hand, if the charges have unlike signs, the potential energy is negative (attraction) and thus, energy is evolved when the distance between them is decreased.

Illustration of Sodium Chloride

The potential energy of interaction of any ion with other ions can be computed if the distances between the ion under investigation with other ions are known. In NaCl crystal, the distance between neighbouring Na^+ and Cl^- ions is 280 pm (henceforth designated as r). With this information, the distance of any Na^+ from others and the corresponding potential energy can be computed.

Consider, for example, the unit cell of NaCl (Fig. 3.19.19). We shall calculate the potential energy of the sodium ion in the centre of the cube.

Length of the cube = $2r$

This big cube may be considered to be made up of eight smaller cubes with each side of length r.

Face diagonal of the small cube = $\sqrt{2}r$

Cross diagonal of the small cube = $\sqrt{3}r$

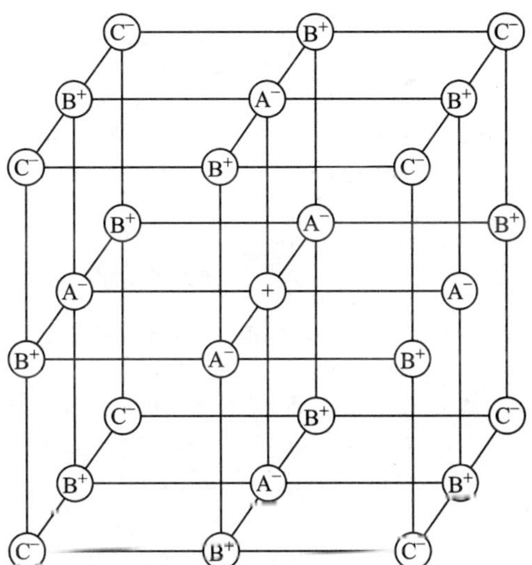

Fig. 3.19.19 A unit cell of sodium chloride crystal (A^-, C^- are the chloride ions, B^+ are the sodium ions)

(1) The nearest neighbours of the reference Na^+ ion are the Cl^- ions (labelled as A^-). There are six Cl^- ions and each is situated at a distance r from the Na^+ ion. Thus, the potential energy (PE) contributed from these six ions is

$$PE_1 = 6\left(\frac{(+e)(-e)}{(4\pi\varepsilon_0)\, r}\right) = -\frac{e^2}{(4\pi\varepsilon_0)\, r}(6)$$

(2) The next nearest neighbours of the central Na^+ ion are the twelve Na^+ ions (labelled as B^+). Each of these ions is situated at a distance $\sqrt{2}r$ from the central Na^+. The contribution from this group is

$$\text{PE}_2 = 12\left(\frac{(e)(e)}{(4\pi\varepsilon_0)\sqrt{2}r}\right) = -\frac{e^2}{r}\left(-\frac{12}{(4\pi\varepsilon_0)\sqrt{2}}\right)$$

(3) Next come, the eight Cl^- ions (labelled as C^-) which are at a distance of $\sqrt{3}r$ from the central ion. Their contribution is

$$\text{PE}_3 = 8\left(\frac{(+e)(-e)}{(4\pi\varepsilon_0)\sqrt{3}r}\right) = -\frac{e^2}{r}\left(\frac{8}{(4\pi\varepsilon_0)\sqrt{3}}\right)$$

If we continue like this for a crystal of a more realistic size, we would find that there are

6 Na^+ ions at $2r$

24 Cl^- ions at $r\sqrt{5}$

24 Na^+ ions at $r\sqrt{6}$, etc.

Adding up all the contributions, we get

$$\text{PE} = -\frac{e^2}{(4\pi\varepsilon_0)r}\left(6 - \frac{12}{\sqrt{2}} + \frac{8}{\sqrt{3}} - \frac{6}{2} + \frac{24}{\sqrt{5}} + \cdots\right)$$

Madelung Constant

The sum of the terms of the infinite series in parentheses is called the *Madelung constant A*. Therefore, we can write

$$\text{PE} = -A\frac{e^2}{(4\pi\varepsilon_0)r}$$

If 1 mole of NaCl is involved, it contains N_A Na^+ ions and N_A Cl^- ions, where N_A is Avogadro constant. The total electrostatic energy for the formation of one mole of NaCl is not $2N_A$ times the above expression, since this procedure would count each interaction twice, but one half of $2N_A$ or N_A times the above expression:

$$\text{PE} = -N_A A\frac{e^2}{(4\pi\varepsilon_0)r}$$

Value of Lattice Energy for Sodium Chloride

For NaCl, $A = 1.748 = 1.75$ and hence the numerical value of Coulomb energy for a single NaCl molecule is

$$\text{PE} = -1.75\frac{e^2}{(4\pi\varepsilon_0)r}$$

$$= -1.75\frac{(1.602\,19\times10^{-19}\text{ C})^2}{4\pi(8.854\times10^{-12}\text{ C}^2\text{ N}^{-1}\text{ m}^{-2})(2.8\times10^{-10}\text{ m})}$$

$$= -1.443\times10^{-18}\text{ N m} = -1.443\times10^{-18}\text{ J}$$

Hence, PE per mole is

$$(-1.443 \times 10^{-18}\ \text{J})(6.022 \times 10^{23}\ \text{mol}^{-1}) = -868\,975\ \text{J mol}^{-1}$$
$$= -869.0\ \text{kJ mol}^{-1}$$

Because of the finite size of the electronic cloud around each ion, there exist repulsive forces between neighbouring ions which we have not included in our calculations. The net effect of these repulsions is to make the actual lattice energy about 10 per cent less negative than the value given by the Coulomb forces alone. Thus, the true lattice energy of NaCl is about -780 kJ mol^{-1}. Comparing this number with the Coulomb energy of a single sodium chloride molecule, we get

$$-\frac{e^2}{(4\pi\varepsilon_0)\,r} = -\frac{(1.602\,19 \times 10^{-19}\ \text{C})^2}{4\pi(8.854 \times 10^{-12}\ \text{C}^2\ \text{N}^{-1}\ \text{m}^{-2})(2.8 \times 10^{-10}\ \text{m})}$$
$$= -8.244 \times 10^{-19}\ \text{J}$$

Hence, the Coulomb energy per mol is

$$-(8.244 \times 10^{-19}\text{J})(6.022 \times 10^{23}\ \text{mol}^{-1}) = -496\,454\ \text{J mol}^{-1}$$
$$= -496.5\ \text{kJ mol}^{-1}$$

which indicates that the crystal is more stable by an approximate amount of 283 kJ mol^{-1}.

General Expression of Potential Energy of Interaction for Symmetrical Salts

The crystal structures of symmetrical salts such as rock salt, CsCl, zinc blende and wurtzite, are all cubic. If we know the distance r between a cation and the nearest anion, then the distance between any two ions in the crystal can be computed from the geometry of the structure. We write the distance between ions i and j as $r_{ij} = \alpha_{ij} r$ in which α_{ij} is a numerical factor obtained from the geometry of the structure. Thus

$$\text{PE} = \frac{(z_i e)(z_j e)}{(4\pi\varepsilon_0)\,r\,\alpha_{ij}}$$

The interaction energy of the ion i with all the others is obtained by summing this expression, that is,

$$\text{PE} = \frac{(z_i e)}{(4\pi\varepsilon_0)\,r} \sum_{i \neq i} \frac{(z_j e)}{\alpha_{ij}}$$

In every term of this sum $z_i = \pm z_j$ so that we can write

$$\text{PE} = \frac{(z_i e)^2}{(4\pi\varepsilon_0)\,r} \sum_{j \neq i} \left(\pm \frac{1}{\alpha_{ij}} \right)$$

Each term in the sum is simply a number, determined by geometry, so the sum is a number which we write as S_i. Thus

$$\text{PE} = E_i = \frac{(z_i e)^2 \, S_i}{(4\pi\varepsilon_0)\, r}$$

Madelung Energy

Sum of the energies of all the ions in the lattice yields the total energy of interaction E_M which is known as *Madelung energy*.

$$E_M = \frac{1}{2}\sum_i E_i = \frac{(ze)^2}{(4\pi\varepsilon_0)\, r}\sum_i \frac{1}{2} S_i; \qquad \left(z_i^2 = z_j^2 = z^2\right)$$

(Note the division by 2 since otherwise the interaction between any pair of ions would have been counted twice.)

The sum $\sum_i S_i/2$ in the above expression is simply a summation of numbers and is found to be negative and proportional to N_A. Thus

$$\sum_i \frac{1}{2} S_i = -AN_A$$

where A is *Madelung constant*. The cohesive energy of the crystal is negative of the interaction energy. Thus

$$E_c = -E_M = N_A A \frac{(ze)^2}{(4\pi\varepsilon_0)\, r}$$

Born Repulsion

As stated earlier, the experimental value of cohesive energy is about 10 per cent smaller than the theoretical value. This is a consequence of neglecting the repulsion which arises at close distances when the electronic cloud of one ion begins to encroach upon another's domain. Its nature is the same as those of neutral atoms and is given by a term b/r^n where b and n are constants, n varying usually from 6 to 12. This form of repulsion energy was first introduced by M. Born, and is called the *Born repulsion*. Thus, the cohesive energy is written as

$$-E_c = -\frac{N_A A (ze)^2}{(4\pi\varepsilon_0) r} + \frac{b}{r^n}$$

Often n can be determined by studying the compressibility coefficient of the crystal. It is given by

$$n = 1 + \frac{9V_0 r_0}{N_A A (ze)^2 \kappa}$$

where $\kappa = -\dfrac{1}{V_0}\left(\dfrac{\partial V}{\partial p}\right)_T$

The term b can be eliminated from the above expression for the lattice energy by recognizing that at the value of r_0 for which the crystal is most stable, E_c is minimum (Fig. 3.19.20) and thus $dE_c/dr = 0$. Therefore, on differentiating we get

$$-\frac{dE_c}{dr} = \frac{N_A A(ze)^2}{(4\pi\varepsilon_0)r^2} - \frac{nb}{r^{n+1}}$$

Setting this equal to zero at r_0, and solving for b, we obtain

$$b = \frac{N_A A(ze)^2 r_0^{n-1}}{(4\pi\varepsilon_0)\, n}$$

Thus, $$-E_c = -\frac{N_A A(ze)^2}{(4\pi\varepsilon_0)\, r_0} + \frac{N_A A(ze)^2 r_0^{n-1}}{(4\pi\varepsilon_0)\, nr_0^n} = -\frac{N_A A(ze)^2}{(4\pi\varepsilon_0)\, r_0}\left(1 - \frac{1}{n}\right)$$

The value of n is approximately 10. Thus, we see that the true cohesive energy is only about 10 per cent different from the value calculated using Coulomb interaction alone.

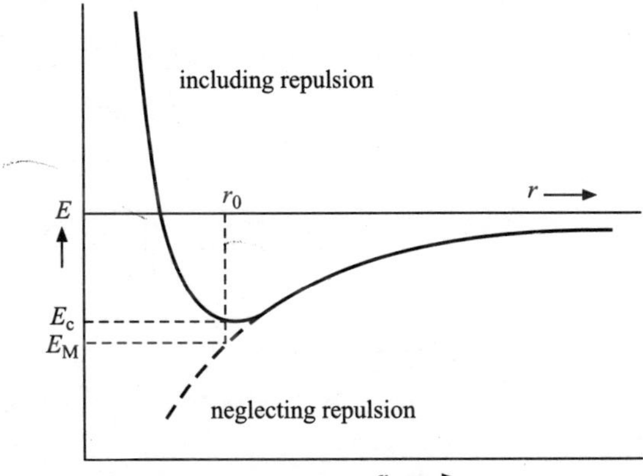

Fig. 3.19.20 Variation of potential energy of interaction with *r*

Example 3.19.10 Calculate the lattice energy of KCl when a_0 = 628 pm, n(exponent of repulsion term) = 9 and A = 1.746.

Solution Substituting the values in the relation

$$-E_c = -\frac{N_A A(ze)^2}{(4\pi\varepsilon_0)\, r_0}\left(1 - \frac{1}{n}\right)$$

we get $$-E_c = -\frac{(6.022 \times 10^{23}\ \text{mol}^{-1})\,(1.746)\,(1 \times 1.602\ 19 \times 10^{-19}\ \text{C})^2}{4\pi(8.84 \times 10^{\ 12}\ \text{C}^2\ \text{N}^{-1}\ \text{m}^{-2})\,(3.14 \times 10^{-10}\ \text{m})} \times \frac{8}{9}$$

or $$E_c = 6.867 \times 10^5\ \text{J mol}^{-1} = 686.7\ \text{kJ mol}^{-1}$$

Covalent Crystals

Maximum stability in covalent crystals is not obtained with the greatest possible number of neighbours, but by forming the allowed number of covalent bonds in the proper directions. We give in the following a few elementary remarks about covalent crystals.

'Few solids are held together exclusively by covalent bonds. The majority of solids incorporating covalent bonds are bound also by either ionic bond or van der Waals forces. The common occurrence is to find distinct molecules held together by covalent bonds and the molecules bound in the crystal by van der Waals forces.'

Diamond Structure

Only those atoms which form four covalent bonds produce a repetitive three-dimensional structure using only covalent bonds, e.g., diamond structure. The latter is based on a face-centred cubic lattice where four out of the eight tetrahedral holes are occupied by carbon atoms. Every atom in this structure is surrounded tetrahedrally by four others. No discrete molecule can be discerned in diamond. The entire crystal is a giant molecule, a unit cell of which is shown in Fig. 3.19.21. Germanium, silicon and grey tin also crystallize in the same way as diamond.

Void Volume in Diamond Crystal

From Fig. 3.19.21, we observe that

C—C bond = $2r_c$, where r_c is the radius of sphere representing carbon

$\angle$C—C—C = 109° 28′

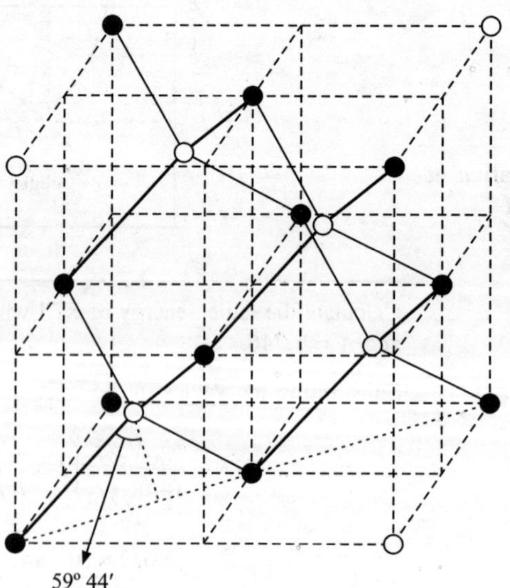

Fig. 3.19.21 A unit cell of diamond crystal

Diagonal face of the cube = $\sqrt{2}a$

This distance will be equal to four times the distance $d_{C—C}$ sin (54° 44′), that is

$$4d_{C—C} \sin (54° 44') = \sqrt{2}a$$

or $$a = \frac{4d_{C—C} \sin (54° 44')}{\sqrt{2}}$$

Area of the cube = $a^3 = \left\{\frac{4d_{C—C} \sin (54° 44')}{\sqrt{2}}\right\}^3$

Total number of carbon atoms/unit cell = 8

Volume occupied by 8 carbons = $8\left(\frac{4}{3}\pi r_C^3\right) = \frac{4}{3}\pi(d_{C—C})^3$

Fraction of volume occupied = $\dfrac{\frac{4}{3}\pi(d_{C—C})^3}{\left\{\frac{4d_{C—C} \sin (54° 44')}{\sqrt{2}}\right\}^3} = 0.34$

Example 3.19.11 The unit cell of diamond structure is face-centred cube with atoms at 0, 0, 0; $\frac{1}{2}, 0, \frac{1}{2}$; $0, \frac{1}{2}, \frac{1}{2}$; $\frac{1}{2}, \frac{1}{2}, 0$; $\frac{1}{4}, \frac{1}{4}, \frac{1}{4}$; $\frac{3}{4}, \frac{3}{4}, \frac{1}{4}$; $\frac{3}{4}, \frac{1}{4}, \frac{3}{4}$; $\frac{1}{4}, \frac{3}{4}, \frac{3}{4}$. The unit cell edge length is 357 pm. Calculate (a) carbon-carbon bond length in diamond; and (b) bond angle ∠CCC.

Solution **(a)** The positions of three carbon atoms are shown in Fig. 3.19.22.

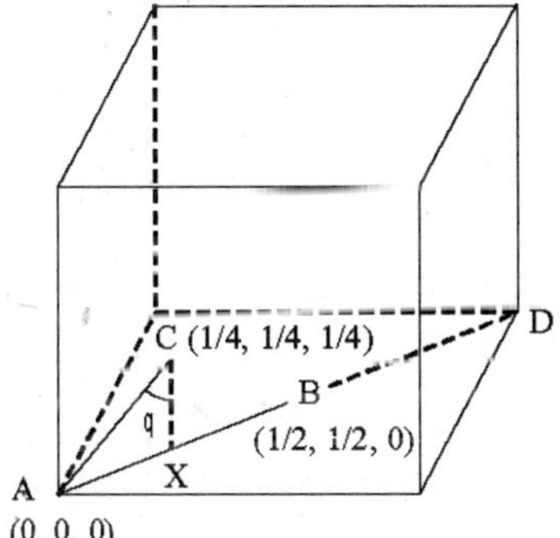

Fig. 3.19.22 Positions of three carbon atoms at (0, 0, 0), $\left(\frac{1}{2}, \frac{1}{2}, 0\right)$ and $\left(\frac{1}{4}, \frac{1}{4}, \frac{1}{4}\right)$ of a cubic unit cell

$$\text{Distance AB} = \sqrt{\left(\frac{1}{2}-0\right)^2 + \left(\frac{1}{2}-0\right)^2 + (0-0)^2} = \frac{1}{\sqrt{2}} \text{ units}$$

$$\text{Distance AC} = \sqrt{\left(\frac{1}{4}-0\right)^2 + \left(\frac{1}{4}-0\right)^2 + \left(\frac{1}{4}-0\right)^2} = \frac{\sqrt{3}}{4} \text{ units}$$

The distance AB is one-half of the face diagonal AD. Thus, the actual distance

$$AB = \frac{\sqrt{2}(357 \text{ pm})}{2}$$

If the unit $1/\sqrt{2}$ represents a distance of $357 \text{ pm}/\sqrt{2}$, the unit $\sqrt{3}/4$ will represent a distance of

$$\left(\frac{357 \text{ pm}}{\sqrt{2}}\right)\left(\frac{\sqrt{2}}{1}\right)\left(\frac{\sqrt{3}}{4}\right) = \frac{(357 \text{ pm})\, 1.732}{4} = 154.58 \text{ pm}$$

Thus carbon-carbon bond length in diamond = 154.58 pm

(b) $$\sin\theta = \frac{AX}{AC} = \frac{357 \text{ pm}/2\sqrt{2}}{154.58 \text{ pm}} = 0.816\,4$$

Hence, $\theta = 54°44'$

$$\angle CCC = 2\theta = 109°28'$$

Example 3.19.12 The edge length of the unit cube of diamond is 356.7 pm and this cube contains 8 carbon atoms. Calculate: (a) the distance $d_{C\text{–}C}$ between carbon atoms, assuming them to be spheres in contact; (b) the fraction of the total volume that is occupied by carbon atoms.

Solution **(a)** Edge length of the cube = 356.7 pm

Face diagonal of the cube = $\sqrt{2}$ (356.7 pm) = 504.3 pm

This distance as may be seen from Fig. 3.19.19 is also given by

$$d = 4(AX) = 4(d_{C\text{–}C} \sin 54°44') = 4d_{C\text{–}C}(0.816\,4)$$

Thus, $$d_{C\text{–}C} = \frac{504.3 \text{ pm}}{4 \times 0.816\,4} = 154.4 \text{ pm}$$

Hence, Radius of carbon = $d_{C\text{–}C}/2$ = 154.4 pm/2 = 77.2 pm

Volume of the unit cell = $(356.7 \text{ pm})^3 = 4.538 \times 10^{-29} \text{ m}^3$

Volume occupied by 8 spheres $= 8\left(\frac{4}{3}\pi(77.2 \text{ pm})^3\right) = \frac{4}{3}\pi(154.4 \times 10^{-12} \text{ m})^3$

$$= 1.542 \times 10^{-29} \text{ m}^3$$

Fraction of the volume occupied by carbon atoms $\dfrac{1.542 \times 10^{-29} \text{ m}^3}{4.538 \times 10^{-29} \text{ m}^3} = 0.34$

Graphite Structure

Graphite is another allotropic form of carbon and is more stable than diamond. This has a layer type structure as shown in Fig. 3.19.23.

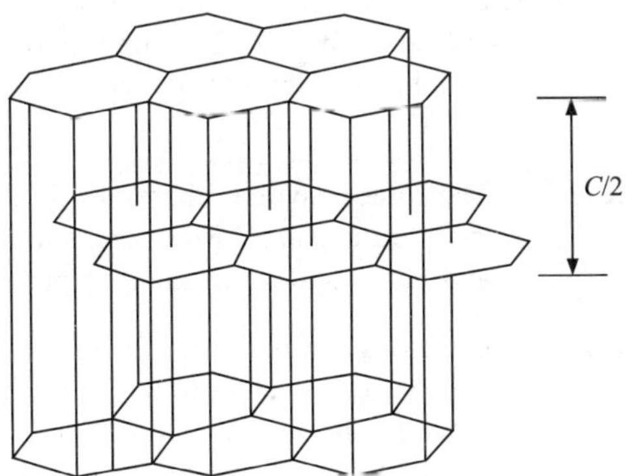

Fig. 3.19.23 Crystal structure of graphite

In each layer, we have carbons attached to each other through the overlapping of sp^2 hybrid orbitals, a stronger bond. Different layers are held by weaker joinings, i.e. the π-bondings. The great difference between graphite and diamond can be understood in terms of the above crystal lattice. The distance between atoms in the plane is 143 pm, but the distance between the two atomic layer planes is 335 pm. In two directions the carbon atoms are tightly held as in the diamond, but in the third direction, the forces of attraction are much less. As a result of this, one layer can slip over another. The crystals are flaky, and yet the material is not wholly disintegrated by shearing action. This planar structure is a part of the explanation of the lubricating action of graphite.

General Characteristics

Atoms with a valency of 2 cannot form isotropic three-dimensional structures. Consequently, we find structures which consist of endless chains of atoms extending through the crystal, the individual chains being held together by much weaker forces. Another example is that of rhombic sulphur. Here there are eight membered rings of sulphur atoms. The bivalence of sulphur is maintained and different molecules are held together by the van der Waals attractions.

Generally, the covalent solids have comparatively low densities as a result of the low coordination numbers. This effect is intensified in those crystals in which covalently bound structural units are held in the crystal by van der Waals forces. The distance between two units held by van der Waals forces is significantly greater than that between units held by covalent, ionic or metallic bonds; these large distances result in a low density of the solid.

Main Characteristics of Different Types of Crystals

Molecular Crystals

Forces which hold the constituents of molecular crystals are of van der Waals type. These are weaker forces because of which molecular crystals are soft and possess comparatively low melting points. Examples are CO_2, $CC1_4$, Ar and most of the organic compounds.

Ionic Crystals

Forces involved here are of electrostatic nature. These are stronger and non-directional type. Therefore, the ionic crystals are strong and likely to be brittle. They have very little elasticity and cannot be easily bent. The melting points are high, which decrease with increasing size of the ions. In ionic crystals, some of the atoms may be held together by covalent bonds to form ions having definite positions and orientations in the crystal lattice. Example: $CaCO_3$.

Covalent Crystals

The forces involved here are of chemical nature (covalent bonds) extended in three dimensions. They are strong, and consequently the crystals are strong and hard, with high melting points. Examples: diamond, silicon, etc.

Metallic Crystals

Electrons are held loosely in these type of crystals. Therefore, they are good conductors of electricity. Metallic crystals can be bent and are also strong. Since the forces have non-directional characteristics, the arrangement of atoms frequently corresponds to the closest-packing of spheres.

3.20 GENERAL DISCUSSION ON STRUCTURE OF LIQUIDS

Characteristics of Liquids

Liquids are neither characterized by the random chaotic motion of molecules, such as in gases, nor by the perfect orderly molecular arrangement in solids. They occupy an intermediary position where molecules are more disorderly than those of a crystalline solid, but much less disorderly than those of a gas. Because of this fact the enthalpy change when a crystal melts is always positive, and the corresponding entropy change is also positive. This implies that there is less of order when a crystal melts. The liquid is thus intermediate between the complete order of the crystalline state and the complete disorder of the gaseous state.

The properties of a substance at temperatures just below and just above its melting point do not show much change. For example, the molar volume of nearly all substances increases by only about 10 per cent on melting. This fact suggests that the difference in structures between crystals and liquids is not very significant and that liquids should have at least some characteristics of the orderly arrangement of crystals.

X-ray Diffraction of Liquids

The above point of view is supported by observations of the diffraction of X-rays from liquids. The diffraction pattern of liquid argon at a series of temperatures is shown in Fig. 3.20.1.

It can be seen that the X-ray diffraction pattern of liquid argon just above its melting point (curve a) shows only very few broad maxima and minima which become broader and more diffused as temperature is raised. The presence of these maxima and minima indicate the presence of some ordered arrangement of molecules as these would not be present if the molecular arrangement in the liquid had been completely random.

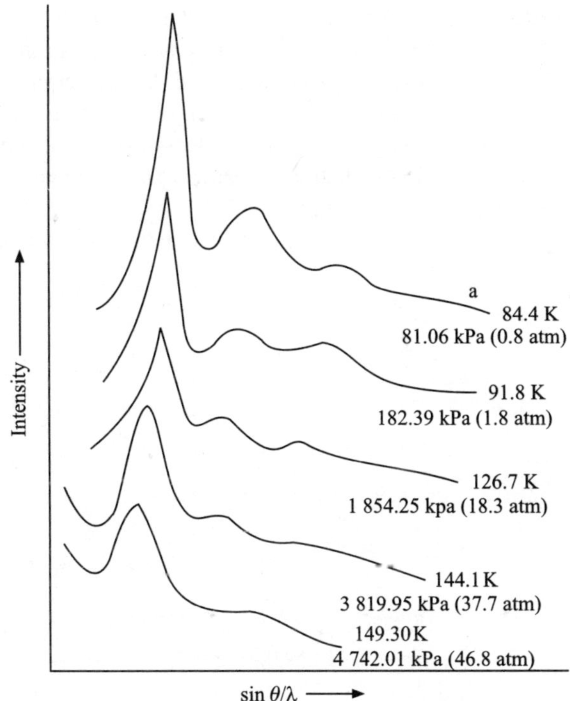

Fig. 3.20.1 X-ray diffraction pattern of liquid argon at a series of temperatures. The ordinate of each curve is shifted relative to the preceding curve

A completely amorphous structure would have exhibited a continuous scattering of X-rays without maxima and minima.* The diffraction pattern of X-rays by liquids show one or more maxima and minima. This indicates that a partially ordered structure exists in the liquid. The range of the order of a typical liquid molecule never exceeds more than a few multiples of the mean intermolecular distance. In other words, there exists a short-range order in the liquids. This way, liquids resemble gases.

The X-ray diffraction pattern of a liquid is not a consequence of any definite arrangement of its constituent atoms (or molecules). Instead, it represents a time average over all of the positions resulting from their translational motion. These pattern can be used for the calculation of the radial distribution function $4\pi r^2 N(r)$ which gives the number of constituent species whose centres are at a distance r from the centre of any chosen constituent species in the liquid. For a solid at 0 K, the radial distribution function would consist of a series of lines at values of r equal to the distances between the nearest neighbours in the crystal lattice, next nearest neighbours, and so on. As the temperature is raised, the lines broaden into narrow bands, with the area under each band proportional to the number of species at that distance.

---

*The width of X-ray diffraction lines in the Debye-Scherrer method depends upon the size of the particles. As the size decreases, the extent of orderly arrangement decreases and the diffraction lines become broadened. With particles around 10 nm in diameter, the lines become diffuse halos, and with still further decrease in particle size, the diffraction maxima become altogether blurred.

Radial Distribution Function

Figure 3.20.2 shows the radial distribution function (as the dotted lines) along with the X-ray diffraction for argon at different temperatures. The positions of the diffraction bands in the crystalline argon is shown by the lines in the bottom (marked as a). Curves b, c and d are those of liquid argon at different temperatures; curve b is just above the melting point of argon and the curve d is just below its critical state. Curve e is that of gaseous argon at just above its critical state.

The following conclusions can be drawn from Fig. 3.20.2.

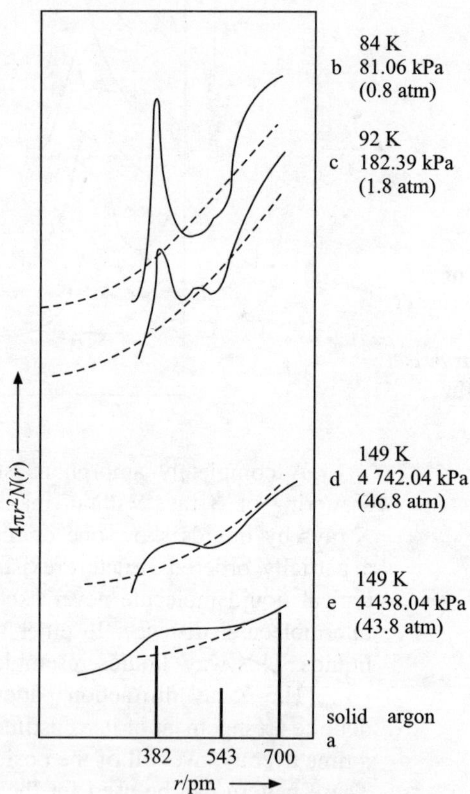

Fig. 3.20.2 Graphs of the radial distribution function of argon at a series of temperatures. (i) Rising dotted lines are the graph of $4\pi r^2$. (ii) Curve *a* shows the position of the diffraction bands in crystalline argon. (iii) Curves *b*, *c* and *d* refer to liquid argon. (iv) Curve *e* refers to gaseous argon

(i) For crystalline argon, a series of bands are observed at r = 382 pm, 543 pm, etc. Since the area under each band is proportional to the number of species present at that distance, the areas under different bands of crystalline argon indicate the presence of twelve nearest neighbours at a distance 382 pm and six next nearest neighbours at a distance 543 pm.

(ii) For liquid argon, the positions of diffraction bands are shifted slightly to larger values of r and even become broader and shallower as the temperature is raised. Moreover, the area under the maximum decreases, indicating the presence of fewer number of ordered atoms. For example, the first maximum of the curve b has an area corresponding to about 10.5 species. These effects indicate that the decrease in degree of ordered arrangement in liquids with rise in temperature.

(iii) For gaseous argon at temperature just above its critical state (curve e), the diffraction pattern is similar to the curve which represents diffraction pattern of liquid argon just below its critical state. This suggests that the two states show very close resemblance with each other near the critical state. This fact is in agreement with the principle of continuity of state (discussed in Chapter 1) according to which the liquid state close to its critical isotherm may be regarded as the compressed gaseous state, since one state can be converted into another without any discontinuity.

The Hole Theory of Liquids

The X-ray diffraction of liquid argon revealed that the distance between the nearest neighbours is of the same order as that of crystalline argon. This is, in fact, true for most liquids. The decrease in density when a solid melts can be explained on the basis that the additional 'free volume' is built up as vacancies in the more or less close-packed liquid lattice. This means that the coordination number of a liquid molecule must decrease as the liquid expands on warming, and this is exactly the inference we draw from the area under the nearest neighbour hump in the radial distribution function of argon at a series of temperatures. It is found there that the coordination number of liquid argon is about 10-11 just below its melting point. The comparison of this with the coordination number of 12 in crystalline argon very nicely explains why liquid argon is about 12 per cent less dense than solid argon. As the liquid is warmed, its density decreases and the apparent coordination number, as determined by the X-ray data, decreases to about 4 just below its critical temperature. However, in all this range the nearest neighbour argon-argon distance is nearly unchanged.

The above information, leads us to a representation of a liquid structure very similar to that shown in Fig. 3.20.3. In the crystalline state, any given molecule has its maximum possible number of nearest neighbours and hence has maximum intermolecular attractions with the largest possible stabilization. If, however, we warm such a system, the molecules acquire, on an average, the amount kT of kinetic energy. When this energy becomes high enough, some molecules can migrate through the lattice forcing other molecules aside and leaving behind a vacancy. It thus produces an abnormal coordination number It is known that even a single abnormal coordination among a few hundred atoms is sufficient to produce a long-range disorder, which is typical of the liquid state. The thermal motions in one region of a crystal is sufficient to destroy its regular structure, the irregularity rapidly spreads throughout the entire specimen. As the temperature of the liquid rises, its volume increases and the average coordination number becomes smaller and smaller. Thus, we can view a liquid as containing some molecules that can manifest their kinetic energy by travelling through it, and others, that are confined by their nearest neighbours and can have only vibrational energy.

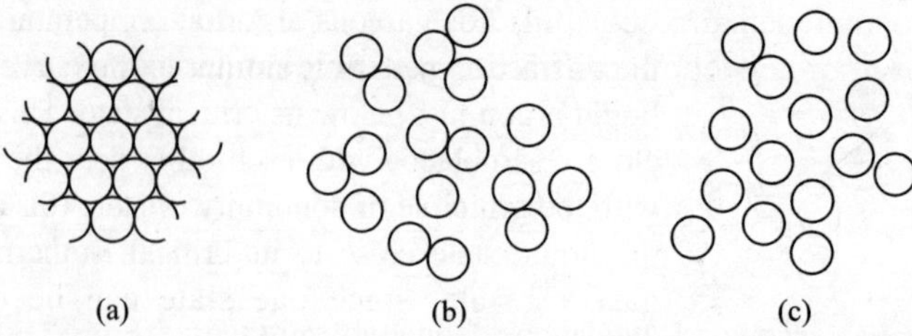

Fig. 3.20.3 Structures of (a) crystalline state, (b) liquid at melting point, (c) liquid near critical temperature

Basis of Hole Theory

Based on the above facts, the *Eyring hole theory of liquids* was proposed by Henry Eyring and his co-workers. According to this, a liquid may be considered to be a composite system of two significant structures, gases on one side and solids on the other side. In a liquid some of the molecules have gas-like character and the others have solid-like character. If the 'free volume' in a liquid is assumed to be arranged in molecule-sized vacancies, then any molecule next to a vacancy can move into it, acts as a gas molecule. On the other hand, any molecule that is completely surrounded and has no vacancy next to it, can only vibrate and thus acts as a solid molecule.

If V_1 and V_s are the respective volumes occupied by a liquid and the same amount of a solid, then the extra volume $(V_1 - V_s)$ will appear as the vacancies of molecular size and if we assume that these vacancies are randomly distributed in the liquid, the mole fraction of molecules behaving as gas is $(V_1 - V_s)/V_1$. The remaining mole fraction V_s/V_1 represents solid-like molecules.

Any physical property of a liquid may thus be considered to be the sum of a contribution from solid-like molecules and a contribution from gas-like molecules. For example, suppose we wish to compute the molar heat capacity of liquid argon at different temperatures. The molar heat capacity of argon gas (predicted on the basis of the law of equipartition of energy) is $3R/2$ or 12.47 J K^{-1} mol^{-1}. For an idealized solid lattice of argon, the molar heat capacity of solid argon is $3R$ or 24.952 J K^{-1} mol^{-1}. Then, the molar heat capacity of liquid argon at any given temperature will be given by the expression.

$$C_p/\text{J K}^{-1}\text{mol}^{-1} = 24.952 \times (\text{solid-like fraction}) + 12.47 \times (\text{gas-like fraction})$$

$$= 24.952\left(\frac{V_s}{V_1}\right) + 12.47\left(\frac{V_1 - V_s}{V_1}\right)$$

Figure 3.21.3 shows the result of this approach. The agreement between experimental and predicted values is excellent.

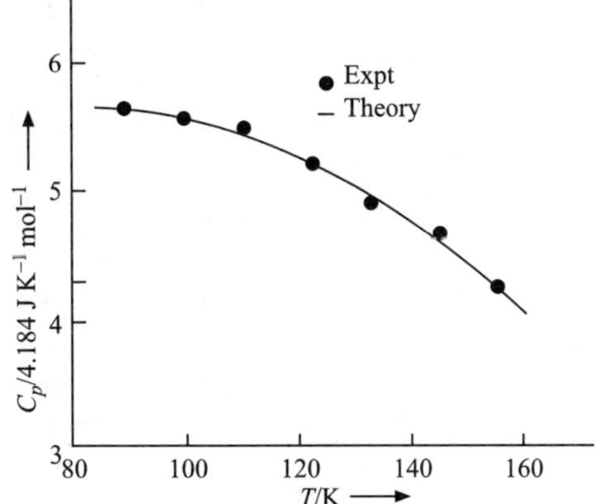

Fig. 3.20.4 Comparison of experimental values of molar heat capacity of argon with theoretical values on the basis of hole theory

REVISIONARY PROBLEMS

3.1 Explain the following:
(a) First law of crystallography.
(b) Haüy's idea of unit cell.
(c) Space lattice.
(d) Unit cell (primitive and nonprimitive).
(e) Law of rational indices.

3.2 Comment upon the following:
(a) Crystal symmetry generates unit cell but not vice versa.
(b) There exists long-range order in crystals.
(c) Lattice points need not be occupied by atoms or ions. The only condition is that the lattice points must have identical environments.
(d) Normal developed faces of a crystal are those which have maximum site densities.
(e) The unit cell has orientation but no position.

3.3 (a) What do you understand by symmetry elements and symmetry operations? Explain the following symmetry elements and the associated operations:

(i) Proper rotation axis,
(ii) Mirror plane (horizontal and vertical),
(iii) Improper rotation axis.

(b) Point out the symmetry elements present in the following compounds: *cis* and *trans* $C_2H_2Cl_2$, BCl_3, CH_4, SF_6, B_2Cl_6, H_2O and NH_3.
(c) How many symmetry axes and planes are present in a simple cube and in a tetrahedral arrangement?
(d) The allowed axes of rotations in a crystal are those which are consistent with the translational periodicity; on this basis, explain why five-fold axis of rotation cannot exist in crystals.
(e) If a crystal has more than one triad axis, it necessarily must have four triad axes.

3.4 (a) How many unit cells are possible in a planar lattice?
(b) Sketch a two-dimensional lattice of closest-packed identical circles, indicating a suitable unit cell.

3.5 (a) Explain how the classification of crystals into seven crystal classes are done on the basis of minimum symmetry elements of rotational axis (proper or improper).
(b) The dimensions of these seven crystals correspond to maximum possible symmetry of their respective systems. Outline the crystal-dimensions and the symmetry elements which they possess.

3.6 What are 14 Bravais lattices? Explain why face-centred and end-centred Bravais lattice are excluded in the tetragonal system.

3.7 What are point groups and space groups? How many point and space groups are theoretically possible?

3.8 (a) What are the Miller indices for sets of similar planes?
(b) Compute the Miller indices for a face having intercepts on the three axes as

(i) $a : \frac{1}{3}b : \frac{1}{2}c$ (ii) $\frac{1}{2}a : \frac{1}{4}b : \infty\, c$

(iii) $a : b : \infty\, c$ (iv) $2a : 2b : 3c$

(c) Show, with the help of diagrams, the planes in cubic unit cells having the following Miller indices:
(i) Primitive cubic cell (100), (110) and (111)
(ii) Face-centred cubic cell (200), (220) and (111)
(iii) Body-centred cubic cell (200), (100) and (222)

(d) Show qualitatively or otherwise that the interplanar distance in a cube having Miller indices (*hkl*) is given by

$$d_{hkl} = \frac{a}{(h^2 + k^2 + l^2)^{1/2}}$$

where a is the edge length of the cube.
(e) Calculate the ratio of d_{hkl} for a given set of planes of cube having the Miller indices as given in part (c).

3.9 (a) Derive the Bragg's equation

$$n\lambda = 2d \sin \theta$$

for the reflection of X-rays from the faces of a crystal and show that it can be written as

$$\lambda = 2d_{hkl} \sin \theta$$

(b) Explain why one should employ monochromatic X-rays in studying the reflections from the faces of the crystals. What would happen if, instead, polychromatic X-rays are employed?
(c) On what factors does the intensity of the diffracted beam from different sets of planes depend?

3.10 There are three principal methods for observing diffraction. These are:
(i) *Laüe method* Single crystal sample, fixed incident angle, continuous wavelength range (white radiation). The diffracted beams are observed (usually with photographic film) in directions where the Bragg equation is satisfied for some λ and n for some set of planes.
(ii) *Rotating crystal method* Single crystal, rotation through all incident angles, fixed wavelength (monochromatic radiation). The diffracted beam is observed at the instant where the crystal is correctly oriented to satisfy the Bragg condition for certain set of planes.

(iii) *Debye-Scherrer method* Powder (or polycrystalline) sample, fixed incident angle, fixed wavelength. The Bragg condition is satisfied for different planes by some of the randomly oriented grains in the sample.
Describe the diffraction pattern in each case and indicate the usefulness and weakness of each method.

3.11 Answer the following:
(a) Explain, why it is not possible to deduce the position of hydrogen atom from X-ray diffraction. How can this difficulty be overcome by using the neutron diffraction method?
(b) Explain, why it is not possible to distinguish by X-ray diffraction, two different atoms which differ only by the possession of one additional electron.
(c) Explain, why in the powder method, the crystallites in the powder crystal must have an average dimension of a few microns. What would happen if the size is too large or too small?

3.12 (a) Show that reflections of X-rays from the planes of a cube are given by

$$\sin^2 \theta_{hkl} = K(h^2 + k^2 + l^2)$$

where K is a constant, θ is the incident angle and hkl are the Miller indices. Hence, show that the reflections for which $\sin^2 \theta_{hkl} = 7K$, $15K$, $23K$ and so on are not observed.
(b) Explain qualitatively, why the reflections present (i) in body-centred cubic crystals are from those planes for which $(h + k + l)$ is even and (ii) in face-centred cubic crystals are from those planes for which hkl are either all odd or all even.
(c) Sketch out the typical reflection patterns that would be obtained for the primitive, body-centred and face-centred cubic systems and on this basis justify that: 'From missing reflections in the X-ray diffraction pattern, it is possible to distinguish between different cubic lattices.'
(d) LiBr, NaBr, KBr and RbBr all have the same crystal structure. X-ray diffraction, however, indicates RbBr to be a simple cubic while the other three have face-centred cubic lattices. Explain.

3.13 Explain the following:
(a) A useful check on the correctness of a proposed crystal structure is to compare the crystal density computed from its structural lattice with that of the experiment.
(b) How does the relative intensities of the reflection maxima of the different orders and planes help in deciding that NaCl crystal essentially consists of two interpenetrating face-centred cubic lattices, one composed entirely of Na^+ ions and the other of Cl^- ions?

3.14 How many points per unit cell are there in:
(i) Primitive cubic lattice,
(ii) Body centred cubic lattice,
(iii) Face-centred cubic lattice.
How are these points designated in crystallography?

3.15 Show that the maximum proportion of the available volume which may be filled by hard spheres in various structures are

$$\text{Simple cubic: } \frac{\pi}{6} (= 0.52)$$

$$\text{Body-centred cubic (BCC): } \frac{\pi\sqrt{3}}{8} (= 0.68)$$

Face-centred cubic (FCC): $\frac{\pi\sqrt{3}}{6}(= 0.74)$

Hexagonal closest-packed (HCP): $\frac{\pi\sqrt{3}}{6}(= 0.74)$

Diamond: $\frac{\pi\sqrt{3}}{16}(= 0.34)$

What is the percentage of void space in each structure?

3.16 (a) What do you understand by the stacking sequences

ABAB . . .

ABCABC . . .

What kind of lattices do these sequences lead to?

(b) How many octahedral and tetrahedral holes are present in, FCC and HCP structures, respectively?

(c) Is it possible to have patterns like ABCACB. . ., ABAC. . ., etc.?

3.17 (a) The arrangement of ions in crystals is primarily determined by the radius ratio of cation to anion. Calculate the limiting radius ratio of cation to that of anion in each of the following arrangements:

(i) Body-centred cubical structure, (ii) Octahedral arrangement,
(iii) Tetrahedral arrangement, (iv) Square planar arrangement,
(v) Triangular arrangement, (vi) Linear arrangement.

(b) What is the coordination number of cations and anions in each of the above arrangements?

(c) Only metallic, never ionic or covalent substances, form crystals exhibiting the maximum coordination number of twelve among like-sized atoms. Account for this effect.

(d) LiCl, KCl and RbCl have structures of face-centred cubic crystal while CsCl has the structure of body-centred cubic crystal. Explain.

3.18 Show that in a closest-packed structure:

(a) The number of octahedral holes is equal to the number of atoms present and these holes are located midway between the two closest-packed layers.

(b) The number of tetrahedral holes is twice the number of atoms present and these are located above each atom in the first layer and below each atom in the second layer.

3.19 (a) Compute the molecular formula of the ionic oxides having the following structures:

(i) *Rock salt* Anions form cubical-closest packing and cations occupying octahedral holes.

(ii) *Anti-fluorite* Anions form cubical-closest packing and cations occupying tetrahedral holes.

(iii) *Fluorite* Cations form cubical-closest packing and anions occupying tetrahedral holes.

(iv) *Zinc blende* Anions are in FCC positions and half of the tetrahedral holes are occupied.

(v) *Corundum* Anions form hexagonal-closest packing and cations occupying only two-thirds of octahedral holes.

(vi) *Spinel Structure* Oxide ions form cubical closest packing, 1/8th of tetrahedral holes occupied by A^{2+} and 1/2 of octahedral holes occupied by B^{3+}.

(b) Compute the void space per unit volume of unit cell of the structures given in part (a).

3.20 Discuss briefly the classification and corresponding properties of crystals based on the bond type.

3.21 (a) Show that the cohesive energy of an ionic crystal is given by

$$-E_c = -\frac{N_A A(ze)^2}{r_0}\left(1 - \frac{1}{n}\right)$$

where the symbols have their usual meanings.
(b) How is the cohesive energy determined experimentally using Born-Haber cycle?

3.22 (a) Draw the structures of diamond and graphite. What differences are essentially present in these structures and how do they lead to different physical properties of these two allotropic forms?
(b) Compute the void volume present in the diamond structure.

3.23 How and why does the X-ray diffraction pattern obtained with liquids differ from those obtained with solids? What is the significance of radial distribution function in liquids?

3.24 Discuss, in brief, the Eyring hole theory of liquids.

TRY YOURSELF PROBLEMS

3.1 Explain why a crystal cannot have an axis of greater than six-fold symmetry.

3.2 Show that the face-centred tetragonal is really body-centred tetragonal.

3.3 Show that a centre of symmetry can be represented by a rotation-inversion axis.

3.4 In studying the diffraction by crystals, one often employs X-rays. Why cannot ultraviolet light of wavelength, say 10 nm, be used?

3.5 Show that a face-centred rhombohedral lattice can also be represented as a rhombohedral lattice. Calculate the rhombohedral angle. (Ans. 60°)

3.6 Why is 'white' X-radiation (a broad range of wavelengths) necessary for the Laüe method, while monochromatic X-radiation (a single-wavelength or very narrow range) is necessary for the powder and rotating-crystal methods?

3.7 List all the symmetry elements of each of the following molecules: CHDFCl, CH_2Cl_2 (tetrahedral), OCS, *Trans*-ClBrHC—CHBrCl, Eclipsed form of ethane, Boat form of cyclohexane.

3.8 What is the plane of closet packing in following?
(a) Face-centred cubic structure.
(b) Body-centred cubic structure

NUMERICAL PROBLEMS

Miller indices

3.1 Calculate the Miller indices of the faces having the following intercepts with the three axes perpendicular to each other.

(i) $\frac{1}{2}a, 2b, \infty c$ (Ans. 410)

(ii) $\frac{3}{2}a, 2b, 1c$ (Ans. 436)

(iii) $\frac{5}{2}a, \frac{5}{2}b, 3c$ (Ans. 665)

(iv) $2a, 3b, 4c$ (Ans. 643)

X-ray Diffraction

3.2 The first order reflection of a beam of X-rays from a given crystal occurs at 5° 15′. At what angle will be the third order reflection? (Ans. 15° 56′)

3.3 The first order reflections from (100), (110) and (111) planes of a given cubic crystal occur at angles 7° 10′, 10° 12′ and 12° 30′, respectively. To what type of cubic lattice does the crystal belong? (Ans. Primitive)

Density

3.4 The density of CaF_2 is 3.18 g cm^{-3} at 20 °C. Calculate the dimensions of a unit cube of the substance containing 4 Ca^{2+} and 8 F^- ions. (Ans. 546 pm)

3.5 The effective radius of an iron atom is 124 pm. Iron occurs both in a BCC structure and a FCC structure. Calculate the density of each in g/cm^3.
(Ans. BCC 7.9 g/cm^3, FCC 8.6 g/cm^3)

3.6 Calculate the density of diamond from the fact that it has a face-centred cubic structure with two atoms per lattice point and a unit cell edge of 356.9 pm.
(Ans. 3.509 g/cm^3)

3.7 Caesium chloride, bromide and iodide form interpenetrating simple cubic crystals rather than interpenetrating face-centred cubic crystals like the other alkali halides. The length of the side of the unit cell of CsCl is 412.1 ± 0.3 pm. Determine:

(a) Density of CsCl.

(b) Ion radius of Cs^+, assuming that the ions touch along a diagonal through the unit cell and that the ionic radius of Cl^- is 181 pm.

(Ans. (a) 3.99 g/cm^3, (b) 176 pm)

Type of Crystal Structure

3.8 Metallic aluminium, with one atom per lattice point and a density of 2.70 g/cm^3 at 25 °C, exhibits the first four X-ray reflections at θ = 19.25°, 22.38°, 32.58° and 39.15°. If the wavelength of radiation is 154.2 pm, determine:
(a) The type and size of cubic unit cell.
(b) The atomic mass of Al.
(c) The distance between closest Al atoms.
(d) The angle θ at which (400) reflection occurs.

(Ans. (a) face-centred, edge length 405.03 pm
(b) 27 g mol^{-1}, (c) 286.4 pm, (d) 49.59°)

3.9 The X-ray spectrum of a cubic metal using radiation of λ = 154.18 pm gave lines at the following angles:

21.8°, 25.4°, 37.2°, 45.4°. 47.8°, 58.8°, 68.6° and 72.7°

(a) Index the lines.
(b) Calculate the unit cell edge length.
(c) Identify the type of unit cell.

(Ans. (a) 111, 200, 220, 311, 222, 400, 331, 420
(b) 360 pm, (c) face-centred])

3.10 Titanium has a hexagonal closest-packed structure containing 6 atoms in the unit cell: $a = b$ = 295.3 pm, c = 472.9 pm. Calculate its density. (Ans. 4.442 g cm^{-3})

3.11 White tin crystallizes in tetragonal system with $a = b$ = 582 pm and c = 317.5 pm. Its density is 7.29 g/cm^3. Determine the number of atoms in the unit cell. (Ans. 4)

3.12 A certain compound, whose density is 4.56 g cm^{-3}, crystallizes in the tetragonal system with unit lattice distances of $a = b$ = 658 pm and c = 593 pm. If the unit lattice contains four molecules, calculate the molar mass of the compound.

(Ans. 176.29 g mol^{-1})

Debye Powder Method

3.13 The Debye-Scherrer picture of a cubic crystal with X-rays of $\lambda = 153.9$ pm displayed lines at the following scattering angles:

No. of lines	1	2	3	4	5	6	7	8	9
θ/deg	13.70	15.89	22.75	26.91	28.25	33.15	37.00	37.60	41.95
Intensity	w	vs	s	vs	m	w	w	m	m

Index these lines. Calculate a_0 for the crystal. Identify the crystal.

(Ans. face-centred, 562.8 pm)

Ionic Compounds

3.14 Determine the simplest formula of an ionic compound in which:

(a) The unit cell consists of a cube such that there are cations (C) at each corner and an anion (A) at the centre of the cube;

(b) the unit cell consists of a cube in which there are cations (C) at each corner and anions (A) at the centre of each face. (Ans. (a) CA, (b) CA_3)

3.15 (a) A spinel is an important class of oxides consisting of two types of metal ions with the oxide ions arranged in CCP layers. The normal spinel has one-eighth of the tetrahedral holes occupied by one type of metal ion and one-half of the octahedral holes occupied by another type of metal ion. Such a spinel is formed by Zn^{2+}, Al^{3+} and O^{2-} with Zn^{2+} in the tetrahedral holes. Give formula of the spinel. (Ans. $ZnAl_2O_4$)

(b) If all the species in Problem 3.15a touch each other, determine the fraction of the volume occupied in the unit cell. (Ans. 0.62)

3.16 Determine the location and size of the largest sphere that can exist among identical spheres in a body-centred structure without disturbing that structure.

(Ans. 0.252 a at $(\frac{1}{2}, \frac{1}{4}, 0)$)

3.17 An oxide of copper crystallizes in a cubic unit cell with oxide atoms at 0, 0, 0 and $\frac{1}{2}, \frac{1}{2}, \frac{1}{2}$. Copper atoms are located at $\frac{1}{4}, \frac{1}{4}, \frac{1}{4}$; $\frac{1}{4}, \frac{3}{4}, \frac{3}{4}$; $\frac{3}{4}, \frac{1}{4}, \frac{3}{4}$; and $\frac{3}{4}, \frac{3}{4}, \frac{1}{4}$. Draw the unit cell showing the positions of all atoms in the cell. How many copper atoms are there in the unit cell? How many oxide atoms? What is the simplest formula of the oxide? (Ans. 4, 2, Cu_2O)

3.18 Caesium bromide crystallizes in the cubic system. Its unit cell has a Cs^+ ion at the body centre and Br^- ion at each corner. Its density is 4.44 g cm^{-3}. Determine:

(a) edge length of the unit cell (b) the d_{200} distance, and (c) fraction of the volume per unit cell occupied. (Ans. (a) 430 pm, (b) 215 pm, (c) 0.77)

3.19 In the book 'X-ray Crystal Systems' Braggs give an example of X-ray analysis as follows. The first order reflection from (100) planes of KCl occurs at 5° 23′ but for NaCl it occurs at 6° 0′ for the rays of same X-rays. If the side of NaCl unit cell is 564 pm, what is the size of KCl unit cell? The densities of NaCl and KCl are 2.17 g cm^{-3} and 1.99 g cm^{-3}, respectively. Do these values support the X-ray analysis?

(Ans. 628.38 pm, yes)

3.20 KCl crystallizes in the same type of lattice as does NaCl. The ionic radius of Na^+ is 0.5 of that of Cl^-, and is 0.7 of that of K^+ Calculate (a) the ratio of the side of the unit cell for KCl to that for NaCl and (b) the ratio of the density of NaCl to that of KCl. (Molar masses, K = 39 g mol^{-1}; Na = 23 g mol^{-1} and Cl = 35.5 g mol^{-1}).

(Ans. (a) 1.143, (b) 1.172)

3.21 NH_4Cl crystallizes in a body-centred cubic lattice with a unit distance of 387 pm. Calculate (a) the distance between the oppositively charged ions in the lattice and (b) the radius for the NH_4^+ ion, the radius for the Cl^- ion being 181 pm.

(Ans. (a) 335.15 pm, (b) 154.15 pm)

Annexure I Symmetry Elements and Symmetry Operations

The study of symmetry of a molecule is often useful in making its theoretical and experimental treatments simpler. A point in the molecule is chosen and the symmetry with respect to lines and planes passing through this point is studied. Before going into the detail, it is worthwhile to make distinction between the symmetry element and the associated symmetry operation.

The symmetry element is a geometrical entity such as a line, a plane or a point with respect to which the symmetry operation is carried out. The associated symmetry operation is actually carrying out the operation on the object. If on carrying out some operation (rotation, reflection, etc.), the object is thrown into a new configuration which is completely indistinguishable from the previous one, then the object is said to possess the above symmetry operation and the corresponding symmetry element.

Five types of symmetry elements are used to describe the symmetry of a molecule. Table AI.1 records these symmetry elements and the corresponding symmetry operations.

Symmetry Elements In Crystallography

In crystallography, the first three symmetry elements shown in Table AI.1 are usually employed in studying the symmetry of crystals.

Symmetry Elements of a Cube

As an illustration, we give below a few symmetry elements possessed by a cubic crystal (Fig. AI.1).

Proper-rotation axis Three types of rotation axes are present in a cubic crystal. These are:

(a) *Tetrad axes* There are *three* such axes, each passing through the centres of the two opposite faces.

(b) *Triad axes* There are *four* such axes, each passing through the opposite cross-corners of the cube.

(c) *Diad axes* There are *six* such axes, each passing through the centre of the two opposite edges of two opposite faces.

Planes of symmetry There are two types of planes present in a cubic crystal. These are:

(a) *Parallel planes* There are *three* such planes, each passing through the middle of the two opposite faces and being parallel to the sides of the cube.

(b) *Diagonal planes* There are *six* such planes, each passing through the diagonals of the two opposite faces.

Centre of symmetry One centre of symmetry is present at the centre of the cube.

Thus in all, we have illustrated 13 proper rotation axes, 9 planes of symmetry and 1 centre of symmetry, making a total of 23 symmetry elements. These are shown in Fig. AI.1.

Table AI.1 Symmetry Elements and Associated Symmetry Operations

Symmetry element	Operation	Symbol used: Harmann-Maugin notation*	Symbol used: Schoenflies notation**
(1) Axis of symmetry	Rotation around the axis by an angle 360°/n where n is the order of rotation axis.	n	C_n
(2) Plane of symmetry	Reflection through the plane.	m	σ
Three different types of planes are ususally observed. These are			
(a) Horizontal mirror plane	Reflection through the plane perpendicular to the principal axis (axis of highest symmetry).	$/m$	σ_h
(b) Vertical mirror plane	Reflection through the plane containing the principal axis.	m	σ_v
(c) Diagonal mirror plane	Reflection through the plane containing the principal axis and bisecting the angle formed by two horizontal C_2 axes which are perpendicular to the principal axis.		σ_d
(3) Rotatory-inversion axis	Rotation about the axis by an angle 360°/n followed by an inversion across a centre	$\bar{n}$	
(4) Rotation-reflection axis (Improper axis)	Rotation about the axis by an angle 360°/n followed by reflection in a plane perpendicular to the axis.		S_n
(5) Inversion-centre or centre of symmetry	A projection through the inversion centre to an equal distance on the other side from the centre.		i

*Symbols used in crystallography.

**Symbols used in studying the symmetry of molecules.

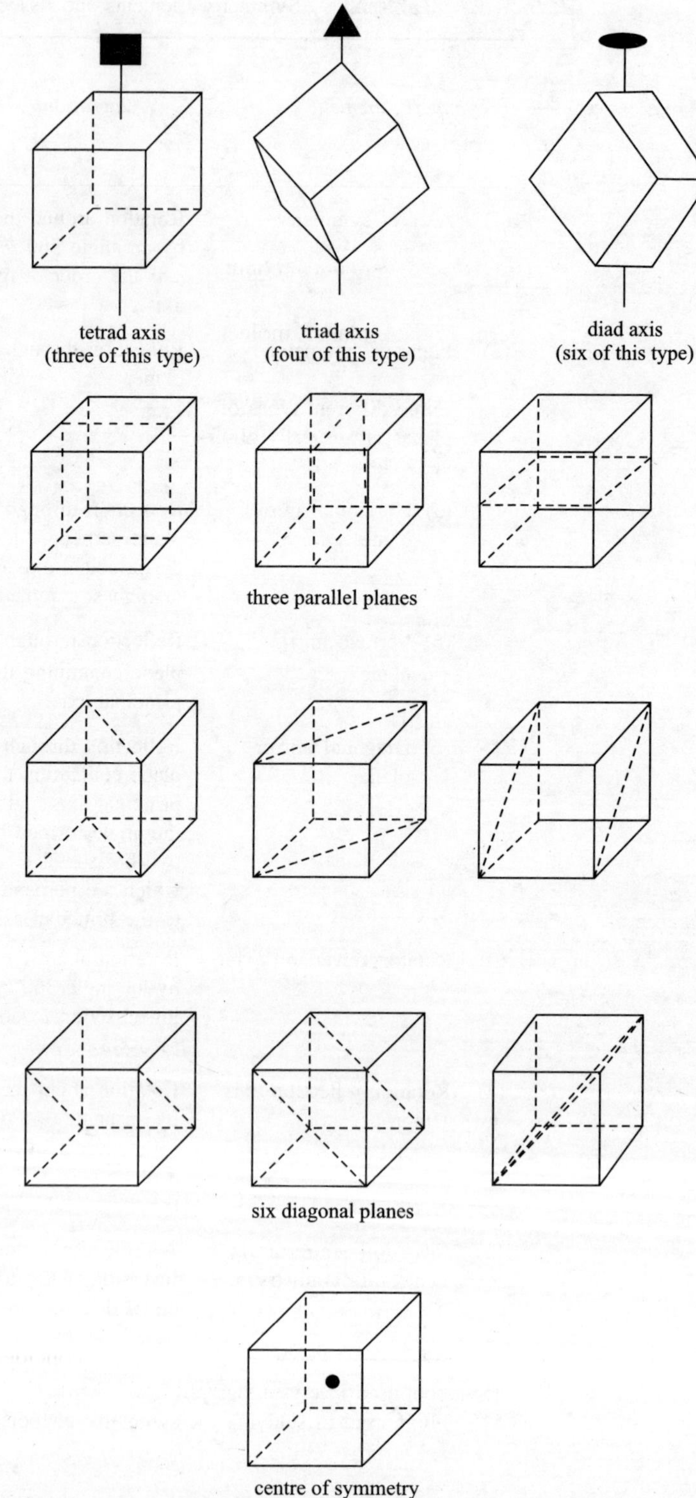

Fig. AI.1 Symmetry elements of a cube

A Few Facts Regarding the Symmetry Elements

A few facts regarding the symmetry elements and operations may be described here.

1. A symmetry element may generate more than one symmetry operation. For example, the symmetry element triad axis of rotation generates three distinct operations of rotation around the same rotation axis. These are rotations by angles 120° (= 1 × 360°/3), 240° (= 2 × 360°/3) and 360° (= 3 × 360°/3) and are represented by the symbols C_3, C_3^2 and C_3^3, respectively. The first two operations lead to the indistinguishable equivalent configurations whereas the last one leads to the completely identical configuration with the original one. The operation of the latter type is also known as *identity operation* and is represented by the symbol *E*.

2. A planar molecule has at least one symmetry element which is the molecular plane of symmetry.

3. All atoms which do not lie on a plane of symmetry must occur in even number as each one must occupy an identical point which lies on the other side of the plane. For example in H_2O molecule, a plane of symmetry passing through O atom and bisecting the bond angle exists. This does not include H atoms.

4. If a molecule has more than one symmetry plane and if it has only one atom of one kind, then this atom must lie on the line of intersection between two or more planes or at the point of intersection of three or more planes. The examples include C of CH_4, N of NH_3 and B of BCl_3.

5. A linear molecule has infinite number of symmetry planes and a ∞-fold axis of rotation collinear with the molecular axis.

6. It one atom of a certain kind lies off a C_n axis, then there must be $(n - 1)$ more or a total of n such atoms. This follows from the fact that the application of C_n successively n times, the first atom is moved to a total of n different equivalent points. One of the examples is BCl_3 where C_3 axis perpendicular to the molecular plane and passing through the atom B exists. Each of Cl atoms is thrown into the equivalent position when a rotation of 360°/3 is carried out and thus there exists 3 Cl atoms.

7. The existence of the C_n axis and one C_2 axis perpendicular to C_n axis means that the other C_2 axes at angles 1 × 360°/*n*, 2 × 360°/*n*, ..., must exist. The number of such C_2 axis is equal to the order n of rotation axis C_n. This follows from the fact that the application of C_n once generates the second C_2 axis from the first and application of C_n^2 generates the third C_2 axis from the first and so on. Examples include BCl_3 and benzene.

8. In a molecule if C_n and a perpendicular plane of symmetry exist then a rotation-reflection axis of the order n (i.e. S_n) also exists. The reverse of this fact might not be true, i.e. there may exist S_n without the existence of either C_n or a perpendicular plane of symmetry. One of the examples is staggered ethane in which S_6 (axis passes through C—C) exists whereas neither C_6 nor a perpendicular plane of symmetry exists.

9. The improper axis may coincide with a proper axis of rotation. Example is a regular tetrahedron where S_4 axis concides with the C_2 axis.

10. The application of two or more operations one after the other may produce a net effect which can be produced by the application of a third single operation.

A Few Examples

Finally, a few examples with their symmetry elements and operations are listed in Table AI.2

Table AI.2 Symmetry Elements and Operations of Some Molecules

Molecule	*Symmetry elements*	*Symmetry operations*
H_2O	(i) A diad axis in the molecular plane passing through O bisecting HOH angle	C_2, C_2^2 ($\equiv E$)
	(ii) Two vertical planes of symmetry: One includes molecular plane and the other perpendicular to it passing through O atom	σ_v, σ_v'
NH_3	(i) A triad axis passing through N	C_3, C_3^2, C_3^3 ($\equiv E$)
	(ii) Three vertical planes of symmetry, each includes N and one of the H atoms	σ_v, σ_v', σ_v''
BCl_3 (planar molecule)	(i) A triad axis passing through B and perpendicular to molecular plane	C_3, C_3^2, C_3^3 ($\equiv E$)
	(ii) Three diad axes, each in the molecular plane and includes B and one of Cl atoms	C_2, C_2', C_2''
	(iii) Three vertical planes of symmetry, each includes C_3 axis, atom B and one of Cl atoms	σ_v, σ_v', σ_v''
	(iv) One horizontal plane of symmetry, It is perpendicular to C_3 axis and includes molecular plane	σ_h
	(v) One S_3 axis coincidence with C_3	S_3, S_3^2, S_3^3 ($\equiv E$)
CH_4, CCl_4 (tetrahedron)†	(i) Three diad axes coinciding with *x*, *y*, and *z* axes	$3C_2$, $3C_2^2$ ($\equiv E$)
	(ii) Four triad axes, each passes through one apex and the centre of the opposite face	$4C_3$, $4C_3^2$, $4C_3^3$ ($\equiv E$)
	(iii) Six diagonal planes of symmetry	$6\sigma_d$
	(iv) Three S_4 axes coinciding with *x*, *y*, and *z* axes	$3S_4$, $3S_4{}^2$ ($\equiv C_2$), $3S_4^3$, $3S_4{}^4$ ($\equiv E$)
	(There are a total of 24 distinct operations. These are E, $8C_3$, $3C_2$, $6S_4$ and $6\ \sigma_d$.)	
SF_6 (octahedron)*	(i) Three S_4 axes, each passing through a pair of opposite apices	$3S_4$, $3S_4^2$ ($\equiv C_2$) $3S_4^3$, $3S_4^4$ ($\equiv E$)
	(ii) Three C_2 axes collinear with the S_4's	already accounted for under (i)
	(iii) Three C_4 axes collinear with S_4s and C_2s.	$3C_4$, $3C_4^2$ ($\equiv C_2$) $3C_4^3$, $3C_4^4$ ($\equiv E$)
	(iv) Six C_2' axes bisecting opposite edges	$6C_2'$, $6C_2'^2$ ($\equiv E$)
	(v) Four S_6 axes, each passing through the centres of a pair of opposite triangular faces	$4S_6$, $4S_6^2$ ($\equiv C_3$) $4S_6^3$ ($\equiv i$), $4S_6^4$ ($\equiv C_3^2$) $4S_6^5$, $4S_6^6$ ($\equiv E$)

(Contd.)

Molecule	*Symmetry elements*	*Symmetry operations*
	(vi) Four C_3 axes colinear with the S_6's	already accounted for under (v)
	(vii) A centre of symmetry	already accounted for under (v)
	(viii) Three horizontal planes of symmetry which pass through four of the six apices	$3\sigma_h$
	(ix) Six diagonal planes of symmetry which pass through two apices and bisect two opposite edges	$6\sigma_d$
	(There are a total of 48 operations. These are E, $8C_3$, $6C_4$, $6C_2$, $3C_2$ (= C_4^2), i, $6S_4$, $8S_6$, $3\sigma_h$ and $6\sigma_d$.)	

†(tetrahedron) (A tetrahedron can be drawn within a cube. We choose the rectangular axes with origin at the centre of cube and axes parallel to the sides of cube.)

* (Octahedron) (An octahedron can be drawn within the cube. The symmetry operations of a cube are the same as those of octahedron.)

Annexure II Supplementary Materials

The Vector H_{hkl}

The vector $\boldsymbol{H}_{hkl}$ is defined as

$$\boldsymbol{H}_{hkl} = h\boldsymbol{b}_1 + k\boldsymbol{b}_2 + l\boldsymbol{b}_3 \qquad \text{(AII.1)}$$

where $\boldsymbol{b}_1$, $\boldsymbol{b}_2$ and $\boldsymbol{b}_3$, known as *reciprocal vectors*, are defined in terms of the primary vectors $\boldsymbol{a}_1$, $\boldsymbol{a}_2$ and $\boldsymbol{a}_3$ of the three crystallographic axes by the expressions

$$\boldsymbol{b}_1 = \frac{\boldsymbol{a}_2 \times \boldsymbol{a}_3}{\boldsymbol{a}_1 \cdot \boldsymbol{a}_2 \times \boldsymbol{a}_3} \qquad \text{(AII.2)}$$

$$\boldsymbol{b}_2 = \frac{\boldsymbol{a}_3 \times \boldsymbol{a}_1}{\boldsymbol{a}_1 \cdot \boldsymbol{a}_2 \times \boldsymbol{a}_3} \qquad \text{(AII.3)}$$

$$\boldsymbol{b}_3 = \frac{\boldsymbol{a}_1 \times \boldsymbol{a}_2}{\boldsymbol{a}_1 \cdot \boldsymbol{a}_2 \times \boldsymbol{a}_3} \qquad \text{(AII.4)}$$

(Note that the indices of the reciprocal vector and the primary vectors in the numerator run cyclically and the denominator in all the reciprocal vectors represents the volume of a unit cell with side lengths a_1, a_2 and a_3.)

The symbols h, k and l in Eq. (AII.1) represents Miller indices which define a set of parallel equidistant planes, one of which passes through the origin and the next makes intercepts a_1/h, a_2/k and a_3/l on the three crystallographic axes (Fig. 3.10.1).

The primary vectors $\boldsymbol{a}_1$, $\boldsymbol{a}_2$ and $\boldsymbol{a}_3$ and the reciprocal vectors $\boldsymbol{b}_1$, $\boldsymbol{b}_2$ and $\boldsymbol{b}_3$ satisfy the normal and orthogonal conditions given by the expression

$$\boldsymbol{a}_i \times \boldsymbol{b}_j = \delta_{ij} \qquad \text{(AII.5)}$$

where δ_{ij} is Kronecker delta (it is equal to one when $i = j$ and is zero when $i \neq j$).

For examples,

$$\boldsymbol{a}_1 \cdot \boldsymbol{b}_1 = \boldsymbol{a}_1 \cdot \frac{\boldsymbol{a}_2 \times \boldsymbol{a}_3}{\boldsymbol{a}_1 \cdot \boldsymbol{a}_2 \times \boldsymbol{a}_3} = 1 \qquad \text{(AII.6)}$$

$$\boldsymbol{a}_1 \cdot \boldsymbol{b}_2 = \boldsymbol{a}_1 \cdot \frac{\boldsymbol{a}_3 \times \boldsymbol{a}_1}{\boldsymbol{a}_1 \cdot \boldsymbol{a}_2 \times \boldsymbol{a}_3} = 0 \qquad \text{(AII.7)}$$

The second expression is zero because the vector $\boldsymbol{a}_3 \times \boldsymbol{a}_1$ is perpendicular to $\boldsymbol{a}_1$ and hence the scalar product of $\boldsymbol{a}_1$ and $\boldsymbol{a}_3 \times \boldsymbol{a}_1$ will be equal to zero.

Characteristics of the Vector H_{hkl}

The vector $\boldsymbol{H}_{hkl}$ has the following characteristics.

1. Its orientation is perpendicular to the planes *hkl*.
2. Its magnitude is equal to the reciprocal of the interplanar distance d_{hkl}.

Proof of first characteristic Consider the vectors $\boldsymbol{a}_1/h - \boldsymbol{a}_2/k$ and $\boldsymbol{a}_2/k - \boldsymbol{a}_3/l$ which have orientations parallel to the *hkl* planes (Fig. AII.1).

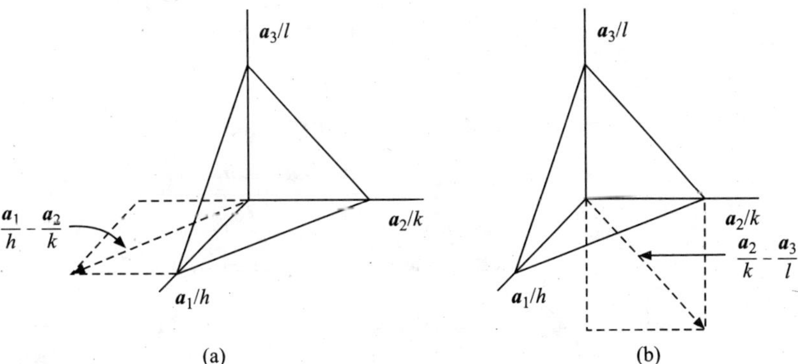

Fig. AII.1

The scalar product of these vectors with $\boldsymbol{H}_{hkl}$ are as follows.

$$\left(\frac{\boldsymbol{a}_1}{h}-\frac{\boldsymbol{a}_2}{k}\right)\cdot\boldsymbol{H}_{hkl}=\left(\frac{\boldsymbol{a}_1}{h}-\frac{\boldsymbol{a}_2}{k}\right)\cdot(h\boldsymbol{b}_1+k\boldsymbol{b}_2+l\boldsymbol{b}_3)$$

$$=\boldsymbol{a}_1\cdot\boldsymbol{b}_1+\frac{k}{h}\boldsymbol{a}_1\cdot\boldsymbol{b}_2+\frac{l}{h}\boldsymbol{a}_1\cdot\boldsymbol{b}_3-\frac{h}{k}\boldsymbol{a}_2\cdot\boldsymbol{b}_1-\boldsymbol{a}_2\cdot\boldsymbol{b}_2-\frac{l}{k}\boldsymbol{a}_2\cdot\boldsymbol{b}_3$$

$$=1+0+0-0-1-0 \qquad \text{(using Eq. AII.5)}$$

$$=0$$

$$\left(\frac{\boldsymbol{a}_2}{k}-\frac{\boldsymbol{a}_3}{l}\right)\cdot\boldsymbol{H}_{hkl}=\left(\frac{\boldsymbol{a}_2}{k}-\frac{\boldsymbol{a}_3}{l}\right)\cdot(h\boldsymbol{b}_1+k\boldsymbol{b}_2+l\boldsymbol{b}_3)$$

$$=\frac{h}{k}\boldsymbol{a}_2\cdot\boldsymbol{b}_1+\boldsymbol{a}_2\cdot\boldsymbol{b}_2+\frac{l}{k}\boldsymbol{a}_2\cdot\boldsymbol{b}_3-\frac{h}{l}\boldsymbol{a}_3\cdot\boldsymbol{b}_1-\frac{k}{l}\boldsymbol{a}_3\cdot\boldsymbol{b}_2-\boldsymbol{a}_3\cdot\boldsymbol{b}_3$$

$$=0+1+0-0-0-1$$

$$=0$$

These scalar products indicate that the vector $\boldsymbol{H}_{hkl}$ is perpendicular to the both the vectors $(\boldsymbol{a}_1/h-\boldsymbol{a}_2/k)$ and $(\boldsymbol{a}_2/k-\boldsymbol{a}_3/l)$. Since both the latter vectors are parallel to the *hkl* planes, it follows that the vector $\boldsymbol{H}_{hkl}$ is perpendicular to the *hkl* planes.

Proof of second characteristic To show that the magnitude $|\boldsymbol{H}_{hkl}|$ is equal to $1/d_{hkl}$, we consider a unit vector $\boldsymbol{n}$ perpendicular to the planes (Fig. 3.10.1). If ϕ is the angle between the a_1 axis and the normal vector $\boldsymbol{n}$, it follows that the distance d_{hkl} between the two successive planes will be given by

$$d_{hkl}=\frac{|\boldsymbol{a}_1|}{h}\cos\phi=\frac{\boldsymbol{a}_1}{h}\cdot\boldsymbol{n} \qquad \text{(AII.8)}$$

Since the vector $\boldsymbol{H}_{hkl}$ is perpendicular to the *hkl* planes, we can define the unit vector $\boldsymbol{n}$ as

$$\boldsymbol{n} = \frac{\boldsymbol{H}_{hkl}}{|\boldsymbol{H}_{hkl}|} = \frac{h_1\boldsymbol{b}_1 + k\boldsymbol{b}_2 + l\boldsymbol{b}_3}{|\boldsymbol{H}_{hkl}|} \tag{AII.9}$$

With this, Eq. (AII.8) becomes

$$d_{hkl} = \frac{\boldsymbol{a}_1}{h} \cdot \frac{h\boldsymbol{b}_1 + k\boldsymbol{b}_2 + l\boldsymbol{b}_3}{|\boldsymbol{H}_{hkl}|} \tag{AII.10}$$

which in view of Eq. (AII.5) becomes

$$d_{hkl} = \frac{1}{|\boldsymbol{H}_{hkl}|} \quad \text{i.e.} \quad |\boldsymbol{H}_{hkl}| = \frac{1}{d_{hkl}} \tag{AII.11}$$

Reciprocal Lattice

The reciprocal lattice is obtained by the repetition of reciprocal vectors $\boldsymbol{b}_1$, $\boldsymbol{b}_2$ and $\boldsymbol{b}_3$. The vector from the origin to any point *hkl* of the lattice represents the vector $\boldsymbol{H}_{hkl}$ which has orientation perpendicular to the *hkl* planes and magnitude equal to the reciprocal of the interplanar distance d_{hkl}.

Bragg Law in the Vector Form

Consider the reflection of X-rays from the *hkl* planes as shown in Fig. AII.2.

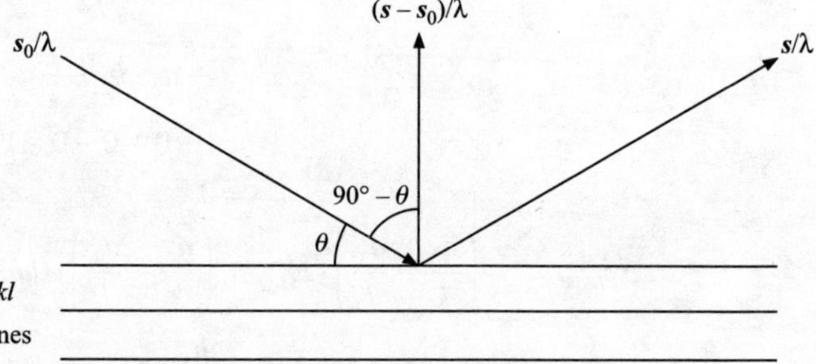

Fig. AII.2 Reflection of X-rays from *hkl* planes

Both the incident X-ray (represented by the unit vector $\boldsymbol{s}_0$) and the reflected X-ray (represented by the unit vector $\boldsymbol{s}$) make equal angle θ with the *hkl* planes. We consider the vectors $\boldsymbol{s}_0/\lambda$ and $\boldsymbol{s}/\lambda$ as shown in Fig. AII.2. From Fig. AII.2, it follows that the vector $(\boldsymbol{s} - \boldsymbol{s}_0)/\lambda$ has a direction perpendicular to the *hkl* planes and its magnitude is given by

$$\left|\frac{\boldsymbol{s} - \boldsymbol{s}_0}{\lambda}\right| = \frac{2\cos(90° - \theta)}{\lambda} = \frac{2\sin\theta}{\lambda} \tag{AII.12}$$

From Bragg law,

$$\lambda = 2d_{hkl}\sin\theta$$

we get
$$\frac{2\sin\theta}{\lambda} = \frac{1}{d_{hkl}} \tag{AII.13}$$

Since $|\boldsymbol{H}_{hkl}| = 1/d_{hkl}$, we write the above expression as

$$\frac{2\sin\theta}{\lambda} = |\boldsymbol{H}_{hkl}| \tag{AII.14}$$

Comparing Eq. (AII.12) with Eq. (AII.14), we get

$$\left| \frac{s - s_0}{\lambda} \right| = | \boldsymbol{H}_{hkl} | \tag{AII.15}$$

Since both the above vectors have orientations perpendicular to the *hkl* planes, we can write

$$\frac{s - s_0}{\lambda} = \boldsymbol{H}_{hkl} \tag{AII.16}$$

which is the vector form of Bragg law.

Expression of Interplanar Distances

Since $| \boldsymbol{H}_{hkl} | = 1/d_{hkl}$, we write

$$\frac{1}{d_{hkl}^2} = | \boldsymbol{H}_{hkl} |^2 = \boldsymbol{H}_{hkl} \cdot \boldsymbol{H}_{hkl}$$

Since $| \boldsymbol{H}_{hkl} | = h\boldsymbol{b}_1 + k\boldsymbol{b}_2 + l\boldsymbol{b}_3$, we get

$$\begin{aligned}\frac{1}{d_{hkl}^2} &= (h\boldsymbol{b}_1 + k\boldsymbol{b}_2 + l\boldsymbol{b}_3) \cdot (h\boldsymbol{b}_1 + k\boldsymbol{b}_2 + l\boldsymbol{b}_3) \\ &= h^2 (\boldsymbol{b}_1 \cdot \boldsymbol{b}_1) + k^2 (\boldsymbol{b}_2 \cdot \boldsymbol{b}_2) + l^2 (\boldsymbol{b}_3 \cdot \boldsymbol{b}_3) + 2hk (\boldsymbol{b}_1 \cdot \boldsymbol{b}_2) \\ &\quad + 2kl (\boldsymbol{b}_2 \cdot \boldsymbol{b}_3) + 2hl (\boldsymbol{b}_3 \cdot \boldsymbol{b}_1)\end{aligned}$$

From the definitions of reciprocal vectors (Eqs AII.2 to AII.4), we can write the above expression as

$$\begin{aligned}\frac{1}{d_{hkl}^2} = \frac{1}{(\boldsymbol{a}_1 \cdot \boldsymbol{a}_2 \times \boldsymbol{a}_3)^2} &[h^2 (\boldsymbol{a}_2 \times \boldsymbol{a}_3) \cdot (\boldsymbol{a}_2 \times \boldsymbol{a}_3) \\ &+ k^2 (\boldsymbol{a}_3 \times \boldsymbol{a}_1) \cdot (\boldsymbol{a}_3 \times \boldsymbol{a}_1) + l^2 (\boldsymbol{a}_1 \times \boldsymbol{a}_2) \cdot (\boldsymbol{a}_1 \times \boldsymbol{a}_2) \\ &+ 2hk (\boldsymbol{a}_2 \times \boldsymbol{a}_3) \cdot (\boldsymbol{a}_3 \times \boldsymbol{a}_1) + 2kl (\boldsymbol{a}_3 \times \boldsymbol{a}_1) \cdot (\boldsymbol{a}_1 \times \boldsymbol{a}_2) \\ &+ 2lh (\boldsymbol{a}_1 \times \boldsymbol{a}_2) \cdot (\boldsymbol{a}_2 \times \boldsymbol{a}_3)]\end{aligned}$$

Using the expression

$$(\boldsymbol{a} \times \boldsymbol{b}) \cdot (\boldsymbol{c} \times \boldsymbol{d}) = (\boldsymbol{a} \cdot \boldsymbol{c}) (\boldsymbol{b} \cdot \boldsymbol{d}) - (\boldsymbol{a} \cdot \boldsymbol{d}) (\boldsymbol{b} \cdot \boldsymbol{c})$$

the above expression becomes

$$\begin{aligned}\frac{1}{d_{hkl}^2} = \frac{1}{(\boldsymbol{a}_1 \cdot \boldsymbol{a}_2 \times \boldsymbol{a}_3)^2} &[h^2 \{(\boldsymbol{a}_2 \cdot \boldsymbol{a}_2) (\boldsymbol{a}_3 \cdot \boldsymbol{a}_3) - (\boldsymbol{a}_2 \cdot \boldsymbol{a}_3) (\boldsymbol{a}_3 \cdot \boldsymbol{a}_2)\} \\ &+ k^2 \{(\boldsymbol{a}_3 \cdot \boldsymbol{a}_3) (\boldsymbol{a}_1 \cdot \boldsymbol{a}_1) - (\boldsymbol{a}_3 \cdot \boldsymbol{a}_1) (\boldsymbol{a}_1 \cdot \boldsymbol{a}_3)\} \\ &+ l^2 \{(\boldsymbol{a}_1 \cdot \boldsymbol{a}_1) (\boldsymbol{a}_2 \cdot \boldsymbol{a}_2) - (\boldsymbol{a}_1 \cdot \boldsymbol{a}_2) (\boldsymbol{a}_2 \cdot \boldsymbol{a}_1)\} \\ &+ 2hk \{(\boldsymbol{a}_2 \cdot \boldsymbol{a}_3) (\boldsymbol{a}_3 \cdot \boldsymbol{a}_1) - (\boldsymbol{a}_2 \cdot \boldsymbol{a}_1) (\boldsymbol{a}_3 \cdot \boldsymbol{a}_3)\} \\ &+ 2kl \{(\boldsymbol{a}_3 \cdot \boldsymbol{a}_1) (\boldsymbol{a}_1 \cdot \boldsymbol{a}_2) - (\boldsymbol{a}_3 \cdot \boldsymbol{a}_2) (\boldsymbol{a}_1 \cdot \boldsymbol{a}_1)\} \\ &+ 2lh \{(\boldsymbol{a}_1 \cdot \boldsymbol{a}_2) (\boldsymbol{a}_2 \cdot \boldsymbol{a}_3) - (\boldsymbol{a}_1 \cdot \boldsymbol{a}_3) (\boldsymbol{a}_2 \cdot \boldsymbol{a}_2)\}]\end{aligned}$$

$$= \frac{1}{(\boldsymbol{a}_1 \cdot \boldsymbol{a}_2 \times \boldsymbol{a}_3)^2} [h^2 \{a_2^2 a_3^2 - (a_2 a_3 \cos \alpha_{23})^2$$
$$+ k^2 \{a_3^2 a_1^2 - (a_3 a_1 \cos \alpha_{31})^2\} + l^2 \{a_1^2 a_2^2 - (a_1 a_2 \cos \alpha_{12})^2\}$$
$$+ 2hk \{(a_2 a_3 \cos \alpha_{23})(a_3 a_1 \cos \alpha_{31}) - (a_1 a_2 \cos \alpha_{12}) a_3^2\}$$
$$+ 2kl \{(a_3 a_1 \cos \alpha_{13})(a_1 a_2 \cos \alpha_{12}) - (a_2 a_3 \cos \alpha_{23}) a_1^2\}$$
$$+ 2lh \{(a_1 a_2 \cos \alpha_{12})(a_2 a_3 \cos \alpha_{23}) - (a_3 a_1 \cos \alpha_{31}) a_2^2\}]$$
$$= \frac{a_1^2 a_2^2 a_3^2}{(\boldsymbol{a}_1 \cdot \boldsymbol{a}_2 \times \boldsymbol{a}_3)^2} \left[\frac{h^2 \sin^2 \alpha_{23}}{a_1^2} + \frac{k^2 \sin^2 \alpha_{31}}{a_2^2} + \frac{l^2 \sin^2 \alpha_{12}}{a_3^2} \right.$$
$$+ \frac{2hk}{a_1 a_2} (\cos \alpha_{23} \cos \alpha_{31} - \cos \alpha_{12})$$
$$+ \frac{2kl}{a_2 a_3} (\cos \alpha_{31} \cos \alpha_{12} - \cos \alpha_{23})$$
$$\left. + \frac{2lk}{a_1 a_3} (\cos \alpha_{12} \cos \alpha_{23} - \cos \alpha_{31}) \right] \qquad \text{(AII.17)}$$

The expression $\boldsymbol{a}_1 \cdot \boldsymbol{a}_2 \times \boldsymbol{a}_3$ represents the volume of unit cell. By definition, we have

$$(\boldsymbol{a}_1 \cdot \boldsymbol{a}_2 \times \boldsymbol{a}_3)^2 = \begin{vmatrix} \boldsymbol{a}_1 \cdot \boldsymbol{a}_1 & \boldsymbol{a}_1 \cdot \boldsymbol{a}_2 & \boldsymbol{a}_1 \cdot \boldsymbol{a}_3 \\ \boldsymbol{a}_2 \cdot \boldsymbol{a}_1 & \boldsymbol{a}_2 \cdot \boldsymbol{a}_2 & \boldsymbol{a}_2 \cdot \boldsymbol{a}_3 \\ \boldsymbol{a}_3 \cdot \boldsymbol{a}_1 & \boldsymbol{a}_3 \cdot \boldsymbol{a}_2 & \boldsymbol{a}_3 \cdot \boldsymbol{a}_3 \end{vmatrix}$$

$$= a_1^2 (a_2^2 a_3^2 - a_2^2 a_3^2 \cos^2 \alpha_{23})$$
$$- (a_1 a_2 \cos \alpha_{12}) [(a_1 a_2 \cos \alpha_{12}) a_3^2 - (a_2 a_3 \cos \alpha_{23})(a_3 a_1 \cos \alpha_{31})]$$
$$+ (a_3 a_1 \cos \alpha_{31}) [(a_1 a_2 \cos \alpha_{12})(a_2 a_3 \cos \alpha_{23}) - a_2^2 (a_3 a_1 \cos \alpha_{31})]$$
$$= a_1^2 a_2^2 a_3^2 [1 + 2 \cos \alpha_{12} \cos \alpha_{23} \cos \alpha_{31} - \cos^2 \alpha_{12} - \cos^2 \alpha_{23} - \cos^2 \alpha_{31}]$$

Substituting the above expression in Eq. (AII.17) and using the conventional crystallographic notations

$a_1 = a$; $\alpha_{23} = \alpha$

$a_2 = b$; $\alpha_{31} = \beta$

$a_3 = c$; $\alpha_{12} = \gamma$

we get

$$\frac{1}{d_{hkl}^2} = \left[\frac{1}{1 + 2\cos\alpha\cos\beta\cos\gamma - \cos^2\alpha - \cos^2\beta - \cos^2\gamma}\right]$$

$$\left[\frac{h^2\sin^2\alpha}{a^2} + \frac{k^2\sin^2\beta}{b^2} + \frac{l^2\sin^2\gamma}{c^2} + \frac{2hk}{ab}(\cos\alpha\cos\beta - \cos\gamma) + \frac{2kl}{bc}(\cos\beta\cos\gamma - \cos\alpha) + \frac{2lh}{ac}(\cos\gamma\cos\alpha - \cos\beta)\right] \quad \text{(AII.18)}$$

Equation (AII.18) is applicable to all the seven crystal systems. For triclinic system ($a \neq b \neq c$, $\alpha \neq \beta \neq \gamma \neq 90°$), Eq. (AII.18) is to be used as written, but for other crystal systems, simplified expressions deduced by using the corresponding crystal parameters can be used. These are described in the following.

Rhombohedral crystal: $a = b = c$; $\alpha = \beta = \gamma \neq 90°$

$$\frac{1}{d_{hkl}^2} = \frac{(h^2 + k^2 + l^2)\sin^2\alpha + 2(hk + kl + lh)(\cos^2\alpha - \cos\alpha)}{a^2(1 + 2\cos^3\alpha - 3\cos^2\alpha)} \quad \text{(AII.19)}$$

Hexagonal crystal: $a = b \neq c$; $\alpha = \beta = 90°$, $\gamma = 120°$

$$\frac{1}{d_{hkl}^2} = \frac{4}{3}\left(\frac{h^2 + hk + k^2}{a^2}\right) + \frac{l^2}{c^2} \quad \text{(AII.20)}$$

Monoclinic crystal: $a \neq b \neq c$; $\alpha = \beta = 90° \neq \gamma$

$$\frac{1}{d_{hkl}^2} = \frac{1}{\sin^2\gamma}\left(\frac{h^2}{a^2} + \frac{k^2}{b^2} + \frac{l^2\sin^2\gamma}{c^2} - \frac{2hk\cos\gamma}{ab}\right) \quad \text{(AII.21)}$$

Orthorhombic crystal: $a \neq b \neq c$; $\alpha = \beta = \gamma = 90°$

$$\frac{1}{d_{hkl}^2} = \frac{h^2}{a^2} + \frac{k^2}{b^2} + \frac{l^2}{c^2} \quad \text{(AII.22)}$$

Tetragonal crystal: $a = b \neq c$; $\alpha = \beta = \gamma = 90°$

$$\frac{1}{d_{hkl}^2} = \frac{h^2 + k^2}{a^2} + \frac{l^2}{c^2} \quad \text{(AII.23)}$$

Cubic crystal: $a = b = c$; $\alpha = \beta = \gamma = 90°$

$$\frac{1}{d_{hkl}^2} = \frac{h^2 + k^2 + l^2}{a^2} \quad \text{(AII.24)}$$

Scattering of X-Rays by Crystals

Crystals exhibit the phenomenon of diffraction when they are exposed to X-rays. The electron present in atoms/molecules of a crystal under the influence of electric component of X-rays experiences acceleration, and according to the classical electromagnetic theory, they emit radiations (know as scattered radiations) in all directions with the frequency identical to that of the incident X-rays.

Scattering by an Electron

For a beam of polarized X-rays, the intensity of scattered X-rays at a distance R from an electron (mass: m_e and charge: e) is given by the expression

$$I = I_0\left[\frac{e^4}{(4\pi\varepsilon_0)^2 m_e^2 c^4 R^2}\right] \qquad \text{(AII.25)}$$

where I_0 is the intensity of incident X-rays and c is the speed of light.

Scattering by an Atom

The scattering of X-rays by an atom is due to its electronic charge. Assuming charge distribution in an atom to be spherical symmetrical, the intensity of scattered X-rays at a distance R from the atom is given by

$$I = I_0\left[\frac{e^4}{(4\pi\varepsilon_0)^2 m_e^2 c^4 R^2}\left(\sum_n f_{en}^2\right)\right] \qquad \text{(AII.26)}$$

where the summation is over the number of electrons in the atom and f_e represents the scattering factor of an electron and is given by the expression

$$f_e = \int_{r=0}^{\infty} 4\pi r^2 \rho_r \left(\frac{\sin kr}{kr}\right) \mathrm{d}r \qquad \text{(AII.27)}$$

In this expression, ρ_r is the charge density at a distance r from the nucleus of the atom and $k = (4\pi \sin \theta)/\lambda$, where θ is half of the angle between the incident and reflected X-rays.

Scattering from a Small Crystal

The intensity of scattered X-rays from a small crystal depends upon the number, nature and placement of its constituent atoms in the crystal lattice. The expression of intensity of reflected X-rays is given by

$$I = I_0\left[\frac{e^4}{(4\pi\varepsilon_0)^2 m_e^2 c^4 R^2}\right] FF^* \left(\frac{\sin^2 N_1x_1}{\sin^2 x_1}\right)\left(\frac{\sin^2 N_2x_2}{\sin^2 x_2}\right)\left(\frac{\sin^2 N_3x_3}{\sin^2 x_3}\right) \qquad \text{(AII.28)}$$

The number N_1, N_2 and N_3 in the above expression when multiplied by the magnitude of the respective unit vectors $\boldsymbol{a}_1$, $\boldsymbol{a}_2$ and $\boldsymbol{a}_3$ give the edges of crystal along the three crystallographic axes. The symbols x_1, x_2 and x_3 are given by the expression

$$\left.\begin{aligned} (\pi/\lambda)(\boldsymbol{s} - \boldsymbol{s}_0)\cdot \boldsymbol{a}_1 &= x_1 \\ (\pi/\lambda)(\boldsymbol{s} - \boldsymbol{s}_0)\cdot \boldsymbol{a}_2 &= x_2 \\ (\pi/\lambda)(\boldsymbol{s} - \boldsymbol{s}_0)\cdot \boldsymbol{a}_3 &= x_3 \end{aligned}\right\} \qquad \text{(AII.29)}$$

where $\boldsymbol{s}_0$ and $\boldsymbol{s}$ are the unit vectors pointing towards the incident and reflected X-rays, respectively.

Structure Factor of a Unit Cell

The term F in Eq. (AII.28) is known as the structure factor of the unit cell of the crystal and is given by the expression

$$F = \sum_n f_n \exp\{(2\pi\,i/\lambda)(\boldsymbol{s} - \boldsymbol{s}_0)\cdot \boldsymbol{r}_n\} \qquad \text{(AII.30)}$$

where $\boldsymbol{r}_n$ is the vector representing the placement of atom in the unit cell of the crystal. Since

$$\lim_{x\to n\pi} \frac{\sin Nx}{\sin x} = \lim_{x\to n\pi} \frac{N\cos Nx}{\cos x} = \pm N$$

the term $\sin^2 Nx/\sin^2 x$ has a value of N^2 near the vicinity of $x = n\pi$ while at other values of x, the term has a small value near to zero (Fig. AII.3).

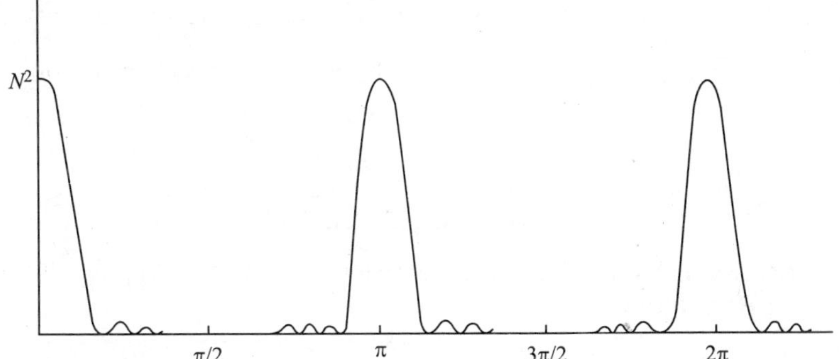

Fig. AII.3 The plot of function $(\sin Nx/\sin x)^2$ versus x

Hence, the intensity of the scattered X-rays as given by Eq. (AII.28) will have smaller values (near to zero) unless the three quotients are simultaneously close to their maximum values for which we must have

$$x_1 = h'\pi; \quad x_2 = k'\pi \quad \text{and} \quad x_3 = l'\pi \qquad \text{(AII.31)}$$

where h', k' and l' are integers. Since for an average crystal the integrals N_1, N_2 and N_3 have large values, the intensities of the scattered X-rays under the conditions given by Eq. (AII.31) will be much larger than those prevailing at other values of x_1, x_2 and x_3.

The Three Laüe Equations

Substituting Eq. (AII.31) in Eq. (AII.29), we get

$$(\pi/\lambda)\,(\boldsymbol{s} - \boldsymbol{s}_0)\cdot \boldsymbol{a}_1 = h'\pi \quad \text{i.e.} \quad (\boldsymbol{s} - \boldsymbol{s}_0)\cdot \boldsymbol{a}_1 = h'\lambda$$

$$(\pi/\lambda)\,(\boldsymbol{s} - \boldsymbol{s}_0)\cdot \boldsymbol{a}_2 = k'\pi \quad \text{i.e.} \quad (\boldsymbol{s} - \boldsymbol{s}_0)\cdot \boldsymbol{a}_2 = k'\lambda$$

$$(\pi/\lambda)\,(\boldsymbol{s} - \boldsymbol{s}_0)\cdot \boldsymbol{a}_3 = l'\pi \quad \text{i.e.} \quad (\boldsymbol{s} - \boldsymbol{s}_0)\cdot \boldsymbol{a}_3 = l'\lambda \qquad \text{(AII.32)}$$

These three expressions are known as the 3 Laüe equations.

Equivalency of the Three Laüe Equations with Bragg Law

Under the conditions given by Eq. (AII.31), it can be shown that the three expressions of Eq. (AII.32) are equivalent to Bragg's law.

Let any arbitrary vector $\boldsymbol{r}$ be written in terms of reciprocal vectors as

$$\boldsymbol{r} = p_1\boldsymbol{b}_1 + p_2\boldsymbol{b}_2 + p_3\boldsymbol{b}_3 \qquad \text{(AII.33)}$$

Taking the scalar product of Eq. (AII.33) with the primary vector $\boldsymbol{a}_1$, we get

$$\boldsymbol{r}\cdot\boldsymbol{a}_1 = (p_1\boldsymbol{b}_1 + p_2\boldsymbol{b}_2 + p_3\boldsymbol{b}_3)\cdot\boldsymbol{a}_1$$

$$= p_1 \qquad (\text{as } \boldsymbol{a}_i\cdot\boldsymbol{b}_j = \delta_{ij})$$

Similarly, it can be shown that

$$\boldsymbol{r}\cdot\boldsymbol{a}_2 = p_2 \quad \text{and} \quad \boldsymbol{r}\cdot\boldsymbol{a}_3 = p_3$$

With these expressions of p_1, p_2 and p_3, Eq. (AII.33) becomes

$$\boldsymbol{r} = (\boldsymbol{r}\cdot\boldsymbol{a}_1)\,\boldsymbol{b}_1 + (\boldsymbol{r}\cdot\boldsymbol{a}_2)\,\boldsymbol{b}_2 + (\boldsymbol{r}\cdot\boldsymbol{a}_3)\,\boldsymbol{b}_3 \qquad (\text{AII.34})$$

Let $\boldsymbol{r}$ be the vector $\boldsymbol{s} - \boldsymbol{s}_0$. Hence

$$\boldsymbol{s} - \boldsymbol{s}_0 = \{(\boldsymbol{s}-\boldsymbol{s}_0)\cdot\boldsymbol{a}_1\}\,\boldsymbol{b}_1 + \{(\boldsymbol{s}-\boldsymbol{s}_0).\boldsymbol{a}_2\}\,\boldsymbol{b}_2 + \{(\boldsymbol{s}-\boldsymbol{s}_0)\cdot\boldsymbol{a}_3)\}\,\boldsymbol{b}_3$$

which on using the 3 Laüe equations becomes

$$\boldsymbol{s} - \boldsymbol{s}_0 = (h'\lambda)\,\boldsymbol{b}_1 + (k'\lambda)\,\boldsymbol{b}_2 + (l'\lambda)\,\boldsymbol{b}_3$$

$$= \lambda(h'\boldsymbol{b}_1 + h'\boldsymbol{b}_2 + l'\boldsymbol{b}_3)$$

$$= \lambda\,\boldsymbol{H}_{h'h'l'} \qquad (\text{AII.35})$$

which is Bragg's law for the reflection from $h'k'l'$ planes (Eq. AII.16).

Structure Factor for a Bragg Reflection

The structure factor of a unit cell (Eq. AII.30) is

$$F = \sum_n f_n \exp\,[(2\pi\,\mathrm{i}/\lambda)\,(\boldsymbol{s}-\boldsymbol{s}_0)\cdot\boldsymbol{r}_n] \qquad (\text{AII.36})$$

where f_n is the atomic scattering factor, $\boldsymbol{s}$ and $\boldsymbol{s}_0$ are the respective unit vectors directed along the incident and scattered X-rays and $\boldsymbol{r}_n$ is the position vector of the atom in the unit cell.

According to Bragg law (Eq. AII.16), we have

$$\boldsymbol{s} - \boldsymbol{s}_0 = \lambda\,\boldsymbol{H}_{hkl} = \lambda(h\boldsymbol{b}_1 + k\boldsymbol{b}_2 + l\boldsymbol{b}_3) \qquad (\text{AII.37})$$

The vector $\boldsymbol{r}_n$ in terms of components along the crystallographic axes $\boldsymbol{a}_1$, $\boldsymbol{a}_2$ and $\boldsymbol{a}_3$, and the fractional coordinates x_n, y_n and z_n as shown in Fig. AII.4 is

$$\boldsymbol{r}_n = x_n\boldsymbol{a}_1 + y_n\boldsymbol{a}_2 + z_n\boldsymbol{a}_3 \qquad (\text{AII.38})$$

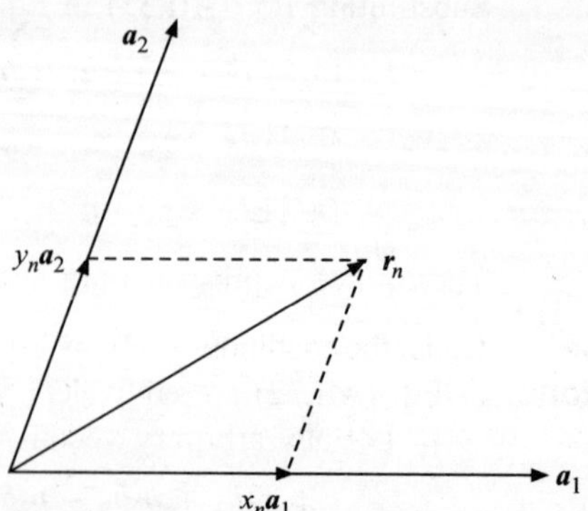

Fig. AII.4 The vector $\boldsymbol{r}_n$ in terms of the fractional coordinates

From Eqs (AII.37) and (AII.38), we can write

$$(\boldsymbol{s} - \boldsymbol{s}_0) \cdot \boldsymbol{r}_n = \lambda\ (h\boldsymbol{b}_1 + k\boldsymbol{b}_2 + l\boldsymbol{b}_3) \cdot (x_n\boldsymbol{a}_1 + y_n\boldsymbol{a}_2 + z_n\boldsymbol{a}_3)$$

which on using the fact $\boldsymbol{a}_i \cdot \boldsymbol{b}_j = \delta_{ij}$, we get

$$(\boldsymbol{s} - \boldsymbol{s}_0) \cdot \boldsymbol{r}_n = \lambda\ (hx_n + ky_n + lz_n) \tag{AII.39}$$

Substituting this in Eq. (AII.36), we get

$$F = \sum_n f_n \exp[(2\pi \mathrm{i})(hx_n + ky_n + lz_n)] \tag{AII.40}$$

Equation (AII.40) provides the structure factor for a reflection from *hkl* planes in a crystal. Its application to a few unit cells is given below.

Face-Centred Cubic Bravais Lattice

For a face-centred Bravais lattice, one needs to specify the locations of four points in its unit cell, conventionally, these are taken as

$$(0, 0, 0);\quad \left(\tfrac{1}{2}, \tfrac{1}{2}, 0\right);\quad \left(\tfrac{1}{2}, 0, \tfrac{1}{2}\right);\quad \text{and}\quad \left(0, \tfrac{1}{2}, \tfrac{1}{2}\right)$$

The other points of the lattice can be generated by translational symmetry of the four points, i.e. by increasing the coordinates of each by unity each time

The structural factor (Eq. AII.40) becomes

$$F = f[1 + \exp\{\pi\ \mathrm{i}\ (h + k) + \exp\{\pi\ \mathrm{i}\ (h + l)\} + \exp\{\pi\ \mathrm{i}\ (k + l)] \tag{AII.41}$$

Since $\exp(\mathrm{i}\ \pi\ m) = \cos\ (\pi\ m) + \mathrm{i}\ \sin\ (\pi\ m)$, the value of $\exp(\mathrm{i}\ \pi\ m)$ will be equal to $(-1)^m$ for an integral value of m. In Eq. (AII.41), $m = h + k$ or $h + l$ or $k + l$, i.e. m involves the sum of two Miller indices at one time. The value of m may be even or odd depending upon the values of *hkl* indices. The following cases may be distinguished.

Case 1 The indices *hkl* are all either odd or even.
In this case, sum of two indices will always be a even number. Hence

$$F_{hkl} = f(1 + 1 + 1 + 1) = 4\,f \tag{AII.42}$$

Case 2 The indices *hkl* have mixed even and odd values.
In this case, two of the three sums of two indices will have odd value while the third sum will have even value. For the former, $(-1)^m = -1$ and for the latter $(-1)^m = +1$. Hence

$$F_{hkl} = f(1 - 1 - 1 + 1) = 0 \tag{AII.43}$$

Intensity of scattered X-rays Since the intensity of scattered X-ray depends on the square of F_{hkl}, the reflections from planes having mixed *hkl* will be missing (as their intensities are zero).

Body-Centred Cubic Bravais Lattice

For a body-centred Bravais lattice, one needs to specify the locations of two points in its unit cell. Conventionally, these are taken as

$$(0, 0, 0)\quad \text{and}\quad \left(\tfrac{1}{2}, \tfrac{1}{2}, \tfrac{1}{2}\right)$$

Substituting these in Eq. (AII.40), we get

$$f_{hkl} = f\,[1 + \exp\{(\pi\ \mathrm{i})\ (h + k + l)\}] \tag{AII.44}$$

Two cases may be distinguished.

Case 1 The sum of *h*, *k* and *l* indices is even.

In this case, $\exp\{\pi\ \mathrm{i}\ (h + k + \lambda)\} = +\,1$. Hence

$$F_{hkl} = f\ (1 + 1) = 2f \tag{AII.45}$$

Case 2 The sum of *h*, *k*, and *l* indices is odd.

In this case, $\exp\{\pi\ \mathrm{i}(h + k + l)\} = -1$ and hence

$$F_{hkl} = f\ (1 - 1) = 0 \tag{AII.46}$$

Intensity of scattered X-rays Since the intensity of scattered X-rays depends on the value of F^2, the refections from planes for which the sum of indices is odd will be missing.

Rock-Salt Structure

The rock-salt structure has face-centred cubic Bravais lattice. In the unit of sodium chloride, there are four Na^+ ions and four Cl^- ions. Conventionally, their locations are taken as follows.

$$\mathrm{Cl}\ \ (0, 0, 0);\quad \left(\tfrac{1}{2}, \tfrac{1}{2}, 0\right);\quad \left(\tfrac{1}{2}, 0, \tfrac{1}{2}\right);\quad \left(0, \tfrac{1}{2}, \tfrac{1}{2}\right)$$

$$\mathrm{Na}^+\ \left(\tfrac{1}{2}, 0, 0\right);\quad \left(0, \tfrac{1}{2}, 0\right);\quad \left(0, 0, \tfrac{1}{2}\right);\quad \left(\tfrac{1}{2}, \tfrac{1}{2}, \tfrac{1}{2}\right)$$

The structural factor in this case is

$$\begin{aligned} F_{hkl} &= f_{\mathrm{Cl}}[1 + \exp\{\pi\,\mathrm{i}\,(h + k)\} + \exp\{\pi\,\mathrm{i}(h + l)\} + \exp\{\pi\,\mathrm{i}(k + l)\}] \\ &\quad + f_{\mathrm{Na}}[\exp(\pi\,\mathrm{i}h) + \exp(\pi\,\mathrm{i}k) + \exp(\pi\,\mathrm{i}l) + \exp\{\pi\,\mathrm{i}(h + k + l)\}] \end{aligned} \tag{AII.47}$$

Three cases may be distinguished.

Case 1 The indices *hkl* are all even.
In this case

$$\begin{aligned} F_{hkl} &= f_{\mathrm{Cl}}(1 + 1 + 1 + 1) + f_{\mathrm{Na}}(1 + 1 + 1 + 1) \\ &= 4(f_{\mathrm{Cl}} + f_{\mathrm{Na}}) \end{aligned} \tag{AII.48}$$

Case 2 The indices *hkl* are all odd.
In this case

$$\begin{aligned} F_{hkl} &= f_{\mathrm{Cl}}(1 + 1 + 1 + 1) + f_{\mathrm{Na}}(-1\ -1\ -1\ -1) \\ &= 4(f_{\mathrm{Cl}} - f_{\mathrm{Na}}) \end{aligned} \tag{AII.49}$$

Case 3 The indices *hkl* have mixed even and odd values.
In this case, either two of the three indices have even or odd values. Hence

$$F_{hkl} = 0$$

Intensity of scattered X-rays Since the intensity of scattered X-rays depends on the value of F^2, We conclude that

(i) The refection from the planes for which the indices *hkl* have mixed values is missing.

(ii) The reflection from the planes having all the three indices even has intensity larger than those from the planes having all the three indices odd.

Caesium Chloride

Caesium chloride has body-centred cubic Bravais lattice with Cs^+ and Cl^- ions occupying the locations

$$Cs^+\ (0, 0, 0) \quad \text{and} \quad Cl\ \left(\tfrac{1}{2}, \tfrac{1}{2}, \tfrac{1}{2}\right)$$

The structural factor takes the form

$$F = f_{Cs} + f_{Cl} \exp\{\pi\, i\, (h + k + l)\} \tag{AII.50}$$

The following cases may be distinguished

Case 1 When $h + k + l$ = even

Here $$F = f_{Cs} + f_{Cl} \tag{AII.51}$$

Case 2 When $h + k + l$ = odd

Here $$F = f_{Cs} - f_{Cl} \tag{AII.52}$$

Zinc-Blende Structure

The zinc-blende structure has a face-centred cubic Bravais lattice with the following locations of Zn^{2+} and S^{2-} ions in its unit cell.

$$Zn^{2+}\ (0, 0, 0);\ \left(\tfrac{1}{2}, \tfrac{1}{2}, 0\right);\ \left(\tfrac{1}{2}, 0, \tfrac{1}{2}\right);\ \left(0, \tfrac{1}{2}, \tfrac{1}{2}\right)$$
$$S^2\ \left(\tfrac{1}{4}, \tfrac{1}{4}, \tfrac{1}{4}\right);\ \left(\tfrac{3}{4}, \tfrac{3}{4}, \tfrac{1}{4}\right);\ \left(\tfrac{3}{4}, \tfrac{1}{4}, \tfrac{3}{4}\right);\ \left(\tfrac{1}{4}, \tfrac{3}{4}, \tfrac{3}{4}\right)$$

The structure factor in this case becomes

$$\begin{aligned} F_{hkl} &= f_{Zn}[1 + \exp\{\pi\, i\, (h + k)\} + \exp\{\pi\, i\, (h + l)\} + \exp\{\pi\, i\, (k + l)\}] \\ &\quad + f_S[\exp\{\pi\, i\, (h + k + l)/2\} + \exp\{\pi\, i\, (3h + 3k + l)/2\} \\ &\quad + \exp\{\pi\, i\, (3h + k + 3l)/2\} + \exp\{\pi\, i\, (h + 3k + 3l)/2\}] \\ &= [1 + \exp\{\pi\, i\, (h + k)\} + \exp\{\pi\, i\, (h + l)\} + \exp\{\pi\, i\, (k + l)\}] \\ &\quad [f_{Zn} + f_S \exp\{\pi\, i\, (h + k + l)/2\}] \end{aligned} \tag{AII.53}$$

The following cases may be distinguished.

Case 1 The indices *hkl* have mixed even and odd values.
In this case, two of the three sums of two indices have odd values while the third sum has an even value. With these, the first term in Eq. (AII.53) is zero. Hence,

$$F_{hkl} = 0 \tag{AII.54}$$

Case 2 The indices *hkl* are all either even or odd.
In this case, sum of the two indices will always be even and hence each exponential in the first term of Eq. (AII.53) is equal to + 1. Hence

$$F_{hkl} = 4\ [f_{Zn} + f_S \exp\{\pi\ i\ (h + k + l)/2\}]$$

Since F_{hkl} involves imaginary quantity and the intensity of the scattered X-rays involves FF^*, we write

$$FF^* = 16\ [f_{Zn} + f_S \exp\{\pi\, i\, (h + k + 1)/2\}][f_{Zn} + f_S \exp\{-\pi\, i\, (h + k + 1)/2\}]$$

$$= 16\ [f_{Zn}^2 + f_S^2 + 2 f_{Zn} f_S \cos\{\pi (h + k + l)/2\}] \qquad \text{(AII.55)}$$

The following three subcases may be considered.

(i) When $h + k + l = 4n$

Here $FF^* = 16\ (f_{Zn} + f_S)^2$ (AII.56)

(ii) When $h + k + l = 4n + 2$

Here $FF^* = 16\ (f_{Zn} - f_S)^2$ (AII.57)

(iii) When the indices *hkl* are all odd

Here $FF^* = 16\ (f_{Zn}^2 + f_S^2)$ (AII.58)

Note: The structure of graphite is similar to that of zinc-blende structure. The structure factors of unit cell can be obtained from those of zinc-blende structure by replacing each of f_{Zn} and f_S by f_C.

4 Ionic Equilibria

4.1 THE DISSOLUTION PROCESS

General Principle of Solubility

Most chemical reactions occur in solutions. The study of such solutions constitutes one of the most important branches of physical chemistry. In general, if the solubility of solutes in various solvents is analyzed, it is observed that the polar solutes are more soluble in polar solvents whereas nonpolar solutes are more soluble in nonpolar solvents. This is the general principle of solubility, i.e. *like dissolves like.* For example, sodium chloride is soluble in water whereas it is insoluble in carbon tetrachloride. The high dielectric constant and polar nature of water, makes it one of the most important solvents for ionic solutes. The higher dielectric constant weakens the forces of attraction between the oppositely charged ions of the ionic crystals and its polar character generates the ion-dipole interactions in which the positive ion is attracted by the negative end of the water dipole, whereas the negative ion is attracted by the positive end of the dipole as shown in Fig. 4.1.1. The consequence of this is that the ions are pulled out of the crystal lattice and are drifted into the liquid phase. Ions move in the solution in the hydrated forms. Certain covalent molecules with relatively high dipole moment can also dissolve in water to produce an ionic solution because of the stronger ion-dipole interactions (e.g., hydrochloric acid).

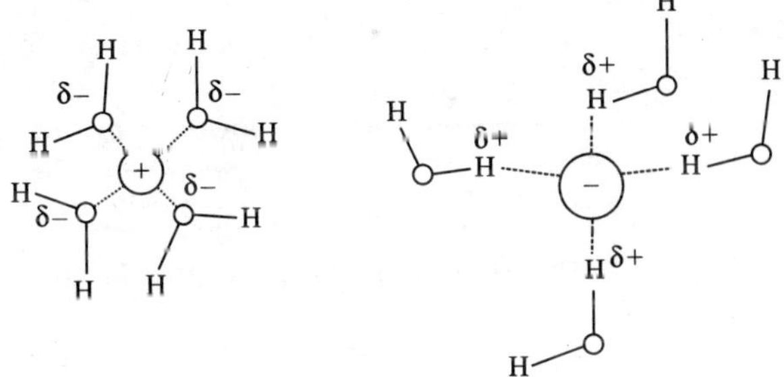

Fig. 4.1.1 Ion-dipole interactions

Ionic and Molecular Solutions

A given substance on dissolution in a solvent (e.g., water), in general, yields either an ionic solution or a molecular solution. In the former, the substance splits up into ions, whereas in the latter, it is present as such. The formation of these two types of solutions can be represented as follows:

$$MX(s) + aq \rightarrow M^+(aq) + X^-(aq) \qquad (4.1.1a)$$

$$MX(s) + aq \rightarrow MX(aq) \qquad (4.1.1b)$$

The formation of these two solutions may be considered through the following steps.

Formation of Ionic Solutions

Steps involved in the formation of ionic solutions are given below

Vaporization of the substance to form gaseous molecules

$$MX(s) \xrightarrow{\Delta_{vap}H} MX(g)$$

Dissociation of gaseous molecules into atoms

$$MX(g) \xrightarrow{\Delta_{diss}H} M(g) + X(g)$$

Formation of gaseous ions from these atoms

$$M(g) \xrightarrow{\Delta_{ioniz}H} M^+(g) + e^-$$

$$X(g) + e^- \xrightarrow{\Delta_{EA}H} X^-(g)$$

Solvation of these gaseous ions

$$M^+(g) + aq \xrightarrow{\Delta_{h1}H} M^+(aq)$$

$$X^-(g) + aq \xrightarrow{\Delta_{h2}H} X^-(aq)$$

The enthalpy change in the formation of an ionic solution is equal to the sum of the above changes, i.e.

$$\Delta H = \Delta_{vap}H + \Delta_{diss}H + \Delta_{ioniz}H + \Delta_{EA}H + \Delta_{h1}H + \Delta_{h2}H = \Delta H_1 + \Delta H_2$$

where $\Delta H_1 = \Delta_{vap}H + \Delta_{diss}H + \Delta_{ioniz}H + \Delta_{EA}H$

$$\Delta H_2 = \Delta_{h1}H + \Delta_{h2}H$$

Formation of Molecular Solution

Steps involved in the formation of molecular solutions are given below

Vaporization of the substance to form gaseous molecules

$$MX(s) \xrightarrow{\Delta_{vap}H} MX(g)$$

***Dissolution of* MX(g) *to give* MX(aq)**

$$MX(g) \xrightarrow{\Delta_{solv}H} MX(aq)$$

with a total enthalpy change $\Delta H = \Delta_{vap}H + \Delta_{solv}H$, which corresponds to the enthalpy change in the formation of a molecular solution.

Illustration

The question whether the given substance is soluble or not and whether on dissolution it forms an ionic or a molecular solution may be answered only from the consideration of the enthalpies involved in the formation of the solution. This is illustrated by taking the typical examples of $CaCl_2$ and $HgCl_2$. The enthalpies involved in various steps of formation of the solution are listed in Table 4.1.1.

Table 4.1.1 Enthalpies Involved in the Formation of Ionic and Molecular Solutions of $CaCl_2$ and $HgCl_2$

	Step	$CaCl_2$	$HgCl_2$
I	*For an ionic solution*	ΔH/kJ mol^{-1}	ΔH/kJ mol^{-1}
	(a) Sublimation	209.2	83.78
	(b) Bond breaking	1 004.2	460.2
	(c) Ionization	1 715.4	2 815.8
	Electron affinity	2 (– 359.8)	2 (– 359.8)
	(d) Cation hydration	– 1 598.3	– 1 845.1
	Anion hydration	2 (– 355.6)	2 (– 355.6)
	Total	– 100.3	83.8
II	*For a molecular solution*		
	(a) Sublimation	209.2	83.78
	(b) Dissolution of gaseous molecules	– 33.5	– 66.9
	Total	175.7	16.8

Comparison of the total enthalpy involved in the formation of the ionic solution indicates that this type of solution is more likely to be formed by $CaCl_2$ than by $HgCl_2$. Analysis of the individual enthalpies indicates that though steps (a) and (b) are more favourable to $HgCl_2$ than to $CaCl_2$, the subsequent step (c), namely, the cation formation, is highly unfavourable to Hg, with the result that $HgCl_2$ does not form an ionic solution. Similar comparison of the total enthalpy involved in the case of molecular solution indicates that the formation of such type of solution is *very* unfavourable for $CaCl_2$ and *slightly* unfavourable for $HgCl_2$.

More precisely, the formation of a solution (a spontaneous process) should be decided by the change in the value of Gibbs free energy ΔG given as

$$\Delta G = \Delta H - T\,\Delta S$$

where ΔH and ΔS are the respective enthalpy change and entropy change of the process. The former represents change in the value of heat content at constant pressure and the latter represents change in the extent of disorderlines of the system. Since the formation of a solution is always accompained by the increase in entropy, the factor $T\,\Delta S$ is always positive. For a spontaneous dissolution, ΔG should be negative. Bóth ΔH and ΔS favour this for an exothermic reaction, whereas, for an endothermic reaction, the entropy factor has to outweigh the enthalpy change. However, this term is usually not large, and does not contribute more than 30 kJ to the overall free energy change. Nevertheless, it becomes very important for an endothermic reaction with a very small value of ΔH as in the case of $HgCl_2$.

4.2 CLASSIFICATION OF SUBSTANCES

Classification of Electrolytes

Based on the relative values of conductivities of aqueous solutions, the dissolved substance can be classified into any one of the following categories.

(i) Strong electrolyte: high conducting
(ii) Weak electrolyte: low conducting
(iii) Nonelectrolyte: nonconducting

Table 4.2.1 records a few typical examples of strong, weak and nonelectrolytes.

Table 4.2.1 A Few Typical Examples of Strong, Weak and Nonelectrolytes

Compounds	*Crystal type*	*Solutions*
Halides, hydroxides and acetates of Gp. 1 and Gp. 2 elements	Ionic	Strong electrolytes
Nitrate, chlorate and sulphates of M^+ and M^{2+} cations	Ionic	Strong electrolytes
$PbBr_2$, $PbCl_2$, $PbAc_2$, $HgCl_2$, $CuCl_2$	Ionic to molecular	Weak electrolytes
HCl, HBr, HI	Molecular	Strong electrolytes
H_2SO_4, $HClO_4$, HNO_3,	Molecular (H bonding)	Strong electrolytes
RCOOH, H_2CO_3	Molecular	Weak electrolytes
ROH, HCN, other organic compounds	Molecular	Nonelectrolytes

Limitation of Classification

The classification of compounds in terms of strong and weak electrolytes is based on their behaviour in a particular solvent, namely, water. However, such classification suffers from a great disadvantage in the sense that a particular electrolyte, though weak in water, might behave as a strong one when dissolved in some other solvent or vice versa. For example, sodium chloride behaves as a strong electrolyte and acetic acid as a weak electrolyte when dissolved in water. However, when acetic acid and sodium chloride are dissolved in ammonia, their conductivity values are comparable, indicating a strong electrolytic behaviour for acetic acid (Table 4.2.2).

Table 4.2.2 Molar Conductivity Values

Solute	*Solvent*	
	Water Λ/S cm^2 mol^{-1}	*Ammonia* Λ/S cm^2 mol^{-1}
Sodium chloride	106.7(s)	284.0(s)
Acetic acid	4.7(w)	216.0(s)

Thus, the above classification depends upon the solvent used.

True and Potential Electrolytes

Another classification which is largely based on the characteristics of the solute and not on that of the solvent, is to label them as the *true electrolyte* and the *potential electrolyte.* The essential characteristics of true electrolyte is that even in the pure liquid state it is an ionic conductor. In dissolution process, all that a polar solvent does is that it uses ion-dipole forces to disengage ions from their lattice sites, solvates them and disperses them into the solution. Examples are NaCl, KCl, etc. The potential electrolyte, however, does not conduct electricity in the pure liquid state, though it provides a conducting solution on dissolution in an ionic solvent. Examples are hydrochloric acid, acetic acid, etc.

4.3 THE ARRHENIUS THEORY OF DISSOCIATION

The increase in molar conductivity with decreasing concentration observed in dilute solutions of all electrolytes led Arrhenius to postulate that a chemical equilibrium exists between the molecule of undissociated electrolyte and the ions that result from dissociation

$$AB \rightleftharpoons A^+ + B \tag{4.3.1}$$

On dilution, more of AB dissociates to give A^+ and B^-, which accounts for the increase in molar conductivity. In dilute solutions, it is known today that the above equilibrium is valid only for weak electrolytes. Strong electrolytes are already present in the form of ions in the solid state. Evidence for the existence of equilibrium in weak electrolytes can be seen from the study of colligative properties (properties which depend only on the number of species and not on their nature). Such properties are osmotic pressure, relative lowering of vapour pressure, elevation of boiling point and depression of freezing point.

For example, if we have 0.01 mol kg^{-1} solutions of CH_3OH and NaCl, the depression of freezing point in the latter is double that of the former. It is because of the fact that solution of NaCl would be 0.01 mol kg^{-1} with respect to Na^+ and 0.01 mol kg^{-1} with respect to Cl^- and that the total concentration of the species in solution is 0.02 mol kg^{-1}. Thus:

Depression of freezing point in 0.01 mol kg^{-1} CH_3OH = 0.018 6 °C

Depression of freezing point in 0.01 mol kg^{-1} NaCl = 0.037 2 °C

Similarly,

Depression of freezing point in 0.01 mol kg^{-1} $Al_2(SO_4)_3$ = 0.093 0 °C

Dissociation of a Weak Electrolyte

The depression of freezing point in case of a weak electrolyte AB (0.01 mol kg^{-1}) is in between the values of 0.018 6 °C and 0.037 2 °C. Thus, the total concentration of species in the solution is greater than 0.01 mol kg^{-1} but less than 0.02 mol kg^{-1}.

Dissociation in Terms of Extent of Reaction

In general, if ξ (known as extent of reaction) is the amount of AB that has dissociated, then the amounts of various species in solution are

$$\underset{0.01\ \text{mol} - \xi}{AB} \rightleftharpoons \underset{\xi}{A^+} + \underset{\xi}{B^-} \tag{4.3.2}$$

Total amount of species in the solution is (0.01 mol + ξ) and, therefore, the depression of freezing point will be equal to (0.01 + ξ/mol) (1.86 °C).

Dissociation in Terms of Degree of Dissociation

The extent of dissociation of a substance can also be expressed in terms of *degree of dissociation,* which is, by definition, equal to the fraction of the total substance that is present in the form of ions. If α is the degree of dissociation, it means that the amount α mole out of 1 mol of the solute is present in the form of ions and thus the remaining amount of the undissociated species is (1 – α) (1 mol). If c is the concentration of the solute AB, then the concentrations of various species in solution are as follows:

$$\underset{c(1-\alpha)}{AB} \rightleftharpoons \underset{c\alpha}{A^+} + \underset{c\alpha}{B^-} \qquad (4.3.3)$$

Similarly, for the electrolyte A_2B (assuming single-step dissociation):

$$\underset{c(1-\alpha)}{A_2B} \rightleftharpoons \underset{c(2\alpha)}{2A^+} + \underset{c\alpha}{B^{2-}} \qquad (4.3.4)$$

In general, $$\underset{c(1-\alpha)}{A_xB_y} \rightleftharpoons \underset{c(x\alpha)}{xA^{y+}} + \underset{c(y\alpha)}{yB^{x-}} \qquad (4.3.5)$$

Expression of Equilibrium Constant

A chemical equilibrium is a dynamic equilibrium and can be characterized by an equilibrium constant,* which by definition is

$$K_{eq} = \frac{\text{Product of concentrations of species appearing on the right side of equilibrium, each raised to the corresponding stoichiometric number}}{\text{Product of conentrations of species appearing on the left side of equilibrium, each raised to the corresponding stoichiometric number}} \qquad (4.3.6)^\dagger$$

In the above examples, K_{eq}s are

$$K_{eq}(AB) = \frac{[A^+][B^-]}{[AB]} \qquad (4.3.7)$$

---

*Concentrations are to be expressed in mol dm^{-3}. By convention the ions are written on the right side of the dissociation reaction.

†In general, a chemical reaction is written as

$$0 = \Sigma_B\ \nu_B\ B$$

where ν_B, the stoichiometric number, is positive for products and negative for reactants. The expression of equilibrium constant is written as

$$K_{eq} = \Pi_B\ [B]^{\nu_B}$$

Taking the example of dissociation of A_2B, we have

Equilibrium reaction: $0 \rightleftharpoons 2A^+ + B^- - A_2B$ i.e. $A_2B \rightleftharpoons 2A^+ + B^-$

Equilibrium constant: $K_{eq} = [A^+]^2\,[B^-][A_2B]^{-1}$ i.e. $K_{eq} = \dfrac{[A^+]^2\,[B^-]}{[A_2B]}$

Through out the treatment of ionic equilibria, we write equilibrium reaction and its equilibrium constant the way these are written at the end of the above two expressions.

$$K_{eq}(A_2B) = \frac{[A^+]^2\,[B^{2-}]}{[A_2B]} \tag{4.3.8}$$

In general, $$K_{eq}(A_xB_y) = \frac{[A^{y+}]^x\,[B^{x-}]^y}{[A_xB_y]} \tag{4.3.9}$$

The value of the equilibrium constant is a characteristic of a given weak electrolyte and depends only on the temperature. It is independent of the individual concentrations of AB, A^+ and B^-. If a strong electrolyte containing either A^+ or B^- is added to the solution of a weak electrolyte AB, even then the above expression for the equilibrium constant holds good. The effect of a strong electrolyte is to suppress the extent of dissociation of the weak electrolyte, i.e. the degree of dissociation of the weak electrolyte is decreased.

4.4 EFFECT OF DILUTION ON DEGREE OF DISSOCIATION

Ostwald Dilution Law

We write the equilibrium for a weak electrolyte AB as

$$AB + H_2O \rightleftharpoons A^+(aq) + B^-(aq) \tag{4.4.1}$$

If α is the degree of dissociation at a given concentration c of AB, then the concentrations of various species in solution are

$$[AB] = c(1 - \alpha); \qquad [A^+(aq)] = c\alpha; \qquad [B^-(aq)] = c\alpha$$

Substituting these in the expression of equilibrium constant, we have

$$K_{eq} = \frac{[A^+][B^-]}{[AB][H_2O]} = \frac{[A^+][B^-]}{[AB](1\,000\text{ g dm}^{-3}/18\text{ g mol}^{-1})}$$

$$= \frac{c\alpha^2}{(1-\alpha)(55.56\text{ M})} \tag{4.4.2}$$

The water concentration will practically remain the same (i.e. 55.56 M) since only very small quantity of this will combine with A^+ and B^-. Combining this concentration with K_{eq} gives another constant K_{diss} which is called the *dissociation constant* or *the ionization constant*. Thus

$$K_{diss} = K_{eq} \times (55.56\text{ M}) = \frac{c\alpha^2}{(1-\alpha)} \tag{4.4.3}$$

Since α is usually a very small quantity, it is, therefore, negligible in comparison to unity, i.e. $(1 - \alpha) \approx 1$. Thus

$$K_{diss} = \frac{c\alpha^2}{1}$$

or $$\alpha = \sqrt{\frac{K_{diss}}{c}} = \sqrt{K_{diss}V_m} \tag{4.4.4}$$

where V_m is the volume containing 1 mol of the solute. Its unit is taken as $dm^3\ mol^{-1}$.

It follows from Eq. (4.4.4) that as c decreases (dilution), α increases. In the limit when $c \to 0$, α will approach 1, i.e. at *infinite dilution, the whole of the weak electrolyte gets ionized*. This is the *Ostwald dilution law*.

The expression

$$\alpha = \sqrt{K_{\text{diss}} V_{\text{m}}}$$

can be used to determine the K_{diss}, if the value of α is known at a given concentration. The value of α can be determined by using any other physicochemical technique such as molar conductivity, colligative properties, etc. A plot of α^2 versus V_{m} will be a straight line; the slope of the resulting plot gives the value of K_{diss}.

Example 4.4.1 At 25 °C, acid dissociation constant of HCN is 4.9×10^{-10} M. Calculate the degree of dissociation of HCN, if its concentrations are (i) 0.1 M and (ii) 0.01 M.

Solution If α is the degree of dissociation of HCN, then the concentrations of various species in solution are

$$\underset{c(1-\alpha)}{HCN} + H_2O \rightleftharpoons \underset{c\alpha}{H_3O^+} + \underset{c\alpha}{CN}$$

Substituting these in the dissociation expression, we have

$$K_{\text{diss}} = \frac{[H_3O^+][CN^-]}{[HCN]} = \frac{(c\,\alpha)(c\,\alpha)}{c\,(1-\alpha)} = \frac{c\,\alpha^2}{1-\alpha} \approx c\,\alpha^2$$

Thus (i) $\alpha = \sqrt{\dfrac{K_{\text{diss}}}{c}} = \sqrt{\dfrac{(4.9 \times 10^{-10}\ \text{M})}{0.1\ \text{M}}} = 7 \times 10^{-5}$

(ii) $\alpha = \sqrt{\dfrac{(4.9 \times 10^{-10}\ \text{M})}{0.01\ \text{M}}} = 2.21 \times 10^{-4}$

4.5 DISSOCIATION OF PURE WATER

Equilibrium Constant of Water

Pure water is itself a very weak electrolyte and ionizes according to the equation

$$H_2O + H_2O \rightleftharpoons H_3O^+ + OH^- \tag{4.5.1}$$

The equilibrium constant of the reaction is

$$K_{\text{eq}} = \frac{[H_3O^+][OH^-]}{[H_2O]^2} \tag{4.5.2}$$

Ionization Constant of Water

The ionization of water may be written as

$$H_2O \rightleftharpoons H^+ + OH^- \tag{4.5.3}$$

for which the ionization constant is given by

$$K_i = \frac{[H^+][OH^-]}{[H_2O]} \quad (4.5.4)$$

It is obvious that $K_i = K_{eq}\,[H_2O]$. (4.5.5)

Ionic Product of Water

Since water is found to be poorly ionized (degree of dissociation is 1.8×10^{-9} at 25 °C), concentration of water remains practically the same $\{(1\ 000\ g\ dm^{-3})/(18\ g\ mol^{-1}) = 55.56\ M\}$. Its concentration can be combined with the ionization constant K_i to give a new constant, known as the *ionic product* of water, K_w. From Eq. (4.5.4), we get

$$K_w = K_i\,[H_2O] = [H^+]\,[OH^-] \quad (4.5.6)$$

The concentration of OH^- in pure water will be the same as that of H^+; therefore

$$K_w = [H^+]^2 \quad (4.5.7)$$

The value of $[H^+]$ in water at 25 °C is found to be 1.0×10^{-7} M. The value of ionic product at 25 °C is thus equal to

$$K_w = (1.0 \times 10^{-7}\ M)\ (1.0 \times 10^{-7}\ M)$$

$$= 1.0 \times 10^{-14}\ M^2 \quad (4.5.8)$$

Because of equal concentrations of hydrogen and hydroxyl ions in pure water, the latter is neutral in its behaviour.

Nature of Solution

Acidity or alkalinity of a solution depends upon the concentration of hydrogen ions relative to that of hydroxyl ions. In any aqueous solution, both hydrogen and hydroxyl ions coexist in accordance with Eq. (4.5.3). The product of hydrogen and hydroxyl ion concentrations is given by Eq. (4.5.6), the value of which depends only on the temperature and not on the individual ionic concentrations. *If the concentration of hydrogen ions exceeds that of the hydroxyl ions, the solution is said to be acidic; whereas, if concentration of hydroxyl ions exceeds that of the hydrogen ions, the solution is said to be alkaline*. Taking into account Eq. (4.5.6), it amounts to

For neutral solution	$[H^+] = [OH^-] = \sqrt{K_w}$		
For acidic solution	$[H^+] > [OH^-]$	or	$[H^+] > \sqrt{K_w}$
For alkaline solution	$[H^+] < [OH^-]$	or	$[H^+] < \sqrt{K_w}$

At 25 °C, these reduce to

Neutral solution	$[H^+] = 10^{-7}$ M
Acidic solution	$[H^+] > 10^{-7}$ M
Alkaline solution	$[H^+] < 10^{-7}$ M

4.6 THE pH-SCALE

Definitions of pH and pOH

Since hydrogen-ion concentrations commonly met within solutions vary considerably over the range 10^{-14} to 1 M, Sorenson introduced a logarithmic scale for the sake of convenience, and gave it a symbol pH. It is expressed as

$$\text{pH} = -\log_{10}\{[\text{H}^+]/\text{M}\} = \log\left\{\frac{1}{[\text{H}^+]/\text{M}}\right\} \tag{4.6.1}$$

Thus, it is equal to the logarithm of the reciprocal of $[\text{H}^+]/\text{M}$. For neutral water at 25 °C, pH is given by

$$\text{pH} = -\log(1.0 \times 10^{-7}) = -(-7) = 7 \tag{4.6.2}$$

The pH corresponding to the acidic and alkaline solutions at 25 °C will be less than and greater than seven, respectively.

In a similar manner, we can define a pOH scale as the negative logarithm of numerical value of the hydroxyl-ion concentration. However, the acidity or alkalinity of a solution is often expressed in terms of pH of a solution. Both pH and pOH are related to each other through the expression

$$\text{pH} + \text{pOH} = \text{p}K_\text{w}^\circ \qquad (4.6.3)^\dagger$$

where $\text{p}K_\text{w}^\circ$, like pH and pOH, is equal to $-\log\{K_\text{w}/\text{M}^2\}$. Its value at 25 °C is equal to 14.

Example 4.6.1 The values of ionic product of water at various temperatures are given below.

$\theta_c/°\text{C}$	0	10	25	40	50
$K_\text{w} \times 10^{14}/\text{M}^2$	0.114	0.292	1.008	2.919	5.474

What are the pH values of the pure water at these temperatures?

Solution Since $[\text{H}^+] = \sqrt{K_\text{w}}$, therefore

$$\text{pH} = -\log\{[\text{H}^+]/\text{M}\} = -\frac{1}{2}\log\{K_\text{w}/\text{M}^2\}$$

Thus, the calculated values of pH at the given temperatures are as follows.

Temperature	pH
0 °C	$-\frac{1}{2}\log(0.114 \times 10^{-14}) = 7.472$
10 °C	$-\frac{1}{2}\log(0.292 \times 10^{-14}) = 7.267$
25 °C	$-\frac{1}{2}\log(1.008 \times 10^{-14}) = 7.002$
40 °C	$-\frac{1}{2}\log(2.919 \times 10^{-14}) = 6.767$
50 °C	$-\frac{1}{2}\log(5.474 \times 10^{-14}) = 6.631$

---

† Throughout, the equilibrium constant, K, carries the unit of $(\text{mol dm}^{-3})^{\Sigma\nu}$. The expression $K/(\text{mol dm}^{-3})^{\Sigma\nu}$ is written as K° and is spelled as standard equilibrium constant. It is a unitless quantity.

Example 4.6.2 At 25 °C, the degree of ionization of water was found to be 1.8×10^{-9}. Calculate the ionization constant and the ionic product of water at this temperature.

Solution If α is the degree of dissociation of water, then we have

$$H_2O \rightleftharpoons H^+ + OH$$

$$[H^+] = [OH^-] = c\alpha$$

If mass of 1 dm^3 water is taken as 1000 g, than

$$c = \frac{n}{V} = \frac{m/M}{V} = \frac{(1000\ g)/(18\ g\ mol^{-1})}{1\ dm^3} = 55.56\ M$$

Thus $$K_i = \frac{[H^+][OH^-]}{[H_2O]} = \frac{(c\alpha)^2}{c(1-\alpha)} \approx c\alpha^2 \qquad \text{(assuming } \alpha << 1)$$

$$= (55.56\ M)(1.8 \times 10^{-9})^2 = 1.8 \times 10^{-16}\ M$$

and $$K_w = [H^+][OH^-] = (c\alpha)^2 = \{(55.56\ M)(1.8 \times 10^{-9})\}^2$$

$$= 1.0 \times 10^{-14}\ M^2$$

Example 4.6.3 What is the pH at 25 °C, if a solution which is twice as alkaline (i.e. which contains twice as many hydroxide ions) as pure water?

Solution For a solution to have twice alkalinity, we have

$$[OH^-] = 2.0 \times 10^{-7}\ M$$

Thus $$pOH = -\log\{[OH^-]/M\} = -\log(2.0 \times 10^{-7}) = 7 - 0.30 = 6.70$$

and hence pH = 14 – pOH = 7.30

Example 4.6.4 The ionic product of water at 100 °C is 55 times than that at 25 °C. (i) Calculate the value of pH of water at 100 °C. (ii) A given solution at 100 °C has a pH value 5.0. Indicate whether the solution is acidic or alkaline or neutral.

Solution **(i)** Given that

$$K_w(100\ °C) = 55 \times K_w(25\ °C) = 55 \times (1.0 \times 10^{-14}\ M^2)$$

Thus $$pH(100\ °C) = -\frac{1}{2}\log\{K_w/M^2\} = -\frac{1}{2}\log(55 \times 10^{-14}) = 6.13$$

(ii) Since for a given solution, pH equal to 5.0 is less than the corresponding pH of pure water at 100 °C, the solution is acidic.

4.7 CLASSIFICATION OF ACIDS AND BASES IN WATER

Strong and Weak Acids

A given acid can be classified into a strong acid or a weak acid depending upon the value of its dissociation constant. For strong acids, K_a is much larger than 1 indicating that they are present in solution in an almost ionized form. On the other hand, K_a is much smaller than 1 for weak acids. This implies that weak acids are feebly ionized. Table 4.7.1 gives the values of ionization constants of some of the common acids.

Strong and Weak Bases

Similarly a given base can be classified into a stronger base ($K_b >> 1$) or a weak base ($K_b << 1$). Table 4.7.2 records K_b for some of the common weak bases.

4.8 EXACT TREATMENT FOR IONIZATION OF A MONOPROTIC ACID

Computing Concentrations of Species in Solution

Before attempting the actual computation of concentrations of various species in a solution of an acid (or a base), the problem may be analyzed from a purely mathematical viewpoint. In order to compute the concentrations of various species present in solution, we must have equations at least equal to the number of the unknowns. These equations are obtained from the equilibrium expressions and from the equations specifying the conservation of charge and mass. The mathematical solution of these equations is, in general, very complicated and the exact solution could only be achieved through the use of a digital computer. However, under certain conditions, calculations can be simplified through certain approximations without affecting the results significantly.

Derivation of Exact Expression to Compute [H$_3$O$^+$]

Equilibria Existing in Solution

Various equilibria existing in an aqueous solution of a monoprotic acid are

$$\mathrm{HA + H_2O \rightleftharpoons H_3O^+ + A^-}, \qquad K_a = \frac{[\mathrm{H_3O^+}][\mathrm{A^-}]}{[\mathrm{HA}]} \tag{4.8.1}$$

$$\mathrm{H_2O + H_2O \rightleftharpoons H_3O^+ + OH^-}, \qquad K_w = [\mathrm{H_3O^+}][\mathrm{OH^-}] \tag{4.8.2}$$

Condition of Mass Balance

Mass-balance equation is

$$[\mathrm{HA}]_0 = [\mathrm{HA}] + [\mathrm{A^-}] \tag{4.8.3}$$

where $[\mathrm{HA}]_0$ is the total concentration of the acid and [HA] and [A$^-$] are the concentrations of undissociated and dissociated forms of the acid, respectively.

Table 4.7.1 Ionization Constants of Common Acids at 25 °C

Acid	*Equilibria*	K_a/M
Nitric	$HNO_3 \rightarrow H^+ + NO_3^-$	$\geq 10^2$
Hydrochloric	$HCl \rightarrow H^+ + Cl^-$	$\geq 10^7$
Hydrobromic	$HBr \rightarrow H^+ + Br^-$	$\geq 10^9$
Hydroiodic	$HI \rightarrow H^+ + I^-$	$\geq 10^{11}$
Sulphuric	$H_2SO_4 \rightarrow H^+ + HSO_4^-$	$\geq 10^{10}$
Acetic	$HAc \rightleftharpoons H^+ + Ac^-$	1.8×10^{-5}
Benzoic	$C_7H_5O_2H \rightleftharpoons H^+ + C_7H_5O_2^-$	6.0×10^{-5}
Chlorous	$HClO_2 \rightleftharpoons H^+ + ClO_2^-$	1.1×10^{-2}
Formic	$HCO_2H \rightleftharpoons H^+ + HCO_2^-$	1.8×10^{-4}
Hydrocyanic	$HCN \rightleftharpoons H^+ + CN^-$	4.0×10^{-10}
Hydrofluoric	$HF \rightleftharpoons H^+ + F^-$	6.7×10^{-4}
Hypobromous	$HOBr \rightleftharpoons H^+ + OBr^-$	2.1×10^{-9}
Hypochlorous	$HOCl \rightleftharpoons H^+ + OCl^-$	3.2×10^{-8}
Nitrous	$HNO_2 \rightleftharpoons H^+ + NO_2^-$	4.5×10^{-4}
Arsenic	$H_3AsO_4 \rightleftharpoons H^+ + H_2AsO_4^-$	2.5×10^{-4}
	$H_2AsO_4^- \rightleftharpoons H^+ + HAsO_4^{2-}$	5.6×10^{-8}
	$HAsO_4^{2-} \rightleftharpoons H^+ + AsO_4^{3-}$	3×10^{-13}
Carbonic	$CO_2 + H_2O \rightleftharpoons H^+ + HCO_3^-$	4.2×10^{-7}
	$HCO_3^- \rightleftharpoons H^+ + CO_3^{2-}$	4.8×10^{-11}
Hydrosulphuric	$H_2S \rightleftharpoons H^+ + HS^-$	1.1×10^{-7}
	$HS^- \rightleftharpoons H^+ + S^{2-}$	1.0×10^{-14}
Oxalic	$H_2C_2O_4 \rightleftharpoons H^+ + HC_2O_4^-$	5.9×10^{-2}
	$HC_2O_4^- \rightleftharpoons H^+ + C_2O_4^{2-}$	6.4×10^{-5}
Phosphoric	$H_3PO_4 \rightleftharpoons H^+ + H_2PO_4^-$	7.5×10^{-5}
	$H_2PO_4^- \rightleftharpoons H^+ + HPO_4^{2-}$	6.2×10^{-8}
	$HPO_4^{2-} \rightleftharpoons H^+ + PO_4^{3-}$	1×10^{-12}
Phosphorous	$H_3PO_3 \rightleftharpoons H^+ + H_2PO_3^-$	1.6×10^{-2}
	$H_2PO_3^- \rightleftharpoons H^+ + HPO_3^{2-}$	7×10^{-7}
Sulphuric	$H_2SO_4 \rightarrow H^+ + HSO_4^-$	strong
	$HSO_4^- \rightleftharpoons H^+ + SO_4^{2-}$	1.3×10^{-2}
Sulphurous	$SO_2 + H_2O \rightleftharpoons H^+ + HSO_3^-$	1.3×10^{-2}
	$HSO_3^- \rightleftharpoons H^+ + SO_3^{2-}$	5.6×10^{-8}
Succinic	$(CH_2COOH)_2 \rightleftharpoons HOOC(CH_2)_2COO^- + H^+$	6.17×10^{-5}
	$HOOC(CH_2)_2COO^- \rightleftharpoons (CH_2COO^-)_2 + H^+$	2.29×10^{-6}

Table 4.7.2 Ionization Constants of Common Bases at 25 °C

Base	*Equilibria*	K_b/M
Ammonia	$NH_3 + H_2O \rightleftharpoons NH_4^+ + OH^-$	1.8×10^{-5}
Aniline	$C_6H_5NH_2 + H_2O \rightleftharpoons C_6H_5NH_3^+ + OH^-$	4.6×10^{-10}
Dimethylamine	$(CH_3)_2NH + H_2O \rightleftharpoons (CH_3)_2NH_2^+ + OH^-$	7.4×10^{-4}
Hydrazine	$N_2H_4 + H_2O \rightleftharpoons N_2H_5^+ + OH^-$	9.8×10^{-7}
Methylamine	$CH_3NH_2 + H_2O \rightleftharpoons CH_3NH_3^+ + OH^-$	5.0×10^{-4}
Pyridine	$C_5H_5N + H_2O \rightleftharpoons C_5H_5NH^+ + OH^-$	1.5×10^{-9}
Trimethylamine	$(CH_3)_3N + H_2O \rightleftharpoons (CH_3)_3NH^+ + OH^-$	7.4×10^{-5}

Condition of Charge Balance

Since the solution is electrically neutral, sum of concentrations of all the positive charges must be equal to that of all the negative charges, i.e. the charge-balance condition gives

$$[H_3O^+] = [OH^-] + [A^-] \tag{4.8.4}$$

Derivation of Exact Expression

From Eq. (4.8.2), we have

$$[OH^-] = \frac{K_w}{[H_3O^+]}$$

Substituting this in Eq. (4.8.4) and rearranging the resultant expression, we obtain

$$[A^-] = [H_3O^+] - \frac{K_w}{[H_3O^+]} \tag{4.8.5}$$

Rewriting Eq. (4.8.3), we have

$$[HA] = [HA]_0 - [A^-]$$

which on using Eq. (4.8.5) becomes

$$[HA] = [HA]_0 - [H_3O^+] + \frac{K_w}{[H_3O^+]} \tag{4.8.6}$$

Substituting $[A^-]$ and [HA] from Eqs (4.8.5) and (4.8.6) in Eq. (4.8.1), we get

$$K_a = \frac{[H_3O^+]\left\{[H_3O^+] - \dfrac{K_w}{[H_3O^+]}\right\}}{\left\{[HA]_0 - [H_3O^+] + \dfrac{K_w}{[H_3O^+]}\right\}} \tag{4.8.7}$$

or $$K_a = \frac{[H_3O^+]^3 - [H_3O^+]\,K_w}{[H_3O^+][HA]_0 - [H_3O^+]^2 + K_w} \tag{4.8.8}$$

This is a cubic equation in $[H_3O^+]$ and can be solved for $[H_3O^+]$. The concentrations of all other species, e.g., OH^-, A^-, HA can be expressed in terms of $[H_3O^+]$ (Eqs (4.8.2), (4.8.5) and (4.8.6), respectively), and thus their values can be determined.

Simplification of Exact Expression

Since the exact solution of a cubic equation (or higher degree) involves too much of mathematical operations, it would be of interest to determine the conditions under which such higher order equation could be simplified.

For Strong Acids (a Typical Example of HCl)

The following equilibrium exists for a strong acid, HCl

$$H_2O + HCl \rightleftharpoons H_3O^+ + Cl^-$$

with a very large value of the dissociation constant ($\cong 10^7$). Equation (4.8.7) for this condition is

$$K_a = \frac{[H_3O^+]\left\{[H_3O^+] \quad \dfrac{K_w}{[H_3O^+]}\right\}}{\left\{[HCl]_0 - [H_3O^+] + \dfrac{K_w}{[H_3O^+]}\right\}}$$

Rearranging this, we get

$$[HCl]_0 = \left\{[H_3O^+] - \frac{K_w}{[H_3O^+]}\right\}\left\{1 + \frac{[H_3O^+]}{K_a}\right\} \qquad (4.8.9)$$

Since K_a is very large, the term $[H_3O^+]/K_a$ can be neglected in comparison to 1. Thus, the above expression becomes

$$[HA] = [HA]_0 - [H_3O^+] + \frac{K_w}{[H_3O^+]} \qquad (4.8.10)$$

This approximation amounts to the fact that HCl is completely dissociated in the solution. This expression is to be used whenever the concentration of H_3O^+ in the solution is less than 10^{-6} M. In this case, the contribution of H_3O^+ obtained from the dissociation of the acid is comparable to that obtained from the dissociation of water. Thus, we have two sources of contribution of H_3O^+, namely (i) the dissociation of the acid, and (ii) the dissociation of water.

A Direct Method to Derive Eq. (4.8.10)

Equation (4.8.10) can also be obtained through the charge-balance expression:

$$[H_3O^+] = [OH^-] + [Cl^-]$$

But $\quad [Cl^-] = [HCl]_0 \quad$ and $\quad [OH^-] = \dfrac{K_w}{[H_3O^+]}$

Thus, $$[H_3O^+] = \frac{K_w}{[H_3O^+]} + [HCl]_0 \tag{4.8.10}$$

If the $[H_3O^+]$ of the solution $\geq 10^{-6}$ M, the term $K_w/[H_3O^+] \leq 10^{-8}$ M, and it can be neglected in comparison to $[H_3O^+]$. Thus

$$[HCl]_0 \approx [H_3O^+] \tag{4.8.11}$$

This amounts to the fact that if the concentration of HCl $\geq 10^{-6}$ M, the H_3O^+ contribution due to the dissociation of water will be negligible in comparison to the amount of H_3O^+ contributed by the acid. It is, thus clear that for a very dilute solution (concentration $\leq 10^{-6}$ M), Eq. (4.8.11) will not be applicable. In such cases, we must use Eq. (4.8.10) to calculate the H_3O^+ concentration.

Example 4.8.1 Calculate at 25 °C the pH of a solution of (a) 0.01 M HCl and (b) 10^{-7} M HCl.

Solution **(a)** Since the concentration of HCl > 10^{-6} M, Eq. (4.8.11) can be employed to evaluate pH. We have

$$[H_3O^+] = 0.01 \text{ M}$$

Hence, $$\text{pH} = -\log\{[H_3O^+]/\text{M}\} = -\log(0.01) = -(-2) = 2$$

(b) Since the concentration of HCl < 10^{-6} M, therefore, we have to use Eq. (4.8.10) to get the value of H_3O^+ concentration:

$$[H_3O^+]^2 - [HCl]_0\,[H_3O^+] - K_w = 0$$

$$[H_3O^+] = \frac{[HCl]_0 \pm \sqrt{[HCl]_0^2 + 4\,K_w}}{2}$$

$$= \frac{(10^{-7}\text{ M}) \pm \sqrt{(10^{-7}\text{ M})^2 + 4.0 \times 10^{-14}\text{ M}^2}}{2}$$

Two solutions are

$$[H_3O^+] = 1.62 \times 10^{-7}\text{ M} \quad \text{and} \quad [H_3O^+] = -0.62 \times 10^{-7}\text{ M}$$

Since $[H_3O^+]$ cannot be negative, thus

$$[H_3O^+] = 1.62 \times 10^{-7}\text{ M}$$

Now, for pH, we write

$$\text{pH} = -\log\{[H_3O^+]/\text{M}\} = -\log(1.62 \times 10^{-7})$$

$$= -\{\log(1.62) + \log(10^{-7})\}$$

$$= -(0.21 - 7) = 6.79$$

Comment The contribution of water increases the $[H_3O^+]$ above that due to the acid alone. No matter how dilute the solution is, the pH of an acid solution would always be less than 7 at 25 °C.

Example 4.8.2 At 25 °C, the degree of ionization for water is 1.8 × 10^{-9}. How is this affected in the presence of 10^{-8} M HCl (assume complete ionization for the acid)? Calculate the total hydrogen-ion concentration in the presence of the acid. What is the pH of the solution?

Solution In order to compute hydrogen-ion concentration in 10^{-8} M HCl, we will have to consider the contribution of H_3O^+ from the dissociation of water. In such a case, the concentration of H_3O^+ is given by

$$[H_3O^+] = \frac{K_w}{[H_3O^+]} + [HCl]_0$$

$$[H_3O^+]^2 - [HCl]_0\,[H_3O^+] - K_w = 0$$

$$[H_3O^+]^2 - (10^{-8}\ M)[H_3O^+] - 10^{-14}\ M^2 = 0$$

$$[H_3O^+] = \frac{(10^{-8}\ M) + \sqrt{10^{-16}\ M^2 + 4.0 \times 10^{-14}\ M^2}}{2}$$

$$[H_3O^+] = 1.05 \times 10^{-7}\ M$$

This is the total concentration of H_3O^+, out of which 1.0 × 10^{-8} M comes from the dissociation of HCl and the remaining (1.05 × 10^{-7} M – 1.0 × 10^{-8} M = 9.5 × 10^{-8} M) from the dissociation of water.

Thus,
$$\underset{\substack{55.56\ M \\ c}}{2H_2O} \rightleftharpoons \underset{\substack{9.5\times10^{-8}\ M \\ c\alpha}}{H_3O^+} + \underset{\substack{9.5\times10^{-8}\ M \\ c\alpha}}{OH^-}$$

$$\text{Degree of ionization of } H_2O = \frac{[H_3O^+]}{c} = \frac{(9.5\times10^{-8}\ M)}{c} = \frac{(9.5\times10^{-8}\ M)}{(55.56\ M)}$$
$$= 1.7\times10^{-9}$$

Thus, the degree of ionization of water decreases from 1.8 × 10^{-9} to 1.7 × 10^{-9}.

$$\text{pH of the solution} = -\log\{[H_3O^+]/M\}$$
$$= -\log(1.05 \times 10^{-7}) = -\log(1.05) - \log(10^{-7})$$
$$= -0.021\ 2 + 7 \approx 6.98$$

For Weak Acids

First approximation If $[H_3O^+] \geq 10^{-6}$ M, then

$$\frac{K_w}{[H_3O^+]} \leq 10^{-8}\ M,$$

the term $\{[H_3O^+] - K_w/[H_3O^+]\}$ could be approximated to $[H_3O^+]$. Using this approximation, Eq. (4.8.7) simplifies to

$$K_a = \frac{[H_3O^+]^2}{[HA]_0 - [H_3O^+]} \tag{4.8.12}$$

This approximation leads to the assumption that the amount of hydronium ions coming from water is negligible in comparison to that contributed by the acid. Equation (4.8.12) can be solved for $[H_3O^+]$ by using either of the following two methods.

(a) *As a quadratic equation* Expression (4.8.12) could be arranged to give a quadratic equation

$$[H_3O^+]^2 + K_a [H_3O^+] - [HA]_0 K_a = 0 \tag{4.8.13}$$

and since $[H_3O^+]$ cannot be negative, the only solution of the equation is

$$[H_3O^+] = \frac{-K_a + \sqrt{K_a^2 + 4[HA]_0 K_a}}{2} \tag{4.8.14}$$

(b) *Method of successive approximation* In this method, H_3O^+ concentration is calculated in an iterative manner. Equation (4.8.13) can be rewritten as

$$[H_3O^+]_{i+1} = \sqrt{K_a([HA]_0 - [H_3O^+]_i)} \tag{4.8.15}$$

We assume an approximate value of $[H_3O^+]$ (i.e. $[H_3O^+]_i$) which on substituting into Eq. (4.8.15), gives a refind value of $[H_3O^+]$ (i.e. $[H_3O^+]_{i+1}$). Then this refined value of $[H_3O^+]$ is used to compute a more refined value, and so on. Such an iteration is carried out until the two consecutive values of $[H_3O^+]$ agree within a reasonable limit. This is illustrated below:

$$[H_3O^+]_1 = \sqrt{K_a([HA]_0 - [H_3O^+]_0)}$$

$$[H_3O^+]_2 = \sqrt{K_a([HA]_0 - [H_3O^+]_1)}$$

....

$$[H_3O^+]_n = \sqrt{K_a([HA]_0 - [H_3O^+]_{n-1})}$$

To start the operation, we can assign $[H_3O^+]$ equal to zero or 10^{-7} M (for pure water) or any other reasonable value.

Second approximation If the solution is fairly concentrated and the value of the equilibrium constant K_a is small, i.e. the extent of dissociation is negligibly small relative to the amount of acid dissolved, then

$$[HA]_0 >> [H_3O^+]$$

With this, Eq. (4.8.12) reduces to

$$K_a = \frac{[H_3O^+]^2}{[HA]_0}$$

Thus, $$[H_3O^+] = \sqrt{K_a[HA]_0}$$

Example 4.8.3 What is the pH of (a) 0.1 M and (b) 0.001 M solution of acetic acid; $K_a = 1.85 \times 10^{-5}$ M.

Solution The type of approximations valid for a given problem can be decided from the values of α and $[H_3O^+]$ obtained by using the simplest expression, in which, it is assumed that $\alpha << 1$ and that there is no contribution of H_3O^+ from water.

Thus, we have

$$\underset{c(1-\alpha)}{\text{HAc}} + \text{H}_2\text{O} \rightleftharpoons \underset{c\alpha}{\text{H}_3\text{O}^+} + \underset{c\alpha}{\text{Ac}^-} \qquad \text{(where Ac}^- \text{ stands for acetate ion)}$$

$$K_a = \frac{[\text{H}_3\text{O}^+][\text{Ac}^-]}{[\text{HAc}]} = \frac{(c\alpha)(c\alpha)}{c(1-\alpha)} = \frac{c\alpha^2}{1-\alpha} \approx c\alpha^2$$

or
$$\alpha = \sqrt{\frac{K_a}{c}} \quad \text{and} \quad [\text{H}_3\text{O}^+] = c\alpha$$

(a) 0.1 M solution. In this case, we have

$$\alpha = \sqrt{\frac{(1.85 \times 10^{-5}\ \text{M})}{(0.1\ \text{M})}} = \sqrt{1.85 \times 10^{-4}} = 1.36 \times 10^{-2}$$

$$[\text{H}_3\text{O}^+] = (1.36 \times 10^{-2})\ (0.1\ \text{M}) = 1.36 \times 10^{-3}\ \text{M}$$

Thus, it can be seen that for this solution $\alpha \ll 1$ and $[H_3O^+] \gg 10^{-6}$ M and hence the use of above expressions are justified. Hence, pH of the solution is

$$\text{pH} = -\log\ (1.36 \times 10^{-3}) = -\ 0.133\ 5 + 3 = 2.87$$

(b) 0.001 M solution. Using the simplest expression of α, we have

$$\alpha = \sqrt{\frac{(1.85 \times 10^{-5}\ \text{M})}{(0.001\ \text{M})}} = 0.136$$

Here $\alpha < 1$, a more correct value can be obtained by solving the following quadratic equation in α.

$$K_a = \frac{c\alpha^2}{(1-\alpha)}$$

$$c\alpha^2 + K_a\alpha - K_a = 0$$

$$\alpha = \frac{-K_a + \sqrt{K_a^2 + 4K_a c}}{2c}$$

or
$$= \frac{(-1.85 \times 10^{-5}\ \text{M}) + \sqrt{\{(1.85 \times 10^{-5})^2 + 4\,(1.85 \times 10^{-5})\,(10^{-3})\}\text{M}^2}}{2(10^{-3}\ \text{M})}$$

$$= \frac{-1.85 \times 10^{-5}\ \text{M} + \sqrt{7.433\,5 \times 10^{-8}\ \text{M}^2}}{2\,(10^{-3}\ \text{M})} = \frac{2.541 \times 10^{-4}\ \text{M}}{2 \times 10^{-3}\ \text{M}}$$

$$= 0.127$$

Therefore,

$$[\text{H}_3\text{O}^+] = c\alpha = (10^{-3}\ \text{M})\ (0.127) = 1.27 \times 10^{-4}\ \text{M}$$

$$\text{pH} = -\log\ \{[\text{H}_3\text{O}^+]/\text{M}\} = -\log\ (1.27 \times 10^{-4})$$

$$= -\ 0.103\ 8 + 4 \approx 3.90$$

Example 4.8.4

Given a solution that is 0.5 M CH_3COOH. To what volume at 25 °C must one dm^3 of this solution be diluted in order to (a) double the pH; (b) double the hydroxide-ion concentration? Given that $K_a = 1.8 \times 10^{-5}$ M.

Solution If α is the degree of dissociation of acetic acid of concentration c then the concentrations of various species in the solution are

$$\underset{c(1-\alpha)}{CH_3COOH} + H_2O \rightleftharpoons \underset{c\alpha}{CH_3COO^-} + \underset{c\alpha}{H_3O^+}$$

With these concentrations, the equilibrium constant becomes

$$K_a = \frac{[CH_3COO^-][H_3O^+]}{[CH_3COOH]} = \frac{(c\alpha)(c\alpha)}{c(1-\alpha)} \approx c\alpha^2$$

or $$\alpha = \sqrt{\frac{K_a}{c}}$$

Substituting the values, we have

$$\alpha = \sqrt{\frac{(1.8 \times 10^{-5}\ M)}{(0.5\ M)}} = 6 \times 10^{-3}$$

The concentration of hydrogen ions is given as

$$[H_3O^+] = c\alpha = (0.5\ M)(6 \times 10^{-3}) = 3 \times 10^{-3}\ M$$

Hence, $pH = -\log\{[H_3O^+]/M\} = -\log(3 \times 10^{-3}) = 2.52$

(a) *To double the* pH

Thus, $pH = 5.04$

Since $pH = -\log\{[H_3O^+]/M\}$, therefore $[H_3O^+]/M = 10^{-pH}$. Substituting the value of pH, we have

$$[H_3O^+]/M = 10^{-5.04} = 9.12 \times 10^{-6}$$

Thus, $c_1\alpha = 9.12 \times 10^{-6}\ M$

In dilution, α will increase, and its value will not be negligible in comparison to one. Thus, we shall have to use the expression

$$K_a = \frac{(c_1\alpha)^2}{c_1(1-\alpha)} = \frac{c_1\alpha^2}{1-\alpha} = \frac{(c_1\alpha)\alpha}{1-\alpha} = \frac{(9.12 \times 10^{-6}\ M)\alpha}{1-\alpha}$$

or $(1.8 \times 10^{-5}\ M)(1-\alpha) = (9.12 \times 10^{-6}\ M)\ \alpha$

which gives

$$(9.12 \times 10^{-6}\ M + 1.8 \times 10^{-5}\ M)\ \alpha = 1.8 \times 10^{-5}\ M$$

or $$\alpha = \frac{1.8 \times 10^{-5}\ M}{27.12 \times 10^{-6}\ M} = 0.663\,7$$

Since $c_1\alpha = 9.12 \times 10^{-6}\ M$ and $\alpha = 0.663\,7$, therefore,

$$c_1 = \frac{9.12 \times 10^{-6}\ M}{0.663\,7} = 1.374 \times 10^{-5}\ M$$

Volume to which the solution should be diluted

$$= \frac{cV}{c_1} = \frac{(0.5\ M)(1\ dm^3)}{(1.374 \times 10^{-5}\ M)} = 3.369 \times 10^4\ dm^3$$

(b) *To double the hydroxyl-ion concentration*

Since $[H_3O^+]$ in 0.5 M acetic acid is 3×10^{-3} M, therefore

$$[OH^-] = \frac{(1.0 \times 10^{-14}\ M^2)}{(3 \times 10^{-3}\ M)}$$

In the present case, the concentration of hydroxyl ion becomes

$$[OH^-] = \frac{2(1.0 \times 10^{-14}\ M^2)}{(3 \times 10^{-3}\ M)}$$

which gives $[H_3O^+] = \dfrac{(3 \times 10^{-3}\ M)}{2} = 1.5 \times 10^{-3}$ M

For this concentration, we can use

$$K_a = c_2\alpha^2 = (c_2\alpha)\ (\alpha)$$

or $$\alpha = \frac{K_a}{(c_2\alpha)} = \frac{(1.8 \times 10^{-5}\ M)}{(1.5 \times 10^{-3}\ M)} = 1.2 \times 10^{-2}$$

Thus, $$c_2 = \frac{(1.5 \times 10^{-3}\ M)}{1.2 \times 10^{-2}} = 0.125\ M$$

Volume to which the solution should be diluted

$$= \frac{cV}{c_2} = \frac{(0.5\ M)\ (1\ dm^3)}{(0.125\ M)} = 4\ dm^3$$

Example 4.8.5 For propanoic acid $K_a = 1.34 \times 10^{-5}$ M at 25 °C. Find for a 0.01 M solution of the acid:

(a) The degree of dissociation.
(b) Hydrogen-ion concentration.
(c) pH and pOH.

Solution **(a)** If α is the degree of dissociation of propanoic acid of concentration c, then the concentrations of various species in solution are

$$\underset{c(1-\alpha)}{CH_3CH_2COOH} + H_2O \rightleftharpoons \underset{c\alpha}{CH_3CH_2COO^-} + \underset{c\alpha}{H_3O^+}$$

With these concentrations, the equilibrium constant becomes

$$K_a = \frac{[CH_3CH_2COO^-][H_3O^+]}{[CH_3CH_2COOH]} = \frac{(c\alpha)\,(c\alpha)}{c(1-\alpha)} = \frac{c\alpha^2}{1-\alpha} \approx c\alpha^2$$

or $$\alpha = \sqrt{\frac{K_a}{c}} = \sqrt{\frac{(1.34 \times 10^{-5}\ M)}{(0.01\ M)}} = \sqrt{1.34 \times 10^{-3}}$$

Thus, $\alpha = 3.66 \times 10^{-2}$

(b) $[H_3O^+] = \alpha c = (3.66 \times 10^{-2})\ (0.01\ M) = 3.66 \times 10^{-4}$ M

(c) $pH = -\log\{[H_3O^+]/M\} = -\log\ (3.66 \times 10^{-4}) = 4 - 0.563 = 3.44$

$pOH = 14 - pH = 10.56$

Example 4.8.6 Glycine $HC_2H_4O_2N(HG)$ is one of the amino acids that make up body proteins. Its ionization constant at 25 °C is 1.7×10^{-10} M. What is the value of $[H_3O^+]$ in a 10^{-4} M solution of glycine?

Solution If we ignore the dissociation of water and assume that $[H_3O^+]$ of the solution is very much smaller than the concentration of HG, its $[H_3O^+]$ can then be calculated using the expression

$$[H_3O^+] = \sqrt{K_a[HG]} = \sqrt{(1.7 \times 10^{-10}\ M)(10^{-4}\ M)}$$
$$= 1.3 \times 10^{-7}\ M$$

This concentration of hydronium ion is smaller than 10^{-6} M and, moreover, it is of the same order as that of pure water.

Consequently, we cannot ignore the hydrogen ions from the dissociation of water. For a correct solution, we have to use the following expression:

$$K_a = \frac{[H_3O^+]\left\{[H_3O^+] - \dfrac{K_w}{[H_3O^+]}\right\}}{\left\{[HG]_0 - [H_3O^+] + \dfrac{K_w}{[H_3O^+]}\right\}} \tag{4.8.17}$$

Since the expected value of $[H_3O^+] << [HG]_0$, therefore, the denominator in the above expression can be approximated to $[HG]_0$. Hence

$$K_a = \frac{[H_3O^+]\left\{[H_3O^+] - \dfrac{K_w}{[H_3O^+]}\right\}}{[HG]_0}$$

Thus, $\quad [H_3O^+] = \sqrt{K_w + K_a\,[HG]_0}$

Substituting the values, we get

$$[H_3O^+] = \sqrt{(10^{-14}\ M^2) + (1.7 \times 10^{-10}\ M)(10^{-4}\ M)}$$
$$= \sqrt{2.7 \times 10^{-14}\ M^2} = 1.643 \times 10^{-7}\ M$$

Note: Without prior knowledge about the type of approximations to be used to simplify the expression, we can obtain $[H_3O^+]$ from Eq. (4.8.17), which is a cubic expression in $[H_3O^+]$. The solution to the problem can be carried out by using the method of successive approximation. The above expression can be written as

$$[H_3O^+]_{i+1} = ([H_3O^+]_i\ \{K_w + K_a\ [HG]_0\} - [H_3O^+]_i^2\ K_a + K_wK_a)^{1/3}$$

To start with, i.e. for $i = 0$, $[H_3O^+]$ is taken to be zero. The first computed value of $[H_3O^+]$ is then substituted in the right side of the above expression and a more refined value of $[H_3O^+]$ is calculated. This process is repeated till the two consecutive values of $[H_3O^+]$ agree within a reasonable limit. A few computed values are as follows:

Cycle (i)	$[H_3O^+]$/M	*Cycle (i)*	$[H_3O^+]$/M
0	0	5	1.591×10^{-7}
1	1.193×10^{-8}	6	1.626×10^{-7}
2	6.815×10^{-8}	7	1.638×10^{-7}
3	1.230×10^{-7}	8	1.642×10^{-7}
4	1.493×10^{-7}	9	1.643×10^{-7}
		10	1.643×10^{-7}

4.9 EXACT TREATMENT FOR IONIZATION OF A BASE

Derivation of Exact Expression to Compute $[OH^-]$

Equilibria Existing in Solution

Various equilibria existing in an aqueous solution of a base are

$$\mathrm{BOH} \rightleftharpoons \mathrm{B^+ + OH^-}, \qquad K_b = \frac{[\mathrm{B^+}][\mathrm{OH^-}]}{[\mathrm{BOH}]} \tag{4.9.1}$$

$$\mathrm{H_2O + H_2O} \rightleftharpoons \mathrm{H_3O^+ + OH^-}, \qquad K_w = [\mathrm{H_3O^+}][\mathrm{OH^-}] \tag{4.9.2}$$

Condition of Mass Balance

Mass-balance equation is

$$[\mathrm{BOH}]_0 = [\mathrm{BOH}] + [\mathrm{B^+}] \tag{4.9.3}$$

where $[\mathrm{BOH}]_0$ is the total concentration of a base and [BOH] and $[\mathrm{B^+}]$ are the respective concentrations of the undissociated and dissociated forms of the base.

Condition of Charge Balance

Electroneutrality (i.e. charge-balance) condition gives

$$[\mathrm{B^+}] + [\mathrm{H_3O^+}] = [\mathrm{OH^-}] \tag{4.9.4}$$

or

$$[\mathrm{B^+}] = [\mathrm{OH^-}] - [\mathrm{H_3O^+}] = [\mathrm{OH^-}] - \frac{K_w}{[\mathrm{OH^-}]} \tag{4.9.5}$$

Derivation of Exact Expression

Substituting for $[\mathrm{B^+}]$ in Eq. (4.9.3), we get

$$[\mathrm{BOH}] = [\mathrm{BOH}]_0 - [\mathrm{OH^-}] + \frac{K_w}{[\mathrm{OH^-}]} \tag{4.9.6}$$

Thus, the equilibrium constant of the base (Eq. 4.9.1) becomes

$$K_b = \frac{\left\{[\mathrm{OH^-}] - \dfrac{K_w}{[\mathrm{OH^-}]}\right\}[\mathrm{OH^-}]}{\left\{[\mathrm{BOH}]_0 - [\mathrm{OH^-}] + \dfrac{K_w}{[\mathrm{OH^-}]}\right\}} \tag{4.9.7}$$

Alternatively, we have

$$K_b = \frac{\left\{\dfrac{K_w}{[H_3O^+]} - [H_3O^+]\right\}\dfrac{K_w}{[H_3O^+]}}{\left\{[BOH]_0 - \dfrac{K_w}{[H_3O^+]} + [H_3O^+]\right\}} \tag{4.9.8}$$

Equation (4.9.7) (or 4.9.8) is a cubic expression in $[OH^-]$ (or $[H_3O^+]$) and could be solved for $[OH^-]$ (or $[H_3O^+]$). The concentrations of other species, viz., B^+ and BOH can be calculated using Eqs (4.9.5) and (4.9.6), respectively.

Simplification of Exact Expression

For Strong Bases (a Typical Example of NaOH)

The following equilibrium exists for a strong base NaOH

$$NaOH \rightleftharpoons Na^+ + OH^-$$

with a very large value of the dissociation constant. Equation (4.9.7) can be rewritten as

$$[NaOH]_0 = \left\{[OH^-] - \frac{K_w}{[OH^-]}\right\}\left\{1 + \frac{[OH^-]}{K_b}\right\}$$

Since K_b is very large, the term $[OH^-]/K_b$ can be neglected in comparison to 1; thus the above expression becomes

$$[NaOH]_0 = [OH^-] - \frac{K_w}{[OH^-]} \tag{4.9.9}$$

The above approximation amounts to the fact that NaOH is completely dissociated in the solution. This expression is to be used when the concentration of OH^- in the solution is less than 10^{-6} M. In this case, the contribution of OH^- obtained from the dissociation of the base is comparable to that obtained from the dissociation of water. Thus, we have two sources of contribution of OH^-, i.e. from the dissociation of (i) the base and (ii) water.

A Direct Method to Derive Eq. (4.9.9)

Equation (4.9.9) can also be obtained by using the charge-balance expression:

$$[OH^-] = [Na^+] + [H_3O^+]$$

But $[Na^+] = [NaOH]_0$ and $[H_3O^+] = \dfrac{K_w}{[OH^-]}$

Therefore,

$$[OH^-] = [NaOH]_0 + \frac{K_w}{[OH^-]} \qquad \text{(Eq. 4.9.9)}$$

Simplification

If the $[OH^-]$ of the solution $\geq 10^{-6}$ M, the term $K_w/[OH^-] \leq 10^{-8}$ M, and can be neglected in comparison to $[OH^-]$. Therefore,

$$[OH^-] = [NaOH]_0 \tag{4.9.10}$$

This amounts to the fact that if concentration of NaOH $\geq 10^{-6}$ M, then the amount of OH^- due to dissociation of water will be negligible in comparison to the amount contributed by the alkali. It is thus clear that for a very dilute solution (concentration $\leq 10^{-6}$ M), Eq. (4.9.10) will not be applicable and in such cases, one must use Eq. (4.9.9) to calculate the OH^- concentration.

Example 4.9.1 Calculate at 25 °C the pH of a solution of (a) 0.01 M NaOH, and (b) 10^{-7} M NaOH.

Solution **(a)** Since the concentration of NaOH $\geq 10^{-6}$ M, therefore, we have

$$[OH^-] = [NaOH]_0 = 10^{-2}\ M$$

Therefore, pOH = – log {$[OH^-]/M$} = 2

and $\quad$ pH = 14 – 2 = 12

(b) Since the concentration of NaOH < 10^{-6} M, therefore, we have

$$[OH^-]^2 - [NaOH]_0\,[OH^-] - K_w = 0$$

i.e.,

$$[OH^-] = \frac{[NaOH]_0 + \sqrt{[NaOH]_0^2 + 4K_w}}{2}$$

Substituting the values, we will get

$$[OH^-] = 1.62 \times 10^{-7}\ M$$

Therefore, pOH = – log {$[OH^-]/M$} = – log (1.62×10^{-7}) = 6.79

and $\quad$ pH = 14 – 6.79 = 7.21

For Weak Bases

First approximation If $[OH^-] \geq 10^{-6}$ M, then $K_w/[OH^-] \leq 10^{-8}$ M. The term $[OH^-] - K_w/[OH^-]$ could be approximated to $[OH^-]$. This amounts to the fact that $[OH^-]$ coming from the dissociation of water is negligible relative to the $[OH^-]$ due to the dissociation of the base. Under this condition, Eq. (4.9.7) yields

$$K_b = \frac{[OH^-][OH^-]}{[BOH]_0 - [OH^-]} \quad \text{or} \quad K_b = \frac{\left(\dfrac{K_w}{[H_3O^+]}\right)^2}{[BOH]_0 - \dfrac{K_w}{[H_3O^+]}} \qquad (4.9.11)$$

Equation (4.9.11) is quadratic in $[OH^-]$ which could be solved for $[OH^-]$ either by direct method or by using the method of successive approximations.

Second approximation For a fairly concentrated solution of a relatively weak base (i.e. when K_b is small), we have

$$[BOH]_0 \geq \frac{K_w}{[H_3O^+]} \quad \text{or} \quad [BOH]_0 \geq [OH^-]$$

Hence, Eq. (4.9.11) modifies to

$$K_b = \frac{[OH^-]^2}{[BOH]_0} \quad \text{or} \quad [OH^-] = \sqrt{K_b\,[BOH]_0}$$

In terms of $[H_3O^+]$, we have

$$K_b = \frac{(K_w/[H_3O^+])^2}{[BOH]_0} \quad \text{or} \quad [H_3O^+] = \frac{K_w}{\sqrt{K_b\,[BOH]_0}} \tag{4.9.12}$$

In terms of degree of dissociation, we have

$$\underset{c(1-\alpha)}{BOH} \rightleftharpoons \underset{c\alpha}{B^+} + \underset{c\alpha}{OH^-}$$

$$K_b = \frac{[B^+][OH^-]}{[BOH]} = \frac{(c\alpha)(c\alpha)}{c(1-\alpha)} = \frac{c\alpha^2}{1-\alpha}$$

Thus $\quad c\alpha^2 + K_b\alpha - K_b = 0$

If $\alpha << 1$, then $K_b = c\alpha^2$ or $\alpha = \sqrt{K_b/c}$

Example 4.9.2 What is the pH of 10^{-3} M solution of NH_4OH ($K_b = 1.85 \times 10^{-5}$ M) at 25 °C?

Solution Using the simplest expression for α, we have

$$\alpha = \sqrt{\frac{K_b}{c}} = \left(\frac{1.85 \times 10^{-5}\ \text{M}}{0.001\ \text{M}}\right)^{1/2} = 0.136$$

Here $\alpha < 1$. A more correct value can be obtained by solving the quadratic expression

$$c\alpha^2 + K_b\,\alpha - K_b = 0$$

that is, $$\alpha = \frac{-K_b + \sqrt{K_b^2 + 4K_b\,c}}{2c}$$

$$= \frac{-(1.85 \times 10^{-5}\ \text{M}) + \sqrt{(1.85 \times 10^{-5}\ \text{M})^2 + 4\,(1.85 \times 10^{-5}\ \text{M})\,(10^{-3}\ \text{M})}}{2\,(10^{-3}\ \text{M})}$$

$$= 0.127$$

Hence, we have

$$[OH^-] = c\alpha = (10^{-3}\ \text{M})\,(0.127) = 1.27 \times 10^{-4}\ \text{M}$$

$$\text{pOH} = -\log\{[OH^-]/\text{M}\} = -\log(1.27 \times 10^{-4}) = 3.90$$

$$\text{pH} = 14 - \text{pOH} = 10.10$$

4.10 EXACT TREATMENT FOR IONIZATION OF A DIPROTIC ACID

Derivation of Exact Expression

Equilibria Existing in Solution In an aqueous solution of a diprotic acid H_2A the following equilibria exist:

$$H_2A + H_2O \rightleftharpoons H_3O^+ + HA^-; \quad K_1 = \frac{[HA^-][H_3O^+]}{[H_2A]} \tag{4.10.1}$$

$$\mathrm{HA^- + H_2O \rightleftharpoons H_3O^+ + A^{2-};} \quad K_2 = \frac{[\mathrm{A^{2-}}][\mathrm{H_3O^+}]}{[\mathrm{HA^-}]} \tag{4.10.2}$$

and

$$\mathrm{H_2O + H_2O \rightleftharpoons H_3O^+ + OH^-;} \quad K_w = [\mathrm{H_3O^+}][\mathrm{OH^-}] \tag{4.10.3}$$

Condition of Mass Balance

Mass-balance condition gives

$$[\mathrm{H_2A}]_0 = [\mathrm{H_2A}] + [\mathrm{HA^-}] + [\mathrm{A^{2-}}] \tag{4.10.4}$$

Condition of Charge Balance

The restriction regarding electroneutrality gives

$$[\mathrm{H_3O^+}] = [\mathrm{HA^-}] + 2[\mathrm{A^{2-}}] + [\mathrm{OH^-}] \tag{4.10.5}$$

The second term of the right hand side is multiplied by two, since the anion A^{2-} carries two negative charges.

Derivation of Exact Expression

With the use of Eqs (4.10.1), (4.10.2) and (4.10.3), Eq. (4.10.5) becomes:

$$[\mathrm{H_3O^+}] = \frac{K_1[\mathrm{H_2A}]}{[\mathrm{H_3O^+}]} + \frac{2K_1K_2[\mathrm{H_2A}]}{[\mathrm{H_3O^+}]^2} + \frac{K_w}{[\mathrm{H_3O^+}]}$$

Rearranging the above equation, we get

$$[\mathrm{H_3O^+}] = \frac{[\mathrm{H_2A}]}{[\mathrm{H_3O^+}]}K_1\left[1 + \frac{2K_2}{[\mathrm{H_3O^+}]}\right] + \frac{K_w}{[\mathrm{H_3O^+}]} \tag{4.10.6}$$

The concentration of H_2A in terms of the known parameter $[H_2A]_0$ can be obtained from the mass-balance expression (Eq. 4.10.4) which on using Eqs (4.10.1) and (4.10.2), becomes

$$[\mathrm{H_2A}]_0 = [\mathrm{H_2A}] + \frac{K_1[\mathrm{H_2A}]}{[\mathrm{H_3O^+}]} + \frac{K_1K_2[\mathrm{H_2A}]}{[\mathrm{H_3O^+}]^2}$$

or

$$[\mathrm{H_2A}] = \frac{[\mathrm{H_2A}]_0}{1 + \dfrac{K_1}{[\mathrm{H_3O^+}]} + \dfrac{K_1K_2}{[\mathrm{H_3O^+}]^2}} \tag{4.10.7}$$

Substituting $[H_2A]$ in Eq. (4.10.6), we have

$$[\mathrm{H_3O^+}] = \frac{K_1}{[\mathrm{H_3O^+}]}\left\{\frac{[\mathrm{H_2A}]_0}{1 + \dfrac{K_1}{[\mathrm{H_3O^+}]} + \dfrac{K_1K_2}{[\mathrm{H_3O^+}]^2}}\right\}\left\{1 + \frac{2K_2}{[\mathrm{H_3O^+}]}\right\} + \frac{K_w}{[\mathrm{H_3O^+}]} \tag{4.10.8}$$

Simplification of Exact Expression

Equation (4.10.8) is a fourth power expression in $[H_3O^+]$ which can, however, be simplified under certain approximations, as discussed in the following.

First approximation If $[H_3O^+]$ of the solution is greater than or equal to 10^{-6} M, it can then be assumed that the ionization of water is not an important source of H_3O^+, i.e. the last term $K_w/[H_3O^+]$ is negligible in comparison to $[H_3O^+]$. Thus, Eq. (4.10.8) is simplified to

$$[H_3O^+] = \left\{ \frac{[H_2A]_0\, K_1}{[H_3O^+] + K_1 + \dfrac{K_1K_2}{[H_3O^+]}} \right\} \left\{ 1 + \frac{2\,K_2}{[H_3O^+]} \right\} \tag{4.10.9}$$

Second approximation For a diprotic acid, in general $K_2 << K_1$ and K_2 itself is very small, the term $K_2/[H_3O^+] << 1$ and $K_1K_2/[H_3O^+] << ([H_3O^+] + K_1)$. Therefore, Eq. (4.10.9) gets reduced to

$$[H_3O^+] = \frac{[H_2A]_0\, K_1}{[H_3O^+] + K_1}$$

or

$$K_1 = \frac{[H_3O^+]^2}{[H_2A]_0 - [H_3O^+]} \tag{4.10.10}$$

which is an expression derived for a monoprotic acid under the assumption that water does not contribute significantly towards total $[H_3O^+]$. Hence, even for a diprotic acid (or, in general, for a polyprotic acid) the hydronium ion concentration can be calculated from its first equilibrium constant alone, provided $K_2 << K_1$.

Third approximation If K_1 is extremely small (of the order of 10^{-6} M or less) and if the concentration of the acid H_2A is not too low, the amount of acid that undergoes ionization is negligible in comparison to the original concentration of acid. Hence, the term $[H_2A]_0 - [H_3O^+] \approx [H_2A]_0$. With this approximation, expression (4.10.10) reduces to

$$[H_3O^+] = \sqrt{K_1\,[H_2A]_0} \tag{4.10.11}$$

Concentrations of Other Species

Concentrations of other species present in the solution can be obtained from the expressions obtained by combining mass-balance expression (Eq. 4.10.4) with the dissociation constants as given by Eqs. (4.10.1) and (4.10.2). These are

$$[H_2A] = \frac{[H_2A]_0}{1 + \dfrac{K_1}{[H_3O^+]} + \dfrac{K_1K_2}{[H_3O^+]^2}} \tag{4.10.12}$$

$$[HA^-] = \frac{[H_2A]_0}{\dfrac{[H_3O^+]}{K_1} + 1 + \dfrac{K_2}{[H_3O^+]}} \tag{4.10.13}$$

$$[A^{2-}] = \frac{[H_2A]_0}{\dfrac{[H_3O^+]^2}{K_1K_2} + \dfrac{[H_3O^+]}{K_2} + 1} \tag{4.10.14}$$

Simplification of Expressions (4.10.12), (4.10.13) and (4.10.14)

(i) If K_2 is very small, the term containing K_2 in the denomerator will have a negligible value in comparison to other terms. Therefore,

$$[H_2A] = \frac{[H_2A]_0}{1 + \dfrac{K_1}{[H_3O^+]}} \tag{4.10.15}$$

$$[HA^-] = \frac{[H_2A]_0}{\dfrac{[H_3O^+]}{K_1} + 1} \tag{4.10.16}$$

$$[A^{2-}] = \frac{[H_2A]_0}{\dfrac{[H_3O^+]^2}{K_1K_2} + \dfrac{[H_3O^+]}{K_2}} \tag{4.10.17}$$

(ii) If K_1 is very small in comparison to $[H_3O^+]$ of the solution, the concentrations of various species can then be calculated by using the following expressions:

$$[H_2A] = [H_2A]_0 \tag{4.10.18}$$

$$[HA^-] = [H_2A]_0 \frac{K_1}{[H_3O^+]} \tag{4.10.19}$$

$$[A^{2-}] = [H_2A]_0 \frac{K_1K_2}{[H_3O^+]^2} \tag{4.10.20}$$

Under these conditions, $[H_3O^+]$ is given by

$$[H_3O^+] = \sqrt{K_1\,[H_2A]_0} \qquad \text{(Eq. 4.10.11)}$$

Substituting this in (4.10.19), we have

$$[HA^-] = \frac{[H_2A]_0\,K_1}{\sqrt{K_1[H_2A]_0}} = \sqrt{K_1[H_2A]_0} = [H_3O^+] \tag{4.10.21}$$

Thus, $$[A^{2-}] = \frac{[H_2A]_0\,K_1K_2}{[H_3O^+]^2} = K_2 \tag{4.10.22}$$

Example 4.10.1 What are the concentrations of H^+, $H_2C_2O_4$, $HC_2O_4^-$ and $C_2O_4^{2-}$ in a 0.1 M solution of oxalic acid? ($K_1 = 5.9 \times 10^{-2}$ M and $K_2 = 6.4 \times 10^{-5}$ M.)

Solution Since the solution is fairly concentrated and $K_2/K_1 \simeq 10^{-3}$ (i.e. $K_2 << K_1$), we can use the expression:

$$K_1 = \frac{[H_3O^+]^2}{[H_2A]_0 - [H_3O^+]} \qquad \text{(Eq. 4.10.10)}$$

which gives

$$[H_3O^+] = \frac{-K_1 + \sqrt{K_1^2 + 4[H_2A]_0\, K_1}}{2}$$

Substituting the given values of K_1 and $[H_2A]_0$ in the above expression, we get

$$[H_3O^+] = \frac{-(5.9\times10^{-2}\ \text{M}) + \sqrt{(5.9\times10^{-2}\ \text{M})^2 + 4(0.1\ \text{M})(5.9\times10^{-2}\ \text{M})}}{2}$$

$$= \frac{-(5.9\times10^{-2}\ \text{M}) + (1.645\times10^{-1}\ \text{M})}{2}$$

$$= 0.052\,8\ \text{M}$$

We can obtain the concentrations of $H_2C_2O_4$, $HC_2O_4^-$ and $C_2O_4^{2-}$ in 0.1 M solution of oxalic acid from the following relations:

$$[H_2C_2O_4] = \frac{[H_2A]_0}{1 + \dfrac{K_1}{[H_3O^+]}} = \frac{0.1\ \text{M}}{1 + \left(\dfrac{5.9\times10^{-2}\ \text{M}}{5.28\times10^{-2}\ \text{M}}\right)} = \frac{0.1\ \text{M}}{1 + 1.118}$$

$$= 0.047\,2\ \text{M}$$

$$[HC_2O_4^-] = \frac{[H_2A]_0}{\dfrac{[H_3O^+]}{K_1} + 1} = \frac{0.1\ \text{M}}{\left(\dfrac{5.28\times10^{-2}\ \text{M}}{5.90\times10^{-2}\ \text{M}}\right) + 1} = \frac{0.1\ \text{M}}{0.894\,9 + 1}$$

$$= 0.052\,8\ \text{M}$$

$$[C_2O_4^{2-}] = \frac{[H_2A]_0}{\dfrac{[H_3O^+]^2}{K_1K_2} + \dfrac{[H_3O^+]}{K_2}}$$

$$= \frac{0.1\ \text{M}}{\dfrac{(5.28\times10^{-2}\ \text{M})^2}{(5.9\times10^{-2}\ \text{M})(6.4\times10^{-5}\ \text{M})} + \dfrac{5.28\times10^{-2}\ \text{M}}{6.4\times10^{-5}\ \text{M}}} = \frac{0.1\ \text{M}}{738.3 + 825.0}$$

$$= 0.000\,064\ \text{M}$$

Example 4.10.2 What are the concentrations of H_3O^+, HS^-, S^{2-} and H_2S in a 0.1 M solution of hydrogen sulphide? (K_1 = 1.1 × 10^{-7} M and K_2 = 1.0 × 10^{-14} M.)

Solution Since here K_1 itself is small and the solution is fairly concentrated, we can use the simplest expression to compute $[H_3O^+]$, i.e.

$$[H_3O^+] = \sqrt{K_1[H_2A]_0} = \sqrt{(1.10\times10^{-7}\ \text{M})(0.1\ \text{M})}$$

$$= 1.0\times10^{-4}\ \text{M}$$

As $K_1 << [H_3O^+]$, we can use the simplest expressions to compute concentrations of various other species in the solution, i.e.

$[H_2A] = [H_2A]_0$ = 0.1 M

$[HA^-] = [H_3O^+]$ = 1.0 × 10^{-4} M

$[A^{2-}] = K_2$ = 1.0 × 10^{-14} M

Direct Solution Since K_1 for H_2S is 1.1×10^{-7} M, the small amount of H_2S that ionizes is negligible in comparison to the original concentration of H_2S. In addition, the concentrations of H^+ and HS^- are not significantly altered by the secondary ionization ($K_2 = 1.0 \times 10^{-14}$ M). Thus,

$$\underset{0.10\ \text{M}}{H_2S} \rightleftharpoons \underset{x}{H^+} + \underset{x}{HS^-}$$

$$K_1 = \frac{[H^+][HS^-]}{[H_2S]} = 1.1 \times 10^{-7}\ \text{M}$$

$$= \frac{x^2}{0.10\ \text{M}} = 1.1 \times 10^{-7}\ \text{M}$$

$$x = [H^+] = [HS^-] = 1.0 \times 10^{-4}\ \text{M}$$

These concentrations are also applicable to the secondary ionization.

$$\underset{1.0\times10^{-4}\ \text{M}}{HS^-} \rightleftharpoons \underset{1.0\times10^{-4}\ \text{M}}{H^+} + \underset{?}{S^{2-}}$$

Hence, $$K_2 = \frac{[H^+][S^{2-}]}{[HS^-]} = 1.0 \times 10^{-14}\ \text{M}$$

which gives $[S^{2-}] = 1.0 \times 10^{-14}$ M

Example 4.10.3

What are the concentrations of H^+, HSO_4^-, SO_4^{2-} and H_2SO_4 in a 0.20 M solution of sulphuric acid? Given,

$$H_2SO_4 \longrightarrow H^+ + HSO_4^-; \qquad \text{Strong}$$

$$HSO_4^- \rightleftharpoons H^+ + SO_4^{2-}; \qquad K_2 = 1.3 \times 10^{-2}\ \text{M}$$

Solution

Since the first dissociation is strong, therefore, the $[H^+]$ due to this dissociation is 0.20 M. Let x be the amount of H^+ produced in the second dissociation. Hence,

$$[H^+]_{\text{total}} = 0.20\ \text{M} + x$$

This in equilibrium gives

$$\underset{(0.20\ \text{M} - x)}{HSO_4^-} \rightleftharpoons \underset{(0.20\ \text{M} + x)}{H^+} + \underset{x}{SO_4^{2-}}$$

$$K_2 = \frac{[H^+][SO_4^{2-}]}{[HSO_4^-]} = \frac{(0.20\ \text{M} + x)(x)}{(0.20\ \text{M} - x)} = 1.3 \times 10^{-2}\ \text{M}$$

or $$(0.20\ \text{M} + x)(x) = (1.3 \times 10^{-2}\ \text{M})(0.20\ \text{M} - x)$$

or $$x^2 + (0.20\ \text{M} + 1.3 \times 10^{-2}\ \text{M})\,x - (1.3 \times 10^{-2}\ \text{M})(0.20\ \text{M}) = 0$$

This is a quadratic equation in x, which gives

$$x = \frac{-0.213\ \text{M} + \sqrt{\{0.213^2 + 4\,(1.3\times10^{-2})\,(0.20)\}\ \text{M}^2}}{2}$$

$$= \frac{-0.213\ \text{M} + 0.236\,2\ \text{M}}{2} = 0.011\,6\ \text{M}$$

Thus, $[H^+]_{total} = 0.2\ \text{M} + x = 0.211\ 6\ \text{M}$

$[HSO_4^-] = 0.2\ \text{M} - x = 0.188\ 4\ \text{M}$

$[SO_4^{2-}] = x = 0.011\ 6\ \text{M}$

$[H_2SO_4] = 0$

Example 4.10.4 The equilibrium constants for amino acids are given in terms of the successive ionization constants of the protonated form. The equilibrium constants for glycine are

$$\underset{(H_2G^+)}{H_3N^+CH_2COOH} \rightleftharpoons H^+ + \underset{(HG)}{H_2NCH_2COOH}$$

$$\underset{(HG)}{H_2NCH_2COOH} \rightleftharpoons H^+ + \underset{(G^-)}{H_2NCH_2COO^-}$$

The numerical values of the dissociation constants for the above reactions are

$$K_{a1} = \frac{[H^+][HG]}{[H_2G^+]} = 4.5\times10^{-3}\ \text{M}$$

$$K_{a2} = \frac{[H^+][G^-]}{[HG]} = 1.7\times10^{-10}\ \text{M}$$

(i) Calculate the pH of 0.01 M glycine in pure water.
(ii) Show that the pH at the isoelectric point is

$$\text{pH} = \frac{1}{2}(\text{p}K_{a1}^\circ + \text{p}K_{a2}^\circ)$$

Solution **(i)** Since the solution is fairly concentrated, pH can be calculated using the formula:

$$[H^+] = \sqrt{K_{a1}\,[H_2G^+]_0} = \{(4.5\times10^{-3}\ \text{M})\,(10^{-2}\ \text{M})\}^{1/2}$$

$$= 6.7\times10^{-3}\ \text{M}$$

$$\text{pH} = -\log\,(6.7\times10^{-3}) = 2.17$$

(ii) At the isoelectric point $[H_2G^+] = [G^-]$. Therefore,

$$K_{a1}K_{a2} = \frac{[H^+][HG]}{[H_2G^+]}\frac{[H^+][G^-]}{[HG]} = \frac{[H^+]^2\,[G^-]}{[H_2G^+]} = [H^+]^2$$

i.e. $(K_{a1}/\text{M})(K_{a2}/\text{M}) = ([H^+]/\text{M})^2$

Taking logarithm and multiplying with –1, we have

$$\text{pH} = \frac{1}{2}\,[\text{p}K_{a1}^\circ + \text{p}K_{a2}^\circ]$$

4.11 DISSOCIATION OF POLYPROTIC ACID

Step-wise Ionization

Polyprotic acids are those acids which contain more than one acid hydrogen per molecule. Examples include sulphuric acid, oxalic acid, phosphoric acid, etc. These acids ionize in steps and each step is characterized by the corresponding ionization constant. Thus, for a triprotic acid, H_3A, the dissociation could be written as

$$H_3A \rightleftharpoons H_2A^- + H^+ \qquad K_1 = \frac{[H_2A^-][H^+]}{[H_3A]} \tag{4.11.1}$$

$$H_2A^- \rightleftharpoons HA^{2-} + H^+ \qquad K_2 = \frac{[HA^{2-}][H^+]}{[H_2A^-]} \tag{4.11.2}$$

$$HA^{2-} \rightleftharpoons A^{3-} + H^+ \qquad K_3 = \frac{[A^{3-}][H^+]}{[HA^{2-}]} \tag{4.11.3}$$

In general, for all polyprotic acids the primary ionization (K_1) is stronger than the secondary (K_2), which is, in turn, stronger than the tertiary (K_3), i.e. $K_3 << K_2 < K_1$. This trend in the values of ionization constants is consistent with the nature of species that ionizes in each step. One would predict that a proton would be released more readily by an uncharged molecule than by a uni-negative ion and more readily by a uni-negative ion than by a bi-negative ion. Thus, most of the $[H_3O^+]$ in the solution will be due to the primary dissociation. The $[H_3O^+]$ coming from the subsequent dissociations will be negligibly small in comparison. This is justified because of two facts: (i) $K_3 << K_2 < K_1$, and (ii) the hydronium-ion concentration from the first dissociation will suppress the subsequent dissociations due to the common-ion effect.

From Eqs (4.11.1) to (4.11.3), we can write

$$K_1 = \frac{[H_2A^-][H^+]}{[H_3A]} \tag{4.11.4}$$

$$K_1K_2 = \frac{[HA^{2-}][H^+]^2}{[H_3A]} \tag{4.11.5}$$

$$K_1K_2K_3 = \frac{[A^{3-}][H^+]^3}{[H_3A]} \tag{4.11.6}$$

Distribution Functions

The mass-balance condition for the acid is

$$[H_3A]_0 = [H_3A] + [H_2A^-] + [HA^{2-}] + [A^{3-}]$$

Making use of Eqs (4.11.4) to (4.11.6), we get

$$[H_3A]_0 = [H_3A] + \frac{K_1[H_3A]}{[H^+]} + \frac{K_1K_2[H_3A]}{[H^+]^2} + \frac{K_1K_2K_3[H_3A]}{[H^+]^3}$$

Hence, $$\frac{[H_3A]}{[H_3A]_0} = \frac{1}{1 + (K_1/[H^+]) + (K_1K_2/[H^+]^2) + (K_1K_2K_3/[H^+]^3)}$$

The ratio $[H_3A]/[H_3A]_0$ is represented by the symbol α_3. Note that the subscript 3 represents the number of hydrogens attached to A in H_3A. Hence

$$\alpha_3 = \left(1 + \frac{K_1}{[H^+]} + \frac{K_1K_2}{[H^+]^2} + \frac{K_1K_2K_3}{[H^+]^3}\right)^{-1} \tag{4.11.7}$$

Working similarly, we find

$$\alpha_2 = \frac{[H_2A^-]}{[H_3A]_0} = \frac{[H_3A]}{[H_3A]_0}\frac{[H_2A^-]}{[H_3A]} = \alpha_3\left(\frac{K_1}{[H^+]}\right) \tag{4.11.8}$$

$$\alpha_1 = \frac{[HA^{2-}]}{[H_3A]_0} = \frac{[H_3A]}{[H_3A]_0}\frac{[HA^{2-}]}{[H_3A]} = \alpha_3\left(\frac{K_1K_2}{[H^+]^2}\right) \tag{4.11.9}$$

$$\alpha_0 = \frac{[A^{3-}]}{[H_3A]_0} = \frac{[H_3A]}{[H_3A]_0}\frac{[A^{3-}]}{[H_3A]} = \alpha_3\left(\frac{K_1K_2K_3}{[H^+]^3}\right) \tag{4.11.10}$$

Simplification of Distribution Functions

Since $K_3 << K_2 < K_1$ and assuming that K_1 is also small, we can simplify Eqs (4.11.7) to (4.11.10).

(i) From Eq. (4.11.7), we get

$$\alpha_3 = 1 \quad \text{i.e.} \quad [H_3A] = [H_3A]_0 \tag{4.11.11}$$

(ii) From Eq. (4.11.8), we get

$$\alpha_2 = \frac{K_1}{[H^+]} \quad \text{i.e.} \quad [H_2A^-] = [H_3A]_0\frac{K_1}{[H^+]} \tag{4.11.12}$$

From the primary ionization (Eq. 4.11.1), we can write

$$[H_2A^-] = [H^+]$$ (As the ionization of H_2A^- will be negligibly small as $K_3 << K_2 < K_1$)

Hence, Eq. (4.11.12) becomes

$$[H_2A^-]\,[H^+] = K_1\,[H_3A]_0$$

or $$[H_2A^-] = [H^+] = (K_1\,[H_3A]_0)^{1/2} \tag{4.11.13}$$

(iii) From Eq. (4.11.9), we get

$$\alpha_1 = \frac{K_1K_2}{[H^+]^2} \quad \text{i.e.} \quad [HA^{2-}] = [H_3A]_0\frac{K_1K_2}{[H^+]^2}$$

Using Eq. (4.11.13), we get

$$[HA^{2-}] = K_2 \tag{4.11.14}$$

(iv) From Eq. (4.11.10), we get

$$\alpha_0 = \frac{K_1K_2K_3}{[H^+]^3} \quad \text{i.e.} \quad [A^{3-}] = [H_3A]_0 \frac{K_1K_2K_3}{[H^+]^3}$$

Using Eq. (4.11.13), we get

$$[A^{3-}] = \frac{K_2K_3}{[H^+]} = \frac{K_2K_3}{(K_1[H_3A]_0)^{1/2}}$$

Example 4.11.1 Calculate H^+, $H_2PO_4^-$, HPO_4^{2-}, PO_4^{3-} and H_3PO_4 in a 0.10 M solution of phosphoric acid ($K_1 = 7.5 \times 10^{-3}$ M, $K_2 = 6.2 \times 10^{-8}$ M and $K_3 = 1.0 \times 10^{-12}$ M).

Solution The principal source of H^+ is the primary ionization because the H^+ produced by the other ionization sources as well as those from the ionization of water are negligible in comparison. Furthermore $H_2PO_4^-$ derived from the primary ionization is not significantly diminished by the secondary ionization. Thus, we can write

$$\underset{(0.1\,\text{M}-x)}{H_3PO_4} \rightleftharpoons \underset{x}{H^+} + \underset{x}{H_2PO_4^-}$$

$$K_1 = \frac{[H^+][H_2PO_4^-]}{[H_3PO_4]} = \frac{x^2}{(0.1\,\text{M} - x)} = 7.5 \times 10^{-3}\ \text{M}$$

Thus solving the quadratic equation for x, we get

$$x = [H^+] = [H_2PO_4^-] = 2.4 \times 10^{-2}\ \text{M}$$

Therefore, $[H_3PO_4] = 0.1\,\text{M} - x = 7.6 \times 10^{-2}$ M

These $[H^+]$ and $[H_2PO_4^-]$ are applicable to the secondary ionization.

Hence, $$\underset{2.4\times10^{-2}\,\text{M}}{H_2PO_4^-} \rightleftharpoons \underset{2.4\times10^{-2}\,\text{M}}{H^+} + \underset{?}{HPO_4^{2-}}$$

$$K_2 = \frac{[H^+][HPO_4^{2-}]}{[H_2PO_4^-]} = [HPO_4^{2-}] = 6.2 \times 10^{-8}\ \text{M}$$

For tertiary ionization

$$\underset{6.2\times10^{-8}\,\text{M}}{HPO_4^{2-}} \rightleftharpoons \underset{2.4\times10^{-2}\,\text{M}}{H^+} + \underset{?}{PO_4^{3-}}$$

$$K_3 = \frac{[H^+][PO_4^{3-}]}{[HPO_4^{2-}]}$$

$$[PO_4^{3-}] = K_3\left(\frac{[HPO_4^{2-}]}{[H^+]}\right) = (1 \times 10^{-12}\ \text{M})\left(\frac{6.2 \times 10^{-8}\ \text{M}}{2.4 \times 10^{-2}\ \text{M}}\right)$$

$$= 2.58 \times 10^{-18}\ \text{M}$$

Example 4.11.2 Calculate the concentrations of H^+, $H_2AsO_4^-$, $HAsO_4^{2-}$, AsO_4^{3-} and H_3AsO_4 in a 0.50 M solution of arsenic acid. Given $K_1 = 2.5 \times 10^{-4}$ M, $K_2 = 5.6 \times 10^{-8}$ M and $K_3 = 3 \times 10^{-13}$ M.

Solution The various equilibria are

$$H_3AsO_4 \rightleftharpoons H^+ + H_2AsO_4^-; \qquad K_1 = [H^+][H_2AsO_4^-]/[H_3AsO_4]$$

$$H_2AsO_4^- \rightleftharpoons H^+ + HAsO_4^{2-}; \qquad K_2 = [H^+][HAsO_4^{2-}]/[H_2AsO_4^-]$$

$$HAsO_4^{2-} \rightleftharpoons H^+ + AsO_4^{3-}; \qquad K_3 = [H^+][AsO_4^{3-}]/[HAsO_4^{2-}]$$

Let $[H_3O^+] = [H_2AsO_4^-] = x$. Therefore,

$$K_1 = \frac{x^2}{0.5\ \text{M} - x} = 2.5 \times 10^{-4}\ \text{M}$$

Solving the quadratic equation for x, we have

$$x = 1.105 \times 10^{-2}\ \text{M}$$

$$[H_3AsO_4] = 0.5\ \text{M} - 0.011\ 05\ \text{M} \approx 0.489\ \text{M}$$

Secondary dissociation

$$\underset{1.105\times10^{-2}\ \text{M}}{H_2AsO_4^-} \rightleftharpoons \underset{1.105\times10^{-2}\ \text{M}}{H^+} + \underset{?}{HAsO_4^{2-}}$$

$$[HAsO_4^{2-}] = K_2 = 5.6 \times 10^{-8}\ \text{M}$$

Tertiary dissociation

$$\underset{5.6\times10^{-8}\ \text{M}}{HAsO_4^{2-}} \rightleftharpoons \underset{1.105\times10^{-2}\ \text{M}}{H^+} + \underset{?}{AsO_4^{3-}}$$

$$K_3 = \frac{[H^+][AsO_4^{3-}]}{[HAsO_4^{2-}]} = 3 \times 10^{-13}\ \text{M}$$

$$[AsO_4^{3-}] = \frac{(3 \times 10^{-13}\ \text{M})(5.6 \times 10^{-8}\ \text{M})}{(1.105 \times 10^{-2}\ \text{M})} = 15.21 \times 10^{-19}\ \text{M}$$

4.12 SOLUTIONS OF SALTS IN WATER: HYDROLYSIS

Phenomenon of Hydrolysis A given salt on dissolving in water may produce acidic, neutral or alkaline solution depending upon its ions. This is due to the fact that certain ions can react with water and thereby produce either an acidic or an alkaline solution according to the following reactions:

$$A^- + H_2O \rightleftharpoons HA + OH^- \qquad (4.12.1)$$

$$B^+ + 2H_2O \rightleftharpoons BOH + H_3O^+ \qquad (4.12.2)$$

This phenomenon is known as *hydrolysis*.*

---

*See also Section 4.35.

Such hydrolysis equilibria are characterized by the following hydrolysis constants;

$$K_{ha} = \frac{[HA][OH^-]}{[A^-]} \qquad (4.12.3)^*$$

$$K_{hc} = \frac{[BOH][H_3O^+]}{[B^+]} \qquad (4.12.4)^\dagger$$

Hydrolysis of Anions

The anion A^- which is a weaker base than OH^- and which has its conjugate acid HA stronger than water but weaker than H_3O^+ shows the phenomenon of hydrolysis.‡ Examples include CH_3COO^-, CN^-, NO_2^-, S^{2-}, etc. The hydrolysis constant in such a case is given by

$$K_h = \frac{[HA][OH^-]}{[A^-]}$$

Multiplying the numerator and denominator by $[H_3O^+]$, we have

$$K_h = \frac{[HA]}{[A^-][H_3O^+]}[OH^-][H_3O^+]$$

The first term is the reciprocal of the dissociation constant K_a of the conjugate acid HA of A^- and the second term is the ionic product of water. Thus,

$$K_h = \frac{K_w}{K_a} \qquad (4.12.5)$$

Hence, the value of hydrolysis constant depends upon the value of K_a. Smaller the value of K_a, larger the value of K_h, i.e. the weaker the conjugate acid, more extensive hydrolysis of the anion. For example, the dissociation constants of HCN and CH_3COOH are 4×10^{-10} M and 1.8×10^{-5} M, respectively. Since the former is smaller than the latter, it is expected that CN^- will be hydrolyzed to a greater extent than CH_3COO^-. Consequently, the pH of a solution containing NaCN would be larger than the solution containing the same amount of CH_3COONa. This fact is in agreement with the observation that pH of 0.1 M solution of NaCN is 11.2 whereas that of 0.1 M solution of CH_3COONa is 8.9.

Hydrolysis of Cations

The cation B^+ which is a weaker acid than H_3O^+ and which has its conjugate base BOH stronger than water but weaker than OH^- shows the phenomenon of hydrolysis.† Examples include NH_4^+, $C_6H_5NH_3^+$, $C_5H_5NH^+$, $N_2H_5^+$, etc.

---

*Since the anion acts as a base, the hydrolysis constant K_h is often represented as ionization constant K_b.

†Since the cation acts as an acid, the hydrolysis constant K_h is often represented as the ionization constant K_a.

‡These characteristics follow from the fact that the position of equilibrium in Eqs (4.12.1) and (4.12.2) favours the formation of a weak acid and a weak base (see, p. 370, 374 and 375).

The hydrolysis constant in such a case is given by

$$K_h = \frac{[BOH][H_3O^+]}{[B^+]}$$

Multiplying the numerator and denominator by $[OH^-]$, we have

$$K_h = \frac{[BOH]}{[B^+][OH^-]}[H_3O^+][OH^-]$$

The first term is a reciprocal of the dissociation constant K_b of the conjugate base BOH of B^+ and the second term is the ionic product of water. Thus,

$$K_h = \frac{K_w}{K_b} \tag{4.12.6}$$

Hence, the value of hydrolysis constant depends upon the value of K_b; smaller the value of K_b, larger the value of K_h, i.e. weaker the conjugate base, more extensive hydrolysis of the cation. For example, the dissociation constants of aniline and NH_4OH are 4.0×10^{-10} M and 1.8×10^{-5} M, respectively. Since the former is smaller than the latter, it is expected that anilinium ion will be hydrolyzed to a greater extent than ammonium ion. Consequently, the pH of a solution containing anilinium chloride would be smaller than the solution containing the same amount of ammonium chloride. This fact is in agreement with the observation that pH of 0.1 M solution of anilinium chloride is 2.8 whereas that of 0.1 M ammonium chloride is 5.1.

Comment on Hydrolysis

Whenever a salt is dissolved in water, the nature of the resulting solution depends upon the extent to which either one or both the ions interact with water. In the subsequent sections, we discuss the exact treatment to compute $[H_3O^+]$ of an aqueous solution of a salt.

Neither of the two ions of a salt formed from a strong acid and a strong base undergo hydrolysis. Therefore, whenever such a salt is dissolved in water, its pH remains the same, i.e. the solution remains neutral. Examples include NaCl, KNO_3, etc.

From Eqs (4.12.5) and (4.12.6), it follows that the product K_aK_h or K_bK_h is always equal to K_w. Since K_h may be written as K_b or K_a, we can express this product as $K_aK_b = K_w$, i.e. the product of ionization constants of an acid (or base) and its conjugate base (or acid) is always equal K_w.

4.13 EXACT TREATMENT OF HYDROLYSIS OF SALT FORMED FROM A WEAK ACID AND A STRONG BASE

Only the anion of such a salt will undergo hydrolysis and will produce free hydroxyl ions thereby rendering the solution alkaline, i.e. pH > 7 at 25 °C. Examples include sodium acetate, sodium cyanide, potassium nitrite, etc.

Equilibria Existing in the Solution

Let *c* be the concentration of such a salt (say, NaA) in the solution. The following equilibria exist in the solution:

$$A^- + H_2O \rightleftharpoons HA + OH^- \tag{4.13.1}$$

$$H_2O + H_2O \rightleftharpoons H_3O^+ + OH^- \tag{4.13.2}$$

These are characterized by the following equilibrium constants.

$$K_h = \frac{[HA][OH^-]}{[A^-]} = \frac{[HA][OH^-]}{[A^-]}\frac{[H_3O^+]}{[H_3O^+]} = \frac{K_w}{K_a} \tag{4.13.3}$$

and $$K_w = [H_3O^+]\,[OH^-] \tag{4.13.4}$$

Condition of Mass Balance

The mass-balance condition gives

$$[NaA]_0 = [A^-] + [HA] = c \tag{4.13.5}$$

Condition of Charge Balance

The charge-balance condition gives

$$[Na^+] + [H_3O^+] = [OH^-] + [A^-] \tag{4.13.6}$$

Derivation of Exact Expression

Since the salt is completely dissociated, therefore

$$[Na^+] = c$$

Hence, Eq. (4.13.6) reduces to

$$[A^-] = c + [H_3O^+] - \frac{K_w}{[H_3O^+]} \tag{4.13.7}$$

With this, Eq. (4.13.5) reduces to

$$[HA] = \frac{K_w}{[H_3O^+]} - [H_3O^+] \tag{4.13.8}$$

Substituting $[A^-]$ and [HA] in Eq. (4.13.3), we get

$$\frac{K_w}{K_a} = \frac{\left\{\dfrac{K_w}{[H_3O^+]} - [H_3O^+]\right\}\dfrac{K_w}{[H_3O^+]}}{\left\{c + [H_3O^+] - \dfrac{K_w}{[H_3O^+]}\right\}} \tag{4.13.9}$$

or $$[H_3O^+]^3 + (c + K_a)\,[H_3O^+]^2 - K_w[H_3O^+] - K_aK_w = 0 \tag{4.13.10}$$

Simplification of Exact Expression

Equation (4.13.10) is a cubical expression in $[H_3O^+]$. A simpler expression can be used to compute $[H_3O^+]$ under the following approximations:

(i) If in the solution $[H_3O^+] \leq 10^{-8}$ M, then $K_w/[H_3O^+] \geq 10^{-6}$ M. In such cases, we can assume

$$\frac{K_w}{[H_3O^+]} - [H_3O^+] \approx \frac{K_w}{[H_3O^+]}$$

and $[H_3O^+]$ can be neglected in comparison to $\{c - K_w/\{H_3O^+]\}$.

(ii) If the solution is fairly concentrated and the value of K_h is very small (which is so in most of the cases), then the term $K_w/\{H_3O^+\}$ will be negligible in comparison to c.

With these approximations, Eq. (4.13.9) is reduced to

$$\frac{K_w}{K_a} = \frac{K_w^2}{c[H_3O^+]^2}$$

or $$[H_3O^+] = \sqrt{\frac{K_w K_a}{c}} \tag{4.13.11}$$

To express Eq. (4.13.11) in pH form, we write it as

$$\frac{[H_3O^+]}{M} = \sqrt{\frac{(K_w/M^2)(K_a/M)}{(c/c^\circ)}}; \qquad (\text{where } c^\circ = 1\ M = 1\ mol\ dm^{-3})$$

or $$-\log\left(\frac{[H_3O^+]}{M}\right) = -\frac{1}{2}\log\left(\frac{K_w}{M^2}\right) - \frac{1}{2}\log\left(\frac{K_a}{M}\right) + \frac{1}{2}\log\left(\frac{c}{c^\circ}\right)$$

or $$pH = \frac{1}{2}pK_w^\circ + \frac{1}{2}pK_a^\circ + \frac{1}{2}\log(c/c^\circ) \tag{4.13.12}$$

It can be seen that the pH of the solution, besides depending upon pK_a°, would also vary with the solution concentration.

Direct Approach to Compute pH

The above two approximations are equivalent to the following two statements:

(i) The amount of OH^- coming from the dissociation of water is negligible in comparison to that produced due to the hydrolysis of the ion. Thus from Eq. (4.13.1), we have

$$[OH^-] = [HA]$$

(ii) Usually K_h is very small. Therefore, the amount of ions that is actually reacted with water and converted to HA is negligible in comparison to the total concentration of the ions dissolved. Thus, the equilibrium concentration of A^- can be approximated to the initial concentration c.

Consider the hydrolysis equilibrium

$$A^- + H_2O \rightleftharpoons HA + OH^-$$

we have $$K_h = \frac{K_w}{K_a} = \frac{[HA][OH^-]}{[A^-]} = \frac{[OH^-]^2}{c}$$

since $[HA] = [OH^-]$ and $[A^-] = c$. Rearranging the above expression, we get

$$[OH^-]^2 = K_h \times c = \frac{K_w}{K_a}c \quad \text{or} \quad \frac{K_w^2}{[H_3O^+]^2} = \frac{K_w}{K_a}c$$

or $$[H_3O^+] = \sqrt{\frac{K_w K_a}{c}}$$

which gives $pH = \frac{1}{2}pK_w^\circ + \frac{1}{2}pK_a^\circ + \frac{1}{2}\log(c/c^\circ)$

an expression identical to Eq. (4.13.12).

Expression of Degree of Hydrolysis

The degree of hydrolysis can be computed as follows:

$$\underset{c(1-\alpha)}{A^-} + H_2O \rightleftharpoons \underset{c\alpha}{HA} + \underset{c\alpha}{OH^-}$$

$$K_h = \frac{K_w}{K_a} = \frac{[HA][OH^-]}{[A^-]} = \frac{(c\alpha)(c\alpha)}{c(1-\alpha)} = \frac{c\alpha^2}{1-\alpha}$$

Under the approximation (ii) stated above, $(1 - \alpha) \approx 1$.

Therefore, $K_h = c\alpha^2$

or $$\alpha = \sqrt{\frac{K_h}{c}} = \sqrt{\frac{K_w}{K_a c}} \qquad (4.13.13)$$

It can be seen from this expression that the degree of hydrolysis increases with dilution.

Example 4.13.1

Calculate at 25 °C the hydrolysis constant and its degree of hydrolysis in 0.10 M solution of: (a) sodium acetate, and (b) sodium carbonate. What will be the pH values? (Given: $K_a(HAc) = 1.8 \times 10^{-5}$ M and $K_a(HCO_3^-) = 4.7 \times 10^{-11}$ M.)

Solution

(a) Hydrolysis of acetate ion (Ac^-) is

$$Ac^- + H_2O \rightleftharpoons HAc + OH^-$$

If it be assumed that the decrease in the concentration of Ac^- ions because of hydrolysis is negligible in comparison to the original concentration of Ac^-, and also, if the contribution of OH^- from the ionization of water is negligible in comparison to that derived from hydrolysis, then we have

$$[HAc] = [OH^-] \quad \text{and} \quad [Ac] = [Ac^-]_0 = 0.1\ M$$

$$K_h = \frac{K_w}{K_a} = \frac{(1.0 \times 10^{-14}\ M)}{(1.8 \times 10^{-5}\ M)}$$

Also $$K_h = \frac{[HAc][OH^-]}{[Ac^-]} \simeq \frac{[OH^-]^2}{[Ac^-]_0} = \frac{[OH^-]^2}{0.1\ M}$$

Hence $$[OH^-] = \sqrt{\frac{(0.1\ M)(1.0 \times 10^{-14}\ M)}{(1.8 \times 10^{-5}\ M)}} = 7.5 \times 10^{-6}\ M$$

Thus, we see that the decrease in the concentration of Ac^- (= $[OH^-]$ = 7.5×10^{-6} M) due to hydrolysis is negligible as compared to the concentration of Ac^-(0.1 M) and that the concentration of OH^- due to the dissociation of water is just 1.3% {= $(10^{-7}\ \text{M}/7.5 \times 10^{-6}\ \text{M}) \times 100$} of the concentration of OH^- derived above, justifying the above procedure to compute the concentration of OH^-. Now the concentration of H_3O^+ would be

$$[H_3O^+] = \frac{K_w}{[OH^-]} = \frac{1.0 \times 10^{-14}\ \text{M}^2}{7.5 \times 10^{-6}\ \text{M}} = 1.3 \times 10^{-9}\ \text{M}$$

Thus $\text{pH} = -\log\{[H^+]/\text{M}\} = -\log(1.3 \times 10^{-9}) = 8.9$

and $$\alpha = \frac{[OH^-]}{c} = \frac{7.5 \times 10^{-6}\ \text{M}}{0.1\ \text{M}} = 7.5 \times 10^{-5}$$

(b) Hydrolysis of CO_3^{2-} is represented as

$$CO_3^{2-} + H_2O \rightleftharpoons HCO_3^- + OH^-$$

Here HCO_3^- is an amphiprotic anion, i.e. both a weak acid and a weak base, and its hydrolysis is negligible.

The hydrolysis constant of the above reaction is

$$K_h = \frac{[HCO_3^-][OH^-]}{[CO_3^{2-}]}$$

Multiplying numerator and denominator by $[H_3O^+]$, we have

$$K_h = \frac{[HCO_3^-][OH^-]}{[CO_3^{2-}]} \frac{[H_3O^+]}{[H_3O^+]} = \frac{K_w}{K_{a2}}$$

where K_{a2} is the dissociation constant of the reaction

$$HCO_3^- + H_2O \rightleftharpoons H_3O^+ + CO_3^{2-}$$

The numerical value of K_h is

$$K_h = \frac{(1.0 \times 10^{-14}\ \text{M}^2)}{(4.7 \times 10^{-11}\ \text{M})} = 2.127 \times 10^{-4}\ \text{M}$$

Since $[HCO_3^-] = [OH^-]$ and $[OH^-] << [CO_3^{2-}]$, therefore

$$K_h = \frac{[OH^-]^2}{[CO_3^{2-}]_0}$$

or $$[OH^-] = \sqrt{K_h[CO_3^{2-}]_0} = \sqrt{(2.127 \times 10^{-4}\ \text{M})(0.1\ \text{M})}$$

$$= 4.61 \times 10^{-3}\ \text{M}$$

Now $\text{pOH} = -\log\{[OH^-]/\text{M}\} = -\log(4.61 \times 10^{-3}) = 2.34$

$\text{pH} = 14 - \text{pOH} = 11.66$

and $$\alpha = \frac{[OH^-]}{c} = \frac{(4.61 \times 10^{-3}\ \text{M})}{(0.1\ \text{M})} = 4.61 \times 10^{-2}$$

Example 4.13.2

It is found that 0.1 M solution of three sodium salts NaX, NaY and NaZ at 25 °C have pHs 7.0, 9.0 and 11.0, respectively. Arrange the acids HX, HY and HZ in order of increasing strength. Where possible, calculate the ionization constants of the acids.

Solution

Hydrolysis of the anions are

$$X^- + H_2O \rightleftharpoons HX + OH^-; \qquad pH = 7.0$$

$$Y^- + H_2O \rightleftharpoons HY + OH^-; \qquad pH = 9.0$$

$$Z^- + H_2O \rightleftharpoons HZ + OH^-; \qquad pH = 11.0$$

Using the expression

$$pH = \frac{1}{2}pK_w^\circ + \frac{1}{2}pK_a^\circ + \frac{1}{2}\log(c/c^\circ)$$

we can calculate pK_a° of the conjugate acid as follows:

(i) Solution of pH = 7. No Hydrolysis, therefore X^- must have conjugate acid HX strong.

(ii) Solution of pH = 9. Using $pK_a^\circ = 2pH - pK_w^\circ - \log(c/c^\circ)$, we have

$$pK_a^\circ = 18 - 14 + 1 = 5$$

(iii) Solution of pH = 11

$$pK_a^\circ = 22 - 14 + 1 = 9$$

Thus, $K_a(HY) = 10^{-5}$ M and $K_a(HZ) = 10^{-9}$ M

Acid HZ is weaker than acid HY.

Example 4.13.3

Calculate the hydrolytic constant, degree of hydrolysis and pH of 0.25 M NaCN solution at 25 °C ($K_a(HCN) = 4.8 \times 10^{-10}$ M).

Solution

The hydrolysis reaction of CN^- is

$$CN^- + H_2O \rightleftharpoons HCN + OH^-$$

The value of hydrolysis constant is

$$K_h = \frac{[HCN][OH^-]}{[CN^-]} = \frac{K_w}{K_a} = \frac{(1.0 \times 10^{-14}\ M^2)}{(4.8 \times 10^{-10}\ M)} = 2.08 \times 10^{-5}\ M$$

Its degree of hydrolysis is

$$\alpha = \sqrt{\frac{K_h}{c}} = \sqrt{\frac{(2.08 \times 10^{-5}\ M)}{(0.25\ M)}} = 9.12 \times 10^{-3}$$

Thus, the concentration of OH^- in the solution is

$$[OH^-] = c\alpha = (0.25\ M)(9.12 \times 10^{-3}) = 2.28 \times 10^{-3}\ M$$

Thus, $pOH = -\log\{[OH^-]/M\} = 2.64$

and $pH = 14 - pOH = 11.36$

4.14 EXACT TREATMENT OF HYDROLYSIS OF SALT FORMED FROM A STRONG ACID AND A WEAK BASE

The cation of such a salt will interact with water to produce H_3O^+ in solution. Therefore, an aqueous solution of such a salt will be acidic (i.e. pH < 7 at 25 °C). Examples include salts like ammonium chloride, anilinium chloride, etc.

Equilibria Existing in Solution

Let c be the concentration of such a salt (BCl) in the solution. The following equilibria will exist in the solution.

$$B^+ + 2H_2O \rightleftharpoons BOH + H_3O^+ \tag{4.14.1}$$

$$H_2O + H_2O \rightleftharpoons H_3O^+ + OH^-$$

These are characterized by the following equilibrium constants.

$$K_h = \frac{[BOH][H_3O^+]}{[B^+]} = \frac{[BOH][H_3O^+]}{[B^+]}\frac{[OH^-]}{[OH^-]} = \frac{K_w}{K_b} \tag{4.14.2}$$

and $$K_w = [H_3O^+][OH^-] \tag{4.14.3}$$

Condition of Mass Balance

The mass-balance condition gives

$$[BCl]_0 = [B^+] + [BOH] \tag{4.14.4}$$

Condition of Charge Balance

The charge-neutrality condition gives

$$[B^+] + [H_3O^+] = [Cl^-] + [OH^-] \tag{4.14.5}$$

Derivation of Exact Expression

Since the salt is completely dissociated, we have

$$[Cl^-] = c$$

Hence, Eq. (4.14.5) reduces to

$$[B^+] = c + \frac{K_w}{[H_3O^+]} - [H_3O^+] \tag{4.14.6}$$

With this, Eq. (4.14.4) reduces to

$$[BOH] = [H_3O^+] - \frac{K_w}{[H_3O^+]} \tag{4.14.7}$$

Substituting $[B^+]$ and [BOH] in Eq. (4.14.2), we get

$$\frac{K_w}{K_b} = \frac{\left\{[H_3O^+] - \dfrac{K_w}{[H_3O^+]}\right\}[H_3O^+]}{\left\{c + \dfrac{K_w}{[H_3O^+]} - [H_3O^+]\right\}} \tag{4.14.8}$$

or $$[H_3O^+]^3K_b + [H_3O^+]^2K_w - (K_b + c)\,K_w[H_3O^+] - K_w^2 = 0 \tag{4.14.9}$$

Simplification of Exact Expression

Equation (4.14.9) is a cubical expression in $[H_3O^+]$. A simpler expression can be used to compute the $[H_3O^+]$ under the following approximations:

(i) If in the resultant solution $[H_3O^+] \geq 10^{-6}$ M, then $K_w/[H_3O^+] \leq 10^{-8}$ M. With this

$$[H_3O^+] - \frac{K_w}{[H_3O^+]} \approx [H_3O^+]$$

and
$$\left\{c - [H_3O^+] + \frac{K_w}{[H_3O^+]}\right\} \approx c - [H_3O^+]$$

(ii) If the solution is fairly concentrated and the value of K_h is small, then

$$c - [H_3O^+] \approx c$$

With these approximations, Eq. (4.14.8) is reduced to

$$\frac{K_w}{K_b} = \frac{[H_3O^+]^2}{c}$$

or
$$[H_3O^+] = \sqrt{\frac{K_w c}{K_b}} \tag{4.14.10}$$

i.e.
$$\text{pH} = \frac{1}{2}\text{p}K_w^\circ - \frac{1}{2}\text{p}K_b^\circ - \frac{1}{2}\log(c/c^\circ) \tag{4.14.11}$$

It can be seen that the pH of the solution, besides depending on K_b, also depends on the concentration of the salt in solution.

Direct Approach to Compute pH

The above two approximations are equivalent to the following two statements:

(i) The amount of H_3O^+ coming from the dissociation of water is negligible in comparison to that produced due to the hydrolysis of the ion, thus making $[BOH] = [H_3O^+]$.

(ii) Usually K_h is very small and, therefore, the concentration of ions that actually reacted with water and converted into BOH is negligible in comparison to the total concentration of the ions dissolved. Thus, the equilibrium concentration of B^+ can be approximated to the initial concentration c.

Considering the hydrolysis equilibrium

$$B^+ + 2H_2O \rightleftharpoons BOH + H_3O^+$$

we have
$$K_h = \frac{K_w}{K_b} = \frac{[BOH][H_3O^+]}{[B^+]} = \frac{[H_3O^+]^2}{c}$$

since $[BOH] = [H_3O^+]$ and $[B^+] = c$.

Rearranging, we get

$$[H_3O^+] = \sqrt{\frac{K_w c}{K_b}}$$

or $$pH = \frac{1}{2}pK_w^\circ - \frac{1}{2}pK_b^\circ - \frac{1}{2}\log(c/c^\circ)$$

an expression identical to Eq. (4.14.11).

Expression of Degree of Hydrolysis

The degree of hydrolysis can be computed as follows:

$$\underset{c(1-\alpha)}{B^+} + 2H_2O \rightleftharpoons \underset{c\alpha}{BOH} + \underset{c\alpha}{H_3O^+}$$

$$K_h = \frac{K_w}{K_b} = \frac{[BOH][H_3O^+]}{[B^+]} = \frac{(c\alpha)(c\alpha)}{c(1-\alpha)} \approx c\alpha^2$$

i.e. $$\alpha = \sqrt{\frac{K_w}{K_b c}} \tag{4.14.12}$$

We can conclude that the degree of hydrolysis increases with dilution.

Example 4.14.1

Calculate the hydrolytic constant of urea hydrochloride salt. Given: K_b(urea) = 1.5×10^{-14} M.

Solution

The hydrolysis reaction of the cation $NH_2CONH_3^+$ is

$$NH_2CONH_3^+ + H_2O \rightleftharpoons NH_2CONH_2 + H_3O^+$$

Thus, $$K_h = \frac{[NH_2CONH_2][H_3O^+]}{[NH_2CONH_3^+]}$$

or $$K_h = \frac{[NH_2CONH_2]}{[NH_2CONH_3^+][OH^-]}[H_3O^+][OH^-] = \frac{K_w}{K_b}$$

Substituting the values, we have

$$K_h = \frac{1.0 \times 10^{-14}\ \text{M}}{1.5 \times 10^{-14}\ \text{M}} = 0.666\ \text{M}$$

Example 4.14.2

Calculate the hydrolysis constant and its degree of hydrolysis in 10^{-2} M solution of NH_4Cl. What will be its pH value? Given: $K_b(NH_4OH)$ = 1.8×10^{-5} M.

Solution

If α is the degree of hydrolysis of NH_4^+, the concentrations of species involved in the hydrolysis reaction are

$$\underset{c(1-\alpha)}{NH_4^+} + 2H_2O \rightleftharpoons \underset{c\alpha}{NH_4OH} + \underset{c\alpha}{H_3O^+}$$

The $[H_3O^+]$ can be computed from the expression

$$K_h = \frac{[NH_4OH][H_3O^+]}{[NH_4^+]} = \frac{[H_3O^+]^2}{[NH_4^+]_0}$$

or $$[H_3O^+] = \sqrt{K_h[NH_4^+]_0}$$

Now $$K_h = \frac{K_w}{K_b} = \frac{(1.0 \times 10^{-14}\ M^2)}{(1.8 \times 10^{-5}\ M)} = 5.6 \times 10^{-10}\ M$$

Thus, $$[H_3O^+] = \sqrt{(5.6 \times 10^{-10}\ M)(10^{-2}\ M)} = 2.39 \times 10^{-6}\ M$$

and $$pH = -\log(2.39 \times 10^{-6}) = 5.63$$

Since $[H_3O^+] = c\alpha$, the degree of hydrolysis of NH_4^+ is given as

$$\alpha = \frac{[H_3O^+]}{c} = \frac{(2.39 \times 10^{-6}\ M)}{(10^{-2}\ M)} = 2.39 \times 10^{-4}$$

4.15 EXACT TREATMENT OF HYDROLYSIS OF SALT FORMED FROM A WEAK ACID AND A WEAK BASE

Both the cation and the anion of such a salt undergo hydrolysis, in general, to different extents. An aqueous solution of such a salt may be neutral, acidic or alkaline depending upon the relative strengths of the conjugate acid and base. Examples include salts such as ammonium acetate, ammonium cyanide, etc.

Equilibria Existing in Solution

Let c be the concentration of such a salt (say, B^+A^-) in the solution. The following equilibria will exist in the solution:

$$B^+ + 2H_2O \rightleftharpoons BOH + H_3O^+;\ K_{hc} = \frac{K_w}{K_b} = \frac{[BOH][H_3O^+]}{[B^+]} \tag{4.15.1}$$

$$A^- + H_2O \rightleftharpoons HA + OH^-, \quad K_{ha} = \frac{K_w}{K_a} = \frac{[HA][OH^-]}{[A^-]} \tag{4.15.2}$$

$$H_2O + H_2O \rightleftharpoons H_3O^+ + OH^-;\ K_w = [H_3O^+][OH^-] \tag{4.15.3}$$

Expression of Overall Hydrolysis

The overall hydrolysis is

$$B^+ + A^- + H_2O \rightleftharpoons BOH + HA$$

The hydrolysis constant of this reaction is given as

$$K_h = \frac{[BOH][HA]}{[B^+][A^-]} = \frac{[HA]}{[A^-][H_3O^+]}\frac{[BOH]}{[B^+][OH^-]}[H_3O^+][OH^-]$$

or $$K_h = \frac{K_w}{K_a K_b} \tag{4.15.4}$$

That is, the hydrolysis constant depends upon both K_a and K_b.

Condition of Mass Balance

The mass-balance condition gives

$$[B^+] = [B^+]_0 - [BOH] = c - [BOH] \tag{4.15.5}$$

$$[A^-] = [A^-]_0 - [HA] = c - [HA] \tag{4.15.6}$$

Condition of Charge Balance

The condition of electroneutrality gives

$$[B^+] + [H_3O^+] = [A^-] + [OH^-] \tag{4.15.7}$$

Derivation of Exact Expression

Substituting for [BOH] from Eq. (4.15.1) into Eq. (4.15.5), we get

$$[B^+] = c - \frac{K_w[B^+]}{K_b[H_3O^+]}$$

or $$[B^+] = \frac{c}{1 + \dfrac{K_w}{K_b[H_3O^+]}} \tag{4.15.8}$$

Similarly, substituting for [HA] from Eq. (4.15.2) into Eq. (4.15.6), we get

$$[A^-] = c - \frac{K_w[A^-]}{K_a[OH^-]} = c - \frac{[H_3O^+][A^-]}{K_a}$$

or $$[A^-] = \frac{c}{1 + \dfrac{[H_3O^+]}{K_a}} \tag{4.15.9}$$

Now substituting Eqs (4.15.8) and (4.15.9) in Eq. (4.15.7), we have

$$\frac{c}{1 + \dfrac{K_w}{K_b[H_3O^+]}} + [H_3O^+] = \frac{c}{1 + \dfrac{[H_3O^+]}{K_a}} + \frac{K_w}{[H_3O^+]} \tag{4.15.10}$$

Simplification of Exact Expression

Equation (4.15.10) is a fourth power in $[H_3O^+]$ and, in principle, can be solved for the given values of c, K_a and K_b. However, a simpler expression can be used under the approximation that H_3O^+ and OH^- concentrations of the resultant solution are negligible in comparison to the unhydrolyzed concentrations of B^+ and A^-, respectively. This approximation holds good if the amount of the salt dissolved is quite large and only a small fraction of the ions is hydrolyzed. With this, Eq. (4.15.7) reduces to

$$[B^+] = [A^-]$$

which according to Eqs (4.15.8) and (4.15.9) becomes

$$\frac{c}{1+\dfrac{K_w}{K_b[H_3O^+]}} = \frac{c}{1+\dfrac{[H_3O^+]}{K_a}}$$

or

$$1+\frac{[H_3O^+]}{K_a} = 1+\frac{K_w}{K_b[H_3O^+]}$$

or

$$[H_3O^+]^2 = \frac{K_wK_a}{K_b} \quad \text{or} \quad [H_3O^+] = \sqrt{\frac{K_wK_a}{K_b}} \tag{4.15.11}$$

or

$$\text{pH} = \frac{1}{2}(\text{p}K_w^\circ + \text{p}K_a^\circ - \text{p}K_b^\circ) \tag{4.15.12}$$

It can be seen that Eq. (4.15.11) is independent of concentration of the salt. We can draw the following conclusions regarding the nature of the solution at 25 °C:

(i) If $K_a = K_b$ then $[H_3O^+] = 10^{-7}$ M; neutral solution.
(ii) If $K_a > K_b$ then $[H_3O^+] > 10^{-7}$ M; acidic solution.
(iii) If $K_a < K_b$ then $[H_3O^+] < 10^{-7}$ M; alkaline solution.

Direct Approach to Compute pH

In the hydrolysis of the salt of a weak acid and a weak base such as ammonium acetate, anilinium acetate, ammonium cyanide, etc., both the ions are hydrolyzed. If we assume that $K_a \approx K_b$, then the hydrolysis of the cation and anion of the salt occur approximately to equal extent.

For a salt such as NH_4CN, $K_a < K_b$, it would be expected at a first glance that CN^- ions hydrolyze to a much greater extent than NH_4^+ ions. However, the hydrolysis of CN^- ions produces OH^- ions according to the equation

$$CN^- + H_2O \rightleftharpoons HCN + OH^-$$

which can react with NH_4^+ ions as

$$OH^- + NH_4^+ \rightleftharpoons NH_4OH$$

This latter reaction causes equilibrium in the former reaction to be displaced to the right, because OH^- ions are removed from the solution. Also the production of OH^- by the former reaction displaces the latter reaction to the right. Therefore, the hydrolysis of one ion drags the hydrolysis of the other ion along so that both hydrolysis are fairly extensive and not too far in extent from each other. So it is fairly safe to assume that [HCN] = [NH_4OH], even in the case of the salt where $K_a \neq K_b$.

Expression of Degree of Hydrolysis

Consider the following reaction:

$$\underset{c(1-\alpha)}{NH_4^+} + \underset{c(1-\alpha)}{CN^-} + H_2O \rightleftharpoons \underset{c\alpha}{NH_4OH} + \underset{c\alpha}{HCN}$$

Thus $$K_h = \frac{[NH_4OH][HCN]}{[NH_4^+][CN^-]} = \frac{(c\alpha)(c\alpha)}{c(1-\alpha)\,c(1-\alpha)} \qquad (4.15.13)$$

or $$\frac{\alpha^2}{(1-\alpha)^2} = K_h \quad \text{or} \quad \frac{\alpha}{1-\alpha} = \sqrt{K_h}$$

This gives $$\alpha = \frac{\sqrt{K_h}}{1+\sqrt{K_h}}$$

Since [HCN] = [NH_4OH], [NH_4^+] = [CN^-], and $K_h = K_w/K_aK_b$, Eq. (4.15.13) may be written as

$$\frac{[HCN]^2}{[CN^-]^2} = \frac{K_w}{K_aK_b} \quad \text{or} \quad \frac{[HCN]}{[CN^-]} = \sqrt{\frac{K_w}{K_aK_b}}$$

or $$\frac{[H_3O^+]}{K_a} = \sqrt{\frac{K_w}{K_aK_b}} \quad \text{or} \quad [H_3O^+] = \sqrt{\frac{K_wK_a}{K_b}}$$

or $$pH = \frac{1}{2}pK_w^\circ + \frac{1}{2}pK_a^\circ - \frac{1}{2}pK_b^\circ$$

Example 4.15.1 Calculate the value of K_h, α_h and pH of the following solutions at 25 °C.

(i) 0.1 M ammonium acetate $\quad K_a = K_b = 1.8 \times 10^{-5}$ M

(ii) 0.1 M anilinium acetate $\quad K_a = 1.8 \times 10^{-5}$ M; $K_b = 4.6 \times 10^{-10}$ M

(iii) 0.1 M ammonium carbonate $\quad K_{a1} = 4.5 \times 10^{-7}$ M; $K_{a2} = 4.7 \times 10^{-11}$ M

Solution (i) 0.1 M ammonium acetate (NH_4Ac). Hydrolysis reaction is

$$\underset{c(1-\alpha)}{NH_4^+} + \underset{c(1-\alpha)}{Ac^-} + H_2O \rightleftharpoons \underset{c\alpha}{NH_4OH} + \underset{c\alpha}{HAc}$$

$$K_h = \frac{[NH_4OH][HAc]}{[NH_4^+][Ac^-]} = \frac{K_w}{K_aK_b}$$

Substituting the concentrations of the species, we get

$$K_h = \frac{(c\alpha)(c\alpha)}{c(1-\alpha)\,c(1-\alpha)} = \frac{\alpha^2}{(1-\alpha)^2} = \frac{K_w}{K_aK_b}$$

or $$\frac{\alpha}{1-\alpha} = \sqrt{\frac{K_w}{K_a K_b}}$$

Substituting the values of K_w, K_a and K_b, we get

$$\frac{\alpha}{1-\alpha} = \sqrt{K_h} = \sqrt{\frac{(1.0 \times 10^{-14}\ \text{M}^2)}{(1.8 \times 10^{-5}\ \text{M})(1.8 \times 10^{-5}\ \text{M})}}$$
$$= 0.555\,5 \times 10^{-2}$$

Hence, $$\alpha = \frac{0.555\,5 \times 10^{-2}}{1.005\,5} = 0.553 \times 10^{-2}$$

and $$K_h \simeq \frac{\alpha^2}{(1-\alpha)^2} = (0.555\,5 \times 10^{-2})^2 = 0.308\,6 \times 10^{-4}$$

Since $[NH_4OH]$ = [HAc] and $[NH_4^+]$ = $[Ac^-]$, therefore

$$K_h = \frac{[\text{HAc}]^2}{[\text{Ac}^-]^2}$$

Making use of the expression $K_a = \dfrac{[H_3O^+][Ac^-]}{[HAc]}$, we get

$$K_h = \frac{[H_3O^+]^2}{K_a^2}$$

This on rearranging gives $[H_3O^+] = K_a\sqrt{K_h}$

Substituting the values of K_a and $\sqrt{K_h}$, we get

$$[H_3O^+] = (1.8 \times 10^{-5}\ \text{M})\ (0.555\,5 \times 10^{-2}) = 1.0 \times 10^{-7}\ \text{M}$$

Thus, $\text{pH} = -\log\{[H_3O^+]/\text{M}\} = -\log(10^{-7}) = 7$

(ii) 0.1 M anilinium acetate. Since

$$\frac{\alpha}{1-\alpha} = \sqrt{K_h} = \sqrt{\frac{K_w}{K_a K_b}} = \left(\frac{1.0 \times 10^{-14}\ \text{M}^2}{(1.8 \times 10^{-5}\ \text{M})(4.6 \times 10^{-10}\ \text{M})}\right)^{1/2}$$
$$= (1.208)^{1/2} = 1.099$$

Hence, $$\alpha = \frac{1.099}{2.099} = 0.52 \quad \text{and} \quad K_h = 1.208$$

and $$[H_3O^+] = K_a\sqrt{K_h} = (1.8 \times 10^{-5}\ \text{M})(1.099) = 1.98 \times 10^{-5}\ \text{M}$$

Thus, $\text{pH} = -\log\{[H_3O^+]/\text{M}\} = -\log(1.98 \times 10^{-5}) = 4.71$

(iii) 0.1 M $(NH_4)_2CO_3$. Since

$$\frac{\alpha}{1-\alpha} = \sqrt{K_h} = \sqrt{\frac{K_w}{K_{a2}\,K_b}} = \sqrt{\frac{1.0\times10^{-14}\ \text{M}}{(4.7\times10^{-11}\ \text{M})\,(1.8\times10^{-5}\ \text{M})}}$$
$$= \sqrt{11.82} = 3.438$$

Therefore, $\alpha = \dfrac{3.438}{4.438} = 0.775$ and $K_h = 11.82$

and $\quad [H_3O^+] = K_{a2}\sqrt{K_h} = (4.7\times10^{-11}\ \text{M})\,(3.438) = 16.16\times10^{-11}\ \text{M}$

Thus, $\quad \text{pH} = -\log\,(16.16\times10^{-11}) = 9.79$

Example 4.15.2

Using the appropriate acidity constants, calculate the equilibrium constant for the following reaction at 25 °C.

$$CN^- + HAc \rightleftharpoons HCN + Ac^-$$

where Ac^- stands for acetate anion.

Solution

For the given reaction, the equilibrium constant is given as

$$K = \frac{[HCN]\,[Ac^-]}{[CN^-]\,[HAc]}$$

Multiplying and dividing by $[H_3O^+]$, we get

$$K = \frac{[HCN]\,[Ac^-]\,[H_3O^+]}{[CN^-]\,[HAc]\,[H_3O^+]} = \frac{[H_3O^+]\,[Ac^-]}{[HAc]}\ \frac{[HCN]}{[H_3O^+]\,[CN^-]}$$
$$= \frac{K_a(HAc)}{K_a(HCN)}$$

Substituting the values of $K_a(HAc)$ and $K_a(HCN)$, we get

$$K = \frac{(1.8\times10^{-5}\ \text{M})}{(4.9\times10^{-10}\ \text{M})} = 3.674\times10^{4}$$

4.16 EXACT TREATMENT OF HYDROLYSIS OF SALT INVOLVING WEAK CONJUGATE CATION AND AN AMPHIPROTIC ANION

When a salt like $NaHCO_3$, NaHS, NaH_2PO_4, Na_2HPO_4, etc., is dissolved in water, the amphiprotic anion can either accept a proton from water or can donate a proton to water.

Equilibria Existing in Solution

Consider a solution of such a salt (say, NaHA) with concentration equal to c. The following equilibria will exist in the solution.

$$HA^- + H_2O \rightleftharpoons H_3O^+ + A^{2-}; \quad K_{a2} = \frac{[H_3O^+][A^{2-}]}{[HA^-]} \tag{4.16.1}$$

$$HA^- + H_2O \rightleftharpoons H_2A + OH^-; \quad K_h = \frac{[H_2A][OH^-]}{[HA^-]} \tag{4.16.2}$$

$$H_2O + H_2O \rightleftharpoons H_3O^+ + OH^-; \quad K_w = [H_3O^+][OH^-] \tag{4.16.3}$$

$$H_2A + H_2O \rightleftharpoons HA^- + H_3O^+; \quad K_{a1} = \frac{[HA^-][H_3O^+]}{[H_2A]} \tag{4.16.4}$$

Condition of Charge Balance

Electroneutrality condition gives

$$[Na^+] + [H_3O^+] = [HA^-] + 2[A^{2-}] + [OH^-] \tag{4.16.5}$$

Condition of Mass Balance

Mass-balance condition gives

$$[NaHA]_0 = [H_2A] + [HA^-] + [A^{2-}] \tag{4.16.6}$$

Derivation of Exact Expression

Subtracting Eq. (4.16.6) from Eq. (4.16.5), we have

$$[Na^+] + [H_3O^+] - [NaHA]_0 = [HA^-] + 2[A^{2-}] + [OH^-] - [H_2A] - [HA^-] - [A^{2-}]$$

or

$$[H_3O^+] = [A^{2-}] - [H_2A] + [OH^-] \tag{4.16.7}$$

as $[Na^+] = [NaHA]_0$, since the salt is completely ionized.

Substituting for $[A^{2-}]$, $[H_2A]$ and $[OH^-]$ from the equilibrium expressions (4.16.1), (4.16.4) and (4.16.3), respectively, we have

$$[H_3O^+] = \frac{K_{a2}[HA^-]}{[H_3O^+]} - \frac{[H_3O^+][HA^-]}{K_{a1}} + \frac{K_w}{[H_3O^+]}$$

or

$$[H_3O^+]^2 K_{a1} = K_{a1} K_{a2} [HA^-] - [HA^-] [H_3O^+]^2 + K_{a1} K_w$$

or

$$[H_3O^+]^2 \{K_{a1} + [HA^-]\} = K_{a1} K_{a2} [HA^-] + K_{a1} K_w$$

or

$$[H_3O^+] = \left\{ \frac{K_{a1}K_{a2}[HA^-] + K_{a1}K_w}{K_{a1} + [HA^-]} \right\}^{1/2} \tag{4.16.8}$$

Simplification of Exact Expression

A simpler expression can be used to compute $[H_3O^+]$ under the following approximations:

(i) The dissociation constant K_{a1} is negligible when compared to $[HA^-]$, i.e.

$$K_{a1} + [HA^-] \approx [HA^-]$$

(ii) $K_{a1}K_w$ is negligible in comparison to $K_{a1}K_{a2}$ $[HA^-]$, i.e.

$$K_{a1}K_{a2} [HA^-] + K_{a1}K_w \approx K_{a1}K_{a2} [HA^-]$$

With these, Eq (4.16.8) reduces to

$$[H_3O^+] = \left\{\frac{K_{a1}\,K_{a2}\,[HA^-]}{[HA^-]}\right\}^{1/2} = \sqrt{K_{a1}\,K_{a2}} \tag{4.16.9}$$

or

$$pH = \frac{1}{2}(pK_{a1}^\circ + pK_{a2}^\circ) \tag{4.16.10}$$

Direct Approach to Compute pH

The hydrogen-ion concentration of the solution as given by Eq. (4.16.1) is

$$[H_3O^+] = \frac{K_{a2}\,[HA^-]}{[A^{2-}]}$$

Substituting $[HA^-]$ from Eq. (4.16.2), we get

$$[H_3O^+] = K_{a2}\frac{[H_2A][OH^-]}{K_h}\cdot\frac{1}{[A^{2-}]}$$

Replacing $[OH^-]$ by $K_w/[H_3O^+]$ and $K_h = K_w/K_{a1}$, and rearranging, we get

$$[H_3O^+]^2 = K_{a1}\,K_{a2}\frac{[H_2A]}{[A^{2-}]}$$

Since both K_h and K_{a2} are usually very small, one may assume $[H_2A] \approx [A^{2-}]$ and hence the above expression modifies to

$$[H_3O^+]^2 = K_{a1}\,K_{a2}$$

or

$$pH = \frac{1}{2}(pK_{a1}^\circ + pK_{a2}^\circ)$$

Example 4.16.1

Calculate the pH of 0.1 M solution of (i) $NaHCO_3$, (ii) Na_2HPO_4 and (iii) NaH_2PO_4. Given that:

$CO_2 + H_2O \rightleftharpoons H^+ + HCO_3^-$; $K_1 = 4.2 \times 10^{-7}$ M

$HCO_3^- \rightleftharpoons H^+ + CO_3^{2-}$; $K_2 = 4.8 \times 10^{-11}$ M

$H_3PO_4 \rightleftharpoons H^+ + H_2PO_4^-$; $K_1 = 7.5 \times 10^{-3}$ M

$H_2PO_4^- \rightleftharpoons H^+ + HPO_4^{2-}$; $K_2 = 6.2 \times 10^{-8}$ M

$HPO_4^{2-} \rightleftharpoons H^+ + PO_4^{3-}$; $K_3 = 1.0 \times 10^{-12}$ M

Solution

In an aqueous solution of the salt containing an amphiprotic anion, pH of the solution is given by

$$pH = \frac{1}{2}[pK_1^\circ + pK_2^\circ]$$

Thus, we have

(i) pH of $NaHCO_3$ solution

$$pH = \frac{1}{2}[-\log(4.2\times10^{-7}) - \log(4.8\times10^{-11})] = 8.35$$

(ii) pH of Na_2HPO_4 solution

$$\mathrm{pH} = \frac{1}{2}[\mathrm{p}K_2^\circ + \mathrm{p}K_3^\circ] = \frac{1}{2}[-\log(6.2\times10^{-8}) - \log(1.0\times10^{-12})] = 9.60$$

(iii) pH of NaH_2PO_4 solution

$$\mathrm{pH} = \frac{1}{2}[\mathrm{p}K_1^\circ + \mathrm{p}K_2^\circ] = \frac{1}{2}[-\log(7.5\times10^{-3}) - \log(6.2\times10^{-8})] = 4.66$$

4.17 EXACT TREATMENT OF HYDROLYSIS OF SALT INVOLVING STRONG CONJUGATE CATION AND AMPHIPROTIC ANION

Equilibria Existing in Solution

Examples of such salts are $NH_4^+HCO_3^-$, $NH_4^+HS^-$, etc. Consider a solution of concentration c of such a salt (say B^+HA^-). The following equilibria will exist in the solution.

$$\underset{\text{(hydrolysis of cation)}}{B^+ + 2H_2O \rightleftharpoons BOH + H_3O^+}; \quad K_{hc} = \frac{K_w}{K_b} = \frac{[BOH][H_3O^+]}{[B^+]} \tag{4.17.1}$$

$$\underset{\text{(ionization of anion)}}{HA^- + H_2O \rightleftharpoons H_3O^+ + A^{2-}}; \quad K_{a2} = \frac{[H_3O^+][A^{2-}]}{[HA^-]} \tag{4.17.2}$$

$$\underset{\text{(hydrolysis of anion)}}{HA^- + H_2O \rightleftharpoons H_2A + OH^-}; \quad K_{ha} = \frac{K_w}{K_{a1}} = \frac{[H_2A][OH^-]}{[HA^-]} \tag{4.17.3}$$

$$H_2O + H_2O \rightleftharpoons H_3O^+ + OH^-; \quad K_w = [H_3O^+][OH^-] \tag{4.17.4}$$

Condition of Mass Balance

The mass-balance conditions are

$$[B^+]_0 = [B^+] + [BOH] = c \tag{4.17.5}$$

and

$$[HA^-]_0 = [HA^-] + [H_2A] + [A^{2-}] = c \tag{4.17.6}$$

Condition of Charge Balance

The condition of electroneutrality gives

$$[B^+] + [H_3O^+] = [HA^-] + 2[A^{2-}] + [OH^-] \tag{4.17.7}$$

Derivation of Exact Expression

Substituting for [BOH] from Eq. (4.17.1) into Eq. (4.17.5), we have

$$[B^+]_0 = c = [B^+] + \frac{K_w}{K_b}\frac{[B^+]}{[H_3O^+]} = [B^+]\left[1 + \frac{K_w}{K_b[H_3O^+]}\right]$$

or

$$[B^+] = \frac{c}{1 + \dfrac{K_w}{K_b[H_3O^+]}} \tag{4.17.8}$$

Similarly, substituting for $[H_2A]$ and $[A^{2-}]$, respectively, from Eq. (4.17.3) and Eq. (4.17.2) into Eq. (4.17.6), we get

$$[\text{HA}^-]_0 = c = [\text{HA}^-] + \frac{K_\text{w}}{K_\text{a1}}\frac{[\text{HA}^-]}{[\text{OH}^-]} + K_\text{a2}\frac{[\text{HA}^-]}{[\text{H}_3\text{O}^+]}$$

or

$$[\text{HA}^-] = \frac{c}{1 + \dfrac{K_\text{w}}{K_\text{a1}[\text{OH}^-]} + \dfrac{K_\text{a2}}{[\text{H}_3\text{O}^+]}} = \frac{c}{1 + \dfrac{[\text{H}_3\text{O}^+]}{K_\text{a1}} + \dfrac{K_\text{a2}}{[\text{H}_3\text{O}^+]}} \tag{4.17.9}$$

The concentration of A^{2-} in terms of c can be written using Eqs (4.17.2) and (4.17.9), Thus, we have

$$[\text{A}^{2-}] = K_\text{a2}\frac{[\text{HA}^-]}{[\text{H}_3\text{O}^+]} = \frac{K_\text{a2}}{[\text{H}_3\text{O}^+]}\left[\frac{c}{1 + \dfrac{[\text{H}_3\text{O}^+]}{K_\text{a1}} + \dfrac{K_\text{a2}}{[\text{H}_3\text{O}^+]}}\right] \tag{4.17.10}$$

Substituting Eqs (4.17.8), (4.17.9) and (4.17.10) in Eq. (4.17.7), we get

$$\frac{c}{1 + \dfrac{K_\text{w}}{K_\text{b}[\text{H}_3\text{O}^+]}} + [\text{H}_3\text{O}^+] = \frac{c}{1 + \dfrac{[\text{H}_3\text{O}^+]}{K_\text{a1}} + \dfrac{K_\text{a2}}{[\text{H}_3\text{O}^+]}}$$

$$+ 2\frac{K_\text{a2}}{[\text{H}_3\text{O}^+]}\left[\frac{c}{1 + \dfrac{[\text{H}_3\text{O}^+]}{K_\text{a1}} + \dfrac{K_\text{a2}}{[\text{H}_3\text{O}^+]}}\right] + \frac{K_\text{w}}{[\text{H}_3\text{O}^+]} \tag{4.17.11}$$

Simplification of Exact Expression

Equation (4.17.11) can be solved for $[H_3O^+]$ for the given values of c, K_{a1}, K_{a2}, and K_b. However, a simpler expression can be used under the approximation that H_3O^+ and OH^- concentrations of the resultant solution are negligible in comparison to the concentrations of unhydrolyzed B^+ and HA^-, respectively. This approximation holds good if the amount of the salt dissolved is quite large and only a small fraction of the ions are hydrolyzed. With this, Eq. (4.17.7) reduces to

$$[\text{B}^+] = [\text{HA}^-]$$

Substituting for $[B^+]$ and $[HA^-]$ from Eqs (4.17.8) and (4.17.9), respectively, we have

$$\frac{c}{1 + \dfrac{K_\text{w}}{K_\text{b}[\text{H}_3\text{O}^+]}} = \frac{c}{1 + \dfrac{[\text{H}_3\text{O}^+]}{K_\text{a1}} + \dfrac{K_\text{a2}}{[\text{H}_3\text{O}^+]}}$$

or

$$1 + \frac{[\text{H}_3\text{O}^+]}{K_\text{a1}} + \frac{K_\text{a2}}{[\text{H}_3\text{O}^+]} = 1 + \frac{K_\text{w}}{K_\text{b}[\text{H}_3\text{O}^+]}$$

or $$[H_3O^+]^2 + K_{a1}\,K_{a2} = \frac{K_w\,K_{a1}}{K_b}$$

or $$[H_3O^+] = \sqrt{\frac{K_{a1}\,K_w}{K_b} - K_{a1}\,K_{a2}} \qquad (4.17.12)$$

Direct Approach to Compute pH

The hydrogen-ion concentration as given by Eq. (4.17.1) is

$$[H_3O^+] = \frac{K_w}{K_b}\frac{[B^+]}{[BOH]}$$

If we assume that the concentrations of H_3O^+ and OH^- of the solution are negligible in comparison to the concentrations of unhydrolyzed B^+ and HA^-, respectively, we may write

$$[B^+] \approx [HA^-]$$

and $$[BOH] = [A^{2-}] + [H_2A]$$

Substituting the above equations in the previous expression, we get

$$[H_3O^+] = \frac{K_w}{K_b}\frac{[HA^-]}{[A^{2-}]+[H_2A]}$$

Taking the inverse, we get

$$\frac{1}{[H_3O^+]} = \frac{K_b}{K_w}\frac{[A^{2-}]}{[HA^-]} + \frac{K_b}{K_w}\frac{[H_2A]}{[HA^-]}$$

Making use of the expressions of K_{a1} and K_{a2}, we get

$$\frac{1}{[H_3O^+]} = \frac{K_b}{K_w}\frac{K_{a2}}{[H_3O^+]} + \frac{K_b}{K_w}\frac{[H_3O^+]}{K_{a1}}$$

or $$[H_3O^+]^2 = \frac{K_{a1}\,K_w}{K_b}\left[1 - \frac{K_b\,K_{a2}}{K_w}\right]$$

or $$[H_3O^+] = \sqrt{\frac{K_{a1}\,K_w}{K_b} - K_{a1}\,K_{a2}}$$

Example 4.17.1 Calculate pH of 0.1 M (NH_4^+) (HCO_3^-), given $K_b(NH_4OH) = 1.75 \times 10^{-5}$ M, K_{a1} (H_2CO_3) = 4.2×10^{-7} M and $K_{a2}(HCO_3^-) = 4.8 \times 10^{-11}$ M.

Solution $[H_3O^+]$ of an aqueous solution containing such a salt is given by

$$[H_3O^+] = \sqrt{K_{a1}\left(\frac{K_w}{K_b} - K_{a2}\right)}$$

Substituting the values of K_{a1}, K_{a2}, K_w and K_b, we get

$$[H_3O^+] = \left[(4.2 \times 10^{-7}\ M)\left(\frac{1.0 \times 10^{-14}\ M^2}{1.75 \times 10^{-5}\ M} - 4.8 \times 10^{-11}\ M\right)\right]^{1/2}$$

$$= 1.48 \times 10^{-8}\ M$$

Therefore,

$$pH = -\log\{[H_3O^+]/M]\} = 7.83$$

4.18 HYDROLYSIS OF A SALT CONTAINING MULTIVALENT CATION OR ANION

Hydrolysis of multivalent cations and anions takes place in a stepwise manner and more than one hydrolytic products are formed. This can be illustrated through the following examples:

Hydrolysis of Fe^{2+} The hydrolysis reactions of Fe^{2+} are

$$Fe^{2+} + 2H_2O \rightleftharpoons Fe(OH)^+ + H_3O^+$$

and

$$Fe(OH)^+ + 2H_2O \rightleftharpoons Fe(OH)_2(aq) + H_3O^+$$

The hydrolysis constants K_{h1} and K_{h2} are

$$K_{h1} = \frac{[Fe(OH)^+][H_3O^+]}{[Fe^{2+}]} = \frac{[Fe(OH)^+]}{[Fe^{2+}][OH^-]}[H_3O^+][OH^-] = \frac{K_w}{K_{b2}} \tag{4.18.1}$$

where K_{b2} is the equilibrium constant of the reaction

$$Fe(OH)^+ \rightleftharpoons Fe^{2+} + OH^-$$

and

$$K_{h2} = \frac{[Fe(OH)_2\ (aq)][H_3O^+]}{[Fe(OH)^+]} = \frac{[Fe(OH)_2\ (aq)]}{[Fe(OH)^+][OH\ ]}[H_3O^+][OH^-]$$

$$= \frac{K_w}{K_{b1}} \tag{4.18.2}$$

where K_{bl} is the equilibrium constant of the reaction

$$Fe(OH)_2\ (aq) \rightleftharpoons Fe(OH)^+ + OH^-$$

The values of K_{bl} and K_{b2} are 1.0×10^{-4} M and 2.5×10^{-6} M, respectively. Therefore, the values of hydrolysis constants are

$$K_{h1} = \frac{K_w}{K_{b2}} = \frac{1.0 \times 10^{-14}\ M^2}{2.5 \times 10^{-6}\ M} = 4.0 \times 10^{-9}\ M$$

$$K_{h2} = \frac{K_w}{K_{b1}} = \frac{1.0 \times 10^{-14}\ M^2}{1.0 \times 10^{-4}\ M} = 1.0 \times 10^{-10}\ M$$

Thus $K_{hl} >> K_{h2}$. Hence the hydrolysis of $Fe(OH)^+$ in comparison to Fe^{2+} can be neglected.

Hydrolysis of S^{2-}

The hydrolysis reactions of S^{2-} are

$$S^{2-} + H_2O \rightleftharpoons HS^- + OH^-$$

$$HS^- + H_2O \rightleftharpoons H_2S + OH^-$$

The hydrolysis constants K_{h1} and K_{h2} are

$$K_{h1} = \frac{[HS^-][OH^-]}{[S^{2-}]} = \frac{[HS^-]}{[S^{2-}][H_3O^+]}[OH^-][H_3O^+] = \frac{K_w}{K_{a2}} \qquad (4.18.3)$$

where K_{a2} is the second dissociation constant of H_2S, i.e.

$$HS^- + H_2O \rightleftharpoons S^{2-} + H_3O^+$$

and $$K_{h2} = \frac{[H_2S][OH^-]}{[HS^-]} = \frac{[H_2S]}{[HS^-][H_3O^+]}[OH^-][H_3O^+] = \frac{K_w}{K_{a1}} \qquad (4.18.4)$$

where K_{a1} is the first ionization constant of acid H_2S, i.e.

$$H_2S + H_2O \rightleftharpoons HS^- + H_3O^+$$

The values of K_{a1} and K_{a2} are 1.1×10^{-7} M and 1.0×10^{-14} M, respectively. Therefore, the values of K_{h1} and K_{h2} are

$$K_{h1} = \frac{K_w}{K_{a2}} = \frac{1.0 \times 10^{-14}\ M^2}{1.0 \times 10^{-14}\ M} = 1.0\ M$$

and $$K_{h2} = \frac{K_w}{K_{a1}} = \frac{1.0 \times 10^{-14}\ M^2}{1.1 \times 10^{-7}\ M} = 9.1 \times 10^{-8}\ M$$

Thus $K_{h1} >> K_{h2}$, i.e. the hydrolysis of HS^- in comparison to S^{2-} can be neglected.

Exact Treatment of Hydrolysis of Fe^{2+}

We proceed as follows.

Material-balance condition gives

$$[Fe^{2+}]_0 = c = [Fe^{2+}] + [Fe(OH)^+] + [Fe(OH)_2(aq)] \qquad (4.18.5)$$

Charge-balance condition gives

$$2\,[Fe^{2+}] + [Fe(OH)^+] + [H_3O^+] = [OH^-] + 2[X^{2-}] \qquad (4.18.6)$$

where X^{2-} is the anion associated with Fe^{2+}.

Following the procedure given earlier, we can express $[Fe^{2+}]$ and $[Fe(OH)^+]$ in terms of $[Fe^{2+}]_0$ (= c). The expressions are

$$[Fe^{2+}] = \frac{c}{1 + \dfrac{K_{h1}}{[H_3O^+]} + \dfrac{K_{h1}\,K_{h2}}{[H_3O^+]^2}} \qquad (4.18.7)$$

$$[\mathrm{Fe(OH)^+}] = \frac{K_{h1}}{[\mathrm{H_3O^+}]}\left[\frac{c}{1+\dfrac{K_{h1}}{[\mathrm{H_3O^+}]}+\dfrac{K_{h1}\,K_{h2}}{[\mathrm{H_3O^+}]^2}}\right] \tag{4.18.8}$$

These on substituting in Eq. (4.18.6) gives

$$2\left[\frac{c}{1+\dfrac{K_{h1}}{[\mathrm{H_3O^+}]}+\dfrac{K_{h1}\,K_{h2}}{[\mathrm{H_3O^+}]^2}}\right]+\frac{K_{h1}}{[\mathrm{H_3O^+}]}\left[\frac{c}{1+\dfrac{K_{h1}}{[\mathrm{H_3O^+}]}+\dfrac{K_{h1}\,K_{h2}}{[\mathrm{H_3O^+}]^2}}\right]+[\mathrm{H_3O^+}] = \frac{K_w}{[\mathrm{H_3O^+}]}+2c \tag{4.18.9}$$

Simpler Expressions

Equation (4.18.9) can be solved for $[\mathrm{H_3O^+}]$. However, it is convenient to use a simpler expression without any loss of accuracy under the following approximations:

(i) The contribution of H_3O^+ due to the dissociation of water is negligible in comparison to the contribution of H_3O^+ due to the hydrolysis of Fe^{2+}.

(ii) The contribution of H_3O^+ due to the hydrolysis of Fe^{2+} is fairly larger than the contribution of H_3O^+ due to the hydrolysis of $Fe(OH)^+$. This is justified since the hydrolysis constant of Fe^{2+} is itself small and, therefore, its hydrolysis will proceed to a limited extent and the concentration of $Fe(OH)^+$ will be very low. The subsequent hydrolysis of $Fe(OH)^+$ (which occurs to a still lesser extent, since $K_{h2} << K_{hl}$) is, therefore, negligible and can thus be ignored. With this approximation we can write

$$[\mathrm{Fe(OH)^+}] = [\mathrm{H_3O^+}] \tag{4.18.10}$$

and Eq. (4.18.7) as

$$[\mathrm{Fe^{2+}}] = \frac{c}{1+K_{h1}/[\mathrm{H_3O^+}]} \tag{4.18.11}$$

With the above two approximations, Eq. (4.18.5) can be written as

$$[\mathrm{Fe^{2+}}] + [\mathrm{H_3O^+}] = c$$

Substituting for $[\mathrm{Fe^{2+}}]$ from Eq. (4.18.11), we have

$$\frac{c}{1+K_{h1}/[\mathrm{H_3O^+}]} + [\mathrm{H_3O^+}] = c$$

or $$c + [\mathrm{H_3O^+}] + K_{h1} = c + \frac{c\,K_{h1}}{[\mathrm{H_3O^+}]}$$

Ignoring K_{hl} in comparison with $[\mathrm{H_3O^+}]$, we get

$$[\mathrm{H_3O^+}] = c\frac{K_{h1}}{[\mathrm{H_3O^+}]} \quad \text{or} \quad [\mathrm{H_3O^+}] = \sqrt{c\,K_{h1}} \tag{4.18.12}$$

Direct Approach to Compute $[H_3O^+]$

Equation (4.18.12) can be derived directly from the first hydrolysis expression

$$Fe^{2+} + 2H_2O \rightleftharpoons Fe(OH)^+ + H_3O^+$$

If the concentration x of Fe^{2+} undergoes hydrolysis, then we have

$$[Fe^{2+}] = c - x$$

$$[Fe(OH)^+] = [H_3O^+] = x$$

and hence

$$K_{h1} = \frac{[Fe(OH)^+][H_3O^+]}{[Fe^{2+}]} = \frac{(x)(x)}{c - x}$$

Assuming the solution to be fairly concentrated and the concentration x to be negligible in comparison to c, we can write

$$c - x \simeq c$$

Thus, $$K_{h1} = \frac{x^2}{c}$$

or $$x = [H_3O^+] = \sqrt{K_{h1}\, c}$$

The above procedure of computing $[H_3O^+]$ from the first hydrolysis expression alone is applicable to any multivalent cation. For example, the $[H_3O^+]$ of a solution containing Fe^{3+} can be computed considering only the following hydrolysis expression

$$Fe^{3+} + 2H_2O \rightleftharpoons Fe(OH)^{2+} + H_3O^+$$

Mathematical treatment of hydrolysis of S^{2-} can be carried out following the procedure given above for Fe^{2+}. In fact, the entire procedure can be rewritten by replacing Fe^{2+}, $Fe(OH)^+$, $Fe(OH)_2(aq)$ and H_3O^+ by S^{2-}, HS^-, H_2S and OH^-, respectively.

Example 4.18.1

Calculate at 25 °C the pH of a

(a) 0.10 M solution of $Fe(NO_3)_2$, given that

$$Fe(OH)_2 \rightleftharpoons Fe(OH)^+ + OH^-; \qquad K_1 = 1.0 \times 10^{-4}\ M$$

$$Fe(OH)^+ \rightleftharpoons Fe^{2+} + OH^-; \qquad K_2 = 2.5 \times 10^{-6}\ M$$

(b) 0.10 M solution of Na_2S, given that

$$H_2S \rightleftharpoons H^+ + HS^-; \qquad K_1 = 1.1 \times 10^{-7}\ \text{M}$$

$$HS^- \rightleftharpoons H^+ + S^{2-}; \qquad K_2 = 1.0 \times 10^{-14}\ \text{M}$$

Solution **(a)** 0.10 M solution of $Fe(NO_3)_2$. We have

$$\underset{0.1\,\text{M} - x}{Fe^{2+}} + 2H_2O \rightleftharpoons \underset{x}{Fe(OH)^+} + \underset{x}{H_3O^+}$$

Thus, $$K_h = \frac{[Fe(OH)^+][H_3O^+]}{[Fe^{2+}]} = \frac{x^2}{0.1\,\text{M} - x}$$

Also $$K_h = \frac{K_w}{K_2} = \frac{(1.0 \times 10^{-14}\ \text{M}^2)}{(2.5 \times 10^{-6}\ \text{M})} = 4.0 \times 10^{-9}\ \text{M}$$

Hence, $$\frac{x^2}{0.1\,\text{M} - x} = 4.0 \times 10^{-9}\ \text{M}$$

Ignoring x in comparison to 0.1 M, we have

$$x = \sqrt{(0.1\,\text{M})\,(4.0 \times 10^{-9}\ \text{M})} = 2 \times 10^{-5}\ \text{M}$$

Hence, $\text{pH} = -\log\{[H_3O^+]/\text{M}\} = -\log(2 \times 10^{-5}) = 4.70$

(b) 0.1 M solution of Na_2S, we have

$$\underset{0.1\,\text{M} - x}{S^{2-}} + H_2O \rightleftharpoons \underset{x}{HS^-} + \underset{x}{OH^-}$$

Thus, $$K_h = \frac{K_w}{K_2} = \frac{(1.0 \times 10^{-14}\ \text{M}^2)}{(1.0 \times 10^{-14}\ \text{M})} = 1.0\ \text{M}$$

$$= \frac{x^2}{0.1\,\text{M} - x}$$

or $$x^2 + x(1.0\ \text{M}) - 0.1\ \text{M}^2 = 0$$

Solving for x, we get

$$x = [OH^-] = 9.16 \times 10^{-2}\ \text{M}$$

Hence, $$[H_3O^+] = \frac{K_w}{[OH^-]} = \frac{(1.0 \times 10^{-14}\ \text{M}^2)}{(9.16 \times 10^{-2}\ \text{M})} = 1.1 \times 10^{-13}\ \text{M}$$

$$\text{pH} = -\log\{[H_3O^+]/\text{M}\} = -\log(1.1 \times 10^{-13}) = 12.96$$

Example 4.18.2

Potassium alum is $KA1(SO_4)_2 \cdot 12H_2O$. As a strong electrolyte, it is considered to be 100% dissociated into K^+, Al^{3+}, and SO_4^{2-}. The solution is acidic because of the hydrolysis of Al^{3+}, but not so acidic as might be expected, because the SO_4^{2-} can sponge up some of the H_3O^+ by forming HSO_4^-. Given a solution made by dissolving 11.4 g of $KA1(SO_4)_2 \cdot 12H_2O$ in enough water to make 0.10 dm^3 of solution, calculate its $[H_3O^+]$:

(a) Considering the hydrolysis

$$Al^{3+} + 2H_2O \rightleftharpoons Al(OH)^{2+} + H_3O^+ \quad \text{with} \quad K_h = 1.4 \times 10^{-5}\ \text{M}$$

(b) Allowing also for the equilibrium

$$HSO_4^- + H_2O \rightleftharpoons H_3O^+ + SO_4^{2-} \quad \text{with} \quad K_2 = 1.26 \times 10^{-2}\ \text{M}$$

Solution

(a) Amount of alum $= \dfrac{11.4\ \text{g}}{474.38\ \text{g mol}^{-1}} = 0.024\ \text{mol}$

Molarity of the prepared solution $= \dfrac{0.024\ \text{mol}}{0.1\ \text{dm}^3} = 0.24\ \text{M}$

Hydrolysis of $A1^{3+}$ is

$$Al^{3+} + 2H_2O \rightleftharpoons Al(OH)^{2+} + H_3O^+$$

$$K_h = \frac{[Al(OH)^{2+}][H_3O^+]}{[Al^{3+}]}$$

If x is the concentration of $A1^{3+}$ that has hydrolyzed, we have

$$K_h = \frac{(x)(x)}{0.24\ \text{M} - x} = 1.4 \times 10^{-5}\ \text{M}$$

Solving for x, we get

$$[H_3O^+] = x = 1.82 \times 10^{-3}\ \text{M}$$

(b) We will have to consider the following equilibria.

$$Al^{3+} + 2H_2O \rightleftharpoons Al(OH)^{2+} + H_3O^+$$

$$H_3O^+ + SO_4^{2-} \rightleftharpoons HSO_4^- + H_2O$$

Let z be the concentration of SO_4^{2-} that combines with H_3O^+ and y be the net concentration of H_3O^+ that is present in the solution. Since the concentration z of SO_4^{2-} combines with the concentration z of H_3O^+, it is obvious that the net concentration of H_3O^+ produced in the hydrolysis reaction of $A1^{3+}$ is $(y + z)$. Thus, the concentration $(y + z)$ of $A1^{3+}$ out of 0.24 M hydrolyzes in the solution. With these, the concentrations of various species in the solution are

$$\underset{0.24\,\text{M}-y-z}{\text{Al}^{3+}} + 2\text{H}_2\text{O} \rightleftharpoons \underset{y+z}{\text{Al(OH)}^{2+}} + \underset{y}{\text{H}_3\text{O}^+}$$

$$\underset{y}{\text{H}_3\text{O}^+} + \underset{0.48\,\text{M}-z}{\text{SO}_4^{2-}} \rightleftharpoons \underset{z}{\text{HSO}_4^-} + \text{H}_2\text{O}$$

Thus, $$K_h = \frac{(y+z)(y)}{(0.24\,\text{M} - y - z)} = 1.4 \times 10^{-5}\,\text{M} \qquad \text{(i)}$$

$$K_2 = \frac{z}{y(0.48\,\text{M} - z)} = \frac{1}{1.26 \times 10^{-2}\,\text{M}} \qquad \text{(ii)}$$

From Eq. (ii), we get

$$z = \frac{(0.48\,\text{M})\,y}{(1.26 \times 10^{-2}\,\text{M}) + y}$$

Substituting this in Eq. (i), we get

$$\frac{\left(y + \dfrac{(0.48\,\text{M})\,y}{(1.26 \times 10^{-2}\,\text{M}) + y}\right)y}{\left(0.24\,\text{M} - y - \dfrac{(0.48\,\text{M})\,y}{(1.26 \times 10^{-2}\,\text{M}) + y}\right)} = 1.4 \times 10^{-5}$$

Making an assumption that $y \ll 1.26 \times 10^{-2}$ M, and then solving for y, we get

$$[\text{H}_3\text{O}^+] = y = 2.932 \times 10^{-4}\,\text{M}$$

4.19 EXACT TREATMENT OF A MIXTURE OF A WEAK ACID (HA) AND SALT OF ITS CONJUGATE BASE (MA)

The exact expression for the computation of $[\text{H}_3\text{O}^+]$ present in an aqueous solution of a mixture of weak acid and the salt of its conjugate base can be derived as follows:

Equilibria Existing in Solution

Various equilibria are

(i) $\text{HA} + \text{H}_2\text{O} \rightleftharpoons \text{H}_3\text{O}^+ + \text{A}^-$; $\quad K_a = \dfrac{[\text{H}_3\text{O}^+][\text{A}^-]}{[\text{HA}]}$ (4.19.1)

(ii) $\text{MA} \rightarrow \text{M}^+ + \text{A}^-$; $\quad K_d$ = Very large value, almost completely dissociated (4.19.2)

(iii) $\text{H}_2\text{O} + \text{H}_2\text{O} \rightleftharpoons \text{H}_3\text{O}^+ + \text{OH}^-$; $\quad K_w = [\text{H}_3\text{O}^+]\,[\text{OH}^-]$ (4.19.3)

Condition of Mass Balance

The material-balance condition gives

$$[HA]_0 + [MA]_0 = [HA] + [A^-] \tag{4.19.4}$$

where $[HA]_0$ and $[MA]_0$ are the respective concentrations of acid and salt in the solution.

Condition of Charge Balance

The condition of electroneutrality gives

$$[M^+] + [H_3O^+] = [A^-] + [OH^-] \tag{4.19.5}$$

Derivation of Exact Expression

Making use of Eq. (4.19.3) and on rearranging the above expression, we get

$$[A^-] = [M^+] + [H_3O^+] - \frac{K_w}{[H_3O^+]} \tag{4.19.6}$$

Substituting this concentration of A^- in Eq. (4.19.4) and rearranging, we have

$$[HA] = [HA]_0 + [MA]_0 - [M^+] - [H_3O^+] + \frac{K_w}{[H_3O^+]} \tag{4.19.7}$$

Since MA is almost completely dissociated,

$$[MA]_0 = [M^+]$$

Hence, Eq. (4.19.7) reduces to

$$[HA] = [HA]_0 - [H_3O^+] + \frac{K_w}{[H_3O^+]} \tag{4.19.8}$$

Substituting [HA] and $[A^-]$ in Eq. (4.19.1), we get

$$K_a = \frac{[H_3O^+]\left\{[M^+] + [H_3O^+] - \dfrac{K_w}{[H_3O^+]}\right\}}{\left\{[HA]_0 - [H_3O^+] + \dfrac{K_w}{[H_3O^+]}\right\}} \tag{4.19.9}$$

Simplification of Exact Expression

Equation (4.19.9) is a cubical expression in $[H_3O^+]$. A simpler expression can be used to compute the $[H_3O^+]$ under the following approximations.

(i) If $[H_3O^+]$ of the solution > 10^{-6} M, then $K_W/[H_3O^+] \le 10^{-8}$ M. With this

$$[M^+] + [H_3O^+] - \frac{K_w}{[H_3O^+]} \approx [M^+] + [H_3O^+]$$

and $$[HA]_0 - [H_3O^+] + \frac{K_w}{[H_3O^+]} \approx [HA]_0 - [H_3O^+]$$

(ii) If the solution is moderately concentrated then $[MA]_0$ (or $[M^+]$) and $[HA]_0$ are much larger than $[H_3O^+]$. With this

$$[M^+] + [H_3O^+] \approx [M^+] \quad \text{and} \quad [HA]_0 - [H_3O^+] \approx [HA]_0$$

Thus, Eq. (4.19.9) reduces to

$$K_a = \frac{[H_3O^+][M^+]}{[HA]_0} \tag{4.19.10}$$

The above two approximations are equivalent to the following two statements:

(i) The $[H_3O^+]$ due to the dissociation of water is negligible when compared to the $[H_3O^+]$ due to the dissociation of the acid.

(ii) The amount of acid HA present in the ionized form is negligible in comparison to the total concentration of HA. The latter assumption is justified because of the presence of the common ion A^- from the salt MA.

Direct Approach to Compute pH

Equation (4.19.10) can be derived from the dissociation expression of the acid alone:

$$HA + H_2O \rightleftharpoons H_3O^+ + A^-$$

Since $[A^-]$ due to the above reaction will be small, therefore, the concentration of A^- in solution will primarily be due to salt. Thus

$[A^-]$ = concentration of the salt

Writing the expression of dissociation constant, we have

$$K_a = \frac{[H_3O^+][A^-]}{[HA]} = \frac{[H_3O^+][M^+]}{[HA]_0}$$

Writing this expression in terms of pH, we have

$$pH = pK_a^\circ + \log \frac{[M^+]}{[HA]_0}$$

or

$$pH = pK_a^\circ + \log \frac{[\text{salt}]}{[\text{acid}]} \tag{4.19.11}$$

The expression (4.19.11) is known as *Henderson–Hasselbalch equation* and can be used to compute pH of a solution containing large amounts of a weak acid and a salt of its conjugate base.

Example 4.19.1

A solution at 25 °C is 0.01 M in acetic acid and 0.025 M in sodium acetate. Determine for this solution (a) the degree of ionization of the acid and (b) pOH and pH. Given: $K_a = 1.8 \times 10^{-5}$ M.

Solution

Using the Henderson's equation, we have

$$pH = pK_a^\circ + \log \frac{[\text{salt}]}{[\text{acid}]} = -\log (K_a/\text{M}) + \log \frac{[\text{salt}]}{[\text{acid}]}$$

$$= -\log(1.8 \times 10^{-5}) + \log\left(\frac{0.025\text{ M}}{0.01\text{ M}}\right) = 4.745 + 0.398 \simeq 5.14$$

Thus, $[H_3O^+] = \text{antilog}(-5.14) = \text{antilog}(\bar{6}.86)$

i.e. $[H_3O^+] = 7.24 \times 10^{-6}$ M

$\text{pOH} = 14 - \text{pH} = 14 - 5.14 = 8.86$

Now $[H_3O^+] = c\alpha$

Hence, $$\alpha = \frac{[H_3O^+]}{c} = \frac{7.24 \times 10^{-6}\text{ M}}{0.01\text{ M}} = 7.24 \times 10^{-4}$$

4.20 EXACT TREATMENT OF A MIXTURE OF A WEAK BASE (BOH) AND SALT OF ITS CONJUGATE ACID (BA)

The exact expression for the computation of $[OH^-]$ (or $[H_3O^+]$) present in an aqueous solution of a mixture of a weak base and the salt of its conjugate acid can be derived as follows:

Equilibria Existing in Solution

Various equilibria are

$$\text{BOH} \rightleftharpoons \text{B}^+ + \text{OH}^-; \qquad K_b = \frac{[\text{B}^+][\text{OH}^-]}{[\text{BOH}]} \tag{4.20.1}$$

$$\text{BA} \rightarrow \text{B}^+ + \text{A}^-; \qquad K_d = \text{very large value} \tag{4.20.2}$$

$$H_2O + H_2O \rightleftharpoons H_3O^+ + OH^-; \quad K_w = [H_3O^+][OH^-] \tag{4.20.3}$$

Condition of Mass Balance

Material-balance condition gives

$$[\text{BOH}]_0 + [\text{BA}]_0 = [\text{BOH}] + [\text{B}^+] \tag{4.20.4}$$

where $[\text{BOH}]_0$ and $[\text{BA}]_0$ are the concentrations of the base and the salt in the solution, respectively.

Condition of Charge Balance

Condition of electroneutrality gives

$$[\text{B}^+] + [H_3O^+] = [\text{A}^-] + [OH^-] \tag{4.20.5}$$

Derivation of Exact Expression

Making use of Eq. (4.20.3) and on rearranging Eq. (4.20.5), we get

$$[\text{B}^+] = [\text{A}^-] + [OH^-] - \frac{K_w}{[OH^-]} \tag{4.20.6}$$

Substituting this concentration of B^+ in Eq. (4.20.4), we get

$$[\text{BOH}] = [\text{BOH}]_0 + [\text{BA}]_0 - [\text{A}^-] - [OH^-] + \frac{K_w}{[OH^-]} \tag{4.20.7}$$

Since BA is almost completely dissociated, therefore

$$[\text{BA}]_0 = [\text{A}^-]$$

With this, Eq. (4.20.7) reduces to

$$[\mathrm{BOH}] = [\mathrm{BOH}]_0 - [\mathrm{OH^-}] + \frac{K_\mathrm{w}}{[\mathrm{OH^-}]} \tag{4.20.8}$$

Substituting $[B^+]$ and [BOH] in Eq. (4.20.1), we get

$$K_\mathrm{b} = \frac{\left\{[\mathrm{A^-}] + [\mathrm{OH^-}] - \dfrac{K_\mathrm{w}}{[\mathrm{OH^-}]}\right\}[\mathrm{OH^-}]}{\left\{[\mathrm{BOH}]_0 - [\mathrm{OH^-}] + \dfrac{K_\mathrm{w}}{[\mathrm{OH^-}]}\right\}} \tag{4.20.9}$$

Simplification of Exact Expression

Equation (4.20.9) is a cubical expression in $[OH^-]$. A simpler expression can be used to compute the $[OH^-]$ under the following approximations:

(i) If for the solution $[\mathrm{OH^-}] \geq 10^{-6}$ M, then $K_\mathrm{w}/[\mathrm{OH^-}] \leq 10^{-8}$ M. With this

$$[\mathrm{A^-}] + [\mathrm{OH^-}] - \frac{K_\mathrm{w}}{[\mathrm{OH^-}]} \approx [\mathrm{A^-}] + [\mathrm{OH^-}]$$

and $$[\mathrm{BOH}]_0 - [\mathrm{OH^-}] + \frac{K_\mathrm{w}}{[\mathrm{OH^-}]} \approx [\mathrm{BOH}]_0 - [\mathrm{OH^-}]$$

(ii) If the solution is fairly concentrated, then $[A^-]$ and $[\mathrm{BOH}]_0$ are larger than $[OH^-]$. With this

$$[\mathrm{A^-}] + [\mathrm{OH^-}] \approx [\mathrm{A^-}]_0 \quad \text{and} \quad [\mathrm{BOH}]_0 - [\mathrm{OH^-}] \approx [\mathrm{BOH}]_0$$

Thus, Eq. (4.20.9) reduces to

$$K_\mathrm{b} = \frac{[\mathrm{A^-}]_0\,[\mathrm{OH^-}]}{[\mathrm{BOH}]_0} = \frac{[\mathrm{B^+}]_0\,[\mathrm{OH^-}]}{[\mathrm{BOH}]_0} \tag{4.20.10}$$

The above two approximations are equivalent to the following two statements.

(i) The $[OH^-]$ due to the dissociation of water is negligible in comparison to the $[OH^-]$ due to the dissociation of the base.

(ii) The amount of the base BOH present in the ionized form is negligible in comparison to the total concentration of BOH. The latter is very much justified because of the common ion B^+ from the salt BA.

Direct Approach to Compute pH

Equation (4.20.10) can be derived from the dissociation expression of the base alone:

$$\mathrm{BOH} \rightleftharpoons \mathrm{B^+} + \mathrm{OH^-}$$

Since B^+ due to the above reaction will be small, therefore, the concentration of B^+ in solution will primarily be due to the salt. Thus,

$[B^+]$ = concentration of the salt

Writing the expression of dissociation constant, we have

$$K_\mathrm{b} = \frac{[\mathrm{B^+}][\mathrm{OH^-}]}{[\mathrm{BOH}]} = \frac{[\mathrm{B^+}]_0\,[\mathrm{OH^-}]}{[\mathrm{BOH}]_0}$$

Writing this expression in terms of pOH, we have

$$pOH = pK_b^o + \log \frac{[B^+]_0}{[BOH]_0}$$

or
$$pOH = pK_b^o + \log \frac{[salt]}{[base]}$$

Hence,
$$pH = 14 - pOH = 14 - \left(pK_b + \log \frac{[salt]}{[base]} \right) \qquad (4.20.11)$$

Equation (4.20.11) is Henderson–Hasselbalch equation as applicable to a solution containing a fairly large concentration of a weak base and a salt of its conjugate acid.

4.21 EXACT TREATMENT OF A MIXTURE OF A STRONG ACID AND A WEAK ACID

The exact expression for the computation of $[H_3O^+]$ of a solution containing a weak acid HA and a strong acid HB can be derived as follows:

Equilibria Existing in Solution

The various equilibria are

$$HA + H_2O \rightleftharpoons H_3O^+ + A^-; \qquad K_a = \frac{[H_3O^+][A^-]}{[HA]} \qquad (4.21.1)$$

$$HB + H_2O \rightarrow H_3O^+ + B^-; \qquad \text{completely dissociated}$$

$$H_2O + H_2O \rightleftharpoons H_3O^+ + OH^-; \qquad K_w = [H_3O^+][OH^-] \qquad (4.21.2)$$

Condition of Mass Balance

Material-balance expressions are

$$[HA]_0 = [HA] + [A^-] = c_1 \qquad (4.21.3)$$

$$[HB]_0 = [B^-] = c_2 \qquad (4.21.4)$$

where c_1 and c_2 are the concentrations of HA and HB, respectively.

Condition of Charge Balance

Charge-balance expression is

$$[H_3O^+] = [A^-] + [B^-] + [OH^-] \qquad (4.21.5)$$

Derivation of Exact Expression

From Eq. (4.21.3), we get

$$[A^-] = c_1 - [HA]$$

which on using Eq. (4.21.1) becomes

$$[A^-] = c_1 - \frac{[H_3O^+][A^-]}{K_a} \quad \text{or} \quad [A^-] = \frac{c_1}{1 + [H_3O^+]/K_a}$$

Substitution of this in Eq. (4.21.5) gives

$$[H_3O^+] = \frac{c_1}{1 + [H_3O^+]/K_a} + c_2 + \frac{K_w}{[H_3O^+]} \qquad (4.21.6)$$

Simplification of Exact Expression

Equation (4.21.6) can be solved for $[H_3O^+]$. However, a simpler expression can be used if we assume:

(i) The contribution of H_3O^+ from the dissociation of water is negligible. In this case, we will get

$$[H_3O^+] = \frac{c_1}{1 + [H_3O^+]/K_a} + c_2$$

i.e. $$[H_3O^+]^2 + (K_a - c_2)[H_3O^+] - K_a(c_1 + c_2) = 0$$

or $$[H_3O^+] = \frac{1}{2}[(c_2 - K_a) + \{(K_a - c_2)^2 + 4K_a(c_1 + c_2)\}^{1/2}]$$

or $$[H_3O^+] = \frac{1}{2}[(c_2 - K_a) + \{(K_a + c_2)^2 + 4K_a c_1)\}^{1/2}] \qquad (4.21.7)$$

(ii) The value of K_a is negligible in comparison to the concentration c_2. With this, Eq. (4.21.7) reduces to

$$[H_3O^+] = \frac{c_2 + \sqrt{c_2^2 + 4K_a c_1}}{2} \qquad (4.21.8)$$

Direct Method to Compute $[H_3O^+]$

Let the concentration x out of the concentration c_1 of weak acid dissociate in the solution. If c_2 is the concentration of the monoprotic strong acid, then the concentrations of various species in the solution are

$$\underset{c_1 - x}{HA} + H_2O \rightleftharpoons \underset{x + c_2}{H_3O^+} + \underset{x}{A^-}$$

Substituting these in the equilibrium expression, we get

$$K_a = \frac{[H_3O^+][A^-]}{[HA]} = \frac{(x + c_2)(x)}{c_1 - x}$$

or $$x^2 + x(c_2 + K_a) - K_a c_1 = 0$$

Solving for x, we get

$$x = \frac{1}{2}[-(c_2 + K_a) + \{(c_2 + K_a)^2 + 4K_a c_1\}^{1/2}\}$$

Since $[H_3O^+] = x + c_2$, we have

$$[H_3O^+] = \frac{1}{2}[(c_2 - K_a) + \{(c_2 + K_a)^2 + 4K_a c_1\}^{1/2}\}$$

This expression of $[H_3O^+]$ is identical with that of Eq. (4.21.7).

A simpler expression can be used if K_a is assumed to be negligible in comparison to c_2. Thus, we have

$$[H_3O^+] = \frac{c_2 + \sqrt{c_2^2 + 4K_a\, c_1}}{2}$$

Example 4.21.1 Calculate the pH of a solution containing 1.0 M acetic acid (HAc) and 0.01 M HCl, given that $K_a(\text{HAc}) = 1.8 \times 10^{-5}$ M.

Solution Making use of the expression

$$[\text{H}_3\text{O}^+] = \frac{c_2 + \sqrt{c_2^2 + 4K_a\, c_1}}{2}; \qquad \text{(where } c_2 = [\text{HCl}],\ c_1 = [\text{HAc}])$$

we get
$$[\text{H}_3\text{O}^+]/\text{M} = \frac{10^{-2} + \sqrt{10^{-4} + (4 \times 1.8 \times 10^{-5} \times 1.0)}}{2}$$

$$= \frac{10^{-2} + \sqrt{1.72 \times 10^{-4}}}{2} = \frac{10^{-2} + 1.31 \times 10^{-2}}{2}$$

i.e. $[\text{H}_3\text{O}^+] = 1.16 \times 10^{-2}$ M

Hence, $\text{pH} = -\log\{[\text{H}_3\text{O}^+]/\text{M}\}$

$= -\log(1.16 \times 10^{-2}) = 1.94$

4.22 EXACT TREATMENT OF A MIXTURE OF TWO WEAK ACIDS

The exact expression for the computation of $[\text{H}_3\text{O}^+]$ in a solution containing two weak acids HA and HB can be derived as follows.

Equilibria Existing in Solution The various equilibria are

$$\text{HA} + \text{H}_2\text{O} \rightleftharpoons \text{H}_3\text{O}^+ + \text{A}^-; \qquad K_1 = \frac{[\text{H}_3\text{O}^+][\text{A}^-]}{[\text{HA}]} \qquad (4.22.1)$$

$$\text{HB} + \text{H}_2\text{O} \rightleftharpoons \text{H}_3\text{O}^+ + \text{B}^-; \qquad K_2 = \frac{[\text{H}_3\text{O}^+][\text{B}^-]}{[\text{HB}]} \qquad (4.22.2)$$

$$\text{H}_2\text{O} + \text{H}_2\text{O} \rightleftharpoons \text{H}_3\text{O}^+ + \text{OH}^-; \qquad K_w = [\text{H}_3\text{O}^+][\text{OH}^-] \qquad (4.22.3)$$

Condition of Mass Balance The material-balance expressions are

$$[\text{HA}]_0 = [\text{HA}] + [\text{A}^-] = c_1 \qquad (4.22.4)$$

$$[\text{HB}]_0 = [\text{HB}] + [\text{B}^-] = c_2 \qquad (4.22.5)$$

where c_1 and c_2 are the concentrations of HA and HB, respectively.

Condition of Charge Balance The condition of electroneutrality gives

$$[\text{H}_3\text{O}^+] = [\text{A}^-] + [\text{B}^-] + [\text{OH}^-] \qquad (4.22.6)$$

Derivation of Exact Expression From Eq. (4.22.4), we get

$$[\text{A}^-] = c_1 - [\text{HA}]$$

which on using Eq. (4.22.1) becomes

$$[\text{A}^-] = c_1 - \frac{[\text{H}_3\text{O}^+][\text{A}^-]}{K_1}$$

or $$[\mathrm{A}^-] = \frac{c_1}{1 + [\mathrm{H_3O^+}]/K_1} \tag{4.22.7}$$

Similarly from Eq. (4.22.5), we get

$$[\mathrm{B}^-] = \frac{c_2}{1 + [\mathrm{H_3O^+}]/K_2} \tag{4.22.8}$$

Substituting Eqs (4.22.7) and (4.22.8) in Eq. (4.22.6), we get

$$[\mathrm{H_3O^+}] = \frac{c_1}{1 + [\mathrm{H_3O^+}]/K_1} + \frac{c_2}{1 + [\mathrm{H_3O^+}]/K_2} + \frac{K_\mathrm{w}}{[\mathrm{H_3O^+}]}$$

or $$[\mathrm{H_3O^+}] = \frac{K_1 c_1}{K_1 + [\mathrm{H_3O^+}]} + \frac{K_2 c_2}{K_2 + [\mathrm{H_3O^+}]} + \frac{K_\mathrm{w}}{[\mathrm{H_3O^+}]} \tag{4.22.9}$$

Simplification of Exact Expression

Equation (4.22.9) can be solved for $[\mathrm{H_3O^+}]$. However, a simpler expression can be used under the following approximations:

(i) The contribution of $\mathrm{H_3O^+}$ from the dissociation of water is negligible in comparison to those available from the dissociation of acids.

(ii) The values of K_1 and K_2 are negligible in comparison to $[\mathrm{H_3O^+}]$. Thus, we have

$$[\mathrm{H_3O^+}] = \frac{K_1 c_1}{[\mathrm{H_3O^+}]} + \frac{K_2 c_2}{[\mathrm{H_3O^+}]}$$

or $$[\mathrm{H_3O^+}] = \sqrt{K_1 c_1 + K_2 c_2} \tag{4.22.10}$$

The concentrations of A^- and B^- in the solution can be calculated using Eqs (4.22.7) and (4.22.8). Thus, we have

$$[\mathrm{A}^-] = \frac{c_1}{1 + [\mathrm{H_3O^+}]/K_1} = \frac{K_1 c_1}{K_1 + [\mathrm{H_3O^+}]} \approx \frac{K_1 c_1}{[\mathrm{H_3O^+}]} \tag{4.22.11}$$

$$[\mathrm{B}^-] = \frac{c_2}{1 + [\mathrm{H_3O^+}]/K_2} = \frac{K_2 c_2}{K_2 + [\mathrm{H_3O^+}]} \approx \frac{K_2 c_2}{[\mathrm{H_3O^+}]} \tag{4.22.12}$$

An Alternative Approach

Let x be the concentration of HA and y be the concentration of HB that dissociate, then the concentrations of various species in the solution will be

$[\mathrm{HA}] = c_1 - x$, $[\mathrm{HB}] = c_2 - y$

$[\mathrm{A}^-] = x$, $[\mathrm{B}^-] = y$, $[\mathrm{H_3O^+}] = x + y$

Substituting these in the equilibrium expressions, we get

$$K_1 = \frac{[\mathrm{H_3O^+}][\mathrm{A}^-]}{[\mathrm{HA}]} = \frac{(x+y)(x)}{c_1 - x} \tag{4.22.13}$$

$$K_2 = \frac{[\mathrm{H_3O^+}][\mathrm{B}^-]}{[\mathrm{HB}]} = \frac{(x+y)\,y}{c_2 - y} \tag{4.22.14}$$

If we assume that x and y are negligible in comparison to c_1 and c_2, then the above expressions simplify to

$$K_1 = \frac{(x+y)(x)}{c_1} \quad \text{and} \quad K_2 = \frac{(x+y)(y)}{c_2} \tag{4.22.15}$$

or
$$x = \frac{K_1 c_1}{(x+y)} \quad \text{and} \quad y = \frac{K_2 c_2}{(x+y)} \tag{4.22.16}$$

Adding the two expressions of Eq. (4.22.16), we get

$$(x+y) = \frac{K_1 c_1}{x+y} + \frac{K_2 c_2}{x+y} = \frac{K_1 c_1 + K_2 c_2}{x+y}$$

or
$$(x+y)^2 = K_1 c_1 + K_2 c_2$$

or
$$(x+y) = \sqrt{K_1 c_1 + K_2 c_2} \tag{4.22.17}$$

Substituting Eq. (4.22.17) in Eqs (4.22.15) and (4.22.16), we get

$$x = [A^-] = \frac{K_1 c_1}{\sqrt{K_1 c_1 + K_2 c_2}} \tag{4.22.18}$$

$$y = [B^-] = \frac{K_2 c_2}{\sqrt{K_1 c_1 + K_2 c_2}} \tag{4.22.19}$$

Thus, the total concentration of H_3O^+ in the solution is given by

$$[H_3O^+] = x + y = \sqrt{K_1 c_1 + K_2 c_2}$$

The values of x and y can be improved further by using the actual values of denominators in the expressions given by Eq. (4.22.15), (i.e. $c_1 - x$ and $c_2 - y$, where x and y are the values that are obtained above), instead of c_1 and c_2, respectively. The numerators are kept as such and then from these two equations, x and y are solved following the procedure adopted above. This process is repeated till self-consistent values of x and y are obtained.

Example 4.22.1

A solution is prepared by mixing one mole of HA with one mole of HB, diluting to a total volume of 1 dm^3 with water. Both HA and HB are weak acids which dissociate according to the following reversible equations.

$$HA + H_2O \rightleftharpoons H_3O^+ + A^-; \qquad K_1 = 1.0 \times 10^{-6}\ M$$

$$HB + H_2O \rightleftharpoons H_3O^+ + B^-; \qquad K_2 = 1.0 \times 10^{-6}\ M$$

(i) Calculate the equilibrium concentrations of H_3O^+, A^- and B^-.
(ii) How does the presence of HB affect the dissociation of HA?

Solution

(i) Making use of the expressions

$$[H_3O^+] = \sqrt{K_1c_1 + K_2c_2}$$

$$[A^-] = \frac{K_1c_1}{[H_3O^+]}$$

and

$$[B^-] = \frac{K_2c_2}{[H_3O^+]}$$

we get $[H_3O^+]/M = \sqrt{(1.0 \times 10^{-6})(1) + (1.0 \times 10^{-6})(1)} = \sqrt{2.0 \times 10^{-6}}$

i.e. $[H_3O^+] = 1.414 \times 10^{-3}$ M

$$[A^-] = \frac{(1.0 \times 10^{-6}\ M)(1\ M)}{(1.414 \times 10^{-3}\ M)} = 0.707 \times 10^{-3}\ M$$

$$[B^-] = \frac{(1.0 \times 10^{-6}\ M)(1\ M)}{(1.414 \times 10^{-3}\ M)} = 0.707 \times 10^{-3}\ M$$

(ii) The concentration of H_3O^+ present in a solution containing only HA would be

$$K_1 = \frac{[A^-][H_3O^+]}{[HA]} \approx \frac{[H_3O^+]^2}{[HA]_0}$$

$$[H_3O^+] = \sqrt{K_1\,[HA]_0} = \sqrt{1.0 \times 10^{-6}\ M^2} = 1.0 \times 10^{-3}\ M$$

The concentration of H_3O^+ due to the dissociation of HA in the solution containing both the weak acids will be equal to $[A^-]$, i.e.

$$[H_3O^+] = 0.707 \times 10^{-3}\ M$$

Since this value is smaller than 1.0×10^{-3} M, it may be concluded that the presence of HB suppresses the dissociation of HA.

Example 4.22.2

The dissociation constants of HF and HNO_2 are 6.71×10^{-4} M and 4.5×10^{-4} M, respectively. Calculate the pH and the ratio of $[F^-]$ to $[NO_2^-]$ in a solution that is simultaneously 0.5 M in HF and 0.5 M in HNO_2.

Solution

Making use of the expression

$$[H_3O^+] = \sqrt{K_1c_1 + K_2c_2}$$

we get $[H_3O^+]/M = \sqrt{6.71 \times 10^{-4} \times 0.5 + 4.5 \times 10^{-4} \times 0.5} = \sqrt{5.605 \times 10^{-4}}$

i.e. $[H_3O^+] = 2.367\,5 \times 10^{-2}$ M

Hence, $pH = -\log(2.367\,5 \times 10^{-2}) = 1.63$

The concentrations of F^- and NO_2^- in the solution are

$$[F^-] = \frac{K_1c_1}{[H_3O^+]} = \frac{(6.71 \times 10^{-4}\ M)(0.5\ M)}{(2.367\,5 \times 10^{-2}\ M)}$$

$$= 1.417 \times 10^{-2}\ M$$

$$[NO_2^-] = \frac{K_2 c_2}{[H_3O^+]} = \frac{(4.5 \times 10^{-4}\ M)(0.5\ M)}{(2.367\,5 \times 10^{-2}\ M)}$$

$$= 9.50 \times 10^{-3}\ M$$

Thus, their ratio is

$$\frac{[F^-]}{[NO_2^-]} = \frac{(1.417 \times 10^{-2}\ M)}{(9.5 \times 10^{-3}\ M)} = 1.49$$

Alternative Method The two equilibria are

$$HF + H_2O \rightleftharpoons H_3O^+ + F^-; \qquad K_1 = \frac{[H_3O^+][F^-]}{[HF]} = 6.71 \times 10^{-4}\ M$$

and $$HNO_2 + H_2O \rightleftharpoons H_3O^+ + NO_2^-; \quad K_2 = \frac{[H_3O^+][NO_2^-]}{[HNO_2]} = 4.5 \times 10^{-4}\ M$$

Let the concentration x of HF and the concentration y of HNO_2 dissociate, then the concentrations of various species will be

$[HF] = 0.5\ M - x, \qquad [F^-] = x$

$[HNO_2] = 0.5\ M - y, \qquad [NO_2^-] = y, \qquad [H_3O^+] = x + y$

Thus, we have

$$K_1 = \frac{(x + y)(x)}{(0.5\ M - x)} = 6.71 \times 10^{-4}\ M$$

and $$K_2 = \frac{(x + y)(y)}{(0.5\ M - y)} = 4.5 \times 10^{-4}\ M \qquad (1)$$

We can start with the assumption that x and y are much less than 0.5 M; thus we have

$$\frac{(x + y)(x)}{0.5\ M} = 6.71 \times 10^{-4}\ M$$

and $$\frac{(x + y)(y)}{0.5\ M} = 4.5 \times 10^{-4}\ M \qquad (2)$$

or $\quad x(x + y) = 3.355 \times 10^{-4}\ M^2$ and $y(x + y) = 2.25 \times 10^{-4}\ M^2$

Dividing the two, we have

$$\frac{x}{y} = \frac{3.355}{2.25} = 1.489 \text{ or } x = (1.489)\,y$$

Substituting $x = 1.489\,y$ in $x(x + y) = 3.355 \times 10^{-4}\ M^2$, we get

$$(1.489\,y)(1.489\,y + y) = 3.355 \times 10^{-4}\ M^2$$

or $$y^2 = \frac{3.355 \times 10^{-4}\ M^2}{1.489 \times 2.489} = 0.000\,090\,36\ M^2$$

or $$y = 0.009\,506\ M$$

and hence,

$$x = (1.489)\ (0.009\ 506\ \text{M}) = 0.014\ 15\ \text{M}$$

Thus, $[H_3O^+] = 0.014\ 15\ \text{M} + 0.009\ 506\ \text{M} = 0.023\ 66\ \text{M}$

$$\text{pH} = -\log\{[H_3O^+]/\text{M}\} = 1.63$$

4.23 BUFFER SOLUTIONS

Introduction

Aqueous solutions of a salt of a strong acid and a strong base, e.g., sodium chloride, and salt of weak acid and a weak base, e.g., ammonium acetate, both have a pH of 7. But the addition of 1 cm^3 of 0.1 M hydrochloric acid to 1 dm^3 of the solution alters the pH to 4 in the case of the former, while it hardly affects the pH of the latter. The addition of an equivalent quantity of sodium hydroxide would likewise change the pH of the sodium chloride solution from 7 to 10, but it would not appreciably alter that of ammonium acetate solution. The solution of ammonium acetate thus has the property of resisting change in pH when an acid or an alkali is added to it and this property of solution is known as *buffer action.*

Definition

In general, a buffer solution is one which is resistant to change in pH upon the addition of a *small amount* of acid or alkali. Such solutions usually consist of a mixture of a weak acid and salt of its conjugate base, e.g., HAc and NaAc (called *acidic buffers*), or of a weak base and salt of its conjugate acid, e.g., NH_4OH and NH_4Cl (called *basic buffers*). A salt of a weak acid and a weak base, e.g. NH_4Ac, also has a buffer action, as indicated above.

Mechanism of Buffer Action

The buffer action of a solution of a weak acid HA and its highly ionized salt could be explained as follows:

$$\left.\begin{aligned} H_3O^+ + A^- &\rightleftharpoons H_2O + HA \\ \text{and} \quad OH^- + HA &\longrightarrow H_2O + A^- \end{aligned}\right\} \qquad (4.23.1)$$

that is, the addition of acid causes the former reaction in which equivalent amount of A^- on combination with added H_3O^+ gives rise to the undissociated weak acid HA and the addition of base causes the latter reaction in which an equivalent amount of the undissociated acid is replaced by the salt. Thus, the solution does not contain excess of H_3O^+ or OH^- and, therefore, its pH value remains the same.

The buffer action of a weak base and its salt can be explained similarly through the following reactions:

$$\left.\begin{aligned} H_3O^+ + BOH &\longrightarrow B^+ + 2H_2O \\ OH^- + B^+ &\rightleftharpoons BOH \end{aligned}\right\} \qquad (4.23.2)$$

Buffer solutions usually contain relatively larger amounts of a weak acid and salt of its conjugate base or a weak base and salt of its conjugate acid.

pH of a Buffer Solution

The pH of a buffer solution is given by Henderson–Hasselbalch equation;

Acidic buffer

$$\text{pH} = \text{p}K_a^\circ + \log \frac{[\text{salt}]}{[\text{acid}]} \qquad (4.23.3)$$

Basic buffer $$\text{pOH} = 14 - \text{pH} = \text{p}K_\text{b}^\circ + \log \frac{[\text{salt}]}{[\text{base}]} \tag{4.23.4}$$

It can be seen that the pH of a buffer solution depends only on the ratio of concentrations of salt and acid (or salt and base), and not on the individual concentrations. On dilution, the ratio of concentrations of salt and acid (or salt and base) will still be the same and thus pH will remain unchanged.

The addition of H^+ or OH^- changes the ratio of [salt]/[acid] (or [salt]/[base]), which causes a change in its pH value. The change in ratio (and hence pH) will be larger for a buffer solution containing relatively lesser concentrations than a more concentrated buffer solution, i.e. the lesser concentrated buffer solutions will have lower capacity to withstand change in pH on addition of H^+ or OH^- ions.

Expression of Buffer Index

Buffer capacity of a solution is defined in terms of buffer index which is the change in the concentration of buffer acid (or base) required for changing its pH value by one, keeping $(c_\text{acid} + c_\text{salt})$ or $(c_\text{base} + c_\text{salt})$ constant. Thus mathematically, it is given as

$$-\partial c_\text{HA}/\partial(\text{pH}) \quad \text{or} \quad \partial c_\text{BOH}/\partial(\text{pH})$$

Let we have an acid buffer containing the amount a of acid and the amount s of the salt. Its pH is given by

$$(\text{pH})_1 = \text{p}K_\text{a}^\circ + \log\left(\frac{s}{a}\right)$$

To this buffer solution, let a little amount Δb of a strong base is added. Since the addition of base converts weak acid into equivalent amount of salt, we will have

$$(\text{pH})_2 = \text{p}K_\text{a}^\circ + \log\left(\frac{s + \Delta b}{a - \Delta b}\right)$$

The change in pH is given by

$$\Delta\text{pH} = (\text{pH})_2 - (\text{pH})_1 = \log\left(\frac{s + \Delta b}{a - \Delta b}\right) - \log\left(\frac{s}{a}\right)$$

$$= \log\left[\frac{s + \Delta b}{a - \Delta b} \cdot \frac{a}{s}\right] = \log\left(\frac{1 + \Delta b/s}{1 - \Delta b/a}\right) = \frac{1}{2.303} \ln\left(\frac{1 + \Delta b/s}{1 - \Delta b/a}\right)$$

Since $\Delta b/s << 1$ and $\Delta b/a << 1$,[†] we write

$$\Delta\text{pH} = \frac{1}{2.303}\left(\frac{\Delta b}{s} + \frac{\Delta b}{a}\right) = \frac{\Delta b}{2.303}\left(\frac{a + s}{a\,s}\right)$$

For infinitesimal value of Δb, we can write

$$\frac{\partial b}{\partial \text{pH}} = 2.303\left(\frac{a\,s}{a + s}\right) \tag{4.23.5}$$

[†] $\ln(1 + x) \simeq x$ for $x << 1$.

Alternative Method

Alternatively, Eq. (4.23.5) may be derived by differentiating Eq. (4.23.3) with respect to [salt]. If the solution contains the total amount of x of acid and salt and if it contains the amount s of the salt, we can write Eq. (4.23.3) as

$$\mathrm{pH} = \mathrm{p}K_\mathrm{a}^\circ + \log\left(\frac{s}{x-s}\right) = \mathrm{p}K_\mathrm{a}^\circ + \frac{1}{2.303}\ln\left(\frac{s}{x-s}\right)$$

Differentiating with respect to s, we get

$$\frac{\partial \mathrm{pH}}{\partial s} = \frac{1}{2.303}\left[\frac{1}{s} + \frac{1}{x-s}\right] = \frac{1}{2.303}\frac{x}{s(x-s)}$$

or
$$\frac{\partial s}{\partial \mathrm{pH}} = 2.303\frac{s(x-s)}{x} = 2.303\left(\frac{sa}{s+a}\right) \qquad \text{(Eq. 4.23.5)}$$

where a is the amount of acid in the solution.

Table 4.23.1 gives the values of $\partial b/\partial(\mathrm{pH})$ for an acid buffer solution containing x = 1.0 M at different values of s.

Table 4.23.1 Buffer Capacity of a Buffer Solution containing Acid and Salt

$\frac{[salt]}{\mathrm{M}}$	$\frac{[salt]}{[acid]}$	$\frac{1}{2.303}\left(\frac{\partial s}{\partial(\mathrm{pH})}\right)$
0.2	0.25	0.16
0.3	0.43	0.21
0.4	0.67	0.24
0.5	1.00	0.25
0.6	1.25	0.24
0.7	2.33	0.21
0.8	4.00	0.16

The same results are plotted in Fig. 4.22.1. It can be seen from Fig. 4.22.1 that the buffer index has a maximum value when the ratio of [salt]/[acid] is one† and it falls off as the ratio of [salt] to [acid] changes in either direction.

Buffer Range

Although it is difficult to give an exact limit up to which a buffer can be used, it is generally accepted that a solution has useful buffer capacity provided that the value of [salt]/[acid] lies within the range of 10 to 0.1, i.e. ten parts of salt to one

† The same results could be obtained by differentiating expression of Eq. (4.23.5) with respect to s and setting the resultant expression to zero, i.e.

$$\frac{\partial}{\partial s}\left\{\frac{\partial s}{\partial \mathrm{pH}}\right\} = \frac{2.303}{x}(x-2s) = 0$$

Solving this we have $s = x/2$.

Thus,
$$\frac{[\mathrm{salt}]}{[\mathrm{acid}]} = \frac{s}{x-s} = \frac{x/2}{x-x/2} = 1$$

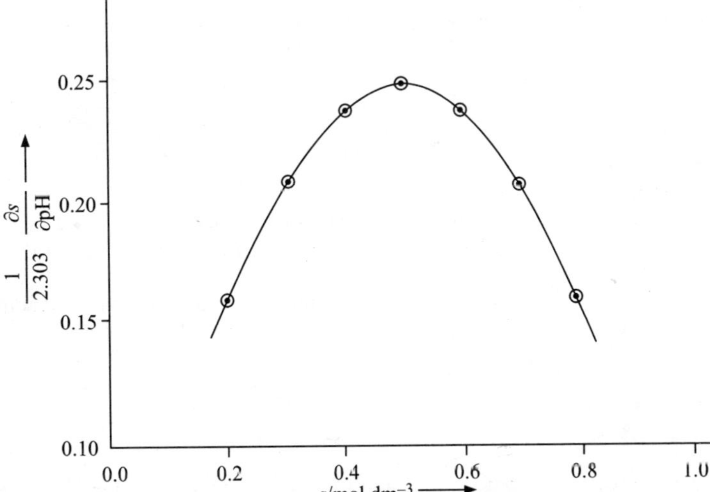

Fig. 4.23.1 Plot of (1/2.303) ($\partial s/\partial$pH) versus salt concentration s

of acid at one extreme, to one part of salt to ten of acid, at the other extreme. It follows, therefore, from Eq. (4.23.3) that a particular weak acid can be employed for making useful buffer solution of pH lying within the range of (pK_a° – 1) to (pK_a° + 1). Acetic acid, for example, has pK_a° of 4.75 at 25 °C; hence mixtures of acetic acid and sodium acetate can be used for preparing buffer solutions whose pHs are roughly in the range of 3.75 to 5.75. Outside this range the buffer capacity of acetic acid-sodium acetate is too small to be of any practical application.

Preparation of Buffer Solution

To make a buffer solution of a given pH, first it is necessary to choose a weak acid (or base) with its pK° value as near as possible to the required pH. Actual ratio of acid to salt (or base to salt) can then be determined by using Henderson's equation. In practice, the solutions employed are usually about 0.1 M to 0.2 M. Table 4.23.2 records some of the recommended mixtures together with their effective pH ranges.

Table 4.23.2 Some of the Recommended Buffers and their pH Ranges

Constituents	pH-*range*
Glycine and glycine hydrochloride	1.0–3.7
Phthalic acid and potassium acid phthalate	2.2–3.8
Acetic acid and sodium acetate	3.7–5.6
Disodium citrate and trisodium citrate	5.0–6.3
Monosodium phosphate and disodium phosphate	5.8–8.0
Boric acid and borax	6.8–9.2
Borax and sodium hydroxide	9.2–11.0
Disodium phosphate and trisodium phosphate	11.0–12.0

Applications of Buffer Solutions

Buffers are an important part of many industrial processes, e.g. electroplating, leather, photographic material and dyeing. In bacteriological research, it is very essential to maintain pH required for the growth of certain bacteria. Human blood is buffered to a pH of 7.3–7.4 by means of HCO_3^-, PO_4^{3-} and other nitrogenous bases.

Problem 4.23.1

Show that the exact expression of buffer index for a buffer containing a weak acid and its sodium salt is given by

$$\left(\frac{\partial[\mathrm{Na^+}]}{\partial \mathrm{pH}}\right)_{[\mathrm{A^-}]} = 2.303\left(\frac{[\mathrm{H_3O^+}][\mathrm{A^-}]K_a}{\{[\mathrm{H_3O^+}]+K_a\}^2} + \frac{K_w}{[\mathrm{H_3O^+}]} + \mathrm{H_3O^+}]\right)$$

Also show that this exact expression is reduced to Eq. (4.23.5) if the solution contains appreciable amounts of acids and its sodium salt.

Solution

In a solution containing a weak acid and its sodium salt, the equilibria to be considered are

$$\mathrm{HA + H_2O \rightleftharpoons H_3O^+ + A^-} \qquad K_a = \frac{[\mathrm{H_3O^+}][\mathrm{A^-}]}{[\mathrm{HA}]} \tag{4.23.6}$$

$$\mathrm{H_2O + H_2O \rightleftharpoons H_3O^+ + OH^-} \qquad K_w = [\mathrm{H_3O^+}][\mathrm{OH^-}] \tag{4.23.7}$$

The mass-balance expression is

$$[\mathrm{A^-}]_{\mathrm{total}} = [\mathrm{HA}] + [\mathrm{A^-}] \tag{4.23.8}$$

The charge-balance expression is

$$[\mathrm{Na^+}] + [\mathrm{H_3O^+}] = [\mathrm{A^-}] + [\mathrm{OH^-}] \tag{4.23.9}$$

From Eqs (4.23.6) and (4.23.8), we get

$$[\mathrm{A^-}]_{\mathrm{total}} = \frac{[\mathrm{H_3O^+}][\mathrm{A^-}]}{K_a} + [\mathrm{A^-}] \Rightarrow [\mathrm{A^-}] = \frac{[\mathrm{A^-}]_{\mathrm{total}}K_a}{[\mathrm{H_3O^+}]+K_a} \tag{4.23.10}$$

$$[\mathrm{A^-}]_{\mathrm{total}} = [\mathrm{HA}] + \frac{K_a[\mathrm{HA}]}{[\mathrm{H_3O^+}]} \Rightarrow [\mathrm{HA}] = \frac{[\mathrm{A^-}]_{\mathrm{total}}[\mathrm{H_3O^+}]}{[\mathrm{H_3O^+}]+K_a} \tag{4.23.11}$$

Substituting Eq. (4.23.10) in Eq. (4.23.9), we get

$$[\mathrm{Na^+}] + [\mathrm{H_3O^+}] = \frac{[\mathrm{A^-}]_{\mathrm{total}}K_a}{[\mathrm{H_3O^+}]+K_a} + [\mathrm{OH^-}]$$

Replacing $[\mathrm{OH^-}]$ by $[\mathrm{H_3O^+}]$, we get

$$[\mathrm{Na^+}] = \frac{[\mathrm{A^-}]_{\mathrm{total}}K_a}{[\mathrm{H_3O^+}]+K_a} + \frac{K_w}{[\mathrm{H_3O^+}]} - [\mathrm{H_3O^+}] \tag{4.23.12}$$

Differentiating Eq. (4.23.12) with respect to $[\mathrm{H_3O^+}]$ keeping $[\mathrm{A^-}]_{\mathrm{total}}$ constant, we get

$$\left(\frac{\partial[\mathrm{Na^+}]}{\partial[\mathrm{H_3O^+}]}\right)_{[\mathrm{A^-}]} = -\frac{[\mathrm{A^-}]_{\mathrm{total}}K_a}{\{[\mathrm{H_3O^+}]+K_a\}^2} - \frac{K_w}{[\mathrm{H_3O^+}]^2} - 1$$

Using the fact

$$\left(\frac{\partial[\mathrm{Na}^+]}{\partial \mathrm{pH}}\right)_{[\mathrm{A}^-]} = \left(\frac{\partial[\mathrm{Na}^+]}{\partial(-\log\{[\mathrm{H_3O^+}]/\mathrm{M}\})}\right)_{[\mathrm{A}^-]} = \left(\frac{\partial[\mathrm{Na}^+]}{-\dfrac{1}{2.303}\partial \ln\{[\mathrm{H_3O^+}]/\mathrm{M}\}}\right)_{[\mathrm{A}^-]}$$

$$= -2.303\,[\mathrm{H_3O^+}]\left(\frac{\partial[\mathrm{Na}^+]}{\partial[\mathrm{H_3O^+}]}\right)_{[\mathrm{A}^-]}$$

we get $$\left(\frac{\partial[\mathrm{Na}^+]}{\partial \mathrm{pH}}\right)_{[\mathrm{A}^-]} = 2.303\left[\frac{[\mathrm{H_3O^+}][\mathrm{A}^-]_{\mathrm{total}}\,K_\mathrm{a}}{\{[\mathrm{H_3O^+}]+K_\mathrm{a}\}^2} + \frac{K_\mathrm{w}}{[\mathrm{H_3O^+}]} + [\mathrm{H_3O^+}]\right] \tag{4.23.13}$$

For a solution containing appreciable amounts of acids and its salt, the above expression reduces to

$$\left(\frac{\partial[\mathrm{Na}^+]}{\partial \mathrm{pH}}\right)_{[\mathrm{A}^-]} = 2.303\left[\frac{[\mathrm{H_3O^+}][\mathrm{A}^-]_{\mathrm{total}}\,K_\mathrm{a}}{\{[\mathrm{H_3O^+}]+K_\mathrm{a}\}^2}\right]$$

which on using Eqs (4.23.10) and (4.23.11) becomes

$$\left(\frac{\partial[\mathrm{Na}^+]}{\partial \mathrm{pH}}\right)_{[\mathrm{A}^-]} = 2.303\,\frac{[\mathrm{HA}][\mathrm{A}^-]}{[\mathrm{A}^-]_{\mathrm{total}}}$$

Using Eq. (4.23.8), this becomes

$$\left(\frac{\partial[\mathrm{Na}^+]}{\partial \mathrm{pH}}\right)_{[\mathrm{A}^-]} = 2.303\,\frac{[\mathrm{HA}][\mathrm{A}^-]}{[\mathrm{HA}]+[\mathrm{A}^-]} = 2.303\,\frac{(\text{acid})\,(\text{salt})}{(\text{acid}+\text{salt})} \tag{4.23.14}$$

Example 4.23.1

Calculate the pH of a solution made by adding 0.001 mole of NaOH to 100 cm^3 of a solution which is 0.50 M in CH_3COOH and 0.50 M in CH_3COONa.

Solution

Amount of CH_3COOH in 100 cm^3 of the solution $= \dfrac{(0.50\ \text{mol})}{(1\,000\ \text{cm}^3)}(100\ \text{cm}^3) = 0.05\ \text{mol}$

Amount of CH_3COONa in 100 cm^3 of the solution $= \dfrac{(0.50\ \text{mol})}{(1\,000\ \text{cm}^3)}(100\ \text{cm}^3) = 0.05\ \text{mol}$

When alkali is added, the acid is converted into salt. Thus, we have

Amount of the acid after the addition of 0.001 mol of the alkali = 0.049 mol

Amount of the salt = 0.051 mol

Since $$\mathrm{pH} = \mathrm{p}K_\mathrm{a}^\circ + \log\left(\frac{[\text{salt}]}{[\text{acid}]}\right)$$

therefore,

$$\mathrm{pH} = -\log(1.8\times10^{-5}) + \log\left(\frac{0.051}{0.049}\right) = 4.744\,7 + 0.017\,4 \approx 4.76$$

Example 4.23.2 (a) Calculate the pH of a buffer A which is 0.01 M in acetic acid and 0.01 M in sodium acetate and a buffer B which is 0.1 M in acetic acid and 0.1 M in sodium acetate, (b) Calculate the pH when 1.0×10^{-3} mole of HCl is added to one dm^3 of the buffer A and the same quantity of HCl is added to one dm^3 of buffer B.

Solution **(a)** Substituting the data in Henderson–Hasselbalch equation, we have

$$\textit{Buffer}\ \text{A:}\quad \text{pH} = \text{p}K_a^\circ + \log\left(\frac{0.01}{0.01}\right) = \text{p}K_a^\circ = 4.75$$

$$\textit{Buffer}\ \text{B:}\quad \text{pH} = \text{p}K_a^\circ + \log\left(\frac{0.1}{0.1}\right) = \text{p}K_a^\circ = 4.75$$

(b) The addition of 1.0×10^{-3} mole of HCl in the buffer will convert 10^{-3} mole of NaAc into HAc according to the reaction

$$Na^+ + Ac^- + H_3O^+ + Cl^- \rightarrow Na^+ + Cl^- + H_2O + HAc$$

Therefore, pHs of the buffers are

$$\textit{Buffer}\ \text{A:}\quad \text{pH} = 4.75 + \log\left(\frac{0.01 - 0.001}{0.01 + 0.001}\right) = 4.66$$

$$\textit{Buffer}\ \text{B:}\quad \text{pH} = 4.75 + \log\left(\frac{0.1 - 0.001}{0.1 + 0.001}\right) = 4.74$$

Example 4.23.3 Calculate the pH of buffer A given in Example 4.23.2 when 0.01 mole of HCl is added to 1 dm^3 of this buffer solution. Compare this value with the value obtained in a solution which is 0.02 M in acetic acid and 0.01 M in sodium chloride.

Solution The addition of 1.0×10^{-2} mole of HCl when added to the buffer A will convert the whole of 0.01 M sodium acetate to 0.01 M acetic acid, according to the reaction

$$Na^+ + Ac^- + H_3O^+ + Cl^- \rightarrow Na^+ + Cl^- + H_2O + HAc$$

The resultant solution will contain 0.02 M HAc and 0.01 M NaCl. This solution will no more be a buffer solution and therefore its pH cannot be calculated using Henderson equation. However, its pH can be calculated using the dissociation constant of HAc.

$$\underset{(0.02\ \text{M})(1-\alpha)}{\text{HAc}} \rightleftharpoons \underset{(0.02\ \text{M})\,\alpha}{\text{H}^+} + \underset{(0.02\ \text{M})\,\alpha}{\text{Ac}^-}$$

Therefore, $$K_a = \frac{(0.02\ \text{M})^2\,\alpha^2}{(0.02\ \text{M})\,(1-\alpha)} = 1.8 \times 10^{-5}\ \text{M}$$

i.e. $$0.02\,\alpha^2 \approx 1.8 \times 10^{-5}$$

or $$\alpha = \sqrt{\frac{1.8 \times 10^{-5}}{0.02}} = 3 \times 10^{-2}$$

Hence, $$[\text{H}^+] = (0.02\ \text{M})\,\alpha = 6 \times 10^{-4}\ \text{M}$$

$$\text{pH} = -\log\,(6 \times 10^{-4}) = 3.22$$

4.24 ACID–BASE INDICATORS

Indicators, in general, are either organic weak acids or weak bases with a characteristic of having different colours in the ionized and unionized forms. For example, phenolphthalein is a weak acid (ionized form is pink and unionized form is colourless) and methyl orange is a weak base (ionized form is red and unionized form is yellow).

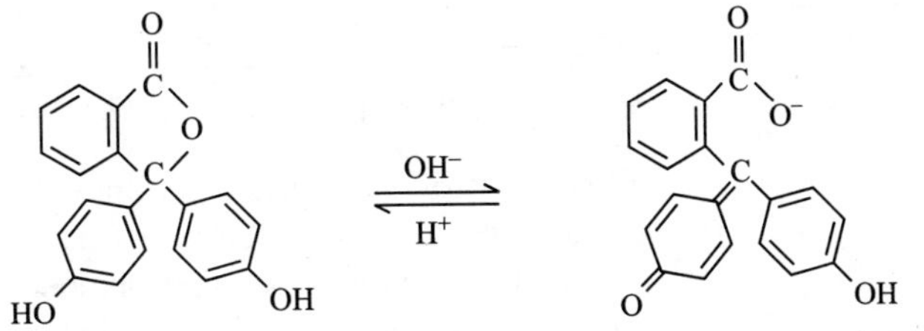

benzenoid form (colourless) $\underset{H^+}{\overset{OH^-}{\rightleftharpoons}}$ quinonoid form (pink)

phenolphthalein

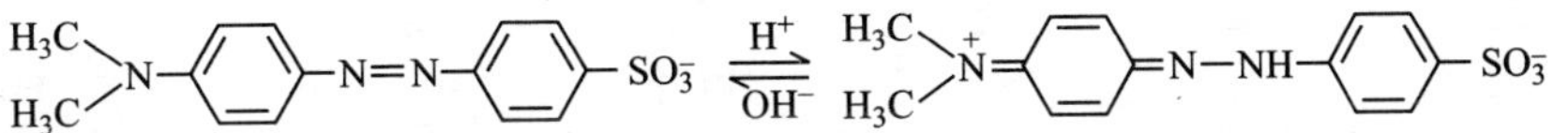

benzenoid form (yellow) $\underset{OH^-}{\overset{H^+}{\rightleftharpoons}}$ quinonoid form (red)

methyl orange

Indicator Constant

The equilibrium between ionized and unionized forms of an indicator can be treated like that of any other weak electrolyte. Considering an acid indicator HIn, the equilibrium involving it and its conjugate base In^- can be represented as

$$HIn \rightleftharpoons H^+ + In^-$$

Its equilibrium constant K_{In} is given by

$$K_{In} = \frac{[H^+][In^-]}{[HIn]} \qquad (4.24.1)$$

This K_{In} is known as the *indicator constant* and like any other dissociation constant, its value depends only on the temperature and not on individual concentrations of H^+, In^- and HIn. These are usually tabulated as pK^{o}_{In} $\{= -\log (K_{In}/M)\}$.

Indicator Range

Rewriting Eq. (4.24.1), we get

$$[H^+] = K_{In}\frac{[HIn]}{[In^-]} \quad \text{or} \quad pH = pK^{o}_{In} + \log\frac{[In^-]}{[HIn^-]} \qquad (4.24.2)$$

In general, the intensity of a coloured solution depends on the concentration of the colour imparting species. If the solution contains two coloured species, say In^- and HIn, the colour of the solution depends on the relative amounts of the two species. On an average, the solution acquires a distinct colour characteristic of In^- if the concentration of the latter is approximately 10 (or more) times greater than that of HIn and vice versa. Thus, we have

(a) At pH = $pK^\circ_{In} - 1$, the ratio $[In^-]/[HIn]$ in the solution is (1/10). For this pH, the percentage ionization of the indicator is

$$\frac{[In^-]}{[In^-]+[HIn]}\times 100 = \frac{1}{(1+10)}\times 100 = \frac{100}{11} \approx 9.1$$

Thus, most of the indicator is present in the unionized acid form HIn and the solution gets the colour characteristic of this form. In fact, pH = $pK^\circ_{In} - 1$ is the maximum pH up to which the solution has a distinct colour characteristic of HIn. At pH smaller than this value, still more of the indicator is present in the unionized form. Thus at pH $\leq pK^\circ_{In} - 1$, the solution has a colour characteristic of HIn.

(b) At pH = $pK^\circ_{In} + 1$, the ratio $[In^-]/[HIn]$ in the solution is (10/1). For this pH, the percentage ionization of the indicator is

$$\frac{[In^-]}{[In^-]+[HIn]}\times 100 = \frac{10}{(10+1)}\times 100 = \frac{1\,000}{11} \approx 91$$

Thus, most of the indicator is present in the ionized base form In^- and the solution gets the colour characteristic of this form. In fact, pH = $pK^\circ_{In} + 1$ is the minimum pH up to which the solution has a distinct colour characteristic of In^-. At pH greater than this value, still more of the indicator is present in the ionized form. Thus at pH $\geq pK^\circ_{In} + 1$, the solution has a colour characteristic of In^-.

In between the pH range $pK^\circ_{In} - 1$ to $pK^\circ_{In} + 1$, transition of colour takes place. A list of indicators is given in Table 4.24.1 which also includes the colours of acid and base forms of each indicator and the pH value at which one or the other form is sufficiently predominant to impart its colour to the solution.

Selection of Indicators

Near the equivalence point of acid-base titration, the pH value of the solution changes steeply and the pH range of this steep change depends on the nature of the acid-base pair being titrated. The centre of this steep change lies at pH = 7 only when a strong acid is titrated with a strong base. For any other titration involving either weak acid or base, the salt formed undergoes hydrolysis and, therefore, gives a pH value either greater than or less than 7. The selected indicator should be such that it offers a colour change near the vicinity of the pH of the solution at the equivalence point. Thus, the selection of indicator must satisfy the following two points:

(i) The steep section of the titration curve at the equivalence point must encompass an interval of pH values at least as large as the pH transition range of an indicator.

(ii) The pH transition range of the indicator must coincide with the steep portion of the titration curve.

Table 4.24.1 Acid-Base Indicators

Indicator	*Colour change*		*pH transition range*	
	Acid form	*Base form*	*Acid form predominate at* pH	*Base form predominate at* pH
Picric acid	Colourless	Yellow	0.0	1.2
Malachite green	Yellow	Green	0.0	2.0
Methyl violet	Yellow	Violet	0.1	3.2
m-Cresol purple	Red	Yellow	1.2	2.8
Thymol blue	Red	Yellow	1.2	2.8
Bromophenol blue	Yellow	Blue	3.0	4.6
Congo red	Blue	Red	3.0	5.0
Methyl orange	Red	Yellow	3.1	4.4
Bromocresol green	Yellow	Blue	3.8	5.4
Methyl red	Red	Yellow	4.2	6.3
Litmus	Red	Blue	4.5	8.3
Propyl red	Red	Yellow	4.6	6.4
Chlorophenol red	Yellow	Red	4.8	6.4
Hematoxylin	Yellow	Red	5.0	6.0
p-Nitrophenol	Colourless	Yellow	5.0	7.0
Bromocresol purple	Yellow	Purple	5.2	6.8
Bromothymol blue	Yellow	Blue	6.0	7.6
Phenol red	Yellow	Red	6.8	8.4
m-Cresol purple	Yellow	Purple	7.4	9.0
Thymol blue	Yellow	Blue	8.0	9.6
Phenolphthalein	Colourless	Red	8.3	10.0
Thymolphthalein	Colourless	Blue	9.3	10.5
Alizarin yellow R	Yellow	Lavender	10.0	12.1
Alizarin blue S	Green	Blue	11.0	13.0
Malachite green	Green	Colourless	11.4	13.0
Trinitrobenzene	Colourless	Orange	12.0	14.0

Indicators in Some Typical Titrations

The following is a brief summary of the choice of indicators employed in different types of titrations:

Strong acid and strong base The pH at the equivalence point is 7 and slightly before and after the equivalence point, it encompasses an interval from pH = 3 to pH = 11 (Fig. 4.24.1). Therefore, any indicator whose pH range is within this interval can be used. Any of the indicators from bromophenol blue to thymolphthalein listed in Table 4.24.1 can be used. However, *the range of the steep change of* pH *at the equivalence point depends upon the concentrations of acid and base being titrated.* For dilute solutions, the corresponding pH change is less as compared to that of concentrated solutions. Hence, the most ideal indicator is that which changes its colour (base form) near pH = 7. Methyl orange, methyl red and phenolphthalein, the common laboratory indicators, can be used for this purpose. The usual choice, however, is phenolphthalein because the colour change in this case is from colourless to a slight pink colour—a change which can be easily detected.

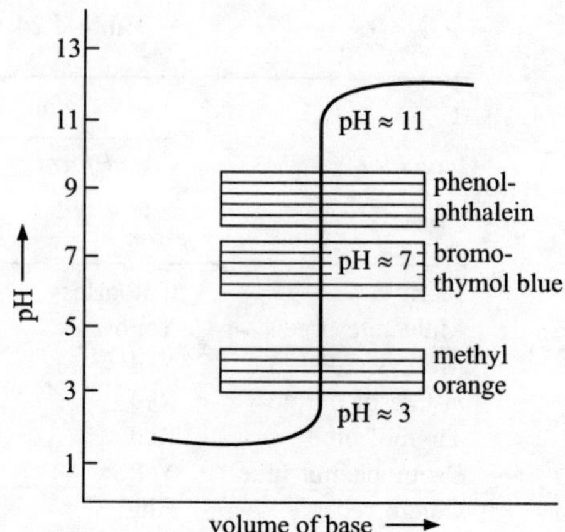

Fig. 4.24.1 Titration curve (only schematic) of a strong acid with a strong base

Weak base and strong acid The pH of the solution at the equivalence point lies slightly in the acidic range. For example, pH of the solution at equivalence point when 0.1 M NH_4OH is titrated with 0.1 M HCl is 5.3; the presence of a slight excess of the acid lowers the pH further down to between 4 and 3 (Fig. 4.24.2). Obviously the use of phenolphthalein as an indicator is ruled out. Methyl red or methyl orange can be used as an indicator for this titration.

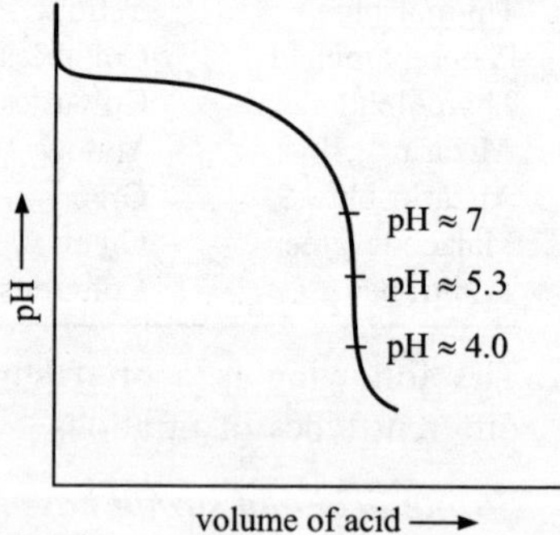

Fig. 4.24.2 Titration curve (only schematic) of a weak base (NH_4OH) with a strong acid (HCl)

Weak acid and strong base The pH of the solution at the equivalence point lies on the alkaline side. For example, when 0.1 M HAc is titrated with 0.1 M NaOH, its pH is 8.7 (Fig. 4.24.3). A slight excess of base brings the pH up to 9 or 10 and hence phenolphthalein can be used as the indicator in this titration.

Weak acid and weak base In this particular case, neither the steep rise of pH near the equivalence point occurs nor the rise of pH encompasses an interval equal to the pH transition range of any of the indicators. Hence, no suitable indicator can be employed to detect the end point as it will cause much error.

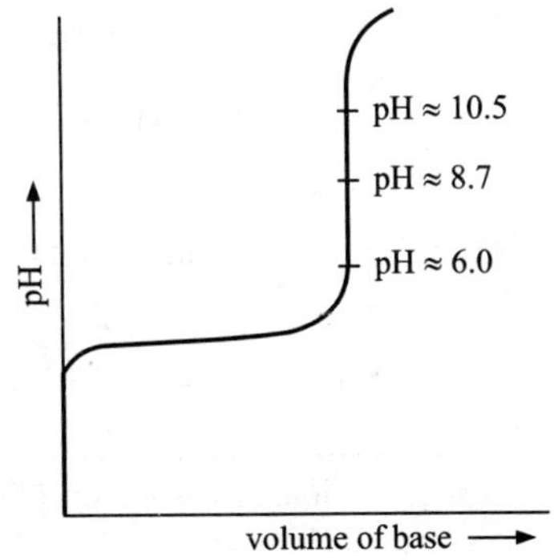

Fig. 4.24.3 Titration curve (only a schematic) of a weak acid (HAc) with a strong base (NaOH)

pH ≈ 10.5

pH ≈ 8.7

pH ≈ 6.0

pH →

volume of base →

Example 4.24.1

For the indicator thymol blue, the value of pH is 2.0 when half of the indicator is present in the unionized form. Calculate the percentage of the indicator in the unionized form in a solution of 4.0×10^{-3} M hydrogen ion concentration.

Solution

Since at pH = 2.0, half of the indicator is present in the unionized form, therefore

$$[\text{HIn}] = [\text{In}^-]$$

Using $\quad \text{pH} = \text{p}K^\circ_{\text{In}} + \log \dfrac{[\text{In}^-]}{[\text{HIn}]}, \quad$ we have

$$\text{p}K^\circ_{\text{In}} = \text{pH} = 2.0$$

pH of the solution containing 4.0×10^{-3} M of H^+ is

$$\text{pH} = -\log(4.0 \times 10^{-3}) = 2.4$$

Thus, $\quad \log\left(\dfrac{[\text{In}^-]}{[\text{HIn}]}\right) = \text{pH} - \text{p}K^\circ_{\text{In}} = 2.4 - 2.0 = 0.4$

or $\quad \dfrac{[\text{In}^-]}{[\text{HIn}]} = 2.5$

Adding 1 on both sides, we get

$$\frac{[\text{In}^-] + [\text{HIn}]}{[\text{HIn}]} = 3.5 \quad \text{or} \quad \frac{[\text{HIn}]}{[\text{In}^-] + [\text{HIn}]} = \frac{1}{3.5} = 0.286$$

Thus, the percentage of indicator in the unionized form = 28.6

Example 4.24.2

An indicator is a weak acid and the pH range of its colour is 3.1 to 4.5. If the neutral point of the indicator lies in the centre of the hydrogen ion concentrations corresponding to given pH range, calculate the ionization constant of the indicator.

Solution

The hydrogen ion concentrations of the given pH range are

$$\text{pH} = 3.1 = -\log\{[\text{H}^+]/\text{M}\} \quad \text{i.e.} \quad \log\{[\text{H}^+]/\text{M}\} = -3.1 = \bar{4}.9$$

Hence, $\quad [\text{H}^+] = 7.9 \times 10^{-4}$ M

$$\text{pH} = 4.5 \quad \text{i.e.} \quad \log\{[\text{H}^+]/\text{M}\} = -4.5 = \bar{5}.5$$

Hence, $\quad [\text{H}^+] = 3.2 \times 10^{-5}$ M

The average of these two hydrogen ion concentrations is

$$\frac{7.9\times10^{-4}\ \text{M}+3.2\times10^{-5}\ \text{M}}{2}=4.11\times10^{-4}\ \text{M}$$

At this H^+ concentration, $[In^-] = [HIn]$. Therefore,

$$\text{pH} = \text{p}K^{\circ}_{\text{HIn}}$$

or $$[H^+] = K_{\text{HIn}} = 4.11 \times 10^{-4}\ \text{M}$$

Example 4.24.3 In the titration of a solution of a weak acid, HX with NaOH, the pH is 5.8 after the addition of 10.0 cm^3 of NaOH solution and is 6.4 after the addition of 20.0 cm^3 of NaOH solution. What is the ionization constant of the HX?

Solution Let x be the initial amount of the acid HX and let y be the amount of base NaOH added per dm^3 of the solution. Let the volume of the acid being titrated be V.

(i) *After the addition of 10.0* cm^3 *of the base* (pH = 5.8):

Amount of salt = Amount of NaOH added

$$=\left(\frac{y}{1\ \text{dm}^3}\right)(10\ \text{cm}^3)=\left(\frac{y}{1\ \text{dm}^3}\right)(10^{-2}\ \text{dm}^3)=y\times10^{-2}$$

$$\text{Amount of acid left} = \left(\frac{x}{1\ \text{dm}^3}\right)(V)-y\times10^{-2}$$

(ii) *After the addition of* 20.0 cm^3 *of the base* (pH = 6.4):

$$\text{Amount of salt} = \left(\frac{y}{1\ \text{dm}^3}\right)(2\times10^{-2}\ \text{dm}^3)=2y\times10^{-2}$$

$$\text{Amount of acid left} = \left(\frac{x}{1\ \text{dm}^3}\right)(V)-2y\times10^{-2}$$

Using the Henderson's equation, we have

$$5.8=\text{p}K^{\circ}+\log\left(\frac{y\times10^{-2}}{x\,(V/\text{dm}^3)-y\times10^{-2}}\right) \tag{1}$$

$$6.4=\text{p}K^{\circ}+\log\left(\frac{2y\times10^{-2}}{x\,(V/\text{dm}^3)-2y\times10^{-2}}\right) \tag{2}$$

Subtracting Eq. (1) from Eq. (2), we get

$$0.6=\log\left\{\frac{2y\times10^{-2}}{x(V/\text{dm}^3)-2y\times10^{-2}}\times\frac{x(V/\text{dm}^3)-y\times10^{-2}}{y\times10^{-2}}\right\}$$

Simplifying, we get

$$\frac{x(V/\text{dm}^3)}{y}=3\times10^{-2}$$

Substituting this in either Eq. (1) (or Eq. 2), we get

$$5.8 = \mathrm{p}K^\circ + \log\left\{\frac{y\times 10^{-2}}{3\times 10^{-2}\ y - y\times 10^{-2}}\right\}$$

i.e. $$5.8 = \mathrm{p}K^\circ + \log\left(\frac{1}{2}\right) \quad \text{or} \quad \mathrm{p}K^\circ = 5.8 - \log\left(\frac{1}{2}\right) = 6.1$$

Universal Indicators

A mixture of the following indicators can be used to determine the approximate pH of the solution in the range 4 to 11.

Methyl red, α-naphtholphthalein, thymolphthalein, phenolphthalein and bromothymol blue.

This mixture gives different colours at different pH values of the solution. These are

pH	4	5	6	7	8	9	10	11
Colour	Red	Orange red	Yellow	Green yellow	Green	Blue green	Blue violet	Red violet

4.25 TITRATION OF A STRONG MONOPROTIC ACID WITH A STRONG BASE

At the Start of Titration

If n_0 is the amount of acid present in the volume V_a of the solution, then

$$[\mathrm{H_3O^+}] = \frac{n_0}{V_a}$$

Before the Equivalence Point

If c_a and c_b are the concentrations of acid and base, respectively, then after the addition of the volume V_b of the base, the amount of H_3O^+ left is

$$n_a = n_0 - c_bV_b \tag{4.25.1}$$

Total volume of the solution = $V_a + V_b$

Hence, the concentration of remaining H_3O^+ in solution is

$$[\mathrm{H_3O^+}] = \frac{n_0 - c_bV_b}{V_a + V_b} \tag{4.25.2}$$

This expression will not be applicable very near to the equivalence point since the concentration of H_3O^+ from the remaining acid becomes comparable with the H_3O^+ concentration due to the dissociation of water. In this case, we use the expression

$$\begin{aligned}[\mathrm{H_3O^+}]_{\mathrm{total}} &= [\mathrm{H_3O^+}]_{\mathrm{acid}} + [\mathrm{H_3O^+}]_{\mathrm{water}} \\ &= [\mathrm{H_3O^+}]_{\mathrm{acid}} + [\mathrm{OH^-}]_{\mathrm{water}} \\ &= \frac{n_0 - c_bV_b}{(V_a + V_b)} + \frac{K_w}{[\mathrm{H_3O^+}]_{\mathrm{total}}}\end{aligned} \tag{4.25.3}$$

At the Equivalence Point Here the concentration of H_3O^+ will be equal to $\sqrt{K_w}$ (i.e. 10^{-7} M at 25 °C) since the salt formed will not undergo hydrolysis.

After the Equivalence Point Here only excess of the base will be present. The total hydroxyl-ion concentration in the solution can be calculated as follows:

(a) *Near the equivalence point* Here $[OH^-]$ coming from both the sources (added NaOH solution and water) will be comparable; therefore

$$[OH^-]_{total} = [OH^-]_{base} + [OH^-]_{water}$$

$$= \frac{\text{Amount of excess base}}{\text{Total volume of the solution}} + [H_3O^+]_{water}$$

$$= \frac{\text{Amount of excess base}}{\text{Total volume of the solution}} + \frac{K_w}{[OH^-]_{total}} \qquad (4.25.4)$$

(b) *When enough base has been added* Here OH^- contribution due to the dissociation of water will be negligible; therefore

$$[OH^-] = \frac{\text{Amount of excess base}}{\text{Total volume of the solution}} \qquad (4.25.5)$$

Illustration Let us illustrate the method of calculations and the titration curve when 50 cm^3 of 0.1 M HCl is titrated against 0.1 M NaOH at 25 °C. Here,

$$V_a = \frac{(50 \text{ cm}^3)}{(1\,000 \text{ cm}^3/1 \text{ dm}^3)} = 0.05 \text{ dm}^3$$

$$n_0 = \frac{0.1 \text{ mol}}{1 \text{ dm}^3} \times 0.05 \text{ dm}^3 = 0.005 \text{ mol}$$

At the start of titration

$$[H_3O^+] = \frac{0.005 \text{ mol}}{(50/1\,000) \text{ dm}^3} = 0.10 \text{ M}; \qquad \text{pH} = 1.0$$

Before the equivalence point The pH of the solution after the addition of 10 cm^3 of the base can be computed as follows:

$$n_a = 0.005 \text{ mol} - (0.1 \text{ M})\ (10 \times 10^{-3} \text{ dm}^3) = 0.004 \text{ mol}$$

$$[H_3O^+] = \frac{0.004 \text{ mol}}{0.05 \text{ dm}^3 + 0.01 \text{ dm}^3} = 6.67 \times 10^{-2} \text{ M}$$

$$\text{pH} = 1.18$$

Proceeding similarly, we can find $[H_3O^+]$ and pH of the solution at the different stages of addition of base. Table 4.25.1 includes a few such calculated values.

At the equivalence point At this point, we have only sodium chloride solution which has pH of 7.

Beyond the equivalence point The pH of the solution at different volume of the base added are given in Table 4.25.2.

Table 4.25.1 pH Values before the Equivalence Point

Volume of base added, V_b/cm^3	$[H_3O^+]$/M *Eqs (4.25.2) and (4.25.3)*	pH
10.0	6.67×10^{-2}	1.18
20.0	4.29×10^{-2}	1.37
30.0	2.50×10^{-2}	1.60
40.0	1.11×10^{-2}	1.95
49.0	1.01×10^{-3}	2.99
49.9	1.00×10^{-4}	4.00
49.99	1.00×10^{-5}	5.00
49.999	1.00×10^{-6}	6.00
49.9999	1.02×10^{-7}	6.99

Table 4.25.2 pH Values after the Equivalence Point

Volume of base added, V_b/cm^3	$[OH^-]$/M *Eq. (4.25.5)*	pH
51.0	9.93×10^{-4}	10.99
52.0	1.96×10^{-3}	11.29
53.0	2.91×10^{-3}	11.46
54.0	3.85×10^{-3}	11.58

Titration Curve

The titration curve of HCl with NaOH is shown in Fig. 4.25.1. The outstanding feature of this type of titration curve is that near the equivalence point, there is an abrupt change in pH value. For example, in the above titration of 50.0 cm^3 of 0.1 M HCl against 0.1 M NaOH, there is a change of pH value by nearly 6 units, as V_b varies from 49.9 cm^3 to 50.1 cm^3. This abrupt change, however, depends upon the concentrations of the acid and the base chosen. For dilute solutions, the corresponding pH change is less as compared to the more concentrated solution.

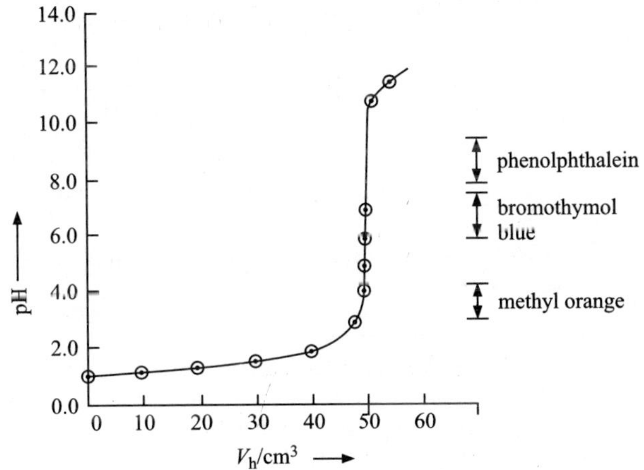

Fig. 4.25.1 Titration curve of a strong acid (HCl) with a strong base (NaOH)

4.26 TITRATION OF A WEAK MONOPROTIC ACID WITH A STRONG BASE

The $[H_3O^+]$ (or pH) of the solution at various stages of titration can be computed as follows.

At the Start of Titration

The concentration of H_3O^+ can be computed using only the expression for ionization constant of the acid, provided the hydronium ion contribution due to the dissociation of water is negligible. We have

$$HA + H_2O \rightleftharpoons H_3O^+ + A^-$$

$$K_a = \frac{[H_3O^+][A^-]}{[HA]}$$

Since $[H_3O^+] = [A^-]$ and $[HA] = [HA]_0 - [H_3O^+]$, therefore

$$K_a = \frac{[H_3O^+]^2}{[HA]_0 - [H_3O^+]} \tag{4.26.1}$$

or $$[H_3O^+]^2 + K_a[H_3O^+] - K_a[HA]_0 = 0$$

Solution of this quadratic expression gives the value of hydrogen-ion concentration. We can use the simpler expression provided the solution is fairly concentrated and the dissociation constant is very small, such that

$$[H_3O^+] = \sqrt{K_a[HA]_0} \tag{4.26.2}$$

If n_0 is the amount of the acid present in the volume V_a of the acid solution, then

$$[HA]_0 = \frac{n_0}{V_a} \tag{4.26.3}$$

Before the Equivalence Point

In this range, the solution contains a weak acid and the salt of its conjugate base. In other words, it behaves like a buffer solution. Hydrogen-ion concentration of the solution can be computed using the Henderson's expression*

$$[H_3O^+] = \frac{[HA]}{[A^-]}K_a$$

or $$\mathrm{pH} = \mathrm{p}K_a^\circ + \log\left(\frac{[A^-]}{[HA]}\right) \tag{4.26.4}$$

If f is the fraction of the acid that has been neutralized by the addition of the volume V_b of the base to the volume V_a of the acid, then

$$[A^-] = \frac{n_0 f}{(V_a + V_b)} \quad \text{and} \quad [HA] = \frac{n_0(1-f)}{(V_a + V_b)}$$

*However, calculation of $[H_3O^+]$ through this equation very near to the equivalence point is not justified because of very low concentration of the weak acid. Moreover at this stage, the contribution of water to the total hydronium ions becomes comparable to that provided by the acid.

Substituting these in Eq. (4.26.4), we get

$$[H_3O^+] = \frac{n_0(1-f)/(V_a+V_b)}{n_0 f/(V_a+V_b)} K_a = \frac{(1-f)}{f} K_a$$

or
$$pH = pK_a^\circ + \log\left(\frac{f}{1-f}\right) \qquad (4.26.5)$$

Thus, $[H_3O^+]$ (or pH) would depend only upon K_a and the fraction of the acid that has been neutralized and is independent of the initial concentration of the acid.

At the Equivalence Point

The acid has been completely neutralized and the solution contains only salt, the anion of which undergoes hydrolysis according to the equation

$$A^- + H_2O \rightleftharpoons HA + OH^-$$

Thus,
$$K_h = \frac{[HA][OH^-]}{[A^-]}$$

If ionization of water is neglected, then

$$[HA] = [OH^-] \quad \text{and} \quad [A^-] = [A^-]_0 - [OH^-]$$

If the solution is fairly concentrated and the value of K_h is very small, then

$$[A^-] \approx [A^-]_0 = \frac{n_0}{(V_a+V_b)}$$

Therefore,
$$[OH^-] = \sqrt{K_h \frac{n_0}{(V_a+V_b)}}$$

or
$$[H_3O^+] = \frac{K_w}{[OH^-]} = \frac{K_w}{\sqrt{\dfrac{K_w}{K_a}\dfrac{n_0}{(V_a+V_b)}}} = \sqrt{\frac{K_w K_a}{n_0/(V_a+V_b)}} \qquad (4.26.6)$$

or
$$pH = \frac{1}{2}pK_w^\circ + \frac{1}{2}pK_a^\circ + \frac{1}{2}\log\left(\frac{n_0/\text{mol L}^{-1}}{V_a+V_b}\right) \qquad (4.26.7)$$

Beyond the Equivalence Point

At this stage, the solution will contain the salt and the excess of free base. If we assume that the $[OH^-]$ coming from the hydrolysis of the salt is negligible in comparison to the concentration of hydroxide ion coming from the base (which is justified because of common ion effect), the concentration of OH^- is given by

$$[OH^-] = \frac{\text{Amount of free base}}{\text{Total volume of the solution}}$$

and $$[H_3O^+] = \frac{K_w}{[OH^-]} \tag{4.26.8}$$

Illustration

The computational procedure is illustrated by evaluating the values of $[H_3O^+]$ (and pH) when 25.0 cm^3 of 0.1 M acetic acid is titrated against 0.1 M NaOH solution.

At the start of titration Here

$$K_a \approx 1.75 \times 10^{-5} \text{ M}$$

and $$[HA]_0 = 0.1 \text{ M}$$

Using the expression $[H_3O^+] = \sqrt{K_a\,[HA]_0}$, we get

$$[H_3O^+] = 1.32 \times 10^{-3} \text{ M}$$

and $$\text{pH} = 2.88$$

Before the equivalence point The value of $[H_3O^+]$ and pH at different stages of the addition of the base are given in Table 4.26.1.

Table 4.26.1 pH Values before the Equivalence Point

Volume of base added, (V_b/cm^3)	$[H_3O^+]$/M *Eq.* (4.26.5)	pH
2.5	15.75 × 10^{-5}	3.80
5.0	7.00 × 10^{-5}	4.15
10.0	2.62 × 10^{-5}	4.58
15.0	1.167 × 10^{-5}	4.93
20.0	0.437 × 10^{-5}	5.36
22.5	0.194 × 10^{-5}	5.70
23.75	0.092 1 × 10^{-5}	6.03
24.00	0.072 9 × 10^{-5}	6.14

At the equivalence point From Eq. (4.26.6), we have

$$[H_3O^+] = \sqrt{\frac{(1.0 \times 10^{-14}\ \text{M}^2)(1.75 \times 10^{-5}\ \text{M})}{0.05\ \text{M}}}$$

$$= \sqrt{\frac{1.75 \times 10^{-19}\ \text{M}^2}{0.05}} = 1.87 \times 10^{-9}\ \text{M}$$

and $$\text{pH} = 8.73$$

After the equivalence point The computed values of $[H_3O^+]$ and pH beyond the equivalence point are given in Table 4.26.2.

Titration Curve

These values of pH when plotted against V_b (the volume of base added) gives the titration curve as shown in Fig. (4.26.1).

Table 4.26.2 pH Values after the Equivalence Point

Total volume of base added, (V_b/cm^3)	*Amount of free base*/mol	$[H_3O^+]$/M *Eq.* (4.26.8)	pH
26.0	0.000 1	5.1×10^{-12}	11.29
28.0	0.000 3	1.77×10^{-12}	11.75
30.0	0.000 5	1.6×10^{-12}	11.80
35.0	0.001 0	8.5×10^{-13}	12.07
40.0	0.001 5	6.0×10^{-13}	12.22
45.0	0.002 0	2.25×10^{-13}	12.65

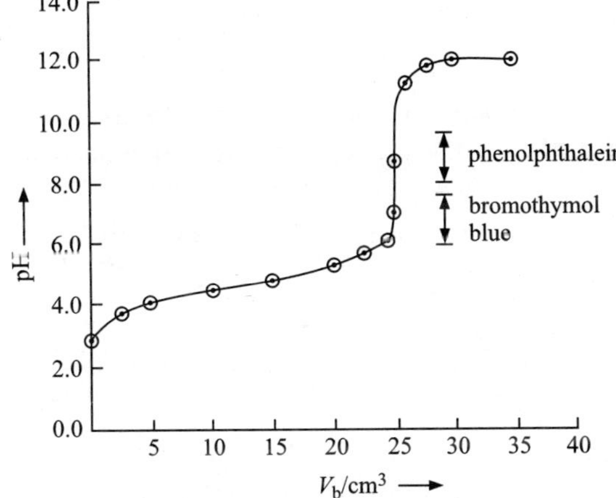

Fig. 4.26.1 Titration curve of a weak acid (HAc) with a strong base (NaOH)

4.27 TITRATION OF A WEAK BASE WITH A STRONG MONOPROTIC ACID

The process of neutralization of a weak base with a strong acid could be studied through a number of steps discussed below.

At the Start of Titration

The $[OH^-]$ can be computed using only the expression of ionization constant of the base, provided the hydroxyl-ion contribution due to the dissociation of water is negligible:

$$BOH \rightleftharpoons B^+ + OH^-$$

$$K_b = \frac{[B^+][OH^-]}{[BOH]}$$

Since $[B^+] = [OH^-]$ and $[BOH] = [BOH]_0 - [OH^-]$, therefore

$$K_b = \frac{[OH^-]^2}{[BOH]_0 - [OH^-]}$$

or $$[OH^-]^2 + K_b[OH^-] - K_b[BOH]_0 = 0 \qquad (4.27.1)$$

We get the value of OH^- ion concentration by solving this quadratic expression. We can use the simpler expression provided the solution is fairly concentrated and the dissociation constant is very small, such that

$$[OH^-] = \sqrt{K_b\,[BOH]_0} \qquad (4.27.2)$$

If n_0 is the amount of the base that is present in volume V_b of the solution, then

$$[BOH]_0 = \frac{n_0}{V_b}$$

Knowing $[OH^-]$, hydronium ion concentration can be calculated by using

$$[H_3O^+] = \frac{K_w}{[OH^-]} = \frac{K_w}{\sqrt{K_b\,[BOH]_0}} \qquad (4.27.3)$$

Before the Equivalence Point

In this range, the solution contains a weak base and its salt with strong acid, thus acting as a buffer solution. The $[OH^-]$ in the solution can be computed by using Henderson's equation

$$[OH^-] = \frac{[BOH]}{[B^+]} K_b$$

i.e. $$pOH = pK_b^\circ + \log \frac{[B^+]}{[BOH]}$$

If f is the fraction of the base neutralized by adding the volume V_a of the acid to the volume V_b of the base, then

Concentration of the salt formed, $[B^+] = \dfrac{n_0 f}{(V_a + V_b)}$

Concentration of the base left, $[BOH] = \dfrac{n_0\,(1-f)}{(V_a + V_b)}$

Substituting these in the above expression of $[OH^-]$, we get

$$[OH^-] = \frac{n_0\,(1-f)/(V_a+V_b)}{n_0 f/(V_a+V_b)} K_b = \frac{1-f}{f} K_b$$

Thus, $$[H_3O^+] = \frac{K_w}{K_b}\left(\frac{f}{1-f}\right)$$

or $$pH = pK_w^\circ - pK_b^\circ - \log\left(\frac{f}{1-f}\right) \qquad (4.27.4)$$

Thus, $[H_3O^+]$ (or pH) would depend only upon K_b and the fraction of the base neutralized and is independent of the initial concentration of base.

At the Equivalence Point

At this stage, the base has been completely neutralized and the solution contains only salt, the cation of which undergoes hydrolysis according to the equation

$$B^+ + 2H_2O \rightleftharpoons BOH + H_3O^+$$

Thus, $$K_h = \frac{[BOH][H_3O^+]}{[B^+]}$$

If ionization of water is neglected, then

$$[BOH] = [H_3O^+]$$

$$[B^+] = [B^+]_0 - [H_3O^+]$$

If the solution is fairly concentrated and the value of K_h is very small, then

$$[B^+] = [B^+]_0 = \frac{n_0}{(V_a + V_b)}$$

$$[H_3O^+] = \sqrt{K_h [B^+]_0} = \sqrt{\frac{K_w}{K_b}\frac{n_0}{(V_a + V_b)}} \tag{4.27.5}$$

or $$pH = \frac{1}{2}pK_w^\circ - \frac{1}{2}pK_b^\circ - \frac{1}{2}\log\left\{\left(\frac{n_0}{V_a + V_b}\right)/\text{mol L}^{-1}\right\}$$

Beyond the Equivalence Point

Now the solution will contain the salt and the excess of free acid. If it is assumed that the $[H_3O^+]$ coming from the hydrolysis of salt is negligible relative to the concentration of hydronium ion coming from the acid (this assumption is justified because of common ion effect), then the concentration of hydronium ion can be computed through the use of the expression

$$[H_3O^+] = \frac{\text{Amount of free acid}}{\text{Total volume of the solution}}$$

Illustration

Following are the values of $[H_3O^+]$ and pH at various stages when 25.0 cm^3 of 0.1 M NH_4OH is titrated against 0.1 M HCl.

At the start of titration

$$K_b = 1.8 \times 10^{-5}\text{ M}$$

$$[BOH]_0 = 0.1\text{ M}$$

Thus, $$[H_3O^+] = \frac{K_w}{\sqrt{K_b[BOH]_0}} = \frac{1.0 \times 10^{-14}\text{ M}^2}{\sqrt{(1.8 \times 10^{-5}\text{ M})(0.1\text{ M})}}$$

$$= 7.45 \times 10^{-12}\text{ M}$$

Hence, pH = 11.13

Before the equivalence point The $[H_3O^+]$ and the corresponding pH of the solution at various stages of the addition of acid are given in Table 4.27.1.

At the equivalence point From Eq. (4.27.5), we get

$$[H_3O^+] = \sqrt{\frac{(1.0 \times 10^{-14}\text{ M}^2)(0.002\,5\text{ mol})}{(1.8 \times 10^{-5}\text{ M})(0.05\text{ dm}^3)}} = 5.27 \times 10^{-6}\text{ mol dm}^{-3}$$

pH = 5.27

Table 4.27.1 pH Values before the Equivalence Point

Volume of acid added, V_a/cm^3	*f*	$[H_3O^+]$/M Eq. (4.27.4)	pH
2.5	0.1	6.17×10^{-11}	10.21
5.0	0.2	1.38×10^{-10}	9.86
10.0	0.4	3.80×10^{-10}	9.42
15.0	0.6	9.55×10^{-10}	9.02
20.0	0.8	2.19×10^{-9}	8.66
22.5	0.9	4.90×10^{-9}	8.31
23.75	0.95	5.25×10^{-8}	7.28
24.88	0.9952	1.12×10^{-7}	6.95

Beyond the equivalence point The values of $[H_3O^+]$ and the corresponding values of pH are given in Table 4.27.2.

Table 4.27.2 pH Values after the Equivalence Point

Total volume of acid added, V_a/cm^3	*Amount of free acid/mol*	$[H_3O^+]$/M	pH
26.0	0.000 1	1.02×10^{-4}	3.99
28.0	0.000 3	3.02×10^{-4}	3.52
30.0	0.000 5	4.90×10^{-4}	3.31
35.0	0.001 0	1.00×10^{-3}	3.0

Titration Curve

The values of pH when plotted against V_a (the volume of acid added) gives the titration curve as shown in Fig. 4.27.1.

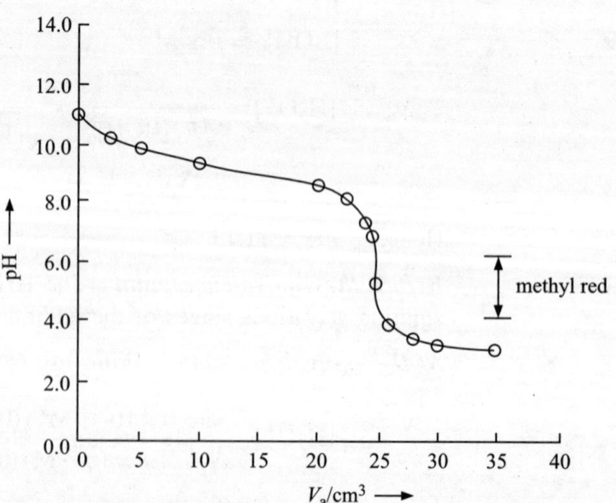

Fig. 4.27.1 Titration curve of a weak base (NH_4OH) with a strong acid (HCl)

4.28 GENERAL TREATMENT OF TITRATION OF AN ACID WITH A STRONG BASE

Basic Equilibrium Reactions

In the titration of an acid (HA) with a strong base (BOH), the equilibria to be studied are

$$HA + H_2O \rightleftharpoons H_3O^+ + A^-; \qquad K_a = \frac{[H_3O^+][A^-]}{[HA]} \tag{4.28.1}$$

$$H_2O + H_2O \rightleftharpoons H_3O^+ + OH^-; \quad K_w = [H_3O^+][OH^-] \tag{4.28.2}$$

Condition of Mass Balance

The material-balance expressions are

$$[HA]_0 = [HA] + [A^-] \tag{4.28.3}$$

$$[B^+]_0 = [B^+] \tag{4.28.4}$$

Condition of Charge Balance

The charge-balance expression is

$$[H_3O^+] + [B^+] = [A^-] + [OH^-] \tag{4.28.5}$$

Derivation of Exact Expression

Eliminating [HA] in Eq. (4.28.3) by using Eq. (4.28.1), we get

$$[HA]_0 = \frac{[H_3O^+][A^-]}{K_a} + [A^-] = \left(\frac{[H_3O^+] + K_a}{K_a}\right)[A^-]$$

or
$$[A^-] = \frac{[HA]_0\, K_a}{[H_3O^+] + K_a} \tag{4.28.6}$$

Using Eqs (4.28.2) and (4.28.6) in Eq. (4.28.5), we get

$$[H_3O^+] + [B^+] = \frac{[HA]_0\, K_a}{[H_3O^+] + K_a} + \frac{K_w}{[H_3O^+]}$$

which on rearranging gives

$$[H_3O^+]^3 + ([B^+] + K_a)\,[H_3O^+]^2 + (K_a\,[B^+] - K_a\,[HA]_0 - K_w)\,[H_3O^+] - K_wK_a = 0 \tag{4.28.7}$$

At any stage of titration, we will have

$$[HA]_0 = \frac{V_aM_a}{V_a + V_b} \quad \text{and} \quad [B^+] = \frac{V_bM_b}{V_a + V_b}$$

where V_b is the volume of base (BOH of molarity M_b) added to the volume V_a of acid (HA of molarity M_a). Substituting these in Eq. (4.28.7), we get

$$[H_3O^+]^3 + \left(\frac{V_bM_b}{V_a + V_b} + K_a\right)[H_3O^+]^2 + \left(K_a\frac{V_bM_b}{V_a + V_b} - K_a\frac{V_aM_a}{V_a + V_b} - K_w\right)[H_3O^+] - K_wK_a = 0 \tag{4.28.8}$$

or $$[H_3O^+]^3 (V_a + V_b) + [V_bM_b + K_a (V_a + V_b)] [H_3O^+]^2$$
$$+ [K_aV_bM_b - K_aV_aM_a - K_w (V_a + V_b)] [H_3O^+] - K_wK_a (V_a + V_b) = 0$$

Collecting the terms of V_a and V_b separately, we get

$$([H_3O^+]^3 + K_a[H_3O^+]^2 - (K_aM_a + K_w) [H_3O^+] - K_wK_a)V_a$$
$$+ ([H_3O^+]^3 + (M_b + K_a)[H_3O^+]^2 + (K_aM_b - K_w)[H_3O^+] - K_wK_a) V_b = 0 \qquad (4.28.9)$$

which gives

$$V_b = -V_a \left[\frac{[H_3O^+]^3 + K_a[H_3O^+]^2 - (K_aM_a + K_w)[H_3O^+] - K_wK_a}{[H_3O^+]^3 + (M_b + K_a)[H_3O^+]^2 + (K_aM_b - K_w)[H_3O^+] - K_wK_a} \right] \qquad (4.28.10)$$

From Eq. (4.28.10), one can compute the volume V_b of base to be added in the volume V_a of acid to obtain the known value of concentration of H_3O^+ which lies within the range accessible in the titration. For a concentration of H_3O^+ outside the accessible range, Eq. (4.28.10) gives negative value of V_b which should be ignored.

For the titration involving a strong acid, all the terms except those containing K_a in Eq. (4.28.10) may be ignored. Cancelling K_a in the numerator and denominator of the resultant expression gives

$$V_b = -V_a \left[\frac{[H_3O^+]^2 - M_a[H_3O^+] - K_w}{[H_3O^+]^2 + M_b[H_3O^+] - K_w} \right] \qquad (4.28.11)$$

Comment

At the equivalence point, we will have

$$V_aM_a = V_bM_b$$

With this, Eq. (4.28.8) is reduced to

$$[H_3O^+]^3 + \left(\frac{V_bM_b}{V_a + V_b} + K_a \right) [H_3O^+]^2 - K_w [H_3O^+] - K_wK_a = 0 \qquad (4.28.12)$$

At the equivalence point, the solution contains a salt (B^+A^-). For a weak acid, Eq. (4.28.12) is found to be identical with Eq. (4.13.10) which was derived earlier for the hydrolysis of salt formed from a weak acid and a strong base.

Buffer Effect of Titrating Solution

The plot of the derivative $(\partial V_b/\partial \text{pH})_{Va,M's}$ versus pH depicts the buffer effect of the titrating solution at different stages of titration. The expression of $\partial V_b/\partial\text{pH}$ may be obtained by differentiating Eq. (4.28.9) with respect to $[H_3O^+]$, we get

$$[3 [H_3O^+]^2 + 2K_a[H_3O^+] - (K_aM_a + K_w)] V_a + [3 [H_3O^+]^2 + 2(M_b + K_a)[H_3O^+]$$
$$+ (K_aM_b - K_w)]V_b + ([H_3O^+]^3 + (M_b + K_a) [H_3O^+]^2$$
$$+ (K_aM_b - K_w) [H_3O^+] - K_wK_a) \left(\frac{\partial V_b}{\partial [H_3O^+]} \right)_{V_a,\, M's} = 0$$

which gives

$$\left(\frac{\partial V_b}{\partial[H_3O^+]}\right)_{V_a,\,M's} = \frac{\alpha}{\beta}$$

where $\alpha = 3\,[H_3O^+]^2\,(V_b + V_a) + 2[H_3O^+]\,(V_bK_a + V_bM_b + V_aK_a)$

$$+ V_bM_bK_a - V_aM_aK_a - V_bK_w - V_aK_w$$

$$\beta = [H_3O^+]^3 + [H_3O^+]^2\,(K_a + M_b) + [H_3O^+]\,(K_aM_b - K_w) - K_wK_a$$

Using the fact that

$$\frac{\partial V_b}{\partial \text{pH}} = \frac{\partial V_b}{\partial\{-\log([H_3O^+]/M)\}} = -2.303\frac{\partial V_b}{\partial\{\ln([H_3O^+]/M)\}}$$

$$= -2.303\,[H_3O^+]\frac{\partial V_b}{\partial[H_3O^+]}$$

we get

$$\frac{\partial V_b}{\partial \text{pH}} = -2.303\,[H_3O^+]\left(\frac{\alpha}{\beta}\right) \tag{4.28.13}$$

Titration Curves

Figure 4.28.1 illustrates the plots of pH versus V_b and the corresponding $\partial V_b/\partial$pH versus V_b plots of a typical titration between a weak acid (say, 25 mL of 0.1 M acetic acid) and a strong base (say, 0.1 M NaOH).

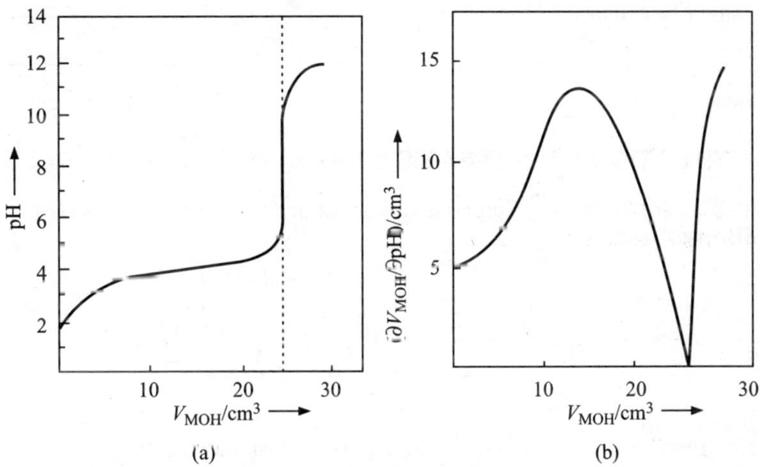

Fig. 4.28.1 (a) Plot of pH versus V_{MOH}, and (b) plot of $\partial V_{MOH}/\partial$pH versus V_{MOH} for a typical titration between a weak acid and a strong base

The derivative plot has a sharp minimum at the equivalence point where buffer capacity has a minimum value.

In the intermediate region of the titration, the buffer capacity has a maximum value as the solution contains both acid and its conjugate base. The addition of a few drops of the base does not cause significant qualitative changes in the composition of the resultant solution. The derivative plot also has a low value at the beginning of titration where small addition of base convert an essentially undissociated acid into a mixture of acid and its conjugate salt. After the equivalence point, buffer capacity goes on increasing as the addition of base into basic solution will have no measureable effect on the pH of the solution.

The buffer capacity of a solution is dependent on its total volume.

The typical plots of distribution functions α_0 (i.e. $[A^-]/[HA]_0 = 1/(1 + [H_3O^+]/K_a)$) and α_1 (i.e. $[HA]/[HA]_0 = 1/(1 + K_a/[H_3O^+])$) versus pH of the solution along with the buffer index (Eq. 4.28.13) is shown in Fig. 4.28.2.

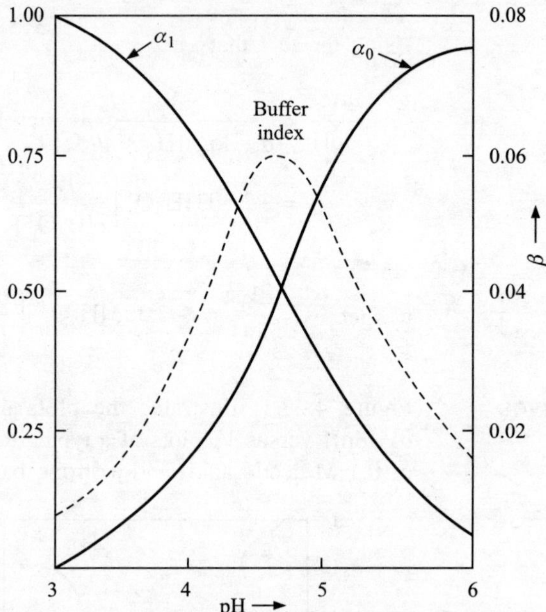

Fig. 4.28.2 Plots of the distribution functions and buffer index versus pH of the titrating solution.

4.29 TITRATION OF A DIBASIC ACID WITH A STRONG BASE

Basic Equilibrium Reactions

Stepwise dissociation of a dibasic acid is represented by the following reactions:

$$H_2A \rightleftharpoons HA^- + H^+; \qquad K_{a1} = \frac{[HA^-][H^+]}{[H_2A]}$$

$$HA^- \rightleftharpoons A^{2-} + H^+; \qquad K_{a2} = \frac{[A^{2-}][H^+]}{[HA^-]}$$

Distribution Functions

The shape and nature of the titration curve for dibasic acid depends upon the relative concentrations of various species present in the solution during the course of titration. The concentrations of these species in solution are conveniently expressed in terms of the following *distribution functions*.

$$\frac{[H_2A]}{[H_2A]_0}, \quad \frac{[HA^-]}{[H_2A]_0} \quad \text{and} \quad \frac{[A^{2-}]}{[H_2A]_0}$$

The expressions relating the distribution functions to $[H_3O^+]$ and equilibrium constants K_{a1} and K_{a2}, as derived earlier, are

$$\frac{[H_2A]}{[H_2A]_0} = \frac{1}{1 + \dfrac{K_{a1}}{[H_3O^+]} + \dfrac{K_{a1}K_{a2}}{[H_3O^+]^2}} \tag{4.29.1}$$

$$\frac{[HA^-]}{[H_2A]_0} = \frac{1}{\dfrac{[H_3O^+]}{K_{a1}} + 1 + \dfrac{K_{a2}}{[H_3O^+]}} \tag{4.29.2}$$

and

$$\frac{[A^{2-}]}{[H_2A]_0} = \frac{1}{\dfrac{[H_3O^+]^2}{K_{a1}K_{a2}} + \dfrac{[H_3O^+]}{K_{a2}} + 1} \tag{4.29.3}$$

These distribution functions provide an excellent means for visualizing how the chemical composition of solution changes as the value of $[H_3O^+]$ is varied. In general, we observe:

(i) The fraction $[H_2A]/[H_2A]_0$ is a monotonically increasing function of $[H_3O^+]$ and will reach its highest value at the lowest accessible values of pH.

(ii) The fraction $[A^{2-}]/[H_2A]_0$ is a monotonically decreasing function of $[H_3O^+]$ and will therefore take on its highest value at the highest pH value.

(iii) The fraction $[HA^-]/[H_2A]_0$ increases first in the lower range of pH and after attaining a maximum, decreases at higher values of pH.

Distribution functions for H_2SO_3 and succinic acid are shown in Figs (4.29.1) and (4.29.2), respectively.

These two distribution diagrams reveal that the maximum value of $[HA^-]/[H_2A]_0$ is about 0.75 for succinic acid and is about 1.0 for sulphurous acid. This amounts to the fact that when the pH of the solution is increased (as during the course of titration), there is a complete conversion of H_2A to HA^- only for sulphurous acid and when this conversion is complete, the subsequent increase in pH causes the gradual change of HA^- to A^{2-}. This is exhibited in the titration curves where two inflexions are observed when sulphurous acid is titrated against a strong base and only one inflexion is observed when succinic acid is titrated against a strong base.

Essential Criterion of Stepwise Neutralization

Hydrogen-ion concentration at which $[HA^-]/[H_2A]_0$ attains a maximum value can be obtained by differentiating Eq. (4.29.2) with respect to $[H_3O^+]$ and setting the resulting expression to zero. Substituting this value of $[H_3O^+]$ in Eq. (4.29.2), we get the maximum value of $[HA^-]/[H_2A]_0$ in the solution. Thus differentiating Eq. (4.29.2), we get

$$\frac{d\left\{\dfrac{[HA^-]}{[H_2A]_0}\right\}}{d([H_3O^+])} = \left\{\frac{K_{a2}}{[H_3O^+]^2} - \frac{1}{K_{a1}}\right\}\left\{\frac{[HA^-]}{[H_2A]_0}\right\}^2 = 0$$

This gives

$$[H_3O^+]_{\text{for }\{[HA^-]/[H_2A]_0\}_{\max}} = \sqrt{K_{a1}K_{a2}}$$

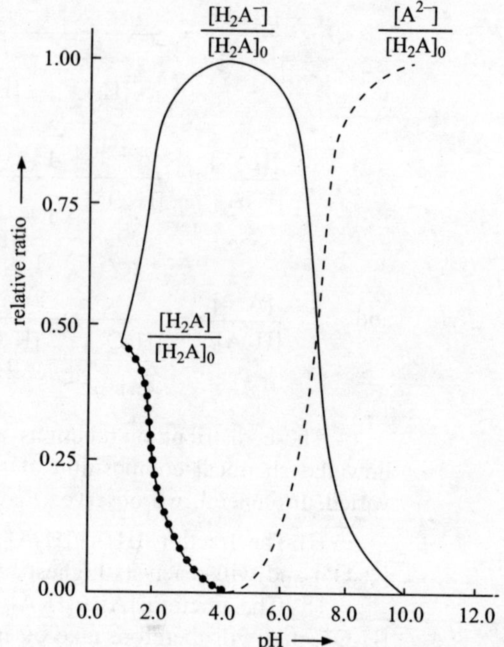

Fig. 4.29.1 Distribution functions of sulphurous acid

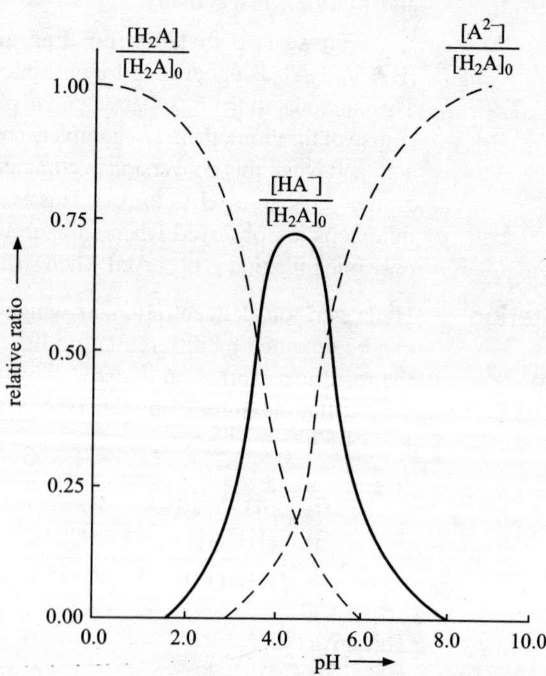

Fig. 4.29.2 Distribution function of succinic acid

This on substitution in Eq. (4.29.2) gives

$$\left\{\frac{[HA^-]}{[H_2A]_0}\right\}_{max} = \left(\frac{1}{1+2\sqrt{K_{a2}/K_{a1}}}\right) \tag{4.29.4}$$

Thus, the maximum value of $[HA^-]/[H_2A]_0$ is a function of the ratio of the two equilibrium constants. Some calculated values are shown in Table 4.29.1.

Table 4.29.1 Maximum Value of $[HA^-]/[H_2A]_0$ for a few Ratios K_{a2}/K_{a1}

K_{a2}/K_{a1}	$[HA^-]/[H_2A]_0$
1/4	0.50
1/9	0.60
1/16	0.67
1/25	0.71
1/36	0.75
$1/10^2$	0.83
$1/10^3$	0.94
$1/10^4$	0.98
$1/10^6$	0.998

It can be concluded from the above table that the essential condition for the stepwise neutralization of a diprotic acid, and thus getting two inflexion points, is that the ratio K_{a1}/K_{a2} must be equal to or greater than 10^3.

Method of Computing pH at Different Stages of Titration

In case of a diprotic acid, $[H_3O^+]$ and hence pH, at various stages of titration can be computed as follows.

At the Start of Titration

If it is assumed that the $[H_3O^+]$ due to the dissociation of the acid HA^- and from the dissociation of water is negligible in comparison to that coming from the dissociation of H_2A (which is justified for $K_{a1} >> K_{a2}$ or $K_{a1}/K_{a2} > 10^3$), then the $[H_3O^+]$ can be computed considering only the primary dissociation:

$$\underset{c(1-\alpha)}{H_2A} + H_2O \rightleftharpoons \underset{c\alpha}{H_3O^+} + \underset{c\alpha}{HA^-}$$

$$K_{a1} = \frac{[H_3O^+][HA^-]}{[H_2A]} = \frac{c\alpha^2}{1-\alpha}$$

or $$c\alpha^2 + K_{a1}\alpha - K_{a1} = 0 \tag{4.29.5}$$

Solving this quadratic equation gives the degree of dissociation of the acid. The $[H_3O^+]$ is equal to $c\alpha$.

If the acid is very weak and the solution is fairly concentrated then the $[H_3O^+]$ can be computed using a simple expression

$$[H_3O^+] = \sqrt{K_{a1}\,[H_2A]_0} \tag{4.29.6}$$

Before the First Equivalence Point

If f' represents the fraction of the acid H_2A that has been converted into HA^- then

$$[H_2A] = c(1 - f')$$

and $$[HA^-] = cf'$$

The concentrations of various species in equilibrium are

$$\underset{c(1-f')(1-\alpha)}{H_2A} + H_2O \rightleftharpoons \underset{c(1-f')\alpha}{H_3O^+} + \underset{c(1-f')\alpha + cf'}{HA^-}$$

Now $$K_{a1} = \frac{[H_3O^+][HA^-]}{[H_2A]} = \frac{\{c\,(1-f')\alpha\}\{c\,(1-f')\,\alpha + cf'\}}{c(1-f')(1-\alpha)}$$

$$= \frac{\alpha\{c\,(1-f')\,\alpha + cf'\}}{(1-\alpha)}$$

or $$c\,(1 - f')\,\alpha^2 + \alpha\,(cf' + K_{a1}) - K_{a1} = 0$$

Solving for α, $[H_3O^+]$ can be calculated from

$$[H_3O^+] = c(1 - f')\,\alpha \tag{4.29.7}$$

If the solution is fairly concentrated, the pH can be calculated using Henderson's equation since the solution will contain a weak acid and the salt of its corresponding conjugate base HA^-; we have

$$\text{pH} = \text{p}K^\circ_{a1} + \log\left(\frac{f'}{1-f'}\right) \tag{4.29.8}$$

At the First Equivalence Point

The solution contains the salt involving HA^-. The HA^- ion acts as an amphiprotic anion. The pH value of the solution, as seen earlier, can be computed through the expression

$$\text{pH} = \frac{1}{2}(\text{p}K^\circ_{a1} + \text{p}K^\circ_{a2}) \tag{4.29.9}$$

After the First Equivalence Point

The addition of alkali causes the following reaction

$$HA^- + BOH \rightarrow B^+ + A^{2-} + H_2O$$

The solution contains a weak acid HA^- and the salt of its corresponding conjugate base A^{2-}. Thus, it behaves like a buffer solution. Depending upon the second dissociation constant, the value of $[H_3O^+]$ and pH can be calculated using Eqs (4.29.7) and (4.29.8) by replacing K_{a1} by K_{a2}, such that

$$[H_3O^+] = c\,(1 - f')\,\alpha \tag{4.29.10a}$$

*All concentration terms include dilution factors, if any.

and $$\mathrm{pH} = \mathrm{p}K^\circ_{a2} + \log\left(\frac{f'}{1-f'}\right) \quad (4.29.10b)$$

At the Second Equivalence Point

Here $[H_3O^+]$ can be calculated using the equation which involves the hydrolysis of A^{2-}. Thus, we have

$$A^{2-} + H_2O \rightleftharpoons HA^- + OH^-$$

$$K_{h1} = \frac{K_w}{K_{a2}} = \frac{[HA^-][OH^-]}{[A^{2-}]}$$

Since $[HA^-] = [OH^-]$, therefore

$$[OH^-] = \sqrt{K_{h1}[A^{2-}]} = \sqrt{\frac{K_w}{K_{a2}}[A^{2-}]}$$

Hence, $$[H_3O^+] = \frac{K_w}{[OH^-]} = \frac{K_w}{\sqrt{\dfrac{K_w}{K_{a2}}[A^{2-}]}} \approx \sqrt{\frac{K_w K_{a2}}{[A^{2-}]_0}} \quad (4.29.11a)$$

or $$\mathrm{pH} = \frac{1}{2}\mathrm{p}K^\circ_w + \frac{1}{2}\mathrm{p}K^\circ_{a2} + \frac{1}{2}\log\{[A^{2-}]_0/\mathrm{mol\ L^{-1}}\} \quad (4.29.11b)$$

After the Second Equivalence Point

The solution will contain the excess of base; its OH^- can be calculated from the knowledge of the excess of base and the total volume of the solution, i.e.

$$[OH^-] = \frac{\text{Amount of free base}}{\text{Total volume of the solution}} \quad (4.29.12)$$

Illustration

Oxalic acid is a dibasic acid with dissociation constants K_{a1} and K_{a2} equal to 5.9×10^{-2} M and 6.4×10^{-5} M, respectively. Its first dissociation constant is fairly large and the ratio of K_{a1} and K_{a2} is of the order of 10^3. Thus, the titration curve of oxalic acid (say, 25.0 cm^3 of 0.05 M acid) with NaOH (0.1 M) can be obtained from the following calculated values of pH.

At the start of titration Using Eq. (4.29.5), we get

$$\alpha = 0.646$$

$$[H_3O^+] = c\alpha = (0.05\ \mathrm{M})\ (0.646) = 0.032\ 3\ \mathrm{M}$$

$$\mathrm{pH} = 1.49$$

Before the first equivalence point Using Eq. (4.29.7), the values of $[H_3O^+]$ and pH at various values of f' are given in Table 4.29.2.

Table 4.29.2 pH Values before the Equivalence Point

f'	$\frac{V_{base}}{cm^3}$	$\frac{V_{total}}{cm^3}$	$\frac{[H_2A]\times 10^2}{M}$	$\frac{[HA^-]\times 10^2}{M}$	α	$\frac{[H_3O^+]\times 10^2}{M}$	pH[†]
0.1	1.25	26.25	4.28	0.476	0.646	2.76	1.56
0.2	2.5	27.5	3.60	0.909	0.646	2.33	1.63
0.4	5.0	30.0	2.50	1.670	0.643	1.61	1.79
0.6	7.5	32.5	1.54	2.30	0.642	0.99	2.00
0.8	10.0	35.0	0.714	2.86	0.640	0.46	2.34
0.9	11.25	36.25	0.345	3.10	0.640	0.22	2.66

[†]Equation (4.29.8) will not give correct pH as $[H_3O^+]$ is comparable with $[H_2A]$.

At the first equivalence point Here pH can be calculated by using Eq. (4.29.9); that is,

$$\text{pH} = \frac{1}{2}(\text{p}K_{a1}^{\circ} + \text{p}K_{a2}^{\circ}) = \frac{1}{2}[-\log(K_{a1}/\text{M}) - \log(K_{a2}/\text{M})]$$

$$= \frac{1}{2}[-\log(5.9\times 10^{-2}) - \log(6.4\times 10^{-5})]$$

$$= \frac{1}{2}[1.23 + 4.19] = 2.68$$

After the first equivalence point The value of K_{a2} is fairly small. The values of pH in this range can be calculated using the simplified expression of Eq. (4.29.10). A few values are shown in Table 4.29.3.

Table 4.29.3 pH Values after the First Equivalence Point

f'	$\frac{V_{base}}{cm^3}$	$\frac{V_{total}}{cm^3}$	$\frac{[HA^-]\times 10^2}{M}$	$\frac{[A^{2-}]\times 10^2}{M}$	α	$\frac{[H_3O^+]\times 10^5}{M}$	pH[†]
0.1	13.75	38.75	2.90	0.322	0.017	49.1	3.31
0.2	15.0	40.0	2.50	0.624	0.010	24.4	3.61
0.4	17.5	42.5	1.760	1.175	0.005	9.45	4.02
0.6	20.0	45.0	1.111	1.665	0.004	4.24	4.37
0.8	22.5	47.5	0.526	2.103	0.003	1.594	4.80
0.9	23.75	48.75	0.256	2.305	0.003	0.709	5.15

[†]Both Eqs (4.29.10a) and (4.29.10b) will give identical values of pH as $[H_3O^+] \ll [HA^-]$.

At the second equivalence point Using expression (4.29.11), we get

$$[H_3O^+] = \left\{\frac{(1.0\times 10^{-14}\ \text{M}^2)(6.4\times 10^{-5}\ \text{M})}{0.025\ \text{M}}\right\}^{1/2}$$

$$= 5.06\times 10^{-9}\ \text{M}$$

$$\text{pH} = 8.30$$

After the second equivalence point According to Eq. (4.29.11) some of the values of pH of the solution are given in Table 4.29.4.

Table 4.29.4 pH Values after the Second Equivalence Point

Amount of base/mol	V_{base}/cm^3	V_{total}/cm^3	$[H_3O^+]/M$	pH
0.000 1	26.0	51.0	5.1×10^{-12}	11.29
0.000 3	28.0	53.0	1.77×10^{-12}	11.75
0.000 5	30.0	55.0	1.12×10^{-12}	11.95
0.001 0	35.0	60.0	6.03×10^{-13}	12.22
0.001 5	40.0	65.0	4.37×10^{-13}	12.36
0.002 0	45.0	70.0	3.47×10^{-13}	12.46

Titration Curve

The plot of pH versus V_b obtained from the data calculated above is shown in Fig. 4.29.3

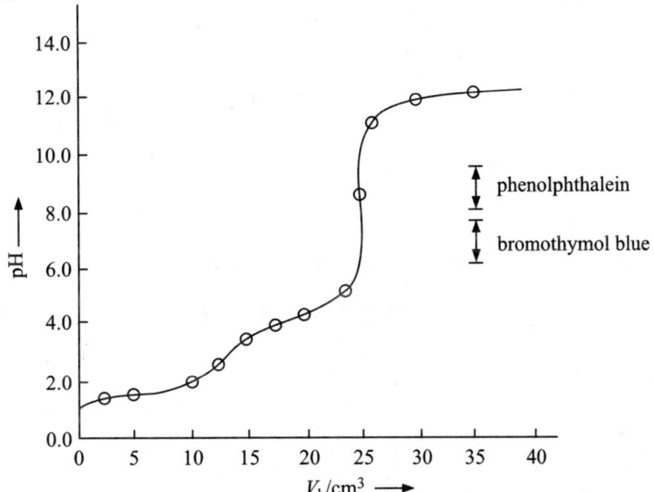

Fig. 4.29.3 Titration curve of oxalic acid with sodium hydroxide

4.30 GENERAL TREATMENT OF TITRATION OF A DIPROTIC ACID WITH A STRONG BASE

Basic Equilibrium Reactions

In the titration of a diprotic acid (H_2A) with a strong base (BOH), the equilibria involved are

$$H_2A + H_2O \rightleftharpoons HA^- + H_3O^+; \quad K_{a1} = \frac{[HA^-][H_3O^+]}{[H_2A]} \qquad (4.30.1)$$

$$HA^- + H_2O \rightleftharpoons A^{2-} + H_3O^+; \quad K_{a2} = \frac{[A^{2-}][H_3O^+]}{[HA^-]} \qquad (4.30.2)$$

$$H_2O + H_2O \rightleftharpoons H_3O^+ + OH^-; \quad K_w = [H_3O^+][OH^-] \qquad (4.30.3)$$

Condition of Mass Balance

The material-balance expressions are

$$[H_2A]_0 = [H_2A] + [HA^-] + [A^{2-}] \qquad (4.30.4)$$

$$[B^+]_0 = [B] \qquad (4.30.5)$$

Condition of Charge Balance

The charge-balance expression is

$$[H_3O^+] + [B^+] = 2[A^{2-}] + [HA^-] + [OH^-] \tag{4.30.6}$$

Derivation of Exact Expression

Using Eqs (4.30.2) and (4.30.3) in Eq. (4.30.6), we get

$$[H_3O^+] + [B^+] = 2\,\frac{K_{a2}\,[HA^-]}{[H_3O^+]} + [HA^-] + \frac{K_w}{[H_3O^+]} \tag{4.30.7}$$

Expressing $[H_2A]$ and $[A^{2-}]$ in Eq. (4.30.4) in terms of $[HA^-]$ by using Eqs (4.30.1) and (4.30.2), we get

$$[H_2A]_0 = \frac{[HA^-][H_3O^+]}{K_{a1}} + [HA^-] + \frac{K_{a2}\,[HA^-]}{[H_3O^+]}$$

i.e.

$$[HA^-] = \frac{[H_2A]_0}{([H_3O^+]/K_{a1}) + 1 + (K_{a2}/[H_3O^+])} \tag{4.30.8}$$

Substituting this in Eq. (4.30.7), we get

$$[H_3O^+] + [B^+] = \left(\frac{[H_2A]_0}{([H_3O^+]/K_{a1}) + 1 + (K_{a2}/[H_3O^+])}\right) \times \left(\frac{2\,K_{a2}}{[H_3O^+]} + 1\right) + \frac{K_w}{[H_3O^+]}$$

which on rewritting is given by

$$[H_3O^+]^4 + [H_3O^+]^3(K_{a1} + [B]) + [H_3O^+]^2(K_{a1}K_{a2} + K_{a1}[B] - K_{a1}\,[H_2A]_0 - K_w) + [H_3O^+]\,K_{a1}\,(K_{a2}\,[B] - 2\,K_{a2}\,[H_2A]_0 - K_w) - K_{a1}\,K_{a2}\,K_w = 0 \tag{4.30.9}$$

At any stage of titration, we can write

$$[H_2A]_0 = \frac{V_a M_a}{V_a + V_b} \tag{4.30.10}$$

$$[B] = \frac{V_b M_b}{V_a + V_b} \tag{4.30.11}$$

where V_b is the volume of base (BOH of molarity M_b) added to the volume V_a of acid (H_2A of molarity M_a). Substituting these in Eq. (4.30.9), we get

$$[H_3O^+]^4 + [H_3O^+]^3\left(K_{a1} + \frac{V_b M_b}{V_a + V_b}\right) + [H_3O^+]^2\left(K_{a1}K_{a2} + K_{a1}\frac{V_b M_b}{V_a + V_b} - K_{a1}\frac{V_a M_a}{V_a + V_b} - K_w\right) + [H_3O^+]\,K_{a1}\left(K_{a2}\frac{V_b M_b}{V_a + V_b} - 2K_{a2}\frac{V_a M_a}{V_a + V_b} - K_w\right) - K_{a1}K_{a2}K_w = 0 \tag{4.30.12}$$

which gives

$$V_b = -V_a \left(\frac{\alpha}{\beta}\right) \tag{4.30.13}$$

where

$$\alpha = [H_3O^+]^4 + [H_3O^+]^3\, K_{a1} + [H_3O^+]^2\,(K_{a1}K_{a2} - K_{a1}M_a - K_w) + [H_3O^+]\,(-2K_{a1}K_{a2}M_a - K_wK_{a1}) - K_{a1}K_{a2}K_w \tag{4.30.14}$$

$$\beta = [H_3O^+]^4 + [H_3O^+]^3\,(K_{a1} + M_b) + [H_3O^+]^2\,(K_{a1}K_{a2} + K_{a1}M_b - K_w) + [H_3O^+]\,(K_{a1}K_{a2}M_b - K_{a1}K_w) - K_{a1}K_{a2}K_w \tag{4.30.15}$$

From Eq. (4.30.13), one can compute the volume V_b of base to be added in the volume V_a of acid to obtain the known value of concentration of H_3O^+ which lies within the range accessible in the titration. For a concentration of H_3O^+ outside the accessible range, Eq. (4.30.13) gives negative value of V_b which should be ignored.

4.31 TITRATION OF SODIUM CARBONATE SOLUTION WITH HYDROCHLORIC ACID

Basic Equilibrium Reactions

Following equilibrium exists in a solution of CO_2 in water:

$$CO_2(aq) + H_2O \rightleftharpoons H_2CO_3$$

with $$K' = \frac{[H_2CO_3]}{[CO_2(aq)]} = 2.6 \times 10^{-3}$$

Carbonic acid is a diprotic acid which undergoes ionization in two stages:

$$H_2CO_3 \rightleftharpoons H^+ + HCO_3^-; \quad K_{a1} = \frac{[H^+][HCO_3^-]}{[H_2CO_3]} = 1.72 \times 10^{-4}\ \text{M}$$

and $$HCO_3^- \rightleftharpoons H^+ + CO_3^-; \quad K_{a2} = \frac{[H^+][CO_3^{2-}]}{[HCO_3^-]} = 4.68 \times 10^{-11}\ \text{M}$$

For the purpose of calculations of pHs and the construction of a titration curve, it is convenient to work with the total analytical concentration of carbon dioxide present as CO_2(aq) and H_2CO_3. For this, we use the value of K_1 given by

$$\begin{aligned} K_1 &= \frac{[H^+][HCO_3^-]}{([CO_2(aq)] + [H_2CO_3])} \approx K_{a1}K' \\ &= (1.72 \times 10^{-4}\ \text{M})(2.6 \times 10^{-3}) \\ &= 4.47 \times 10^{-7}\ \text{M} \end{aligned}$$

Since the value of $K_1/K_{a2} > 10^3$, it can be considered that the reaction of sodium carbonate with hydrochloric acid takes place in stages. First, there is a complete conversion of CO_3^{2-} to HCO_3^- followed by the conversion of HCO_3^- to carbonic acid.

The computation procedure for the calculation of $[H_3O^+]$ in solution at various stages of titration is described below.

Method of Computing pH at Different Stages of Titration

At the Start of Titration

The pH of the solution can be determined from the following principal equilibrium involving the hydrolysis of CO_3^{2-}.

$$CO_3^{2-} + H_2O \rightleftharpoons HCO_3^- + OH^-$$

$$K_h = \frac{K_w}{K_{a2}} = \frac{[HCO_3^-][OH^-]}{[CO_3^{2-}]}$$

Since $[HCO_3^-] = [OH^-]$ and $[CO_3^{2-}] = [CO_3^{2-}]_0 - [OH^-]$, therefore, we have

$$K_h = \frac{K_w}{K_{a2}} = \frac{[OH^-]^2}{[CO_3^{2-}]_0 - [OH^-]}$$

or $$[OH^-]^2 + \frac{K_w}{K_{a2}}\{[OH^-] - [CO_3^{2-}]_0\} = 0 \qquad (4.31.1a)$$

Solving this quadratic expression, we get $[OH^-]$. If the solution is fairly concentrated and the value of K_h is small, we can use the following simple expression to compute $[OH^-]$.

$$[OH^-]^2 = \sqrt{\frac{K_w}{K_{a2}}[CO_3^{2-}]_0} \qquad (4.31.1b)$$

Knowing $[OH^-]$, hydronium-ion concentration can be computed from $K_w/[OH^-]$, from which pH can be calculated.

Before the First Equivalence Point

At this stage of titration, we have the reaction

$$CO_3^{2-} + H^+ \rightleftharpoons HCO_3^-$$

Solution contains CO_3^{2-} and HCO_3^-—a buffer solution. Hence, the pH of the solution can be computed from Henderson's equation

$$pH = pK_{a2}^\circ + \log\frac{[CO_3^{2-}]}{[HCO_3^-]}$$

or $$pH = pK_{a2}^\circ + \log\left(\frac{1-f'}{f'}\right) \qquad (4.31.2)$$

where f' is the fraction of CO_3^{2-} that has been converted into HCO_3^-.

At the First Equivalence Point

The solution contains only HCO_3^-, an amphiprotic anion. Therefore, the pH value of the solution, as seen earlier, is given by

$$pH = \frac{1}{2}(pK_1^\circ + pK_{a2}^\circ) \qquad (4.31.3)$$

Between the First and Second Equivalence Points

The addition of acid causes the following reaction.

$$HCO_3^- + H^+ \rightleftharpoons H_2CO_3 \rightleftharpoons CO_2(aq) + H_2O$$

Using the overall dissociation constant of carbonic acid, we have

$$\{H_2CO_3 + CO_2(aq)\} + H_2O \rightleftharpoons H_3O^+ + HCO_3^-$$

$$K_1 = \frac{[H_3O^+][HCO_3^-]}{[H_2CO_3 + CO_2(aq)]} \tag{4.31.4}$$

The hydrogen-ion concentration is given by

$$[H_3O^+] = K_1 \frac{[H_2CO_3]}{[HCO_3^-]}$$

where $[H_2CO_3]$ is the total analytical concentration of dissolved carbon dioxide in its two forms. Thus, the pH of the solution is given by

$$pH = pK_1^\circ + \log\left(\frac{1-f''}{f''}\right) \tag{4.31.5}$$

where f'' is the fraction of HCO_3^- that has been converted into H_2CO_3.

At the Second Equivalence Point

In Eq. (4.31.4),

$$[H_3O^+] = [HCO_3^-]$$

$$[H_2CO_3] = [H_2CO_3]_0 - [H_3O^+] \simeq [H_2CO_3]_0$$

Hence, $K_1 = \dfrac{[H_3O^+]^2}{[H_2CO_3]_0}$ or $[H_3O^+] = \sqrt{K_1[H_2CO_3]_0}$ (4.31.6)

After the Second Equivalence Point

The solution contains excess of HCl, the $[H_3O^+]$ of which can be determined using

$$[H_3O^+] = \frac{\text{Amount of } H_3O^+}{\text{Total volume of the solution}} \tag{4.31.7}$$

Illustration

The following are the calculated values of pH of the solution at various stages of titration when 0.05 M of sodium carbonate (50.0 cm^3) is titrated against 0.1 M of hydrochloric acid.

At the start of titration The Eq. (4.31.1a) becomes

$$[OH^-]^2 + \frac{1.0 \times 10^{-14}\ M^2}{4.68 \times 10^{-11}\ M}\{[OH^-] - 0.05\ M\} = 0$$

which gives

$$[OH^-] = 3.16 \times 10^{-3}\ M; \quad pOH = 2.5 \quad \text{and} \quad pH = 11.50$$

Before the first equivalence point The values of pH at different values of f' are given in Table 4.31.1.

At the first equivalence point Employing Eq. (4.31.3) we have

$$pH = \frac{1}{2}[pK_1^\circ + pK_{a2}^\circ]$$

$$= \frac{1}{2}[-\log(4.47 \times 10^{-7}) - \log(4.68 \times 10^{-11})] = 8.34$$

Table 4.31.1 pH Values before the First Equivalence Point

Volume of acid added, V_a/cm^3	f'	pH *using* Eq. (4.31.2)
5.0	0.2	10.93
10.0	0.4	10.51
15.0	0.6	10.15
20.0	0.8	9.73
22.5	0.9	9.38

Between the first and second equivalence points The values of pH at various values of f'' are given in Table 4.31.2.

Table 4.31.2 pH Values Between the First and Second Equivalence Points

Volume of acid added, V_a/cm^3	f''	pH *using* Eq. (4.31.4)
30.0	0.2	6.95
35.0	0.4	6.53
40.0	0.6	6.17
45.0	0.8	5.75
47.5	0.9	5.40

At the second equivalence point Here

$$K_1 = \frac{[H^+][HCO_3^-]}{[H_2CO_3]} = \frac{[H^+]^2}{0.025\ M - [H^+]} \approx \frac{[H^+]^2}{0.025\ M}$$

Also $K_1 = 4.47 \times 10^{-7}$ M

Hence, $[H^+] = \sqrt{(0.025\ M)(4.47 \times 10^{-7}\ M)} = 1.06 \times 10^{-4}$ M

$pH = 3.97$

Beyond the second equivalence point Some of the values of pH after the second equivalence point are recorded in Table 4.31.3.

Table 4.31.3 pH Values After the Second Equivalence Point

Volume of HCl *added,* V_a/cm^3	pH
52.5	2.61
55.0	2.32
60.0	2.04

Titration Curve

The above values of pH when plotted against volume of acid added gives the plot as shown in Fig. (4.31.1).

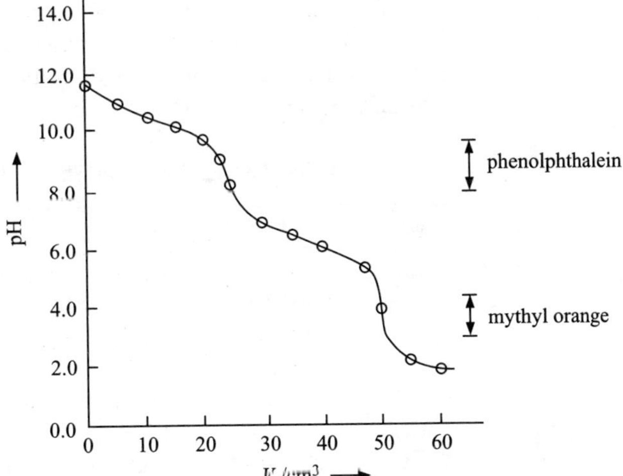

Fig. 4.31.1 Titration curve of sodium carbonate wtih hydrochloric acid

4.32 SOLUBILITY PRODUCT

Definition

If a slightly soluble salt is placed in water, a dynamic equilibrium is established where the rate of dissolution of ions from the solid equals the rate of precipitation of ions from the saturated solution. For example, in AgCl, we have the following equilibrium

$$AgCl(s) \rightleftharpoons Ag^+(aq) + Cl^-(aq)$$

with the equilibrium constant

$$K' = \frac{[Ag^+][Cl^-]}{[AgCl]}$$

Since the concentration of a pure solid is a constant, [AgCl] may be combined with K' to give a new constant, called the *solubility product*:

$$K_s = K' [AgCl] = [Ag^+] [Cl^-]$$

Thus, the solubility product is equal to the product of the ionic concentrations present in a saturated solution. When the salt dissolves to give unequal numbers of positive and negative ions, each concentration term must be raised to a power equal to the coefficient of the ion in the chemical equation. For example:

Salt	*Reaction*	*Solubility product expression*
CaF_2	$CaF_2(s) \rightleftharpoons Ca^{2+} + 2F^-$	$K_s(CaF_2) = [Ca^+] [F^-]^2$
PbI_2	$PbI_2(s) \rightleftharpoons Pb^{2+} + 2I^-$	$K_s(PbI_2) = [Pb^{2+}] [I^-]^2$
Hg_2Cl_2	$Hg_2Cl_2(s) \rightleftharpoons Hg_2^{2+} + 2Cl^-$	$K_s(Hg_2Cl_2) = [Hg_2^{2+}] [Cl^-]^2$
$Al(OH)_3$	$Al(OH)_3(s) \rightleftharpoons Al^{3+} + 3OH^-$	$K_s\{Al(OH)_3\} = [Al^{3+}] [OH^-]^3$
$Ag_3(PO_4)$	$Ag_3(PO_4)(s) \rightleftharpoons 3Ag^+ + PO_4^{3-}$	$K_s\{Ag_3(PO_4)\} = [Ag^+]^3 [PO_4^{3-}]$
$Ca_3(PO_4)_2$	$Ca_3(PO_4)_2(s) \rightleftharpoons 3Ca^{2+} + 2PO_4^{3-}$	$K_s\{Ca_3(PO_4)_2\} = [Ca^{2+}]^3 [PO_4^{3-}]^2$
$KAl(SO_4)_2$	$KAl(SO_4)_2 \rightleftharpoons Al^{3+} + K^+ + 2SO_4^{2-}$	$K_s\{KAl(SO_4)_2\} = [K^+] [Al^{3+}] [SO_4^{2-}]^2$

Since the solubility of a salt varies with temperature, the numerical value of K_s for a salt changes with temperature; values are usually recorded at 25 °C. The numerical values of standard solubility product for some of the common sparingly soluble salts are listed in Table 4.32.1.

Table 4.32.1 Values of Standard Solubility Products at 25 °C†

Salt	K_s°	*Salt*	K_s°	*Salt*	K_s°
$PbBr_2$	4.6×10^{-6}	HgS	3.0×10^{-53}	$Pb(OH)_2$	4.2×10^{-15}
Hg_2Br_2	1.3×10^{-22}	CuS	4.0×10^{-38}	$Mg(OH)_2$	8.9×10^{-13}
AgBr	5.0×10^{-13}	PbS	1.0×10^{-29}	$Mn(OH)_2$	2×10^{-13}
$BaCO_3$	1.6×10^{-9}	SnS	8.0×10^{-29}	$Ni(OH)_2$	1.6×10^{-16}
$CdCO_3$	5.2×10^{-12}	Sb_2S_3	10^{-80}	$Sr(OH)_2$	3.2×10^{-4}
$CaCO_3$	4.7×10^{-9}	Bi_2S_3	1.6×10^{-72}	$Sn(OH)_2$	3×10^{-27}
$CuCO_3$	2.5×10^{-10}	CdS	1.0×10^{-28}	$Zn(OH)_2$	4.5×10^{-17}
$FeCO_3$	2.1×10^{-11}	ZnS	2.5×10^{-22}	PbI_2	8.3×10^{-9}
$PbCO_3$	1.5×10^{-15}	CoS	7×10^{-23}	Hg_2I_2	4.5×10^{-29}
$MgCO_3$	1×10^{-5}	NiS	3×10^{-21}	AgI	8.5×10^{-17}
$MnCO_3$	8.8×10^{-11}	MnS	5.6×10^{-16}	AgCN	1.6×10^{-14}
Hg_2CO_3	9.0×10^{-17}	FeS	1.0×10^{-19}	AgCNS	1.0×10^{-12}
$NiCO_3$	1.4×10^{-7}				
Ag_2CO_3	8.2×10^{-12}	BaF_2	2.4×10^{-5}	BaC_2O_4	1.5×10^{-8}
$SrCO_3$	7×10^{-10}	CaF_2	3.9×10^{-11}	CaC_2O_4	1.3×10^{-9}
$ZnCO_3$	2×10^{-10}	PbF_2	4×10^{-8}	PbC_2O_4	8.3×10^{-12}
		MgF_2	8×10^{-8}	MgC_2O_4	8.6×10^{-5}
$PbCl_2$	1.6×10^{-5}	SrF_2	7.9×10^{-11}	$Ag_2C_2O_4$	1.1×10^{-11}
Hg_2Cl_2	1.1×10^{-18}			SrC_2O_4	5.6×10^{-8}
AgCl	1.7×10^{-10}	$Al(OH)_3$	5×10^{-33}	$Ba_3(PO_4)_2$	6×10^{-39}
		$Ba(OH)_2$	5×10^{-3}	$Ca_3(PO_4)_2$	1.3×10^{-32}
$BaCrO_4$	8.5×10^{-11}	$Cd(OH)_2$	2×10^{-14}	$Pb_3(PO_4)_2$	1×10^{-54}
$PbCrO_4$	2×10^{-16}	$Ca(OH)_2$	1.3×10^{-6}	$Ag_3(PO_4)$	1.8×10^{-18}
Hg_2CrO	2×10^{-9}	$Cr(OH)_3$	6.7×10^{-31}	$Sr_3(PO_4)_2$	1×10^{-31}
Ag_2CrO_4	1.9×10^{-12}	$Co(OH)_2$	2.5×10^{-16}		
$SrCrO_4$	3.6×10^{-5}	$Co(OH)_3$	2.5×10^{-43}	$BaSO_4$	1.5×10^{-9}
		$Cu(OH)_2$	1.6×10^{-19}	$CaSO_4$	2.4×10^{-5}
		$Fe(OH)_2$	1.8×10^{-15}	$PbSO_4$	1.3×10^{-8}
		$Fe(OH)_3$	6×10^{-38}	Ag_2SO_4	1.2×10^{-5}
				$SrSO_4$	7.6×10^{-7}

† In the expression of standard solubility product, c is replaced by c/c°, where c° is the standard concentration of 1 M.

Solubility and Solubility Product

The numerical value of the solubility product for a salt can be determined from its molar solubility and vice versa. The following problems illustrate the procedure.

Example 4.32.1

The solubility of strontium oxalate at 20 °C is 0.000 54 M. Calculate the solubility product.

Solution

Since SrOx $\rightleftharpoons$ $Sr^{2+} + Ox^{2-}$, therefore

$$[Sr^{2+}] = [Ox^{2-}] = 0.000\ 54\ M = 5.4 \times 10^{-4}\ M$$

Solubility product is

$$K_s(SrOx) = [Sr^{2+}]\ [Ox^{2-}] = (5.4 \times 10^{-4}\ M)^2 = 2.92 \times 10^{-7}\ M^2$$

Example 4.32.2

The solubility of CaF_2 in water at 20 °C is 15.6 mg per dm^3 of solution. Calculate the solubility product of CaF_2.

Solution

$$\text{Solubility in moles per dm}^3 = \frac{(15.6 \times 10^{-3}\ g\ dm^{-3})}{(78\ g\ mol^{-1})} = 2.0 \times 10^{-4}\ M$$

Since $CaF_2 \rightleftharpoons Ca^{2+} + 2F^-$, therefore

$$[Ca^{2+}] = 2.0 \times 10^{-4}\ M \quad \text{and} \quad [F^-] = 2 \times 2.0 \times 10^{-4}\ M$$

Hence, the solubility product is

$$K_s = [Ca^{2+}]\ [F^-]^2 = (2.0 \times 10^{-4}\ M)\ (4.0 \times 10^{-4}\ M)^2 = 32 \times 10^{-12}\ M^3$$

Example 4.32.3

What is the solubility of $Ag_2(CrO_4)$ in water if the value of solubility product is $K_s = 1.3 \times 10^{-11}\ M^3$?

Solution

Since $Ag_2CrO_4 \rightleftharpoons 2Ag^+ + CrO_4^{2-}$, therefore

$$[Ag^+] = 2[CrO_4^{2-}]$$

Let s be the solubility of Ag_2CrO_4, then

$$[Ag^+] = 2s \quad \text{and} \quad [CrO_4^{2-}] = s$$

Hence, $K_s = [Ag^+]^2\ [CrO_4^{2-}] = (2s)^2\ (s)$ i.e. $4s^3 = 1.3 \times 10^{-11}\ M^3$

Therefore, $s = \{3.25 \times 10^{-12}\ M^3\}^{1/3} = 1.48 \times 10^{-4}\ M$

Example 4.32.4

A mixture of solid $SrSO_4$ and solid $BaSO_4$ is shaken up with water until saturation equilibrium is established. Given that $K_s(SrSO_4) = 7.6 \times 10^{-7}\ M^2$ and $K_s(BaSO_4) = 1.5 \times 10^{-9}\ M^2$, calculate the concentrations of Sr^{2+}, Ba^{2+} and SO_4^{2-}.

Solution

If x and y are the respective concentrations of $SrSO_4$ and $BaSO_4$ in the solution, then

$$[Sr^{2+}] = x; \qquad [Ba^{2+}] = y; \qquad [SO_4^{2-}] = x + y$$

Thus, $K_s(SrSO_4) = [Sr^{2+}]\ [SO_4^{2-}] = x(x + y) = 7.6 \times 10^{-7}\ M^2$

$$K_s(BaSO_4) = [Ba^{2+}]\ [SO_4^{2-}] = y(x + y) = 1.5 \times 10^{-9}\ M^2$$

Solving for x and y, we get

$$x = [Sr^{2+}] = 8.7 \times 10^{-4}\ M \quad \text{and} \quad y = [Ba^{2+}] = 1.7 \times 10^{-6}\ M$$

$$x + y = [SO_4^{2-}] = 8.717 \times 10^{-4}\ M$$

Factors Affecting Solubility of Salt

There is evidence that some salts ionize in steps. For example, lead chloride

$$PbCl_2(s) \rightleftharpoons Pb^{2+}(aq) + 2Cl^-(aq) \qquad K_s = [Pb^{2+}][Cl^-]^2$$

is thought to ionize according to the expressions

$$PbCl_2(s) \rightleftharpoons PbCl_2(aq) \qquad K_1 = [PbCl_2(aq)]$$

$$PbCl_2(aq) \rightleftharpoons PbCl^+(aq) + Cl^-(aq) \qquad K_2 = \frac{[PbCl^+][Cl^-]}{[PbCl_2(aq)]}$$

$$PbCl^+(aq) \rightleftharpoons Pb^{2+}(aq) + Cl^-(aq) \qquad K_3 = \frac{[Pb^{2+}][Cl^-]}{[PbCl^+]}$$

The product of the constants K_1, K_2 and K_3 is

$$K_1K_2K_3 = [PbCl_2(aq)]\left(\frac{[PbCl^+][Cl^-]}{[PbCl_2(aq)]}\right)\left(\frac{[Pb^{2+}][Cl^-]}{[PbCl^+]}\right)$$

$$= [Pb^{2+}][Cl^-]^2 = K_s$$

In these expressions, $PbCl_2(aq)$ is an undissociated molecule in solution.

Thus, the principle of solubility product applies to all solutions of slightly soluble materials whether they dissociate in steps or not. However, the K_ss of salts that dissociate in steps must be carefully interpreted. For example, in a saturated solution of $PbCl_2$, the concentration of Cl^- is not twice the concentration of Pb^{2+} as the expression for the K_s might lead one to expect; such an erroneous deduction ignores the dissociation of the salt in steps. The value of K_s applies only if the actual concentrations of Pb^{2+} and Cl^- are employed, and the stepwise mechanism is taken into account in order to deduce the correct concentration terms.

Other factors such as complex formation, hydrolysis, etc., introduce errors into solubility calculations for certain salts. For example, the solubility of lead chloride is enhanced in a moderately concentrated solution of chloride ions because of the formation of the complex ion $PbCl_3^-$, so that

$$PbCl_2(s) + Cl^-(aq) \rightleftharpoons PbCl_3^-(aq)$$

Also, the hydrolysis of the Pb^{2+} ions according to the reaction

$$Pb^{2+}(aq) + 2H_2O \rightleftharpoons Pb(OH)^+(aq) + H_3O^+(aq)$$

reduces the concentration of Pb^{2+} ions so that the solubility of $PbCl_2$ is actually higher than the value obtained from the calculation that ignores hydrolysis.

Example 4.32.5

What is the solubility of PbS; (a) ignoring the hydrolysis of ions, and (b) including the hydrolysis of ions (assume pH of the solution to be equal to 7)? Given that $K_s(\text{PbS}) = 7.0 \times 10^{-29}\ \text{M}^2$, $K_{b2}(\text{Pb(OH)}^+) = 1.5 \times 10^{-8}$ M, $K_{a1}(\text{H}_2\text{S}) = 1.0 \times 10^{-7}$ M and $K_{a2}(\text{HS}^-) = 1.0 \times 10^{-14}$ M.

Solution

Ignoring hydrolysis Let s be the concentration of PbS in the solution, then

$$K_s(\text{PbS}) = [\text{Pb}^{2+}]\,[\text{S}^{2-}] = 7.0 \times 10^{-29}\ \text{M}^2$$

Hence, $(s)\,(s) = 7.0 \times 10^{-29}\ \text{M}^2$ or $s = 8.4 \times 10^{-15}$ M

Including hydrolysis The given assumption that pH of the solution to be taken equal to 7 is justified on the basis that any H^+ or OH^- obtained through the hydrolysis of very low concentrations of Pb^{2+} and S^{2-} (of the order of 8.4×10^{-15} M) may be ignored in comparison to that derived from the ionization of water. The solubility of PbS including hydrolysis of ions may be calculated as follows:

The various equilibria are

$$\text{Pb}^{2+} + 2\text{H}_2\text{O} \rightleftharpoons \text{Pb(OH)}^+ + \text{H}_3\text{O}^+ \qquad K_{h1} = \frac{[\text{Pb(OH)}^+][\text{H}_3\text{O}^+]}{[\text{Pb}^{2+}]} \tag{1}$$

$$\text{S}^{2-} + \text{H}_2\text{O} \rightleftharpoons \text{HS}^- + \text{OH}^- \qquad K_{h2} = \frac{[\text{HS}^-][\text{OH}^-]}{[\text{S}^{2-}]} \tag{2}$$

$$\text{HS}^- + \text{H}_2\text{O} \rightleftharpoons \text{H}_2\text{S} + \text{OH}^- \qquad K_{h3} = \frac{[\text{H}_2\text{S}][\text{OH}^-]}{[\text{HS}^-]} \tag{3}$$

The values of K_{h1}, K_{h2} and K_{h3} are

$$K_{h1} = \frac{[\text{Pb(OH)}^+]}{[\text{Pb}^{2+}][\text{OH}^-]}[\text{H}_3\text{O}^+][\text{OH}^-]$$

$$= \frac{K_w}{K_{b2}(\text{Pb(OH}^+))} = \frac{1.0 \times 10^{-14}\ \text{M}^2}{1.5 \times 10^{-8}\ \text{M}} = 6.7 \times 10^{-7}\ \text{M} \tag{4}$$

$$K_{h2} = \frac{[\text{HS}^-]}{[\text{S}^{2-}][\text{H}_3\text{O}^+]}[\text{OH}^-][\text{H}_3\text{O}^+]$$

$$= \frac{K_w}{K_{a2}} = \frac{1.0 \times 10^{-14}\ \text{M}^2}{1.0 \times 10^{-14}\ \text{M}} = 1.0\ \text{M} \tag{5}$$

$$K_{h3} = \frac{[\text{H}_2\text{S}]}{[\text{HS}^-][\text{H}_3\text{O}^+]}[\text{OH}^-][\text{H}_3\text{O}^+]$$

$$= \frac{K_w}{K_{a1}} = \frac{1.0 \times 10^{-14}\ \text{M}^2}{1.1 \times 10^{-7}\ \text{M}} = 9.1 \times 10^{-8}\ \text{M} \tag{6}$$

Mass-balance expressions are

$$[\text{Pb}^{2+}]_0 = [\text{Pb}^{2+}] + [\text{Pb(OH)}^+] \tag{7}$$

$$[\text{S}^{2-}]_0 = [\text{S}^{2-}] + [\text{HS}^-] + [\text{H}_2\text{S}] \tag{8}$$

Making use of Eq. (1) in Eq. (7), we have

$$[\text{Pb}^{2+}]_0 = [\text{Pb}^{2+}] + \frac{K_{h1}\,[\text{Pb}^{2+}]}{[\text{H}_3\text{O}^+]}$$

or

$$[\text{Pb}^{2+}] = \frac{[\text{Pb}^{2+}]_0}{1 + \dfrac{K_{h1}}{[\text{H}_3\text{O}^+]}} \tag{9}$$

Making use of Eqs (2) and (3) in Eq. (8), we have

$$[S^{2-}]_0 = [S^{2-}] + K_{h2}\frac{[S^{2-}]}{[OH^-]} + K_{h2}\,K_{h3}\frac{[S^{2-}]}{[OH^-]^2}$$

or
$$[S^{2-}] = \frac{[S^{2-}]_0}{1 + \dfrac{K_{h2}}{[OH^-]} + \dfrac{K_{h2}K_{h3}}{[OH^-]^2}} \tag{10}$$

Now the expression of solubility product of PbS is

$$K_s(PbS) = [Pb^{2+}]\,[S^{2-}]$$

Substituting $[Pb^{2+}]$ and $[S^{2-}]$ from Eqs (9) and (10), we get

$$K_s(PbS) = \frac{[Pb^{2+}]_0\,[S^{2-}]_0}{\left(1 + \dfrac{K_{h1}}{[H_3O^+]}\right)\left(1 + \dfrac{K_{h2}}{[OH^-]} + \dfrac{K_{h2}K_{h3}}{[OH^-]^2}\right)}$$

If y is the concentration of PbS in the solution, then it is obvious that

$$[Pb^{2+}]_0 = [S^{2-}]_0 = y$$

With this, the previous expression gives

$$y^2 = K_s(PbS)\left(1 + \frac{K_{h1}}{[H_3O^+]}\right)\left(1 + \frac{K_{h2}}{[OH^-]} + \frac{K_{h2}K_{h3}}{[OH^-]^2}\right)$$

Substituting the values of $K_s(PbS)$, K_{h1}, K_{h2}, K_{h3}, $[H_3O^+]$ and $[OH^-]$, we have

$$y^2 = (7.0 \times 10^{-29}\ M^2)\left(1 + \frac{6.7 \times 10^{-7}\ M}{10^{-7}\ M}\right)\left(1 + \frac{1.0\ M}{10^{-7}\ M} + \frac{9.1 \times 10^{-8}\ M^2}{(10^{-7}\ M)^2}\right)$$

$$= (7.0 \times 10^{-29}\ M^2)\,(7.7)\,(1.91 \times 10^7) = 1.029\,5 \times 10^{-20}\ M^2$$

Hence, $y = 1.014\,6 \times 10^{-10}$ M

Thus, the solubility of PbS including hydrolysis is $1.014\,6 \times 10^{-10}$ M. Comparing this with that obtained ignoring hydrolysis indicates that the solubility has been increased by a factor

$$\frac{1.014\,6 \times 10^{-10}}{8.4 \times 10^{-15}} = 12\,078$$

Example 4.32.6 Calculate the molar solubility of AgCN in water. Given:

$$K_s(AgCN) = 2.3 \times 10^{-16}\ M^2 \text{ and } K_b(CN^-) = 1.7 \times 10^{-5}\ M.$$

Solution Material-balance expression of the dissolved salt is

$$[Ag^+] = [CN^-] + [HCN] \tag{1}$$

Charge-balance expression is

$$[Ag^+] = [CN^-] + [OH^-] \tag{2}$$

From these two expressions, we get

$$[OH^-] = [HCN] \tag{3}$$

The base reaction of CN^- is

$$CN^- + H_2O \rightleftharpoons HCN + OH^-$$

Hence, $$K_b = \frac{[HCN][OH^-]}{[CN^-]}$$

Using Eq. (3), we get

$$K_b = \frac{[HCN]^2}{[CN^-]} \quad \text{or} \quad [HCN] = \sqrt{K_b[CN^-]}$$

Substituting this in Eq. (1), we get

$$[Ag^+] = [CN^-] + \sqrt{K_b[CN^-]}$$

With the use of solubility-product expression of AgCN, this becomes

$$[Ag^+] = \frac{K_s}{[Ag^+]} + \left(\frac{K_b\,K_s}{[Ag^+]}\right)^{1/2}$$

or $$[Ag^+]^2 = K_s + (K_b\,K_s\,[Ag^+])^{1/2} \tag{4}$$

Assuming $K_s << (K_b\,K_s\,[Ag^+])^{1/2}$, we get

$$[Ag^+]^2 = (K_b K_s\,[Ag^+])^{1/2}$$

or $$[Ag^+]^4 = K_b K_s\,[Ag^+]$$

or $$[Ag^+] = (K_b K_s)^{1/3}$$

Substituting the values of K_b and K_s, we get

$$[Ag^+] = [(1.7 \times 10^{-5}\ M)\ (2.3 \times 10^{-16}\ M^2)]^{1/3}$$
$$= 1.58 \times 10^{-7}\ M$$

In order to check the assumption we made, let us calculate $\sqrt{K_b K_s[Ag^+]}$ and then compare it with K_s

$$\sqrt{K_b K_s[Ag^+]} = [(1.7 \times 10^{-5}\ M)\,(2.3 \times 10^{-16}\ M^2)\,(1.58 \times 10^{-7}\ M)]^{1/2}$$
$$= 2.48 \times 10^{-14}\ M^2$$

which is about 100 times greater than K_s justifying the assumption made to simplify the computations.

(Alternatively, Eq. (4) may be solved for $[Ag^+]$ by the successive approximation method with initial concentration of Ag^+ equal to $\sqrt{K_s}$.)

Common-ion Effect on Solubility

The common-ion present in the solution decreases the solubility of a given compound, e.g., the solubility of $BaSO_4$ in Na_2SO_4 solution is smaller than that in an aqueous solution.

Example 4.32.7 At 25 °C, a saturated solution of $BaSO_4$ is 3.9×10^{-5} M. What is its solubility in 0.1 M Na_2SO_4 solution?

Solution In an aqueous solution $[Ba^{2+}] = [SO_4^{2-}] = 3.9 \times 10^{-5}$ M

$$K_s(BaSO_4) = [Ba^{2+}]\,[SO_4^{2-}] = (3.9 \times 10^{-5}\ M)^2 = 1.5 \times 10^{-9}\ M^2$$

In 0.1 M Na_2SO_4 solution, the concentration of SO_4^{2-} = 0.1 M. Hence,

$$[Ba^{2+}] = \frac{K_s(BaSO_4)}{[SO_4^{2-}]} = \frac{1.5 \times 10^{-9}\ M^2}{(0.1\ M)} = 1.5 \times 10^{-8}\ M$$

Thus, the solubility of $BaSO_4$ is reduced from 3.9×10^{-5} M to 1.5×10^{-8} M by the common-ion effect.

Example 4.32.8 Freshly precipitated aluminium and magnesium hydroxides are stirred vigorously in a buffer solution containing 0.25 mol L^{-1} of ammonium chloride and 0.05 mol L^{-1} of ammonium hydroxide. Calculate the concentration of aluminium and magnesium ions in solution. Given: $K_b^\circ(NH_4OH) = 1.8 \times 10^{-5}$, $K_s^\circ(Mg(OH)_2) = 8.9 \times 10^{-12}$ and $K_s^\circ(Al(OH)_3) = 6 \times 10^{-32}$.

Solution To start with, we have a buffer solution containing 0.25 mol L^{-1} of NH_4Cl and 0.05 mol L^{-1} of NH_4OH. In the equilibrium reaction

$$NH_4OH \rightleftharpoons NH_4^+ + OH^-; \qquad K_b = \frac{[NH_4^+][OH^-]}{[NH_4OH]}$$

we have $[NH_4OH] = 0.05$ mol L^{-1} and $[NH_4^+] = 0.25$ mol L^{-1}

The dissolution of $Mg(OH)_2$ and $Al(OH)_3$ generates extra OH^- ions which causes the replacement of NH_4^+ by NH_4OH in the buffer solution. In the final solution, we will have

$$\text{(i)}\ \underset{c_1+\alpha}{NH_4OH} \rightleftharpoons \underset{c_2-\alpha}{NH_4^+} + \underset{z}{OH^-}; \qquad K_b = 1.8 \times 10^{-5}\ M$$

where $c_1 = 0.05$ mol L^{-1} and $c_2 = 0.25$ mol L^{-1}.

The concentration of NH_4^+ ions replaced by NH_4OH is α and z is the final concentration of OH^- in the solution. Besides this equilibrium, we also have

$$\text{(ii)}\ Mg(OH)_2 \rightleftharpoons \underset{x}{Mg^{2+}} + \underset{z}{2OH^-}; \qquad K_{s1} = 8.9 \times 10^{-12}\ M^3$$

$$\text{(iii)}\ Al(OH)_3 \rightleftharpoons \underset{y}{Al^{3+}} + \underset{z}{3OH^-}; \qquad K_{s2} = 6 \times 10^{-32}\ M^4$$

From reactions (i) to (iii), we have

$$K_b = \frac{[NH_4^+][OH^-]}{[NH_4OH]} = \frac{(c_2 - \alpha)z}{(c_1 + \alpha)} \tag{1}$$

$$K_{s1} = [Mg^{2+}][OH^-]^2 = xz^2 \tag{2}$$

$$K_{s2} = [Al^{3+}][OH^-]^3 = yz^3 \tag{3}$$

The condition of electrical neutrality gives

$$[NH_4^+] + 2[Mg^{2+}] + 3[Al^{3+}] = [OH^-] + [Cl^-]$$

i.e. $\quad (c_2 - \alpha) + 2x + 3y = z + c_2$

or $\quad 2x + 3y - \alpha - z \qquad (4)$

In the above expression, x, y and α can be replaced in terms of z by using Eqs (1) to (3). Thus, we have

$$2\left(\frac{K_{s1}}{z^2}\right) + 3\left(\frac{K_{s2}}{z^3}\right) - \left(\frac{c_2 z - K_b c_1}{K_b + z}\right) = z \tag{5}$$

Rearranging this expression, we get

$$z^4 = \frac{1}{(K_b + c_2)}[(K_b c_1)z^3 + (2K_{s1})z^2 + (2K_{s1}K_b + 3K_{s2})z + 3K_{s2}K_b - z^5] \tag{6}$$

For the given values of K_b, K_{s1}, K_{s2}, c_1 and c_2, the root z of the above expression can be obtained by the method of successive approximation. We may start with $z = 10^{-5}$ M (or any other value) and substitute this in the right hand side of the above expression and then taken $\sqrt[4]{\ }$ to get the new refined value of z. This refined value is again substituted in the right hand side and get the new refined value of z. This process is repeated till the two successive values of z agree within the allowed accuracy.

Since $K_b << c_2$, K_{s2} has a small value and z^5 is also expected to have a small value, we may use the following simpler expression to compute the root z.

$$z^3 = \frac{1}{c_2}[(K_b\, c_1)\, z^2 + (2\, K_{s1})\, z + 2\, K_{s1} K_b]$$

The results obtained in the iteration are as follows.

Cycle No.	$[OH^-]$/M
1	1.0×10^{-5}
2	1.33×10^{-5}
3	1.42×10^{-5}
4	1.45×10^{-5}
5	1.45×10^{-5}

Finally, $$[\mathrm{Mg^{2+}}] = \frac{K_s(\mathrm{Mg(OH)_2})}{[\mathrm{OH^-}]^2} = \frac{8.9\times10^{-12}\ \mathrm{M^3}}{(1.45\times10^{-5}\ \mathrm{M})^2} = 0.0423\ \mathrm{M}$$

$$[\mathrm{Al^{3+}}] = \frac{K_s(\mathrm{Al(OH)_3})}{[\mathrm{OH^-}]^3} = \frac{6\times10^{-32}\ \mathrm{M^4}}{(1.45\times10^{-5}\ \mathrm{M})^3} = 1.97\times10^{-17}\ \mathrm{M}$$

Precipitation and Solubility Product

The numerical value of solubility product of a salt provides its maximum solubility in water. *For a solution of a salt at a specified concentration, the product of the concentrations of the ions, each raised to the proper power, is called the ionic product. Thus, for a saturated solution in equilibrium with excess solid, the ionic product is equal to its solubility product.* If the ionic product of the solution is less than the corresponding solubility product, it means that the solution is unsaturated and thus more salt can dissolve in it. On the other hand, if the ionic product exceeds the solubility product, the solution is holding more salt that can dissolve in it; therefore, precipitation takes place which continues till the ionic product becomes equal to the solubility product.

Example 4.32.9

Will a precipitate form if 20 cm^3 of 0.01 M $AgNO_3$ and 20 cm^3 of 0.000 4 M NaCl are mixed? Given: $K_s(\mathrm{AgCl}) = 1.7 \times 10^{-10}\ \mathrm{M}^2$.

Solution

Total volume of the solution after mixing is 40 cm^3, thus the concentrations of ions in the solution are decreased to half the values, i.e.

$$[\mathrm{Ag^+}] = 5.0 \times 10^{-3}\ \mathrm{M} \quad \text{and} \quad [\mathrm{Cl^-}] = 2.0 \times 10^{-4}\ \mathrm{M}$$

The ionic product is

$$[\mathrm{Ag^+}]\ [\mathrm{Cl^-}] = (5.0 \times 10^{-3}\ \mathrm{M})\ (2.0 \times 10^{-4}\ \mathrm{M}) = 1.0 \times 10^{-6}\ \mathrm{M^2}$$

This ionic product is larger than K_s (= $1.7 \times 10^{-10}\ \mathrm{M}^2$) and thus precipitation of AgCl will occur.

Example 4.32.10

At 25 °C, will a precipitate of $Mg(OH)_2$ form in a 0.000 1 M solution of $Mg(NO_3)_2$ if pH of the solution is adjusted to 9.0? ($K_s(\mathrm{Mg(OH)_2}) = 8.9 \times 10^{-12}\ \mathrm{M}^3$). At what minimum value of pH will precipitation start?

Solution

If pH = 9.0, $[\mathrm{H_3O^+}] = 1.0 \times 10^{-9}$ M, then $[\mathrm{OH^-}] = 1.0 \times 10^{-5}$ M

The ionic product of $Mg(OH)_2$ in the solution

$$= [\mathrm{Mg^{2+}}]\ [\mathrm{OH^-}]^2 = (1.0 \times 10^{-4}\ \mathrm{M})\ (1.0 \times 10^{-5}\ \mathrm{M})^2$$

$$= 1.0 \times 10^{-14}\ \mathrm{M^3}$$

Since this value is smaller than 8.9×10^{-12} M^3, no precipitate will form. The concentration of OH^- required to precipitate Mg^{2+} from the solution is

$$\sqrt{\frac{K_s}{[Mg^{2+}]}} = \sqrt{\frac{8.9 \times 10^{-12}\ M^3}{1.0 \times 10^{-4}\ M}} = \sqrt{8.9 \times 10^{-8}\ M^2}$$

$$= 2.983 \times 10^{-4}\ M$$

Thus, $\quad pOH = -\log\{[OH^-]/M\} = -\log(2.983 \times 10^{-4}) = 3.54$

and hence, $pH = 14 - pOH = 10.46$

Preferential Precipitation of Salts

Frequently, a solution contains more than one ion capable of forming precipitates with another ion which is added to the solution. For example, in a solution containing Cl^-, Br^-, and I^- ions, if Ag^+ ions are added, then out of the three, the least soluble silver salt is precipitated first. If the addition of Ag^+ ions is continued, eventually a stage is reached when the next lesser soluble salt starts precipitating along with the least soluble salt and so on. If the stoichiometry of the precipitated salts is the same, then the salt with the minimum solubility product (and hence also the minimum solubility) will precipitate first followed by the salt of next higher solubility product and so on.

If the stoichiometry of the precipitated salts is not the same, then, from the solubility product data alone, we cannot predict which ion will precipitate first. Take, for example, a solution containing Cl^- and CrO_4^{2-}. Both these ions form precipitates with Ag^+. Though the solubility product of AgCl is larger than that of Ag_2CrO_4, yet it is AgCl (lesser soluble) which precipitates first when Ag^+ ions are added to the solution. In order to predict which ion (Cl^- or CrO_4^{2-}) precipitates first, we have to calculate the concentration of Ag^+ ions needed to start the precipitation through the solubility product data and the given concentration of Cl^- or CrO_4^{2-}. Since square root is involved in the expression for computing Ag^+ for silver chromate, the quantity of Ag^+ needed to start the precipitation of CrO_4^{2-} is larger than that for Cl^-. Hence, as $AgNO_3$ is added to the solution, the minimum of the two concentrations of Ag^+ to start the precipitation will be reached first and thus the corresponding ion (Cl^- in this case) will be precipitated in preference to the other. During the course of precipitation, concentration of Cl^- decreases and the corresponding concentration of Ag^+ to start the precipitation increases. Its concentration eventually becomes equal to the value required for the precipitation of CrO_4^{2-}. At this stage, practically whole of the Cl^- ions has been precipitated. The addition of more of $AgNO_3$ causes the precipitation of both the ions together.

The following problems illustrate the principle of preferential precipitation.

Example 4.32.11 A solution is 0.1 M in Cl^-, 0.1 M in Br^- and 0.1 M in I^-. Solid $AgNO_3$ is gradually added to this solution. Assuming that the addition of $AgNO_3$ does not change the volume, answer the following:

(a) What concentration of Ag^+ ions will be required to start precipitation of each of the three ions?

(b) Which ion will precipitate first?

(c) What will be the concentration of this ion when the second ion starts precipitating?

(d) What will be the concentration of both the ions when the third ion starts precipitating? Given:

$$K_s(\text{AgCl}) = 1.7 \times 10^{-10}\ \text{M}^2;$$

$$K_s(\text{AgBr}) = 5.0 \times 10^{-13}\ \text{M}^2$$

and $$K_s(\text{AgI}) = 8.5 \times 10^{-17}\ \text{M}^2$$

Solution **(a)** The concentration of Ag^+ ions to start the precipitation can be computed from the solubility product data.

Since $K_s = [\text{Ag}^+]\,[\text{X}^-]$, therefore (where $X^- \equiv Cl^-$, Br^- or I^-)

$$[\text{Ag}^+] = \frac{K_s}{[\text{X}^-]}$$

For chloride: $$[\text{Ag}^+] = \frac{1.7 \times 10^{-10}\ \text{M}^2}{(0.1\ \text{M})} = 1.7 \times 10^{-9}\ \text{M}$$

For bromide: $$[\text{Ag}^+] = \frac{5.0 \times 10^{-13}\ \text{M}^2}{(0.1\ \text{M})} = 5.0 \times 10^{-12}\ \text{M}$$

For iodide: $$[\text{Ag}^+] = \frac{8.5 \times 10^{-17}\ \text{M}^2}{(0.1\ \text{M})} = 8.5 \times 10^{-16}\ \text{M}$$

(b) Since $[Ag^+]$ required to precipitate I^- is minimum, therefore it is AgI which precipitates first. The same conclusion can also be reached from the solubility product data.

(c) The second ion Br^- will be precipitated only when $[Ag^+]$ ion in the solution has reached to 5.0×10^{-12} M. Therefore, the concentration of remaining I^- is

$$[\text{I}^-] = \frac{K_s(\text{AgI})}{[\text{Ag}^+]} = \frac{8.5 \times 10^{-17}\ \text{M}^2}{(5.0 \times 10^{-12}\ \text{M})} = 1.7 \times 10^{-5}\ \text{M}$$

(d) The third ion Cl^- will be precipitated only when $[Ag^+]$ in the solution has reached to 1.7×10^{-9} M. Therefore, the concentrations of remaining I^- and Br^- are:

$$[\text{I}^-] = \frac{K_s(\text{AgI})}{[\text{Ag}^+]} = \frac{8.5 \times 10^{-17}\ \text{M}^2}{(1.7 \times 10^{-9}\ \text{M})} = 5.0 \times 10^{-8}\ \text{M}$$

$$[\text{Br}^-] = \frac{K_s(\text{AgBr})}{[\text{Ag}^+]} = \frac{5.0 \times 10^{-13}\ \text{M}^2}{(1.7 \times 10^{-9}\ \text{M})} = 3.0 \times 10^{-4}\ \text{M}$$

Percentage of I^- remaining in the solution when AgBr starts precipitating

$$= \frac{(1.7 \times 10^{-5}\ \text{M})}{(0.1\ \text{M})} \times 100 = 1.7 \times 10^{-2} = 0.017$$

Percentage of Br^- remaining in the solution when AgCl starts precipitating

$$= \frac{(3.0 \times 10^{-4}\ \text{M})}{(0.1\ \text{M})} \times 100 = 3.0 \times 10^{-1} = 0.3$$

Thus, we see that the precipitation of one ion is almost complete when the second starts precipitating. With a suitable method (*potentiometrically* or *conductometrically*), the concentrations of the individual ions can be determined without involving much error.

Example 4.32.12

A solution is 0.1 M in Cl^- and 0.001 M in CrO_4^{2-}. If solid $AgNO_3$ is gradually added to this solution, which will precipitate first, AgCl or Ag_2CrO_4? Assume that the addition causes no change in volume. Given:

$$K_s(\text{AgCl}) = 1.7 \times 10^{-10}\ \text{M}^2$$

and $$K_s(\text{Ag}_2\text{CrO}_4) = 1.9 \times 10^{-12}\ \text{M}^3$$

Solution

When a precipitate begins to form, the ionic product of the substance just exceeds the corresponding K_s. Therefore, we calculate the $[Ag^+]$ needed to precipitate AgCl and Ag_2CrO_4:

$$\text{AgCl} \rightleftharpoons \text{Ag}^+ + \text{Cl}^-$$

$$K_s = [\text{Ag}^+][\text{Cl}^-] = 1.7 \times 10^{-10}\ \text{M}^2$$

$$[\text{Ag}^+] = \frac{1.7 \times 10^{-10}\ \text{M}^2}{0.1\ \text{M}} = 1.7 \times 10^{-9}\ \text{M}$$

$$\text{Ag}_2\text{CrO}_4 \rightleftharpoons 2\text{Ag}^+ + \text{CrO}_4^{2-}$$

$$K_s = [\text{Ag}^+]^2[\text{CrO}_4^{2-}] = 1.9 \times 10^{-12}\ \text{M}^3$$

$$[\text{Ag}^+] = \sqrt{\frac{1.9 \times 10^{-12}\ \text{M}^3}{0.001\ \text{M}}} = 4.4 \times 10^{-5}\ \text{M}$$

Since the amount of Ag^+ needed to precipitate AgCl is less, this value will, therefore, be reached first in the solution and hence it is the AgCl which will be precipitated first.

Example 4.32.13

In Example 4.32.12, what will be the concentration of Cl^- when Ag_2CrO_4 begins to precipitate? At this point, what percentage of the Cl^- originally present remains in the solution?

Solution

Ag^+ required for the precipitation of Ag_2CrO_4 is 4.4×10^{-5} M. At this point, the concentration of Cl^- ion in the solution can be calculated from K_s(AgCl).

$$K_s = [\text{Ag}^+][\text{Cl}^-] = 1.7 \times 10^{-10}\ \text{M}^2$$

Therefore,

$$[\text{Cl}^-] = \frac{1.7 \times 10^{-10}\ \text{M}^2}{4.4 \times 10^{-5}\ \text{M}} = 3.9 \times 10^{-6}\ \text{M}$$

$$\text{Percentage of Cl}^- \text{ remaining in the solution} = \frac{3.9 \times 10^{-6}\ \text{M}}{0.1\ \text{M}} \times 100 = 0.003\,9$$

From Example 4.32.13 it may be concluded that virtually whole of Cl^- is precipitated when CrO_4^{2-} ion starts precipitating. Thus, the use of CrO_4^{2-} ions as the precipitating indicator in the titration of Cl^- versus Ag^+ is justified. The colour of Ag_2CrO_4 is red and hence the end-point of the above titration is located by the appearance of a red precipitate. In actual practice, the concentration of CrO_4^{2-} used is much less and hence a higher concentration of Ag^+ is required to start precipitation of Ag_2CrO_4. Consequently, much less of the Cl^- will remain untitrated when Ag_2CrO_4 begins to precipitate.

Qualitative Analysis

Qualitative analysis of cations is largely based on the principle of solubility product. Cations are separated into six groups depending upon the solubility of their salts. A brief description of the scheme of qualitative analysis is given below.

Group I

Insoluble chlorides Only Ag^+, Hg_2^{2+} and Pb^{2+} form insoluble chlorides since they have low values of solubility products.

$$K_s(PbCl_2) = 1.6 \times 10^{-5}\ M^3$$

$$K_s(Hg_2Cl_2) = 1.1 \times 10^{-18}\ M^3$$

$$K_s(AgCl) = 1.7 \times 10^{-10}\ M^2$$

Group II

Insoluble sulphides in acidic medium The sulphide-ion concentration in a solution that has been saturated with H_2S is given by

$$[S^{2-}] = \frac{K_1K_2\,[H_2S]}{[H^+]^2} = \frac{1.1\times10^{-21}\ M^2}{[H^+]^2}[H_2S]$$

In an acidic solution, the $[H^+]$ is large and thus $[S^{2-}]$ is low. For example, in 10 cm^3 of a saturated H_2S solution that has been made 0.3 M in H^+ there are approximately seven S^{2-} ions. Thus, when H_2S is passed in acidic medium of a solution containing several cations, only for least soluble sulphides of Hg^{2+}, Bi^{3+}, Pb^{2+}, Cu^{2+}, Cd^{2+}, Sb^{3+}, As^{3+} and Sn^{2+}, the ionic products become greater than the corresponding solubility products which, therefore, are precipitated. For other cations such as Fe^{3+}, Co^{2+}, Ni^{2+}, Zn^{2+}, Cr^{3+}, etc., the ionic products are smaller than their respective solubility products and thus are not precipitated, i.e. their sulphides are soluble.

It is possible to compute the minimum pH of the solution at which a given cation will form an insoluble sulphide when the solution is saturated with H_2S. This can be computed as follows:

The minimum concentration of sulphide ion necessary to start the precipitation of cation M^{2+} is

$$[S^{2-}] = \frac{K_s(MS)}{[M^{2+}]}$$

The hydrogen-ion concentration (maximum) of a solution having the above minimum $[S^{2-}]$ can be computed using the expression

$$[H^+] = \sqrt{\frac{(1.1 \times 10^{-21}\ M^2)[H_2S]}{[S^{2-}]}}$$

In an aqueous solution that is in equilibrium with H_2S gas at 101.325 kPa, $[H_2S] = 0.1$ M; hence

$$[H^+] = \sqrt{\frac{(1.1 \times 10^{-22}\ M^3)}{[S^{2-}]}}$$

$$= \sqrt{\frac{(1.1 \times 10^{-22}\ M^3)[M^{2+}]}{K_s(MS)}}$$

The minimum $[S^{2-}]$ and the maximum $[H^+]$ along with pH values for various sulphides ($[M^{2+}] = 0.01$ M) are being given in Table 4.32.2.

Table 4.32.2 Condition for the Precipitation of Various Sulphides

Solid		K_s°	$[S^{2-}]_{min}/M$	$[H^+]_{max}/M$	*Min* pH
HgS	Gp. II	3.0×10^{-53}	3×10^{-51}	$\approx 2 \times 10^{14}$	≈ -14.3
CuS	Gp. II	4.0×10^{-38}	4×10^{-36}	$\approx 5 \times 10^{6}$	≈ -6.7
PbS	Gp. II	1.0×10^{-29}	1.0×10^{-27}	$\approx 3 \times 10^{2}$	≈ -2.5
SnS	Gp. II	8.0×10^{-29}	8.0×10^{-27}	$\approx 1 \times 10^{2}$	≈ -2.0
Sb_2S_3	Gp. II	10^{-80}	10^{-25}	$\approx 3 \times 10$	≈ -1.5
Bi_2S_3	Gp. II	1.6×10^{-72}	2.5×10^{-23}	≈ 2	≈ -0.3
CdS	Gp. II	1.0×10^{-28}	1.0×10^{-26}	$\approx 1.0 \times 10^{2}$	≈ -2.0
CoS		7×10^{-23}	7×10^{-21}	≈ 0.13	≈ 0.9
NiS		3×10^{-21}	3×10^{-19}	$\approx 2 \times 10^{-2}$	≈ 1.7
ZnS		2.5×10^{-22}	2.5×10^{-18}	$\approx 6 \times 10^{-2}$	≈ 1.2
FeS		1.0×10^{-19}	1.0×10^{-17}	$\approx 3 \times 10^{-3}$	≈ 2.5
MnS		5.6×10^{-16}	5.6×10^{-14}	$\approx 4 \times 10^{-5}$	≈ 4.4

The data in Table 4.32.2 reveal how cations of Group II could be separated from the rest of cations. The concentration of H^+ less than that listed (or pH greater than the listed value) yields sufficient sulphide ions to cause precipitation for each case. Thus, if in an acidic solution containing $[H^+]$ between 0.13 to 2 M (or pH < 0.9) H_2S is passed, only sulphides of Group II are precipitated.

Group III

Insoluble hydroxides in ammoniacal solution containing ammonium chloride The concentration of OH^- in an ammoniacal solution decreases when NH_4Cl is added to it, because of the common-ion effect. Thus only for least soluble hydroxides of Fe^{3+}, Al^{3+} and Cr^{3+}, the ionic products exceed the corresponding solubility products. Therefore, only these three ions are precipitated; for other cations (Co^{2+}, Ni^{2+}, Ba^{2+}, Ca^{2+}, etc.), the respective ionic product is smaller than the corresponding solubility product and thus they remain in the solution. In fact, cations such as Co^{2+}, Ni^{2+}, Mg^{2+}, etc., form complexes with ammonia and thus the amounts of free ions in the solution are decreased to a very large extent.

Consequently, a larger concentrations of OH^- is required to precipitate these ions from the solution as compared to those required for the group III cations (*see* Section 4.33).

Group IV

Insoluble sulphides in ammoniacal medium In ammoniacal solution, the concentration of S^{2-} becomes larger when H_2S is passed through the solution. This is because of the formation of ammonium sulphide. Thus, sulphides of Co^{2+}, Ni^{2+}, Mn^{2+} and Zn^{2+} are precipitated here.

Group V

Insoluble carbonates in ammoniacal solution containing ammonium chloride The carbonates of Ba^{2+}, Ca^{2+} and Sr^{2+} are precipitated here, while $MgCO_3$ does not precipitate due to the low concentration of CO_3^{2-} caused by the shift of equilibrium

$$NH_4^+ + CO_3^{2-} \rightleftharpoons NH_3 + HCO_3^-$$

to the right side in the presence of ammonium salt.

Group VI

This contains the remaining cations, viz., Mg^{2+} and K^+.

Example 4.32.14

A solution that is 0.30 M in H^+, 1.0 M in Pb^{2+} and 1.0 M in Fe^{2+} is saturated with H_2S; should PbS and/or FeS precipitate? The K_s of PbS is 7×10^{-29} M^2 and the K_s of FeS is 4×10^{-19} M^2.

Solution

For any saturated solution of H_2S

$$[H^+]^2 [S^{2-}] = 1.1 \times 10^{-22} \ M^3$$

Since this solution is 0.30 M in H^+, therefore,

$$(0.3 \ M)^2 [S^{2-}] = 1.1 \times 10^{-22} \ M^3$$

$$[S^{2-}] = 1.2 \times 10^{-21} \ M$$

Ionic product of either of the two sulphides

$$= (1.0 \ M) (1.2 \times 10^{-21} \ M) = 1.2 \times 10^{-21} \ M^2$$

This ionic product is greater than the K_s of PbS; therefore PbS will precipitate. However, the ionic product is less than the K_s of FeS; the solubility of FeS has not been exceeded; no FeS precipitate will form.

Example 4.32.15

A solution that is 0.1 M in Cd^{2+} and 0.1 M in H^+ is saturated with H_2S. What concentration of Cd^{2+} remains in solution after CdS has precipitated? The K_s of CdS is 1.0×10^{-28} M^2.

Solution

For each Cd^{2+} ion precipitated, two H^+ ions are added to the solution according to the following reaction:

$$Cd^{2+}(aq) + H_2S(aq) \rightleftharpoons CdS(s) + 2H^+(aq)$$

We shall assume that virtually all Cd^{2+} ions precipitate as CdS. Hence, the precipitation introduces 0.20 mol of H^+ per dm^3 of solution, and the final H^+ is 0.30 M. Therefore,

$$[H^+]^2 [S^{2-}] = 1.1 \times 10^{-22} \ M^3$$

$$[S^{2-}] = \frac{1.1 \times 10^{-22} \ M^3}{[H^+]^2} = \frac{1.1 \times 10^{-22} \ M^3}{(0.30 \ M)^2}$$

$$= 1.22 \times 10^{-21} \ M$$

The concentration of Cd^{2+} after CdS has precipitated may be derived from the K_s of CdS:

$$[Cd^{2+}]\,[S^{2-}] = 1.0 \times 10^{-28}\ M^2$$

or $$[Cd^{2+}] = \frac{1.0 \times 10^{-28}\ M^2}{1.22 \times 10^{-21}\ M} = 8.2 \times 10^{-8}\ M$$

Example 4.32.16

A solution containing zinc and manganese ions each at a concentration of 0.01 M is saturated with H_2S. If the K_s of ZnS is $1.0 \times 10^{-22}\ M^2$ and the K_s of MnS is $5.6 \times 10^{-16}\ M^2$, what is the pH at which the manganous sulphide will form a precipitate? Under these conditions, what will be the concentration of remaining zinc ions? Given: $[H_2S] = 0.10$ M.

Solution

The minimum of concentration S^{2-} ion to start the precipitation is obtained from the K_s with $[Mn^{2+}] = 0.01$ M. Therefore, we have

$$[S^{2-}] = \frac{K_s(MnS)}{[Mn^{2+}]} = \frac{5.6 \times 10^{-16}\ M^2}{(0.01\ M)} = 5.6 \times 10^{-14}\ M$$

The H^+ concentration of the solution having the above $[S^{2-}]$ can be computed from the expression of H_2S equilibrium:

$$\frac{[H^+]^2\,[S^{2-}]}{[H_2S]} = \frac{[H^+]^2\,(5.6 \times 10^{-14}\ M)}{(0.10\ M)} = 1.1 \times 10^{-21}\ M^2$$

This gives

$$[H^+] = 4.43 \times 10^{-5}\ M \quad \text{or} \quad pH = 4.35$$

If the $[H^+] > 4.43 \times 10^{-5}$ M, then the $[S^{2-}]$ will be less than 5.6×10^{-14} M and MnS will no longer precipitate from the solution.

The concentration of Zn^{2+} ion remaining in the solution can be calculated from the solubility product of ZnS:

$$[Zn^{2+}] = \frac{K_s(ZnS)}{[S^{2-}]} = \frac{1.0 \times 10^{-22}\ M^2}{(5.6 \times 10^{-14}\ M)} = 1.79 \times 10^{-9}\ M$$

Thus, by properly adjusting the $[H^+]$ in the solution, it is possible to precipitate effectively all of zinc ions from the solution without precipitating any Mn^{2+} ion.

4.33 EQUILIBRIA INVOLVING COMPLEX IONS

A complex ion is an aggregate consisting of a metal ion surrounded by ligands. The latter may be an ion, molecule or a combination of the two. An essential feature of a ligand is that it must have an unshared pair of electrons with which it can bind to the central ion. A few examples are

$:NH_3$ H—Ö: (with H bonded below O) $[:C{\equiv}N]^-$ $:\ddot{C}l:^-$

The stability of complex ions varies widely. The cations of groups 1 and 2 of the periodic table only form aggregates that are loosely held together by weak ion-dipole or ion-ion electrostatic attractions; these are probably better described as ion-pairs rather than complex ions. At the other extreme are the very stable complex ions such as $Co(NH_3)_6^{3+}$ and $Fe(CN)_6^{4-}$ in which strength of the mutual attractions among central ions and the ligands is of the same order as the strength of the covalent bond. We shall restrict our attention to those complex ions which are formed in definite stoichiometric ratios and which exist in the solution as reversible dissociation entities.

The formation of a complex ion occurs in steps. Taking an example of the formation of $[Ag(NH_3)_2^+]$ ion, we have

$$Ag^+ + NH_3 \rightleftharpoons Ag(NH_3)^+; \quad K_1 = \frac{[Ag(NH_3)^+]}{[Ag^+][NH_3]} \tag{4.33.1}$$

and

$$Ag(NH_3)^+ + NH_3 \rightleftharpoons Ag(NH_3)_2^+; \quad K_2 = \frac{[Ag(NH_3)_2^+]}{[Ag(NH_3)^+][NH_3]} \tag{4.33.2}$$

Formation Constant

These stepwise equilibrium constants are related to the overall equilibrium constant (called the *formation constant*) by the expression

$$Ag^+ + 2NH_3 \rightleftharpoons Ag(NH_3)_2^+; \quad K_{form} = K_1 K_2 = \frac{[Ag(NH_3)_2^+]}{[Ag^+][NH_3]^2} \tag{4.33.3}$$

where K_{form} is the *formation constant.**

Instability Constant

In literature, values of equilibrium constants are given in the form of dissociation of complex ions. This dissociation, like formation, takes place in steps:

$$Ag(NH_3)_2^+ \rightleftharpoons Ag(NH_3)^+ + NH_3; \quad K_1 = \frac{[Ag(NH_3)^+][NH_3]}{[Ag(NH_3)_2^+]} \tag{4.33.4}$$

and

$$Ag(NH_3)^+ \rightleftharpoons Ag^+ + NH_3; \quad K_2 = \frac{[Ag^+][NH_3]}{[Ag(NH_3)^+]} \tag{4.33.5}$$

The product of these two constants gives an overall constant called the instability constant K_{inst}:

$$K_{inst} = \frac{[Ag^+][NH_3]^2}{[Ag(NH_3)_2^+]} \tag{4.33.6}$$

Obviously, K_{inst} is numerically equal to the inverse of K_{form}.

---

*The overall equilibrium constants are often represented by the symbol β. Thus

$$\beta_1 = K_1; \quad \beta_2 = K_1K_2; \quad \beta_n = K_1K_2 \ldots K_n$$

The magnitude of overall instability constant is a measure of the tendency of the complex to dissociate into constituent ions. A very stable complex ion will have very low value of instability constant and hence ordinary ionic reactions are not exhibited by it. For example, $[Fe(CN)_6^{4-}]$ does not give the test for Fe^{2+}.

If a large excess of ligand is employed and if each of the step wise dissociation constant is small, assumption may then be made that the principal complex in solution is the one with the highest number of ligands. The validity of the above assumption may be seen from the following example.

Example 4.33.1 A 1.0×10^{-3} M solution of $AgNO_3$ is made 0.20 M in NH_3. What are the concentrations of Ag^+, $Ag(NH_3)^+$, and $Ag(NH_3)_2^+$ in the resulting solution. Given:

$$Ag(NH_3)_2^+ \rightleftharpoons Ag(NH_3)^+ + NH_3; \qquad K_1 = 1.4 \times 10^{-4}\ \text{M}$$

$$Ag(NH_3)^+ \rightleftharpoons Ag^+ + NH_3; \qquad K_2 = 4.3 \times 10^{-4}\ \text{M}$$

Solution We shall assume that the Ag^+ is completely converted into the higher complex $Ag(NH_3)_2^+$, by this excess of NH_3. Therefore,

$$[Ag(NH_3)_2^+] = 1.0 \times 10^{-3}\ \text{M}$$

$$[NH_3] = 2 \times 10^{-1}\ \text{M} - 2.0 \times 10^{-3}\ \text{M} \approx 2 \times 10^{-1}\ \text{M}$$

From the instability constant

$$\frac{[Ag^+][NH_3]^2}{[Ag(NH_3)_2^+]} = (1.4 \times 10^{-4}\ \text{M})(4.3 \times 10^{-4}\ \text{M}) = 6.02 \times 10^{-8}\ \text{M}^2$$

we have $$[Ag^+] = \{6.02 \times 10^{-8}\ \text{M}^2\}\frac{[Ag(NH_3)_2^+]}{[NH_3]^2} = \frac{\{6.02 \times 10^{-8}\ \text{M}^2\}(1.0 \times 10^{-3}\ \text{M})}{(2 \times 10^{-1}\ \text{M})^2}$$

$$= 1.505 \times 10^{-9}\ \text{M}$$

The concentration of $Ag(NH_3)^+$ may be obtained from either of the stepwise dissociation constants. Thus

$$\frac{[Ag(NH_3)^+][NH_3]}{[Ag(NH_3)_2^+]} = 1.4 \times 10^{-4}\ \text{M}$$

$$[Ag(NH_3)^+] = \frac{(1.4 \times 10^{-4}\ \text{M})[Ag(NH_3)_2^+]}{[NH_3]} = \frac{(1.4 \times 10^{-4}\ \text{M})(1.0 \times 10^{-3}\ \text{M})}{(0.2\ \text{M})}$$

$$= 0.7 \times 10^{-6}\ \text{M}$$

Alternative Method We have

$$K_1 = \frac{[Ag(NH_3)^+][NH_3]}{[Ag(NH_3)_2^+]} = 1.4 \times 10^{-4}\ \text{M} \qquad (4.33.7)$$

$$K_2 = \frac{[Ag^+][NH_3]}{[Ag(NH_3)^+]} = 4.3 \times 10^{-4}\ \text{M} \qquad (4.33.8)$$

$$K_1K_2 = \frac{[\text{Ag}^+][\text{NH}_3]^2}{[\text{Ag(NH}_3)_2^+]} = 6.02 \times 10^{-8}\ \text{M}^2 \tag{4.33.9}$$

The mass-balance condition of Ag^+ gives

$$[\text{Ag}^+]_0 = [\text{Ag}^+] + [\text{Ag(NH}_3)^+] + [\text{Ag(NH}_3)_2^+]$$

Using Eqs (4.33.8) and (4.33.9), we get

$$[\text{Ag}^+]_0 = [\text{Ag}^+] + \frac{[\text{Ag}^+][\text{NH}_3]}{K_2} + \frac{[\text{Ag}^+][\text{NH}_3]^2}{K_1K_2}$$

or
$$s_0 = \frac{[\text{Ag}^+]}{[\text{Ag}^+]_0} = \frac{1}{1 + ([\text{NH}_3]/K_2) + ([\text{NH}_3]^2/K_1K_2)} \tag{4.33.10}$$

$$s_1 = \frac{[\text{Ag(NH}_3)^+]}{[\text{Ag}^+]_0} \equiv \frac{[\text{Ag}^+]}{[\text{Ag}^+]_0}\cdot\frac{[\text{Ag(NH}_3)^+]}{[\text{Ag}^+]} = s_0\,\frac{[\text{NH}_3]}{K_2} \tag{4.33.11}$$

$$s_2 = \frac{[\text{Ag(NH}_3)_2^+]}{[\text{Ag}^+]_0} \equiv \frac{[\text{Ag}^+]}{[\text{Ag}^+]_0}\,\frac{[\text{Ag(NH}_3)_2^+]}{[\text{Ag}^+]} = s_0\,\frac{[\text{NH}_3]^2}{K_1K_2} \tag{4.33.12}$$

Since $[NH_3]$ is present in excess, we assume

$$[\text{NH}_3] = [\text{NH}_3]_0$$

Substituting $[NH_3]_0$ and the values of K_1 and K_2 in Eqs (4.33.10) to (4.33.12), we get

$$s_0 = \frac{1}{1 + (0.2/4.3 \times 10^{-4}) + (0.2^2/6.02 \times 10^{-8})} = 1.50 \times 10^{-6}$$

Hence, $[\text{Ag}^+] = s_0\,[\text{Ag}^+]_0 = (1.50 \times 10^{-6})\,(1 \times 10^{-3}\ \text{M}) = 1.50 \times 10^{-9}\ \text{M}$

$$s_1 = (1.50 \times 10^{-6})\left(\frac{0.2}{4.3 \times 10^{-4}}\right) = 6.98 \times 10^{-4}$$

Hence, $[\text{Ag(NH}_3)^+] = s_1[\text{Ag}^+]_0 = (6.98 \times 10^{-4})(1.0 \times 10^{-3}\ \text{M}) = 6.98 \times 10^{-7}\ \text{M}$

$$s_2 = (1.50 \times 10^{-6})\left(\frac{0.2^2}{6.02 \times 10^{-8}}\right) = 9.97 \times 10^{-1}$$

Hence, $[\text{Ag(NH}_3)_2^+] = s_2[\text{Ag}^+]_0 = (9.97 \times 10^{-1})(1.0 \times 10^{-3}\ \text{M}) = 9.97 \times 10^{-4}\ \text{M}$

Unfortunately, there are many situations in which the above assumption of principal complex being the one with the highest number of ligands may not be made because either or both of criteria (namely, large excess of ligand and each of the dissociation constant is small) are not satisfied. For example, in solving problems concerning the complexes formed by Ag^+ and $S_2O_3^{2-}$ ions, assumption cannot be made that the concentrations of all Ag^+ containing species are negligible except for $Ag(S_2O_3)_3^{5-}$, the highest complex. The equations and corresponding constants for the stepwise dissociation of the $Ag(S_2O_3)_3^{5-}$ ion are

$$Ag(S_2O_2)_3^{5-} \rightleftharpoons Ag(S_2O_3)_2^{3-} + S_2O_3^{2-}; \quad K_1 = 2.0 \times 10^{-1}\ M$$

$$Ag(S_2O_3)_2^{3-} \rightleftharpoons Ag(S_2O_3)^{-} + S_2O_3^{2-}; \quad K_2 = 3.3 \times 10^{-5}\ M$$

$$Ag(S_2O_3)^{-} \rightleftharpoons Ag^{+} + S_2O_3^{2-}; \quad K_3 = 1.5 \times 10^{-9}\ M$$

The constant for the first step of the dissociation is comparatively large. Therefore, even at high concentrations of $S_2O_3^{2-}$, significant concentrations of both $Ag(S_2O_3)_3^{5-}$ and $Ag(S_2O_3)_2^{3-}$ exist. Notice that this conclusion cannot be reached by an examination of the instability constant ($9.9 \times 10^{-15}\ M^3$) alone. In support of the above analysis, we give below an example pertaining to $Ag(S_2O_3)_3^{5-}$.

Example 4.33.2 Determine the concentrations of Ag^+, $Ag(S_2O_3)^-$ and $Ag(S_2O_3)_2^{3-}$, $Ag(S_2O_3)_3^{5-}$ and $S_2O_3^{2-}$ in a solution prepared by dissolving 1.0 mole of $Ag(S_2O_3)_3^{5-}$ in 1.0 dm^{-3} of water.

Solution We have

$$Ag(S_2O_3)_3^{5-} \rightleftharpoons Ag(S_2O_3)_2^{3-} + S_2O_3^{2-}; \quad K_1 = 2.0 \times 10^{-1}\ M$$

$$Ag(S_2O_3)_2^{3-} \rightleftharpoons Ag(S_2O_3)^{-} + S_2O_3^{2-}; \quad K_2 = 3.3 \times 10^{-5}\ M$$

$$Ag(S_2O_3)^{-} \rightleftharpoons Ag^{+} + S_2O_3^{2-}; \quad K_3 = 1.5 \times 10^{-9}\ M$$

Since $K_1 >> K_2 >> K_3$, it can be assumed that $[Ag(S_2O_3)_2^{3-}] = [S_2O_3^{2-}]$, i.e. the primary source of $S_2O_3^{2-}$ is due to the first dissociation of $Ag(S_2O_3)_3^{5-}$. Thus, we have

$$\begin{array}{lccc} & Ag(S_2O_3)_3^{5-} & \rightleftharpoons Ag(S_2O_3)_2^{3-} & + S_2O_3^{2-} \\ t = 0 & 1.0\ M & 0 & 0 \\ t_{equil} & 1.0\ M - x & x & x \end{array}$$

$$\text{Therefore,}\ K_1 = \frac{[Ag(S_2O_3)_2^{3-}][S_2O_3^{2-}]}{[Ag(S_2O_3)_3^{5-}]} = \frac{x^2}{1.0\ M - x}$$

or $\quad x^2 + K_1 x - K_1(1.0\ M) = 0$

which gives

$$x = \frac{-K_1 + \sqrt{K_1^2 + 4K_1\,(1.0\ M)}}{2} = \frac{-0.2\ M + \sqrt{(0.04 + 4 \times 0.2)\ M^2}}{2}$$

$$= \frac{-0.2 + \sqrt{0.84}}{2} \text{M} = \frac{-0.2 + 0.916}{2} \text{M} = 0.358 \text{ M}$$

Hence, $[Ag(S_2O_3)_2^{3-}] = [S_2O_3^{2-}] = 0.358 \text{ M}$

$$[Ag(S_2O_3)_3^{5-}] = 1.0 \text{ M} - 0.358 \text{ M} = 0.642 \text{ M}$$

$$[Ag(S_2O_3)^-] = \frac{K_2[Ag(S_2O_3)_2^{3-}]}{[S_2O_3^{2-}]} = K_2 = 3.3 \times 10^{-5} \text{ M}$$

and $[Ag^+] = \dfrac{K_3[Ag(S_2O_3)^-]}{[S_2O_3^{2-}]} = \dfrac{K_3K_2}{[S_2O_3^{2-}]}$

$$= \frac{(1.5 \times 10^{-9} \text{ M})(3.3 \times 10^{-5} \text{ M})}{0.358 \text{ M}} = 1.38 \times 10^{-13} \text{ M}$$

Solubility of a Precipitate Through Complex Formation

Slightly soluble substances can often be dissolved through the formation of complex ions. The dissolution of AgCl in NH_3 illustrates this technique. Consider the following equilibrium between solid AgCl and its ions:

$$AgCl(s) \rightleftharpoons Ag^+(aq) + Cl^-(aq)$$

This reaction is forced to the right by removal of Ag^+ ion through the formation of the silver-ammonia complex ion and, therefore, AgCl dissolves.

Example 4.33.3

What is the solubility of AgCl in 0.20 M NH_3?

Given: $K_s(AgCl) = 1.7 \times 10^{-10} \text{ M}^2$, $K_1 = [Ag(NH_3)^+]/[Ag^+][NH_3] = 2.33 \times 10^3 \text{ M}^{-1}$ and $K_2 = [Ag(NH_3)_2^+]/[Ag(NH_3)^+][NH_3] = 7.14 \times 10^3 \text{ M}^{-1}$.

Solution

If s be the concentration of AgCl in the solution, then

$$[Cl^-] = s$$

From the K_s for AgCl, we derive

$$[Ag^+] = \frac{K_s}{[Cl^-]} = \frac{1.7 \times 10^{-10} \text{ M}^2}{s}$$

If we assume that the majority of the dissolved Ag^+ goes into solution as $Ag(NH_3)_2^+$ then

$$[Ag(NH_3)_2^+] = s$$

Since two molecules of NH_3 are required for every $Ag(NH_3)_2^+$ ion formed, we have

$$[NH_3] = 0.20 \text{ M} - 2s$$

Therefore,

$$K_{inst} = \frac{[Ag^+][NH_3]^2}{[Ag(NH_3)_2^+]} = \frac{\left(\dfrac{1.7 \times 10^{-10}\ M^2}{s}\right)(0.20\ M - 2s)^2}{s}$$

$$= 6.0 \times 10^{-8}\ M^2$$

From which we derive

$$\frac{(0.20\ M - 2s)^2}{s^2} = \frac{6.0 \times 10^{-8}\ M^2}{1.7 \times 10^{-10}\ M^2} = 3.5 \times 10^2$$

which gives $s = [Ag(NH_3)_2^+] = 9.6 \times 10^{-3}$ M, which is the solubility of AgCl in 0.20 M NH_3.

Alternative Method

We may work out the solubility of silver chloride in 0.20 M NH_3 as follows.

If s is the solubility of AgCl, we will have

$$s = [Ag^+] + [Ag(NH_3)^+] + [Ag(NH_3)_2^+]$$

$$s = [Cl^-] = K_s(AgCl)/[Ag^+]$$

From these, we get

$$[Ag^+] + [Ag(NH_3)^+] + [Ag(NH_3)_2^+] = \frac{K_s}{[Ag^+]}$$

$$[Ag^+] + \beta_1[Ag^+][NH_3] + \beta_2[Ag^+][NH_3]^2 = \frac{K_s}{[Ag^+]}$$

where $$\beta_1 = \frac{[Ag(NH_3)^+]}{[Ag^+][NH_3]} = 2.33 \times 10^3\ M^{-1}$$

$$\beta_2 = \frac{[Ag(NH_3)_2^+]}{[Ag^+][NH_3]^2} = (2.33 \times 10^3)(7.14 \times 10^3)\ M^{-2} = 1.66 \times 10^7\ M^{-2}$$

Rearranging the above expression, we get

$$[Ag^+](1 + \beta_1[NH_3] + \beta_2[NH_3]^2) = \frac{K_s}{[Ag^+]}$$

or $$[Ag^+]^2 = \frac{K_s}{1 + \beta_1[NH_3] + \beta_2[NH_3]^2}$$

$$= \frac{1.7 \times 10^{-10}\ M^2}{1 + (2.33 \times 10^3)(0.20) + (1.66 \times 10^7)(0.20^2)}$$

$$= \frac{1.77 \times 10^{-10}}{6.65 \times 10^5} M^2 = 2.66 \times 10^{-16}\ M^2$$

$$[Ag^+] = 1.63 \times 10^{-8}\ M$$

$$s = [Cl^-] = \frac{K_s}{[Ag^+]} = \frac{1.7 \times 10^{-10}\ M^2}{1.63 \times 10^{-8}\ M} = 1.04 \times 10^{-2}\ M$$

Conditional Stability Constant

The reaction between Mg^{2+} ion and ethylenediamine tetracetate ion (represented as L^{4-}) is

$$Mg^{2+} + L^{4-} \rightleftharpoons MgL^{2-} \tag{4.33.13}$$

Its equilibrium constant is

$$K = \frac{[MgL^{2-}]}{[Mg^{2+}][L^{4-}]} \tag{4.33.14}$$

The use of equilibrium constant in computing the equilibrium concentration (s) of species involved in the reaction is not straightforward if the metal ions react with auxilary complexing agent (say, NH_3) or/and the ligand ions react with H^+ ions present in the solution. The side reactions occurring in the solution are

$$Mg^{2+} + NH_3 \rightleftharpoons Mg(NH_3)^{2+}$$

$$Mg(NH_3)^{2+} + NH_3 \rightleftharpoons Mg(NH_3)_2^{2+}$$

$$Mg(NH_3)_2^{2+} + NH_3 \rightleftharpoons Mg(NH_3)_3^{2+}$$

$$Mg(NH_3)_3^{2+} + NH_3 \rightleftharpoons Mg(NH_3)_4^{2+}$$

$$\cdots \quad \cdots \quad \cdots \quad \cdots \quad \cdots \quad \cdots \quad \cdots \quad \cdots \quad \cdots \quad \cdots$$

$$\cdots \quad \cdots \quad \cdots \quad \cdots \quad \cdots \quad \cdots \quad \cdots \quad \cdots \quad \cdots \quad \cdots \tag{4.33.15}$$

$$L^{4-} + H^+ \rightleftharpoons HL^{3-}$$

$$HL^{3-} + H^+ \rightleftharpoons H_2L^{2-}$$

$$H_2L^{2-} + H^+ \rightleftharpoons H_3L^-$$

$$H_3L^- + H^+ \rightleftharpoons H_4L \tag{4.33.16}$$

The overall equilibrium constants for these reactions are

$$\beta_1 = \frac{[Mg(NH_3)^{2+}]}{[Mg^{2+}][NH_3]}, \qquad \beta_2 = \frac{[Mg(NH_3)_2^{2+}]}{[Mg^{2+}][NH_3]^2}$$

$$\beta_3 = \frac{[Mg(NH_3)_3^{2+}]}{[Mg^{2+}][NH_3]^3}, \qquad \beta_4 = \frac{[Mg(NH_3)_4^{2+}]}{[Mg^{2+}][NH_3]^4} \quad \cdots \tag{4.33.17}$$

$$\gamma_1 = \frac{[HL^{3-}]}{[L^{4-}][H^+]}, \qquad \gamma_2 = \frac{[H_2L^{2-}]}{[L^{4-}][H^+]^2}$$

$$\gamma_3 = \frac{[H_3L^-]}{[L^{4-}][H^+]^3}, \qquad \gamma_4 = \frac{[H_4L]}{[L^{4-}][H^+]^4} \tag{4.33.18}$$

A great simplification in the computation is achieved if one makes use of the conditional (or apparent) equilibrium constant, defined as

$$K' = \frac{[MgL^{2-}]}{[Mg^{2+}]'[L^{4-}]'} \tag{4.33.19}$$

In Eq. (4.33.19), the meaning of the terms $[Mg^{2+}]'$ and $[L^{4-}]'$ are as follows.

1. The term $[Mg^{2+}]'$ represents the concentration of all those Mg^{2+} ions that have not combined with the complexant L^{4-}. It includes the free Mg^{2+} ions and all those Mg^{2+} ions combined with the auxilary complexing agent NH_3. It is given by

$$\begin{aligned}[Mg^{2+}]' &= [Mg^{2+}] + [Mg(NH_3)^{2+}] + [Mg(NH_3)_2^{2+}] \\ &\quad + [Mg(NH_3)_3^{2+}] + [Mg(NH_3)_4^{2+}] + \cdots \\ &= [Mg^{2+}] + \beta_1[Mg^{2+}][NH_3] + \beta_2[Mg^{2+}][NH_3]^2 \\ &\quad + \beta_3[Mg^{2+}][NH_3]^3 + \beta_4[Mg^{2+}][NH_3]^4 + \cdots \\ &= [Mg^{2+}](1 + \beta_1[NH_3] + \beta_2[NH_3]^2 + \beta_3[NH_3]^3 \\ &\quad + \beta_4[NH_3]^4 + \cdots) \\ &= [Mg^{2+}]\,\alpha_{Mg(NH_3)} \end{aligned} \tag{4.33.20}$$

2. The term $[L^{4-}]'$ represents the concentration of all those ligand ions which have not reacted with Mg^{2+} ions. It includes the free L^{4-} ions and all those L^{4-} ions combined with H^+. It is given by

$$\begin{aligned}[L^{4-}]' &= [L^{4-}] + [HL^{3-}] + [H_2L^{2-}] + [H_3L^-] + [H_4L] \\ &= [L^{4-}] + \gamma_1[L^{4-}][H^+] + \gamma_2[L^{4-}][H^+]^2 + \gamma_3[L^{4-}][H^+]^3 \\ &\quad + \gamma_4[L^{4-}][H^+]^4 \\ &= [L^{4-}](1 + \gamma_1[H^+] + \gamma_2[H^+]^2 + \gamma_3[H^+]^3 + \gamma_4[H^+]^4) \\ &= [L^{4-}]\,\alpha_{L(H)} \end{aligned} \tag{4.33.21}$$

Substituting Eqs (4.33.20) and (4.33.21) in Eq. (4.33.19), we get

$$K' = \frac{[MgL^{2-}]}{([Mg^{2+}]\,\alpha_{Mg(NH_3)})\,([L^{4-}]\,\alpha_{L(H)})}$$

$$= \frac{[MgL^{2-}]}{[Mg^{2+}][L^{4-}]} \cdot \frac{1}{\alpha_{Mg(NH_3)}\,\alpha_{L(H)}}$$

$$= \frac{K}{\alpha_{Mg(NH_3)}\,\alpha_{L(H)}} \tag{4.33.22}$$

If the complex MgL^{2-} also exhibits side reactions, the conditional stability constant becomes

$$K' = K\left(\frac{\alpha_{MgL(X)}}{\alpha_{Mg(NH_3)}\,\alpha_{L(H)}}\right) \tag{4.33.23}$$

The conditional equilibrium constant is not true constant for it varies with the pH of the solution as well as with the ammonia concentration. For a series of systems, however, where these two values are unvariable, K' will remain unchanged. During a series of repetitive calculations where only the metal-ion concentration changes, the use of conditional stability constant saves a lot of time.

Example 4.33.4 Determine the conditional equilibrium constant for the formation of the complex of Mg^{2+} with EDTA at pH = 10 in the presence of an 0.1 M $NH_3 - NH_4^+$ buffer.

Given: Stability constant of $MgL^{2-} = 4.89 \times 10^8\ M^{-1}$.

Stability constants of $Mg^{2+} - NH_3$ complexes

$\beta_1 = 1.70\ M^{-1}$; $\beta_2 = 1.20\ M^{-2}$; $\beta_3 = 0.457\ M^{-3}$; $\beta_4 = 0.091\ M^{-4}$

$\beta_5 = 0.010\ M^{-5}$; $\beta_6 = 5.0 \times 10^{-4}\ M^{-6}$

Equilibrium constants of ligand with H^+

$\gamma_1 = 1.82 \times 10^{10}\ M^{-1}$, $\gamma_2 = 2.62 \times 10^{16}\ M^{-2}$, $\gamma_3 = 1.23 \times 10^{19}\ M^{-3}$,

$\gamma_4 = 1.20 \times 10^{21}\ M^{-4}$

Solution The value $\alpha_{Mg(NH_3)}$ is

$$\alpha_{Mg(NH_3)} = 1 + \beta_1[NH_3] + \beta_2[NH_3]^2 + \beta_3[NH_3]^3 + \beta_4[NH_3]^4 + \beta_5[NH_3]^5 + \beta_6[NH_3]^6$$

$$= 1 + (1.70)(0.1) + (1.20)(0.1)^2 + (0.457)(0.1)^3 + (0.091)(0.1)^4 + (0.010)(0.1)^5 + (5.0 \times 10^{-4})(0.1)^6$$

$$= 1.182$$

$$\alpha_{L(H)} = 1 + \gamma_1 [H^+] + \gamma_2 [H^+]^2 + \gamma_3 [H^+]^3 + \gamma_4 [H^+]^4$$

$$= 1 + (1.82 \times 10^{10})(10^{-10}) + (2.62 \times 10^{16})(10^{-10})^2 + (1.23 \times 10^{19})(10^{-10})^3 + (1.20 \times 10^{21})(10^{-10})^4$$

$$= 2.820$$

Hence, $$K' = \frac{K}{\alpha_{Mg(NH_3)}\alpha_{L(H)}} = \frac{4.89 \times 10^8 \text{ M}^{-1}}{(1.182)(2.820)} = 1.467 \times 10^8 \text{ M}^{-1}$$

Example 4.33.5 Calculate the concentration of zinc remaining unchelated by EDTA in 75.0 cm^3 of a solution that is 0.0125 M in EDTA, 0.0125 M in zinc, 0.100 M in NH_3 and has a pH of 10.0

Given: Stability constant of Zn – EDTA = 3.2×10^{16} M^{-1};
Stability constant of Zn – NH_3 complexes

$$\beta_1 = 10^{2.37} \text{ M}^{-1},\ \beta_2 = 10^{4.81} \text{ M}^{-2},\ \beta_3 = 10^{7.31} \text{ M}^{-3} \text{ and } \beta_4 = 10^{9.46} \text{ M}^{-4}$$

Equilibrium constants of EDTA – H^+ reactions

$$\gamma_1 = 1.82 \times 10^{10} \text{ M}^{-1},\ \gamma_2 = 2.62 \times 10^{16} \text{ M}^{-2},\ \gamma_3 = 1.23 \times 10^{19} \text{ M}^{-3} \text{ and}$$

$$\gamma_4 = 1.20 \times 10^{21} \text{ M}^{-4}$$

Solution We have

$$\alpha_{Zn(NH_3)} = 1 + \beta_1 [NH_3] + \beta_2 [NH_3]^2 + \beta_3 [NH_3]^3 + \beta_4 [NH_3]^4$$

$$= 1 + (10^{2.37})(0.1) + (10^{4.81})(0.1)^2 + (10^{7.31})(0.1)^3 + (10^{9.46})(0.1)^4$$

$$= 3.095 \times 10^5$$

$\alpha_{L(H)} = 2.820$ (From Example 4.33.4)

$$K' = \frac{K}{\alpha_{Zn(NH_3)}\ \alpha_{L(H)}} = \frac{3.2 \times 10^{16} \text{ M}^{-1}}{(3.095 \times 10^5)(2.820)}$$

$$= 3.67 \times 10^{11} \text{ M}^{-1}$$

Now $$K' = \frac{[ZnL^{2-}]}{[Zn^{2+}]'\,[L^{4-}]'}$$

In solution $[Zn^{2+}]' = [L^{4-}]'$

Hence, $$[Zn^{2+}]' = \left(\frac{[ZnL^{2-}]}{K'}\right)^{1/2}$$

Since K' is very large, one may assume that $[ZnL^{2-}] \simeq [Zn^{2-}]_0$. Hence

$$[Zn^{2+}]' = \left(\frac{0.012\,5 \text{ M}}{3.67 \times 10^{11} \text{ M}^{-1}}\right)^{1/2} = 1.85 \times 10^{-7} \text{ M}$$

Example 4.33.6 (a) The conditional stability constant for the magnesium complex of EDTA at pH = 9.00 in a 0.1 molar ammonia buffer is 2.4×10^7 M^{-1}. Determine the fraction of Mg present as the EDTA complex if 100 mL of buffered 0.010 0 molar Mg^{2+} is mixed with 100 mL of 0.010 0 molar EDTA.

(b) Calculate the value of stability constant of Mg–EDTA complex and also the concentrations of free Mg^{2+} and EDTA concentrations in solution.

Equilibrium constants of ligand with H^+ are

$$\gamma_1 = 1.82 \times 10^{10}\ \text{M}^{-1},\ \gamma_2 = 2.62 \times 10^{16}\ \text{M}^{-2},\ \gamma_3 = 1.23 \times 10^{19}\ \text{M}^{-3},$$

$$\gamma_4 = 1.20 \times 10^{21}\ \text{M}^{-4}$$

Solution **(a)** On mixing the two solutions, the concentration of each of Mg^{2+} and EDTA becomes half, i.e. 0.005 0 M in Mg^{2+} and 0.005 0 M in EDTA.

The expression of conditional stability constant is

$$K' = \frac{[\text{MgL}^{2-}]}{[\text{Mg}^{2+}]'\,[\text{L}^{4-}]'}$$

Since Mg^{2+} and L^{4-} combine to give 1 : 1 complex, in solution we will have

$$[\text{Mg}^{2+}]' = [\text{L}^{4-}]' = x\ \text{(say)}$$

Then $[\text{MgL}^{2-}] = 0.005\ 0\ \text{M} - x$

With this, we get

$$2.4 \times 10^7\ \text{M}^{-1} = \frac{0.005\ 0\ \text{M} - x}{x^2}$$

Solving for x, we get

$$x = 1.441\ 3 \times 10^{-5}\ \text{M}$$

The concentration of Mg^{2+} that is complexed with EDTA is

$$[\text{MgL}^{2-}] = 0.005\ 0\ \text{M} - 1.441\ 3 \times 10^{-5}\ \text{M} = 4.985\ 6 \times 10^{-3}\ \text{M}$$

Percentage of Mg^{2+} ions undergone complexation is

$$= \frac{4.985\ 6 \times 10^{-3}\ \text{M}}{0.005\ 0\ \text{M}} \times 100 = 99.71\%$$

(b) The relation connecting K and K' is

$$K = K'\,\alpha_{\text{Mg(NH}_3)}\,\alpha_{\text{L(H)}}$$

where $\alpha_{\text{Mg(NH}_3)} = 1 + \beta_1\,[\text{NH}_3] + \beta_2\,[\text{NH}_3]^2 + \cdots$

$= 1.182$ (see Example 4.33.4)

$$\alpha_{\text{L(H)}} = 1 + \gamma_1\,[\text{H}^+] + \gamma_2\,[\text{H}^+]^2 + \gamma_3\,[\text{H}^+]^3 + \gamma_4\,[\text{H}^+]^4$$

$$= 1 + (1.82 \times 10^{10})(10^{-9}) + (2.62 \times 10^{16})(10^{-9})^2$$

$$+ (1.23 \times 10^{19})(10^{-9})^3 + (1.20 \times 10^{21})(10^{-9})^4$$

$$= 19.22$$

Hence, $K = (2.4 \times 10^7\ \text{M}^{-1})\ (1.182)\ (19.22) = 5.45 \times 10^8\ \text{M}^{-1}$

Now $[\text{Mg}^{2+}] = [\text{Mg}^{2+}]'/\alpha_{\text{Mg(NH}_3)} = 1.441\ 3 \times 10^{-5}\ \text{M} / 1.182 = 1.22 \times 10^{-5}\ \text{M}$

$[\text{L}^{4-}] = [\text{L}^{4-}]'/\ \alpha_{\text{L(H)}} = 1.441\ 3 \times 10^{-5}\ \text{M} / 19.22 = 7.5 \times 10^{-7}\ \text{M}$

Some Applications of Complex Formation in Analytical Chemistry

Another important use of complex ion in analytical chemistry is generally based on either separating or estimating one ion in the presence of another ion. This method depends upon the fact that one ion may be transformed into a stable complex ion while the other forms either a weak complex ion or remains in the ionic form as such in the solution. Three typical examples are:

(i) In the separation of Cu^{2+} and Cd^{2+} ions, excess of KCN is added to the solution followed by passing the H_2S gas in the solution when Cd^{2+} is precipitated as CdS whereas Cu^{2+} remains in the solution. This is because of the fact that the cyanide complex of cadmium is less stable than the complex of copper, with the result that the solution contains more free Cd^{2-} than Cu^{2+} and thus only for CdS, the ionic product is exceeded to the corresponding solubility product.

(ii) The formation of a thio-complex ion in an alkaline medium of the metallic sulphide is made use of in separating the cations of IIA group from those of IIB group. The cations of IIB (As^{3+}, Sb^{3+} and Sn^{2+}) form the complex compounds whereas cations of IIA (Hg^{2+}, Pb^{2+}, Cu^{2+}, Bi^{3+} and Cd^{2+}) do not form such compounds.

$$CuS(s) + S^{2-}(aq) \rightarrow \text{No reaction}$$

$$As_2S_3(s) + 3S^{2-}(aq) \rightleftharpoons 2AsS_3^{3-}$$

$$Sb_2S_3(s) + 3S^{2-}(aq) \rightleftharpoons 2SbS_3^{3-}$$

$$SnS_2(s) + S_2^{2-}(aq) \rightleftharpoons SnS_4^{2-}$$

(iii) If a solution containing Al^{3+} and Zn^{2+} is treated with a buffer of NH_3 at a controlled alkaline pH, $Al(OH)_3$ is precipitated, whereas Zn^{2+} stays in solution as $Zn(NH_3)_4^{2+}$. The precipitation of $Zn(OH)_2$ is prevented by the formation of the complex ion. Al^{3+} does not form an ammonia complex:

$$Al^{3+}(aq) + 3OH^-(aq) \rightleftharpoons Al(OH)_3(s)$$

$$Zn^{2+}(aq) + 4NH_3(aq) \rightleftharpoons Zn(NH_3)_4^{2+}(aq)$$

Example 4.33.7

H_2S is bubbled into a 0.2 M NaCN solution which is 0.02 M in each $Ag(CN)_2^-$ and $Cd(CN)_4{}^{2-}$. Determine which sulphide precipitates first.

Given: $K_s(Ag_2S) = 1.0 \times 10^{-50}\ M^3$; $K_{inst}(Ag(CN)_2^-) = 1.0 \times 10^{-20}\ M^2$

$K_s(CdS) = 7.1 \times 10^{-28}\ M^2$; $K_{inst}(Cd(CN)_4^{2-}) = 7.8 \times 10^{-18}\ M^4$

Solution

We have

$$Ag(CN)_2^- \rightleftharpoons Ag^+ + 2CN^-; \quad K_{inst} = 1.0 \times 10^{-20}\ M^2$$

$$Cd(CN)_4^{2-} \rightleftharpoons Cd^{2+} + 4CN^-; \quad K_{inst} = 7.8 \times 10^{-18}\ M^4$$

Concentration of Ag^+ in solution

$$= \frac{K_{inst}\,[Ag(CN)_2^-]}{[CN^-]^2} = \frac{\{1.0 \times 10^{-20}\ M^2\}\,(0.02\ M)}{(0.2\ M)^2}$$

$$= 5.0 \times 10^{-21}\ M$$

Concentration of Cd^{2+} in solution

$$= \frac{K_{inst}\,[Cd(CN)_4^{2-}]}{[CN^-]^4} = \frac{\{7.8 \times 10^{-18}\ M^4\}\,(0.02\ M)}{(0.2\ M)^4}$$

$$= 9.75 \times 10^{-17}\ M$$

$[S^{2-}]$ needed to precipitate Ag^+

$$= \frac{K_s(Ag_2S)}{[Ag^+]^2} = \frac{1.0 \times 10^{-50}\ M^3}{(5 \times 10^{-21}\ M)^2} = 4 \times 10^{-10}\ M$$

$[S^{2-}]$ needed to precipitate Cd^{2+}

$$= \frac{K_s(CdS)}{[Cd^{2+}]} = \frac{7.1 \times 10^{-28}\ M^2}{9.75 \times 10^{-17}\ M} = 7.203 \times 10^{-12}\ M$$

Since $[S^{2-}]$ needed to precipitate Cd^{2+} ion is smaller than that of Ag^+ ion, therefore, Cd^{2+} will be precipitated first.

Example 4.33.8 Given 0.001 mol of Cu^{2+}, Co^{2+} and Ag^+, by using instability constants which are equal to $1.0 \times 10^{-12}\ M^4$, $4.0 \times 10^{-5}\ M^6$ and $6.0 \times 10^{-8}\ M^2$, respectively, calculate the amount of NH_3 required to be added to one dm^3 of solution of each in order to lower the concentration of uncomplexed metal ion to 1.0×10^{-4} M.

Given:

$$Cu(NH_3)_4^{2+} \rightleftharpoons Cu^{2+} + 4NH_3; \qquad K_{inst} = 1.0 \times 10^{-12}\ M^4$$

$$Co(NH_3)_6^{2+} \rightleftharpoons Co^{2+} + 6NH_3; \qquad K_{inst} = 4.0 \times 10^{-5}\ M^6$$

$$Ag(NH_3)_2^{+} \rightleftharpoons Ag^{+} + 2NH_3; \qquad K_{inst} = 6.0 \times 10^{-8}\ M^2$$

Solution After the addition of NH_3 in each case, we have

$$[Cu^{2+}] = [Co^{2+}] = [Ag^+] = 10^{-4}\ M$$

$$[Cu(NH_3)_4^{2+}] = [Co(NH_3)_6^{2+}] = [Ag(NH_3)_2^{+}] = (10^{-3} - 10^{-4})\ M = 9.0 \times 10^{-4}\ M$$

Concentration of ammonia present in each of the solution can be calculated from the corresponding instability constant. Thus

Concentration of NH_3 in Cu^{2+} solution

$$= \left\{\frac{K_{inst}\,[Cu(NH_3)_4^{2+}]}{[Cu^{2+}]}\right\}^{1/4} = \left(\frac{\{1.0 \times 10^{-12}\ M^4\}\,(9.0 \times 10^{-4}\ M)}{10^{-4}\ M}\right)^{1/4}$$

$$= 1.73 \times 10^{-3}\ M$$

Concentration of NH_3 in Co^{2+} solution

$$= \left\{\frac{K_{inst}\,[Co(NH_3)_6^{2+}]}{[Co^{2+}]}\right\}^{1/6} = \left(\frac{\{4.0 \times 10^{-5}\ M^6\}\,(9.0 \times 10^{-4}\ M)}{10^{-4}\ M}\right)^{1/6}$$

$$= 2.67 \times 10^{-1}\ M$$

Concentration of NH_3 in Ag^+ solution

$$= \left\{\frac{K_{inst}\,[Ag(NH_3)_2^{+}]}{[Ag^{+}]}\right\}^{1/2} = \left(\frac{\{6.0 \times 10^{-8}\ M^2\}\,(9.0 \times 10^{-4}\ M)}{10^{-4}\ M}\right)^{1/2}$$

$$= 7.35 \times 10^{-4}\ M$$

Actual amount of NH_3 to be added = Free ammonia + combined ammonia
Thus,

NH_3 to be added in Cu^{2+} solution

$$= 1.73 \times 10^{-3}\ M + 4 \times 9.0 \times 10^{-4}\ M = 5.33 \times 10^{-3}\ M$$

NH_3 to be added in Co^{2+} solution

$$= 2.67 \times 10^{-1}\ M + 6 \times 9.0 \times 10^{-4}\ M = 2.72 \times 10^{-1}\ M$$

NH_3 to be added in Ag^+ solution

$$= 7.35 \times 10^{-4}\ M + 2 \times 9.0 \times 10^{-4}\ M = 25.35 \times 10^{-4}\ M$$

Example 4.33.9 Predict whether or not AgCl will be precipitated from a solution which is 0.02 M in NaCl and 0.05 M in $KAg(CN)_2$. Given $K_{inst}(Ag(CN)_2^-) = 4.0 \times 10^{-19}\ M^2$ and $K_s(AgCl) = 2.8 \times 10^{-10}\ M^2$.

Solution Let x be the concentration of Ag^+ in the 0.05 M solution of $Ag(CN)_2^-$. Then

$$\underset{0.05\,M - x}{Ag(CN)_2^-} \rightleftharpoons \underset{x}{Ag^+} + \underset{2x}{2CN^-}$$

$$K_{inst} = \frac{[Ag^+][CN^-]^2}{[Ag(CN)_2^-]} = 4.0 \times 10^{-19}\ M^2$$

$$= \frac{(x)(2x)^2}{(0.05\ M - x)} \approx \frac{4x^3}{0.05\ M} = 4.0 \times 10^{-19}\ M^2$$

Thus, $x = 1.7 \times 10^{-7}\ M$

Ionic product of AgCl in the solution

$$= (1.7 \times 10^{-7}\ M)\,(0.02\ M) = 3.42 \times 10^{-9}\ M^2$$

Since this value exceeds the solubility product of AgCl, therefore precipitation will occur.

4.34 AMPHOTERISM

The hydroxides of certain metals can function as both bases and acids. Examples include $Al(OH)_3$, $Zn(OH)_2$, $Sn(OH)_2$, $Cr(OH)_3$, Sb_2O_3 and As_2O_3.

Amphoteric behaviour of these compounds is due to the formation of complex ions with hydroxyl ions. For example, $Zn(OH)_2$ is a sparingly soluble amphoteric solid that dissolves via a series of steps, the first of which is

$$Zn(OH)_2(s) \rightleftharpoons Zn(OH)_2(aq)$$

Molecular $Zn(OH)_2$ further dissociates according to the equations

$$Zn(OH)_2(aq) \rightleftharpoons Zn(OH)^+ + OH^-$$

$$Zn(OH)^+ \rightleftharpoons Zn^{2+} + OH^-$$

In addition to this, zinc ion can also form complexes $Zn(OH)_3^-$ and $Zn(OH)_4^{2-}$. The formation of these species can be represented by the equations

$$Zn(OH)_2(aq) + OH^- \rightleftharpoons Zn(OH)_3^-$$

$$Zn(OH)_3^- + OH^- \rightleftharpoons Zn(OH)_4^{2-}$$

If the hydroxide concentration is very small, then Zn^{2+} and $Zn(OH)^+$ are present in high concentrations. On the other hand, the species $Zn(OH)_3^-$ and $Zn(OH)_4^{2-}$ are present at high hydroxide concentrations. Thus in an acidic solution, the concentrations of cationic species increase whereas in alkaline solution the concentrations of anionic species increase. In other words, the solubility of $Zn(OH)_2$ is greater in an acidic or basic solution than in neutral water.

Example 4.34.1 Calculate the solubility of solid zinc hydroxide at a pH of 5, 9 and 13. Given:

$$Zn(OH)_2(s) \rightleftharpoons Zn(OH)_2(aq) \qquad K_1 = 10^{-6}\ M \qquad (1)$$

$$Zn(OH)_2(aq) \rightleftharpoons Zn(OH)^+ + OH^- \qquad K_2 = 10^{-7}\ M \qquad (2)$$

$$Zn(OH)^+ \rightleftharpoons Zn^{2+} + OH^- \qquad K_3 = 10^{-4}\ M \qquad (3)$$

$$Zn(OH)_2(aq) + OH^- \rightleftharpoons Zn(OH)_3^- \qquad K_4 = 10^3\ M^{-1} \qquad (4)$$

$$Zn(OH)_3^- + OH^- \rightleftharpoons Zn(OH)_4^{2-} \qquad K_5 = 10\ M^{-1} \qquad (5)$$

Solution The total concentration of zinc hydroxide that dissolves in water is given by:

$$\text{Dissolved } Zn(OH)_2 = [Zn^{2+}] + [Zn(OH)^+] + [Zn(OH)_2(aq)] + [Zn(OH)_3^-] + [Zn(OH)_4^{2-}]$$

According to reaction (1), the concentration of molecular zinc hydroxide is always the same provided it is in contact with solid zinc hydroxide; thus we write

$$[Zn(OH)_2(aq)] = 10^{-6}\ M$$

The concentrations of the rest of the species can be written in terms of $[OH^-]$ as given below.

$$[Zn(OH)^+] = \frac{[Zn(OH)_2(aq)]}{[OH^-]} K_2 = \frac{(10^{-6}\ M)(10^{-7}\ M)}{[OH^-]} = \frac{10^{-13}\ M^2}{[OH^-]}$$

$$[Zn^{2+}] = \frac{[Zn(OH)^+]}{[OH^-]} K_3 = \frac{10^{-17}\ M^3}{[OH^-]^2}$$

$$[Zn(OH)_3^-] = K_4\,[Zn(OH)_2(aq)]\,[OH^-] = 10^{-3}\,[OH^-]$$

$$[Zn(OH)_4^{2-}] = K_5\,[Zn(OH)_3^-]\,[OH^-] = (10^{-2}\ M^{-1})\,[OH^-]^2$$

Thus, the solubility is given by

$$\begin{aligned}\text{Dissolved } Zn(OH)_2 &= [Zn^{2+}] + [Zn(OH)^+] + [Zn(OH)_2(aq)] + [Zn(OH)_3^-] + [Zn(OH)_4^{2-}]\\ &= \frac{10^{-17}\ M^3}{[OH^-]^2} + \frac{10^{-13}\ M^2}{[OH^-]} + (10^{-6}\ M) + 10^{-3}\,[OH^-]\\ &\quad + (10^{-2}\ M^{-1})\,[OH^-]^2\end{aligned}$$

From this general expression, the solubility at various pH values can be calculated.

$\text{pH} = 13, \qquad [OH^-] = 0.10\ M$

$\text{Dissolved } Zn(OH)_2 = (10^{-15} + 10^{-12} + 10^{-6} + 10^{-4} + 10^{-4})\ M = 2 \times 10^{-4}\ M$

$\text{pH} = 9, \qquad [OH^-] = 10^{-5}\ M$

$\text{Dissolved } Zn(OH)_2 = (10^{-7} + 10^{-8} + 10^{-6} + 10^{-8} + 10^{-12})\ M = 1.12 \times 10^{-6}\ M$

$\text{pH} = 5, \qquad [OH^-] = 10^{-9}\ M$

$\text{Dissolved } Zn(OH)_2 = (10 + 10^{-4} + 10^{-6} + 10^{-12} + 10^{-20})\ M = 10\ M$

Example 4.34.2 (a) At 25 °C what minimum pH will 1.0×10^{-3} mol of $Al(OH)_3$ go into solution (1 dm^3) as $Al(OH)_4^-$? (b) At 25 °C what maximum pH will 1.0×10^{-3} mol of $Al(OH)_3$ go into solution (1 dm^3) as Al^{3+}? Given:

$$K_s(Al(OH)_3) = 5.0 \times 10^{-33}\ M^4 \quad \text{and} \quad K_{inst} \text{ of } Al(OH)_4^- = 1.3 \times 10^{-34}\ M^4$$

Solution (a) At the point at which $Al(OH)_3$ just dissolves

$$K_s = [Al^{3+}]\,[OH^-]^3 = 5.0 \times 10^{-33}\ M^4 \qquad (1)$$

If the $Al(OH)_3$ is completely converted into $Al(OH)_4^-$, then

$$[Al(OH)_4^-] = 1.0 \times 10^{-3}\ M$$

Thus, $$\frac{[Al^{3+}][OH^-]^4}{[Al(OH)_4^-]} = 1.3 \times 10^{-34}\ M^4$$

or $$[Al^{3+}]\,[OH^-]^4 = 1.3 \times 10^{-37}\ M^5 \qquad (2)$$

If we divide Eq. (2) by Eq. (1), we get

$$[OH^-] = \frac{1.3 \times 10^{-37}\ M^5}{5.0 \times 10^{-33}\ M^4} = 2.6 \times 10^{-5}\ M$$

$$pOH = -\log\{[OH^-]/M\} = -\log(2.6 \times 10^{-5}) = 4.6$$

or $$pH = 14 - pOH = 9.4$$

(b) Here $Al(OH)_3$ will be completely converted into Al^{3+}. Thus

$$Al^{3+} = 1.0 \times 10^{-3}\ M$$

and $$[Al^{3+}]\,[OH^-]^3 = 5.0 \times 10^{-33}\ M^4$$

which gives

$$[OH^-] = \left(\frac{5.0 \times 10^{-33}\ M^4}{1.0 \times 10^{-3}\ M}\right)^{1/3} = 1.71 \times 10^{-10}\ M$$

Thus $pOH = -\log\{[OH^-]/M\} = -(1.71 \times 10^{-10}) = 9.78$ and hence

$$pH = 14 - pOH = 4.22$$

4.35 SOME CONCEPTS OF ACIDS AND BASES

One of the most important equilibrium in chemistry involves acids and bases. Various attempts have been made to define these acids and bases, starting from the phenomenological basis to the molecular composition and structure of the compound. Three of the currently accepted concepts are discussed in the following.

The Arrhenius Concept

Arrhenius classified a substance into an acid or a base in terms of the characteristic ions of water which it produces in aqueous solution. Thus, *an acid is a substance which ionizes in water to produce* $H^+(aq)$ *or the hydronium ion; a base is a substance which produces hydroxide ion* $OH^-(aq)$. The strength of an acid is defined in terms of the concentration of $H^+(aq)$ that is present in the aqueous solution of a given concentration of the acid. Likewise, the strength of a base depends upon the relative concentration of $OH^-(aq)$ in an aqueous solution of the base.

The neutralization of an acid with base is explained in terms of the equation

$$H^+(aq) + OH^-(aq) \rightarrow H_2O(l).$$

This reaction is exothermic, as 57.32 kJ of heat is liberated when one mole of a strong acid combines with one mole of a strong base. If a weak acid and/or a weak base is involved, then the heat evolved is less than 57.32 kJ.

The principle of Arrhenius-water concept has been extended to incorporate the reactions in nonaqueous media. An acid is a substance that gives the cation characteristic of the solvent and a base that gives the anion characteristic of the solvent. The reaction of an acid and a base yields the solvent as one of its products. Table 4.35.1 includes some of the solvent systems.

Table 4.35.1 Some Solvent Systems

Solvent	*Acid ion*	*Base ion*	*Typical acid*	*Typical base*
H_2O	H_3O^+	OH^-	HCl	NaOH
NH_3	NH_4^+	NH_2^-	NH_4Cl	$NaNH_2$
CH_3COOH	$H_2C_2H_3O_2^+$	$C_2H_3O_2^-$	HCl	CH_3COONa
SO_2	SO^{2+}	SO_3^{2-}	$SOCl_2$	$CaSO_3$
N_2O_4	NO^+	NO_3^-	NOCl	$AgNO_3$
$COCl_2$	$COCl^+$	Cl^-	$(COCl)AlCl_3$	$CaCl_2$

The Brönsted-Lowry Concept

J. Brönsted and T. Lowry independently proposed a broader concept of acids and bases. According to them, *a substance is known as an acid if it can donate a proton and as a base if it can accept a proton.* The substance may be a molecule or an ion. The reaction of an acid with a base constitutes the transfer of a proton from the acid to the base.

Conjugate Acid-Base Pair

The dissolution of ammonia in water may be represented by the equation

$$H_2O + NH_3 \rightleftharpoons NH_4^+ + OH^-$$

In this reaction, H_2O is serving as an acid and is donating a proton to the base NH_3. If a solution of an ammonium salt is made strongly alkaline, ammonia gas is released. The reaction is, therefore, reversible and the system exists in equilibrium.

In the reverse reaction, NH_4^+ is serving as an acid and OH^- as a base. It follows, then, that in this Brönsted acid-base reaction, two acids (H_2O and NH_4^+) and two bases (OH^- and NH_3) are involved. The base NH_3 gains a proton and thereby forms the acid NH_4^+ which on the loss of a proton forms the base NH_3. Such an acid-base pair, related through the loss or gain of a proton, is called a *conjugate pair*. NH_4^+ is the conjugate acid of the base NH_3, and NH_3 is the conjugate base of the acid NH_4^+. In like manner, the acid H_2O and the base OH^- constitute a second conjugate pair in the preceding reaction. We may indicate conjugate relationships by the use of subscripts in the following manner:

(1) $$\underset{\text{Acid}_1}{H_2O} + \underset{\text{Base}_2}{NH_3} \rightleftharpoons \underset{\text{Acid}_2}{NH_4^+} + \underset{\text{Base}_1}{OH^-}$$

A few other examples are

(2) $$\underset{\text{Acid}_1}{CH_3COOH} + \underset{\text{Base}_2}{H_2O} \rightleftharpoons \underset{\text{Acid}_2}{H_3O^+} + \underset{\text{Base}_1}{CH_3COO^-}$$

(3) $$\underset{\text{Acid}_1}{HCl} + \underset{\text{Base}_2}{H_2O} \rightleftharpoons \underset{\text{Acid}_2}{H_3O^+} + \underset{\text{Base}_1}{Cl^-}$$

(4) $$\underset{\text{Acid}_1}{H_2O} + \underset{\text{Base}_2}{CO_3^{2-}} \rightleftharpoons \underset{\text{Acid}_2}{HCO_3^-} + \underset{\text{Base}_1}{OH^-}$$

Similar type of reactions can be written in a nonaqueous medium:

$$\underset{\text{Base}_2}{H^-} + \underset{\text{Acid}_1}{NH_3} \rightleftharpoons \underset{\text{Acid}_2}{H_2} + \underset{\text{Base}_1}{NH_2^-}$$

Amphiprotic Species There are many molecules and ions that can function as acids in certain reactions and as bases in other reactions. Such species are called *amphiprotic.* For example, H_2O acts as an acid in reactions (1) and (4) and as a base in reactions (2) and (3).

Arrhenius acid-base type reactions are also covered by the Brönsted-Lowry reactions:

$$\underset{\text{Acid}_1}{H_3O^+} + \underset{\text{Base}_2}{OH^-} \rightleftharpoons \underset{\text{Acid}_2}{H_2O} + \underset{\text{Base}_1}{H_2O}$$

$$\underset{\text{Acid}_1}{NH_4^+} + \underset{\text{Base}_2}{NH_2^-} \rightleftharpoons \underset{\text{Acid}_2}{NH_3} + \underset{\text{Base}_1}{NH_3}$$

Strength of Brönsted Acids and Bases

The strength of a Brönsted acid is determined by its tendency to donate a proton, and the strength of a Brönsted base is determined by its tendency to accept a proton. Consider the reaction

$$\underset{\text{Acid}_1}{HCl} + \underset{\text{Base}_2}{H_2O} \longrightarrow \underset{\text{Acid}_2}{H_3O^+} + \underset{\text{Base}_1}{Cl^-}$$

This reaction proceeds virtually to completion, i.e. the reactants are completely converted into products. It can be concluded that:

(i) The acid HCl is stronger than H_3O^+ ion, since HCl has a greater tendency to lose a proton than the H_3O^+ ion.

(ii) The base H_2O is stronger than Cl^- ions, since in the competition for holding the proton, H_2O shows larger tendency and thereby is converted to H_3O^+.

Thus, we see that the strong acid HCl has a weak conjugate base Cl^- ion and the strong base H_2O has a weak conjugate acid H_3O^+. The fact that a strong acid has a weak conjugate base and vice versa is always true since a strong acid with a high tendency to lose a proton is necessarily conjugate to a weak base with a small tendency to gain and hold a proton. Similarly, a strong base with a strong tendency to attract a proton is necessarily conjugate to a weak acid with a lower tendency to lose a proton.

An aqueous solution of 0.1 M acetic acid is only 1.33 per cent ionized at 25 °C. The reaction is

$$\underset{\text{Acid}_1}{HC_2H_3O_2} + \underset{\text{Base}_2}{H_2O} \rightleftharpoons \underset{\text{Acid}_2}{H_3O^+} + \underset{\text{Base}_1}{C_2H_3O_2^-}$$

or briefly

$$\underset{\text{Acid}_1}{HAc} + \underset{\text{Base}_2}{H_2O} \rightleftharpoons \underset{\text{Acid}_2}{H_3O^+} + \underset{\text{Base}_1}{Ac^-}$$

which is largely displaced to the left. Once again, we can conclude that:

(i) The H_3O^+ is a stronger acid than HAc since at equilibrium more H_3O^+ ions than HAc molecules have lost protons.

(ii) The Ac^- is a stronger base than water since it shows greater tendency to attract a proton than H_2O, thus converting itself into HAc.

In the above example, we again notice that the stronger acid H_3O^+ is conjugate to the weaker base H_2O and the stronger base Ac^- is conjugate to the weaker acid HAc. From the above two examples, one more conclusion can be drawn. This can be stated as follows: *In a given reaction, the position of equilibrium favours the formation of a weak acid and a weak base.*

Thus, in the reaction between HCl and H_2O, the equilibrium concentrations of the weaker species H_3O^+ and Cl^- are high, whereas in the solution of acetic acid, the equilibrium concentrations of the stronger species H_3O^+ and $C_2H_3O_2^-$ are low.

Arrhenius concept of a strong and a weak acid in aqueous solution can be explained by the Brönsted concept. In Arrhenius theory, a strong acid is virtually 100% ionized in aqueous solution and thus produces very high concentration of H_3O^+. In the Brönsted system, acids are classified according to their ability to donate protons to a specific base under consideration. It is obvious that if water is used as the specific base, then the acid strengths of the Arrhenius concept can be explained in terms of the Brönsted concept. Thus, strong Arrhenius acids are those substances that are stronger acids than H_3O^+ and weak Arrhenius acids are those substances that are weaker acids than H_3O^+. Table 4.35.2 records some common conjugate acid-base pairs showing the interdependent relationship of their strengths.

Levelling Effect

In Brönsted system, it is obvious that the strength of an acid is influenced by the solvent water. All those acids which are stronger than H_3O^+ will essentially be completely ionized in aqueous solution. For Example:

$$HClO_4 + H_2O \rightarrow H_3O^+ + ClO_4^-$$

$$HNO_3 + H_2O \rightarrow H_3O^+ + NO_3^-$$

$$HCl + H_2O \rightarrow H_3O^+ + Cl^-$$

$$HBr + H_2O \rightarrow H_3O^+ + Br^-$$

Thus, the acids $HClO_4$, HNO_3, HCl and HBr of the same concentration appear to have the same acid strength (equally stronger). Water is said to have a levelling effect on acids stronger than H_3O^+. The strongest acid that can exist in water solution is the conjugate acid of water, i.e. H_3O^+ ion.

Acids that are weaker than H_3O^+ are not levelled by water. Thus, the weaker acids such as HAc, H_3PO_4, HNO_2 and H_2S show a wide variation in their degree of ionization.

Table 4.35.2 Common Conjugate Acid-Base Pairs

Acid		*Conjugate base*	
Name	*Formula*	*Formula*	*Name*
Perchloric acid	$HClO_4$	ClO_4^-	Perchlorate ion
Sulphuric acid	H_2SO_4	HSO_4^-	Hydrogen sulphate ion
Hydroiodic acid	HI	I^-	Iodide ion
Hydrobromic acid	HBr	Br^-	Bromide ion
Hydrochloric acid	HCl	Cl^-	Chloride ion
Nitric acid	HNO_3	NO_3^-	Nitrate ion
Hydronium ion	H_3O^+	H_2O	Water
Hydrogen sulphate ion	HSO_4^-	SO_4^{2-}	Sulphate ion
Phosphoric acid	H_3PO_4	$H_2PO_4^-$	Dihydrogen phosphate ion
Nitrous acid	HNO_2	NO_2^-	Nitrite ion
Acetic acid	$HC_2H_3O_2$	$C_2H_3O_2^-$	Acetate ion
Carbonic acid ($CO_2 + H_2O$)	H_2CO_3	HCO_3^-	Bicarbonate ion
Hydrogen sulphide	H_2S	HS^-	Hydrosulphide ion
Ammonium ion	NH_4^+	NH_3	Ammonia
Hydrogen cyanide	HCN	CN^-	Cyanide ion
Water	H_2O	OH^-	Hydroxide ion
Ammonia	NH_3	NH_2^-	Amide ion

Similarly, bases that are stronger than the OH^- ion are levelled by water. For example, H^- and NH_2^- ions are stronger than OH^- ion and therefore they are levelled to OH^- in aqueous solution.

$$H^- + H_2O \rightarrow H_2 + OH^-$$

$$NH_2^- + H_2O \rightarrow NH_3 + OH^-$$

Thus, the above reactions are essentially complete. The bases that are weaker than OH^- are not levelled in water. Examples include NH_3 and anilinium ion.

The levelling effect is basically due to the fact that water acts as too strong a base with respect to those acids which are stronger than H_3O^+ and also too strong an acid with respect to those bases which are stronger than OH^-. If we want to differentiate these acids and bases, we will have to choose a less basic (more acidic) and less acidic (more basic) solvent than water, respectively. For example:

(i) In acetic acid medium, the strength of some of the acids described above were found to be

$$HClO_4 > HBr > H_2SO_4 > HCl > HNO_3$$

(ii) In liquid ammonia, out of H^- and NH_2^- ions, the former is found to be stronger than the latter.

The levelling effect is also observed for solvents other than water. For example, in liquid ammonia solvent, the strongest acid that can exist is the conjugate acid of NH_3, i.e. NH_4^+ ion, and the strongest base that can exist is the conjugate base of NH_3, i.e. NH_2^- ion. Thus, any acid that is stronger than NH_4^+ ion will be levelled and any base that is stronger than NH_2^- will also be levelled. For example, HAc is a stronger acid than NH_4^+ and thus will be levelled, which will give us:

$$HAc + NH_3 \rightarrow NH_4^+ + Ac^-$$

Nitric acid in methanol (CH_3OH) solution is incompletely dissociated since HNO_3 is a weaker acid than the conjugate acid of methanol, $CH_3OH_2^+$:

$$HNO_3 + CH_3OH \rightleftharpoons CH_3OH_2^+ + NO_3^-$$

Hydrolysis of Anions

The reaction of an ion with water in which either H_3O^+ or OH^- is produced is called hydrolysis. Such reactions are Brönsted acid-base reactions. Anions that function as bases in water solution are hydrolyzed as follows:

$$\underset{Acid_1}{H_2O} + \underset{Base_2}{B^-} \rightleftharpoons \underset{Acid_2}{HB} + \underset{Base_1}{OH^-}$$

The extent of hydrolysis of a given anion depends upon the base strength of the anion, and an indication of the degree of hydrolysis is given by the concentration of OH^- relative to the concentration of anion present in the solution.

The following are the factors which determine the extent of hydrolysis of anions:

Complete hydrolysis The anion which is a stronger base than OH^- and the conjugate acid of the anion much weaker acid than H_2O will show complete hydrolysis. This is, in fact, the levelling effect of water on bases stronger than OH^-, e.g. H^- shows complete hydrolysis. For such a reaction, we have

$$H_2O + H^- \rightarrow H_2 + OH^-$$

No hydrolysis The anion which is a weaker base than OH^- and its conjugate acid stronger than both H_2O and H_3O^+ does not hydrolyze. For example Cl^- ion;

$$Cl^- + H_2O \leftarrow HCl + OH^-$$

If fact, HCl is levelled in water to H_3O^+ and, therefore, the HCl molecules do not exist in solution. The Cl^- ion is simply too weak a base to accept a proton from water. Thus, anion that have conjugate acids stronger than H_3O^+ do not hydrolyze.

Hydrolysis to a limited extent The anion which acts as a weaker base than OH^- ion and the conjugated acid stronger than H_2O but weaker than H_3O^+, shows hydrolysis to a limited extent. Example is CH_3COO^- ion:

$$CH_3COO^- + H_2O \rightleftharpoons CH_3COOH + OH^-$$

Other examples are CN^-, NO_2^-, and S^{2-}.

Hydrolysis of Cations

Cations, like anions, also show hydrolysis. The extent of hydrolysis depends upon the following factors:

Complete hydrolysis The cation which is a stronger acid than H_3O^+ and its conjugate base very much weaker than H_2O will show complete hydrolysis. Example is PH_4^+ ion:

$$PH_4^+ + H_2O \rightarrow H_3O^+ + PH_3$$

No hydrolysis The cation which is a weaker acid than H_3O^+ and its conjugate base stronger than both H_2O and OH^- does not hydrolyze. Example is Na^+ ion:

$$Na^+ + 2H_2O \leftarrow NaOH + H_3O^+$$

Hydrolysis to a limited extent The cation which is a weaker acid than H_3O^+ ion and its conjugate base stronger than H_2O but weaker than OH^-, shows hydrolysis to a limited extent. Example is NH_4^+ ion:

$$NH_4^+ + 2H_2O \rightleftharpoons NH_4OH + H_3O^+$$

The Lewis Concept

The Brönsted-Lowry concept of acids and bases which involves the transfer of a proton from an acid to a base was broadened by Lewis to include substances whose reactions do not involve loss or gain of a proton. According to him, *a base has an unshared electron pair which enables it to form a covalent bond with an atom, molecule or ion. An acid is a substance that can form a covalent bond by accepting an electron pair from a base.*

The following are a few examples of Lewis acids:

(i) Molecules or atoms that have incomplete octets:

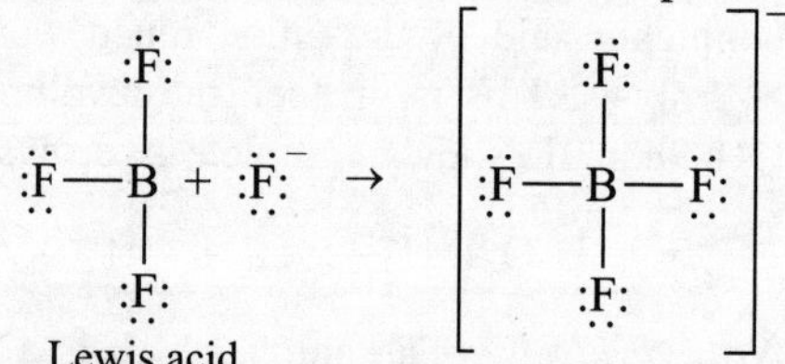

(ii) Simple cations:

$$\underset{\text{Lewis acid}}{Cu^{2+}} + 4\,:NH_3 \rightarrow [Cu(:NH_3)_4]^{2+}$$

$$\underset{\text{Lewis acid}}{Fe^{3+}} + 6\,:CN^- \rightarrow [Fe(:C\equiv N)_6]^{3-}$$

Here, the strength of an acid is determined by the positive charge density on the cation. Thus, the acid strength would increase with increasing charge and decreasing radius.

(iii) Compounds having central atoms capable of expanding their valence shells:

$$\underset{\text{Lewis acid}}{SnCl_2} + 2Cl^- \rightarrow SnCl_4^{2-}$$

$$\underset{\text{Lewis acid}}{SiF_4} + 2F^- \rightarrow SiF_6^{2-}$$

(iv) Molecules in which the central atom contains one or more multiple bonds:

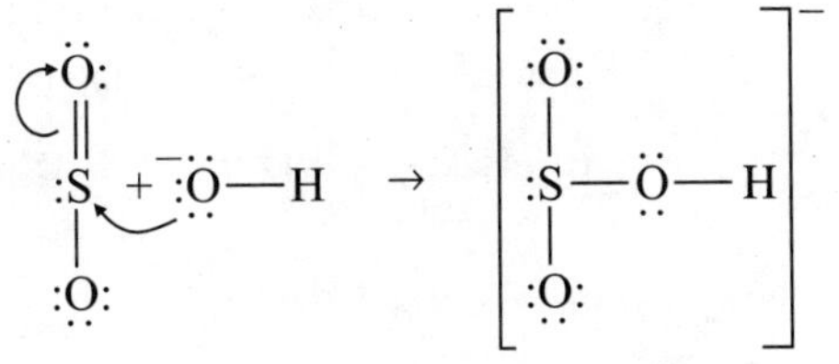

$$:\ddot{O}:\overset{}{\underset{}{=}}S(:\ddot{O}:) + :\ddot{O}—H \rightarrow \left[:\ddot{O}:—S(—\ddot{O}—H)(:\ddot{O}:)\right]^-$$

Examples of Lewis bases will include all negative ions, e.g., OH^-, CN^-, and all molecules possessing one or more lone pair of pairs of electrons, e.g., oxides and ammonia.

The Brönsted acid-base reactions can be interpreted in terms of the Lewis concept by considering a proton as a primary Lewis acid and the Brönsted acid as a complex consisting of this primary Lewis acid and the appropriate Lewis base. For example, the Brönsted acid H_3O^+ is simply a complex formed between the primary Lewis acid H^+ and the Lewis base H_2O. Thus, in the reaction between H_3O^+ and OH^-, we have

$$\underset{\text{Lewis acid}}{H_3O^+} + \underset{\text{Lewis base}}{OH^-} \rightarrow \underset{\text{Lewis acid}}{H_2O} + \underset{\text{Lewis base}}{H_2O}$$

It can be said that the Lewis base OH^- has displaced a weaker base H_2O from the Lewis acid H_3O^+. In the reaction

$$HCl + H_2O \rightarrow H_3O^+ + Cl^-$$

the Lewis base H_2O displaces a weak base Cl^-. Since a base supplies an electron pair to a nucleus, the above type of base displacement reactions are known as nucleophilic displacements. Nucleophilic displacements may be identified among reactions that are not Brönsted acid-base reactions; one of the examples is

$$[Cu(H_2O)_4]^{2+} + 4NH_3 \rightarrow [Cu(NH_3)_4]^{2+} + 4H_2O$$

Lewis acid displacements or electrophilic displacements are not so common as base displacements. A few examples are

(a) $CoCl_2 + AlCl_3 \rightarrow CoCl^+ + AlCl_4^-$

Here $AlCl_3$ displaces the weaker acid $CoCl^+$ from its complex with the base Cl^-.

(b) $SeOCl_2 + BCl_3 \rightarrow SeOCl^+ + BCl_4^-$

Here BCl_3 displaces the weaker acid $SeOCl^+$.

Example 4.35.1

(a) What is the conjugate base of
(i) H_3PO_4, (ii) $H_2PO_4^-$, (iii) NH_3, (iv) HS^- (v) H_3O^+, (vi) HOCl, (vii) $Al(H_2O)_6^{3+}$, (viii) CH_3OH, and (ix) phenol?

(b) What is the conjugate acid of
(i) F^-, (ii) OH^-, (iii) PO_4^{3-}, (iv) NH_3, (v) NH_2^-, (vi) HS^-, (vii) $[Zn(OH)(H_2O)_5]^{2+}$, (viii) CH_3OH, (ix) CH_3OCH_3 and (x) $C_6H_5NH_2$?

Solution **(a)** The following relations give us the conjugate base,

(i) $\underset{\text{Acid}}{H_3PO_4} + H_2O \rightleftharpoons H_3O^+ + \underset{\text{Base}}{H_2PO_4^-}$

(ii) $\underset{\text{Acid}}{H_2PO_4^-} + H_2O \rightleftharpoons H_3O^+ + \underset{\text{Base}}{HPO_4^{2-}}$

(iii) $\underset{\text{Acid}}{NH_3} + H_2O \rightleftharpoons H_3O^+ + \underset{\text{Base}}{NH_2^-}$

(iv) $\underset{\text{Acid}}{HS^-} + H_2O \rightleftharpoons H_3O^+ + \underset{\text{Base}}{S^{2-}}$

(v) $\underset{\text{Acid}}{H_3O^+} + H_2O \rightleftharpoons H_3O^+ + \underset{\text{Base}}{H_2O}$

(vi) $\underset{\text{Acid}}{HOCl} + H_2O \rightleftharpoons H_3O^+ + \underset{\text{Base}}{OCl^-}$

(vii) $\underset{\text{Acid}}{[Al(H_2O)_6^{3+}]} + H_2O \rightleftharpoons H_3O^+ + \underset{\text{Base}}{[Al(H_2O)_5OH]^{2+}}$

(viii) $\underset{\text{Acid}}{CH_3OH} + H_2O \rightleftharpoons H_3O^+ + \underset{\text{Base}}{CH_3O^-}$

(ix) $\underset{\text{Acid}}{C_6H_5OH} + H_2O \rightleftharpoons H_3O^+ + \underset{\text{Base}}{C_6H_5O^-}$

(b) The following relations give us conjugate acid.

(i) $H_2O + \underset{\text{Base}}{F^-} \rightleftharpoons \underset{\text{Acid}}{HF} + OH^-$

(ii) $H_2O + \underset{\text{Base}}{OH^-} \rightleftharpoons \underset{\text{Acid}}{H_2O} + OH^-$

(iii) $H_2O + \underset{\text{Base}}{PO_4^{3-}} \rightleftharpoons \underset{\text{Acid}}{HPO_4^{2-}} + OH^-$

(iv) $H_2O + \underset{\text{Base}}{NH_3} \rightleftharpoons \underset{\text{Acid}}{NH_4^+} + OH^-$

(v) $H_2O + \underset{\text{Base}}{NH_2^-} \rightleftharpoons \underset{\text{Acid}}{NH_3} + OH^-$

(vi) $H_2O + \underset{\text{Base}}{HS^-} \rightleftharpoons \underset{\text{Acid}}{H_2S} + OH^-$

(vii) $H_2O + \underset{\text{Base}}{[Zn(OH)(H_2O)_5]^{2+}} \rightleftharpoons \underset{\text{Acid}}{[Zn(H_2O)_6]^{3+}} + OH^-$

(viii) $H_2O + \underset{\text{Base}}{CH_3OH} \rightleftharpoons \underset{\text{Acid}}{CH_3OH_2^+} + OH^-$

(ix) $H_2O + \underset{\text{Base}}{CH_3OCH_3} \rightleftharpoons \underset{\text{Acid}}{CH_3(OH)^+CH_3} + OH^-$

(x) $H_2O + \underset{\text{Base}}{C_6H_5NH_2} \rightleftharpoons \underset{\text{Acid}}{C_6H_5NH_3^+} + OH^-$

Example 4.35.2

Identify the Brönsted acids and bases in the following:

(i) $NH_3 + HCl \rightarrow NH_4^+ + Cl^-$ (ii) $NH_4^+ + OH^- \rightarrow NH_3 + H_2O$

(iii) $HS^- + OH^- \rightarrow S^{2-} + H_2O$ (iv) $H_3O^+ + HS^- \rightarrow H_2S + H_2O$

(v) $H_2SO_4 + CN^- \rightarrow HCN + HSO_4^-$

Solution

The identifications of Brönsted acids and bases in the given reactions are given below:

(i) $\underset{\text{Base}_1}{NH_3} + \underset{\text{Acid}_2}{HCl} \rightarrow \underset{\text{Acid}_1}{NH_4^+} + \underset{\text{Base}_2}{Cl^-}$ **(ii)** $\underset{\text{Acid}_1}{NH_4^+} + \underset{\text{Base}_2}{OH^-} \rightarrow \underset{\text{Base}_1}{NH_3} + \underset{\text{Acid}_2}{H_2O}$

(iii) $\underset{\text{Acid}_1}{HS^-} + \underset{\text{Base}_2}{OH^-} \rightarrow \underset{\text{Base}_1}{S^{2-}} + \underset{\text{Acid}_2}{H_2O}$ **(iv)** $\underset{\text{Acid}_1}{H_3O^+} + \underset{\text{Base}_2}{HS^-} \rightarrow \underset{\text{Acid}_2}{H_2S} + \underset{\text{Base}_1}{H_2O}$

(v) $\underset{\text{Acid}_1}{H_2SO_4} + \underset{\text{Base}_2}{CN^-} \rightarrow \underset{\text{Acid}_2}{HCN} + \underset{\text{Base}_1}{HSO_4^-}$

Example 4.35.3

Each of the following reactions is displaced to the right. Arrange all the Brönsted acids that appear in these equations in the order of decreasing acid strength. Make a similar list for the Brönsted bases.

$H_3O^+ + H_2PO_4^- \rightleftharpoons H_3PO_4 + H_2O$ $HCN + OH^- \rightleftharpoons H_2O + CN^-$

$H_3PO_4 + CN^- \rightleftharpoons HCN + H_2PO_4^-$ $H_2O + NH_2^- \rightleftharpoons NH_3 + OH^-$

Solution

From the given reactions, we can make a list comparing the acid and base strengths as shown below.

	Acid strength comparison	Base strength comparison
$\underset{\text{Acid}_1}{H_3O^+} + \underset{\text{Base}_2}{H_2PO_4^-} \rightleftharpoons \underset{\text{Acid}_2}{H_3PO_4} + \underset{\text{Base}_1}{H_2O}$	$H_3O^+ > H_3PO_4$	$H_2PO_4^- > H_2O$
$\underset{\text{Acid}_1}{HCN} + \underset{\text{Base}_2}{OH^-} \rightleftharpoons \underset{\text{Acid}_2}{H_2O} + \underset{\text{Base}_1}{CN^-}$	$HCN > H_2O$	$OH^- > CN^-$
$\underset{\text{Acid}_1}{H_3PO_4} + \underset{\text{Base}_2}{CN^-} \rightleftharpoons \underset{\text{Acid}_2}{HCN} + \underset{\text{Base}_2}{H_2PO_4^-}$	$H_3PO_4 > HCN$	$CN^- > H_2PO_4^-$
$\underset{\text{Acid}_1}{H_2O} + \underset{\text{Base}_2}{NH_2^-} \rightleftharpoons \underset{\text{Acid}_2}{NH_3} + \underset{\text{Base}_1}{OH^-}$	$H_2O > NH_3$	$NH_2^- > OH$

Thus, the order of decreasing acid and base strengths are

$$H_3O^+ > H_3PO_4 > HCN > H_2O > NH_3$$

and $NH_2^- > OH^- > CN^- > H_2PO_4^- > H_2O$

Example 4.35.4 Based upon knowledge about the relative strength of acids in Example 4.35.3, would you expect an appreciable reaction (i.e. over 50%) to occur in the following (complete the reaction yourself).

(i) $H_2O + CN^- \rightleftharpoons$ (ii) $NH_3 + CN^- \rightleftharpoons$

(iii) $HCN + H_2PO_4^- \rightleftharpoons$ (iv) $H_3PO_4 + NH_2^- \rightleftharpoons$

Solution The complete reactions along with the designations as acids and bases and the relative strengths of these are as follows.

Reaction	Acid strength comparison	Base strength comparison
(i) $\underset{\text{Acid}_1}{H_2O} + \underset{\text{Base}_2}{CN^-} \rightleftharpoons \underset{\text{Acid}_2}{HCN} + \underset{\text{Base}_1}{OH^-}$	$H_2O < HCN$	$CN^- < OH^-$
(ii) $\underset{\text{Acid}_1}{NH_3} + \underset{\text{Base}_2}{CN^-} \rightleftharpoons \underset{\text{Acid}_2}{HCN} + \underset{\text{Base}_1}{NH_2^-}$	$NH_3 < HCN$	$CN^- < NH_2^-$
(iii) $\underset{\text{Acid}_1}{HCN} + \underset{\text{Base}_2}{H_2PO_4^-} \rightleftharpoons \underset{\text{Acid}_2}{H_3PO_4} + \underset{\text{Base}_1}{CN^-}$	$HCN < H_3PO_4$	$H_2PO_4^- < CN^-$
(iv) $\underset{\text{Acid}_1}{H_3PO_4} + \underset{\text{Base}_2}{NH_2^-} \rightleftharpoons \underset{\text{Acid}_2}{NH_3} + \underset{\text{Base}_1}{H_2PO_4^-}$	$H_3PO_4 > NH_3$	$NH_2^- > H_2PO_4^-$

Only reaction (iv) is expected to proceed in the forward direction since both the reactants are stronger than the products.

4.36 ACID AND BASE STRENGTHS AND STRUCTURE

The correlation between molecular structure and acid or base strength involves many factors. Some of them are listed below.

(1) The acid strength of the hydrogen compounds of the elements in a particular period increases as the atomic number is increased. This is primarily due to the increase in electronegativity of the atom attached to hydrogen. Thus, the decreasing order of acid strength of the last three elements of second period is

$$HF > H_2O > NH_3$$

The conjugate base strength will follow opposite trend, that is

$$NH_2^- > OH^- > F^-$$

Thus, the base strength decreases with increasing electronegativity. Here size difference of the atom is small compared with the electronegativity difference.

(2) There is an increase in acid strength of the hydro-acids of the elements in any group of the periodic classification with increasing atomic size of the electronegative elements. For example, the hydrogen compounds of the Group 16 and 17 elements arranged according to increasing acid strength are

$H_2O < H_2S < H_2Se < H_2Te$

$HF < HCl < HBr < HI$

This is, primarily due to increase in atomic radius of the element in a group.

Trends in a base-strength will be opposite to the above trends. Thus, the decreasing order of base-strength will be

$O^{2-} > S^{2-} > Se^{2-} > Te^{2-}$

$F^- > Cl^- > Br^- > I^-$

that is, for monatomic anions of similar charge, base strength decreases with increasing size.

(3) As the negative charge on the ion decreases, its base strength also decreases. Thus

$N^{3-} > O^{2-} > F^-$

Acid Strength of Oxoacids

The oxoacids have been studied more extensively than any other type of acids.

(1) For acids with the structure

H—O—Z

the acid strength increases with increasing electronegativity of Z. The higher the electronegativity of Z, the more electrons of the bonds are displaced towards Z and thus more readily is the proton removed. For example, in the hypohalous acids, the order of decreasing acidity is

HOCl > HOBr > HOI

(2) In compounds in which additional oxygen atoms are bound to Z, the electron withdrawing power of the group bonded to hydrogen is increased. Thus, the proton is more readily removed. This effect is illustrated by the series in which acidity increases with increasing oxidation state of chlorine (from +1 in HOCl to +7 in $HClO_4$). However, the formal charge of the central atom of an oxoacid is a better indicator of acid strength than oxidation number.

$$H-\ddot{\underset{..}{O}}-\ddot{\underset{..}{Cl}}: < H-\ddot{\underset{..}{O}}-\underset{\oplus}{\ddot{Cl}}-\underset{\ominus}{\ddot{\underset{..}{O}}}: < H-\ddot{\underset{..}{O}}-\underset{(2+)}{\overset{:\ddot{O}:^{\ominus}}{\overset{|}{Cl}}}-\underset{\ominus}{\ddot{\underset{..}{O}}}: < H-\ddot{\underset{..}{O}}:-\underset{\underset{\ominus}{:\underset{..}{O}\cdot}}{\overset{:\ddot{O}:^{\ominus}}{\overset{|\;(3+)}{Cl}}}-\ddot{\underset{..}{O}}:^{\ominus}$$

In the above series, the increasing formal charge on the chlorine atom parallels an increase in the oxidation number of the chlorine atom and serves to indicate increasing acid strength. However, for the oxoacids of phosphorus, oxidation number fails to give a true indication of acid strength. The acids are approximately of equal strength, and in each case, the formal charge on the phosphorus atom is +1.

$$\mathrm{H{-}\ddot{O}{-}\overset{\overset{:\ddot{O}:^{\ominus}}{|}}{\underset{\underset{H}{|}}{P^{\oplus}}}{-}H} \qquad \mathrm{H{-}\ddot{O}{-}\overset{\overset{:\ddot{O}:^{\ominus}}{|}}{\underset{\underset{H}{|}}{P^{\oplus}}}{-}\ddot{O}{-}H} \qquad \mathrm{H{-}\ddot{O}{-}\overset{\overset{:\ddot{O}:^{\ominus}}{|}}{\underset{\underset{\underset{H}{|}}{:\ddot{O}:}}{|}}{P^{\oplus}}}{-}\ddot{O}{-}H}$$

The number of oxygen atoms bonded to the central atom but not bonded to H atoms influence the formal charge on the central atom and, therefore, a qualitative indication of the strength of acids of general formula $(HO)_mZO_n$ is provided by the value of n in the formula. In general, we have

$n = 3$ very strong acid	$HOClO_3$, $HOIO_3$
$n = 2$ strong acid	$HOClO_2$, $(HO)_2SO_2$, $HONO_2$
$n = 1$ weak acid	$HOClO$, $(HO)_3PO$, $HONO$
$n = 0$ very weak acid	$HOCl$, $(HO)_3B$

Within a particular group, acid strength increases with increase in electronegativity of the central atom. Thus,

$$HOClO_2 > HOBrO_2 > HOIO_2$$

This increasing acid strength will show a parallel decreasing base strength of the conjugate base. Thus, for the series

$$HOCl < HOClO < HOClO_2 < HOClO_3$$

the order of the base strength of the anions is

$$OCl^- > ClO_2^- > ClO_3^- > ClO_4^-$$

Example 4.36.1 Which compound in each of the following pairs is the stronger acid and why?
(i) H_3PO_4 or H_3AsO_4, (ii) H_3AsO_3 or H_3AsO_4, (iii) H_2SO_4 or H_2SO_3, (iv) H_3BO_3 or H_2CO_3, (v) H_2Se or HBr.

Solution

(i) $H_3AsO_4 > H_3PO_4$	P and As belong to the same group, but size of As is larger than P.
(ii) $H_3AsO_4 > H_3AsO_3$	More number of oxygen atoms attached to As in H_3AsO_4.
(iii) $H_2SO_4 > H_2SO_3$	More number of oxygen atoms attached to S in H_2SO_4.
(iv) $H_2CO_3 > H_3BO_3$	Same period, electronegativity of C is greater than that of B.
(v) $HBr > H_2Se$	Same period, electronegativity of Br is greater than that of Se.

Example 4.36.2 Which compound in each of the following pairs is the stronger base and why?
(i) P^{3-} or S^{2-} (ii) PH_3 or NH_3, (iii) SiO_3^{2-} or SO_3^{2-}, (iv) NO_2^- or NO_3^-, (v) Br^- or F^-.

Solution

(i) $P^{3-} > S^{2-}$	More negative charge on P.
(ii) $NH_3 > PH_3$	Same group, size of N is smaller than P.
(iii) $SiO_3^{2-} > SO_3^{2-}$	Same period, electronegativity of S is larger than Si.
(iv) $NO_2^- > NO_3^-$	Since HNO_3 is stronger acid than HNO_2, their conjugate base will have opposite order.
(v) $F^- > Br^-$	Same group, size of F^- is smaller than Br^-.

REVISIONARY PROBLEMS

4.1 On dissolving a substance in water, it can produce either an ionic or molecular solution. From the following data, predict what type of solution is likely to be formed by $CaCl_2$ and $HgCl_2$.

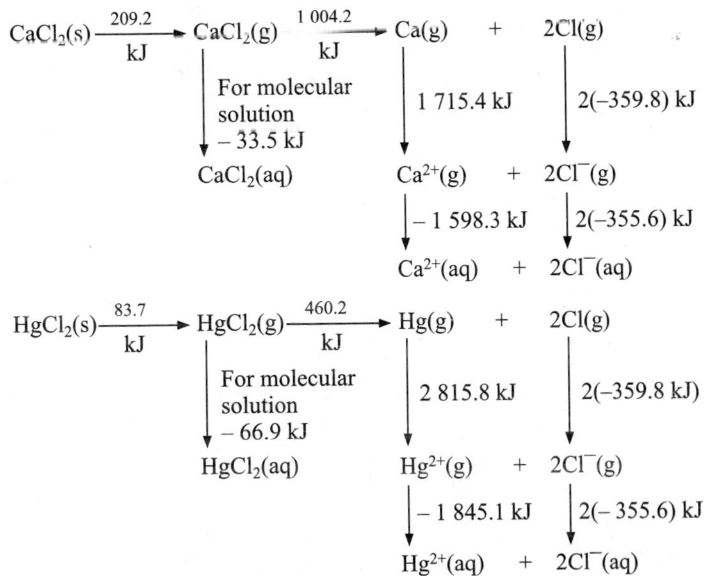

4.2 (a) Explain why the Arrhenius theory of dissociation is only applicable to weak electrolytes.
(b) Write down the expression for the equilibrium constant of a weak electrolyte AB in terms of its degree of dissociation α and concentration c.
(c) Show that for a weak electrolyte $\alpha \rightarrow 1$ as $c \rightarrow 0$ (Ostwald dilution law).

4.3 (a) Discuss the statement:
'Classification of electrolytes into 'strong' and 'weak' is obsolete. It would be more appropriate to classify them as 'true' and 'potential' electrolytes.'
(b) What is the role of a solvent in an ionization process involving (a) strong electrolyte, and (b) weak electrolyte (strong and weak are used in conventional sense).

4.4 (a) What is the ionic product of water? Its value at several temperatures was found to be as follows:

Temperature/°C	0	10	25	40	50
$K_w \times 10^{14}/M^2$	0.114	0.292	1.008	2.919	5.474

If pH of the solution is defined as pH = – log ([H$^+$]/M), calculate the pH of water at the above temperatures.

(b) Comment on the following statements:

(i) pH of water is 7.00 at 25 °C; at other temperatures it may be more or less than 7.00.

(ii) Neutral pH is 7.00 at 25 °C (pure water, by definition, is a neutral solution). The definition of a neutral solution is not that [H$^+$] = 1.0×10^{-7} M, which applies only at 25 °C but rather that

[H$^+$] = [OH$^-$], which applies at any temperature,

(iii) Show that for any aqueous solution

$$\text{pH} + \text{pOH} = \text{p}K_w^{\circ}$$

4.5 (a) Show that the exact concentration of H_3O^+ in an aqueous solution of an acid HA can be computed from the expression

$$K_a = \frac{[H_3O^+]^3 - [H_3O^+]K_w}{[H_3O^+][HA]_0 - [H_3O^+]^2 + K_w}$$

Under what conditions can the following expressions be used:

(i) $K_a = \dfrac{[H_3O^+]^2}{[HA]_0 - [H_3O^+]}$ (ii) $K_a = \dfrac{[H_3O^+]^2}{[HA]_0}$

(iii) $K_a = \dfrac{[H_3O^+]^2 - K_w}{[HA]_0}$

(b) Show that the exact expression given above can be written as

$$[HA]_0 = \left\{[H_3O^+] - \frac{K_w}{[H_3O^+]}\right\}\left\{1 + \frac{[H_3O^+]}{K_a}\right\}$$

and further show that for a strong acid (K_a very large), this expression reduces to

$$[HA]_0 = [H_3O^+] - \frac{K_w}{[H_3O^+]}$$

Under what condition should this equation be used for computing the [H_3O^+] of the solution? What form does it take when the solution is fairly concentrated?

(c) Consider an aqueous solution of a strong acid such as HCl. At high concentration, the acid itself is the only important source of H_3O^+ and thus one can compute H_3O^+ concentration from the equation [H_3O^+] = [HCl]$_0$, but at concentrations near 10^{-6} M and below, the ionization of water contributes appreciably to the concentration of H_3O^+. The expression to be used in such situations can be derived by using the charge-balance expression and the relation [H_3O^+] [OH$^-$] = K_w. Show that the resultant expression is given as

$$[H_3O^+]^2 - [HCl]_0\,[H_3O^+] - K_w = 0$$

(Note that this expression is the same as that given in Part (b).)

4.6 (a) Show that the exact concentration of OH^- in an aqueous solution of a base BOH can be computed from the expression

$$K_b = \frac{[OH^-]^3 - [OH^-]K_w}{[OH^-][BOH]_0 - [OH^-]^2 + K_w}$$

Under what conditions can the following expressions be used?

(i) $K_b = \dfrac{[OH^-]^2}{[BOH]_0 - [OH^-]}$ (ii) $K_b = \dfrac{[OH^-]^2}{[BOH]_0}$

(iii) $K_b = \dfrac{[OH^-]^2 - K_w}{[BOH]_0}$

(b) Show that the exact expression given above can be written as

$$[BOH]_0 = \left\{[OH^-] - \frac{K_w}{[OH^-]}\right\}\left\{1 + \frac{[OH^-]}{K_b}\right\}$$

and further show that for a strong base (K_b very large), this expression reduces to

$$[BOH]_0 = [OH^-] - \frac{K_w}{[OH^-]}$$

Under what condition should this equation be used for computing the $[OH^-]$ of the solution? What form does it take when the solution is fairly concentrated?

(c) Consider an aqueous solution of a strong base such as NaOH. At a high concentration, the base itself is the only important source of OH^- and thus one can compute OH^- concentration from the equation $[OH^-] = [BOH]_0$ but at concentrations near 10^{-6} M and below, the ionization of water contributes appreciably to the concentration of OH^-. The expression to be used in such a situation can be derived by using the charge-balance expression and the relation $[H_3O^+]\,[OH^-] = K_w$. Show that the resultant expression is given as

$$[OH^-]^2 - [NaOH]_0\,[OH^-] - K_w = 0$$

(Note that this expression is the same as that given in Part (b).)

4.7 (a) Show that the exact expression for $[H_3O^+]$ in an aqueous solution of a diprotic acid H_2A is given by

$$[H_3O^+] = \left\{\frac{[H_2A]_0\,K_1}{[H_3O^+] + K_1 + \dfrac{K_1K_2}{[H_3O^+]}}\right\}\left\{1 + \frac{2\,K_2}{[H_3O^+]}\right\} + \frac{K_w}{[H_3O^+]}$$

Under what approximations can the following expressions be used?

(i) $[H_3O^+] = \left\{\dfrac{K_1[H_2A]_0}{[H_3O^+] + K_1 + \dfrac{K_1K_2}{[H_3O^+]}}\right\}\left\{1 + \dfrac{2\,K_2}{[H_3O^+]}\right\}$

(ii) $[H_3O^+] = \dfrac{K_1[H_2A]_0}{[H_3O^+] + K_1}$ or $K_1 = \dfrac{[H_3O^+]^2}{[H_2A]_0 - [H_3O^+]}$

(iii) $[H_3O^+] = \sqrt{K_1[H_2A]_0}$

(b) Show that the expressions (ii) and (iii) in part (a) can be derived directly from the first dissociation constant.

4.8 Show that the concentrations of H_2A, HA^- and A^{2-} in an aqueous solution of a diprotic acid H_2A are given by

$$[H_2A] = \frac{[H_2A]_0}{1 + \dfrac{K_1}{[H_3O^+]} + \dfrac{K_1K_2}{[H_3O^+]^2}}; \qquad [HA^-] = \frac{[H_2A]_0}{\dfrac{[H_3O^+]}{K_1} + 1 + \dfrac{K_2}{[H_3O^+]}}$$

$$[A^{2-}] = \frac{[H_2A]_0}{\dfrac{[H_3O^+]^2}{K_1K_2} + \dfrac{[H_3O^+]}{K_2} + 1}$$

Under what approximations can the following simple forms be used?

(a) $$[H_2A] = \frac{[H_2A]_0}{1 + \dfrac{K_1}{[H_3O^+]}}; \qquad [HA^-] = \frac{[H_2A]_0}{\dfrac{[H_3O^+]}{K_1} + 1}$$

$$[A^{2-}] = \frac{[H_2A]_0}{\dfrac{[H_3O^+]^2}{K_1K_2} + \dfrac{[H_3O^+]}{K_2}}$$

(b) $[H_2A] = [H_2A]_0$; $[HA^-] = [H_3O^+]$; $[A^{2-}] = K_2$

4.9 Show that for a triprotic acid in water,

$$[H_3O^+] = [H_2A^-] = \sqrt{K_1\,[H_3A]_0}; \qquad [HA^{2-}] = K_2$$

$$[A^{3-}] = K_2K_3/\sqrt{K_1[H_3A]_0}$$

Justify the assumptions which lead to these simple expressions.

4.10 What is the phenomenon of hydrolysis? What type of cations and anions will be hydrolyzed in aqueous solution? Show that

$$K_{ha} = \frac{K_w}{K_a} \quad \text{and} \quad K_{hb} = \frac{K_w}{K_b}$$

4.11 (a) Show that the exact expression to compute $[H_3O^+]$ in an aqueous solution of a salt formed from a weak acid and a strong base is given by

$$[H_3O^+]^3 + (c + K_a)\,[H_3O^+]^2 - K_w\,[H_3O^+] - K_aK_w = 0$$

Under what approximations can the following simpler form be used?

$$pH = \frac{1}{2}pK_w^\circ + \frac{1}{2}pK_b^\circ + \frac{1}{2}\log(c/c^\circ)$$

The expression can be derived by considering alone the hydrolysis of the anion A^-. Derive this relation directly.

(b) Show that the degree of hydrolysis of the anion A^- is given by

$$\alpha = \sqrt{\frac{K_h}{c}} = \sqrt{\frac{K_w}{K_a c}}$$

How does α vary with the dilution of the solution?

4.12 (a) Show that the expression to compute $[H_3O^+]$ in an aqueous solution containing a salt formed from a strong acid and a weak base is given by

$$[H_3O^+]^3 K_b + K_w [H_3O^+]^2 - (K_b + c)K_w [H_3O^+] - K_w^2 = 0$$

Under what approximations, the following simpler form can be used?

$$pH = \frac{1}{2}pK_w^\circ - \frac{1}{2}pK_b^\circ - \frac{1}{2}\log(c/c^\circ)$$

This expression can be derived directly by considering alone the hydrolysis of the cation B^+. Derive this relation directly.

(b) Show that the degree of hydrolysis of the cation B^+ is given by

$$\alpha = \sqrt{\frac{K_h}{c}} = \sqrt{\frac{K_w}{K_b c}}$$

How does α vary with dilution of the solution?

4.13 (a) Show that the exact expression to compute $[H_3O^+]$ in an aqueous solution containing a salt of a weak acid and a weak base is given by

$$\frac{c}{1 + \dfrac{K_w}{K_b[H_3O^+]}} + [H_3O^+] = \frac{c}{1 + \dfrac{[H_3O^+]}{K_a}} + \frac{K_w}{[H_3O^+]}$$

Under what approximations can the following simpler form be used?

$$[H_3O^+] = \left(\frac{K_a K_w}{K_b}\right)^{1/2} \quad \text{or} \quad pH = \frac{1}{2}[pK_w^\circ + pK_a^\circ - pK_b^\circ]$$

This expression can be derived by considering hydrolysis of the salt directly. Derive the relation directly.

(b) Show that the degree of hydrolysis of such a salt is given as

$$\alpha = \frac{\sqrt{K_h}}{1 + \sqrt{K_h}}$$

and hence prove that the extent of hydrolysis is independent of concentration of the salt.

4.14 What is an amphiprotic anion? Show that the exact expression to compute $[H_3O^+]$ in an aqueous solution containing amphiprotic anion is given by

$$[H_3O^+] = \left\{\frac{K_{a1}K_{a2}[HA^-] + K_{a1}K_w}{K_{a1}[HA^-]}\right\}^{1/2}$$

where $\quad [HA^-] = [HA^-]_0 \Big/ \left\{\dfrac{[H_3O^+]}{K_{a1}} + 1 + \dfrac{K_{a2}}{[H_3O^+]}\right\}$

Under what approximations can the following simpler expression be used?

$$[H_3O^+] = \sqrt{K_{a1}K_{a2}} \quad \text{or} \quad pH = \frac{1}{2}(pK_{a1}^\circ + pK_{a2}^\circ)$$

4.15 Show that the exact expression to compute $[H_3O^+]$ in an aqueous solution containing a salt of strong conjugate cation and an amphiprotic anion is given by

$$\frac{c}{1 + \dfrac{K_w}{K_b[H_3O^+]}} + [H_3O^+] = \frac{c}{1 + \dfrac{[H_3O^+]}{K_{a1}} + \dfrac{K_{a2}}{[H_3O^+]}} + 2\frac{K_{a2}}{[H_3O^+]}\left[\frac{c}{1 + \dfrac{[H_3O^+]}{K_{a1}} + \dfrac{K_{a2}}{[H_3O^+]}}\right] + \frac{K_w}{[H_3O^+]}$$

Under what approximations can the following simpler form be used?

$$[H_3O^+] = \left\{K_{a1}\left(\frac{K_w}{K_b} - K_{a2}\right)\right\}^{1/2}$$

4.16 Explain why

(a) A solution of Na_2CO_3 is alkaline?

(b) A solution of NH_4Ac is neutral?

(c) A solution of NaCl is neutral?

(d) A solution of $CuSO_4$ is acidic?

4.17 Show that the exact expression to compute $[H_3O^+]$ in an aqueous solution containing a weak acid HA and a salt of its conjugate base MA is given by

$$K_a = \frac{[H_3O^+]\left\{[M^+] + [H_3O^+] - \dfrac{K_w}{[H_3O^+]}\right\}}{\left\{[HA]_0 - [H_3O^+] + \dfrac{K_w}{[H_3O^+]}\right\}}$$

Under what approximations can the following simpler form be used?

$$[H_3O^+] = \frac{K_a[HA]_0}{[M^+]} \quad \text{or} \quad pH = pK_a^\circ + \log\left(\frac{[\text{salt}]}{[\text{acid}]}\right)$$

4.18 Show that the exact expression to compute $[OH^-]$ in an aqueous solution containing a weak base BOH and a salt of its conjugate acid BA is given by

$$K_b = \frac{\left\{[A^-] + [OH^-] - \dfrac{K_w}{[OH^-]}\right\}[OH^-]}{\left\{[BOH]_0 - [OH^-] + \dfrac{K_w}{[OH^-]}\right\}}$$

Under what approximations can the following simpler form be used?

$$[OH^-] = \frac{K_b[BOH]_0}{[A^-]} \quad \text{or} \quad pOH = pK_b^\circ + \log\left(\frac{[A^-]}{[BOH]_0}\right)$$

4.19 Show that the $[H_3O^+]$ of a solution containing a weak monoprotic acid of concentration c_1 and a strong monoprotic acid of concentration c_2 is given by

$$[H_3O^+] = \frac{c_2 + \sqrt{c_2^2 + 4K_a c_1}}{2}$$

4.20 Show that the $[H_3O^+]$ of a solution containing two weak monoprotic acids of concentrations c_1 and c_2 is given by

$$[H_3O^+] = \sqrt{K_1 c_1 + K_2 c_2}$$

4.21 What are buffer solutions? When a small quantity of acid is added to an aqueous solution of ammonium acetate, the pH of the solution remains almost unchanged, whereas when the same acid is added to an aqueous solution of sodium chloride, the pH of the solution changes enormously. Explain why it is so.

4.22 What do you understand by the terms buffer capacity and useful buffer range. Show that the buffer capacity is maximum when the ratio of salt to acid (or base) is one.

4.23 What is an acid-base indicator? How does its colour change with H^+ ion concentration of the solution? What is an indicator range?

4.24 (a) Calculate the pH when 0.0, 10.0, 20.0, 30.0, 40.0, 49.0, 49.9, 49.99, 49.999, 50.0, 50.001, 50.01, 50.1, 51.0, 52.0, 53.0 and 55.0 cm^3 of NaOH of 0.1 M has been added to 50.0 cm^3 of 0.1 M of HCl. Sketch this titration curve.

(b) What is the pH value at the equivalence point? Explain why there is steep rise of pH values near the equivalence point. What is the approximate change of pH on addition of 49.99 cm^3 of NaOH to 50.01 cm^3 of NaOH?

Does this change of pH depend on the concentrations of acid and base being titrated?

(c) From the titration curve justify the use of phenolphthalein indicator.

4.25 (a) Calculate the pH when 0.0, 2.5, 5.0, 10.0, 15.0, 20.0, 22.5, 23.75, 24.88, 25.0, 26.0, 28.0, 30.0 and 35.0 cm^3 of 0.1 M NaOH has been added to 25.0 cm^3 of 0.1 M acetic acid ($K_a^\circ = 1.8 \times 10^{-5}$). Sketch the titration curve.

(b) What is the pH of the solution at the equivalence point? How much change of pH takes place when a slight excess of NaOH (say, 0.01 cm^3) is added at the equivalence point? From this, justify the use of 'phenolphthalein' indicator.

(c) In the titration curve, show the range over which the solution behaves as a buffer solution.

4.26 (a) Calculate the pH when 0.0, 2.5, 5.0, 10.0, 15.0, 20.0, 22.5, 23.75, 24.88, 25.0, 26.0, 28.0, 30.0 and 35.0 cm^3 of 0.1 M HCl has been added to 25.0 cm^3 of 0.1 M NH_4OH ($K_b^\circ = 1.8 \times 10^{-5}$). Sketch the titration curve.

(b) What is the pH at the equivalence point? How much change of pH takes place when a slight excess of HCl (say, 0.01 cm^3) is added at the equivalence point? From this, justify the use of methyl orange indicator in this titration. Why can phenolphthalein not be used as indicator in this titration?

4.27 Explain why the titration of a weak acid with weak base is not carried out using an acid-base indicator.

4.28 The shape and nature of the titration curve for a dibasic acid depends upon the relative concentrations of various species (obtained from the dibasic acid) present in the solution during titration. The concentrations of these species are expressed in terms of various rational functions, called distribution functions. Show that the expressions of the rational functions are given by

$$\frac{[H_2A]}{[H_2A]_0} = \frac{1}{1 + \dfrac{K_1}{[H_3O^+]} + \dfrac{K_1K_2}{[H_3O^+]^2}}; \qquad \frac{[HA^-]}{[H_2A]_0} = \frac{1}{\dfrac{[H_3O^+]}{K_1} + 1 + \dfrac{K_2}{[H_3O^+]}}$$

$$\frac{[A^{2-}]}{[H_2A]_0} = \frac{1}{\dfrac{[H_3O^+]^2}{K_1K_2} + \dfrac{[H_3O^+]}{K_2} + 1}$$

How do these ratios, in general, change with the change in pH of the solution? Show that the maximum value of $[HA^-]/H_2A]_0$ is given by

$$\left(\frac{[HA^-]}{[H_2A]_0}\right)_{max} = \frac{1}{1 + 2\sqrt{K_2/K_1}}$$

From this, justify the statement that: 'The essential condition of the stepwise neutralization of a diprotic acid with a strong base and thus getting two inflexion points is that the ratio of K_1/K_2 must be equal to or greater than 10^3.'

4.29 Calculate the pH values when 0.0, 2.5, 5.0, 7.5, 10.0, 11.25, 12.5, 13.75, 15.0, 17.5, 20.0, 22.5, 23.75, 25.0, 26.0, 28.0, 30.0, 35.0, 40.0 cm^3 of 0.1 M NaOH has been added to 25.0 cm^3 of 0.05 M oxalic acid. Given that $K_1^\circ = 5.9 \times 10^{-2}$ and $K_2^\circ = 6.4 \times 10^{-5}$, sketch the titration curve. Suggest an indicator that can be used in the above titration.

4.30 (a) Calculate the pH when 0.0, 5.0, 10.0, 15.0, 20.0, 22.5, 25.0, 30.0, 35.0, 40.0, 45.0, 47.5, 50.0, 52.5, 55.0 and 60.0 cm^3 of 0.1 M HC1 is added to 50 cm^3 of 0.05 M sodium carbonate solution. Sketch the titration curve.

(b) Justify that: 'When phenolphthalein is used as an indicator, the end point corresponds to the reaction $CO_3^{2-} + H^+ \rightarrow HCO_3^-$ and when methyl-orange is used, it corresponds to $CO_3^{2-} + 2H^+ \rightarrow H_2CO_3 \rightleftharpoons CO_2 + H_2O$.

4.31 Define the term solubility product as applicable to a sparingly soluble salt. How does this help in deciding the precipitation of a salt? Discuss its applications in qualitative and quantitative analysis.

4.32 What are amphoteric hydroxides? Show with a suitable example that the solubility of such hydroxides is greater in an acidic or basic solution than it is in neutral water.

4.33 Discuss in brief (a) the Arrhenius concept, (b) the solvent-system concept, (c) Brönsted-Lowry concept, and (d) the Lewis concept of acids and bases.

4.34 (a) What do you understand by the levelling effect of a solvent? Elaborate on the statement: "The strongest base and the strongest acid that can exist in water are OH^- and H_3O^+, respectively.'

(b) Name the strongest acid and base that can exist in any self-ionizing solvent.

TRY YOURSELF PROBLEMS

4.1 The exact expression for $[H_3O^+]$ in an aqueous solution of a diprotic acid H_2A is given by

$$[H_3O^+] = \left\{\frac{[H_2A]_0\, K_1}{[H_3O^+] + K_1 + \dfrac{K_1K_2}{[H_3O^+]}}\right\}\left\{1 + \frac{2K_2}{[H_3O^+]}\right\} + \frac{K_w}{[H_3O^+]}$$

This expression will be equally applicable to H_2SO_4 with the condition that the value of K_1 is very large (since first dissociation is complete). If the contribution of H_3O^+ from water is neglected, show that the above expression for H_2SO_4 is reduced to

$$[H_3O^+]^2 - [H_3O^+]\,\{[H_2A]_0 - K_2\} - 2[H_2A]_0\,K_2 = 0$$

i.e.
$$[H_3O^+] = \frac{1}{2}\left\{[H_2A]_0 - K_2 + \sqrt{([H_2A]_0 - K_2)^2 + 8[H_2A]_0\,K_2}\right\}$$

Also derive this expression directly by considering the second dissociation constant, charge-balance and material-balance conditions.

4.2 (a) The dissociation of a weak base can be represented as

$$BOH + H_2O \rightleftharpoons B^+(H_2O) + OH^- \qquad (1)$$

The exact expression of $[H_3O^+]$ for the above dissociation is

$$K_b = \frac{\left\{\dfrac{K_w}{[H_3O^+]} - [H_3O^+]\right\}\left\{\dfrac{K_w}{[H_3O^+]}\right\}}{\left\{[BOH]_0 - \dfrac{K_w}{[H_3O^+]} + [H_3O^+]\right\}} \qquad (2)$$

and the approximate expression is

$$[H_3O^+] = \frac{K_w}{\sqrt{K_b\,[BOH]_0}} \qquad (3)$$

The expression showing the hydrolysis of a strong conjugate anion is, in fact, of the same form as that of the dissociation of a weak base

$$A^- + H_2O \rightleftharpoons HA + OH^- \qquad (4)$$

(A^- playing the role of BOH; HA playing the role of $B^+(H_2O)$ and K_h playing the role of K_b.) The exact expression of $[H_3O^+]$ for the reaction (4) is given by the expression (2) with the replacement of K_b with K_h and $[BOH]_0$ with $[A^-]_0$. Thus, we have

$$K_h = \frac{\left\{\dfrac{K_w}{[H_3O^+]} - [H_3O^+]\right\}\left\{\dfrac{K_w}{[H_3O^+]}\right\}}{\left\{[A^-]_0 - \dfrac{K_w}{[H_3O^+]} + [H_3O^+]\right\}} \qquad (5)$$

(Note that this expression is exactly the same as given by Eq. (4.13.9) in the text.) The approximate expression of Eq. (5) will be

$$[H_3O^+] = \frac{K_w}{\sqrt{K_h\,[A^-]_0}}$$

Show that this is equivalent to

$$[H_3O^+] = \sqrt{\frac{K_w K_a}{c}} \quad \text{or} \quad pH = \frac{1}{2}pK_w^\circ + \frac{1}{2}pK_a^\circ + \frac{1}{2}\log(c/c^\circ)$$

(i.e., Eq. (4.13.12) of the text.)

(a) Show similarly that the expressions of $[H_3O^+]$ in the hydrolysis of a strong conjugate cation B^+ can be obtained from the corresponding expressions of a weak acid HA with the replacement of K_a by K_h, and HA by B^+.

4.3 Show that $[OH^-] \approx \sqrt{K_h c}$ in a solution containing a salt formed from a strong base and a weak acid.

4.4 Show that $[H_3O^+] \approx \sqrt{K_h c}$ in a solution containing a salt formed from a weak base and a strong acid.

4.5 (a) Write the complete charge-balance equation for a solution obtained by dissolving NaCl, Na_2HPO_4 $ZnSO_4$ and HCl in water.

(Ans. $[H^+] + [Na^+] + 2[Zn^{2+}] = [OH^-] + [Cl^-] + [HSO_4^-] + 2[SO_4^{2-}] + [H_2PO_4^-] + 2[HPO_4^{2-}] + 3[PO_4^{3-}]$)

(b) Which of the terms in the charge-balance equation can be neglected in good approximation if the concentrations of each of the solutes present is about 1.0 M.

(Ans. $[OH^-]$, $3[PO_4^{3-}]$, $2[H_2PO_4^{2-}]$)

4.6 Show that the solubility of a sparingly soluble strong electrolyte M_mA_a in water is given by the relation

$$s = \left(\frac{K_s}{(m)^m (a)^a} \right)^{1/(m+a)}$$

4.7 Show that the solubility of a sparingly soluble salt MA in which the ion A^- undergoes hydrolysis in the solution is given by

$$s = \sqrt{K_s \left(1 + \frac{[H^+]}{K_a} \right)}$$

[*Hint:* Material-balance expression $[M^+] = [A^-] + [HA^-]$

$$= [A^-]\left[1 + \frac{[H^+]}{K_a}\right] = \frac{K_s}{[M^+]}\left(1 + \frac{[H^+]}{K_a}\right).\Bigg]$$

4.8 Show that the solubility of a sparingly soluble salt $M^{2+}A^{2-}$ in which the ion A^{2-} undergoes hydrolysis is given by

$$s = \sqrt{K_s \left(1 + \frac{[H^+]}{K_{a2}} + \frac{[H^+]^2}{K_{a1}K_{a2}} \right)}$$

4.9 Show that the $[H_3O^+]$ of a solution containing n number of weak acids is given by

$$[H_3O^+] = \sqrt{\sum_{i=1}^{n} K_i c_i}$$

NUMERICAL PROBLEMS

Dissociation of Pure Water

4.1 The pH of pure water at 40 °C is found to be 6.765. What is the value of pK_w°? What will be the nature of the solution at 40 °C having $[H^+] = 1.0 \times 10^{-7}$ M? What will be the pOH of the latter solution? (Ans. $pK_w^\circ = 13.53$, alkaline, pOH = 6.53)

pH Values

4.2 Calculate the pH corresponding to the following conditions in an aqueous solution at room temperature:
(a) $[H^+] = 1.0 \times 10^{-9}$ M (b) $[H^+] = 0.003\ 2$ M
(c) $[OH^-] = 0.012$ M (Ans. (a) 9.0; (b) 2.49; (c) 12.08)

4.3 Convert the following pH values of solutions at 50 °C to the corresponding values of $[H^+]$ and $[OH^-]$. Given that K_w° at 50 °C is 5.474×10^{-14}.
(a) 7.63 (b) 1.48
(Ans. (a) $[H^+] = 2.34 \times 10^{-8}$ M, $[OH^-] = 2.34 \times 10^{-6}$ M
(b) $[H^+] = 3.3 \times 10^{-2}$ M, $[OH^-] = 1.66 \times 10^{-12}$ M)

pH of an Acid Solution

4.4 Calculate the pH of (a) 0.004 M HCl and (b) 2.0×10^{-7} M HCl at 25 °C.
(Ans. (a) 2.40; (b) 6.62)

4.5 Calculate the pH of 0.01 M solution of propanoic acid ($K_a^\circ = 1.34 \times 10^{-5}$). What is its degree of dissociation? (Ans. pH = 3.44, α = 0.036 4)
How much change in pH and degree of dissociation takes place when the above solution of propanoic acid is diluted hundred times? (Ans. pH = 4.52, α = 0.305)

4.6 What is the pH of a 5×10^{-6} M solution of hydrocyanic acid, $K_a^\circ = 5 \times 10^{-10}$?
(Ans. 6.95)

4.7 Given that the standard dissociation constant of HF is 6.71×10^{-4}. How much of water would you need to add to 50.0 cm^3 of 0.27 M HF to cut the H_3O^+ concentration by a factor of 2? Assume additive volumes. (Ans. 150 cm^3)

pH of a Mixture of Acids

4.8 Suppose you have a solution that is simultaneously 0.15 M in HNO_2 ($K_a^\circ = 4.5 \times 10^{-4}$) and 0.2 M in HAc ($K_a^\circ = 1.8 \times 10^{-5}$). Calculate the concentrations of H_3O^+, NO_2^- and Ac^- in the solution.
(Ans. $[H_3O^+] = 8.4 \times 10^{-3}$ M; $[NO_2^-] = 8.0 \times 10^{-3}$ M;
$[Ac^-] = 4.3 \times 10^{-4}$ M)

4.9 What will be the concentrations of all species present in a solution made by mixing 0.20 dm^3 of 4.0 M HAc and 0.60 dm^3 of 0.8 M HCN? Given: K_a°(HAc) $= 1.8 \times 10^{-5}$ and K_a°(HCN) $= 4 \times 10^{-10}$.
(Ans. $[Ac^-] = 4.25 \times 10^{-3}$ M, $[CN^-] = 0.566 \times 10^{-7}$ M)

pH of a Base

4.10 What are the pH of (a) 0.02 M NaOH, and (ii) 1.5×10^{-7} M NaOH at 25°C?
(Ans. (a) 12.30; (b) 7.30)

4.11 (a) What is the pH of 0.015 M solution of ammonium hydroxide ($K_b^\circ = 1.85 \times 10^{-5}$)? What is its degree of dissociation? (Ans. 10.72, 3.51×10^{-2})

How much change in pH and degree of dissociation takes place when the above solution of ammonium hydroxide is diluted hundred times? (Ans. 10.35, 0.294 9)

(b) If $K_b^\circ = 1.81 \times 10^{-5}$ for $NH_3 + H_2O \rightleftharpoons NH_4^+ + OH^-$, what mass of NH_3 would you need to add to 10.0 dm^3 of water to make the $[OH^-]$ equal to 1.5×10^{-3} M?
(Ans. 21.386 g)

(c) How much water do you need to add to 50.0 cm^3 of 6.0 M NH_3 to double the per cent conversion of NH_3 into NH_4^+? (Ans. 150 cm^3)

pH of a Diprotic Acid

4.12 Calculate the $[H_3O^+]$ of 0.01 M HSO_4^-. Given: K_1° complete, $K_2^\circ = 1.1 \times 10^{-2}$.
(Ans. 0.006 34 M)

4.13 Given a solution of 0.15 M H_2SO_4. What are the concentrations of H_3O^+, SO_4^{2-}, and HSO_4^- in this solution? ($K_2^\circ = 1.26 \times 10^{-2}$.)
(Ans. $[H_3O^+] = 0.161$ M; $[SO_4^{2-}] = 0.010\ 9$ M; $[HSO_4^-] = 0.139$ M)

4.14 Given a solution of 0.15 M $NaHSO_4$. What are the concentrations of HSO_4^-, SO_4^{2-} and H_3O^+ in this solution? ($K_2^\circ = 1.26 \times 10^{-2}$)
(Ans. $[H_3O^+] = 3.76 \times 10^{-2}$ M; $[SO_4^{2-}] = 3.76 \times 10^{-2}$ M; $[HSO_4^-] = 0.112$ M)

4.15 Fumaric acid is a diprotic acid with formula $H_2(C_4H_2O_4)$. The ionization constants are $K_1^\circ = 0.9 \times 10^{-4}$ and $K_2^\circ = 4 \times 10^{-5}$. Calculate the pH of 1 M solution of fumaric acid.
(Ans 1.025)

pH of a Triprotic Acid

4.16 Calculate the concentrations of H^+, $H_2AsO_4^{2-}$, AsO_4^{3-} and unionized H_3AsO_4 in a solution of 0.05 M H_3AsO_4. For H_3AsO_4: $K_{a1}^\circ = 5.6 \times 10^{-3}$, $K_{a2}^\circ = 1.7 \times 10^{-7}$ and $K_{a3}^\circ = 3 \times 10^{-12}$.
(Ans. $[H^+] = [H_2AsO_4^-] = 0.014$ M; $[HAsO_4^{2-}] = 1.7 \times 10^{-7}$ M; $[AsO_4^{3-}] = 3.6 \times 10^{-17}$ M; unionized $[H_3AsO_4] = 0.036$ M)

4.17 A typical solution of H_3PO_4 contains the following species at equilibrium: 0.076 M H_3PO_4; 0.023 9 M $H_2PO_4^-$; 6.2×10^{-8} M HPO_4^{2-}; 3×10^{-18} M PO_4^{3-}; 0.023 9 M H_3O^+. Calculate K_{a1}°, K_{a2}° and K_{a3}° for this acid.
(Ans. $K_{a1}^\circ = 7.5 \times 10^{-3}$; $K_{a2}^\circ = 6.2 \times 10^{-8}$ and $K_{a3}^\circ = 1 \times 10^{-12}$)

pH of a Solution Containing a Salt

4.18 The pH of a 0.05 M solution of sodium acetate was found to be 8.70. What is the value of the dissociation constant of acetic acid? (Ans. $K_a^\circ = 2.0 \times 10^{-5}$)

4.19 Calculate the pH of a solution obtained by dissolving 4.92 g of sodium acetate in water to a volume of 400 cm^3. K_a° for acetic acid is 1.8×10^{-5}. (Ans. 8.96)

4.20 The solution of sodium salt of a weak monofunctional acid has a value of 9.0. What will be the pH if the solution is diluted with pure water to twice its volume?
(Ans. 8.85)

4.21 What is the pH of a 0.08 M solution of sodium benzoate if K_a° for benzoic acid is 6.3×10^{-5}?
(Ans. 8.56)

4.22 (a) What is the pH of 0.015 M solution of ammonium chloride if K_b° for ammonia is 1.8×10^{-5}? (b) How much change in pH takes place if this solution is diluted hundred times? (Ans. (a) 5.54 (b) 6.54)

4.23 Calculate the pH value of a 0.04 M solution of anilinium chloride. K_b° for aniline is 4.0×10^{-10}.
(Ans. pH = 3.0)

4.24 Calculate the per cent hydrolysis of acetate anion in 1.0 M NH_4Ac and compare it with the per cent hydrolysis of Ac^- in 1.0 M NaAc. Given: $K_a^\circ(HAc) = K_b^\circ(NH_4OH) = 1.8 \times 10^{-5}$. (Ans. $NH_4Ac = 0.55\%$, $NaAc = 2.4 \times 10^{-3}\%$)

4.25 Calculate the values of K_h°, α_h and pH of the following solutions.

(a) 0.05 M ammonium acetate — $K_a^\circ = K_b^\circ = 1.8 \times 10^{-5}$

(b) 0.05 M anilinium acetate — $K_a^\circ = 1.8 \times 10^{-5}$ and $K_b^\circ = 4.0 \times 10^{-10}$

(c) 0.05 M ammonium carbonate — $K_{a1}^\circ = 4.5 \times 10^{-7}$ and $K_{a2}^\circ = 4.7 \times 10^{-11}$

(d) 0.1 M NH_4CN — $K_b^\circ(NH_3) = 1.8 \times 10^{-5}$ and $K_a^\circ = 4.0 \times 10^{-10}$

(Ans. (a) 3.086×10^{-4}, 0.052 6, 7; (b) 1.388, 0.541, 4.67; (c) 11.8, 0.775, 9.79; (d) 1.11, 0.573, 9.28)

4.26 Calculate the pH of
(a) 0.10 M NaH_2AsO_4 (b) 0.1 M Na_3AsO_4
For H_3AsO_4: $K^\circ_{a1} = 4.9 \times 10^{-3}$, $K^\circ_{a2} = 5.6 \times 10^{-8}$ and $K^\circ_{a3} = 3.0 \times 10^{-12}$.
(Ans. (a) 4.78; (b) 9.39)

4.27 Calculate the per cent hydrolysis of 0.8 M K_2S solution. For H_2S: $K^\circ_{a1} = 1 \times 10^{-7}$ and $K^\circ_{a2} = 1.3 \times 10^{-14}$. (Ans. 61% hydrolyzed)

4.28 Calculate the pH of the solutions containing, (a) 0.1 M $NaHCO_3$ and (b) 0.1 M NH_4HCO_3. Given that:

$$CO_2 + H_2O \rightleftharpoons H^+ + HCO_3^- \qquad K^\circ_{a1} = 4.2 \times 10^{-7}$$
$$HCO_3^- \rightleftharpoons H^+ + CO_3^{2-} \qquad K^\circ_{a2} = 4.8 \times 10^{-11}$$
$$NH_4OH \rightleftharpoons NH_4^+ + OH^- \qquad K^\circ_b = 1.8 \times 10^{-5}$$

(Ans. (a) 8.35, (b) 7.84)

4.29 What are the pH of solutions containing
(i) 0.1 M $CuSO_4$; $K^\circ_h(Cu^{2+}) = 1.0 \times 10^{-8}$
(ii) 0.05 M $Fe(NO_3)_2$; $K^\circ_h(Fe^{2+}) = 5.0 \times 10^{-9}$ (Ans. (i) 4.5, (ii) 4.80)

Buffer Solution

4.30 What minimum volume of a buffer solution which is 0.1 M in both lactic acid ($K^\circ_a = 1.4 \times 10^{-4}$) and sodium lactate must be diluted to exactly 100 cm^3 so that the resulting buffer solution will change its pH by not more than 0.5 pH unit upon the addition of 10.0 cm^3 of 0.01 M HCl? (Ans. 1.9 cm^3)

4.31 A buffer is made by adding 25.0 cm^3 of 0.1 M sodium acetate to 50.0 cm^3 of 0.1 M acetic acid ($K^\circ_a = 1.8 \times 10^{-5}$). What would be the pH if this buffer is added to: (a) 25.0 cm^3 of distilled water; (b) 10.0 cm^3 of 0.1 M HCl, and (c) 10.0 cm^3 of 0.1 M NaOH? (Ans. (a) 4.444, (b) 4.143, (c) 5.347)

4.32 What pH change would be produced on adding 10.0 cm^3 of 0.1 M H_2SO_4 to 0.1 dm^3 of a buffer containing 0.15 M $NaHSO_4$ and 0.15 M Na_2SO_4? Given: $K^\circ_2 = 1.26 \times 10^{-2}$. (Ans. 1.899 6 to 1.890 9)

4.33 (a) Calculate the mass of solid $NaC_2H_3O_2$ that must be dissolved in 250 cm^3 of 0.09 M $HC_2H_3O_2$ ($K^\circ_a = 1.8 \times 10^{-5}$) in order to yield a mixture buffered at 4.7. Assume constant volume during the solubilization process, (b) Calculate also the mass of $NaC_2H_3O_2$ to yield a buffer solution of maximum buffer capacity.
(Ans. (a) 1.66 g, (b) 1.845 g)

4.34 Calculate the mass of NH_4Cl that must be added to each of the following ammonia solutions of specified volume and molarity to establish the desired pH.

Ammonia solution	pH	*Ammonia solution*	pH
(a) 100.0 cm^3 of 0.1 M	9.0	(b) 50.0 cm^3 of 0.12 M	9.25
(c) 30.0 cm^3 of 0.195 M	9.05	(d) 25.0 cm^3 of 0.679 M	10.0

What would be the corresponding masses to get the buffer solution of maximum buffer capacity?
(Ans. (a) 0.953 g, 0.53 g; (b) 0.321 g, 0.318 g; (c) 0.497 g, 0.310 g; (d) 0.162 g, 0.899 7 g)

Effect of Dilution on pH of a Solution

4.35 Calculate the change in pH upon ten–fold dilution of the following solutions:
(a) 0.1 M HCl. (b) 0.1 M acetic acid.
(c) 0.1 M ammonium chloride.
(d) Mixture of 0.1 M acetic acid and 0.1 M sodium acetate.
(Ans. (a) 1 to 2; (b) 2.87 to 3.37; (c) 5.128 to 5.627; (d) unchanged)

Acid-Base Indicator

4.36 An indicator has $pK^\circ_{In} = 5.3$. In a certain solution this indicator is found to be 80% in its acid form. What is the pH of the solution? (Ans. pH = 4.7)

4.37 A weak acid type indicator was found to be 60% dissociated at pH = 9.20. What will be the per cent dissociation at pH = 9.0? (Ans. 48.62)

4.38 An indicator HIn has a standard ionization constant of 9.0×10^{-9}. The acid colour of the indicator is yellow and the alkaline colour is red. The yellow colour is visible when the ratio of yellow form to red form is 30 to 1 and the red colour is predominant when the ratio of red form to yellow form is 2 to 1. What is the pH range of the indicator? (Ans. 6.57 to 8.35)

4.39 A volume of exactly 50 cm^3 of exactly 0.1 M HCl is titrated with exactly 0.1 M NaOH to a certain hue of methyl orange corresponding to exactly pH = 4. What is the titration error? (Ans. – 0.2%)

4.40 A volume of exactly 50 cm^3 of exactly 0.1 M NaOH is titrated with 0.01 M HCl to exactly pH = 4 (methyl orange). What is the titration error? (Ans. 2%)

Solubility Product

4.41 In a saturated solution of calcium phosphate, the concentration of PO_4^{3-} ion is 3.3×10^{-7} M. Calculate the K°_s of $Ca_3(PO_4)_2$. (Ans. 1.321×10^{-32})

4.42 Which oxalate has the lower molar solubility: $Ag_2C_2O_4$($K^\circ_s = 1.1 \times 10^{-11}$) or CaC_2O_4 ($K^\circ_s = 1.3 \times 10^{-9}$)? (Ans. CaC_2O_4)

4.43 Calculate the solubility of calcium oxalate:
(a) In pure water without any hydrolysis.
(b) In pure water involving hydrolysis of the anion.
(c) In aqueous solution buffered to pH = 2.0.
Given: K°_s(CaOx) = 2.3×10^{-9}. For oxalic acid: $K^\circ_{a1} = 5.4 \times 10^{-2}$ and $K^\circ_{a2} = 5.1 \times 10^{-5}$ (Ans. (a) 4.8×10^{-5} M; (b) 4.8×10^{-5} M; (c) 7.32×10^{-4} M)

4.44 What will be the concentrations of Sr^{2+}, F^- and SO_4^{2-} in a solution that is simultaneously in equilibrium with solid SrF_2 and $SrSO_4$. $K^\circ_s(SrF_2) = 7.9 \times 10^{-10}$ and $K^\circ_s(SrSO_4) = 7.6 \times 10^{-7}$.
(Ans. $[Sr^{2+}] = 1.1 \times 10^{-3}$ M; $[SO_4^{2-}] = 6.9 \times 10^{-4}$ M; $[F^-] = 8.5 \times 10^{-4}$ M)

4.45 What is the solubility of CaF_2 in aqueous solution ($K^\circ_s = 3.7 \times 10^{-11}$), (a) in pure water, and (b) in a solution buffered to pH = 3.50? Given: K_a(HF) = 6.7×10^{-4} M. (Ans. (a) 2.099×10^{-4} M; (b) 2.716×10^{-4} M)

(Hint: $[F^-]_0 = [F^-] + [HF] = [F^-]\,(1 + [H^+]/K_a)$

$[Ca^{2+}]_0 = [F^-]_0/2 = [F^-]\,(1 + [H^+]/K_a)/2$

$K_s = [Ca^{2+}]_0\,[F^-]^2 = 4[Ca^{2+}]_0^3/(1 + [H^+]/K_a)^2$

$s = [Ca^{2+}]_0 = [K_s(1 + [H^+]/K_a)^2/4]^{1/3}$.)

4.46 At what pH value will the solubility of CaF_2 ($K^\circ_s = 4.0 \times 10^{-11}$) be 100 times that in pure water? $K^\circ_w = 1.0 \times 10^{-14}$. (Ans. 0.174)

4.47 Calculate the $[Ag^+]$ and $[Hg_2^{2-}]$ when one mole of $AgNO_3$ and one mole of $Hg_2(NO_3)_2$ are added to 1 dm^3 of a solution containing 3 mol of NaCl. Given: K°_s(AgCl) = 1.6×10^{-10} and $K^\circ_s(Hg_2Cl_2) = 2 \times 10^{-18}$.
(Ans. $[Ag^+] = 1.26 \times 10^{-5}$ M; $[Hg_2^{2+}] = 7.94 \times 10^{-7}$ M)

4.48 (a) A solution is 0.15 M in Pb^{2+} and 0.20 M in Ag^+. If solid Na_2SO_4 is very solwly added to this solution, which will precipitate first, $PbSO_4$ or Ag_2SO_4 Neglect volume changes.

(b) The addition of Na_2SO_4 is continued until the second cation just starts to precipitate as the sulphate. What is the concentration of the first cation at this point? Given: $K_s^o(Ag_2SO_4) = 1.2 \times 10^{-5}$ and $K_s^o(PbSO_4) = 1.3 \times 10^{-8}$.

(Ans. (a) $PbSO_4$, (b) 9.03×10^{-7} M)

4.49 What should the H^+ concentration be in a solution that is 0.25 M in Co^{2+} to prevent the precipitation of CoS when the solution is saturated with H_2S? (A saturated solution of H_2S is 0.1 M.) Given: $K_s^o(CoS) = 5 \times 10^{-22}$. (Ans. 0.23 M)

4.50 (a) What concentration of H^+ should be present in a solution that is 0.20 M in Pb^{2+} and 0.20 M in Zn^{2+} so that upon saturation with H_2S, the maximum amount of PbS precipitates but no ZnS precipitates at any stage? (b) What concentration of Pb^{2+} remains in the solution after the PbS has precipitated? (Ignore the hydrolysis of ions, but do not neglect the increase in acidity caused by precipitation.) Given: $K_s^o(PbS) = 7 \times 10^{-29}$ and $K_s^o(ZnS) = 2.5 \times 10^{-22}$.

(Ans. (a) 0.696 6 M, (b) 3.09×10^{-7} M)

Solubility of a Salt

4.51 Compare the molar solubilities of (a) AgCl, (b) AgBr, and (c) AgI in 0.50 M NH_3.

(Ans. (a) 0.024 M, (b) 0.001 435 M; (c) 1.88×10^{-5} M)

4.52 (a) At what maximum concentration of OH^- will 1.0×10^{-3} mol of $Zn(OH)_2$ go into solution (1.0 dm^3) as $Zn(OH)_4^{2-}$?

(b) At what maximum concentration of OH^- will 1.3×10^{-3} mol of $Zn(OH)_2$ go into solution (1.0 dm^3) as Zn^{2+}? (Ans. (a) 9.9×10^{-2} M; (b) 2.1×10^{-7} M)

Concepts of Acids and Bases

4.53 (a) Find out the conjugate bases of the following acids?

$H_2C_2O_4$, H_2SO_3, HS^-, NH_4^+, HSO_3^-, $H_3P_2O_7^-$, HCl, HF, HNO_3, H_2O, H_3O^+, OH^-, NH_3, NH_4^+, HN_2^-, HOCl, $Al(H_2O)_6^{3+}$, $HClO_4$, H_2O_2

(b) What are conjugate acids of the following bases?

Cl^-, O^{2-}, OH^-, H_2O, NH_2^-, CO_3^{2-}, HCO_3^-, F^-, PO_4^{3-}, NH_3, HS^-

4.54 Account for the following observations: 'In water HCl and HI behave as if they are equally strong, although in ether HI acts as a stronger acid than HCl'.

4.55 Identify the Brönsted acids and bases in the following:

$HCO_3^- + HF \rightleftharpoons H_2CO_3 + F^-$ $\qquad$ $HS^- + OH^- \rightleftharpoons S^{2-} + H_2O$

$NH_3 + HCl \rightleftharpoons NH_4^+ + Cl^-$ $\qquad$ $H_3O^+ + HS^- \rightleftharpoons H_2S + H_2O$

$NH_4^+ + OH^- \rightleftharpoons NH_3 + H_2O$ $\qquad$ $HSO_4^- + CN^- \rightleftharpoons HCN + SO_4^{2-}$

$CH_3OH + OH^- \rightleftharpoons CH_3O^- + H_2O$ $\qquad$ $NH_4^+ + H_2O \rightleftharpoons NH_3 + H_3O^+$

4.56 (a) Which compound of each of the following pairs is the stronger acid?

(i) H_3PO_4 or H_3AsO_4 (ii) H_3AsO_3 or H_3AsO_4
(iii) H_2SO_4 or H_2SO_3 (iv) H_3BO_3 or H_2CO_3
(v) H_2Se or HBr

(b) Which compound of each of the following pairs is the stronger base?

(i) P^{3-} or S^{2-} (ii) PH_3 or NH_3
(iii) SiO_3^{2-} or SO_3^{2-} (iv) NO_2^- or NO_3^-
(v) Br^- or F^-

4.57 In each of the following series, state which species is the strongest acid?

(i) HF, HCl, H_2S (ii) PH_4^+, NH_4^+, AsH_4^+
(iii) $HClO_4$, HClO, $HClO_3$, (iv) H_2SO_3, H_2SeO_3, H_2TeO_3
(v) HNO_3, HNO_2, $H_2N_2O_2$ (vi) Na^+, Mg^{2+}, Al^{3+}
(vii) Ca^{2+}, Zn^{2+}, Cu^{2+} (viii) NH_4^+, NH_3 NH_2^-

5 Conductance

5.1 INTRODUCTION

The conducting ability of electrolytic solutions provides a direct proof of the existence of ions in solutions. The experimental determinations of the conducting properties of electrolytic solutions are very important as they can be used to study quantitatively the behaviour of ions in solutions. Moreover, these can be used to determine the values of many physical quantities such as solubilities and solubility products of sparingly soluble salts, ionic products of self-ionizing solvents, hydrolysis constants of salts, dissociation constants of weak acids and bases and to form the basis for conductometric titration methods.

5.2 METALLIC CONDUCTION

The electrons can flow through metals, this property being known as *metallic* (or *electronic*) conduction. The latter has been explained on the basis of mobile electron clouds permeating relatively fixed lattice of positive metal ions. When electrons from an external battery are forced into one end of a metal wire, they displace the electrons of the cloud at the point of entry. The displaced electrons, in turn, assume new positions by pushing neighbouring electrons ahead, and this effect is transmitted right down to the other end of the wire where they are forced out and sent back to the external battery. The rate at which electrons enter from one end of the wire is equal to that of emission from the other end. At any position in the wire, the electrical neutrality is maintained.

The quantity of electric charge is measured in *coulombs*. One coulomb is the charge carried by $6.241\,8 \times 10^{18}$ electrons. The rate of flow of electric charge, i.e. current is given in amperes; 1 ampere is 1 coulomb/second. We can therefore write

$$I = \frac{Q}{t} \quad \text{or} \quad Q = I\,t \tag{5.2.1}$$

where I is the current, Q is the charge transported and t is the time.

The current is forced through the circuit by an electrical potential difference which is measured in volts. It takes 1 joule of work[†] to move 1 coulomb from a

---

[†] In CGS units, the charge and work are defined as follows:

One electrostatic unit of charge (esu) is the charge that will repel another like charge at a distance 1 cm in vacuum by a force of 1 dyne; $2.997\,9 \times 10^9$ esu = 1 coulomb.

1 joule = 10^7 ergs = 2.3905×10^{-1} cal. One erg is the work done when a force of 1 dyn acts through a distance of 1 cm, 1 erg = 1 dyn cm = 1 g cm^2/s^2.

lower to higher potential when the potential difference is 1 volt. Thus,

1 volt = 1 joule/1 coulomb or 1 joule = 1 volt coulomb

i.e. 1 volt coulomb is a unit of energy and equals 1 joule.

Ohm's Law

The current I carried by a wire is directly proportional to electric potential difference ($\Delta\phi$) between its two ends, i.e.

$$I \propto \Delta\phi$$

Removing the proportionality sign, we have

$$I = \frac{\Delta\phi}{R} \tag{5.2.2}$$

where R is a constant of proportionality and is known as the *resistance of the wire*. Equation (5.2.2) is known as Ohm's law. Resistance is measured in ohms: (Symbol : Ω) : 1 volt is required to force a current of 1 ampere through a resistance of 1 ohm.

Effect of Temperature

Resistance to current flow in a metal is caused by the vibration of the metal ions about their lattice positions; such vibrations interfere with the motion of the electrons and consequently retarded the flow of current. On increasing the temperature, the thermal motion of metal ions increases and they offers more resistance against the flow of electrons. Hence, metals become poorer conductors with rise in the temperature.

5.3 ELECTROLYTIC CONDUCTION

Electrolytic conduction, in which charge is carried by ions, will not occur unless ions of the electrolyte are free to move. Hence, electrolytic conduction is exhibited principally by molten salts and by aqueous solutions of electrolytes.

The principles of electrolytic conduction are best illustrated by reference to an electrolytic cell such as that shown in Fig. 5.3.1 for the electrolysis of molten NaCl between inert electrodes. The entire assembly except that of the external battery of Fig. 5.3.1 is known as the cell. The description of a cell is given in Table 5.3.1

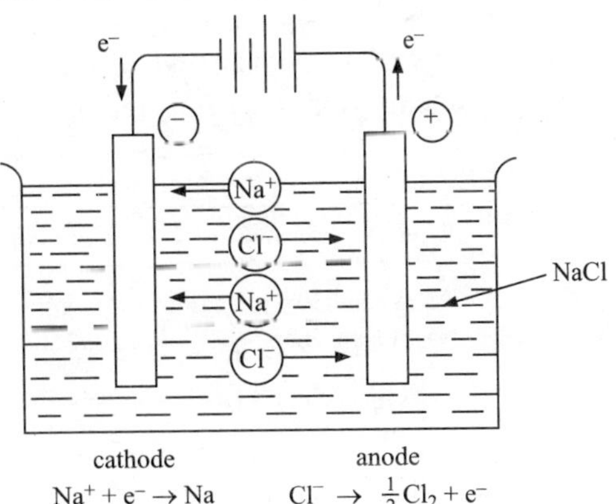

Fig. 5.3.1 Electrolysis of molten sodium chloride

Table 5.3.1 Description of the Cell

	Cathode	*Anode*
Ions attracted	Cations	Anions
Direction of electron movement	Into cell	Out of cell
Half reaction	Reduction	Oxidation
Sign	–ve; since it is attached to the –ve end of battery	+ve; since it is attached to the +ve end of battery

Working of an Electrolytic Cell

The electrons are received from the negative end of the external battery by the negative electrode of the cell. These are used up in the reduction reaction at this electrode. The number of electrons received at the negative electrode are given back to the positive end of the external battery from the positive electrode of the cell where electrons are released as a result of oxidation reaction. Within the cell, current is carried by the movements of ions: cations towards the negative electrode (called the cathode) and anions towards the positive electrode (called the anode).

Factors Affecting Electrolytic Conduction

The movement of ions gives rise to what is known as the electrolytic conduction. The latter, thus, depends on the mobility of ions and anything that inhibits the motion of ions causes resistance to current flow. Factors that influence the electrical conductivity of solutions of electrolytes include interionic attraction, solvation of ions and viscosity of solvent; these factors depend on the attraction of solute-solute, solute-solvent and solvent-solvent, respectively. The average kinetic energy of the solute ions increases as the temperature is raised and, therefore, the resistance of electrolytic conductors generally decreases (i.e. conduction increases) as the temperature is raised.

5.4 ELECTROLYSIS

The electrolysis of molten salts produces ions which are characteristic of the salt. When certain aqueous solutions are electrolyzed, water is, however, involved in the electrode reactions rather than the ions derived from the solute. Hence, the current-carrying ions are not necessarily discharged at the electrodes.

The cathodic and anodic reactions involved in the electrolysis of aqueous solutions of some salts are given in Table 5.4.1.

Table 5.4.1 Electrolysis of Some of the Aqueous Solutions of Salts

Solution	*Electrodes*	*Cathodic reaction*	*Anodic reaction*
NaCl	Inert	$2H_2O + 2e^- \rightarrow H_2(g) + 2OH^-$	$2Cl^- \rightarrow Cl_2(g) + 2e^-$
Na_2SO_4	Inert	$2H_2O + 2e^- \rightarrow H_2(g) + 2OH^-$	$H_2O \rightarrow \frac{1}{2} O_2(g) + 2H^+ + 2e^-$
$CuSO_4$	Inert	$Cu^{2+} + 2e^- \rightarrow Cu$	$H_2O \rightarrow \frac{1}{2} O_2(g) + 2H^+ + 2e^-$
$CuCl_2$	Inert	$Cu^{2+} + 2e^- \rightarrow Cu$	$2Cl^- \rightarrow Cl_2(g) + 2e^-$
$CuSO_4$	Copper	$Cu^{2+} + 2e^- \rightarrow Cu$	$Cu \rightarrow Cu^{2+} + 2e^-$

Faraday's Laws

The quantitative relationships between electricity and chemical change were first described by Michael Faraday. These are:

(1) *The mass of a chemical substance involved at an electrode reaction is directly proportional to the amount of current passed through the cell.*

(2) *The masses of different substances produced by a given amount of current are proportional to the equivalent masses of the substances.*†

These two laws have been shown to hold very rigidly provided the passage of electricity takes place entirely by electrolytic conduction. It applies to molten electrolytes as well as to solutions of electrolytes and is independent of temperature, pressure or the nature of the solvent.

Mathematical Representation

Mathematically, the Faraday's law may be represented as

$$m_B = \frac{It}{F} \frac{M_B}{|\nu_e|/\nu_B}$$

where m_B is the mass of the species B (molar mass: M_B) deposited/liberated on passing a current I for the time t and $|\nu_e|$ and ν_B are the stoichiometric numbers of electrons and the species B, respectively, involved in the reduction/oxidation reaction at the electrode.

Value of Electronic Charge

Faraday's laws are readily interpreted by reference to the electrolysis of molten sodium chloride. The change at the cathode requires one electron for every sodium ion reduced, i.e.

$$Na^+ + e^- \rightarrow Na$$

If the electrons consumed at this electrode is equal to Avogadro constant (i.e. 6.022×10^{23} mol^{-1}), 1 mole of sodium metal (22.989 8 g) is produced. At the same time, 1 mole of electrons* is removed from the anode and 1 mole of Cl^- ions (35.453 g) is discharged, i.e. 0.5 mole of Cl_2 gas is produced.

$$Cl^- \rightarrow \tfrac{1}{2}Cl_2(g) + e^-$$

Thus, 96 487 coulombs of electricity which is necessary to produce 1 equivalent mass of a substance at the electrode will be the total charge carried by 1 mole of electrons. Hence, the negative charge carried by each electron is given by

$$e = \frac{(96\,487 \text{ C mol}^{-1})}{(6.022 \times 10^{23} \text{ mol}^{-1})} = 1.602 \times 10^{-19} \text{ C}$$

Coulometer

The quantity of electricity passed through a circuit can be determined from the chemical changes that are produced when the same amount of current is passed through a suitable electrolytic cell. A cell used for this purpose is called a *coulometer.* The coulometer is placed in the circuit in series with any other

---

†In this chapter, the equivalent mass means the mass corresponding to a total of unit charge on each ion of a single molecule of the substance.

*The quantity of electric charge carried by 1 mole of electrons is conventionally known as 1 faraday of electricity, abbreviated as 1 *F.* This quantity has a value of 96 487 C and in SI units, it is known as faraday constant (1 F = 96 487 C mol^{-1}).

apparatus. The silver coulometer is commonly employed for precise work. This coulometer consists of a platinum dish serving as both cathode and cell vessel, and pure silver as anode. The electrolyte is an aqueous solution of purified silver nitrate. The mass of Ag deposited on the dish is determined when the experiment is over and from this mass, the quantity of electricity passed through the coulometer can be calculated by using Faraday's laws of electrolysis.

5.5 CONDUCTANCE OF SOLUTIONS

Solutions like metallic conductors obey Ohm's law. Thus, the resistance of an electrolytic conductor to current passage can be determined by the application of Ohm's law. However, while dealing with solutions, it is more convenient to speak of the conductance (symbol : G), which is merely the reciprocal of the electrical resistance. The unit of conductance is ohm^{-1}. In SI unit, Ω^{-1} is known as Siemens (symbol : S).

Resistance of a Conductor

The resistance R of a conductor (metallic or electrolytic) is proportional directly to its length l and inversely to its cross-sectional area A. Therefore,

$$R \propto l \quad \text{and} \quad R \propto \frac{1}{A}$$

which give $R \propto \dfrac{l}{A}$ (5.5.1)

Removing the proportionality sign, we have

$$R = \rho\left(\frac{l}{A}\right) \tag{5.5.2}$$

where ρ is constant of proportionality and is called the *resistivity*. It is the resistance offered by a conductor of unit length and unit area of cross-section. The value of ρ depends on, and is characteristic of, the nature of the conductor.

Definition of Conductance

The expression for the corresponding conductance G is

$$G = \frac{1}{R} = \frac{1}{\rho}\left(\frac{A}{l}\right) = \kappa\left(\frac{A}{l}\right) \tag{5.5.3}$$

where κ (= $1/\rho$) is the conductivity of the conductor. This quantity may be considered to be the conductance of a cubic material of edge length unity.

SI Unit of Conductivity

In SI units, κ is expressed as S m^{-1}. It may also be expressed as Ω^{-1} cm^{-1}. The unit Ω^{-1} is also conventionally written as mho, which is ohm spelled backward.

Conductance (or conductivity) is an additive property. For example, in an aqueous solution containing several electrolytes, the total conductance is

$$G_{\text{total}} = \sum_i G_i + G_{\text{water}} \tag{5.5.4}$$

where the summation is to be carried over all the electrolytes present in the solution and G_{water} is the conductance of water which is utilized for making the solution, G_{water} is often negligible in comparison to $\sum_i G_i$ as repeatedly distilled water (known as conductivity water) of very low conductance is employed for making the solution.

Measurement of Resistance

The conductance of a solution can be determined by measuring the resistance offered by the solution contained within the two electrodes of a conductivity cell. For measuring resistance, the Wheatstone bridge (a schematic diagram of which is given in Fig. 5.5.1) is employed.

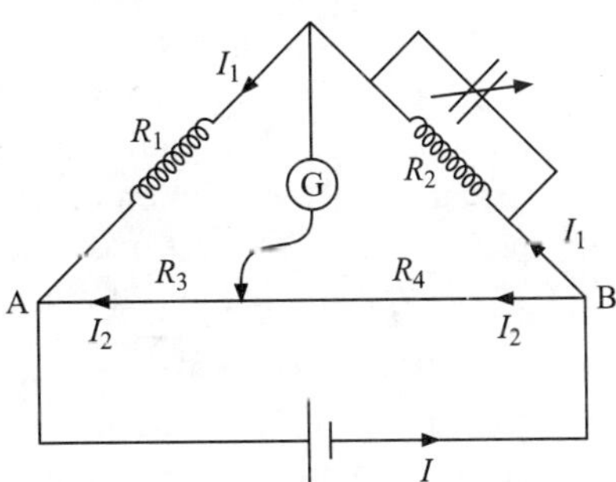

Fig. 5.5.1 Measurement of an unknown resistance

When no current flows through the galvanometer, it means that the potential drop from B (or A) to one terminal of galvanometer is the same as that from B (or A) to the other terminal of galvanometer. These give the following equalities

$$I_1R_2 = I_2R_4 \quad \text{and} \quad I_1R_1 = I_2R_3$$

Dividing these, we get

$$\frac{R_1}{R_2} = \frac{R_3}{R_4} \quad \text{or} \quad R_1 = R_2\left(\frac{R_3}{R_4}\right)$$

Necessary Requirements of the Method

Although the principle of the Wheatstone bridge as given above remains the same, several modifications in technique are necessary before resistance of an electrolytic solution can be measured by this method.

Direct current cannot be used, as this would cause electrolysis and concentration changes at the electrodes. The gas if liberated increases the resistance at the electrodes. These gases also set up a counter emf of polarization which opposes the passage of current. To avoid these, alternating current is employed, usually a frequency of 500-2 000 Hz is supplied by either a vibrating tuning fork or a vacuum-tube oscillator. In alternating current, the direction of the current is reversed about a thousand times per second with the result that the polarization produced by each pulse of the current is neutralized by the next pulse.

Since a current of this frequency is within the range of the human ear, the galvanometer can be replaced by a set of earphones. Passage of the current through these earphones produces a buzzing sound which decreases in intensity as balance is approached and it attains minimum when the balance point is reached. Theoretically, the sound should be zero at the balance point, but due to capacitance introduced by the cell, such an ideal state is not attained. However, by placing a variable condenser across the standard resistance, it is possible to sharpen the balancing by adjusting the condenser to the capacitance of the cell. An alternative detecting device employed at present is the cathode-ray oscilloscope.

Conductivity Cells

The conductivity cells are constructed of glass with electrodes of either platinum or gold. To overcome imperfections in the current and other effects, so as to decrease the effect of polarization of the electrodes, the electrodes are coated with a layer of finely divided platinum black. This is done by electrolysis of 3 per cent solution of chloroplatinic acid containing traces of lead acetate. If the finely divided platinum serves either as a catalyst for decomposition or oxidation of the solution or as an adsorbent for the adsorption of ions, the gray platinum surface electrodes may be employed instead. The latter can be obtained by heating the platinized electrode to red heat.

The distance between the two electrodes of a cell is determined by the conductance of the solution to be measured. For solutions of high conductance the electrodes are widely spaced, while for poorly conducting solutions, the electrodes are mounted close to each other (Fig. 5.5.2).

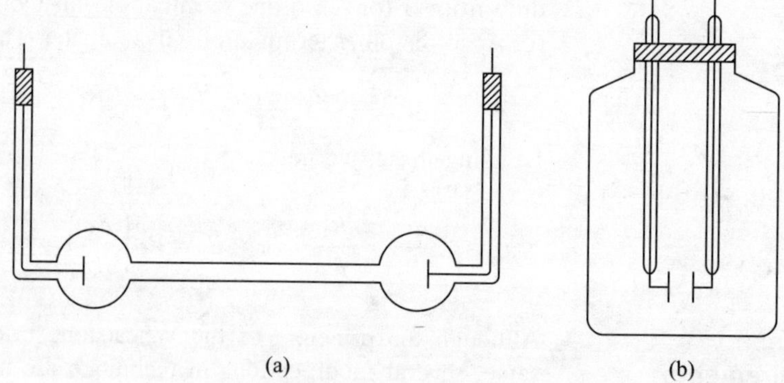

Fig. 5.5.2 Conductivity cell (a) with larger electrode distance, (b) with smaller electrode distance

Cell Constant

According to Eq. (5.5.3), the conductivity of any electrolytic conductor is given

$$\kappa = \left(\frac{l}{A}\right) G = \left(\frac{l}{A}\right) \frac{1}{R}$$

and hence before κ can be calculated from the measured resistance, the ratio (l/A) for the particular cell used, is required. For any given cell, the ratio has a fixed value, called the *cell constant* K_{cell}, which can be obtained without knowing the actual values of l and A. A solution of known κ is placed in the cell, the resistance is measured, and K_{cell} is calculated. Potassium chloride is the accepted standard for which accurately determined values of κ at different concentrations and temperatures in aqueous solution are available.

5.6 EQUIVALENT AND MOLAR CONDUCTIVITIES

Since the charges of solute ions determine the conductance of a solution, the comparison of conductance data is made between values for solutions corresponding to a total of unit charge on each ion of the solute. It is because of this that the equivalent conductivity of the solution is employed for comparison purposes.

Suppose 1 equivalent mass of an electrolyte is dissolved in volume V of the solution. Let this whole solution be placed in a conductivity cell. Multiplying and dividing the right side of Eq. (5.5.3) by the distance l between the two electrodes of the cell, we get

$$G = \kappa \frac{A \times l}{l^2} = \frac{\kappa(V_{eq}/\text{eq}^{-1})}{l^2} \quad \text{or} \quad Gl^2 = \kappa(V_{eq}/\text{eq}^{-1}) \tag{5.6.1}$$

Note: V_{eq} in Eq. (5.6.1) carry the units of $m^3\ eq^{-1}$.

Equivalent Conductivity

The term Gl^2/eq is known as *equivalent conductivity,* abbreviated as Λ_{eq}. Thus, we have

$$\Lambda_{eq} = Gl^2/\text{eq} = \kappa\ V_{eq} \tag{5.6.2}$$

The equivalent conductivity of an electrolyte may be defined as the *conductance of a volume of solution containing one equivalent mass of a dissolved substance when placed between two parallel electrodes which are at a unit distance apart, and large enough to contain between them the whole solution.*

The equivalent conductivity thus gives the conducting power of the ions produced by 1 equivalent mass (i.e. mass corresponding to a total of unit charge on each ion) of an electrolyte at any particular concentration. The unit of Λ_{eq} in CGS units are:

$$\Lambda_{eq} = (\Omega^{-1}\ \text{cm}^{-1})\ (\text{cm}^3\ \text{eq}^{-1}) = \Omega^{-1}\ \text{cm}^2\ \text{eq}^{-1} \equiv \text{S cm}^2\ \text{eq}^{-1}$$

If c_{eq} is the concentration of the solution (in equivalent per unit volume), then V_{eq}(which carry a unit of $m^3\ eq^{-1}$) will be equal to $1/c_{eq}$. Hence, Eq. (5.6.2) may be written as

$$\Lambda_{eq} = \kappa\ (1/c_{eq}) \quad \text{i.e.} \quad \Lambda_{eq} = \kappa/c_{eq} \tag{5.6.3}$$ †

Molar Conductivity

In SI units, c is expressed as the amount of substance per unit volume instead of equivalent of substance per unit volume, and thus one uses the term *molar conductivity* as defined below.

The molar conductivity, Λ_m, of an electrolyte may be defined as the *conductance of a volume of solution containing one mole of a dissolved substance when placed between two parallel electrodes which are at a unit distance apart, and large enough to contain between them the whole solution.* Thus, it gives the conducting power of the ions produced by 1 mole of an electrolyte at any particular concentration. This can be calculated by using the expressions analogous to Eqs (5.6.2) and (5.6.3):

---

† The term 'equivalent conductivity' follows from Eq. (5.6.2) as it is equal to the conductivity divided by equivalent concentration.

$$\Lambda_{\rm m} = \kappa\, V_{\rm m} \tag{5.6.4}$$

$$\Lambda_{\rm m} = \kappa\,(1/c) \quad \text{i.e.} \quad \Lambda_{\rm m} = \frac{\kappa}{c} \tag{5.6.5}$$†

where $V_{\rm m}$ is the volume of the solution containing one mole of the substance and c is the resultant molar concentration. Note $V_{\rm m}$ carries the unit of $m^3\ mol^{-1}$. The unit of $\Lambda_{\rm m}$ will be

$$\Lambda_{\rm m} = (\Omega^{-1}\ {\rm m}^{-1})\ ({\rm m}^3\ {\rm mol}^{-1}) = \Omega^{-1}\ {\rm m}^2\ {\rm mol}^{-1} \equiv {\rm S\ m^2\ mol^{-1}}$$

Relationship Between Molar and Equivalent Conductivities

If $\nu_+ z_+$ (or $\nu_- |z_-|$) is the total charge number of the cations (or magnitude of the charge number of the anions) of a single molecule of an electrolyte, it follows that

$$1\ {\rm mol} \equiv (\nu_+ z_+)\ (1\ {\rm eq})$$

Hence, $$\Lambda_{\rm m} = \frac{\kappa}{c_{\rm m}} \quad \text{and} \quad \Lambda_{\rm eq} = \frac{\kappa}{c_{\rm eq}} = \frac{\kappa}{(\nu_+ z_+\ {\rm eq\ mol^{-1}})\ c_{\rm m}}$$

Thus $$\Lambda_{\rm eq} = \frac{\Lambda_{\rm m}}{\nu_+ z_+\ {\rm eq\ mol^{-1}}} \tag{5.6.6}$$

Comment on Equivalent and Molar Conductivities

The term equivalent conductivity finds no place in IUPAC recommendations. However, these values are tabulated as molar conductivity of the species $\frac{1}{|z_{\rm B}|}{\rm B}^z$, where z is the charge of B and $|z_{\rm B}|$ is the magnitude of charge number. For example, the equivalent conductivity of Mg^{2+} ions is referred to as molar conductivity of $\frac{1}{2}Mg^{2+}$ ions, that is

$$\lambda_{\rm eq}({\rm Mg^{2+}})/{\rm eq^{-1}} = \lambda_{\rm m}\left(\tfrac{1}{2}{\rm Mg^{2+}}\right)/{\rm mol^{-1}}$$

Note that the unit of molar conductivity of any species, such as Mg^{2+}, $\frac{1}{2}Mg^{2+}$, Fe^{3+}, $\frac{1}{3}Fe^{3+}$, SO_4^{2-} and $\frac{1}{2}SO_4^{2-}$, is $S\ m^2\ mol^{-1}$. Also note that

$$\lambda_{\rm m}({\rm B}^z) = |z_{\rm B}|\,\lambda_{\rm m}\left(\frac{1}{|z_{\rm B}|}{\rm B}^z\right) \tag{5.6.7}$$

For example, $\lambda_{\rm m}({\rm Mg^{2+}}) = 2\lambda_{\rm m}\left(\tfrac{1}{2}{\rm Mg^{2+}}\right)$

†The term molar conductivity follows from Eq. (5.6.5) as it is equal to the conductivity divided by amount-of-substance concentration (i.e. molar concentration).

Example 5.6.1 The molar conductivity of 0.05 M solution of $MgCl_2$ is 194.2 S cm^2 mol^{-1} at 25 °C. A cell with electrodes that are 1.50 cm^2 in surface area and 0.50 cm apart is filled with 0.05 M $MgCl_2$ solution. How much current will flow when the potential difference between the two electrodes is 5.0 V?

Solution Molar conductivity of $MgCl_2$ solution, Λ = 194.2 S cm^2 mol^{-1}

Concentration, c = 0.05 M

Therefore,

Conductivity, $\kappa = \Lambda\, c$ = (194.2 S cm^2 mol^{-1}) (0.05 mol dm^{-3})

$$= 9.71 \text{ S (cm}^2\text{ dm}^{-3}) = 9.71 \times 10^{-3} \text{ S cm}^{-1}$$

Conductance of 0.05 M $MgCl_2$ solution,

$$G = \kappa\left(\frac{A}{l}\right) = (9.71 \times 10^{-3} \text{ S cm}^{-1})\left\{\frac{1.50 \text{ cm}^2}{0.50 \text{ cm}}\right\} = 2.913 \times 10^{-2} \text{ S}$$

Resistance offered by the solution,

$$R = \frac{1}{G} = \frac{1}{0.029\,13 \text{ S}} = 34.34\ \Omega$$

The current flowing through the electrodes,

$$I = \frac{(\Delta\phi)}{R} = \frac{(5.0 \text{ V})}{(33.34\ \Omega)} = \frac{(5.0 \text{ V})}{(34.34 \text{ V A}^{-1})} = 0.15 \text{ A}$$

Example 5.6.2 The conductivity of 0.1 M NaOH solution is 0.0221 S cm^{-1}. When an equal volume of 0.1 M HCl solution is added, the conductivity decreases to 0.0056 S cm^{-1}. A further addition of HCl solution, the volume of which is equal to that of the first portion added, the conductivity increases to 0.017 S cm^{-1}. Calculate (a) Λ_m(NaOH), (b) Λ_m(NaCl), (c) Λ_m(HCl).

Solution **(a)** Molar conductivity of 0.1 M NaOH solution,

$$\Lambda_m(\text{NaOH}) = \frac{\kappa}{c} = \frac{(0.022\,1 \text{ S cm}^{-1})}{(0.1 \text{ mol dm}^{-3})} = 0.221 \text{ S cm}^{-1} \text{ dm}^3 \text{ mol}^{-1}$$
$$= 221 \text{ S cm}^2 \text{ mol}^{-1}$$

On adding equal volume of 0.1 M HCl solution, the total volume becomes double and the concentration of the salt formed will be 0.05 M.

(b) Molar conductivity of 0.05 M NaCl solution,

$$\Lambda_m(\text{NaCl}) = \frac{\kappa}{c} = \frac{(0.005\,6 \text{ S cm}^{-1})}{(0.05 \text{ mol dm}^{-3})}$$
$$= 0.112 \text{ S cm}^{-1} \text{ dm}^3 \text{ mol}^{-1} = 112 \text{ S cm}^2 \text{ mol}^{-1}$$

(c) On further addition of equal volume of the acid, the solution will become 0.033 M in NaCl and 0.033 M in HC1. The conductivity of this solution will be given by

$$\kappa = \sum_i c_i \Lambda_i = c(\text{NaCl})\ \Lambda_\text{m}(\text{NaCl}) + c(\text{HCl})\ \Lambda_\text{m}(\text{HCl})$$

Substituting the values, we get

$$0.017 \text{ S cm}^{-1} = (0.033 \times 10^{-3} \text{ mol cm}^{-3})\ (112 \text{ S cm}^2 \text{ mol}^{-1})$$

$$+\ (0.033 \times 10^{-3} \text{ mol cm}^{-3})\ \Lambda_\text{m}(\text{HCl})$$

Solving for Λ_m(HCl), we get

$$\Lambda_\text{m}(\text{HCl}) = 403.15 \text{ S cm}^2 \text{ mol}^{-1}$$

Example 5.6.3

A conductivity cell filled with 0.1 M solution of potassium chloride at 25 °C has a measured resistance of 24.96 Ω. Calculate the cell constant if the conductivity κ for 0.1 M solution of KC1 is 0.011 639 S cm^{-1} and conductivity water with $\kappa = 7.5 \times 10^{-8}$ S cm^{-1} is used to make up the solutions. When the cell is filled with a 0.01 M solution of acetic acid, the cell resistance is 1982 Ω. Calculate the molar conductivity of acetic acid at this concentration.

Solution

Since $\quad R = \rho\left(\dfrac{l}{A}\right) = \rho K_\text{cell}$, therefore,

$$K_\text{cell} = \frac{R}{\rho} = R\kappa = (24.96\ \Omega)\ (0.011\,639 \text{ S cm}^{-1}) = 0.290\,5 \text{ cm}^{-1}$$

Conductivity of 0.01 M of acetic acid solution

$$\kappa = \frac{1}{R} K_\text{cell} = \left(\frac{1}{1982\ \Omega}\right)(0.290\,5 \text{ cm}^{-1}) = 1.465 \times 10^{-4} \text{ S cm}^{-1}$$

Conductivity due to acetic acid alone

$$= 1.465 \times 10^{-4} \text{ S cm}^{-1} - 7.5 \times 10^{-8} \text{ S cm}^{-1} = 1.464 \times 10^{-4} \text{ S cm}^{-1}$$

Molar conductivity of acetic acid at a concentration of 0.01 M

$$= \frac{\kappa}{c} = \left(\frac{1.464 \times 10^{-4} \text{ S cm}^{-1}}{(0.01 \times 10^{-3} \text{ mol cm}^{-3})}\right) = 14.64 \text{ S cm}^2 \text{ mol}^{-1}$$

Example 5.6.4

A sample of water from a large swimming pool has a resistance of 9 200 Ω at 25 °C when placed in a certain conductance cell. When filled with 0.02 M KC1 solution, the cell has a resistance of 85 Ω at 25 °C. Five hundred grams of sodium chloride were dissolved in the pool, which was thoroughly stirred. A sample of this solution gave a resistance of 7 600 Ω. Calculate the volume of water in the pool. Given: Molar conductivity of sodium chloride at that concentration is 126.5 S cm^2 mol^{-1} and molar conductivity of potassium chloride at 0.02 M is 138.3 S cm^2 mol^{-1}.

Solution Molar conductivity of 0.02 M KC1 solution = 138.3 S cm^2 mol^{-1}

Resistance offered by 0.02 M KC1 solution = 85 Ω

Conductivity of 0.02 M KC1 solution,

$$\kappa = \Lambda_m c = (138.3 \text{ S cm}^2 \text{ mol}^{-1})(0.02 \text{ mol dm}^{-3})$$
$$= 2.766 \text{ S cm}^2 \text{ dm}^{-3} = 2.766 \text{ S cm}^2 (10 \text{ cm})^{-3}$$
$$= 0.002\,766 \text{ S cm}^{-1}$$

Cell constant, $K_{cell} = \kappa R = (0.002766 \text{ S cm}^{-1})(85\ \Omega) = 0.235\ 1 \text{ cm}^{-1}$

Conductance of water, $G_1 = \dfrac{1}{R_1} = \dfrac{1}{9\,200\ \Omega}$

Conductance of water after NaCl has been mixed,

$$G_2 = \frac{1}{R_2} = \frac{1}{7\,600\ \Omega}$$

Conductance due to NaCl alone

$$= G_2 - G_1 = \frac{1}{7\,600\ \Omega} - \frac{1}{9\,200\ \Omega} = \frac{(1\,600\ \Omega)}{(7\,600\ \Omega)(9\,200\ \Omega)} = 2.288 \times 10^{-5} \text{ S}$$

Conductivity of sodium chloride

$$= GK_{cell} = (2.288 \times 10^{-5} \text{ S})(0.235\ 1 \text{ cm}^{-1}) = 5.379 \times 10^{-6} \text{ S cm}^{-1}$$

Concentration of NaCl in the swimming pool

$$c = \frac{\kappa}{\Lambda_m} = \frac{(5.379 \times 10^{-6} \text{ S cm}^{-1})}{(126.5 \text{ S cm}^2 \text{ mol}^{-1}}$$
$$= 4.253 \times 10^{-8} \text{ mol cm}^{-3} = 4.253 \times 10^{-5} \text{ mol dm}^{-3}$$

Mass of NaCl in 1 dm^3 solution = $(4.253 \times 10^{-5}$ mol) (58.5 g mol^{-1}) = 0.002 488 g

Total volume of water in the swimming pool containing 500 g NaCl

$$= \left(\frac{1 \text{ dm}^3}{0.002\,488 \text{ g}}\right)(500 \text{ g}) = 2.009 \times 10^5 \text{ dm}^3$$

5.7 VARIATION OF CONDUCTIVITY AND MOLAR CONDUCTIVITY WITH CONCENTRATION

Both the conductivity and molar conductivity of a solution vary with concentration. The conductivity increases with increase in concentration whereas the molar conductivity increases on dilution (i.e. decrease in concentration). For strong electrolytes, conductivity increases sharply with increase in concentration while for weak electrolytes it starts at lower values in dilute solutions and increases much more gradually. In both the cases, this increase is due to the increase in the number of ions per unit volume of the solution. For strong electrolytes, the number of ions per unit volume increases in proportion to the concentration and that is why the increase in conductivity is very rapid. In weak electrolytes, however, the increase in the number of ions is basically due to the change in the partial ionization of the solute, and consequently, the conductivity increases very gradually.

As stated above, molar conductivity Λ_m of both strong and weak electrolytes increases on dilution. The basic reason for this is that the decrease in conductivity is more than compensated by increase in the value of $1/c$ on dilution.

The variation of molar conductivity on dilution for strong and weak electrolytes shows altogether different behaviour as can be seen from Fig. 5.7.1 where Λ_m has been plotted against $\sqrt{c}$. For strong electrolytes, the variation is almost linear in dilute solutions while that for weak electrolytes, the variation is very rapid. As the molar conductivity is a measure of the conducting power of all the ions that are available in 1 mole of a substance, it is, therefore, obvious that the number of ions that are available for conductance increases on dilution. For weak electrolytes, the increase in the number of ions has been explained on the basis of Arrhenius theory of electrolytic dissociation whereas that for strong electrolytes has been explained on the basis of Debye-Hückel-Onsager theory. In brief, the increase in the number of ions in case of weak electrolytes is due to the increase in the degree of ionization of the electrolyte on dilution, whereas in the case of strong electrolytes, this increase is due to the weakening of the ion-ion interactions on dilution.

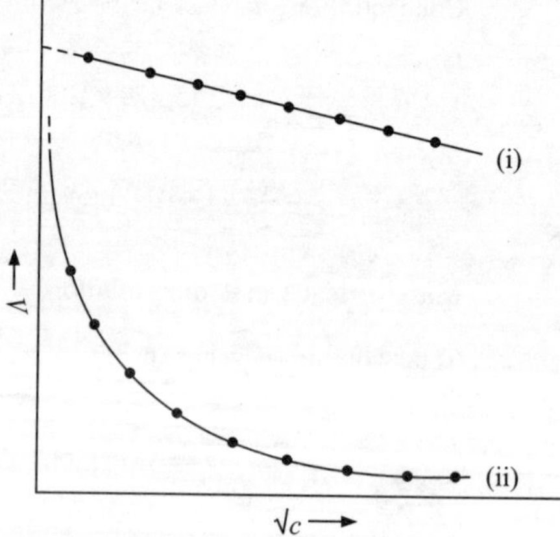

Fig. 5.7.1 Variation of molar conductivity on dilution (i) for strong eiectrolyte and (ii) weak electrolyte

Arrhenius Theory of Electrolytic Dissociation

Ostwald Dilution Law

According to the Arrhenius theory of electrolytic dissociation, there exists an equilibrium between the undissociated molecule AB and the ions A^+ and B^- which result from the dissociation of the molecule. Thus,

$$\mathrm{AB} \rightleftharpoons \mathrm{A^+ + B^-}$$

This equilibrium is characterized by the equilibrium constant, defined as

$$K = \frac{[\mathrm{A^+}][\mathrm{B^-}]}{[\mathrm{AB}]} \tag{5.7.1}$$

If α is the degree of dissociation and c is the initial concentration of the weak electrolyte AB, then concentrations of various species are

$$[\mathrm{A^+}] = [\mathrm{B^-}] = c\alpha \quad \text{and} \quad [\mathrm{AB}] = c(1 - \alpha)$$

Thus, $$K = \frac{(c\alpha)(c\alpha)}{c(1-\alpha)} = \frac{c\alpha^2}{1-\alpha}$$

Since $\alpha \ll 1$, we have $K = c\alpha^2$ or $\alpha = \sqrt{K/c}$. (5.7.2)

The expression $\alpha = \sqrt{K/c}$ is known as the *Ostwald dilution law*, according to which the degree of dissociation increases as c decreases. As α is increased, the concentrations of the ions A^+ and B^- are also increased. Thus, the increase of conductance for a weak electrolyte is primarily due to the increase in the number of ions on dilution.

Expression of Degree of Dissociation

Arrhenius suggested that the degree of dissociation of an electrolyte can be calculated using the expression

$$\frac{\Lambda_c}{\Lambda^\infty} \tag{5.7.3}$$

where Λ_c is the molar conductivity of the solution at a given concentration and Λ^∞ is that at infinite dilution.

Expression of Dissociation Constant

Substituting Eq. (5.7.3) in the expression

$$K = \frac{c\alpha^2}{(1 - \alpha)} \tag{5.7.4}$$

we get $$K = \frac{c\Lambda_c^2}{\Lambda^\infty(\Lambda^\infty - \Lambda_c)}$$

If we plot $\Lambda_c^2/\Lambda^\infty(\Lambda^\infty - \Lambda_c)$ versus $1/c$, we get a straight line with slope equal to K. This gives one of the methods to determine the value of the dissociation constant of a weak electrolyte.

Example 5.7.1 The following are the conductivities of chloroacetic acid in aqueous solution at 25 °C.

mol dm^{-3}/c	16	32	64	128	256	512
Λ_{cm}/S cm^2 mol^{-1}	53.1	72.4	96.8	127.7	164.	205.8

If Λ^∞ for chloroacetic acid be 362 S cm^2 mol^{-1}, are these values in accord with the Ostwald dilution law? If so, calculate K_a.

Solution Since the dissociation constant of a weak electrolyte is given by

$$K_a = \frac{c\Lambda_c^2}{\Lambda^\infty(\Lambda^\infty - \Lambda_c)}$$

we have

$$\frac{\Lambda_c^2}{\Lambda^\infty(\Lambda^\infty - \Lambda_c)} = \frac{K_a}{c}$$

Thus, if on plotting $\Lambda_c^2/\Lambda^\infty(\Lambda^\infty - \Lambda_c)$ versus $1/c$ a straight line is obtained, then the chloroacetic acid is a weak electrolyte: otherwise not. If so, the slope gives the value K_a.

From the given values of Λ_c at various concentrations, we have

mol dm^{-3}/c	$\dfrac{\Lambda_c}{\text{S cm}^2\text{ mol}^{-1}}$	$\dfrac{\Lambda_c^2}{\Lambda^\infty(\Lambda^\infty - \Lambda_c)}$
16	53.1	0.025
32	72.4	0.050
64	96.8	0.090
128	127.7	0.193
256	164	0.375
512	205.8	0.750

The graph between $\Lambda_c^2/\Lambda^\infty(\Lambda^\infty - \Lambda_c)$ and mol dm^{-3}/c was found to be a straight line with a slope of 0.001 44 (Fig. 5.7.1). Thus, K_a = 0.001 44 M.

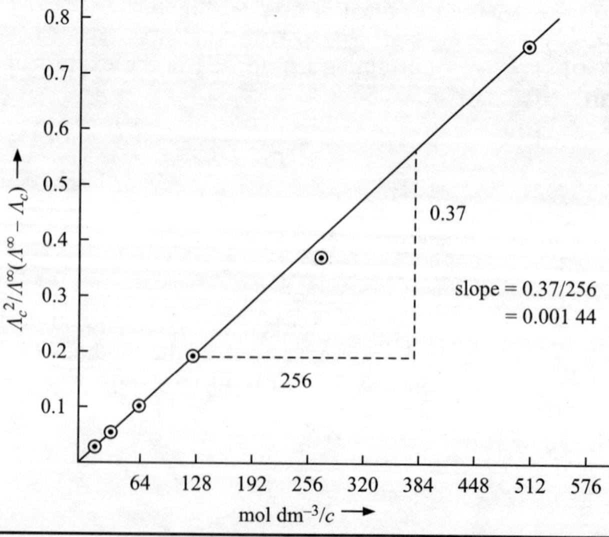

Fig. 5.7.1

Debye-Hückel-Onsager Theory of Ionic Atmosphere

The conductance data of many substances did not conform to Ostwald dilution law. These substances are strong electrolytes which are known to be completely ionized in solution. The variation of molar conductivity of a dilute solution of a strong electrolyte on dilution is explained on the basis of Debye-Hückel-Onsager theory. According to this theory, *each ion in solution is surrounded by an ionic atmosphere consisting of other ions whose net average charge is opposite to that of the central ion.* This ionic atmosphere is spherical and symmetrical in nature in the absence of any disturbing factor such as the application of potential across the two electrodes immersed in the solution during the conductance measurements. However, when the potential is applied, the ions start moving towards the respective electrodes and as a result the ionic atmosphere is distorted which results in a decrease in speeds of the ions. Debye and Hückel showed that these effects are due to two factors, namely, (a) relaxation of ionic atmosphere due to an applied potential or asymmetry effect and (b) the electrophoretic effect.

Asymmetry Effect

This arises from the fact that any central ion and its atmosphere are oppositely charged, i.e. when the central ion is positively charged, the atmosphere is negative, and vice versa. Because of this, the central ion and the atmosphere tend to move in opposite directions as the potential is applied across the electrodes. Thus, a central positive ion will tend to move towards the cathode while its ionic atmosphere will tend towards the anode. This results in the distortion of spherical and symmetrical nature of the ionic atmosphere as shown in Fig. 5.7.2. Consequently, the force exerted by the atmosphere on the central ion is no longer uniform in all directions: it is greater behind the ion than in front of it. Thus, the central ion experiences a retarding force opposite to the direction of its motion with the result that its speed is lowered.

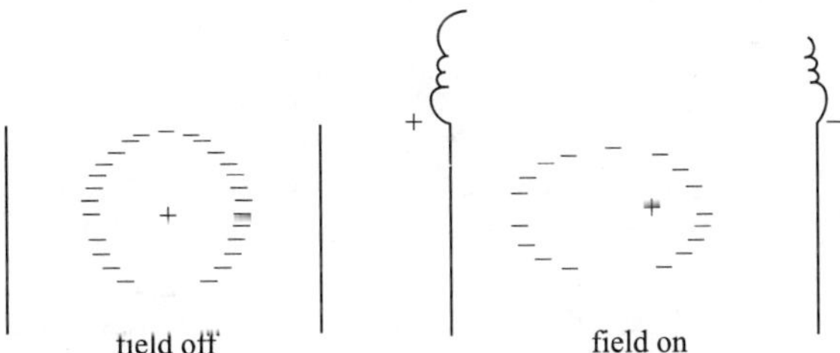

Fig. 5.7.2 Asymmetry effect showing distortion of ionic atmosphere in the presence of an applied potential

Electrophoretic Effect

Ions in solutions are generally solvated and when these move, they carry with them the associated solvent molecules. Since cations and anions move in the opposite directions, it is obvious that any particular ion, say the central ion, does not move through a stationary medium but through a medium in which solvent molecules carried by the ions of the atmosphere move in the opposite direction. Similarly, the ions involved in the ionic atmosphere move in a medium of opposite moving solvent molecules carried by the central ion. Thus, both type of ions, while moving in solution do not travel through a stationary medium but through a medium which moves in the opposite direction. These counter-currents make it more difficult for the ion to move through the solution and thus slow down its motion. This is known as the electrophoretic effect.

Quantitative Expression

Debye and Hückel showed that both of these retarding effects on an ion produce a decrease in molar conductivity. Their mathematical treatment was subsequently extended by Onsager. He derived the following expression for the molar conductivity Λ_m of a very dilute solution of a strong electrolyte;

$$\Lambda_m = \Lambda_m^\infty - \left[\frac{(2.8\times10^6)\, z_+|z_-|q}{\{\varepsilon_r(T/\mathrm{K})\}^{3/2}(1+\sqrt{q})}\Lambda_m^\infty + \frac{(41.25\ \mathrm{S\ cm^2\ mol^{-1}})\,(z_+ + |z_-|)}{(\eta/\eta^\circ)\{\varepsilon_r(T/\mathrm{K})\}^{1/2}}\right] \times \left(\frac{z_+ + |z_-|}{2}\right)^{1/2} (c/c^\circ)^{1/2} \quad (5.7.5)$$

where z_+ and $|z_-|$ are the charge number on cation and magnitude of charge number on anion, respectively, ε_r is the relative permittivity of the solvent, c° is 1 mol dm^{-3}, η is the viscosity of the solvent, η° is unit viscosity and q is given as

$$q = \frac{z_+|z_-|(\lambda_+^\infty + \lambda_-^\infty)}{(z_+ + |z_-|)(z_+\lambda_-^\infty + |z_-|\lambda_+^\infty)} \quad (5.7.6)$$

where λ_+^∞ and λ_-^∞ are the limiting ionic conductivities of cation and anion, respectively.

Equation (5.7.5) may be written as

$$\Lambda_m = \Lambda_m^\infty - (A\Lambda_m^\infty + B)\,(c/c^\circ)^{1/2} \quad (5.7.7)$$

The first term $A\Lambda_m^\infty\sqrt{(c/c^\circ)}$ is the decrease in molar conductivity due to the asymmetric effect. The appearance of ε_r and T in the denominator can be easily interpreted. Since, on increasing ε_r and T, the interionic attraction is decreased, there will, therefore, be less decrease in molar conductivity.

The second term $B\sqrt{(c/c^\circ)}$ is the decrease in molar conductivity due to the electrophoretic effect. Here η also appears in the denominator. This is because of the fact that the motion of both the atmosphere and the ion is slowed down in a very viscous solvent and correspondingly the decrease in conductivity will be less.

It may be mentioned here that Eq. (5.7.5) is applicable only to very dilute solutions and may be considered as essentially a limiting equation for conductivity. For the special case of 1–1 electrolytes (uni-univalent) in water at 25 °C, for which $z_+ = |z_-| = 1$, $\varepsilon_r = 80$ and $\eta = 0.008\,9$ poise, the above equation reduces to

$$\Lambda_m = \Lambda_m^\infty - (0.23\Lambda_m^\infty + 60.22\ \mathrm{S\ cm^2\ mol^{-1}})\sqrt{c/(\mathrm{mol\ dm^{-3}})} \quad (5.7.8)$$

Comparison of the experimental data with the values predicted by the Onsager equation reveals that the agreement is usually excellent in very dilute solutions up to about 0.02 M. In more concentrated solutions, the molar conductivity is usually higher than the value predicted from the Onsager equation.

Example 5.7.2

At 25 °C the constants in the Onsager equation are $A = 0.2289$ and $B = 60.19$ S cm^2 mol^{-1}. Calculate the molar conductivities of 0.01 M HCl solution and 0.01 M KCl solution. Given:

$$\lambda_m^\infty(H^+) = 349.8 \text{ S cm}^2 \text{ mol}^{-1}, \quad \lambda_m^\infty(Cl^-) = 76.35 \text{ S cm}^2 \text{ mol}^{-1},$$

and $\lambda_m^\infty(K^+) = 73.52$ S cm^2 mol^{-1}.

Solution

Molar conductivity of HCl at infinite dilution is

$$\Lambda_m^\infty(HCl) = \lambda_m^\infty(H^+) + \lambda_m^\infty(Cl^-) = (349.8 + 76.35) \text{ S cm}^2 \text{ mol}^{-1}$$

$$= 426.15 \text{ S cm}^2 \text{ mol}^{-1} \qquad (\textit{see}, \text{Eq. 5.9.2})$$

The value of Λ_m(HCl) from Onsager equation is

$$\Lambda_m = \Lambda_m^\infty - (A\Lambda_m^\infty + B)(c/c^\circ)^{1/2}$$

$$= (426.15 \text{ S cm}^2 \text{ mol}^{-1}) - \{(0.228\,9)(426.15 \text{ S cm}^2 \text{ mol}^{-1}) + (60.19 \text{ S cm}^2 \text{ mol}^{-1})\}(0.01)^{1/2}$$

$$= (426.15 - 15.78) \text{ S cm}^2 \text{ mol}^{-1} = 410.37 \text{ S cm}^2 \text{ mol}^{-1}$$

Now, for $\Lambda_m^\infty(KCl) = (73.52 + 76.35)$ S cm^2 mol^{-1} = 149.87 S cm^2 mol^{-1}

$$\Lambda_m(KCl) = \Lambda_m^\infty(KCl) - \{A\Lambda_m^\infty(KCl) + B\}(c/c^\circ)^{1/2}$$

$$= [149.87 - \{0.228\,9 \times 149.87 + 60.19\}\sqrt{0.01}\,] \text{ S cm}^2 \text{ mol}^{-1}$$

$$= (149.87 - 9.45) \text{ S cm}^2 \text{ mol}^{-1} = 140.42 \text{ S cm}^2 \text{ mol}^{-1}$$

5.8 CONDUCTIVITY AT HIGH ELECTRIC FIELDS AND HIGH FREQUENCIES

The evidence that there exists an ionic atmosphere around the ions is shown by the Wien effect and Debye-Falkenhagen effect. These effects are discussed in the following.

Wien Effect

In very high electric fields, $E > 10^5$ V/cm, an increase in conductivity is observed. This results from the fact that a finite time (the relaxation time) is required for the ionic atmosphere to form around an ion. In very high electric fields, the ion moves so rapidly that it effectively loses its atmosphere. As a result there is no time for the formation of atmosphere to slow down the ionic motion.† The asymmetry effect disappears which results in the increase of conductance.

† The relaxation time, i.e. the time required to form an ionic atmosphere is of the order of 10^{-9} s. The average thickness of an ionic atmosphere is of the order of 10^{-8} cm. The ion at the very high electric field moves with a velocity of the order of a metre per second. Thus, time required to pass through the thickness of an ionic atmosphere is of the order of 10^{-10} s. Hence within the relaxation time of 10^{-9} s, the ion covers a distance equal to ten times the thickness of an ionic atmosphere. In other words, the ion moves as if there exists no ionic atmosphere.

Debye-Falkenhagen Effect

Here again the conductivity increases when high frequency AC is used (of the order of 3×10^6 Hz). The ion changes its direction of motion so quickly that the more sluggish atmosphere cannot adjust and follow the motion of the ion. The ion moves as if it had no atmosphere, and thus the conductivity increases. At high frequency, both the asymmetry and electrophoretic effects are absent.

5.9 KOHLRAUSCH'S LAW OF INDEPENDENT MIGRATION OF IONS

Experimental Observations

For a strong electrolyte, the value of Λ_m in a very dilute solution, say in 0.001 or 0.000 1 M solution, is very close to the limiting value of the conductivity Λ_m^∞ at infinite dilution (or at zero concentration obtained by extrapolation). On the other hand, the corresponding value for a weak electrolyte is very far away from the limiting value at zero concentration. For example, at 25 °C for 0.001 M sodium chloride solution Λ_m is 123.7 S cm^2 mol^{-1} as against Λ_m^∞ of 126.5 S cm^2 mol^{-1}. At the same concentration and temperature, the value for acetic acid is 49.2 S cm^2 mol^{-1} as compared to 390.7 S cm^2 mol^{-1} for the value of Λ_m^∞.

Kohlrausch was the first to point out that when Λ_m for a uni-univalent strong electrolyte is plotted against $\sqrt{c}$, the curve approaches linearity in dilute solutions, i.e.

$$\Lambda_m = \Lambda_m^\infty - b\sqrt{c} \tag{5.9.1}$$

where b is constant. The value of Λ_m^∞ can, thus, be obtained by extrapolating the above curve to a value of $\sqrt{c} = 0$.

Statement of Kohlrausch's Law

The extrapolation method cannot be employed for a weak electrolyte as Λ_m versus $\sqrt{c}$ curve does not approach linearity in solutions as dilute as 0.000 1 M. In fact, the variation of Λ_m with dilution is very rapid. However, the value of Λ_m^∞ for a weak electrolyte can be determined by the application of *Kohlrausch's law of independent migration of ions*. This law states that at *infinite dilution, where dissociation for all electrolytes is complete* (including weak electrolytes since $\alpha \rightarrow 1$ as $c \rightarrow 0$; Ostwald dilution law) *and where all interionic effects disappear* (because of larger distance between ions), *each ion migrates independently of its co-ion and contributes to the total molar conductivity of an electrolyte a definite share which depends only on its own nature and not at all on the ion with which it is associated.* Thus, Λ_m^∞ of the electrolyte must be equal to the sum of the molar conductivities of the ions composing it. Thus

$$\Lambda_m^\infty(AB) = \lambda_m^\infty(A^+) + \lambda_m^\infty(B^-) \tag{5.9.2}$$

Experimental Evidences

Evidence for the validity of the above law can be seen from the data of Λ^∞ given in Table 5.9.1.

According to Kohlrausch's law, we should have

$$\begin{aligned}\Lambda^\infty(AB) - \Lambda^\infty(CB) &= \lambda^\infty(A^+) + \lambda^\infty(B^-) - \lambda^\infty(C^+) - \lambda^\infty(B^-) \\ &= \lambda^\infty(A^+) - \lambda^\infty(C^+)\end{aligned}$$

Table 5.9.1 Evidence for Kohlrausch's Law

Electrolyte	Λ^{∞}/S cm² mol⁻¹ at 25 °C	*Difference* $\Delta\Lambda^{\infty}$/S cm² mol⁻¹	*Electrolyte*	Λ^{∞}/S cm² mol⁻¹ at 25 °C	*Difference* $\Delta\Lambda^{\infty}$/S cm² mol⁻¹
KCl	149.9	34.9	HCl	426.2	4.9
LiCl	115.0		HNO_3	421.3	
KNO_3	145.0	34.9	KCl	149.9	4.9
$LiNO_3$	110.1		KNO_3	145.0	
KOH	271.5	34.8	LiCl	115.0	4.9
LiOH	236.7		$LiNO_3$	110.1	

and
$$\Lambda^{\infty}(AB) - \Lambda^{\infty}(AC) = \lambda^{\infty}(A^+) + \lambda^{\infty}(B^-) - \lambda^{\infty}(A^+) - \lambda^{\infty}(C^-)$$
$$= \lambda^{\infty}(B^-) - \lambda^{\infty}(C^-)$$

that is, the difference between the Λ^{∞}s of electrolytes containing a common ion should be equal to the difference in molar conductivities of ions not in common. The difference of Λ^{∞} of different pairs of uni-univalent electrolytes having either cation or anion common and having the same two anions or cations uncommon should yield a constant value. In line with this requirement, we see that irrespective of the nature of the co-ion, the difference between the conductivities of K^+ and Li^+ is constant (= 34.9 S cm² mol⁻¹) and the same is true for the difference in conductivities between Cl^- and NO_3^- ions (= 4.9 S cm² mol⁻¹).

Illustration

Based on Kohlrausch's law, it is possible to calculate the molar conductivity of a weak electrolyte at infinite dilution by simply adding and subtracting Λ^{∞} values of appropriate electrolytes. Taking, for example, acetic acid, we have

$$\Lambda^{\infty}(HAc) = \Lambda^{\infty}(NaAc) + \Lambda^{\infty}(HCl) - \Lambda^{\infty}(NaCl)$$
$$= \lambda^{\infty}(Na^+) + \lambda^{\infty}(Ac^-) + \lambda^{\infty}(H^+) + \lambda^{\infty}(Cl^-) - \lambda^{\infty}(Na^+) - \lambda^{\infty}(Cl^-)$$
$$= \lambda^{\infty}(H^+) + \lambda^{\infty}(Ac^-) = \lambda^{\infty}(HAc)$$

For an electrolyte other than uni-univalent, the molar conductivity can be evaluated by using the expression

$$\Lambda_m^{\infty} = \nu_+ \lambda_+^{\infty} + \nu_- \lambda_-^{\infty}$$

where ν_+ and ν_- are the respective stoichiometric numbers of cation and anion of the electrolyte.

5.10 VALUES OF LIMITING IONIC MOLAR CONDUCTIVITIES

Since the fraction of the total current carried by any ion is given by its transport number (*see* Section 5.11), the latter must also represent the fraction of the total conductivity due to the ion. Consequently λ_+^{∞} and λ_-^{∞} are also related to Λ_m^{∞} by the relations

$$\nu_+ \lambda_+^\infty = t_+^\infty \Lambda_m^\infty \quad \text{and} \quad \nu_- \lambda_-^\infty = t_-^\infty \Lambda_m^\infty \qquad (5.10.1)^\dagger$$

where t_+^∞ and t_-^∞ are the respective transport numbers at infinite dilution as obtained by extrapolation. These equations permit the calculation of the limiting molar ionic conductivities from transport numbers and Λ_m^∞ values of strong electrolytes.

Illustration

For example, for hydrochloric acid at 25 °C Λ_m^∞ is 426.14 S cm² mol⁻¹, while t_+^∞ of the hydrogen ion is 0.820 9. Therefore,

$$\lambda^\infty(H^+) = 0.820\,9 \times 426.14 \text{ S cm}^2 \text{ mol}^{-1} = 349.82 \text{ S cm}^2 \text{ mol}^{-1}$$

$$\lambda^\infty(Cl^-) = 0.179\,1 \times 426.14 \text{ S cm}^2 \text{ mol}^{-1} = 76.32 \text{ S cm}^2 \text{ mol}^{-1}$$

Values of Molar Conductivities

The molar ionic conductivities of other ions have similarly been evaluated. The molar ionic conductivities of ions of weak acids or bases were deduced from the Λ^∞s of their salts, which are strong electrolytes. These values for some of the ions are given in Table 5.10.1.

Λ^∞ for Strong and Weak Electrolytes

The values of Λ^∞ for strong and weak electrolytes can be obtained by adding the appropriate molar ionic conductivities of the cation and anion, e.g., for acetic acid, we have

$$\begin{aligned}\Lambda^\infty(\text{HAc}) &= \lambda^\infty(H^+) + \lambda^\infty(Ac^-)\\ &= 349.8 \text{ S cm}^2 \text{ mol}^{-1} + 40.9 \text{ S cm}^2 \text{ mol}^{-1}\\ &= 390.7 \text{ S cm}^2 \text{ mol}^{-1}\end{aligned}$$

An examination of the values of molar ionic-conductivities as given in Table 5.10.1 brings forward the following two facts:

Comparison of of Alkali-Metal Ions Molar Conductivities

It is interesting to compare the molar conductivities of the alkali-metal ions. The values are

Ion	Li^+	Na^+	K^+	Rb^+	Cs^+
λ_+^∞/S cm² mol⁻¹	38.66	50.11	73.52	77.8	77.3

If we insist on Stokes' law interpretation of these values by the equation

$$\lambda_+^\infty = \frac{Fez_+}{6\pi\eta r_+}$$

---

†In general, we have

$$\lambda_m^{(+)} = z_+ \lambda_{eq}^{(+)} \quad \text{and} \quad \lambda_m^{(-)} = |z_-|\,\lambda_{eq}^{(-)}$$

$$\begin{aligned}\Lambda_m &= \nu_+\lambda_m^{(+)} + \nu_-\lambda_m^{(-)} = \nu_+\,[z_+\lambda_{eq}^{(+)}] + \nu_-\,\{|z_-|\,\lambda_{eq}^{(-)}\}\\ &= \nu_+ z_+\,\{\lambda_{eq}^{(+)} + \lambda_{eq}^{(-)}\} = \nu_+ z_+ \Lambda_{eq}\end{aligned}$$

$$t_+ = \frac{\nu_+\lambda_m^{(+)}}{\Lambda_m} = \frac{\nu_+ z_+ \lambda_{eq}^{(+)}}{\nu_+ z_+ \Lambda_{eq}} = \frac{\lambda_{eq}^{(+)}}{\Lambda_{eq}} \quad \text{and} \quad t_- = \frac{\nu_-\lambda_m^{(-)}}{\Lambda_m} = \frac{\nu_-|z_-|\lambda_{eq}^{(-)}}{\nu_+ z_+ \Lambda_{eq}} = \frac{\lambda_{eq}^{(-)}}{\Lambda_{eq}}$$

Table 5.10.1 Molar Ionic Conductivities at Infinite Dilution at 25 °C†

Ions	λ_+^∞/S cm^2 mol^{-1}	*Ions*	λ_+^∞/S cm^2 mol^{-1}	*Ions*	λ_-^∞/S cm^2 mol^{-1}	*Ions*	λ_-^∞/S cm^2 mol^{-1}
H^+	349.8	$\frac{1}{2}Hg^{2+}$	63.6	OH^-	197.6	$H_2AsO_4^-$	34
Li^+	38.66	$\frac{1}{2}Zn^{2+}$	52.8	F^-	55.4	HS^-	65
Na^+	50.11	$\frac{1}{2}Cd^{2+}$	54	Cl^-	76.34	HSO_3^-	58
K^+	73.52	$\frac{1}{2}Pb^{2+}$	70	Br^-	78.14	HSO_4^-	52
Rb^+	77.8	$\frac{1}{2}Mn^{2+}$	53.5	I^-	76.97	$\frac{1}{2}CO_3^{2-}$	69.3
Cs^+	77.3	$\frac{1}{2}Co^{2+}$	55	CN^-	82	$\frac{1}{2}HPO_4^{2-}$	57
NH_4^+	73.4	$\frac{1}{2}Ni^{2+}$	54	CNS^-	66	$\frac{1}{2}SO_3^{2-}$	72
Ag^+	61.92	$\frac{1}{3}Fe^{3+}$	68	NO_3^-	71.44	$\frac{1}{2}SO_4^{2-}$	80
$\frac{1}{2}Mg^{2+}$	53.06	$\frac{1}{3}Al^{3+}$	63	ClO_2^-	52	$\frac{1}{2}S_2O_3^{2-}$	87.4
$\frac{1}{2}Ca^{2+}$	59.50	$\frac{1}{3}Cr^{3+}$	67	ClO_3^-	64.6	$\frac{1}{2}CrO_4^{2-}$	83
$\frac{1}{2}Sr^{2+}$	59.46	$\frac{1}{3}Co(NH_3)_6^{3+}$	99.2	ClO_4^-	67.4	$\frac{1}{3}PO_4^{3-}$	92.8
$\frac{1}{2}Ba^{2+}$	63.64	$N(CH_3)_4^+$	44.92	BrO_3^-	55.7	$\frac{1}{3}Fe(CN)_6^{3-}$	99.1
$\frac{1}{2}Cu^{2+}$	56.6	$NH(CH_3)_3^+$	42	IO_3^-	40.7	$\frac{1}{4}Fe(CN)_6^{4-}$	111
$\frac{1}{2}Hg_2^{2+}$	68.6			IO_4^-	54.5	Formate$^-$	54.6
				$\frac{1}{2}$ Oxalate^{2-}	74.1	Acetate$^-$	40.9
				$\frac{1}{2}$ Tartrate^{2-}	59.6	Monochloro acetate$^-$	39.8
				$\frac{1}{3}$ Citrate^{3-}	71.5	Dichloro acetate$^-$	38
				MnO_4^-	61	Trichloro acetate$^-$	35
				HCO_3^-	44.5	Benzoate$^-$	32.3
				$H_2PO_4^-$	36	*n*-Propionate$^-$	35.8

†It is a standard practice to report molar conductivity of the entity $\frac{1}{|z_B|}B^z$, where $|z_B|$ is the magnitude of charge number of the species B carrying a charge *z*.

we would be forced to conclude that the radius of Li^+ ion is larger than that of Cs^+ ion. In fact, the crystallographic radius of Li^+ is much smaller than that of Cs^+. However, in solution, these ions are solvated. The number of water molecules attached to any ion depends upon its charge density. Larger the charge density of an ion, larger the number of water molecules attached to the ion. Since the crystallographic radii of the ions follow the order

$$Li^+ < Na^+ < K^+ < Rb^+$$

the charge densities follow the order:

$$Rb^+ < K^+ < Na^+ < Li^+$$

Since Li^+ ion has the maximum charge density, it follows that it will be more extensively hydrated than the rest of the ions, with the result that it has the largest ionic radius. Thus, the actual ionic radii of these ions in solution follow the order

$$Li^+ > Na^+ > K^+ > Rb^+$$

and, therefore, the order of molar ionic conductivities should be

$$Li^+ < Na^+ < K^+ < Rb^+$$

which is the correct order of the observed molar ionic conductivities.

Example 5.10.1 The crystallographic radii of Na^+ and Cl^- are 95 pm and 181 pm. Estimate the molar ionic conductivities using Stokes law, $\lambda = |z|eF/6\pi\eta r$. given: $\eta = 0.89 \times 10^{-2}$ poise. Compare the values obtained with the experimental values of $\lambda(Na^+) = 50.11$ S cm^2 mol^{-1} and $\lambda(Cl^-) = 76.35$ S cm^2 mol^{-1}.

Solution We have $\eta = 0.89 \times 10^{-2}$ poise $= 0.89 \times 10^{-3}$ kg m^{-1} s^{-1}.

Substituting the values of $|z|$, e, F, η and r for the Na^+ in Stokes law

$$\lambda(Na^+) = \frac{|z|eF}{6\pi\eta r}$$

we get

$$\lambda(Na^+) = \frac{(1)(1.602\,19 \times 10^{-19}\ \text{C})(96\,500\ \text{C mol}^{-1})}{6(3.14)(0.89 \times 10^{-3}\ \text{kg m}^{-1}\ \text{s}^{-1})(95 \times 10^{-12}\ \text{m})}$$

$$= 97.06 \times 10^{-4}\ \text{C}^2\ \text{mol}^{-1}\ \text{kg}^{-1}\ \text{s}$$

$$\equiv 97.06 \times 10^{-4}\ \text{S m}^2\ \text{mol}^{-1} \equiv 97.06\ \text{S cm}^2\ \text{mol}^{-1}$$

(Experimental value is 50.11 S cm^2 mol^{-1}.)

and

$$\lambda(Cl^-) = \frac{(1)(1.602\,19 \times 10^{-19}\ \text{C})(96\,500\ \text{C})}{6(3.14)(0.89 \times 10^{-3}\ \text{kg m}^{-1}\ \text{s}^{-1})(181 \times 10^{-12}\ \text{m})}$$

$$= 50.94 \times 10^{-4}\ \text{S m}^2\ \text{mol}^{-1} = 50.94\ \text{S cm}^2\ \text{mol}^{-1}$$

(Experimental value is 76.35 S cm^2 mol^{-1}.)

Abnormally High Molar Conductivities of H^+ and OH^-

The molar conductivities of the hydrogen ion and the hydroxyl ion are much larger than those of other ions. This was first explained by von Grotthus and hence is known as *Grotthus conductance*. It is explained on the basis of a proton jump from one water molecule to another. The process of proton transfer results in a more rapid transfer of positive charge from one region of the solution to another, than would be possible if the ion H_3O^+ has to push its way through the solution as other ions do. The mechanisms of conduction of H^+ ion and OH^- ion are shown below.

This type of mechanism also prevails in any other solvent. Thus, in a given solvent (for example, liquor ammonia) the molar conductivities of its characteristic cation and anion (namely, NH_4^+ and NH_2^-) will have unusually high values than any other cations and anions.

Effect of Temperature and Pressure on Molar Ionic Conductivities

For most of the ions in water, the value of λ^∞ increases with increase in temperature and this increase is about 2 per cent per Celsius degree rise in temperature. The temperature dependence is expressed as

$$\lambda^\infty = \lambda^\infty_{25\,°C}\,[1 + a\{(t/°C) - 25\}]$$

where a is constant (known as the *temperature coefficient*) and t is temperature. For most of the ions, the temperature coefficient has a value of 0.02 with the exceptions of H^+ and OH^- for which the values are 0.014 and 0.016, respectively. This is probably due to the difference in the conduction mechanism. The temperature dependence results from the decrease in viscosity of water with temperature which amounts to about 2 per cent per Celsius degree.

The value of λ^∞ decreases with increase in pressure which again is a consequence of increase in viscosity of water with pressure.

Example 5.10.2

The following data are obtained in an experiment to determine the dissociation constant of the weak acid HA.

Solution	*Conductivity*
0.01 M HA	3.8×10^{-5} S cm^{-1}
100 cm^3 of 0.01 of M HA + 1 cm^3 of 1.0 M NaOH	80.0×10^{-5} S cm^{-1} (neglecting dilution)

Calculate $\Lambda_m(\text{NaA})$, $\lambda_m(\text{A}^-)$ and $\Lambda_m(\text{H}^+\text{A}^-)$ and the degree of dissociation of 0.01 M HA. Given: $\lambda_m(\text{Na}^+) = 50\ \text{S cm}^2\ \text{mol}^{-1}$ and $\lambda_m(\text{H}^+) = 350\ \text{S cm}^2\ \text{mol}^{-1}$.

Solution

On mixing 100 cm^3 of 0.01 M HA solution and 1 cm^3 of 1.0 M NaOH solution, we get the salt formation NaA, the concentration of which is 0.01 M. The molar conductivity of this solution is

$$\Lambda_m(\text{NaA}) = \frac{\kappa}{c} = \frac{(80 \times 10^{-5}\ \text{S cm}^{-1})}{(0.01\ \text{mol dm}^{-3})} = 80 \times 10^{-3}\ \text{S cm}^{-1}\ \text{dm}^3\ \text{mol}^{-1}$$

$$= 80 \times 10^{-3}\ \text{S cm}^{-1}\ (10\ \text{cm})^3\ \text{mol}^{-1} = 80\ \text{S cm}^2\ \text{mol}^{-1}$$

Since the salt formed is a strong electrolyte and the solution is fairly diluted, this value of $\Lambda_m(\text{NaA})$ can be assumed to be $\Lambda_m^\infty(\text{NaA})$.

Hence, according to Kohlrausch law, we have

$$\Lambda_m^\infty(\text{NaA}) = \lambda_m^\infty(\text{Na}^+) + \lambda_m^\infty(\text{A}^-)$$

Since $\lambda_m^\infty(\text{Na}^+) = 50\ \text{S cm}^2\ \text{mol}^{-1}$, therefore, $\lambda_m^\infty(\text{A}^-) = 30\ \text{S cm}^2\ \text{mol}^{-1}$

The molar conductivity at infinite dilution of the weak acid is

$$\Lambda_m^\infty(\text{H}^+\text{A}^-) = \lambda_m^\infty(\text{H}^+) + \lambda_m^\infty(\text{A}^-)$$

$$= 350\ \text{S cm}^2\ \text{mol}^{-1} + 30\ \text{S cm}^2\ \text{mol}^{-1} = 380\ \text{S cm}^2\ \text{mol}^{-1}$$

The apparent molar conductivity of 0.01 M HA will be

$$\Lambda_m(\text{HA}) = \frac{\kappa}{c} = \frac{(3.8 \times 10^{-5}\ \text{S cm}^{-1})}{(0.01\ \text{mol dm}^{-3})}$$

$$= 3.8 \times 10^{-3}\ \text{S cm}^{-1}\ \text{dm}^3\ \text{mol}^{-1} \equiv 3.8\ \text{S cm}^2\ \text{mol}^{-1}$$

$$\alpha = \frac{\Lambda(\text{HA})}{\Lambda^\infty(\text{HA})} = \frac{(3.8\ \text{S cm}^2\ \text{mol}^{-1})}{(380\ \text{S cm}^2\ \text{mol}^{-1})} = 0.01$$

Example 5.10.3

Calculate the molar and equivalent conductivities at infinite dilution of the salt KOOC·COONa. Given the molar ionic conductivities at infinite dilution of Ox^{2-}, K^+ and Na^+ as 148.2, 50.1 and 73.5 S cm^2 mol^{-1}, respectively.

Solution

We have

$$\Lambda_m^\infty(\text{KOOC}\cdot\text{COONa}) = \Lambda_m^\infty(\text{K}^+) + \Lambda_m^\infty(\text{Na}^+) + \Lambda_m^\infty(\text{Ox}^{2-})$$

$$= (50.1 + 73.5 + 148.2)\ \text{S cm}^2\ \text{mol}^{-1}$$

$$= 271.8\ \text{S cm}^2\ \text{mol}^{-1}$$

Since 1 mole of KOOC·COONa contains 2 equivalents of the salt, the equivalent conductivity of this salt will be half of the molar conductivity, i.e.

$$\Lambda_{eq}^\infty(\text{KOOC}\cdot\text{COONa}) = [\tfrac{1}{2}\lambda_m^\infty(\text{K}^+) + \tfrac{1}{2}\lambda_m^\infty(\text{Na}^+) + \tfrac{1}{2}\lambda_m^\infty(\text{Ox}^{2-})]\ \text{mol eq}^{-1}$$

$$= (25.05 + 36.75 + 74.1)\ \text{S cm}^2\ \text{eq}^{-1}$$

$$= 135.9\ \text{S cm}^2\ \text{eq}^{-1}$$

In terms of IUPAC recommendation, it is written as

$$\Lambda_m^\infty(\tfrac{1}{2}\text{KOOC}\cdot\text{COONa}) = 135.9\ \text{S cm}^2\ \text{mol}^{-1}$$

Example 5.10.4

How will you calculate the molar conductivity and equivalent conductivity at infinite dilution of potash alum?

Solution

The formula of potash alum is $K_2SO_4 \cdot Al_2(SO_4)_3 \cdot 24H_2O$. Thus, at infinite dilution 1 mol of potash alum will produce 2 mol of K^+, 2 mol of Al^{3+} and 4 mol of SO_4^{2-}.

Hence, $\Lambda_m^\infty(\text{potash alum}) = 2\lambda_m^\infty(K^+) + 2\lambda_m^\infty(Al^{3+}) + 4\lambda_m^\infty(SO_4^{2-})$

Since total positive (or negative) charge number carried by 1 molecule of potash alum is 8, one equivalent of the salt will be equal to (1/8)th of one mole.

Hence,
$$\Lambda_{eq}^\infty(\text{potash alum}) = (\tfrac{1}{8}\,\text{mol eq}^{-1})\,\Lambda_m^\infty(\text{potash alum})$$
$$= [\tfrac{1}{4}\lambda_m^\infty(K^+) + \tfrac{1}{4}\lambda_m^\infty(Al^{3+}) + \tfrac{1}{2}\lambda_m^\infty(SO_4^{2-})]\,\text{mol eq}^{-1}$$

Since $\lambda_m(SO_4^{2-}) = 2\lambda_m(\tfrac{1}{2}SO_4^{2-})$ and $\lambda_m(Al^{3+}) = 3\lambda_m(\tfrac{1}{3}Al^{3+})$,

we will have

$$\Lambda_{eq}^\infty(\text{potash alum}) = \left[\tfrac{1}{4}\lambda_m^\infty(K^+) + \tfrac{3}{4}\lambda_m^\infty\left(\tfrac{1}{3}Al^{3+}\right) + \lambda_m\left(\tfrac{1}{2}SO_4^{2-}\right)\right]\text{mol eq}^{-1}$$

i.e $\Lambda_{eq}^\infty(\text{potash alum}) = \tfrac{1}{4}\lambda_{eq}^\infty(K^+) + \tfrac{3}{4}\lambda_{eq}^\infty(Al^{3+}) + \lambda_{eq}^\infty(SO_4^{2-})$

Example 5.10.5

The values of Λ_m^∞ for HCl, NaCl and NaAc (sodium acetate) are 420, 126 and 91 S cm^2 mol^{-1}, respectively. The resistance of a conductivity cell is 520 Ω when filled with 0.1 M acetic acid and drops to 122 Ω when enough NaCl is added to make the solution 0.1 M in NaCl as well. Calculate the cell constant and hydrogen-ion concentration of the solution. Given:

$\Lambda_m^\infty(\text{HCl}) = 420\ \text{S cm}^2\ \text{mol}^{-1}$; $\Lambda_m^\infty(\text{NaCl}) = 126\ \text{S cm}^2\ \text{mol}^{-1}$; and $\Lambda_m^\infty(\text{NaAc}) = 91\ \text{S cm}^2\ \text{mol}^{-1}$.

Solution

Resistance of 0.1 M HAc = 520 Ω

Resistance of 0.1 M HAc + 0.1 M NaCl = 122 Ω

Conductance due to 0.1 M NaCl,

$$G = \frac{1}{122\ \Omega} - \frac{1}{520\ \Omega} = 0.006\,27\ \text{S}$$

Conductivity of 0.1 M NaCl solution

$$\kappa = \Lambda_m\, c = (126\ \text{S cm}^2\ \text{mol}^{-1})\,(0.1\ \text{mol dm}^{-3})$$
$$= 12.6\ \text{S cm}^2\ \text{dm}^{-3} = 12.6\ \text{S cm}^2\ (10\ \text{cm})^{-3} = 0.012\,6\ \text{S cm}^{-1}$$

Cell constant, $K_{cell} = \dfrac{\kappa}{G} = \dfrac{(0.012\,6\ \text{S cm}^{-1})}{(0.006\,27\ \text{S})} = 2.01\ \text{cm}^{-1}$

Conductivity of 0.1 M HAc solution

$$\kappa = \frac{K_{\text{cell}}}{R} = \frac{2.01\ \text{cm}^{-1}}{520\ \Omega} = (2.01/520)\ \text{S cm}^{-1}$$

Molar conductivity of 0.1 M HAc solution

$$\Lambda_{\text{m}}(\text{HAc}) = \frac{\kappa}{c} = \frac{(2.01/520)\ \text{S cm}^{-1}}{(0.1\ \text{mol dm}^{-3})}$$

$$= 0.038\,65\ \text{S cm}^{-1}\ \text{dm}^{3}\ \text{mol}^{-1} = 38.65\ \text{S cm}^{2}\ \text{mol}^{-1}$$

According to Kohlrausch law, $\Lambda_{\text{m}}^{\infty}(\text{HAc})$ is given by

$$\Lambda_{\text{m}}^{\infty}(\text{HAc}) = \Lambda_{\text{m}}^{\infty}(\text{HCl}) + \Lambda_{\text{m}}^{\infty}(\text{NaAc}) - \Lambda_{\text{m}}^{\infty}(\text{NaCl})$$

$$= (420 + 91 - 126)\ \text{S cm}^{2}\ \text{mol}^{-1} = 385\ \text{S cm}^{2}\ \text{mol}^{-1}$$

Therefore, the degree of dissociation of acetic acid is given as

$$\alpha = \frac{\Lambda_{\text{m}}}{\Lambda_{\text{m}}^{\infty}} = \frac{(38.65\ \text{S cm}^{2}\ \text{mol}^{-1})}{(385\ \text{S cm}^{2}\ \text{mol}^{-1})} \approx 0.1$$

and the hydrogen-ion concentration of 0.1 M HAc solution is

$$[\text{H}^+] = c\alpha = (0.1\ \text{M})\ (0.1) = 0.01\ \text{M}$$

Thus, its pH is $\text{pH} = -\log\{[\text{H}^+]/\text{M}\} = -\log(0.01) = 2$

5.11 TRANSPORT NUMBERS

Although the current is transported through a solution by the migration of positive and negative ions, the fraction of the total current carried by each ion is not necessarily the same. For example, in dilute solutions of $MgSO_4$, the Mg^{2+} ions carry about 0.38 of the total current, while the SO_4^{2-} ions carry rest of the current, i.e. 0.62. In dilute HNO_3 solution, the nitrate ions carry only 0.16 of the total current while the hydrogen ions carry 0.84. The SO_4^{2-} and H^+ ions transport greater fraction of the total current because in their respective solutions, they move faster than the other ions present. If both ions in a solution move with the same speed, each would transfer the same quantity of electricity in any given time across any fixed plane in the solution. However, when the speeds of two ions are not the same, then in any given period of time, the faster ion will carry a greater fraction of the total current than the slower one.

The quantitative relation between the fraction of the current carried by an ion and its speed can be derived as follows.

Consider two parallel plates at a distance d apart, across which a potential of $\Delta\phi$ is applied and between these two electrodes some volume of an electrolytic solution is placed (Fig. 5.11.1).

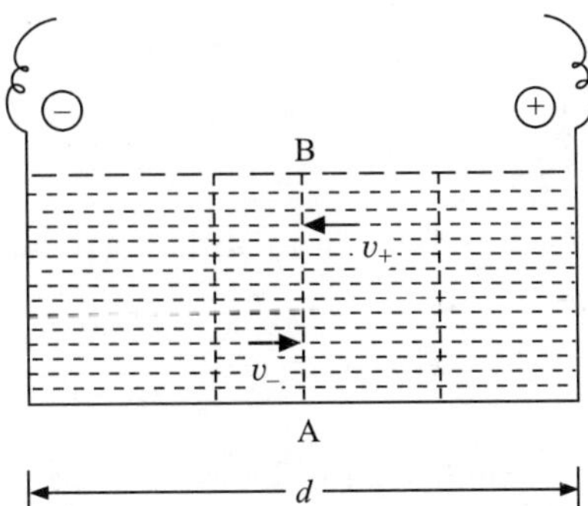

Fig. 5.11.1 Migration of ions across a plane

Let in the solution we have:

Average migration speed of each cation = v_+
Charge number on each cation = z_+
Total number of cations = N_+
Average migration speed of each anion = v_-
Magnitude of charge number on each anion = $|z_-|$
Total number of anions = N_-

Condition for electroneutrality requires that

$$N_+ z_+ = N_- |z_-| \tag{5.11.1}$$

The quantity of electric charge transported by the cations in a unit time across a plane AB will be equal to the total electric charge possessed by all the cations that lie within a distance v_+ from this plane. Thus,

Number of cations within a distance v_+ from this plane = $N_+ \dfrac{v_+}{d}$

Current carried by these ions is

$$I_+ = \frac{N_+ v_+}{d}(z_+ e) \qquad (5.11.2)^\dagger$$

Similarly, the quantity of electric charge transported by the anions in a unit time across the same plane AB will be equal to the total electric charge possessed by all the anions that lie within a distance v_- from this plane. Thus,

Number of anions within a distance v_- from this plane = $N_- \dfrac{v_-}{d}$

Current carried by these ions is

$$I_- = \frac{N_- v_-}{d}(|z_-| e) \qquad (5.11.3)^\dagger$$

The total current carried by these ions is

$$I = I_+ + I_- = \frac{N_+ v_+ z_+ e}{d} + \frac{N_- v_- |z_-| e}{d}$$

† In Eqs (5.11.2) and (5.11.3), e represents the elementary charge (i.e. protonic charge).

Using the condition of electroneutrality as given by Eq. (5.11.1), we have

$$I = \frac{N_+ z_+ e(v_+ + v_-)}{d} \tag{5.11.4}$$

The fraction t_+ of the total current carried by the cations is thus given by

$$t_+ = \frac{I_+}{I} = \frac{N_+ v_+ z_+ e}{N_+ z_+ e(v_+ + v_-)} = \frac{v_+}{(v_+ + v_-)} \tag{5.11.5}$$

Similarly for the anions, we have

$$t_- = \frac{I_-}{I} = \frac{N_- v_- |z_-| e}{N_+ z_+ e(v_+ + v_-)} = \frac{v_-}{(v_+ + v_-)} \tag{5.11.6}$$

Now t_+ and t_- are the transport (or transference) numbers of the cations and anions, respectively. Dividing these two, we get

$$\frac{t_+}{t_-} = \frac{v_+}{v_-} \tag{5.11.7}$$

or $\quad t \propto v$

that is, *the transport numbers of ions are directly proportional to their speeds.* When these are equal, i.e. $t_+ = t_-$, both the ions contribute equally to the transport of the current. However, when v_+ is not equal to v_-, t_+ will not be equal to t_- and the two ions will carry different proportion of the total current.

Since the transport number gives the fraction of electric charges carried by the ions, therefore, it follows that the sum of transport numbers of ions in a solution will be equal to one, i.e. $t_+ + t_- = 1$.

The values of the transport numbers of ions in a solution can be determined experimentally, following either the Hittorf's method or the moving boundary method. These are discussed in the following.

Hittorf's Method for the Determination of Transport Number of Ions

On passing the current through a solution, concentration changes occur in the vicinity of the electrodes. These changes are due to two factors.

Electrode Reactions At cathode, cations accept electrons and are reduced and at anode, anions are oxidized by giving up their electrons. The amount of ions reduced or oxidized on either of the two electrodes is equal to the amount of electrons that is passed through the electrolytic cell divided by the corresponding charge number on the ions.

Migration Effects Within the solution, the charge is carried by the movements of ions. The amount of charge carried by either of the two ions is directly proportional to its ionic speed. Thus, across any plane in the solution, cations move in one direction (towards cathode) and anions move in the opposite direction (towards anode) and the sum of the current carried by the ions is equal to the total amount of the current that is passed through the cell.

In order to understand the nature of the changes caused by the above two factors near the vicinity of electrodes, we consider the following two cases.

Illustration Involving Identical Speeds of Ions

Let the electrolytic cell be divided into the three compartments, namely, the cathodic-, the central- and the anodic-compartment by the two imaginary planes AA′ and BB′ as shown in Fig. 5.11.2.

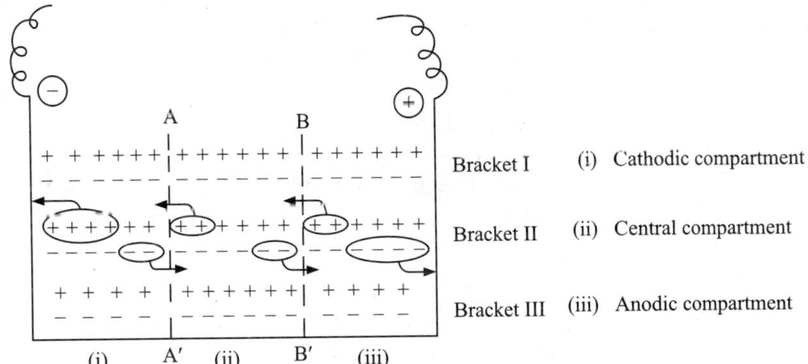

Fig. 5.11.2 Concentration changes when the speeds of ions are same

Let each compartment contain equal amount of uni-univalent electrolyte (say, six mol) and let 4 mol of electrons be passed through the cell. Since the speeds of both the ions in this case are the same, therefore across any plane such as AA′ and BB′, each ion will carry one half of the electric charge, i.e. (2 mol) F of electric charge will be transported by cations from right to left and the equivalent amount will be transported by anions from left to right. The changes that occur after passing 4 mol of electrons are listed below.

Cathodic Compartment

Electrode reaction 4 mol of cations are deposited at the cathode.
Migration effects 2 mol of cations are migrated in from the central compartment and 2 mol of anions are migrated out to the central compartment.

Central Compartment

Electrode reaction Nil.
Migration effects 2 mol of cations are migrated out to the cathodic compartment and the same amount is migrated in from the anodic compartment. Similarly, 2 mol of anions are migrated in from the cathodic compartment and the same amount is migrated out to the anodic compartment.

Anodic Compartment

Electrode reaction 4 mol of anions are deposited at the anode.

Migration effects 2 mol of anions are migrated in from the central compartment and 2 mol of cations are migrated out to the central compartment.

The final result of all these changes are shown in bracket III of Fig. 5.11.2. It can be seen that the concentration in the central compartment remains unchanged while those in the cathodic and anodic compartments suffer changes from six mol to four mol.

Illustration Involving Different Speeds of Ions

Let the speed of the cations be three times larger than that of the anions. It is obvious that the cations will now carry three times as much electric charge as the anions. Thus, out of (4 mol)F of electric charge that is passed, (3 mol)F will be carried by the cations and the remaining (1 mol)F of charge will be carried by the anions. Consequently, across the planes AA′ and BB′, 3 mol of cations will move from right to left and only 1 mol of anions will move from left to right. Brackets II and III of Fig. 5.11.3 depict the changes during and after the passage of (4 mol)F of electric charge. It can be seen that again there is no change in concentration in the central compartment. Moreover, the anodic compartment has suffered a concentration change equal to three times that of the cathodic compartment.

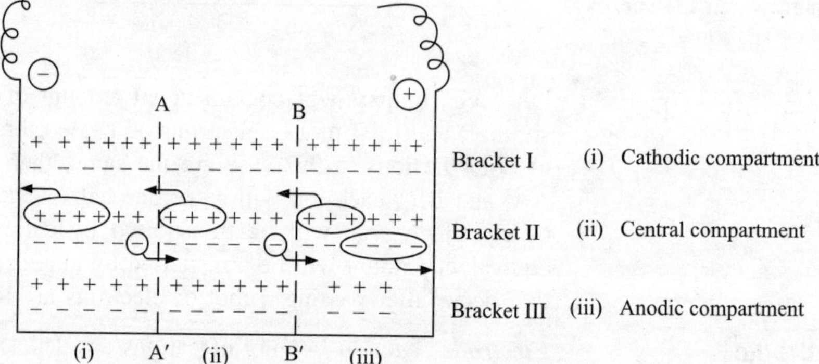

Fig. 5.11.3 Concentration changes when the speed of the cation is three times larger than that of the anion

Conclusion

From the above two examples, the following facts are revealed:

For the case of equal speeds, the loss in concentration of cations due to migration from the anodic compartment is equal to the loss in concentration of anions due to migration from the cathodic compartment. On the other hand, when $v_+ : v_- = 3 : 1$, the loss in concentration of cations from the anodic compartment due to migration is three times that of anions from the cathodic section. This parallelism between concentration loss due to migration and the speed of the ion responsible for it leads to Hittorf's rule, namely,

$$\frac{\text{Loss in the amount of cations at anode due to migration}}{\text{Loss in the amount of anions at cathode due to migration}} = \frac{v_+}{v_-} = \frac{t_+}{t_-}$$

Adding 1 to both the sides, we have

$$\frac{\text{(Loss in the amount of cations at anode due to migration)} + \text{Loss in the amount of anions at cathode due to migration)}}{\text{Loss in the amount of anions at cathode due to migration}} = \frac{t_+ + t_-}{t_-}$$

or

$$\frac{\text{Total amount of electrons passed}}{\text{Loss in the amount of anions at cathode due to migration}} = \frac{1}{t_-}$$

Inverting this, we have

$$\frac{\text{Loss in the amount of anions at cathode due to migration}}{\text{Total amount of electrons passed}} = t_-$$

Similarly,

$$\frac{\text{Loss in the amount of cations at anode due to migration}}{\text{Total amount of electrons passed}} = t_+$$

The above two hypothetical examples can be generalized to include any ionic speed ratio. This is illustrated by the following two examples of the electrolysis of hydrochloric acid and copper sulphate solutions.

ELECTROLYSIS OF HYDROCHLORIC ACID SOLUTION

Consider the electrolytic cell shown in Fig. 5.11.4, divided into the three compartments, namely, the cathodic, the central and the anodic compartments.

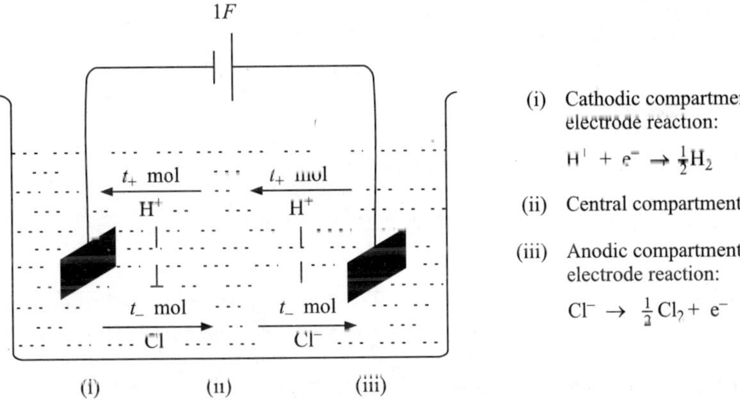

Fig. 5.11.4 Electrolysis of hydrochloric acid solution

The changes that occur in each compartment on passing one mol of electrons are shown in Table 5.11.1.

Table 5.11.1

Cathodic compartment	*Central compartment*	*Anodic compartment*
(i) 1 mol of H^+ removed because of electrode reaction	(i) t_+ mol of H^+ migrated out to the cathodic compartment	(i) 1 mol of Cl^- removed because of electrode reaction
(ii) t_+ mol of H^+ migrated in	(ii) t_+ mol of H^+ migrated in from anodic compartment	(ii) t_+ mol of H^+ migrated out
(iii) t_- mol of Cl^- migrated out	(iii) t_- mol of Cl^- migrated out to the anodic compartment	(iii) t_- mol of Cl^- migrated in
	(iv) t_- mol of Cl^- migrated in from cathodic compartment	

The changes in the three compartments are as follows.

Central Compartment

The concentration in the central compartment remains unchanged by the passage of the current.

Cathodic Compartment

Change in the amount of H^+

$$= \text{Change due to electrode reaction} + \text{Change due to migration effect}$$

$$= -1 \text{ mol} + t_+ \text{ mol} = -t_- \text{ mol}$$

Change in the amount of Cl^-

$$= \text{Change due to electrode reaction} + \text{Change due to migration effect}$$

$$= 0 \text{ mol} - t_- \text{ mol} = -t_- \text{ mol}$$

Thus, the passage of 1 mol of electrons results in the loss of t_- mol of HC1.

Anodic Compartment

Change in the amount of H^+

$$= \text{Change due to electrode reaction} + \text{Change due to migration effect}$$

$$= 0 \text{ mol} - t_+ \text{ mol} = -t_+ \text{ mol}$$

Change in the amount of Cl^-

$$= \text{Change due to electrode reaction} + \text{Change due to migration effect}$$

$$= -1 \text{ mol} + t_- \text{ mol} = -t_+ \text{ mol}$$

This amounts to the loss of t_+ mol of HC1.

Thus, by measuring the change in the concentration of electrolyte either in the cathodic or the anodic compartment, we can find out t_- and t_+ of the ions.

ELECTROLYSIS OF COPPER SULPHATE SOLUTION INVOLVING Pt ELECTRODES

Consider the cell shown in Fig. 5.11.5 divided into the three compartments. On passing 1 mol of electrons, the changes that occur in each of the three compartments are as follows.

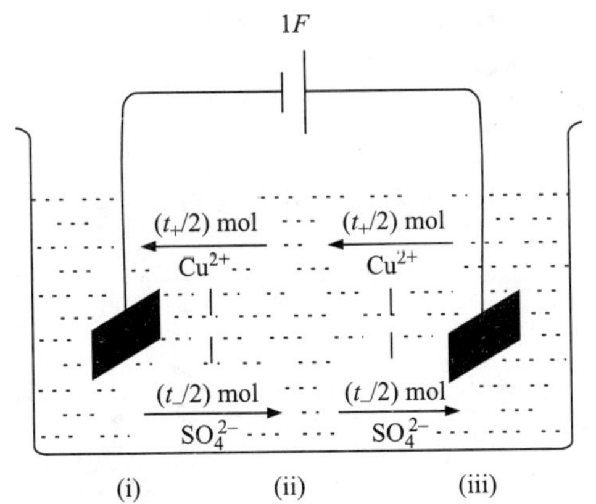

(i) Cathodic compartment electrode reaction:

$\frac{1}{2}Cu^{2+} + e^- \rightarrow \frac{1}{2}Cu$

(ii) Central compartment

(iii) Anodic compartment electrode reaction:

$\frac{1}{2}H_2O \rightarrow \frac{1}{4}O_2 + H^+ + e^-$

Fig. 5.11.5 Electrolysis of copper sulphate solution using Pt-electrodes

Central Compartment No change.

Cathodic Compartment Change in the amount of Cu^{2+}

= Change due to electrode reaction + Change due to migration effect

$= (-1/2)$ mol $+ (t_+/2)$ mol $= -(t_-/2)$ mol

Change in the amount of SO_4^{2-}

= Change due to electrode reaction + Change due to migration effect

$= 0 - (t_-/2)$ mol $= -(t_-/2)$ mol

i.e. there occurs a loss of $(t_-/2)$ mol of $CuSO_4$.

Anodic Compartment Change in the amount of Cu^{2+}

= Change due to electrode reaction + Change due to migration effect

$= 0 - (t_+/2)$ mol $= -(t_+/2)$ mol

Change in the amount of SO_4^{2-}

= Change due to electrode reaction + Change due to migration effect

$= 0 + (t_-/2)$ mol $= (1/2)$ mol $- (t_+/2)$ mol

Change in the amount of H^+

= Change due to electrode reaction + Change due to migration effect

$= 1$ mol $+ 0 = 1$ mol

This amounts to the gain of (1/2) mol of H_2SO_4 and the loss of $(t_+/2)$ mol of $CuSO_4$.

Thus, from the loss of $CuSO_4$ from the cathodic and anodic compartments, the transport numbers can be calculated.

ELECTROLYSIS OF $CuSO_4$ SOLUTION INVOLVING Cu ELECTRODES

When the electrolysis of $CuSO_4$ is done using copper electrodes, the resultant changes in the concentrations of the solution in the vicinity of the electrodes will be different, the analysis of which can, however, be done in a similar manner. Thus, from Fig. 5.11.6, we have

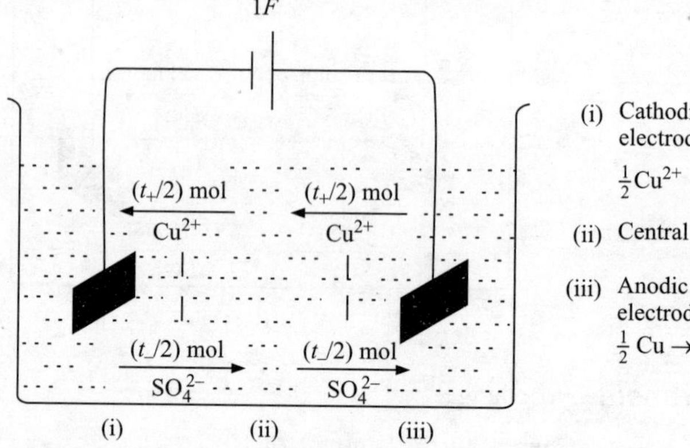

Fig. 5.11.6 Electrolysis of copper sulphate solution using Cu-electrodes

Central Compartment No Change in the concentrations of Cu^{2+} and SO_4^{2-} ions.

Cathodic Compartment Change in the amount of Cu^{2+}

$$= \text{Change due to electrode reaction} + \text{Change due to migration effect}$$

$$= -(1/2)\text{ mol} + (t_+/2)\text{ mol} = -(t_-/2)\text{ mol}$$

Change in the amount of $SO_4^{2-} = -(t_-/2)$ mol

i.e. there occurs a loss of $(t_-/2)$ mol of $CuSO_4$.

Anodic Compartment Change in the amount of Cu^{2+}

$$= \text{Change due to electrode reaction} + \text{Change due to migration effect}$$

$$= +(1/2)\text{ mol} - (t_+/2)\text{ mol} = (t_-/2)\text{ mol}$$

Change in the amount of SO_4^{2-}

$$= \text{Change due to electrode reaction} + \text{Change due to migration effect}$$

$$= 0 + (t_-/2)\text{ mol} = (t_-/2)\text{ mol}$$

i.e. there occurs a gain of $(t_-/2)$ mol of $CuSO_4$.

Thus, from the analysis of the concentrations in either of the compartments, t_- can be found out. The transport number of the cations t_+ can be calculated using the formula $t_+ + t_- = 1$.

Comment It is not necessary to pass only 1 mol of electrons. In fact, we can pass any amount of electrons. If a quantity of electric charge Q (expressed in coulombs) is passed, then the concentration change in either of the electrodes should be divided by Q/F to give the value of transport number.

Conceptual Analysis of Data Sometimes the conceptual analysis of the data is very helpful in determining the transport number of the ion. Consider, for example, the anodic compartment in the electrolysis of copper sulphate solution using copper electrodes. Had there been no transportation, the expected increase of Cu^{2+} ions would have been equal to the half of the amount of electrons passed. But the observed increase of Cu^{2+} ions, which is less than the expected increase, must be due to the migration of some cations from this anodic compartment to the central compartment. Thus, the fraction of the electric charge carried by these migrated cations, which is the transport number of cations, is given by

$$t_+ = \frac{\text{Expected increase} - \text{Actual increase}}{\text{Expected increase}}$$

Experimental Set-up for Determination of Transport Number

Hittorf's Method

The experimental setup for determining the transport numbers of ions by the Hittorf's method is shown in Fig. 5.11.7.

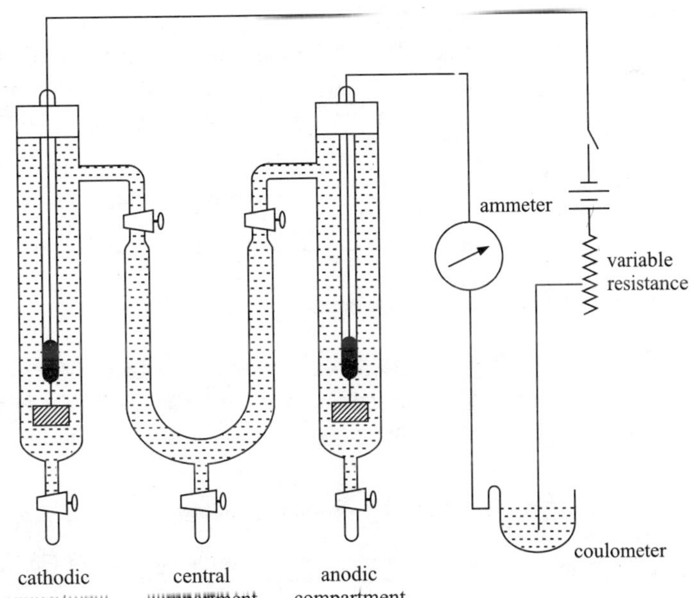

Fig. 5.11.7 Hittorf's apparatus to determine the transport numbers of ions

The three compartments are separated from each other by stopcocks, so that the three portions of the solution can be drawn off separately after the experiment has been performed. The mass and concentration of electrolyte in each portion are measured at the end of the experiment. Knowing the original concentration, we can calculate the change in the amount of electrolyte in each compartment. Analysis of the middle compartment is used as a check to determine if any interfering effects have occurred. The change in the amount of electrolyte in each of the compartments can be related to the transport numbers of the ions by a procedure such as given earlier.

The Hittorf experiment is subject to many practical difficulties. The development of a concentration gradient by the flow of current results in diffusion of the electrolyte from the more concentrated to the less concentrated regions.

This tends to undo the effect to be measured: to minimize diffusion the experiment must not be carried over a long time. On the other hand, if the time is short, the concentration changes are small because a small current must be used. If large currents are used, heating effects occur unevenly and produce convection in the solution; this mixes up the solution again.

Method of Calculations

Since the change in concentration of solution results in change of density, and hence volume, the change of concentration must be determined with reference to a definite mass of the solvent. Therefore, estimations are made with a known mass of solvent and not with a known volume of the solution before and after electrolysis. Take, for example, the cathodic solution of the electrolysis of $CuSO_4$ using Cu-electrodes.

Let the mass a of cathodic solution after electrolysis contain the mass b of $CuSO_4$.

Mass of water in this solution = $(a - b)$

Let the mass c of the solution contain the mass d of $CuSO_4$ before electrolysis. The mass of water in this solution is $(c - d)$. The mass of $CuSO_4$ in the mass $(a - b)$ of water of this solution is given by $d(a - b)/(c - d)$. Thus,

Amount of $CuSO_4$ before electrolysis,

$$x = \frac{d(a-b)}{(c-d)} \frac{1}{\text{Molar mass of CuSO}_4}$$

Amount of $CuSO_4$ after electrolysis,

$$y = \frac{b}{(a-b)} \frac{1}{\text{Molar mass of CuSO}_4}$$

Decrease in the amount of $CuSO_4$ in cathodic compartment = $x - y$

If in the coulometer, the mass m of Ag is deposited, then the total amount of electrons passed

$$= \frac{m}{\text{Molar mass of Ag}} (= z)$$

Hence, $$t\_/2 = \frac{\text{Loss in the amount of CuSO}_4 \text{ in the cathodic compartment}}{\text{Amount of electrons passed}}$$

$$= \frac{(x-y)}{z}$$

Abnormal Transport Number

The experimental values of transport number which at first appear unrealistic may often be explained if one considers how the electrolyte has ionized in the solution. For example, the Cd^{2+} ion shows a transport number of about 0.4 in very dilute solution of CdI_2. With progressively higher electrolytic concentrations, however, the value drops sharply below 0.4, passes through zero and finally becomes negative. At the same time, the transport number of the iodide ion apparently increases beyond 0.6, passes through unity and eventually exceeds it.

The above observations are easily explained when the nature of the ions present in more concentrated solutions of CdI_2 is understood. In such solutions, this electrolyte exists largely as Cd^{2+} and CdI_4^{2-}, i.e. as the simple cadmium ion with a double charge and as a doubly charged anionic complex ion. The latter will obviously migrate in a direction opposite to that of the simple ion. The fact that the measured transport number of Cd^{2+} approaches negative value indicates that the conductance of the complexed form is greater than that of the uncomplexed form so that a net loss of Cd^{2+} occurs at the cathode.

Example 5.11.1

A Hittorf cell fitted with Ag-AgCl electrodes is filled with HC1 solution which contains $0.385\,6 \times 10^{-3}$ g HCl/g water. A current of 2.0 mA is passed for exactly 3 hours. The solutions are withdrawn, weighed and analyzed. The total mass of the cathode solution is 51.743 6 g; it contains 0.026 7 g HCl. The anode solution weighs 52.046 1 g and contains 0.013 3 g of HCl. What is the transport number of H^+ ion?

Solution

Electricity passed = $(2 \times 10^{-3}\ \text{A})\ (3 \times 60 \times 60\ \text{s}) = 21.6\ \text{A s} = 21.6\ \text{C}$

$$\text{Amount of electrons passed} = \frac{21.6\ \text{C}}{(96\,500\ \text{C mol}^{-1})} = 0.000\,224\ \text{mol}$$

Mass of HC1 present initially in (51.743 6 – 0.026 7) g (i.e. 51.716 9 g) of water

$$= \left(\frac{0.3856 \times 10^{-3}\ \text{g}}{1\ \text{g H}_2\text{O}}\right)(51.716\,9\ \text{g H}_2\text{O}) = 0.019\,94\ \text{g}$$

Mass of HCl present after the electrolysis in 51.716 9 g of water = 0.026 7 g

Mass of HCl gained = 0.026 7 g – 0.019 94 = 0.006 76 g

$$\text{Amount of HCl gained} = \frac{0.006\,76\ \text{g}}{(36.5\ \text{g mol}^{-1})} = 0.000\,185\,3\ \text{mol}$$

Had there been no migration, the gain of Cl^- in the cathodic compartment would have been 0.000 224 mol. Actually the gain is only 0.000 185 3 mol. Remaining Cl^- must have migrated to the central compartment. Thus, the transport number of Cl^- is

$$t_{\text{Cl}^-} = \frac{(0.000\,224\ \text{mol} - 0.000\,185\,3\ \text{mol})}{0.000\,224\ \text{mol}} = 0.17$$

and $t_{\text{H}^+} = 1 - 0.17 = 0.83$

Example 5.11.2

After the electrolysis of a 2 mol kg^{-1} solution of $FeCl_3$ employing the platinum electrodes, the cathodic compartment contained 50 g of water and was 1.575 mol kg^{-1} in $FeCl_3$ and 0.50 mol kg^{-1} in $FeCl_2$. Calculate (a) the amount of electrons passed, (b) the transport number of Cl^-, and (c) the equivalent and molar conductivities of Fe^{3+}. Given: $\lambda(\text{Cl}^-) = 76.35$ S cm^2 mol^{-1}.

Solution

The reaction in the cathodic compartment is

$$\text{Fe}^{3+} + \text{e}^- \rightarrow \text{Fe}^{2+}$$

Thus, the given concentration of 0.50 mol kg^{-1} of $FeCl_2$ in the cathodic compartment is due to the above reaction. Therefore,

(a) Amount of electrons passed

$$= \left(\frac{0.50\ \text{mol}}{1\,000\ \text{g H}_2\text{O}}\right)(50\ \text{g H}_2\text{O}) = 0.025\ \text{mol}$$

(b) The amount of Cl^- before the electrolysis

$$= \left(\frac{3 \times 2.0 \text{ mol}}{1\,000 \text{ g } H_2O}\right)(50 \text{ g } H_2O) = 0.30 \text{ mol}$$

The amount of Cl^- after the electrolysis

$$= \left(\frac{3 \times 1.575 \text{ mol} + 2 \times 0.50 \text{ mol}}{1\,000 \text{ g } H_2O}\right)(50 \text{ g } H_2O) = 0.286\,25 \text{ mol}$$

Loss of Cl^- in the cathodic compartment

$$= 0.3 \text{ mol} - 0.286\,25 \text{ mol} = 0.013\,75 \text{ mol}$$

This loss must be due to migration of Cl^-. Thus, the transport number of Cl^- is

$$t(Cl^-) = \frac{0.013\,75 \text{ mol}}{0.025 \text{ mol}} = 0.55 \quad \text{and} \quad t(Fe^{3+}) = 1 - t(Cl^-) = 0.45$$

(c) Since $\frac{t_+}{t_-} = \frac{\lambda_+}{\lambda_-}$; (where λ's are equivalent conductivities)†

$$\text{Therefore, } \lambda(\tfrac{1}{3}Fe^{3+}) = \lambda(Cl^-)\left(\frac{t(Fe^{3+})}{t(Cl^-)}\right) = (76.35 \text{ S cm}^2 \text{ mol}^{-1})\left(\frac{0.45}{0.55}\right)$$

$$= 62.5 \text{ S cm}^2 \text{ mol}^{-1}$$

$$\lambda(Fe^{3+}) = 3 \times 62.5 \text{ S cm}^2 \text{ mol}^{-1}$$

Example 5.11.3 A determination of the transference numbers of Cd^{2+} and I^- by Hittorf method gave the following data.

A stock solution of CdI_2 was prepared and its concentration was (determined by precipitation of iodide as AgI) found to be 0.002 763 g of CdI_2 per gram of solution.

Another sample of the solution was placed in a Hittorf transport number cell and current was passed through the cell. It was found that 0.034 62 g of Cd was deposited at the cathode by the passage of the current. Furthermore, analysis of the anodic-compartment solution, which weighed 152.643 g indicated the presence of 0.371 8 g of CdI_2.

(a) What are the electrode reactions?

(b) Indicate diagrammatically using t_+ and t_- to represent the transport numbers, changes in the amounts of ions in the anodic and cathodic compartments as a result of migration.

(c) What is the amount of electrons passed through the cell in the experiment?

(d) What are the transport numbers of Cd^{2+} and I^- in CdI_2 solution?

(e) What was the change in amount of CdI_2 in the cathodic compartment?

---

† In general $\frac{t_+}{t_-} = \frac{\nu_+ \lambda_+}{\nu_- \lambda_-}$, where λ's are molar conductivities.

Solution The electrolytic cell for the present problem is shown in Fig. 5.11.8.

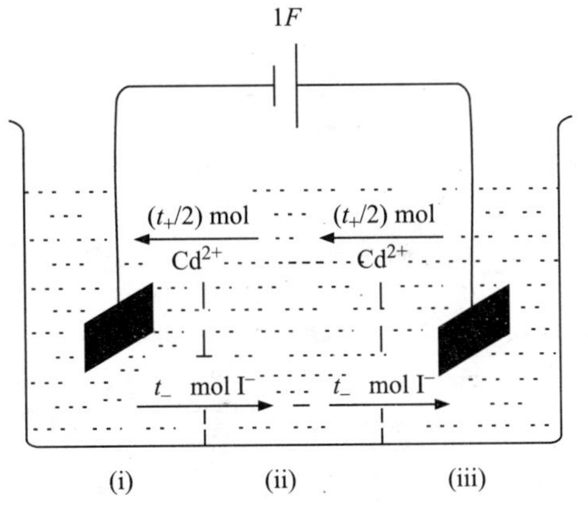

(i) Cathodic compartment electrode reaction:

$\frac{1}{2}Cd^{2+} + e^- \rightarrow \frac{1}{2}Cd$

(ii) Central compartment

(iii) Anodic compartment electrode reaction:

$I^- = \frac{1}{2}I_2 + e^-$

Fig. 5.11.8 Changes in concentrations of ions as a result of migration

(a) The electrode reactions are

At cathode: $\frac{1}{2}Cd^{2+} + e^- \rightarrow \frac{1}{2}Cd$

At anode: $I^- \rightarrow \frac{1}{2}I_2 + e^-$

(b) Let a total of 1 mol of electrons is passed through the cell. Out of this, t_+ mol of charge is carried by cations and t_- mol is carried by anions. Since each Cd^{2+} carries two charges, $(t_+/2)$ mol of Cd^{2+} will be transferred from anodic to centre and from centre to cathodic compartments. Since each anion carries one charge, t_- mol of I^- will be transferred from cathodic to centre and from centre to anodic compartments (Fig. 5.11.8).

(c) Amount of electrons passed through the cell can be determined from the amount of Cd deposited at the cathode.

Molar mass of Cd carries 2 mol of charge. Hence, the charge due to 0.034 62 g Cd deposited will be

$$\left(\frac{2\text{ mol}}{\text{Molar mass of Cd}}\right)(0.034\ 62\text{ g}) = \left(\frac{2\text{ mol}}{112.4\text{ g}}\right)(0.034\ 62\text{ g})$$
$$= 0.000\ 615\ 8\text{ mol}$$

(d) The transport numbers of Cd^{2+} and I^- can be determined from the analysis of anodic compartment. Thus, on passing 1 mol of electrons, we will have

Change in the amount of Cd^{2+}

= Change due to the electrode reaction + Change due to the migration

$$= 0\text{ mol} - \frac{t_+}{2}\text{ mol} = -\frac{t_+}{2}\text{ mol}$$

Change in the amount of I^-

= Change due to the electrode reaction + Change due to the migration

$$= -1\text{ mol} + t_-\text{ mol} = -t_+\text{ mol}$$

That is, there occurs a loss of $(t_+/2)$ mol of CdI_2.

Mass of CdI_2 in (152.643 − 0.371 8) g of water after electrolysis in the anodic compartment = 0.371 8 g

Mass of CdI_2 in (152.643 – 0.371 8) g of water before electrolysis in the anodic compartment

$$= \frac{(0.002\,763\text{ g})\{(152.643 - 0.371\,8)\text{ g H}_2\text{O}\}}{(1.0 - 0.002\,763)\text{ g H}_2\text{O}} = 0.421\,9\text{ g}$$

Mass of CdI_2 lost in the anodic compartment = 0.421 9 g – 0.371 8 g = 0.050 1 g

$$\text{Amount of CdI}_2\text{ lost} = \frac{0.050\,1\text{ g}}{(366.4\text{ g mol}^{-1})} = 0.000\,136\,7\text{ mol}$$

Amount of CdI_2 lost per mol of electrons passed

$$= \left(\frac{0.000\,136\,7\text{ mol}}{0.000\,615\,8\text{ mol of electrons}}\right)(1\text{ mol of electrons}) = 0.222\text{ mol}$$

Thus, $\frac{t_+}{2}\text{ mol} \equiv 0.222\text{ mol}$

$$t_+ = 0.444$$

$$t_- = 1 - t_+ = 0.556$$

Alternatively, we may proceed as follows.

Had there been no migration, 0.000 615 8 mol (≡ amount of electrons passed) of I^- would have lost due to the electrode reaction. The actual lost which is 2 × 0.000 1367 mol (≡ 2 × amount of CdI_2 lost) is less than the expected lost. This is due to the migration which is (0.000 615 8 mol – 0.000 273 4 mol). Hence, the transport number I^- is

$$t(\text{I}^-) = \frac{0.000\,615\,8\text{ mol} - 0.000\,273\,4\text{ mol}}{0.000\,615\,8\text{ mol}} = 0.556$$

(e) In the cathodic compartment, the change in the amount of CdI_2, when 0.000 6158 mol of electrons is passed, can be determined as follows.

On passing 1 mole of electrons, we will have

Change in the amount of Cd^{2+}

= Change due to the electrode reaction + Change due to the migration

$$= -\left(\frac{1}{2}\right)\text{mol} + \left(\frac{t_+}{2}\right)\text{mol} = -\left(\frac{1}{2}\right)t_-\text{ mol}$$

Change in the amount of I^-

= Change due to the electrode reaction + Change due to the migration

$= 0\text{ mol} - t_-\text{ mol} = -t_-\text{ mol}$

Thus, there occurs a net loss $(t_-/2)$ mol of CdI_2 on passing 1 mol of electrons. But, the actual electrons passed is 0.000 615 8 mol

Hence,

Change in the amount of CdI_2 in the cathodic compartment

= (0.556/2) (0.000 615 8 mol) = 0.000 171 3 mol

Example 5.11.4

In a transport experiment with 1 molal CdI_2 solution using inert electrodes, the following results were obtained. While the anodic solution was one molal, the cathodic solution (100 g solvent) was 0.5 molal with respect to CdI_2. The electricity passed as measured by silver coulometer was 9 650 coulombs. Calculate the transport number of Cd^{2+} and I^-. Rationalize the results and suggest a plausible mechanism for the observation.

Solution

The electrolytic cell of the given problem is shown in Fig. 5.11.9.

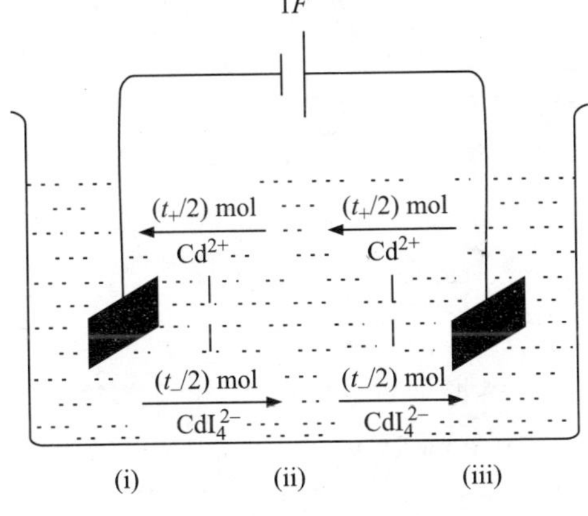

(i) Cathodic compartment electrode reaction:

$\frac{1}{2}Cd^{2+} + e^- \rightarrow \frac{1}{2}Cd$

(ii) Central compartment

(iii) Anodic compartment electrode reaction:

$I^- \rightarrow \frac{1}{2}I_2 + e^-$

Fig. 5.11.9 Electrolysis of concentrated solution of CdI_2

$$\text{Amount of electrons passed} = \frac{(9\,650\ \text{C})}{(96\,500\ \text{C mol}^{-1})} = 0.1\ \text{mol}$$

Amount of CdI_2 before the electrolysis in 100 g of the solvent

$$= \frac{(1\ \text{mol})}{(1\,000\ \text{g H}_2\text{O})}(100\ \text{g H}_2\text{O}) = 0.1\ \text{mol}$$

Amount of CdI_2 in 100 g of solvent in cathodic compartment after the electrolysis

$$= \left(\frac{0.5\ \text{mol}}{1\,000\ \text{g H}_2\text{O}}\right)(100\ \text{g H}_2\text{O}) = 0.05\ \text{mol}$$

Amount of CdI_2 in 100 g of solvent in anodic compartment after the electrolysis

$$= \left(\frac{1\ \text{mol}}{1\,000\ \text{g H}_2\text{O}}\right)(100\ \text{g H}_2\text{O}) = 0.1\ \text{mol}$$

Following the analysis of Example 5.11.3, we find that

$$\frac{t_+}{2} = \text{Amount of CdI}_2 \text{ lost in the anodic compartment} = 0$$

i.e. $\quad t_+ = 0$ and $t_- = 1$

These abnormal transport number can be explained on the basis that in the solution the complex compound $Cd(CdI_4)$ is formed. Thus, it is the Cd^{2+} and CdI_4^{2-} which migrate in the solution as shown in Fig. 5.11.9. Since in the anodic compartment, there occurs no loss of Cd^{2+}, it is obvious that

$$t(Cd^{2+}) = t(CdI_4^{2-}) = 0.5$$

This result may be verified from the following analysis.

The changes in the amounts of Cd^{2+} and I^- when 1 mol of electrons is passed are follows.

Cathodic Compartment Change in the amount of Cd^{2+}

= Change due to electrode reaction + Change due to the migration

$$= -(\tfrac{1}{2})\text{mol} + \{\underset{\text{as } Cd^{2+}}{(t_+/2)\text{ mol}} - \underset{\text{as } CdI_4^{2-}}{(t\_/2)\text{ mol}}\}$$

$= -\ t\_$ mol

Change in the amount of I^-

= Change due to electrode reaction + Change due to the migration

$= 0$ mol + $\{-\ (t\_/2)$ mol of $CdI_4^{2-}\} = -2t\_$ mol

Thus, there occurs a net loss of $t\_$ mol of CdI_2.

Anodic Compartment Change in the amount of Cd^{2+}

= Change due to the electrode reaction + Change due to the migration

$$= 0\text{ mol} + \{\underset{\text{as } Cd^{2+}}{-(t_+/2)\text{ mol}} + \underset{\text{as } CdI_4^{2-}}{(t\_/2)\text{ mol}}\}$$

$$= \tfrac{1}{2}(-1 + 2t\_)\text{ mol}$$

Change in the amount of I^-

= Change due to the electrode reaction + Change due to the migration

$= -1$ mol + $\{(t\_/2)$ mol as $CdI_4^{2-}\} = -1$ mol + $2t\_$ mol

Thus, there occurs a net loss of $\tfrac{1}{2}(1 - 2t\_)$ mole of CdI_2.

Actual Computation On passing 0.1 mole of electrons, we have
Change in the amount of CdI_2 in cathodic compartment

$= 0.1$ mol $-\ 0.05$ mol $= 0.05$ mol

Change in the amount of CdI_2 in anodic compartment

$= 0.1$ mol $-\ 0.1$ mol $= 0$

The expected changes in the amount of CdI_2 when 1 mol of electrons is passed are described in the following:

In cathodic compartment

$$= \left(\frac{0.05\text{ mol}}{0.1\text{ mol of electrons}}\right)(1.0\text{ mol of electrons}) = 0.5\text{ mol}$$

In anodic compartment = 0

Hence, $t\_$ mol $= 0.5$ mol, i.e., $t\_ = 0.5$

$\tfrac{1}{2}(1 - 2t\_)$ mol $= 0$, i.e., $t\_ = 0.5$

Thus, the transport numbers of Cd^{2+} and CdI_4^{2-} in 1 molal solution of CdI_2 are both equal to 0.5.

Example 5.11.5

The transport number of $Co(NH_3)_4Cl_2^+$ in a 0.02 M solution of $[Co(NH_3)_4Cl_2]Cl$ was determined by means of a Hittorf type cell. The anode was silver and the anodic solution was silver nitrate; the middle and the cathodic compartment contained 0.02 M of complex and the cathode was a silver-silver chloride electrode. At the end of the experiment, the silver anode was found to have changed mass by 2 g and the cathodic compartment solution (150 cm^3 by volume) was 0.025 mol dm^{-3}. Calculate the transport numbers.

Solution

The electrolytic cell corresponding to the given problem is shown in Fig. 5.11.10.

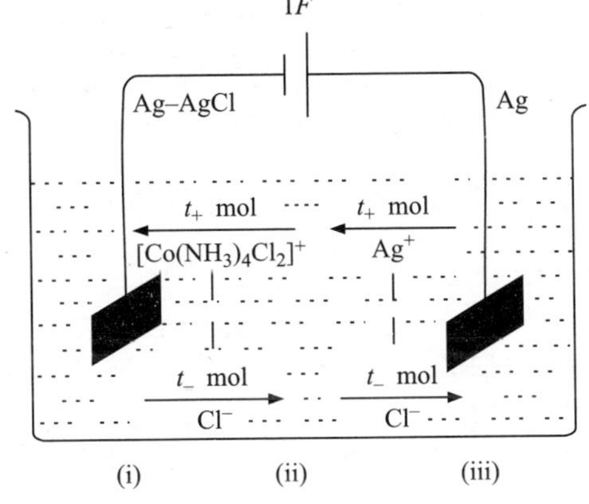

(i) Cathodic compartment electrode reaction:

$AgCl + e^- \rightarrow Ag + Cl^-$

(ii) Central compartment

(iii) Anodic compartment electrode reaction:

$Ag \rightarrow Ag^+ + e^-$

Fig. 5.11.10 Electrolysis of $[Co(NH_3)_4Cl_2]Cl$ solution

Thus, we have

Cathodic reaction: $AgCl + e^- \rightarrow Ag + Cl^-$

Anodic reaction: $Ag \rightarrow Ag^+ + e^-$

Loss in mass of anode = 2 g

$$\text{Amount of electrons passed} = \frac{2\text{ g}}{(107.8\text{ g mol}^{-1})} = 0.018\,55\text{ mol}$$

Changes in amount of cations and anions in the cathodic compartment

= 1 mol of Cl^- from electrode reaction – t_- mol of Cl^- due to the migration out

+ t_+ mol of $[Co(NH_3]_4Cl_2]^+$ due to the migration in

i.e. t_+ mol of complex will be gained on passing 1 mol of electrons.

Amount of the complex in 150 cm^3 of the solution before the electrolysis

$$= \left(\frac{0.02\text{ mol}}{1\,000\text{ cm}^3\text{ solution}}\right)(150\text{ cm}^3\text{ solution}) = 0.003\text{ mol}$$

Amount of the complex in 150 cm^3 of the solution after the electrolysis

$$= \left(\frac{0.025\text{ mol}}{1\,000\text{ cm}^3\text{ solution}}\right)(150\text{ cm}^3\text{ solution}) = 0.003\,75\text{ mol}$$

Gain of complex = 0.000 75 mol

Gain of complex per mole of electrons passed

$$= \left(\frac{0.000\,75\ \text{mol}}{0.018\,55\ \text{mol electrons}}\right)(1\ \text{mol electrons}) = 0.04\ \text{mol}$$

Hence, t_+ mol = 0.04 mol, i.e. $t_+ = 0.04$

$t_- = 1 - t_+ = 0.96$

Example 5.11.6 A solution of 0.01 molal in silver nitrate is used with silver electrodes in the determination of the transport number of Ag^+ by Hittorf method; 32.10 mg of silver is deposited in a silver coulometer in series with the Hittorf cell. At the end of the run, the 20.09 g of solvent in the anodic compartment was found to contain 39.66 mg of Ag; 27.12 g of solution in the cathodic compartment contained 11.14 mg of Ag. Calculate t_+ for Ag^+.

Solution The electrolytic cell showing the electrolysis of $AgNO_3$ solution using Ag electrodes is shown in Fig. 5.11.11.

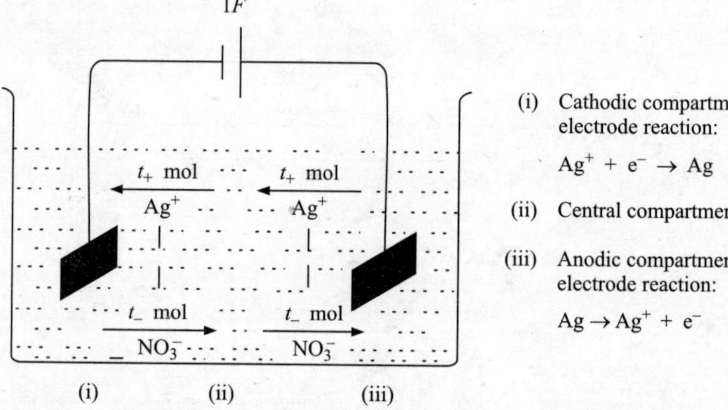

Fig. 5.11.11 Electrolysis of $AgNO_3$ solution using Ag-electrodes.

From the silver deposition in the silver coulometer, we have

$$\text{Amount of electrons passed} = \frac{(32.10 \times 10^{-3}\ \text{g})}{(107.8\ \text{g mol}^{-1})} = 0.000\,297\,7\ \text{mol}$$

The analysis of the anodic compartment is given below.

Electrode reaction: $Ag \rightarrow Ag^+ + e^-$

Change in the amount of Ag^+ on passing 1 mol of electrons

= Change due to the electrode reaction + Change due to the migration

= 1 mol + ($-\,t_+$ mol) = t_- mol

Change in the amount of NO_3^- on passing 1 mol of electrons

= Change due to the migration alone = t_- mol

Thus, there will be a net gain of t_- mol of $AgNO_3$ in the anodic compartment.

The amount of Ag^+ before the electrolysis in 20.09 g of solvent,

$$x = \left(\frac{0.01\ \text{mol}}{1\,000\ \text{g solvent}}\right)(20.09\ \text{g of solvent})$$

$$= 0.000\,2009\ \text{mol}$$

Amount of Ag^+ after electrolysis in 20.09 g of solution

$$= 0.039\ 66\ \text{g}/107.8\ \text{g mol}^{-1} = 0.000\ 367\ 9\ \text{mol}$$

Gain of Ag^+ = 0.000 367 9 mol – 0.000 200 9 mol = 0.000 167 mol

Gain of Ag^+ per mole of electrons passed

$$= \left(\frac{0.000\,167\ \text{mol}}{0.000\,297\,7\ \text{mol electrons}}\right)(1\ \text{mol of electrons}) = 0.56\ \text{mol}$$

Hence, t_- mol = 0.56 mol, i.e. $t_- = 0.56$

$$t_+ = 1 - t_- = 0.44$$

Example 5.11.7

The conductivity of a solution is expressed by $\kappa = \sum_i c_i\lambda_i$, where c_i is the molar concentration of ith ion and λ_i is its molar conductivity, the sum being taken over all the ions in the solution. Calculate the fraction of current carried by each ion (i.e. the transport number) in a solution which is 0.1 M in Na_2SO_4 and 0.01 M in H_2SO_4 (assume complete ionization). Given: $\lambda_m(Na^+) = 50.1\ \text{S cm}^2\,\text{mol}^{-1}$, $\lambda_m(H^+) = 349.8\ \text{S cm}^2\,\text{mol}^{-1}$ and $\lambda_m(SO_4^{2-}) = 160\ \text{S cm}^2\,\text{mol}^{-1}$.

Solution

The transport number of an ion in a solution containing many ions

$$= \frac{\text{Conductivity of the ion in solution}}{\text{Total conductivity of solution}}$$

Now, the conductivity of individual ions and the total conductivity of the solution containing 0.2 mol of Na^+, 0.02 mol of H^+ and 0.11 mol of SO_4^{2-} are given by

Conductivity due to $Na^+ = c(Na^+)\,\lambda_m(Na^+)$

$$= (0.2\ \text{mol dm}^{-3})\ (50.11\ \text{S cm}^2\ \text{mol}^{-1})$$

$$= 10.022\ \text{S dm}^{-3}\ \text{cm}^2 = 0.010\ 022\ \text{S cm}^{-1}$$

Conductivity due to $H^+ = c(H^+)\,\lambda_m(H^+)$

$$= (0.02\ \text{mol dm}^{-3})\ (349.8\ \text{S cm}^2\ \text{mol}^{-1})$$

$$= 6.996\ \text{S dm}^{-3}\ \text{cm}^2 = 0.006\ 996\ \text{S cm}^{-1}$$

Conductivity due to $SO_4^{2-} = c(SO_4^{2-})\,\lambda_m(SO_4^{2-})$

$$= (0.11\ \text{mol dm}^{-3})\ (160\ \text{S cm}^2\ \text{mol}^{-1})$$

$$= 17.6\ \text{S dm}^{-3}\ \text{cm}^2 = 0.017\ 6\ \text{S cm}^{-1}$$

Total conductivity = (0.010 022 + 0.006 996 + 0.017 6) S cm^{-1} = 0.034 618 S cm^{-1}. Therefore,

$$t(Na^+) = \left(\frac{0.010\,022\ \text{S cm}^{-1}}{0.034\,618\ \text{S cm}^{-1}}\right) = 0.29$$

$$t(H^+) = \left(\frac{0.006\,996\ \text{S cm}^{-1}}{0.034\,618\ \text{S cm}^{-1}}\right) = 0.20$$

$$t(SO_4^{2-}) = \left(\frac{0.017\,6\ \text{S cm}^{-1}}{0.034\,618\ \text{S cm}^{-1}}\right) = 0.51$$

Example 5.11.8 What must be the ratio of concentrations of HCl and NaCl in a solution if the transport number of H^+ is 0.5. Given: $\lambda_m(Na^+) = 50.1$ S cm^2 mol^{-1}, $\lambda_m(H^+) = 349.8$ S cm^2 mol^{-1} and $\lambda_m(Cl^-) = 76.35$ S cm^2 mol^{-1}.

Solution Let c_1 and c_2 be the respective concentrations of HCl and NaCl in a solution in which the transport number of H^+ is 0.5. We will have

$$c(H^+) = c_1; \qquad c(Na^+) = c_2 \qquad \text{and} \qquad c(Cl^-) = c_1 + c_2$$

The expression for the transport number of H^+ is given by

$$t(H^+) = \frac{c(H^+)\,\lambda(H^+)}{c(H^+)\,\lambda(H^+) + c(Na^+)\,\lambda(Na^+) + c(Cl^-)\,\lambda(Cl^-)}$$

$$= \frac{c_1\,\lambda(H^+)}{c_1\,\lambda(H^+) + c_2\,\lambda(Na^+) + (c_1 + c_2)\,\lambda(Cl^-)}$$

$$= \frac{(c_1/c_2)\,\lambda(H^+)}{(c_1/c_2)\,\lambda(H^+) + \lambda(Na^+) + \{(c_1/c_2) + 1\}\,\lambda(Cl^-)}$$

where $t(H^+) = 0.5$, Solving for (c_1/c_2), we get

$$\frac{c_1}{c_2} = \frac{\lambda(Na^+) + \lambda(Cl^-)}{\lambda(H^+) - \lambda(Cl^-)}$$

Substituting the values, we get

$$\frac{c_1}{c_2} = \frac{(50.1 + 76.35)\ \text{S cm}^2\ \text{mol}^{-1}}{(349.8 - 76.35)\ \text{S cm}^2\ \text{mol}^{-1}} = 0.463$$

Example 5.11.9 Calculate the transport numbers of K^+, Na^+ and $C_2O_4^{2-}$ in a 10^{-3} mol dm^{-3} solution of KOOC·COONa. The limiting molar conductivities for $C_2O_4^{2-}$, K^+ and Na^+ are 148.2, 50.1 and 73.5 S cm^2 mol^{-1}, respectively.

Solution From the given data, we have

$$[Na^+] = 10^{-3}\ \text{M}; \quad [K^+] = 10^{-3}\ \text{M}; \quad [Ox^{2-}] = 10^{-3}\ \text{M}$$

$$t(Na^+) = \frac{\kappa(Na^+)}{\kappa(Na^+) + \kappa(K^+) + \kappa(Ox^{2-})}$$

where

$$\kappa(Na^+) = c(Na^+)\,\lambda(Na^+) = (10^{-3}\ \text{mol dm}^{-3})\,(73.5\ \text{S cm}^2\ \text{mol}^{-1})$$

$$= 73.5 \times 10^{-6}\ \text{S cm}^{-1}$$

$$\kappa(K^+) = c(K^+)\,\lambda(K^+) = (10^{-3}\ \text{mol dm}^{-3})\,(50.1\ \text{S cm}^2\ \text{mol}^{-1})$$

$$= 50.1 \times 10^{-6}\ \text{S cm}^{-1}$$

$$\kappa(Ox^{2-}) = c(Ox^{2-})\,\lambda(Ox^{2-}) = (10^{-3}\ \text{mol dm}^{-3})\,(148.2\ \text{S cm}^2\ \text{mol}^{-1})$$

$$= 148.2 \times 10^{-6}\ \text{S cm}^{-1}$$

$$\kappa(Na^+) + \kappa(K^+) + \kappa(Ox^{2-}) = (73.5 + 50.1 + 148.2) \times 10^{-6}\ S\ cm^{-1}$$
$$= 271.8 \times 10^{-6}\ S\ cm^{-1}$$

Thus, $$t(Na^+) = \frac{73.5 \times 10^{-6}\ S\ cm^{-1}}{271.8 \times 10^{-6}\ S\ cm^{-1}} = 0.27$$

$$t(K^+) = \frac{50.1 \times 10^{-6}\ S\ cm^{-1}}{271.8 \times 10^{-6}\ S\ cm^{-1}} = 0.18$$

$$t(Ox^{2-}) = \frac{148.2 \times 10^{-6}\ S\ cm^{-1}}{271.8 \times 10^{-6}\ S\ cm^{-1}} = 0.55$$

The Moving Boundary Method

In this method, the motion of ions under the influence of an applied potential is observed directly.

A schematic diagram of the apparatus is shown in Fig. 5.11.12. It consists of a long vertical tube of uniform area of cross-section *A* fitted with two electrodes at the two ends. The electrolyte AX (called the principal electrolyte, the transport number of whose cation A^+ is to be determined) is gently introduced above the solution of another suitable electrolyte BX, called the indicator electrolyte. The latter is subjected to three conditions, namely,

(i) The anion X^- should be common to both the electrolytes.

(ii) The cation B^+ should move slower than the cation A^+.

(iii) The electrolyte BX should be more dense than AX.

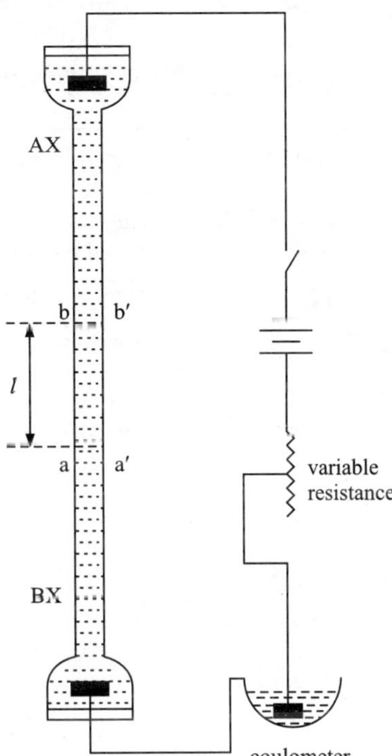

Fig. 5.11.12 Moving boundary method to determine the transport number of ions

The boundary between these two electrolytes is visible because of a difference in refractive indices of the solutions (or in some cases a difference in colour). On passing current between the two electrodes, A^+ and B^+ ions move upward towards the cathode while X^- ions move downward towards the anode. Thus, the boundary between the two solutions is moved up, which can be followed by refractive index measurements. Let the boundary move from aa′ to bb′ (= l) after passing the electric charge Q, then

Volume of the liquid moved = $(A \times l)$

Concentration of AX in the solution = c

Amount of A^+ carried towards the cathode = $(c)\ (Al)$

The corresponding amount of electricity carried by A^+ ions = $(c)\ (Al)$

Amount of electrons passed = Q/F

Transport number of A^+ cations† = $(lAc)/(Q/F)$ (5.11.8)

Since B^+ ions have a lower velocity than A^+ ions, they never overtake the latter and a sharp boundary is always maintained between the two solutions. However, the B^+ ions are never very far behind A^+ ions, for otherwise the solution below the boundary would get diluted and its increasing resistance would cause an increased potential drop, thereby increasing the velocity of B^+ ions. Thus, the boundary moves slowly in the upward direction.

The measurements of the transport number can be made over a range of concentration of the electrolyte; the plot of t versus $\sqrt{c}$ is linear in dilute solution and can be extrapolated to $\sqrt{c} = 0$ so as to obtain the value of the transport number at infinite dilution t^∞.

Example 5.11.10 Calculate the transport numbers of H^+ and Cl^- from the following data obtained by moving boundary method using $CdCl_2$ as the indicator electrolyte; concentration of HC1 solution = 0.1 M, mass of Ag deposited in the coulometer = 0.120 9 g; movement of boundary = 7.50 cm and cross-sectional area of the tube = 1.24 cm^2.

Solution Transport number of H^+ is given by

$$t(\mathrm{H}^+) = \frac{l\,A\,c}{(Q/F)}$$

where $l = 7.50$ cm, $A = 1.24$ cm^2, $c = 0.1$ M

and $(Q/F) = (0.120\ 9\ \mathrm{g}/108\ \mathrm{g\ mol^{-1}})$

Therefore,

---

†For the electrolyte other than uni-univalent, the amount of A^{z+} carried towards the cathode is $(\nu_+ c)\ (Al)$, where ν_+ is the stoichiometric number of A^{z+} in the molecule. The corresponding amount of electricity carried will be $z_+(\nu_+ c)\ (Al)$. Hence,

$$t(\mathrm{A}^{z+}) = \frac{\nu_+ z_+\, lAc}{(Q/F)} \qquad (5.11.9)$$

$$t(H^+) = \frac{(7.50\ cm)\ (1.24\ cm^2)\ (0.1\ mol\ dm^{-3})}{(0.120\ 9/108)\ mol}$$

$$= 830.8\ cm^3\ dm^{-3} = 830.8\ cm^3\ (10\ cm)^{-3} = 0.830\ 8 \simeq 0.83$$

and $t(Cl^-) = 1 - 0.83 = 0.17$

Example 5.11.11 A solution of 0.10 M in LiCl with conductivity $1.06 \times 10^{-2}\ S\ cm^{-1}$ is placed in a moving boundary cell of cross-sectional area $1.17\ cm^2$. It was electrolyzed for 131 minutes with a constant current of 9.42 milliamperes. The Li^+ was observed to move a distance of 2.08 cm. What are the velocity, mobility, transport number and molar conductivity of Li^+ in this solution?

Solution We have $c = 0.1$ M, $A = 1.17\ cm^2$, $l = 2.08$ cm,

$I = 9.42 \times 10^{-3}$ A and $t = 131 \times 60$ s

Amount of electrons passed

$$\frac{Q}{F} = \frac{It}{F} = \frac{(9.42 \times 10^{-3}\ A)\ (131 \times 60\ s)}{(96\ 500\ C\ mol^{-1})} = 0.000\ 767\ 3\ mol$$

Therefore,

$$t_+ = \frac{l\,A\,c}{(Q/F)} = \frac{(2.08\ cm)\ (1.17\ cm^2)\ (0.1\ mol\ dm^{-3})}{(0.000\ 767\ 3\ mol)} = 317.2\ cm^3\ dm^{-3} = 0.317$$

Now $$\Lambda_m(LiCl) = \frac{\kappa}{c} = \left(\frac{1.06 \times 10^{-2}\ S\ cm^{-1}}{0.1\ mol\ dm^{-3}}\right)$$

$$= 1.06 \times 10^{-1}\ S\ cm^{-1}\ dm^{-3}\ mol^{-1} = 106\ S\ cm^2\ mol^{-1}$$

$$\lambda_m(Li) = t(Li^+)\ \Lambda_m(LiCl) = (0.317)\ (106\ S\ cm^2\ mol^{-1})$$

$$= 33.6\ S\ cm^2\ mol^{-1}$$

$$\text{Velocity of } Li^+ = \frac{l}{t} = \frac{(2.08\ cm)}{(131 \times 60\ s)} = 0.000\ 265\ cm\ s^{-1}$$

Mobility of Li^+

$$= \frac{\lambda_m}{z_+ F} = \frac{33.6\ S\ cm^2\ mol^{-1}}{(1)\ (96\ 500\ C\ mol^{-1})} = 0.000\ 348\ S\ cm^2\ C^{-1}$$ (*see*, Eq. 5.12.7)

$$= 0.000\ 348\ (V\ A^{-1})^{-1}\ cm^2\ (A\ s)^{-1} = 0.000\ 348\ cm^2\ V^{-1}\ s^{-1}$$

5.12 IONIC SPEED AND IONIC MOBILITY

Let a volume V_m of the solution containing 1 mol of the substance be placed between two parallel electrodes. Let d be the distance between these two electrodes. The electrodes are large enough to include between them the whole of the solution. In a unit time, as seen earlier, the total current flowing through these electrodes is given by

$$I_m = I_+ + I_- = \frac{N_+ z_+ e v_+}{d} + \frac{N_- |z_-| e v_-}{d}$$

Since 1 mol of the substance is involved, the numbers N_+ and N_- will be related to the Avogadro constant N_A by the expressions

$$N_+ = \nu_+ N_A$$

$$N_- = \nu_- N_A$$

where ν_+ and ν_- are the respective stoichiometric numbers of cations and anions appearing in one molecule of the substance. With these, we have

$$I_m = \frac{N_A e}{d}(\nu_+ z_+ v_+ + \nu_- |z_-| v_-) = \frac{F}{d}(\nu_+ z_+ v_+ + \nu_- |z_-| v_-) \quad (5.12.1)$$

where F is faraday constant and is equal to $(N_A e)$.

The molar conductivity of the solution is given by

$$\Lambda = \frac{\kappa}{c} = \kappa(V_m) = \kappa(d)(A_m) \quad (5.12.2)$$

where A_m is the area of cross section of the electrodes dipping into the solution (which include 1 mol of electrolyte). Its unit is m^2 mol^{-1}.

Conductance of the solution is given by

$$G = \kappa\left(\frac{A}{d}\right)$$

which according to Eq. (5.12.2), becomes

$$G = \frac{\Lambda}{A_m d}\left(\frac{A}{d}\right) = \frac{(\Lambda/\text{mol}^{-1})}{d^2} \quad (5.12.3)$$

According to Ohm's law, the current flowing through the solution (or electrodes) is given by

$$I = \frac{(\Delta\phi)}{R} = (\Delta\phi)G$$

which according to Eq. (5.12.3) becomes

$$I = (\Delta\phi)\frac{(\Lambda/\text{mol}^{-1})}{d^2} \quad (5.12.4)$$

Comparing Eq. (5.12.1) and Eq. (5.12.4), we get

$$I_{\mathrm{m}}/\mathrm{mol}^{-1} = I$$

i.e.
$$\frac{F}{d}(\nu_+ z_+ \upsilon_+ + \nu_- |z_-| \upsilon_-) = \frac{(\Delta\phi)\Lambda}{d^2}$$

If the solution is infinitely diluted, then using Kohlrausch law the above expression becomes

$$\frac{F}{d}(\nu_+ z_+ \upsilon_+^\infty + \nu_- |z_-| \upsilon_-^\infty) = \frac{\Delta\phi}{d^2}(\nu_+ \lambda_+^\infty + \nu_- \lambda_-^\infty)$$

This gives

$$\upsilon_+^\infty = \left(\frac{\Delta\phi}{d}\right)\left(\frac{\lambda_+^\infty}{z_+ F}\right) \quad \text{and} \quad \upsilon_-^\infty = \left(\frac{\Delta\phi}{d}\right)\left(\frac{\lambda_-^\infty}{|z_-| F}\right) \tag{5.12.5}$$

Defining the term ionic mobility u, as the speed of the ion when the potential gradient is unity, we have

$$u_+^\infty = \frac{\upsilon_+^\infty}{(\Delta\phi/d)} = \frac{\lambda_+^\infty}{z_+ F} \quad \text{and} \quad u_-^\infty = \frac{\upsilon_-^\infty}{(\Delta\phi/d)} = \frac{\lambda_-^\infty}{|z_-| F} \tag{5.12.6}$$

Since $\upsilon \propto (\Delta\phi/d)$, the ionic mobility may be interpreted as the proportionality constant between ionic speed and electric field.

At any other concentration, it follows that

$$u_+ = \frac{\lambda_+}{z_+ F} \quad \text{and} \quad u_- = \frac{\lambda_-}{|z_-| F} \tag{5.12.7}$$

Unit of ionic mobility Since

$$u = \frac{\upsilon}{(\Delta\phi/d)} = \frac{\mathrm{m\ s^{-1}}}{\mathrm{V\ m^{-1}}}$$

the unit of u is $\mathrm{m^2\ V^{-1}\ s^{-1}}$.

An Alternative Derivation

The speed of an ion (say, cation) is given by

$$\upsilon_+ = \frac{\mathrm{d}l}{\mathrm{d}t} \tag{5.12.8}$$

where $\mathrm{d}l$ is the length which it covers in time $\mathrm{d}t$.

According to the definition of ionic mobility (speed of the ion per unit potential gradient), we have

$$u = \frac{\text{Speed}}{\text{Potential gradient}} = \frac{\mathrm{d}l/\mathrm{d}t}{\mathrm{d}(\Delta\phi)/\mathrm{d}l} \tag{5.12.9}$$

Using Ohm's law, $I = \Delta\phi/R$, we have

$$\Delta\phi = IR$$

Therefore, $$\frac{\mathrm{d}(\Delta\phi)}{\mathrm{d}l} = I\frac{\mathrm{d}R}{\mathrm{d}l} \tag{5.12.10}$$

Now, the resistance of the electrolytic solution is given by

$$R = \rho\left(\frac{l}{A}\right) = \frac{1}{\kappa}\left(\frac{l}{A}\right)$$

Therefore, $$\frac{\mathrm{d}R}{\mathrm{d}l} = \frac{1}{\kappa}\left(\frac{1}{A}\right)$$

Substituting this in Eq. (5.12.10), we have

$$\frac{\mathrm{d}(\Delta\phi)}{\mathrm{d}l} = I\frac{1}{\kappa A}$$

With this, Eq. (5.12.9) yields

$$u = \left(\frac{\mathrm{d}l}{\mathrm{d}t}\right)\left(\frac{A\kappa}{I}\right) = \frac{\kappa(A\,\mathrm{d}l)}{(I\,\mathrm{d}t)} = \frac{\kappa V}{Q} \tag{5.12.11}$$

where $V = (A\,\mathrm{d}l)$ and $Q = I\,\mathrm{d}t$

The transport number of the cation is given by

$$t_+ = \frac{\nu_+ z_+ lAc}{(Q/F)} \qquad \text{(Eq. 5.11.9)}$$

This, by definition, is also given by

$$t_+ = \frac{\nu_+\lambda_+}{\Lambda} = \frac{\nu_+\lambda_+}{(\kappa/c)}$$

Equating these two, we get

$$\frac{\nu_+ z_+ lAc}{(Q/F)} = \frac{\nu_+\lambda_+}{(\kappa/c)}$$

Rearranging this, we get

$$\frac{(lA)\,\kappa}{Q} = \frac{\lambda_+}{z_+F} \quad \text{or} \quad \frac{V\kappa}{Q} = \frac{\lambda_+}{z_+F}$$

Thus, the ionic mobility as given by Eq. (5.12.11) becomes

$$u_+ = \frac{\lambda_+}{z_+ F}$$

For an anion, the expression is $u_- = \dfrac{\lambda_-}{|z_-|F}$

Example 5.12.1

A potential of 5.60 V is applied to two electrodes placed 9.8 cm apart; how far would an ammonium ion be expected to move in 1 hour in a dilute solution of ammonium salt at 25 °C. Given: $\lambda(NH_4^+) = 73.4$ S cm^2 mol^{-1}.

Solution

$$\text{Ionic mobility} = \frac{\lambda_m(NH_4^+)}{F} = \frac{(73.4\ \text{S cm}^2\ \text{mol}^{-1})}{(96\,500\ \text{C mol}^{-1})} = 7.61 \times 10^{-4}\ \text{cm}^2\ \text{V}^{-1}\ \text{s}^{-1}$$

$$\text{Potential gradient} = \left(\frac{5.60\ \text{V}}{9.80\ \text{cm}}\right) = 0.571\ \text{V cm}^{-1}$$

Actual speed = (Ionic mobility) (potential gradient)

$= (7.61 \times 10^{-4}\ \text{cm}^2\ \text{V}^{-1}\ \text{s}^{-1})\ (0.571\ \text{V cm}^{-1}) = 4.34 \times 10^{-4}\ \text{cm s}^{-1}$

Distance moved in 1 hour = $(14.34 \times 10^{-4}\ \text{cm s}^{-1})\ (60 \times 60\ \text{s}) = 1.56$ cm

---

Example 5.12.2

For H^+ and Na^+, the values of λ_+^∞ are 349.8 and 50.11 S cm^2 mol^{-1}. Calculate the mobilities and velocities of these ions if they are in a cell in which the electrodes are 5 cm apart and to which a potential of 2 V is applied.

Solution

Ionic mobility of H^+

$$= \frac{\lambda_m(H^+)}{F} = \left(\frac{349.8\ \text{S cm}^2\ \text{mol}^{-1}}{96\,500\ \text{C mol}^{-1}}\right) = 000\ 362\ 5\ \text{cm}^2\ \text{V}^{-1}\ \text{s}^{-1}$$

Actual speed with which H^+ moves

= (ionic mobility) (Potential gradient)

$= (0.003\ 625\ \text{cm}^2\ \text{V}^{-1}\ \text{s}^{-1})\ (2\ \text{V}/5\ \text{cm}) = 0.001\ 448\ \text{cm s}^{-1}$

Ionic mobility of Na^+

$$= \frac{(50.11\ \text{S cm}^2\ \text{mol}^{-1})}{(96\,500\ \text{C})} = 0.000\ 519\ 3\ \text{cm}^2\ \text{V}^{-1}\ \text{s}^{-1}$$

Actual speed with which Na^+ moves

$= (0.000\ 519\ 3\ \text{cm}^2\ \text{V}^{-1}\ \text{s}^{-1})\ (2\ \text{V}/5\ \text{cm}) = 0.000\ 207\ 7\ \text{cm s}^{-1}$

5.13 THE WALDEN'S RULE

When an ion in a solution is subjected to an electrical field it moves with a constant speed. The ion attains this constant speed when the accelerating electrical force becomes equal to the retarding frictional force. These two forces are given as

Electrical force acting on the ion = $(|z|e)E$

Frictional force as given by Stokes' law = $6\pi\eta r v$

where (ze) is the charge carried by the ion, r is its radius, v is the speed with it which moves and η is the viscosity of the medium. Equating these two, we get

$$|z|eE = 6\pi\eta r v$$

From this, we obtain the ionic mobility as

$$u = \frac{v}{E} = \frac{|z|e}{6\pi\eta r} \tag{5.13.1}$$

Now the molar conductivity of the solution is given by

$$\Lambda = \nu_+\lambda_+ + \nu_-\lambda_-$$

which on making use of Eq. (5.12.7) becomes

$$\Lambda = \nu_+ z_+ u_+ F + \nu_- |z_-| u_- F$$

Substituting u_+ and u_- from Eq. (5.13.1), we get

$$\Lambda = \frac{Fe}{6\pi\eta}\left[\frac{\nu_+ z_+^2}{r_+} + \frac{\nu_- |z_-|^2}{r_-}\right] \tag{5.13.2}$$

The only quantity on the right hand side of Eq. (5.13.2) which depends on the medium is η. Thus for an electrolyte in different medium, we must have

$$\Lambda\eta = \text{constant} \tag{5.13.3}$$

Equation (5.13.3) is known as the Walden's rule.

When Eq. (5.13.3) is applied to a specific electrolyte in several different solvents, it is found that except for electrolytes containing very large ions, the product of Λ and η is not constant. This arises from the fact that the ions are solvated. An ion is attached to molecules of solvent which are carried along with the ion as it moves. The effective radius of the ion is therefore larger than its crystallographic radius and is different in each solvent. The amount of solvent held to the ion is less with larger ions, so that the effective radius is more nearly the same in various solvents, consequently Walden's rule is more accurate for large ions.

5.14 APPLICATION OF CONDUCTANCE MEASUREMENTS

Degree of Dissociation of a Weak Electrolyte and its Dissociation Constant

Degree of dissociation of a weak electrolyte can be obtained by using Arrhenius equation

$$\alpha = \frac{\Lambda_c}{\Lambda^\infty} \tag{5.14.1}$$

where Λ_c is the molar conductivity at the given concentration and Λ^∞ is the corresponding value at infinite dilution when the weak electrolyte is present in completely ionized form.

Knowing α, numerical value of the dissociation constant can be evaluated using the appropriate expression.

Example 5.14.1

Given are the following molar conductivities at 25 °C: HCl, 426 S cm^2 mol^{-1}; NaCl, 126 S cm^2 mol^{-1}; NaC(sodium crotonate), 83 S cm^2 mol^{-1}. Calculate the degree of dissociation and dissociation constant of crotonic acid. The conductivity of a 0.001 mol dm^{-3} acid solution is 3.83 × 10^{-5} S cm^{-1}. Neglect the variation of conductivity with concentration.

Solution

The molar conductivity of the dissociated form of crotonic acid is

$$\Lambda_m(H^+, C^-) = \Lambda_m(HCl) + \Lambda_m(NaC) - \Lambda_m(NaCl)$$

$$= (426 + 83 \quad 126)\ \text{S cm}^2\ \text{mol}^{-1} = 383\ \text{S cm}^2\ \text{mol}^{-1}$$

The apparent molar conductivity of HC,

$$\Lambda_m(HC) = \frac{\kappa}{c} = \frac{3.83 \times 10^{-5}\ \text{S cm}^{\ 1}}{0.001\ \text{mol dm}^{-3}}$$

$$= 3.83 \times 10^{-2}\ \text{S cm}^{-1}\ \text{dm}^3\ \text{mol}^{-1} = 38.3\ \text{S cm}^2\ \text{mol}^{-1}$$

The degree of dissociation,

$$\alpha = \frac{\Lambda_m(HC)}{\Lambda_m^\infty(H^+C^-)} = \frac{(38.3\ \text{S cm}^2\ \text{mol}^{-1})}{(383\ \text{S cm}^2\ \text{mol}^{-1})} = 0.1$$

$$K_a = \frac{c\,\alpha^2}{1-\alpha} = \frac{(10^{-3}\ \text{M})(0.1)^2}{1-0.1} = 1.11 \times 10^{-5}\ \text{M}$$

Determination of the Ionic Product of Water

The ionic product of water is

$$K_w = [H^+]\,[OH^-] \tag{5.14.2}$$

Since in pure water $[H^+] = [OH^-]$, therefore, it follows that

$$[H^+] = [OH^-] = \sqrt{K_w} \tag{5.14.3}$$

The conductivity κ of pure water is related to its molar concentration by the equation

$$\kappa = [H^+]\,\lambda_{H^+} + [OH^-]\,\lambda_{OH^-}$$

Using Eq. (5.14.3) it becomes

$$\kappa = \sqrt{K_w}\,(\lambda_{H^+} + \lambda_{OH^-})$$

The ions are present in such low concentrations that the molar ionic conductivities may be taken as the ones at infinite dilution, so that $\lambda^{\infty}_{H^+} + \lambda^{\infty}_{OH^-} = 547.6$ S cm^2 mol^{-1} and we obtain for K_w:

$$K_w = \left(\frac{\kappa}{547.6 \text{ S cm}^2 \text{ mol}^{-1}}\right)^2 \tag{5.14.4}$$

Example 5.14.2 The conductivity of water at 298 K is 0.554×10^{-7} S cm^{-1}. Calculate the degree of dissociation and ionic product of water. Given: $\lambda(H^+) = 349.8$ S cm^2 mol^{-1} and $\lambda(OH^-) = 197.8$ S cm^2 mol^{-1}.

Solution Let α be the degree of dissociation. The concentrations of H^+ and OH^- in water become

$$[H^+] = [OH^-] = c\alpha$$

where c is the concentration of water and is equal to 55.56 M.

The conductivity of water is given by

$$\kappa = [H^+]\,\lambda_m(H^+) + [OH^-]\,\lambda_m(OH^-)$$

$$= c\alpha\,[\lambda_m(H^+) + \lambda_m(OH^-)] = c\alpha\Lambda_m(H^+, OH^-)$$

or $$\alpha = \frac{\kappa}{c\Lambda_m\,(H^+, OH^-)}$$

Substituting the data, we get

$$\alpha = \frac{0.554 \times 10^{-7}\text{ S cm}^{-1}}{(55.56\text{ mol dm}^{-3})\{(349.8 + 197.8)\text{ S cm}^2\text{ mol}^{-1}\}}$$

$$= \frac{0.554 \times 10^{-7}\text{ S cm}^{-1}}{(55.56 \times 10^{-3}\text{ mol cm}^{-3})(547.6\text{ S cm}^2\text{ mol}^{-1})} = 1.821 \times 10^{-9}$$

and $$K_w = [H^+]\,[OH^-] = (c\alpha)^2 = (55.56\text{ M})^2\,(1.821 \times 10^{-9})^2$$

$$= 1.023 \times 10^{-14}\text{ M}^2$$

Determination of Solubility and Solubility Product

The solubility of a salt in the solution can be determined from the conductivity measurements. Since the molar conductivity of the solution is given by

$$\Lambda = \frac{\kappa}{c} \tag{5.14.5}$$

therefore, it follows that the concentration of the salt in the solution is given by

$$c = \frac{\kappa}{\Lambda} \tag{5.14.5}$$

In using this expression, κ used should be that of the salt alone, the value of which can be calculated by subtracting κ of pure water from κ of the solution, i.e.

$$\kappa(\text{salt}) = \kappa(\text{soln.}) - \kappa(\text{water})$$

If the salt is sparingly soluble, its solubility product can be determined from its solubility obtained by using Eq. (5.14.5). Here Λ can be replaced by Λ^{∞} as the solution will be fairly diluted, i.e. it will have very low concentration.

Example 5.14.3

Calculate the solubility and solubility product of $Co_2[Fe(CN)_6]$ in water at 25 °C from the following data.

Conductivity of a saturated solution of $Co_2[Fe(CN)_6]$ is 2.06×10^{-6} S cm^{-1} and that of water used is 4.1×10^{-7} S cm^{-1}. The ionic molar conductivities of Co^{2+} and $Fe(CN)_6^{4-}$ are 86.0 S cm^2 mol^{-1} and 444.0 S cm^2 mol^{-1}, respectively.

Solution

Molar conductivity of $Co_2[Fe(CN)_6]$ becomes

$$\Lambda_m = 2\,\lambda_m(Co^{2+}) + \lambda_m(Fe(CN)_6^{4-})$$
$$= 2\,(86.0 \text{ S cm}^2 \text{ mol}^{-1}) + 444.0 \text{ S cm}^2 \text{ mol}^{-1} = 616.0 \text{ S cm}^2 \text{ mol}^{-1}$$

Its conductivity is

$$\kappa(Co_2[Fe(CN)_6]) = \kappa(\text{soln.}) - \kappa(\text{water})$$
$$= 2.06 \times 10^{-6} \text{ S cm}^{-1} - 4.1 \times 10^{-7} \text{ S cm}^{-1}$$
$$= 1.65 \times 10^{-6} \text{ S cm}^{-1}$$

Now using the expression $c = \kappa/\Lambda_m$, we get

$$c = \frac{(1.65 \times 10^{-6} \text{ S cm}^{-1})}{(616.0 \text{ S cm}^2 \text{ mol}^{-1})} = 2.678 \times 10^{-9} \text{ mol cm}^{-3}$$
$$= 2.678 \times 10^{-6} \text{ mol dm}^{-3}$$

and

$$K_s(Co_2[Fe(CN)_6]) = [Co^{2+}]^2\,[Fe(CN)_6^{4-}]$$
$$= (2 \times 2.678 \times 10^{-6} \text{ M})^2\,(2.678 \times 10^{-6} \text{ M})$$
$$= 7.68 \times 10^{-17} \text{ M}^3$$

Example 5.14.4

The conductivity of 0.001 M Na_2SO_4 solution is 2.6×10^{-4} S cm^{-1} and increases to 7.0×10^{-4} S cm^{-1} when the solution is saturated with $CaSO_4$. The molar conductivities of Na^+ and Ca^{2+} are 50 S cm^2 mol^{-1} and 120 S cm^2 mol^{-1}, respectively. Calculate (a) the conductivity of only $CaSO_4$ in the solution, (b) solubility of $CaSO_4$, and (c) solubility product of $CaSO_4$. The conductivity of water used is 0.50×10^{-6} S cm^{-1}.

Solution

(a) Concentration of Na_2SO_4 solution = 0.001 M
Conductivity due to Na_2SO_4 alone in the solution

$$= (2.6 \times 10^{-4} - 0.5 \times 10^{-6}) \text{ S cm}^{-1} = 2.595 \times 10^{-4} \text{ S cm}^{-1}$$

Molar conductivity of Na_2SO_4

$$= \frac{\kappa}{c} = \left(\frac{2.595 \times 10^{-4} \text{ S cm}^{-1}}{0.001 \text{ mol dm}^{-3}}\right) = 2.595 \times 10^{-1} \text{ S cm}^{-1} \text{ dm}^3 \text{ mol}^{-1}$$
$$= 259.5 \text{ S cm}^2 \text{ mol}^{-1}$$

Molar ionic conductivity of SO_4^{2-}

$$= \Lambda_m(Na_2SO_4) - 2\,\lambda_m(Na^+) = (259.5 - 2 \times 50)\ S\ cm^2\ mol^{-1}$$

$$= 159.5\ S\ cm^2\ mol^{-1}$$

Conductivity due to $CaSO_4$ alone in the solution

$$= (7.0 \times 10^{-4} - 2.6 \times 10^{-4})\ S\ cm^{-1} = 4.4 \times 10^{-4}\ S\ cm^{-1}$$

Conductivity of $CaSO_4$ solution

$$= \kappa\ (CaSO_4) + \kappa(\text{water}) = (4.4 \times 10^{-4} + 0.50 \times 10^{-6})\ S\ cm^{-1}$$

$$= 4.405 \times 10^{-4}\ S\ cm^{-1}$$

(b) Molar conductivity of $CaSO_4$

$$= \lambda_m(Ca^{2+}) + \lambda_m(SO_4^{2-}) = (120 + 159.5)\ S\ cm^2\ mol^{-1}$$

$$= 279.5\ S\ cm^2\ mol^{-1}$$

Solubility of $CaSO_4$ in 0.001 M Na_2SO_4 solution

$$= \frac{\kappa}{\Lambda} = \frac{(4.4 \times 10^{-4}\ S\ cm^{-1})}{(279.5\ S\ cm^2\ mol^{-1})}$$

$$= 1.576 \times 10^{-6}\ mol\ cm^{-3} = 1.576 \times 10^{-3}\ mol\ dm^{-3}$$

(c) Solubility product of $CaSO_4$

$$= [Ca^{2+}]\ [SO_4^{2-}] = (0.001\ 576\ M)\ \{(0.001\ 576 + 0.001)\ M\}$$

$$= 4.056 \times 10^{-6}\ M^2$$

Example 5.14.5 The conductivity of a saturated solution of CaF_2 at 18 °C was found to be 4.2×10^{-5} S cm^{-1}. The conductivity of water used for making the solution was 2.0×10^{-6} S cm^{-1}. The molar ionic conductivities at infinite dilution of Ca^{2+} and F^- are 104.0 and 48.0 S cm^2 mol^{-1}, respectively. What is the solubility and solubility product of CaF_2? (Assume that only Ca^{2+} and F^- are present in the solution.)

Solution Conductivity due to CaF_2 alone

$$= \kappa(\text{soln.}) - \kappa(\text{water}) = (4.2 \times 10^{-5} - 2.0 \times 10^{-6})\ S\ cm^{-1}$$

$$= 4.0 \times 10^{-5}\ S\ cm^{-1}$$

Concentration of CaF_2, in the solution

$$= \frac{\kappa}{\Lambda_m} = \frac{\kappa}{\lambda_m(Ca^{2+}) + 2\lambda_m(F^-)} = \frac{(4.0 \times 10^{-5}\ S\ cm^{-1})}{(104 + 2 \times 48)\ S\ cm^2\ mol^{-1}}$$

$$= 2.0 \times 10^{-7}\ mol\ cm^{-3} = 2.0 \times 10^{-4}\ mol\ dm^{-3}$$

$$K_s(CaF_2) = [Ca^{2+}]\ [F^-]^2 = (2.0 \times 10^{-4}\ M)\ (4.0 \times 10^{-4}\ M)^2$$

$$= 3.2 \times 10^{-11}\ M^3$$

Determination of Hydrolysis Constant

Hydrolysis constant of a salt can be determined from conductivity measurements. Let c be the concentration of the salt BH^+Cl^- (formed from a weak base and strong acid) in the solution. The cation of this salt undergoes hydrolysis in the solution as

$$BH^+ + H_2O \rightleftharpoons B + H_3O^+$$

If α is the degree of hydrolysis, then the concentrations of BH^+ and H_3O^+ in the solution will be $c(1-\alpha)$ and $c\alpha$, respectively. The conductivity of this solution is given by

$$\begin{aligned}\kappa_1 &= c(BH^+)\,\lambda(BH^+) + c(H_3O^+)\,\lambda(H_3O^+) + c(Cl^-)\,\lambda(Cl^-)\\ &= c(BH^+)\,\lambda(BH^+) + c(H_3O^+)\,\lambda(H_3O^+) + \{c(BH^+) + c(H_3O^+)\}\,\lambda(Cl^-)\\ &= c(BH^+)\,\{\lambda(BH^+) + \lambda(Cl^-)\} + c(H_3O^+)\,\{\lambda(H_3O^+) + \lambda(Cl^-)\}\\ &= c(1-\alpha)\,\Lambda(BH^+, Cl^-) + c\alpha\,\Lambda(H_3O^+, Cl^-)\end{aligned} \qquad (5.14.6)$$

or $$\kappa_1 = (1-\alpha)\,\kappa_2 + \alpha\kappa_3 \qquad (5.14.7)$$

where, κ_2 is the conductivity of the unhydrolyzed salt of concentration c and κ_3 is the conductivity of the HC1 solution of concentration c. An alternative form of Eq. (5.14.6) is obtained by taking c to the left side, which gives

$$\Lambda_1 = (1-\alpha)\,\Lambda_2 + \alpha\Lambda_3 \qquad (5.14.8)$$

where Λ_1 is the molar conductivity of the solution involving hydrolysis,
Λ_2 is the molar conductivity of the solution of concentration c involving no hydrolysis

and Λ_3 is the molar conductivity of the hydrochloric acid solution of concentration c.

The value of κ_2 or Λ_2 can be determined experimentally by adding a large excess of the base to the solution, which causes the equilibrium to shift to the left, i.e. it minimizes the hydrolysis.

Knowing the value of α from either of the two equations, the hydrolysis constant can be obtained by using the expression

$$K_h = \frac{c\alpha^2}{1-\alpha}$$

Example 5.14.6

The conductivity measurements were made at 25 °C for the following solutions.

Solution	κ/S cm^{-1}
10^{-3} M phenanthrolinium chloride (BHCl)	1.360×10^{-4}
10^{-3} M BHCl plus a large excess of B	1.045×10^{-4}
10^{-3} M HCl	4.210×10^{-4}

Phenanthrolinium chloride is a strong electrolyte, i.e. it exists as BH^+ and Cl^-. Phenanthroline is a nonelectrolyte. Calculate K_a for the acid dissociation $BH^+ \rightleftharpoons B + H^+$.

Solution The degree of dissociation of $BH^+ \rightleftharpoons B + H^+$ can be calculated using the expression

$$\kappa_1 = (1 - \alpha)\,\kappa_2 + \alpha\kappa_3$$

where κ_1, κ_2 and κ_3 are the conductivities of BHCl, BHCl + excess B, and HCl, respectively, Rearranging the above expression, we get

$$\alpha = \frac{\kappa_1 - \kappa_2}{\kappa_3 - \kappa_2}$$

Substituting the values, we get

$$\alpha = \frac{(1.360 \times 10^{-4}\ \text{S cm}^{-1}) - (1.045 \times 10^{-4}\ \text{S cm}^{-1})}{(4.210 \times 10^{-4}\ \text{S cm}^{-1}) - (1.045 \times 10^{-4}\ \text{S cm}^{-1})} = \frac{0.315}{3.165} \approx 0.1$$

$$K_\text{h} = \frac{c\alpha^2}{1-\alpha} = \frac{(10^{-3}\ \text{M})\,(0.1)^2}{1-(0.1)} = 1.1 \times 10^{-5}\ \text{M}$$

Example 5.14.7 An aqueous solution (0.007 8 M) of aniline hydrochloride had molar conductivity equal to 119.4 S cm^2 mol^{-1} which became 103.0 S cm^2 mol^{-1}, when a few drops of aniline were added to the solution. Molar conductivity of hydrochloric acid at the same ionic concentration was found to be 413 S cm^2 mol^{-1}. Calculate the degree of hydrolysis and hydrolysis constant of aniline hydrochloride.

Solution The expression to be used is

$$\Lambda_1 = (1 - \alpha)\,\Lambda_2 + \alpha\Lambda_3$$

where Λ_1, Λ_2 and Λ_3 are molar conductivities of salt, salt + base, and acid, respectively. Rearranging and substituting the values, we get

$$\alpha = \frac{\Lambda_1 - \Lambda_2}{\Lambda_3 - \Lambda_2} = \frac{(119.4\ \text{S cm}^2\ \text{mol}^{-1}) - (103\ \text{S cm}^2\ \text{mol}^{-1})}{(413\ \text{S cm}^2\ \text{mol}^{-1}) - (103\ \text{S cm}^2\ \text{mol}^{-1}} = 0.052\,89$$

Now $$K_\text{h} = \frac{c\alpha^2}{1-\alpha} = \frac{(0.007\,8\ \text{M})\,(0.052\,89)^2}{(1-0.052\,89)} = 2.304 \times 10^{-5}\ \text{M}$$

Conductometric Titrations

Strong Acid Versus Strong Base

The principle of conductometric titrations is based on the fact that during the titration, one of the ions is replaced by the other and invariably these two ions differ in the ionic conductivity with the result that the conductivity of the solution varies during the course of the titration. Take an example of the titration of a strong acid, say HCl, with a strong base, say NaOH, the following changes in conductance is observed during the course of titration.

Before NaOH is added, the conductance of HCl solution has a high value due to the presence of highly mobile hydrogen ions. As NaOH is added, H^+ ions combine with OH^- ions to form undissociated water and thus the faster moving H^+ ions are replaced by relatively slower moving Na^+ ions. Consequently, the conductance of the solution decreases and this continues right up to the equivalence point where the solution contains only NaCl. Beyond the equivalence point, if more of NaOH is added, then the solution contains an excess of the fast moving OH^- ions with the result that its conductance is increased and it continues to increase as more and more of NaOH is added. If we plot the conductance value versus the amount of NaOH added, we get a curve of the type shown in Fig. 5.14.1. The descending portion AB represents the conductances before the equivalence point (solution contains a mixture of acid HCl and the salt NaCl) and the ascending portion CD represents the conductances after the equivalence point (solution contains the salt NaCl and the excess of NaOH). The point E which represents the minimum conductance is due to the solution containing only NaCl with no free acid or alkali and thus represents the equivalence point. This point can, however, be obtained by the extrapolation of the lines AB and. DC, and, therefore, one is not very particular in locating this point experimentally as it is in the case of ordinary acid-base titrations involving the acid-base indicators.

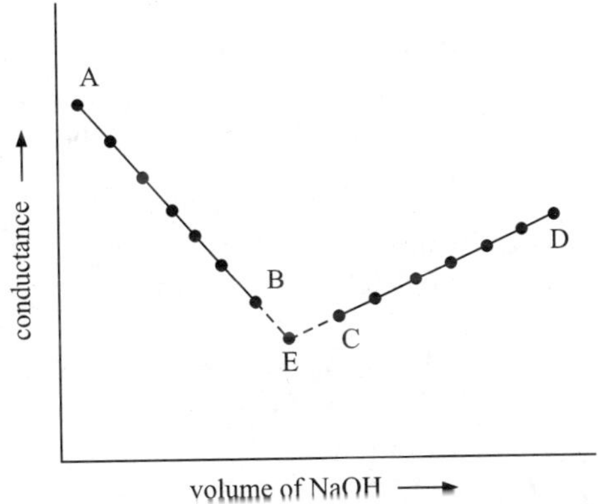

Fig. 5.14.1 Conductometric titration curve of a strong acid (e.g., HCl) versus a strong base (e.g., NaOH)

In general, the conductometric titrations can be performed in solution where the slower moving ion is being replaced by the faster moving ion and vice versa. The equivalence point can be obtained by extrapolating the two lines before and after the equivalence point. A few other typical cases are described below.

Weak Acid versus Strong Base

Let us take the specific example of acetic acid being titrated against NaOH. Before the addition of alkali, the solution shows poor conductance due to feeble ionization of acetic acid. Initially the addition of alkali causes not only the replacement of H^+ by Na^+ but also suppresses the dissociation of acetic acid due to the common ion Ac^- and thus the conductance of the solution decreases in the beginning. But very soon the conductance starts increasing as addition of NaOH neutralizes the undissociated HAc to Na^+Ac^- thus causing the replacement of non-conducting HAc with strong-conducting electrolyte Na^+Ac^-. The increase in conductance continues right up to the equivalence point. Beyond this point, conductance increases more rapidly with the addition of NaOH due to the highly conducting OH^- ions. The graph near the equivalence point is curved due to the hydrolysis of the salt NaAc. The actual equivalence point can, as usual, be obtained by the extrapolation method. The nature of curve obtained is shown in Fig. 5.14.2

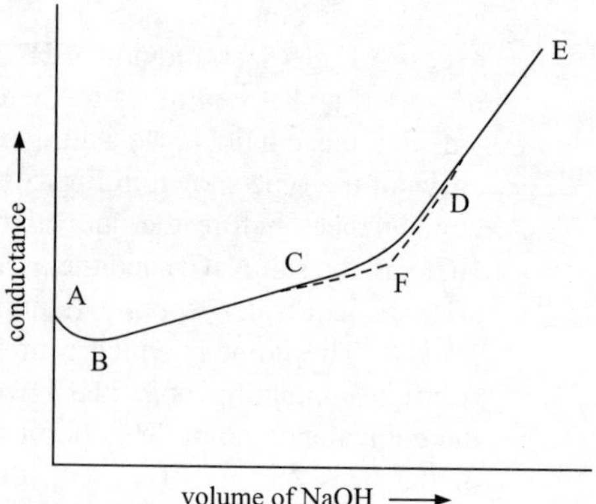

Fig. 5.14.2 Conductometric titration curve of a weak acid (e.g., HAc) versus strong base (e.g., NaOH)

Weak base versus strong acid

The nature of curve is exactly of the type shown in Fig. 5.14.2.

Weak base versus weak acid

The nature of curve before the equivalence point is similar to that shown in Fig. 5.14.2. After the equivalence point, conductance virtually remains the same as the weak base which is being added is feebly ionized and, therefore, is not very much conducting.

Mixture of Weak and Strong Acids versus a Strong Base

In this case, the strong acid is first neutralized and only when this has been completely neutralized, the weak acid starts neutralizing. Thus, we have the neutralization reaction in two stages and the curve contains two breaks as shown in Fig. 5.14.3. The first break corresponds to the neutralization of the strong acid and the second corresponds to that of the weak acid. From these the individual concentrations of the strong acid and the weak acid can be calculated.

Comments

Besides the above titrations, we can employ conductometric titrations involving a mixture of weak and strong bases (e.g., NH_4OH and NaOH versus HCl), diprotic acid (e.g., oxalic acid versus NaOH, stepwise neutralization, two breaks in the curve), precipitation (e.g., NaCl versus $AgNO_3$), oxidation-reduction (e.g., Fe^{2+} versus MnO_4^-), displacement reactions (e.g., sodium acetate versus HCl) and other types of reactions. The method serves as a very important tool where direct titrations using indicators are either difficult or impossible.

In the end it may be mentioned that the conductometric titration curves given in Figs. 15.14.1 to 15.14.3 are actually observed only if the volume of the solution remains unchanged. In practice, this condition is fulfilled by taking the solution in the burette about 10 to 20 times more concentrated than the solution in the titrating flask. A high precision microburette is employed in order to increase the accuracy of the results.

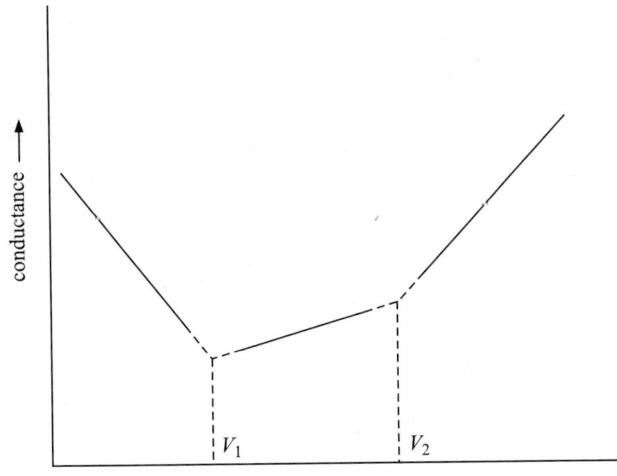

Fig. 5.14.3 Conductometric titration curve of a mixture of a strong acid (e.g., HC1) and a weak acid (e.g., HAc) with a strong base (e.g., NaOH)

REVISIONARY PROBLEMS

5.1 Explain how the mechanism of metallic conduction differs from that of electrolytic conduction and hence explain the following fact:

'Conduction of an electrolytic solution increases with increase in temperature whereas that of metal decreases.'

5.2 Define the term electrolysis. State and explain Faraday's laws of electrolysis. How will you calculate the electronic charge from Faraday's laws of electrolysis?

5.3 Define the terms: resistance, conductance, resistivity, conductivity, equivalent conductivity and molar conductivity. Express equivalent and molar conductivities in terms of conductivity.

5.4 (a) Explain, why conductivity decreases on dilution whereas molar conductivity increases.
(b) Explain, why the variation of molar conductivity on dilution of a strong electrolyte differs from that of a weak electrolyte. Sketch a diagram indicating the typical variations of molar conductivities for these two electrolytes
(c) Explain, why Arrhenius theory is not applicable to strong electrolytes.

5.5 (a) State and explain Kohlrausch's law of independent migration of ions. How does it help in determining the molar conductivity at infinite dilution of a weak electrolyte?
(b) Explain, why conductivities of H^+ and OH^- ions in aqueous medium are rather abnormally high.
(c) Explain, why ionic molar conductivity at infinite dilution increases in the order Li^+, Na^+, K^+, Rb^+, etc.

5.6 (a) Explain the terms: (i) asymmetry effect, and (ii) electrophoretic effect, (b) Discuss qualitatively the Debye-Hückel-Onsager theory as applied to the variation of molar conductivity with dilution of a dilute solution of a strong electrolyte.

5.7 Explain why the conductivity is increased when a very high electrical field ($E \simeq 10^5$ V/cm, Wien effect) or a very high frequency AC($\nu \simeq 3 \times 10^6$ Hz, Debye-Falkenhagen effect) is employed.

5.8 What is the effect of temperature on ionic molar conductivities? How will you account for this change in ionic molar conductivities?

5.9 (a) What is meant by transport number of an ion? Describe, in brief, the Hittorf method and the moving boundary method to determine experimentally the transport number of an ion.
(b) Show that the transport number of an ion in a solution containing many ions is given by

$$t_i = \frac{\text{Conductivity of ion in the solution}}{\text{Total conductivity of the solution}} \quad \text{i.e.} \quad t_i = \frac{c_i\lambda_i}{\sum_i c_i\lambda_i}$$

5.10 (a) Define ionic mobility. What is its unit? How does it differ from ionic speed?
(b) Derive the relation of ionic mobility u_i in terms of ionic molar conductivity λ_i, i.e. show that

$$u_i = \frac{\lambda_i}{|z_i| F}$$

where $|z_i|$ is the magnitude of charge number of the ion and F is Faraday constant.

5.11 What is Walden rule? Explain, why Walden rule is not applicable to cations of small sizes.

5.12 Explain, how the conductometric measurements can be employed for the determination of:
(i) Degree of dissociation of a weak electrolyte and its dissociation constant,
(ii) Ionic product of water.
(iii) Solubility and solubility product of a sparingly soluble salt,
(iv) Hydrolysis constant of a salt.

5.13 Describe briefly the principle involved in the conductometric titrations. Discuss its advantages over the ordinary titrations.

TRY YOURSELF PROBLEMS

5.1 (a) Suppose that a monoprotic acid having a concentration c_a is titrated with a base having concentration c_b. If V_0 is the volume of acid and V is the volume of base added at any stage of the titration, show that conductivity before the equivalence point is reached is given as a function of V by

$$\kappa = \left(\frac{V_0}{V_0 + V}\right)\left[\kappa_a - c_b\left(\frac{V}{V_0}\right)(\lambda_{H^+} - \lambda_{M^+})\right]$$

where κ_a is the conductivity of acid solution before any base has been added. Assume that the values of λ do not change with the volume of solution.
(b) Show that after the equivalence point is passed

$$\kappa = \left(\frac{V_0 + V_e}{V_0 + V}\right)\kappa_e + \left(\frac{V - V_e}{V_0 + V}\right)\kappa_b$$

where κ_b is the conductivity of the basic solution, κ_e that of the solution at the equivalence point, V_e is the volume of base added at the equivalence point.

5.2 Explain why the amide ion in liquid ammonia has abnormally high transport number.

5.3 'Transport number of Cl^- ion in aqueous solution of HCl is 0.16 and it is 0.62 in aqueous solution of NaCl.' Explain the difference.

5.4 The equivalent conductivity of an aqueous solution of $K_2SO_4 \cdot Al_2(SO_4)_3 \cdot 24H_2O$ is x. What is its molar conductivity?

5.5 Discuss the following conductometric-titration graphs (Take necessary data from the text):

(a) HCl versus NH_4OH

(b) $AgNO_3$ versus $BaCl_2$

(c) $Ba(OH)_2$ versus $MgSO_4$

(d) A mixture of HCl and HAc versus NaOH

(e) A mixture of HCl and HAc versus NH_4OH

(f) Oxalic acid versus NaOH

(g) A mixture of NaCl and NaI versus $AgNO_3$

(h) NH_4Cl solution versus NaOH solution

5.6 In a transport experiment with a solution containing platinum chloride and an excess of hydrochloric acid, it is found that platinum migrates exclusively towards the anode. This implies the formation of a complex anion between Pt^{4+} ion and the Cl^- ions. The compound H_2PtCl_6 is in fact known. This compound may dissociate according to the following two schemes:

(i) $H_2PtCl_6 \rightleftharpoons 2H^+ + Pt^{4+} + 6Cl^-$

(ii) $H_2PtCl_6 \rightleftharpoons 2H^+ + PtCl_6^{2-}$

Which of these two schemes is consistent with the above experimental behaviour?

NUMERICAL PROBLEMS

Conductivity and Molar Conductivity

5.1 A particular conductivity cell has a resistance of 468 Ω when filled with 0.001 M HCl, of 1 580 Ω when filled with 0.001 M NaCl, and of 1 650 Ω when filled with 0.001 M $NaNO_3$. The molar conductivity of $NaNO_3$ is 121 S cm^2 mol^{-1}. Neglecting changes in values with concentration, calculate

(a) Conductivity of 0.001 M $NaNO_3$.

(b) Cell constant.

(c) Resistance of the cell when filled with 0.001 M HNO_3 and molar conductivity of HNO_3.

(Ans. (a) 0.121 2 × 10^{-3} S cm^{-1}, (b) 0.199 65 cm^{-1}, (c) 414.4 S cm^2 mol^{-1}, 474 Ω)

5.2 A conductance cell is filled with 0.05 M KCl (conductivity = 6.67 × 10^{-3} S cm^{-1}) at 25 °C and the measured resistance is 243 ohms. When the cell is filled with 0.01 M NaOH, the measured resistance is 681 ohms at 25 °C. What are the conductivity and the molar conductivity of 0.01 M NaOH?

(Ans. 2.38 × 10^{-3} S cm^{-1}, 238 S cm^2 mol^{-1})

5.3 Given the following Λ^∞/S cm^2 mol^{-1} values at 25 °C: HCl, 426.2; KCl, 149.9, $CaCl_2$,271.6; NaCl, 126.5; Na_2SO_4, 259.8. Calculate Λ^∞ of (a) H_2SO_4, (b) K_2SO_4, and (c) $CaSO_4$.

(Ans. (a) 859.2 S cm^2 mol^{-1}, (b) 306.6 S cm^2 mol^{-1}, (c) 278.4 S cm^2 mol^{-1})

5.4 Given the following Λ^∞/S cm^2 mol^{-1} values at 25 °C; $AgNO_3$, 133.4; KCl, 149.9; KNO_3, 144.9. What is the molar conductivity at infinite dilution of AgCl?

(Ans. (a) 138.4 S cm^2 mol^{-1})

5.5 The resistance of a conductance cell containing 0.1 M KC1 was found to be 60.53 ohms at 25 °C. When this cell was filled with a solution containing 2.38 g of $MgCl_2$ per dm^3, its resistance was 151.0 ohms. Calculate (a) cell constant; (b) conductivity of the $MgCl_2$ solution; and (c) molar conductivity of the solution. Given: Λ(0.1 M KC1) = 128.93 S cm^2 mol^{-1}.

(Ans. (a) 0.780 cm^{-1}, (b) 5.16 × 10^{-3} S cm^{-1}, (c) 206.62 S cm^2 mol^{-1})

5.6 (a) Calculate the equivalent conductivity and molar conductivity at infinite dilution of KOOC·COONa (assuming complete ionization) at 25 °C from the following data. Ionic mobilities at 25 °C of Na^+, K^+ and $C_2O_4^{2-}$ are 0.0005 19, 0.000 762 and 0.000 768 cm^2 V^{-1} s^{-1}, respectively.

(Ans. 135.92 S cm^2 mol^{-1}, 271.84 S cm^2 mol^{-1})

(b) The conductivity of 0.01 M KC1 solution is 1.41 × 10^{-3} S cm^{-1}. The measured resistance of a 0.01 M KC1 solution in the conductivity cell was found to be 1 000 ohms. Assuming complete ionization and neglecting inter-ionic effects, show that the resistance in this cell of a 0.01 M KOOC·COONa solution is 518.7 ohms.

Transport Number

5.7 A $AgNO_3$ solution containing 0.007 39 g of $AgNO_3$ per gram of H_2O is electrolyzed between silver electrodes. During the experiment 0.078 g of Ag is deposited at the cathode. At the end of the experiment, the anode portion contains 23.14 g of H_2O and 0.236 g of $AgNO_3$. What are the transport numbers of Ag^+ and NO_3^- ions?

(Ans. t_+ = 0.47)

5.8 A 4 molal solution of $FeCl_3$ is electrolyzed between platinum electrodes. After the electrolysis the cathode portion, weighing 30 g is 3.15 molal in $FeCl_3$ and 1.0 molal in a $FeCl_2$. What are the transport numbers of Fe^{3+} and Cl^- ions?

(Ans. $t(Fe^{3+})$ = 0.45, $t(Cl^-)$ = 0.55)

5.9 An aqueous solution of $K_4Fe(CN)_6$, initially 1.0 molal, was electrolyzed at 25 °C in a Hittorf apparatus with platinum electrodes until 964.87 C of electric charge had passed through. The only anode reaction is $Fe(CN)_6^{4-} \rightarrow Fe(CN)_6^{3-} + e^-$. The solution in the anodic compartment after electrolysis contained 100.0 g H_2O and was 1.015 mol kg^{-1} with respect to total Fe.

(a) How many equivalents are there in a mole of $Fe(CN)_6^{4-}$ (i) from the point of view of conductance, and (ii) from the point of view of the reaction at the electrode?

(b) What is the amount of $Fe(CN)_6^{4-}$ that has reacted at the anode?

(c) What is the amount of $Fe(CN)_6^{4-}$ that has entered or left the anodic compartment?

(d) Calculate the transport number of $Fe(CN)_6^{4-}$ in $K_4Fe(CN)_6$.

(e) Calculate the ionic conductivity of $Fe(CN)_6^{4-}$ at infinite dilutioin. at 25 °C. Given that $\lambda^\infty(K^+)$ = 73.52 S cm^2 mol^{-1}. Point out any appaoximations that you make in this calculation.

(Ans. (a) (i) 4, (ii) 1, (b) 0.01 mol, (c) 0.0015 mol, (d) t_- = 0.60, (e) λ = 110.28 S cm^2 mol^{-1})

5.10 The apparent transport numbers t_+ of the Zn^{2+} ion in ZnI_2 solution were measured and it was found that the transport number varies with molality of ZnI_2 as follows:

m/mol kg^{-1}	0.05	0.1	1.0	2.5	4.0	5.0	10.0
t_+	0.382	0.363	0.291	0.115	–0.05	–0.19	–0.55

How do you account for this variation of t_+?

(Ans. Complex formation $Zn^{2+} + I^- \rightleftharpoons ZnI^+$; $ZnI^+ + I^- \rightleftharpoons ZnI_2$
$ZnI_2 + I^- \rightleftharpoons ZnI_3^-$; $ZnI_3^- + I^- \rightleftharpoons ZnI_4^{2-}$)

5.11 A moving boundary experiment is carried out with a 0.1 M solution of hydrochloric acid at 25 °C ($\kappa = 4.24$ S m^{-1}). Sodium ions are caused to follow the hydrogen ions. Three milliamperes current is passed through the tube of 0.3 cm^2 cross-sectional area and it is observed that the boundary moves 3.08 cm in 1 hour. Calculate (a) the hydrogen-ion mobility, (b) the hydrogen-ion transport number, (c) the chloride-ion mobility, and (d) the electric field strength.

(Ans. (a) 3.63×10^{-7} m^2 V^{-1} s^{-1}, (b) 0.826, (c) 7.64×10^{-8} m^2 V^{-1} s^{-1}, (d) 23.6 V m^{-1})

5.12 The transport numbers of ions in a 1.0 molal solution of $AgNO_3$ were determined by the moving boundary method using a solution of 0.60 mol kg^{-1} $Cd(NO_3)_2$ as the following solution. When a current of 15.0 milliamperes was used, the boundary swept through a volume of 0.145 cm^3 in 33.0 minutes. Calculate the transport numbers of Ag^+ and NO_3^- ions. (Ans. $t_+ = 0.471$)

5.13 At 25 °C the molar conductivity of a 0.02 M $AgNO_3$ solution is 128.71 S cm^2 mol^{-1} while transport number of Ag^+ is 0.477. Calculate the molar ionic conductivities and the ionic mobilities of Ag^+ and NO_3^- in a 0.02 M solution of $AgNO_3$.

(Ans. 61.39 S cm^2 mol^{-1}, 67.32 S cm^2 mol^{-1} 6.36×10^{-4} V^{-1} cm^2 s^{-1}, 6.976×10^{-4} V^{-1} cm^2 s^{-1})

5.14 At 18 °C, the mobility at infinite dilution of ammonium ion is 0.000 66 cm^2 V^{-1} s^{-1} while that of the chlorate ion is 0.000 57 cm^2 V^{-1} s^{-1}. Calculate Λ_m^∞ of ammonium chlorate and the transport numbers of the two ions.

(Ans. 118.69 S cm^2 mol^{-1}, $t_+ = 0.537$, $t_- = 0.463$)

5.15 Calculate the mobility of NO_3^- in a very dilute solution. Given the following molar conductivities at infinite dilution in S cm^2 mol^{-1} at 25 °C:
KC1, 149.9; KNO_3, 144.9; HC1, 426.1 and the transport number of H^+ in HC1 in 0.831. (Ans. 7.38×10^{-4} cm^2 V^{-1} s^{-1})

5.16 The transport number of Na^+ in dilute NaCl solution is 0.4 and $\lambda(C1^-)$ = 75 S cm^2 mol^{-1}. In addition, the following resistances have been measured for a conductivity cell filled successively with the indicated solutions: 0.1 M KC1 solution, 7 000 Ω; a solution of 0.1 M KC1 + 0.2 M in NaCl, 2 600 Ω. Calculate the value of Λ(KCl). (Ans. 147.7 S cm^2 mol^{-1})

5.17 Given that $\lambda(Li^+)$ is 40 S cm^2 mol^{-1}, $\lambda(Na^+)$ is 50 S cm^2 mol^{-1} and $\lambda(NO_3^-)$ is 70 S cm^2 mol^{-1} neglecting changes in values with concentration, calculate for a solution which is 0.1 M in $LiNO_3$ and 0.2 M in $NaNO_3$ (both salts in the same solution):

(a) Conductivity.
(b) Transport number of Li^+.
(c) Distance Li^+ ions would move in 100 seconds if some of the solution placed in a tube of 5 cm^2 cross section and current of 0.1 A were flowing.

(Ans. (a) 0.035 S cm^{-1}, (b) $t(Li^+) = 0.114$, (c) 0.023 6 cm)

Application of Conductance

5.18 Hydrofluoric acid is a weak electrolyte. At 25 °C, the molar conductivity of 0.002 M HF is 176.2 S cm^2 mol^{-1}; Λ^∞(HF) = 405.2 S cm^2 mol^{-1}. What is the degree of dissociation of HF in a 0.002 M solution? (Ans. 0.434 8)

5.19 At 25 °C, the resistance of a cell filled with 0.01 M KC1 solution is 525 ohms. The resistance of the same cell filled with 0.1 M NH_4OH is 2 030 ohms. What is the degree of dissociation of NH_4OH in this solution? (Ans. = 0.014 3)

5.20 The conductivity of a staturated solution of CaF_2 at 18 °C was found to be 4.2×10^{-5} S cm^{-1}. The conductivity of water used for making this solution was 2.8×10^{-6} S cm^{-1}. Given: $\lambda^{\infty}(\frac{1}{2}\,Ca^{2+}) = 52.0$ and $\lambda^{\infty}(F^{-}) = 48.0$ cm^2 S mol^{-1}. What is the solubility product and solubility in g dm^{-3} of CaF_2?

(Ans. 3.01×10^{-11} M^3, 1.529×10^{-2} g)

5.21 The conductivity of a saturated aqueous solution of silver oxalate. ($Ag_2C_2O_4$) is 3.8×10^{-5} S cm^{-1} 25 °C. Calculate the molar conductivity of oxalate ion ($C_2O_4^{2-}$). Given at 25 °C, conductivity of water is 6.2×10^{-8} S cm^{-1}, molar conductivity of Ag^+ ion at infinite dilution is 62 S cm^2 mol^{-1}, standard solubility product of $Ag_2C_2O_4$ is 1.1×10^{-11}.

(Ans. 147 S cm^2 mol^{-1})

5.22 Silver chloride is only slightly soluble in water, and a saturated solution of AgCl may be regarded as infinitely dilute. At 25 °C the conductivity of a saturated solution is 1.80×10^{-6} S cm^{-1} after correction for the conductivity of pure water. What is the solubility and solubility product of AgCl at 25 °C?

(Ans. 1.30×10^{-5} M, 1.69×10^{-10} M^2)

APPENDIX I A Note on Changing Concepts in Physical Chemistry

The scientific presentation of a few concepts in chemistry needs modification in the light of recommendations of the International Union of Pure and Applied Chemistry (IUPAC). A few such concepts are highlighted here.

1 PHYSICAL QUANTITIES

Definition of Physical Quantity

In physical sciences, we commonly deal with quantities such as pressure, volume, mass, temperature, current, etc. These quantities are known as physical quantities. A physical quantity has two components, namely, numerical value and its unit and is written as

Physical quantity = (Numerical value) (Unit)

Seven Base Physical Quantities

The IUPAC has recommended the use of seven physical quantities having their own dimensions. Their dimensions are completely independent of one another and it is for this reason that these are known as dimensionally independent physical quantities. These physical quantities along with their recommended symbols are given in Table AI.1.

Table AI.1 Seven Base Physical Quantities

Physical quantity	*Symbol*
Length	l
Mass	m
Time	t
Electric current	I
Thermodynamic temperature	T
Amount of substance	n
Luminous intensity	I_V

Of the seven physical quantities, luminous intensity is not needed in physical chemistry. It is used in optical photometry and is, therefore, included here only for the sake of completeness.

2 THE INTERNATIONAL SYSTEM OF UNITS (SI UNITS)

In order to have consistency in scientific recording, IUPAC has recommended the use of coherent units, known as the International System of Units, commonly abbreviated as SI. The SI units of seven base quantities are listed in Table AI.2.

Table AI.2 SI units of Seven Base Physical Quantities

Physical quantity	*Name of SI unit*	*Symbol for SI unit*
Length	metre	m
Mass	kilogram	kg
Time	second	s
Electric current	ampere	A
Thermodynamic Temperature	kelvin	K
Amount of substance	mole	mol
Luminous intensity	candela	cd

Definition of Seven Base Physical Quantities

The SI base units stated in Table AI.2 have been precisely defined as follows:

Metre The metre is the length of path travelled by light in vacuum during a time interval of 1/299 792 458 of a second.

Kilogram The kilogram is the unit of mass; it is equal to the mass of the international prototype of the kilogram. This prototype is a polished cylinder of platinum-iridium alloy which was chosen for its durability and resistance to corrosion. The cylinder is kept at the International Bureau of Weights and Measures in a suburb of Paris, France.

Second The second is the duration of 9 192 631 770 periods of the radiation corresponding to the transition between the two hyperfine levels of the ground state of the caesium-133 atom.

Ampere The ampere is that constant current which, if maintained in two straight parallel conductors of infinite length, of negligible cross-section, and placed 1 metre apart in vacuum, would produce between these conductors a force equal to 2×10^{-7} newton per metre of length.

Kelvin The kelvin, unit of thermodynamic temperature, is the fraction 1/273.16 of the thermodynamic temperature of the triple point of water.

Mole The mole is the amount of substance of a system which contains as many elementary entities as there are atoms in 0.012 kilogram of carbon-12.

Candela The candela is the luminous intensity, in a given direction, of a source that emits monochromatic radiation of frequency 540×10^{12} hertz and that has a radiant intensity in that direction of (1/683) watt per steradian.

Comment on Amount of Substance

Of the seven base physical quantities, it is worth highlighting the physical quantity 'amount of substance' along with its unit 'mol'. This is because a basic change in our conventional way of referring to this quantity is required. Suppose we have 0.5 mol of a substance (say, hydrogen). Conventionally, we speak or write it as

Number of moles of hydrogen = 0.5

Now suppose we have 1 kg of hydrogen. We never speak or write it as

Number of kg of hydrogen = 1

since we are not accustomed to speak or write it like this. Instead, we say

Mass of hydrogen = 1 kg

Many such examples can be cited. For example, we never say

Number of kg of apples = 1

Number of dozens of apples = 1

Number of litres of milk = 1

Number of metres of a line = 1

A mole of a substance is a collection of 6.022×10^{23} particles of that substance. It is like a dozen which is a collection of 12 articles. When we do not refer to a collection of 12 articles as

Number of dozens of articles = 1

why should we refer to a collection of 6.022×10^{23} articles as

Number of moles of articles = 1?

So, some inconsistency has arisen in referring to these two quantities which basically belong to the same category. To avoid this, IUPAC has recommended the use of the phrase 'amount of substance' for a physical quantity whose unit is 'mol'. More precisely, the word 'amount' is exclusively reserved whenever the quantity is to be expressed in terms of moles. It is like using the words mass for kg, volume for m^3 (or L) and length for metre. Thus, the use of the phrase 'number of moles equal to 0.5' should be completely replaced by 'amount of substance equal to 0.5 mol'. It may be mentioned that to write

Amount of substance = 0.5 g

is not acceptable as the word 'amount' is exclusively reserved for the unit 'mol' and not for 'g'.

Definition of Amount of Substance

By definition, the amount of substance is proportional to the number of specified entities of that substance. The proportionality factor is the same for all substances and is equal to $1/N_A$, where N_A is Avogadro constant which has a value of

$$N_A = 6.022 \times 10^{23}\ \text{mol}^{-1}$$

(Note: Earlier, N_A was commonly referred to as Avogadro number. But it is not a pure number as it has both a numerical value as well as a unit. So, it is a physical quantity and is called Avogadro constant.)

Mathematically, the amount of substance is given as

$$n = \frac{N}{N_A}$$

where N is the number of specified entities of the substance and N_A is Avogadro constant. For example, 3.011 5 × 10^{23} molecules of hydrogen will contain

$$\text{Amount of substance} = \frac{3.011 \times 10^{23}}{6.022 \times 10^{23}\ \text{mol}^{-1}} = 0.5\ \text{mol}$$

Comment on Specified Entities

A comment on the *specified entities* of the substance may be made here. Specified entities need not be confined to those known or thought to exist as independent entities or to groups containing integral number of atoms. For example, we may have

amount of $\frac{1}{2}H_2O$	or	$n(\frac{1}{2}H_2O)$
amount of $(H_2 + \frac{1}{2}O_2)$	or	$n(H_2 + \frac{1}{2}O_2)$
amount of $(H_2 + 0.234\ O_2)$	or	$n(H_2 + 0.234\ O_2)$
amount of $(0.075\ {}^6Li + 0.925\ {}^7Li)$	or	$n(0.075\ {}^6Li + 0.925\ {}^7Li)$
amount of $Fe_{0.91}S$	or	$n(Fe_{0.91}S)$
amount of $\frac{1}{2}Fe^{2+}$	or	$n(\frac{1}{2}Fe^{2+})$
amount of $\frac{1}{2}Fe^{3+}$	or	$n(\frac{1}{2}Fe^{3+})$
amount of $\frac{1}{3}Fe^{3+}$	or	$n(\frac{1}{3}Fe^{3+})$
amount of $\frac{1}{5}KMnO_4$	or	$n(\frac{1}{5}KMnO_4)$

and so on.

The following equality may also be noted down.

$$n(\tfrac{1}{5}KMnO_4) = 5\,n(KMnO_4)$$

$$n(\tfrac{1}{3}Fe^{3+}) = 3\,n(Fe^{3+})$$

that is, in the same solution of potassium permangate the amount of $\frac{1}{5}KMnO_4$ species will be five times the amount of $KMnO_4$ species and so on.

3 MANIPULATION OF A PHYSICAL QUANTITY

It is advisable to use a physical quantity with its complete definition, i.e. to use unit along with the numerical value. The unit is treated the same way as any other quantity in an algebraic operation is treated. For example, in algebra, each of the following expressions represents one and the same thing.

$$x = 6 \qquad \frac{x}{2} = 3 \qquad x = 1.5 \times 4$$

$$x = 2 \times 3 \qquad \frac{x}{4} = 1.5 \qquad \frac{x}{1.5} = 4$$

and so on.

Similarly, the following expressions represent one and the same thing.

$$V = 25.0\ \text{cm}^3; \qquad \frac{V}{\text{cm}^3} = 25.0; \qquad \frac{V}{25.0} = 1\ \text{cm}^3$$

and so on.

Display of Data in Tabular and Graphical Forms

The way of writing $V/\text{cm}^3 = 25.0$ is especially useful for the headings in tables and as labels on the axes of graphs. For example, the data on variation of rate constant with temperature are shown in Table AI.3.

Table AI.3 Variation of Rate Constant with Temperature*

T/K	k/s^{-1}	K/T**	log (k/s^{-1})
273	7.87×10^{-7}	3.663×10^{-3}	$\bar{7}.896$
293	1.70×10^{-5}	3.413×10^{-3}	$\bar{5}.230$
313	2.57×10^{-4}	3.193×10^{-3}	$\bar{4}.410$
333	1.78×10^{-3}	3.003×10^{-3}	$\bar{3}.250$
353	2.14×10^{-2}	2.833×10^{-3}	$\bar{2}.330$

* The conventional way of writing the table headings is as follows.

T	k	$1/T$	log k
(K)	(s^{-1})	(K^{-1})	

** The data of $1/T$ may be recorded as follows.

T/K	273	293	313	333	353
10^3 K/T (or kK/T)	3.663	3.413	3.193	3.003	2.833

Earlier, the heading 10^3 K/T was used to be written as $(1/T) \times 10^{-3}$. (K^{-1})

Figure AI.1 displays the plot of log (k/s^{-1}) versus 10^3 K/T. Note the way axes have been marked. Both log (k/s^{-1}) and 10^3 K/T are dimensionless quantities. Thus, we are plotting one number against the other. The slope of the obtained straight line plot is also a dimensionless quantity, i.e. it is a number. The obtained slope is

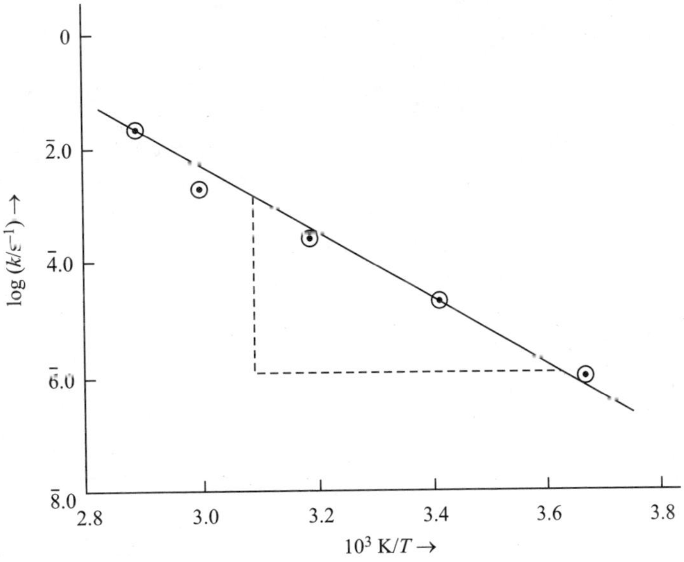

Fig. AI.1 Plot of log (k/s^{-1}) versus 10^3 K/T

$$\text{slope} = \frac{\Delta \log (k/\text{s}^{-1})}{\Delta(10^3 \text{K}/T)} = \frac{\bar{6}.0 - \bar{3}.1}{3.6\,2 - 3.10} = \frac{-3.10}{0.52} = -5.96$$

The expression representing the straight-line plot is

$$\log\left(\frac{k}{k^\circ}\right) = \log\left(\frac{A}{k^\circ}\right) - \frac{E_a}{2.303\,R}\frac{1}{T} \qquad (\text{where } k^\circ = \text{s}^{-1})$$

The slope of this expression is

$$m = \frac{\Delta \log (k/k^\circ)}{\Delta(1/T)} = -\frac{E_a}{2.303\,R}$$

From the slope of the straight-line plot shown in Fig. 1, the value of E_a is calculated as follows.

$$\frac{\Delta \log (k/\text{s}^{-1})}{\Delta(10^3 \text{ K}/T)} = \frac{\Delta \log (k/\text{s}^{-1})}{10^3 \text{ K}\, \Delta(1/T)} = -5.96$$

or
$$\frac{\Delta \log(k/\text{s}^{-1})}{\Delta(1/\text{T})} = -5.96 \times 10^3 \text{ K}$$

Equating this to $-E_a/2.303\,R$, we get

$$-\frac{E_a}{2.303\,R} = -5.96 \times 10^3 \text{ K}$$

or
$$\begin{aligned} E_a &= (2.303\,R)\,(5.96 \times 10^3 \text{ K}) \\ &= (2.303)\,(8.314 \text{ J K}^{-1} \text{ mol}^{-1})\,(5.96 \times 10^3 \text{ K}) \\ &= 1.14 \times 10^5 \text{ J mol}^{-1} \equiv 114 \text{ kJ mol}^{-1} \end{aligned}$$

4 DERIVED PHYSICAL QUANTITIES

Physical quantities other than base physical quantities are known as derived physical quantities. These may be expressed in terms of base physical quantities by multiplication and division. It is not necessary that all the seven will be required to define a particular physical quantity. Some may require only two, other may require three and so on. Some of the derived physical quantities have special names of SI unit (Table AI.4). Table AI.5 records some of the other derived physical quantities.

5 SI PREFIXES

The prefixes shown in Table AI.6 may be used to express multiple and submultiple of SI units.

A Few Comments

- Prefix symbols should be printed in roman type with no space between the prefix and the unit symbol, e.g. ms, pm, kg, and so on.
- A prefix when combined with a unit symbol provides a new symbol that can be raised to any power without the use of parentheses.

Table AI.4 SI Derived Units with Special Names and Symbols

Physical quantity	*Name of SI unit*	*Symbol for SI unit*	*Expression in terms of SI base units*
frequency	hertz	Hz	s^{-1}
force	newton	N	$kg\ m\ s^{-2}$
pressure	pascal	Pa	$N\ m^{-2} \equiv kg\ m^{-1}\ s^{-2}$
energy, work, heat	joule	J	$N\ m \equiv kg\ m^2\ s^{-2}$
power	watt	W	$J\ s^{-1} \equiv kg\ m^2\ s^{-3}$
electric charge	coulomb	C	A s
electric potential, electromotive force	volt	V	$J\ C^{-1} \equiv kg\ m^2\ s^{-3}\ A^{-1}$
electric resistance	ohm	Ω	$V\ A^{-1} \equiv kg\ m^2\ s^{-3}\ A^{-2}$
electric conductance	siemens	S	$\Omega^{-1} \equiv kg^{-1}\ m^{-2}\ s^3\ A^2$
electric capacitance	farad	F	$C\ V^{-1} \equiv kg^{-1}\ m^{-2}\ s^4\ A^2$
magnetic flux density	tesla	T	$V\ s\ m^{-2} \equiv kg\ s^{-2}\ A^{-1}$
magnetic flux	weber	Wb	$V\ s \equiv kg\ m^2\ s^{-2}\ A^{-1}$
inductance	henry	H	$V\ A^{-1}\ s \equiv kg\ m^2\ s^{-2}\ A^{-2}$

Table AI.5 A few Derived Physical Quantities

Physical quantity	*Units*
area	m^2
volume	m^3
speed, velocity	$m\ s^{-1}$
angular velocity	s^{-1}, rad s^{-1}
acceleration	$m\ s^{-2}$
wave number	m^{-1}
density	$kg\ m^{-3}$
specific volume	$m^3\ kg^{-1}$
amount concentration	$mol\ m^{-3}$
molality	$mol\ kg^{-1}$
molar volume	$m^3\ mol^{-1}$
heat capacity, entropy	$J\ K^{-1} \equiv kg\ m^2\ s^{-2}\ K^{-1}$
molar heat capacity	$J\ K^{-1}\ mol^{-1}$
specific heat capacity	$J\ K^{-1}\ kg^{-1} \equiv m^2\ s^{-2}\ K^{-1}$
molar energy	$J\ mol^{-1}$
surface tension	$N\ m^{-1} \equiv J\ m^{-2} \equiv kg\ s^{-2}$
dynamic viscosity	$N\ s\ m^{-2} \equiv Pa\ s \equiv m^{-1}\ kg\ s^{-1}$
conductivity	$S\ m^{-1} \equiv kg^{-1}\ m^{-3}\ s^3\ A^2$
molar conductivity	$S\ m^2\ mol^{-1}$
permittivity	$F\ m^{-1} \equiv kg^{-1}\ m^{-3}\ s^4\ A^2$
permeability	$H\ m^{-1} \equiv kg\ m\ s^{-2}\ A^{-2}$
electric field strength	$V\ m^{-1} \equiv kg\ m\ s^{-3}\ A^{-1}$
magnetic field strength	$A\ m^{-1}$

Table AI.6 Recommended Multiple and Submultiples along with their Symbols

Submultiple	*Prefix*	*Symbol*	*Multiple*	*Prefix*	*Symbol*
10^{-1}	deci	d	10	deca	da
10^{-2}	centi	c	10^2	hecto	h
10^{-3}	milli	m	10^3	kilo	k
10^{-6}	micro	μ	10^6	mega	M
10^{-9}	nano	n	10^9	giga	G
10^{-12}	pico	p	10^{12}	tera	T
10^{-15}	femto	f	10^{15}	peta	P
10^{-18}	atto	a	10^{18}	exa	E

For example,

$$1\ \mathrm{cm^3} \equiv (0.01\ \mathrm{m})^3 \equiv 10^{-6}\ \mathrm{m^3}$$

$$1\ \mu\mathrm{s}^{-1} \equiv (10^{-6}\ \mathrm{s})^{-1} \equiv 10^6\ \mathrm{s}^{-1}$$

$$1\ \mathrm{mmol/dm^3} \equiv (10^{-3}\ \mathrm{mol})/(10^{-1}\ \mathrm{m})^3 \equiv \mathrm{mol\ m^{-3}}$$

$$1\ \mathrm{g\ cm^{-3}} \equiv (10^{-3}\ \mathrm{kg})\ (10^{-2}\ \mathrm{m})^{-3} \equiv 10^3\ \mathrm{kg\ m^{-3}}$$

- Prefixes are not to be combined into compound prefixes. For example 10^{-12} m is not to be written as 1 μμm but as 1 pm.
- The name and symbols of decimal multiples and submultiples of the SI base unit of mass, the kg, which already contains a prefix, are constructed by adding the appropriate prefix to the word gram and symbol g. For example

  1 milligram is written as 1 mg and not as 1 μkg

  1 megagram is written as 1 Mg and not as 1 kkg

6 PRINTING OF SYMBOLS

The symbol for a physical quantity should generally be a single letter of the Latin or Greek alphabet. As far as possible, only the recommended symbols should be used to represent physical quantities (Table AI.7) and these should be printed in italics (sloping type). Capital and lower case (especially for specific quantites) may both be used.

The symbol may be modified by subscript and/or superscripts of specific meaning. Subscripts and superscripts that are themselves symbols for physical quantities or numbers should be printed in italic type; all other should be printed in roman (upright) type.

A few examples are given below.

C_p heat capacity at constant pressure

x_i mole fraction of the ith species

C_{B} heat capacity of substance B

V_{m} molar volume

The meaning of symbols for physical quantities may be further qualified by the use of one or more subscripts or by information contained in round brackets.

Table AI.7 Some of the Recommended Symbols of Physical Quantities

Physical quantity	*Symbol*	*Physical quantity*	*Symbol*
length	l	Bohr radius	a_0
breadth	b	Rydberg constant	R_∞
height	h	Bohr magneton	μ_B
distance, diameter	d	nuclear magneton	μ_N
radius	r	disintegration (rate) constant	λ
area	A	half life	$t_{1/2}$
volume	V	transition frequency	ν
time	t	transition wavenumber	$\tilde{\nu}$
speed	v, u, w, c	wavelength	λ
mass	m	speed of light in vacuum	c_0
reduced mass	μ	in a medium	c
density	ρ	Avogadro constant	N_A, L
momentum	$\boldsymbol{p}$	Boltzmann constant	k, k_B
angular momentum	L	gas constant	R
moment of inertia	I	average speed	$\bar{c}, \bar{u}$
force	$\boldsymbol{F}$		$<c>, <u>$
energy	E	number of entities	N
potential energy	E_p, V	amount of a substance	n
kinetic energy	E_k, T, K	atomic mass	m_a
weight	G, W	molar mass	M
gravitational constant	G	relative molecular or	
dynamic viscosity	η	molar mass	M_r
fluidity	ϕ	molar volume	V_m
power	P	mole fraction	x, y
electric charge	Q	pressure	p, P
electric potential	V, ϕ	partial pressure	p_B
electric potential difference	$U, \Delta V, \Delta\phi$	amount concentration	c
capacitance	C	solubility	s
permittivity	ε	molality	m
permittivity of vacuum	ε_0	surface concentration	Γ
relative permittivity	ε_r	stoichiometric number	ν
electric dipole moment	p, μ	extent of a reaction	ξ
electric current	I	degree of dissociation	α
magnetic flux density	B	heat	q, Q
magnetic field strength	H	work	w, W
permeability	μ	internal energy	U
permeability of vacuum	μ_0	enthalpy	H
electric resistance	R	thermodynamic temperature	T
conductance	G	Celsius temperature	θ_c, t
resistivity	ρ	entropy	S
conductivity	κ	Helmholtz function	A
mass number	A	Gibbs function	G
atomic number	Z	specific quantity X	x
electron rest mass	m_e	Compressibility,	
atomic mass constant	m_u	isothermal	κ_T
elementary charge	e	isentropic	κ_s
Planck constant	h	cubic expansion coefficient	α, α_V

(Contd.)

Physical quantity	*Symbol*	*Physical quantity*	*Symbol*
heat capacity,		partial order of reaction	n_B
at constant pressure	C_p	overall order of reaction	n
at constant volume	C_V	rate constant	k
ratio of heat capacities	γ	Faraday constant	F
Joule-Thomson coefficient	μ, μ_{JT}	charge number of an ion	z
second virial coefficient	B	charge number of electrochemical cell reaction	z, n
compression factor	Z	emf	E
chemical potential	μ	conductivity cell constant	K_{cell}
standard equilibrium constant	K^o	molar conductivity of an electrolyte	Λ
osmotic pressure	Π	ionic molar conductivity	λ
rate of change of quantity X	$\dot{X}$	electric mobility	μ
rate of reaction	$\dot{\xi}$	transport number	t
rate of concentration change	r_B, v_B		
rate of reaction (based on amount concentration)	v, r		

A few examples are listed below.

$C_{p,\,B}$ heat capacity of the substance B

$C_{p,m}(B)$ molar heat capacity of the substance B

$$\mu_i = \left(\frac{\partial G}{\partial n_i}\right)_{T,\,p,\,n_j}$$ chemical potential of ith species

$\Delta_f S^o(HgCl_2, s, 25\ °C) = -\ 154.3\ J\ K^{-1}\ mol^{-1}$.
standard entropy of formation of solid mercuric chloride

7 PRINTING OF UNITS

- The symbol for a unit should be printed in roman (upright) type.
  The symbol for a unit should remain unaltered in the plurals. For example
  5 cms is wrong
  5 cm is correct
- The symbol for a unit should not be followed by a full stop except when it occurs at the end of a sentence in text. For example,
  Take 5 g. of a substance — wrong
  Take 5 g of a substance — correct
  Mass of a substance is 5 g. — correct
- Symbols for prefixes for units should be printed in roman type with no space between the prefix and the unit. For example,
  m s^{-1} stand for metre per second
  ms^{-1} stands for millisecond inverse
- Symbols for units should be printed in the lower case letters, unless they are derived from a personal name when they should begin with a capital letter. For example,

  5 cm, 5 J, 5 kJ

8 SPECIFIC AND MOLAR PHYSICAL QUANTITIES

Specific Physical Quantities

The adjective *specific* before the name of an extensive quantity means the physical quantity per unit mass. When the symbol for the extensive quantity is a capital letter, the symbol used for the specific quantity is often the corresponding lower case letter. For example,

$$\text{Specific volume, } v = \frac{V}{m}$$

$$\text{Specific heat capacity at constant pressure, } c_p = \frac{C_p}{m}$$

A few physical quantity which were formerly named with the adjective specific but do not involve the unit mass have been renamed. These are given below.

Specific conductance	Conductivity
Specific resistance	Resistivity
Specific rate constant	Rate constant
Specific heat	Specific heat capacity

(The physical quantity specific heat is wrongly named as the basic physical quantity is heat capacity and not merely heat.)

Molar Physical Quantities

The adjective *molar* before the name of an extensive quantity generally means divided by amount of substance. The subscript m on the symbol for the extensive quantity denotes the corresponding molar quantity. For example,

Molar volume, $V_{\text{m}} = V/n$

Molar enthalpy, $H_{\text{m}} = H/n$

Molar entropy, $S_{\text{m}} = S/n$

$$\text{Molar heat capacity, } C_{p,\text{m}} = \frac{C_p}{n}$$

There are a few quantities where the adjective molar implies division by amount of substance concentration. For example

$$\text{Molar absorption coefficient, } \varepsilon = \frac{A}{c}$$

$$\text{Molar conductivity, } \Lambda = \frac{\kappa}{c}$$

(The molar conductivity is formerly known as molar conductance. It is called molar conductivity as it is conductivity divided by amount of substance concentration.)

9 DIMENSIONAL ANALYSIS

A physical expression should also be dimensionally correct. For example, the conversion expression of Celsius temperature to kelvin temperature may be written as

$$T = \theta_c + 273.15$$

This expression is numerically correct but not dimensionally as the unit of θ_c is °C and that of T is K. One can add or subtract two physical quantities if they have the same unit. Thus, a correct conversion equation would be

$$T/\mathrm{K} = \theta_c/^\circ\mathrm{C} + 273.15$$

For example, for 25°C, we would have

$$T/\mathrm{K} = 25^\circ\mathrm{C}/^\circ\mathrm{C} + 273.15$$

$$= 25 + 273.15 = 298.15$$

or $$T = 298.15\ \mathrm{K}$$

A few other examples are

$$\log\left(\frac{k}{k^\circ}\right) = \log\left(\frac{A}{k^\circ}\right) - \frac{E_a}{2.303\,RT}$$

$$\ln\left(\frac{\eta}{\eta^\circ}\right) = \ln\left(\frac{A}{\eta^\circ}\right) + \frac{E}{RT}$$

$$\ln\left(\frac{p}{p^\circ}\right) = -\frac{\Delta_{\mathrm{vap}}H}{RT} + C$$

Note that the division by k°, η° and p° (which stand for the corresponding unit physical quantities) make the expression within the logarithm brackets unitless. Other examples are

$$\mathrm{pH} = -\log\{[\mathrm{H^+}]/\mathrm{mol\ dm^{-3}}\}$$

$$\mathrm{p}K_\mathrm{w}^\circ = -\log\{K_\mathrm{w}/(\mathrm{mol\ dm^{-3}})^2\}$$

$$\mathrm{p}K_\mathrm{a}^\circ = -\log\{K_\mathrm{a}/\mathrm{mol\ dm^{-3}}\}$$

The use of standard equilibrium constant K° (= $K/(\mathrm{mol\ dm^{-3}})^{\Sigma\nu}$) would avoid the division by units.

10 EXPRESSIONS OF PHYSICAL QUANTITIES ARE INDEPENDENT OF THE CHOICE OF UNITS

The defining equation of a physical quantity in terms of other physical quantities has nothing to do with the units employed for the latter. For example, earlier molar conductivity was defined as

$$\Lambda = \frac{1\,000\,\kappa}{c}$$

with the understanding that κ is expressed in S cm^{-1} and c in terms of mol L^{-1} and the resultant units of Λ will come out to be S cm^2 mol^{-1}. The above way of defining Λ with the restriction on the choices of units of κ and c, is, in fact, not needed. One can define directly Λ as

$$\Lambda = \frac{\kappa}{c}$$

without the factor of 1000 in the numerator. The unit of Λ can be derived in terms of units in which κ and c are expressed. When we try to multiply or divide the same type of units, the numerical value of 1000 will automatically appear during the conversion of one unit in terms of its equivalent unit. Taking a typical example where

$$\kappa = 2.380 \times 10^{-3}\ \text{S cm}^{-1}$$

$$c = 0.01\ \text{mol dm}^{-3}$$

we will have

$$\begin{aligned}
\Lambda &= \frac{\kappa}{c} \\
&= \frac{2.380 \times 10^{-3}\ \text{S cm}^{-1}}{0.01\ \text{mol dm}^{-3}} = 2.38 \times 10^{-1}\ \text{S cm}^{-1}\ \text{dm}^3\ \text{mol}^{-1} \\
&= 2.38 \times 10^{-1}\ \text{S cm}^{-1}\ (10\ \text{cm})^3\ \text{mol}^{-1} \\
&= 238\ \text{S cm}^2\ \text{mol}^{-1}
\end{aligned}$$

Another example is provided by defining equations of ebullioscopic constant (i.e. boiling point elevation constant) and freezing point depression constant. Earlier, these were defined as

$$K_\text{b} = \frac{RT_\text{b}^{*2} M_1}{1000\ \Delta_\text{vap} H_\text{m}} \quad \text{and} \quad K_\text{f} = \frac{RT_\text{f}^{*2} M_1}{1000\ \Delta_\text{fus} H_\text{m}}$$

The factor of 1000 in the denominators of the above expressions is basically due to defining equation of molality as

$$m = \frac{w_2/M_2}{w_1/1000}$$

with the understanding that w_1 is expressed in grams. However, this is an unnecessary condition and the correct defining equations would be

$$m = \frac{w_2/M_2}{w_1}$$

$$K_\text{b} = \frac{RT_\text{b}^{*2} M_1}{\Delta_\text{vap} H_\text{m}} \quad \text{and} \quad K_\text{f} = \frac{RT_\text{f}^{*2} M_1}{\Delta_\text{fus} H_\text{m}}$$

In case, if it is desired to write the relationship between numerical values, the expression $\Lambda = 1\,000\ \kappa/c$ may be written as

$$\Lambda/(\text{S cm}^2\ \text{mol}^{-1}) = \frac{1\,000\ \kappa/(\text{S cm}^{-1})}{c/(\text{mol dm}^{-3})}$$

Similarly, for molality we can write

$$m/(\text{mol kg}^{-1}) = \frac{(w_2/\text{g})\,(M_2/\text{g mol}^{-1})}{(w_1/\text{g})/1\,000}$$

The definition of a unit physical quantity from its expression involving other physical quantities should not involve any particular choice of units for the latter. For example, conductivity is defined as

$$\kappa = G\left(\frac{l}{A}\right)$$

The following definition of conductivity is wrong.

> The conductivity κ is defined by the relation $\kappa = G(l/A)$ where l is the length in cm, A is the area in cm^2, and G is the conductance in ohm^{-1}.

The correct way would be

> The conductivity κ is defined by the relation $\kappa = G(l/A)$ where l is the length, A is the area and G is the conductance.

Similarly, the following statement is incorrect.

> The conductivity of a conductor is defined as the conductance offered by the conductor having 1 cm length and 1 cm^2 area of cross-section.

Instead, we will have

> The unit conductivity of a conductor is defined as the conductance offered by the conductor having unit length and unit area of cross-section.

11 PROGRESS OF A CHEMICAL REACTION

Extent of Reaction

A chemical reaction, in general, may be written as

$$0 = \sum_{\text{B}} \nu_{\text{B}}\text{B}$$

where ν_{B} is the stoichiometric coefficient (or number) of the species B in the chemical equation. The summation is overall the species of chemical equation. The stoichiometric coefficients have positive values for products and negative for reactants. The change in the amount of a reactant or a product during the progress of a reaction may be represented in terms of the variable ξ (pronounced xi), known as the extent of reaction. By definition, we have

$$n(\text{B}) = n_0(\text{B}) + \nu_{\text{B}}\xi$$

where $n_0(\text{B})$ is the amount of B at $t = 0$ (when $\xi = 0$) and $n(\text{B})$ is the amount when the extent of reaction is ξ. Obviously,

$$\frac{\mathrm{d}n_{\text{B}}}{\mathrm{d}\xi} = \nu_{\text{B}}$$

that is, the change in the amount of a reactant or a product with the change in the extent of reaction is equal to its stoichiometric coefficient in the balanced chemical equation. For a finite change in the extent of reaction, we will have

$$\frac{\Delta n_B}{\Delta \xi} = \nu_B \quad \text{i.e.} \quad \Delta n_B = \nu_B \, \Delta \xi$$

Thus, the change in the extent of reaction when multiplied by the stoichiometric number gives the change in the amount of the corresponding species during the progress of the reaction. Obviously, the unit of $\Delta\xi$ is the same as that of Δn, i.e. mol, as the stoichiometric number is a pure number.

Illustration

Taking a specific example of the decomposition of N_2O_5 to give NO_2 and O_2, we write the reaction as

$$0 = 4NO_2(g) + O_2(g) - 2N_2O_5(g)$$

or

$$2N_2O_5(g) = 4NO_2(g) + O_2(g)$$

If the reaction is started with the amount n_0 of N_2O_5, we will have

	$2N_2O_5(g)$	$\rightarrow$ $4NO_2(g)$	$+ O_2(g)$
$t = 0$	n_0		
t	$n_0 - 2\xi$	4ξ	ξ

At the maximum permitted value of ξ, we will have

$$n_0 - 2\,\xi_{max} = 0$$

$$\xi_{max} = \frac{n_0}{2}$$

For example, for $n_0 = 0.4$ mol, $\xi_{max} = 0.2$ mol, for $n_0 = 2$ mol, $\xi_{max} =$ 1 mol and for $n_0 = 4.0$ mol, $\xi_{max} = 2.0$ mol. Quite often, the equation written above is interpreted as.

2 mol of $N_2O_5(g)$ on decomposition produces 4 mol of $NO_2(g)$ and 1 mol of $O_2(g)$.

This interpretation is true only for unit extent of reaction, i.e. $\xi = 1$ mol. It is not a general statement as we can predict only the relative amounts of the species being formed or consumed from the expression of the chemical equation.

12 THERMODYNAMIC RECOMMENDATIONS

Sign Convention of Heat and Work

Heat absorbed by the system and work done on the system are assigned positive values as these operations increase the internal energy of the system. The first law of thermodynamics as a consequence of this is defined as

$$\Delta U = q + w$$

with $\quad dw = -p\, dV$

Standard State Pressure

Standard state pressure is defined equal to 1 bar (= 10^5 N m^{-2}). The change from 1 atm to 1 bar would cause the following changes in the values of thermodynamic functions.

(a) $\Delta_f H^\circ$ for gaseous and condensed phases remain unaffected.
(b) $\Delta_f S^\circ$ for gaseous phases increase by a factor of

$$R \ln \frac{101\,325 \text{ Pa}}{10^5 \text{ Pa}} = 0.109 \text{ J K}^{-1} \text{ mol}^{-1}$$

(c) $\Delta_r G^\circ$ of a reaction is altered by an amount – (0.109 J K^{-1} mol^{-1}) $T\,\Delta \nu_g$, where $\Delta \nu_g$ (= $\Sigma\, \nu_g$) is the change in the stoichiometric numbers of gaseous species appearing in the chemical equation and is given as

$$\Delta \nu_g = \Sigma\, \nu_g(\text{product}) - \Sigma\, |\, \nu_g(\text{reactant})|$$

(d) Standard equilibrium constant of a reaction having nonzero $\Delta \nu_g$ will be changed by a factor of $(1.013\,25)^{\Delta \nu_g}$.

Relation Between $\Delta_r H$ and $\Delta_r U$

In the expression

$$\Delta_r H = \Delta_r U + (\Delta \nu_g)\, RT$$

$\Delta \nu_g$ is to be interpreted as the change in the stoichiometric number of gaseous species in the chemical equation and not the change in the amount of gaseous species.

Recommended Subscripts

Symbols used as subscripts (in roman) to denote a chemical reaction
The type of reaction is indicated by the following symbols written immediately after Δ (and not after H, U, S and G, as the case may be).

Process	*Subscript*	*Example*
Vaporization	vap	$\Delta_{vap} H$
Sublimation	sub	$\Delta_{sub} H$
Melting, fusion	fus	$\Delta_{fus} H$
Transition (between two phases)	trs	$\Delta_{trs} H$
Mixing of fluids	mix	$\Delta_{mix} H$
Solution (of solute in solvent)	sol	$\Delta_{sol} H$
Dilution (of a solution)	dil	$\Delta_{dil} H$
Adsorption	ads	$\Delta_{ads} H$
Reaction in general	r	$\Delta_r H$
Atomization	at	$\Delta_{at} H$
Combustion reaction	c	$\Delta_c H$
Formation reaction	f	$\Delta_f H$

In general, $\Delta_{vap} H$ (or a similar expression such as $\Delta_{sub} H$, $\Delta_{fus} H$) is regarded as extensive thermodynamic quantity associated with the process. The more useful quantity is $\Delta_{vap} H_m$, which may be called molar enthalpy of vaporization.

Recommended Superscripts

Recommended superscripts

standard	o	ideal	id
pure substance	*	activated complex	‡
infinite dilution	∞	excess quantity	E

Standard or Thermodynamic Equilibrium Constant

For a gaseous reaction

$$0 = \sum_{B} \nu_B B$$

the free energy change with its extent of reaction is given by

$$dG = \sum_{B} \mu_B \, dn_B = \sum_{B} \mu_B (\nu_B \, d\xi) = \sum_{B} (\nu_B \, \mu_B) \, d\xi$$
$$= \Delta_r G \, d\xi$$

or $$\left(\frac{\partial G}{\partial \xi}\right)_{T,p} = \Delta_r G$$

For a reaction at equilibrium

$$\left(\frac{\partial G}{\partial \xi}\right)_{T,p} = 0$$

Hence $\Delta_r G = 0$

Now since $\mu_B = \mu_B^\circ + RT \ln (p_B/p^\circ)$, we get

$$\Delta_r G = \Delta_r G^\circ + RT \sum_{B} \nu_B \ln \frac{p_B}{p^\circ} = \Delta_r G^\circ + RT \ln \left\{ \prod_{B} \left(\frac{p_B}{p^\circ}\right) \right\}^{\nu_B}$$
$$= \Delta_r G^\circ + RT \ln Q_p^\circ$$

where Q_p°, known as standard reaction quotient, is given by

$$Q_p^\circ = \prod_{B} \left(\frac{p_B}{p^\circ}\right)^{\nu_B}$$

Now at equilibrium, $\Delta_r G = 0$. hence

$$\Delta_r G^\circ = - RT \ln K_p^\circ$$

where $K_p^\circ - (Q_p^\circ)_{eq}$

and is known as *standard equilibrium constant.* It is a unitless quantity as it involves the ratio of p_B/p°. Another quantity which does not involve the ratio is equilibrium constant, defined as

$$K_p = \prod_{B} p_B^{\nu_B}$$

It has the SI unit of $Pa^{\Sigma\nu_g}$ (= $Pa^{\Delta\nu_g}$). The expression connecting K_p and K_p° is

$$K_p^\circ = \frac{K_p}{(p^\circ)^{\Delta\nu_g}}$$

The expression connecting K_p and K_c of a reaction is

$$K_p = K_c(RT)^{\Delta\nu_g}$$

The corresponding expression involving the standard equilibrium constants is

$$K_p^\circ = K_c^\circ\,(c^\circ\, RT/p^\circ)^{\Delta\nu_g}$$

13 ENTHALPHY CHANGE OF A REACTION

What is the unit of enthalpy of a reaction? Is it kJ or kJ mol^{-1}? Quite often, if the unit is written as kJ mol^{-1}, we interpret mol^{-1} with the formation of 1 mol of a product or with the consumption of 1 mol of a reactant. For example, for the enthalpy of formation of $H_2O(1)$, we write

$$H_2(g) + \tfrac{1}{2}O_2(g) \rightarrow H_2O(1) \qquad \Delta H^\circ = -\ 285.8\ \text{kJ mol}^{-1}$$

Here, the appearance of mol^{-1} in the unit of ΔH° is explained on the basis that 1 mol of $H_2O(1)$ is formed. Another example is the enthalpy of combustion, say of $CH_4(g)$, we write it as

$$CH_4(g) + 2O_2(g) \rightarrow CO_2(g) + 2H_2O(1) \qquad \Delta H^\circ = -\ 890.4\ \text{kJ mol}^{-1}$$

Here, the appearance of mol^{-1} is interpreted as the combustion of 1 mol of $CH_4(g)$. Now consider the reaction

$$2C_3H_6(g) + 9O_2(g) \rightarrow 6CO_2(g) + 6H_2O(1) \quad \Delta H^\circ = -4\,116\,\text{kJ mol}^{-1}$$

Here, the appearance of mol^{-1} in the unit of ΔH cannot be interpreted likewise as neither 1 mol of any reactant is disappearing nor 1 mol of any product is appearing. So, we argue that the unit of ΔH is kJ and not kJ mol^{-1}. In fact, it is not so. The mol^{-1} in the unit of ΔH is not for the appearance of 1 mol of a product or disappearance of 1 mol of a reactant. It is for the unit extent of the reaction or more precisely, $\Delta_r H$ is reported for a unit extent of reaction. To be specific, we again take the example of

$$2N_2O_5(g) \rightarrow 4NO_2(g) + O_2(g)$$

Let the reaction be started with the amount n_0 of $N_2O_5(g)$.

Enthalpy of the system to start with, where $\xi = 0$, is

$$H^\circ_{\text{initial}} = n_0\, H^\circ_{\text{m}}(N_2O_5,\ g)$$

Now suppose the reaction has proceeded to the extent ξ. At this stage, we will have

$$\underset{n_0-2\xi}{2N_2O_5(g)} \rightarrow \underset{4\xi}{4NO_2(g)} + \underset{\xi}{O_2(g)}$$

Enthalpy of the system at this stage is

$$H^{\circ}_{\text{final}} = (n_0 - 2\xi)\, H^{\circ}_{\text{m}}(N_2O_5, g) + 4\xi\, H^{\circ}_{\text{m}}(NO_2, g) + \xi H^{\circ}_{\text{m}}(O_2, g)$$

Enthalpy change of the system at this stage is

$$\begin{aligned}\Delta H^{\circ} &= H^{\circ}_{\text{final}} - H^{\circ}_{\text{initial}} \\ &= [(n_0 - 2\xi)H^{\circ}_{\text{m}}(N_2O_5, g) + 4\xi H^{\circ}_{\text{m}}(NO_2, g) + \xi H^{\circ}_{\text{m}}(O_2, g)] \\ &\qquad - [n_0\, H^{\circ}_{\text{m}}(N_2O_5, g)] \\ &= -2\xi H^{\circ}_{\text{m}}(N_2O_5, g) + 4\xi H^{\circ}_{\text{m}}(NO_2, g) + \xi H^{\circ}_{\text{m}}(O_2, g)]\end{aligned}$$

The unit of ΔH° is kJ as ξ has the unit of mol and H°_{m} has the unit of kJ mol^{-1}.

Enthalpy of Reaction

By definition, *enthalpy of reaction* is given by

$$\begin{aligned}\Delta_r H^{\circ} &= \frac{\Delta H^{\circ}}{\xi} \\ &= -2\, H^{\circ}_{\text{m}}(N_2O_5, g) + 4H^{\circ}_{\text{m}}(NO_2, g) + H^{\circ}_{\text{m}}(O_2, g)\end{aligned}$$

Obviously, the unit of $\Delta_r H^{\circ}$ is kJ mol^{-1} and not merely kJ. So here mol^{-1} is meant for the unit extent of reaction and not for one mol of a particular reactant disappearing or product appearing.

Thus, the enthalpy of a reaction is related to the entire chemical equation and not for one particular species of the equation. For example, for the reaction

$$H_2(g) + \tfrac{1}{2}O_2(g) \rightarrow H_2O(l) \qquad \Delta_r H^{\circ} = -285.8 \text{ kJ mol}^{-1}$$

$\Delta_r H^{\circ}$ should be interpreted as the enthalpy change for the unit extent of reaction, i.e. when 1 mol of H_2 combines with (1/2) mol of $O_2(g)$ to give 1 mol of $H_2O(l)$, the associated enthalpy change is – 285.8 kJ. Similarly, the reaction

$$2C_3H_6(g) + 9O_2(g) \rightarrow 6CO_2(g) + 6H_2O(l) \quad \Delta_r H^{\circ} = -4\,116 \text{ kJ mol}^{-1}$$

should be interpreted as the enthalpy change for the unit extent of reaction, i.e. when 2 mol of $C_3H_6(g)$ combines with 9 mol of $O_2(g)$ to give 6 mol of $CO_2(g)$ and 6 mol of $H_2O(l)$, the associated enthalpy change is – 4 116 kJ.

Alternatively, the above mentioned particular examples may be generalized as shown in the following.

The general expression for a chemical equation is

$$0 = \sum_{B} \nu_B B$$

where ν_B is the stoichiometric coefficient (or number) of the species B in the chemical equation. The stoichiometric coefficient has positive values for products and negative for reactants.

The enthalpy change of the reaction as it progresses is given by

$$\Delta H^\circ = \sum_B (\nu_B \xi) H^\circ_m(B)$$

where ξ is the extent of reaction. The enthalpy of the reaction is given by

$$\Delta_r H^\circ = \left(\frac{\partial(\Delta H^\circ)}{\partial \xi}\right)_{T,p} = \left(\frac{\partial\{\sum_B (\nu_B \xi) H^\circ_m(B)\}}{\partial \xi}\right)_{T,p}$$

$$= \sum_B \nu_B H^\circ_m(B)$$

In a similar manner, we will have

$$\Delta_r U^\circ = \left\{\frac{\partial(\Delta U^\circ)}{\partial \xi}\right\}_{T,V} = \sum_B \nu_B U^\circ_m(B)$$

$$\Delta_r S^\circ = \left\{\frac{\partial(\Delta S^\circ)}{\partial \xi}\right\}_{T,P} = \sum_B \nu_B S^\circ_m(B)$$

$$\Delta_r G^\circ = \left\{\frac{\partial(\Delta G^\circ)}{\partial \xi}\right\}_{T,p} = \sum_B \nu_B G^\circ_m(B)$$

The units of $\Delta_r U^\circ$, $\Delta_r S^\circ$ and $\Delta_r G^\circ$ would be J mol^{-1}, J K^{-1}mol^{-1} and J mol^{-1}, respectively. Note that all the reaction quantities mentioned above are intensive quantities.

Relationship of $\Delta_r H$ with $\Delta_r U$

We have $\Delta_r H = \Sigma_B \nu_B H_m(B)$

Now since $H_m = U_m + pV_m$, we get

$$\Delta_r H = \Sigma_B \nu_B (U_m + pV_m)_B$$

$$= \Sigma_B \nu_B U_m(B) + \Sigma_B \nu_B (pV_m)_B$$

Now using the fact that for one mole of a substance

$pV \approx 0$ for a condensed (solid or liquid) phase

$pV_m = RT$ for a gaseous phase

the above expression becomes

$$\Delta_r H = \Delta_r U + \Sigma_B (\nu_B)_g\, RT$$

$$= \Delta_r U + (\Delta \nu_g)\, RT$$

where $\Delta\nu_g$ is the change in the stoichiometric number of gaseous species involved in the chemical equation and is given by

$$\Delta \nu_g = \Sigma_B(\nu_B)_g$$
$$= \Sigma_B \nu_g(\text{product}) - \Sigma_B \mid \nu_g(\text{reactant}) \mid$$

It may be emphasized here that $\Delta \nu_g$ in the above expression represents change in the stoichiometric number of gaseous species and not change in the amount of gaseous species.

Alternatively, we may proceed as follows.

If ξ is the extent of reaction for the reaction

$$0 = \sum_B \nu_B B$$

then at this stage

Enthalpy change, $\Delta H^\circ = \sum_B (\nu_B \xi)\, H^\circ_m(B)$

Energy change, $\Delta U^\circ = \sum_B (\nu_B \xi)\, U^\circ_m(B)$

pV change, $\Delta(pV) = \Delta(n_g RT) = (\Delta n_g)\, RT$

$$= \sum_B (\nu_B \xi)_g\, RT$$

Now since $\Delta H^\circ = \Delta U^\circ + \Delta(pV)$, we get

$$\sum_B (\nu_B \xi) H^\circ_m(B) = \sum_B (\nu_B \xi) U^\circ_m(B) + \sum_B (\nu_B \xi)_g\, RT$$

Dividing throughout by ξ, we get

$$\sum_B \nu_B H^\circ_m(B) = \sum_B \nu_B U^\circ_m(B) + \sum_B (\nu_B)_g\, RT$$

or $\quad \Delta_r H^\circ = \Delta_r U^\circ + \sum_B (\nu_B)_g\, RT$

or $\quad \Delta_r H^\circ = \Delta_r U^\circ + (\Delta \nu_g)\, RT$

14 RATE OF A REACTION

If a reaction does not involve the condition of constant volume it is obvious that the concentration of a substance involved in the reaction can vary not only due to the reaction in progress, but also due to the change in volume. In order to avoid such difficulties, IUPAC has recommended that the rate of a reaction be defined in terms of change of amount of substance with time. For example, for the reaction

$$2N_2O_5(g) \rightarrow 4NO_2(g) + O_2(g)$$

$$\text{rate} = -\frac{1}{2}\frac{d\{n(N_2O_5)\}}{dt} = \frac{1}{4}\frac{d\{n(NO_2)\}}{dt} = \frac{d\{n(O_2)\}}{dt}$$

If the reaction is started with the amount n_0, we will have

	$2N_2O_5(g) \rightarrow$	$4NO_2(g) +$	$O_2(g)$
$t = 0$	n_0	0	0
t	$n_0 - 2\xi$	4ξ	ξ

Hence,
$$\text{rate} = -\frac{1}{2}\frac{dn(N_2O_5)}{dt} = -\frac{1}{2}\frac{d(n_0 - 2\xi)}{dt} = \frac{d\xi}{dt}$$
$$= \frac{1}{4}\frac{dn(N_2O_5)}{dt} = \frac{1}{4}\frac{d(4\xi)}{dt} = \frac{d\xi}{dt}$$
$$= \frac{dn(O_2)}{dt} = \frac{d\xi}{dt}$$

that is, the rate of reaction is equal to the change of extent of reaction with time.

15 WEIGHT VERSUS MASS

The word weight means gravitational force ($F = mg$, where m is the mass and g is the acceleration due to gravity). It is because of this, that the terms atomic and molecular weights are misleading, though in chemistry, these have been precisely defined. The IUPAC has recommended the use of following terms while dealing with atomic and molecular masses.

Relative Atomic Mass of an Element

The relative atomic mass of an element is the ratio of the average mass per atom of a specified composition (natural isotopic composition is assumed unless some other composition is specified) of an element to 1/12 of the mass of an atom of the nuclide carbon-12, i.e.

$$A_r = \frac{\text{mass of an atom}}{(1/12)\text{ mass of an atom of }^{12}\text{C}} \qquad (1)$$

Relative Molecular Mass of a Compound

The relative molecular mass of a compound is the ratio of the average mass per molecule of a specified isotopic composition of a compound to 1/12 of the mass of an atom of nuclide carbon-12, i.e.

$$M_r = \frac{\text{mass of a molecule}}{(1/12)\text{ mass of an atom of }^{12}\text{C}} \qquad (2)$$

The quantities A_r and M_r are formerly known as atomic weight and molecular weight, respectively. It may be noted that A_r and M_r carry no units as these are simply the ratio of two masses.

Unified Atomic Mass Unit

The quantity (1/12) mass of an atom ^{12}C is known as the unified atomic mass unit (symbol: m_u, unit: u, also commonly abbreviated as amu). Hence

$$m_u = 1\text{ u}$$
$$= \frac{m_a(^{12}\text{C})}{12} = \frac{(0.012\text{ kg mol}^{-1})/(6.022 \times 10^{23}\text{ mol}^{-1})}{12}$$
$$= 1.661 \times 10^{-27}\text{ kg} \equiv 1.661 \times 10^{-24}\text{ g}$$

Atomic Mass

The average mass per atom of a specified isotopic composition of an element is known as the atomic mass (symbol: m_a). It is simply a mass and thus has the unit of mass (i.e. g or kg). From Eq. (1), it follows that

$$\text{mass of an atom, } m_a = A_r\, m_u$$

that is, the atomic mass is equal to the relative atomic mass times the atomic mass unit.

Molecular Mass

The average mass per molecule of a specified isotopic composition of a compound is known as molecular mass. It is simply a mass and thus carries the unit of mass (i.e. g or kg). From Eq. (2), it follows that

$$\text{mass of a molecule, } m_f = M_r\, m_u$$

that is, the molecular mass is equal to the relative molecular mass times the atomic mass unit.

Molar Mass

The average mass per unit amount of a substance (element or compound) of a specific isotopic composition is known as molar mass. Mathematically, we write it as

$$M = \frac{m}{n}$$

Since the unit of mass (m) is kg or g, and that of amount of substance (n) is mol, it follows that the unit of molar mass is kg mol^{-1} or g mol^{-1}.

Relation Between Molar Mass and Atomic or Molecular Mass

If a system has N entities (atoms or molecules), we will have

For elements	*For compounds*
$m = N\, m_a$	$m = N\, m_f$
$n = N/N_A$	$n = N/N_A$
$M_m = \dfrac{m}{n}$	$M_m = \dfrac{m}{n}$
$= \dfrac{N\, m_a}{N/N_A}$	$= \dfrac{N\, m_f}{N/N_A}$
$= N_A m_a$	$= N_A m_f$
$= N_A(A_r\, m_u)$	$= N_A(M_r\, m_u)$
$= A_r\,(N_A\, m_u)$	$= M_r(N_A\, m_u)$
$= A_r\,(1 \text{ g mol}^{-1})$	$= M_r\,(1 \text{ g mol}^{-1})$

In words, the relative atomic (or molecular) mass is the numerical value of the molar mass expressed in g mol^{-1}.

Illustration

The above terms on masses may be illustrated by taking specific examples of sodium and carbon dioxide. We will have

Relative atomic mass of sodium = 23

Atomic mass of sodium = 23 u = 23 (1.66×10^{-27} kg)

$= 3.82 \times 10^{-26}$ kg

Molar mass of sodium = 23 g mol^{-1}

Relative molecular mass of carbon dioxide = 44

Molecular mass of carbon dioxide = 44 u = 44 (1.66×10^{-27} kg)

= 7.30×10^{-26} kg

Molar mass of carbon dioxide = 44 g mol^{-1}

Quite often, we do not specify the units of atomic, molecular and molar masses. Hence, our statements are not exact and precise. In other words, a mere replacement of the term weight by mass without carrying the associated unit does not complete our adoption of IUPAC recommendations. So, whenever we write or speak of atomic, molecular and molar masses, we must state the associated units.

It may be noted that the terms gram atomic weight, gram molecular weight, gram formular weight, gram atom, gram molecule, etc., are obsolete terms and should thus be abandoned.

16 CONCEPT OF EQUALITY OF AMOUNTS OF SUBSTANCES

The concepts of equivalent mass and normality of a solution which are commonly used in the volumetric analysis find no place in the IUPAC recommendations. The computation of these quantities very much depends upon the chemical reaction one is actually dealing with. Sometimes, this important point is not given due attention and thus eventually may lead to the wrong computations. One of the examples is provided by the equivalent mass of $S_2O_3^{2-}$ ions in the titration of thiosulphate ions and iodine. Since each $S_2O_3^{2-}$ ion carries two negative charge, its equivalent mass (without actually referring to the reaction) would be

$$\text{Equivalent mass} = \frac{\text{Molar mass}}{2 \text{ eq mol}^{-1}}$$

The result is obviously wrong since the basic reaction of $S_2O_3^{2-}$ is

$$2S_2O_3^{2-} \rightarrow S_4O_6^{2-} + 2e^-$$

Hence, its

$$\text{Equivalent mass} = \frac{2 \text{ Molar mass}}{2 \text{ eq mol}^{-1}} = \frac{\text{Molar mass}}{\text{eq mol}^{-1}}$$

The concepts of equivalent mass and normality are basically not required because we can derive this information directly by referring to the chemical equation. For example, in the volumetric titration of MnO_4^- and Fe^{2+} ions, the required equation is

$$MnO_4^- + 5Fe^{2+} + 8H^+ \rightarrow 5Fe^{3+} + Mn^{2+} + 4H_2O$$

Hence, $n(MnO_4^-) \equiv n(5Fe^{2+})$ or $n(MnO_4^-) \equiv \frac{1}{5} n(Fe^{2+})$

In terms of concentrations, we would have

$$[MnO_4^-] \equiv [5Fe^{2+}] \quad \text{or} \quad [MnO_4^-] \equiv \frac{1}{5}[Fe^{2+}]$$

Now if V_1 and V_2 are the equivalent volumes of MnO_4^- and Fe^{2+} solutions, then we will have

$$[MnO_4^-]\ V_1 \equiv [5Fe^{2+}]\ V_2 \quad \text{or} \quad [MnO_4^-]V_1 \equiv \frac{1}{5}[Fe^{2+}]V_2$$

or $$[Fe^{2+}] = 5\,[MnO_4^-]\left(\frac{V_1}{V_2}\right)$$

Alternatively, a redox reaction may be written in such a manner that the absolute value of the stoichiometric coefficient for the electrons transferred (which are normally omitted from the overall equation) is equal to one. In the above example, we would have

$$\frac{1}{5}MnO_4^- + Fe^{2+} + \frac{8}{5}H^+ \rightarrow \frac{1}{5}Mn^{2+} + Fe^{3+} + \frac{4}{5}H_2O$$

Hence, $[\frac{1}{5}MnO_4^-] \equiv [Fe^{2+}]$ or $5\ [MnO_4^-] \equiv [Fe^{2+}]$

Using the expression of equality of amounts, we get

$$[\tfrac{1}{5}MnO_4^-]V_1 = [Fe^{2+}]V_2$$

or $$[Fe^{2+}] = [\tfrac{1}{5}MnO_4^-]\left(\frac{V_1}{V_2}\right) = 5\,[MnO_4^-]\left(\frac{V_1}{V_2}\right)$$

17 ELECTROLYSIS

Let the chemical reaction occurring at the electrode be represented as

$$0 = \sum_B \nu_B B$$

One of the constituents of the reaction is electron (e^-) with stoichiometric coefficient ν_e. The change in the amount of any other constituent with the absorption or emission of the amount n_e of electron will be given as

$$\frac{\Delta n_B}{\nu_B} = \frac{n_e}{|\nu_e|} \quad \text{that is} \quad \Delta n_B = \frac{\nu_B}{|\nu_e|}n_e$$

We take the magnitude of ν_e as the electron may appear as a reactant or a product depending upon the type of reaction (reduction or oxidation) occurring at the electrode. Now since

$$n_e = \frac{N_e}{N_A} \quad \text{and} \quad \Delta n_B = \frac{\Delta m_B}{M_B}$$

we have

$$\frac{\Delta m_B}{M_B} = \frac{\nu_B}{|\nu_e|}\frac{N_e}{N_A} \quad \text{or} \quad \Delta m_B = M_B \frac{\nu_B}{|\nu_e|}\frac{N_e}{N_A}$$

Multiplying and dividing the right side of the above expression by the charge of a proton (e), we get

$$\Delta m_B = M_B \frac{\nu_B}{|\nu_e|}\frac{(N_e e)}{(N_A e)} = M_B \frac{\nu_B}{|\nu_e|}\frac{Q}{F} = \frac{M_B}{|\nu_e|/\nu_B}\frac{Q}{F} \tag{AI.1}$$

where Q is the quantity of electricity (= It) passed and F is the Faraday constant (= 96487 C mol^{-1} $\simeq$ 96500 C mol^{-1}).

Equation (AI.1) is the quantitative expression of Faraday's law of electrolysis.

(Note: There is no need of expressing the Faraday's law of electrolysis in terms of electrochemical equivalent of the substance.)

Illustrations

Equation (AI.1) may be illustrated by taking the following reaction taking place at the electrode.

1. Cathodic reduction of Ag^+ to Ag

$$Ag^+ + e^- = Ag \qquad \nu_{Ag} = +1; \qquad |\nu_e| = 1; \qquad \Delta m_{Ag} = \left(\frac{M_{Ag}}{1}\right)\left(\frac{Q}{F}\right)$$

2. Cathodic reduction of Au^{3+} to Au

$$Au^{3+} + 3e^- = Au \qquad \nu_{Au} = +1; \qquad |\nu_e| = 3; \qquad \Delta m_{Au} = \left(\frac{M_{Au}}{3}\right)\left(\frac{Q}{F}\right)$$

3. Anodic oxidation of Sn to Sn^{2+}

$$Sn = Sn^{2+} + 2e^- \qquad \nu_{Sn} = -1; \qquad |\nu_e| = 2; \qquad \Delta m_{Sn} = -\left(\frac{M_{Sn}}{2}\right)\left(\frac{Q}{F}\right)$$

4. Anodic oxidation of Cl^- to ClO_3^-

$$Cl^- + 3H_2O \rightarrow ClO_3^- + 6H^+ + 6e^-$$

$$\nu_{Cl^-} = -1; \qquad |\nu_e| = 6 \qquad \Delta m_{Cl^-} = -\left(\frac{M_{Cl^-}}{6}\right)\left(\frac{Q}{F}\right);$$

$$\nu_{ClO_3^-} = +1; \qquad |\nu_e| = 6 \qquad \Delta m_{ClO_3^-} = \left(\frac{M_{ClO_3^-}}{6}\right)\left(\frac{Q}{F}\right)$$

APPENDIX II Units and Conversion Factors

CGS Units vis-á-vis SI Units

Physical quantity	*CGS units*		*SI units*	
	Name	*Symbol*	*Name*	*Symbol*
length	centimetre	cm	metre	m
	Angstrom (10^{-8} cm)	Å		
mass	gram	g	kilogram	kg
time	second	s	second	s
temperature	Celsius	°C	kelvin	K
	kelvin	°K		
energy	calorie	cal	joule	J
	kilocalorie	kcal	kilojoule	kJ
	litre-atmosphere	lit-atm		
	ergs	erg		
electric current	ampere	A	ampere	A

Conversion of CGS Units to SI Units

Quantity	*Unit*	*Equivalent*
length	Å	10^{-10} m = 10^{-1} nm = 10^{2} pm
volume	litre	1 L = 1 dm^3
force	dyn	10^{-5} N
energy	erg	10^{-7} J
	cal	4.184 8 J
	eV	1.602 1 × 10^{-19} J
	eV/mole	98.484 kJ mol^{-1}
pressure	atmosphere	101.325 kN m^{-2}
	mmHg (or Torr)	133.322 N m^{-2}
	bar (10^6 dyn/cm^2)	10^5 N m^{-2}
viscosity	poise	10^{-1} kg m^{-1} s^{-1}

Values of Some Physico-Chemical Constants

Constant	*SI units*
Acceleration of gravity, g	$9.806\ 65\ \text{m s}^{-2}$
Avogadro constant, N_A	$6.022\ 05 \times 10^{23}\ \text{mol}^{-1}$
Bohr magneton, μ_B	$9.274\ 09 \times 10^{-24}\ \text{J T}^{-1}$
Bohr radius, a_0	$5.291\ 77 \times 10^{-11}\ \text{m}$
Boltzmann constant, k	$1.380\ 66 \times 10^{-23}\ \text{J K}^{-1}$
Elementary charge, e	$1.602\ 19 \times 10^{-19}\ \text{C}$
Electronic rest mass, m_e	$9.109\ 53 \times 10^{-31}\ \text{kg}$
Faraday constant, F	$9.648\ 46 \times 10^{4}\ \text{C mol}^{-1}$
Gas constant, R	$8.205\ 75 \times 10^{-2}\ \text{dm}^3\ \text{atm K}^{-1}\ \text{mol}^{-1}$
	$8.314\ 41\ \text{J K}^{-1}\ \text{mol}^{-1}$
Molar volume of ideal gas at 0 °C and 1 atm, V_m	$2.241\ 4 \times 10^{-2}\ \text{m}^3\ \text{mol}^{-1}$
Planck's constant, h	$6.626\ 18 \times 10^{-34}\ \text{J s}$
Proton rest mass, m_p	$1.672\ 65 \times 10^{-27}\ \text{kg}$
Vacuum speed of light, c_0	$2.997\ 925 \times 10^{8}\ \text{m s}^{-1}$
Standard atmospheric pressure	101.325 kPa
	1.013 25 bar

Index

Absolute temperature, 3
Acid, 284, 420
Acid-base,
    indicators, 355
    strength and structure, 430
    titration curves, 358, 359
Acid strength,
    and structure, 430
    of oxoacids, 431
Activation energy of viscous flow, 157
Amagat's law, 12
Amorphous solids, 171
Amount of substance, 518
Amphiprotic, 324, 422
Amphoterism, 417
Angle of contact, 161
Antifluorite unit cell, 225
Arrhenius,
    concepts of acids and bases, 420
    theory of electrolytic dissociation, 277, 450
Asymmetry effect, 463
Atomic mass, 538
Average energy,
    rotational motion, 85
    translational motion, 85
    vibrational motion, 86
Average kinetic energy, 19
Average speed, 56
Avogadro,
    constant, 9
    law, 9

Barometric distribution law, 62
Base, 284, 420
Base physical quantities, 517
Berthelot's equation of state, 29, 114, 136
Bimolecular collisions, 68
Body-centred cubic lattice, 177
Boiling point, 152
Boltzmann,
    constant, 20
    distribution law, 64
Born repulsion, 236
Boyle's law, 2
Boyle temperature, 33
Bragg equation, 192
Bravais lattice, 176
Bronsted-Lowry theory of acids, 421
Buffer,
    action, 348
    index, 349
    preparation of, 351
    range, 350
    solution, 348
Cailletal-Mathias law, 38
Capillary action, 162
Capillary rise method for surface tension, 162
Cell constant, 454
Centre of symmetry, 254
Charles law, 2
Clapeyron equation, 147
Classical energy associated with,
    rotation motion, 85
    translational motion, 85
    vibrational motion, 86
Clausius-Clapeyron equation, 147
Closest-packing,
    of ions, 224
    of spheres, 212
Coefficient of diffusion, 77
Coefficient of thermal conductivity, 81
Coefficient of viscosity, 74, 153
Cohesive energy of crystals, 232
Collision,
    cross-section of a molecule, 68
    diameter, 68
    rate, bimolecular, 69
Common-ion effect, 393
Complex ions, 403
Compressibility, coefficient, 236
Compression factor, 21, 107, 133
Concepts of acids and bases, 420
Conditional stability constant, 410

Conductance,
  and its measurement, 453
  application of, 502
  definition, 452
  Grotthus, 470
Conduction,
  electrolytic, 449
  metallic, 408
Conductivity,
  cell, 454
  equivalent, 455
  molar, 455
  variation on dilution, 459
Conjugate acid and base, 421
Contact angle, 161
Continuity of state, 39
Corresponding states, 44
Corundum structure, 230
Coulometer, 451
Covalent crystals, 238, 242
Co-volume, 24
Critical,
  constants, 36
  density, 38
  isotherm, 37
  pressure, 37
  ratio of ionic radii, 222
  state, 37
  temperature, 36
  volume, 37
Crystal,
  classification on bond type, 210
  cleavage, 182
  covalent, 238
  density, 206
  edges, 171
  faces, 171, 182
  interplanar distances, 263
  lattice, 172
  planes, 184
  structure of potassium chloride, 205
  structure of sodium chloride, 202
  symmetry, 254
  systems, 173
Cubic,
  interplanar distances, 187
  lattices, 186
  closest-packing, 212
  system, 175
  system, diffraction pattern of, 197
  system, interplanar distances, 187

Dalton's law of partial pressures, 12
Debye-Falkenhagen effect, 466
Debye-Hiickel-Onsager theory, 463
Debye-Scherrer powder method, 195
Degree of dissociation, 279, 502
Degrees of freedom, 86
Density,
  critical, 38
  of cubic crystals, 206
  orthobaric, 38
Derived physical quantities, 522
Designation of Crystal planes, 184
Diamond structure, 238
Dieterici's equation of state, 29, 119, 138
Diffraction,
  of radiation, 189
  of X-rays, 191
  pattern of cubic system, 199
Diffusion coefficient, 77
Dimensional analysis, 528
Dissociation,
  constant, of a weak acid, 284
  degree of, 279
  of polyprotic acid, 305
  of pure water, 280
Distribution of molecular speeds, 52
Drop number method, 164
Drow weight method, 164

Effect of impurities on
  boiling point, 150
Effect of temperature on,
  surface tension, 164
  vapour pressure, 148
  viscosity, 76, 153
Electrical conductivity, 83
  electrolytic, 449
  metallic, 448
Electrolysis, 450, 541

Electrolyte,
classification, 276
potential, 277
true, 277
weak, 277
Electrolytic cell, 449
Electrophoretic effect, 463
Enthalpy change of a reaction, 534
Equation of state, 5
Equilibrium constant, 278
Equivalent conductivity, 455
Excluded volume, 24
Extensive property, 6
Exact treatment of a,
base, 295
diprotic acid, 298
mixture of BOH and BA, 339
mixture of HA and MA, 336
mixture of strong and weak acids, 341
monoprotic acid, 284
Extent of reaction, 530

Face-centred cubic lattice, 177
Faraday,
constant, 451
laws, 451
First law of crystallography, 172
Formation constant of a complex ion, 404
Formation of a bubble, 160
Fluidity, 157
Fluorite unit cell, 228

Gas collision, 68
Gas constant, 6
Gaseous state, 1
Gay-Lussac's law, 4
Glide plane, 181
Graham's law of diffusion, 18
Graphite structure, 241
Grotthus Conductance, 470

Harmann-Maugin notations, 255
Haüy's idea and space lattice, 172
Heat capacities of gases, 88
Henderson-Hasselbalch equation, 348
Hexagonal system, 175
Hittorf's method, 465
Hole theory of liquids, 245
Holes,
octahedral, 224
tetrahedral, 224
Hydrolysis, 308
Hydrolysis,
constant, determination, 507
of anions, 309, 425
of cations, 309, 426
of Fe^{2+}, 330
of S^{2-}, 331
Hydrolysis of salt formed from,
a strong acid and a strong base, 310
a strong acid and a weak base, 316
a weak acid and a strong base, 310
a weak acid and a weak base, 319
Hydrolysis of salt involving,
multivalent cation or anion, 330
strong conjugate cation and an amphiprotic anion, 327
weak conjugate cation and an amphiprotic anion, 324

Ideal gas, 7
Indicator, 355
constant, 355
range, 355
selection of, 356
Instability constant, 404
Intensive property, 6
Inerfacial,
angle, 171
tension, 165
Ionic,
atmosphere, 463
conductivity, 467
crystals, 242
crystals and closest packing, 224
equilibria, 273
migration, 466
mobility, 497
product of water, 281, 503
solutions, 275
velocity, 497

Ionization,
constant of water, 280
of weak acids, 285
of weak bases, 286
Isoteniscope method, 150

Kelvin temperature scale, 3
Kinetic,
gas equation, 14
theory of gases, 14
Knudsen method, 79
Kohlrausch's law, 466

Laminar flow, 153
Lattice energy, 218
Lattice planes, 186
Laüe,
equations, 267
method, 191
Law of,
corresponding state, 44
equipartition of energy, 84
rational indices, 184
rectilinear diametery, 176
Levelling effect, 423
Lewis acid, 426
Limiting densities method, 106
Liquid,
state, 1, 242
structure of, 242
surface tension of, 158
viscosity of, 153
X-ray diffraction pattern of, 242

Madelung constant, 234
Maxwell distribution law, 52
Mean,
free path, 70
speed, 56
square speed, 19
Measurement of,
surface tension, 162
vapour pressure, 150
viscosity, 154
Metallic,
conduction, 448
crystals, 242
Miller indices, 184
Molar conductivity, 455
Molar mass, 539
Molecular,
beam method, 54
collision diameter, 68
collision in a gas, 68
crystals, 211, 241
mass, 538
solutions, 273
Mole fraction, 12
Monoclinic system, 175
Moving boundary method, 495
Most probable speed, 53

Nonprimitive unit cell, 173

Octahedral holes, 224
Ohm's law, 449
Orthobaric densities, 38
Orthorhombic system, 175
Ostwald dilution law, 279
Ostwald viscometer, 154
Packing of spheres, 212
Parachor, 167
Partial,
pressure, 12
volume, 12
pH-scale, 282
Physical properties of liquids, 144
Physical quantities,
base, 517
definition, 518
derived, 522
molar, 527
specific, 527
Point groups, 180
Poise (unit), 154
Poiseuille's equation, 154
Powder method, 195
Preferential precipitation of salts, 397
Pressure,
critical, 37
from kinetic theory, 14
reduced, 44
Primitive unit cell, 173

Radius ratio, in ionic crystals, 222
Rate of reaction, 537
Rational indices, law of, 180
Real gases, 20
Reciprocal,
  lattice, 262
  vectors, 260
Rectilinear diameter, law of, 38
Reduced,
  equation of state, 44
  pressure, 44
  temperature, 44
  variables, 133
  volume, 44
Redlich-Kwong equation, 29, 126, 141
Relative,
  atomic mass, 538
  molecular mass, 538
Relaxation time, 463
Resistance, 452
Rhombohedral system, 175
Rock-salt structure, 224
Root-mean-square speed, 19
Rotation axis, 254

Schoenflies symbols, 255
Scattering process,
  by an atom, 266
  by an electron, 266
  by a small crystal, 266
  by a unit cell, 267
Screw axis, 181
Second virial coefficient, 32
Self diffusion, 77
SI prefixes, 502
SI units, 517
Solid state, 1, 171
Solubility product, 387
  and precipitation, 396
  and qualitative analysis, 400
  determination, 504
Space,
  groups, 181
  lattice, 172
Specific physical quantity, 527
Speed,
  average, 56
  distribution of, 52
  most probable, 53
  root mean square, 19
Standard state pressure, 531
State,
  critical, 37
  equation of, 5
  gaseous, 1
  liquid, 1
  solid, 1
Stalagmometer, 163
States of matter, 1
Structure factor, 267
Structure of,
  common ionic oxides, 224
  diamond, 238
  graphite, 241
  liquids, 242
Stokes law, 155
Surface active materials, 166
Surface energy, 158
Surface tension, 158
  and its measurement, 162
  of mixtures, 166
Symbols of physical quantities, 523
Symmetry,
  elements, 254
  operations, 254

Temperature,
  absolute, 3
  critical, 36
  inversion, 95
  reduced, 44
Tetrahedral holes, 224
Tetragonal system, 175
Thermal conductivity, 81
Thermodynamic recommendations, 531
Titrations,
  acid-base, 355
  conductometric, 508
Titration curve of,
  a dibasic acid with a strong base, 374

a strong acid with a strong base, 361
a weak acid with a strong base, 364
a weak base with a strong acid, 367
sodium carbonate with HCl, 383
Transport number, 474
Triclinic system, 175
Trigonal system, 175
Triple point, 147
Trouton's rule, 152

Unit cell, 172
Unified atomic mass unit, 538
Universal indicators, 361

Van der Waals,
constants, 26
crystals, 211
equation, 23, 107, 134
Vapour pressure,
and boiling point, 150
definition, 146
and its measurement, 150
Variation of conductivity with concentration, 449
Virial equation of state, 30
Viscometer, 154
Viscosity of,
gas, 73
liquid, 153
Void volume,
in body-centred cubic lattice, 212
in closest packing, 215
in simple cubic lattice, 211
in diamond structure, 239
Volume,
critical, 37
excluded, 24
reduced, 44

Walden's rule, 501
Weiss indices, 184
Wheatstone bridge, 453
Wien effect, 465

X-ray,
diffraction of liquids, 242
powder method, 195

Zinc blende structure, 229